Fuel Cell Fundamentals and Applications

Yinshi Li

Fuel Cell Fundamentals and Applications

Springer

Yinshi Li
Xi'an Jiaotong University
Xi'an, China

ISBN 978-981-96-9844-8 ISBN 978-981-96-9845-5 (eBook)
https://doi.org/10.1007/978-981-96-9845-5

This Springer imprint is published by the registered company Springer Nature Singapore Pte Ltd.
The registered company address is: 152 Beach Road, #21-01/04 Gateway East, Singapore 189721, Singapore

Preface

The contemporary global landscape confronts two existential imperatives: accelerating the energy transition and mitigating the climate crisis. Overreliance on conventional fossil fuels has precipitated severe environmental degradation while jeopardizing the long-term viability of socioeconomic systems worldwide. Within this critical context, hydrogen energy emerges as a transformative solution, distinguished by its wide availability, exceptional energy density, and carbon-neutral emissions.

Fuel cells, serving as electrochemical conversion systems, epitomize this paradigm shift. By enabling direct transformation of chemical energy into electricity with high thermodynamic efficiency and minimal environmental footprint, fuel cells are redefining hydrogen utilization across diverse applications—from zero-emission transportation and industrial manufacturing to building energy supply and decentralized energy networks. Their proliferating adoption underscores a systemic reshaping of the world's energy landscape.

However, the large-scale commercial application of fuel cells still faces numerous challenges. Persistent bottlenecks, including difficulties in hydrogen storage and transportation, insufficient catalyst material activity, and short system lifetime, demand rigorous scientific inquiry and engineering innovation. Addressing these barriers necessitates unprecedented interdisciplinary synergy across academia, industry, and governmental institutions to advance materials science, system integration, and policy frameworks.

This book provides a comprehensive overview of the state of the art in fuel cell technology, and the major features are summarized as follows:

Chapter 1 systematically reviews hydrogen storage technologies (compressed gas, cryogenic liquid, organic/inorganic carriers, solid materials) and utilization in industries, transport, medicine, and agriculture. The analysis contextualizes global hydrogen strategies, China's policies, challenges in storage/transportation, and fuel cells' role in enabling a carbon-neutral "hydrogen society" paradigm.

Chapter 2 delineates fuel cell fundamentals, including major types (AFCs: Alkaline Fuel Cells, PEMFCs: Proton Exchange Membrane Fuel Cells, PAFCs: Phosphoric Acid Fuel Cells, MCFCs: Molten Carbonate Fuel Cells, SOFCs: Solid Oxide

Fuel Cells), thermodynamics (Gibbs free energy, reversible potentials), kinetics (activation energy, Butler-Volmer equation), and mass/charge transport phenomena. It highlights PEMFCs' dominance in transportation via high power density while noting challenges: polarization losses (activation, ohmic, concentration), cost barriers, and hydrogen infrastructure needs. Innovations in materials and system integration are emphasized as pivotal for scalable decarbonization.

Chapter 3 summarizes fuel cell characterization techniques, including electrochemical methods (CV: Cyclic Voltammetry, EIS: Electrochemical Impedance Spectroscopy, polarization curves), material analysis (SEM: Scanning Electron Microscope, TEM: Transmission Electron Microscope, XRD: Diffraction of X-rays), system mapping (current, temperature, water visualization), and in situ techniques (CT: Computed Tomography, neutron imaging). These multiscale diagnostics are critically evaluated for performance optimization and durability enhancement.

Chapter 4 reviews the fundamental principles, key components (such as membranes, catalyst layers, and bipolar plates), and system architecture of PEMFCs. Applications are rigorously analyzed across vehicles, submarines, aircraft, and trains, with emphasis on efficiency benchmarks and operational constraints.

Chapter 5 focuses on the highly efficient electrocatalysts for PEMFCs, primarily Pt-based materials, to drive the hydrogen oxidation reaction and oxygen reduction reaction. Research focuses on reducing Pt usage via alloys, nano-structuring, and non-precious alternatives like M-N-C catalysts. Challenges include cost, durability, and activity optimization. Advances in single-atom catalysts, transition metal oxides, and metal-free carbon materials show promise for sustainable, high-performance fuel cells.

Chapter 6 analyzes PEMFCs modeling, including empirical correlations, mechanistic models, and data-driven approaches. Oxygen transport in cathodes and two-phase flow (VOF, M2) are highlighted as a focus in fuel cell simulation. Data-driven models, as a special class of simulation models, do not require the governing equations of the internal mechanism and are characterized by efficient and fast computation.

Chapter 7 details PEMFCs micromodeling strategies, covering microstructure reconstruction of gas diffusion layers (GDLs), microporous layers (MPLs), and catalyst layers (CLs) using stochastic and imaging techniques. Pore-scale simulations via Lattice Boltzmann, Monte Carlo, density functional theory (DFT), and molecular dynamics are evaluated for transport/reaction optimization and catalyst design.

Chapter 8 catalogs PEMFCs degradation mechanisms, including membrane mechanical/thermal/chemical decay, Pt catalyst dissolution, carbon support corrosion, GDL degradation, bipolar plate corrosion, and impurity effects. And the mitigation strategies are also provided for the degradation of each component at the same time.

Chapter 9 explores degradation experiments and numerical simulations in PEMFCs, covering experimental methods (steady-state durability tests, accelerated stress tests) and numerical simulations (mechanistic, data-driven, hybrid models). It details degradation mechanisms in components like membranes, catalysts, and electrodes and discusses challenges and future directions for improving durability predictions and mitigation strategies.

Chapter 10 discusses Anion Exchange Membrane Fuel Cells (AEMFCs), focusing on AEM properties, stability challenges, catalysts for methanol/ethanol/formate oxidation, and prospects as low-cost, durable alternatives to PEMFCs.

Chapter 11 examines SOFCs, covering their working principles, key materials (electrolytes, cathodes, anodes), system designs (planar, tubular), applications in transportation and stationary power, challenges (high costs, thermal management), and prospects for clean energy generation.

This book serves as an interdisciplinary resource in fuel cells. It is intended as an introduction to the topic for students, practitioners, and academics specializing in energy systems, advanced materials, and industrial chemistry, while concurrently providing policymakers, corporate strategists, and venture capital stakeholders with an analytical framework to inform decarbonization investments. We extend profound gratitude to all contributing authors for sharing cutting-edge insights and empirical advancements that collectively elevate the scholarly rigor of this book. Additionally, we acknowledge the professional staff from the Springer Nature editorial team for their meticulous support throughout the publication process.

Xi'an, China Yinshi Li

Contents

Contributors

Chapter 1: Hydrogen Storage and Utilization

Mingsheng Hao School of Energy and Power Engineering, Xi'an Jiaotong University, Xi'an, China
Qiangsheng Sun National Innovation Platform (Center) for Industry-Education Integration of Energy Storage Technology, Xi'an Jiaotong University, Xi'an, China
Rui Wang School of Chemical Engineering and Technology, Xi'an Jiaotong University, Xi'an, China
Yinshi Li School of Energy and Power Engineering, Xi'an Jiaotong University, Xi'an, China

Chapter 2: Fuel Cells

Xianda Sun Institute for Advanced Technology, Shandong University, Jinan, China
Yinshi Li School of Energy and Power Engineering, Xi'an Jiaotong University, Xi'an, China

Chapter 3: Fuel Cell Diagnostic Tools

Mingsheng Hao School of Energy and Power Engineering, Xi'an Jiaotong University, Xi'an, China
Kang Wang School of Energy and Power Engineering, Xi'an Jiaotong University, Xi'an, China
Yinshi Li School of Energy and Power Engineering, Xi'an Jiaotong University, Xi'an, China

Chapter 4: Proton Exchange Membrane Fuel Cells: Fundamentals

Shipei Deng School of Energy and Power Engineering, Xi'an Jiaotong University, Xi'an, China
Yuxuan Lou School of Energy and Power Engineering, Xi'an Jiaotong University, Xi'an, China
Mingsheng Hao School of Energy and Power Engineering, Xi'an Jiaotong University, Xi'an, China
Qiangsheng Sun National Innovation Platform (Center) for Industry-Education Integration of Energy Storage Technology, Xi'an Jiaotong University, Xi'an, China
Yinshi Li School of Energy and Power Engineering, Xi'an Jiaotong University, Xi'an, China

Chapter 5: Proton Exchange Membrane Fuel Cells: Electrocatalysts

Na Zhou School of Energy and Power Engineering, Xi'an Jiaotong University, Xi'an, China
Jinling Xue School of Energy and Power Engineering, Xi'an Jiaotong University, Xi'an, China
Yibin Fan School of Future Technology, Xi'an Jiaotong University, Xi'an, China
Rui Wang School of Chemical Engineering and Technology, Xi'an Jiaotong University, Xi'an, China
Yinshi Li School of Energy and Power Engineering, Xi'an Jiaotong University, Xi'an, China

Chapter 6: Proton Exchange Membrane Fuel Cells: Macro-model

Shipei Deng School of Energy and Power Engineering, Xi'an Jiaotong University, Xi'an, China
Ruijin Zhang School of Future Technology, Xi'an Jiaotong University, Xi'an, China
Shengyuan Chen School of Energy and Power Engineering, Xi'an Jiaotong University, Xi'an, China
Yubo Hu School of Energy and Power Engineering, Xi'an Jiaotong University, Xi'an, China
Yinshi Li School of Energy and Power Engineering, Xi'an Jiaotong University, Xi'an, China

Chapter 7: Proton Exchange Membrane Fuel Cells: Micro-model

Qinrui Tang School of Energy and Power Engineering, Xi'an Jiaotong University, Xi'an, China
Shengyuan Chen School of Energy and Power Engineering, Xi'an Jiaotong University, Xi'an, China
Yubo Hu School of Energy and Power Engineering, Xi'an Jiaotong University, Xi'an, China
Mingsheng Hao School of Energy and Power Engineering, Xi'an Jiaotong University, Xi'an, China
Yinshi Li School of Energy and Power Engineering, Xi'an Jiaotong University, Xi'an, China

Chapter 8: Proton Exchange Membrane Fuel Cells: Degradation Mechanism

Shengyuan Chen School of Energy and Power Engineering, Xi'an Jiaotong University, Xi'an, China
Yubo Hu School of Energy and Power Engineering, Xi'an Jiaotong University, Xi'an, China
Zongyao Ding School of Energy and Power Engineering, Xi'an Jiaotong University, Xi'an, China
Rui Wang School of Chemical Engineering and Technology, Xi'an Jiaotong University, Xi'an, China
Yinshi Li School of Energy and Power Engineering, Xi'an Jiaotong University, Xi'an, China

Chapter 9: Proton Exchange Membrane Fuel Cells: Degradation Experiments and Numerical Simulations

Yubo Hu School of Energy and Power Engineering, Xi'an Jiaotong University, Xi'an, China
Shengyuan Chen School of Energy and Power Engineering, Xi'an Jiaotong University, Xi'an, China
Mingsheng Hao School of Energy and Power Engineering, Xi'an Jiaotong University, Xi'an, China
Yinshi Li School of Energy and Power Engineering, Xi'an Jiaotong University, Xi'an, China

Chapter 10: Anion Exchange Membrane Fuel Cells

Xue Zhou School of Energy and Power Engineering, Xi'an Jiaotong University, Xi'an, China
Jinling Xue School of Energy and Power Engineering, Xi'an Jiaotong University, Xi'an, China
Na Zhou School of Energy and Power Engineering, Xi'an Jiaotong University, Xi'an, China
Zhipeng Liu School of Energy and Power Engineering, Xi'an Jiaotong University, Xi'an, China
Ruijin Zhang School of Future Technology, Xi'an Jiaotong University, Xi'an, China
Yinshi Li School of Energy and Power Engineering, Xi'an Jiaotong University, Xi'an, China

Chapter 11: Solid Oxide Fuel Cells

Jinling Xue School of Energy and Power Engineering, Xi'an Jiaotong University, Xi'an, China
Zhihao Chen School of Future Technology, Xi'an Jiaotong University, Xi'an, China
Tian Xie School of Energy and Power Engineering, Xi'an Jiaotong University, Xi'an, China
Xue Zhou School of Energy and Power Engineering, Xi'an Jiaotong University, Xi'an, China
Mingsheng Hao School of Energy and Power Engineering, Xi'an Jiaotong University, Xi'an, China
Qinrui Tang School of Energy and Power Engineering, Xi'an Jiaotong University, Xi'an, China
Ruijin Zhang School of Future Technology, Xi'an Jiaotong University, Xi'an, China
Yinshi Li School of Energy and Power Engineering, Xi'an Jiaotong University, Xi'an, China

Chapter 1
Hydrogen Storage and Utilization

1.1 Introduction

Solar fuels are chemical fuels produced using solar energy and include mainly formic acid, formate, hydrogen, and methanol. These solar fuels can be used as energy carriers. They can all be produced in places where solar energy resources are abundant, and then transported to places where energy is scarce. Therefore, they can alleviate energy shortages and reduce environmental pollution. Compared to other solar fuels, hydrogen has many interesting properties, making it the most studied and promising solar fuel.

First, hydrogen energy offers significant benefits, emitting only water during combustion and producing no carbon emissions, aiding carbon reduction goals. As the most abundant element in the universe, hydrogen has the potential to be low-cost and widely available for energy storage. However, most of the hydrogen is in the form of water molecules, and pure hydrogen cannot be found in nature. This implies that if we want to use hydrogen, we must first develop efficient and rapid hydrogen production technologies. Fortunately, there are a variety of technologies for preparing hydrogen, mainly thermochemical hydrogen production and water electrolysis, which will not be repeated here. While for the storage and utilization of hydrogen, there are still a lot of problems to be solved, and they will be the rest of the chapter.

The discovery of hydrogen dates back to the sixteenth century. Dr. Paracelsus discovered that when the metal was dissolved in an acid, a flammable gas was produced. However, he regarded the gas as hydrocarbons and other flammable gases such as carbon monoxide. In 1781, Cavendish's experiments demonstrated that water is produced when hydrogen is burned. Then, in 1929, Karl Friedrich Bonhoeffer and Paul Harteck reported, based on the experimental studies mentioned above, that ordinary hydrogen is a mixture of two types of molecules, ortho-hydrogen and para-hydrogen. Because of its simple structure, hydrogen allows relatively easy theoretical

Y. Li, *Fuel Cell Fundamentals and Applications*,
https://doi.org/10.1007/978-981-96-9845-5_1

calculations of its properties. Therefore, hydrogen is often used as a theoretical model for more complex atoms.

Before introducing hydrogen storage, transportation, and utilization technologies, it is necessary to briefly review modern strategies for the development of hydrogen energy. Up to now, more than 20 countries, accounting for 70% of the world's GDP, have formulated hydrogen energy development strategies, and more than 50 policies directly support the deployment of hydrogen energy worldwide. The most positive responses come from Japan, the European Union, the United States, and South Korea. Among them, the United States plays a leading role in the development of hydrogen energy and fuel cells. As early as 1990, the United States promulgated the relevant laws and regulations on hydrogen energy research and development. Japan is strongly dependent on imported fossil fuels, which account for more than 80% of the primary energy supply [1]. In 2017, Japan unveiled its Hydrogen Strategy, which aims to create a "hydrogen society". The main objectives of the strategy are to achieve cost parity between hydrogen and other fuels, to achieve hydrogen substitution for conventional energy sources in automobiles (including trucks and forklifts) and power generation, and to develop combined heat and power fuel cell systems for households. In 2019, the U.S. Hydrogen Economy Roadmap was released. Hydrogen demand is expected to exceed 17 million tons by 2030, with 5.3 million FCEVs on U.S. roads and 5600 hydrogen fueling stations nationwide. In January of that year, the European region released the European Hydrogen Energy Roadmap, which aims to provide 24% of Europe's final energy needs with hydrogen energy by 2050. In 2008, Korea issued a low-carbon and green growth strategy and implemented demonstration projects such as one million green homes and a green hydrogen city. By 2020, the world's first hydrogen law was promulgated, legally boosting industry development. That same year, "A Hydrogen Strategy for a Climate-Neutral Europe" was released, kicking off the top-level design for green and digital transformation. The Clean Hydrogen Alliance Energy was founded to speed up industrial decarbonization and maintain its lead. In 2020, the "EU Hydrogen strategies on hydrogen and the energy system integration strategy" were published, with the aim of setting a new clean energy investment agenda for the EU to achieve carbon neutrality by 2050 [2].

To achieve the goal of carbon peak and neutrality, China also focuses on hydrogen development. In 2014, the "Strategic Action Plan for Energy Development (2014–2020)" officially designated hydrogen energy and fuel cells as the direction of energy technology innovation. In 2016, China's National Development and Reform Commission (NDRC) and the National Energy Administration (NEA) formulated a roadmap for the development of hydrogen energy, taking the development of hydrogen energy and fuel cell technology as a key task. In March 2019, hydrogen energy was included in the Government Work Report for the first time, making it clear that the construction of hydrogenation and other facilities will be promoted. At the end of 2019, the Energy Statistical Statement System included hydrogen in energy statistics for the first time. In 2022, the NDRC and the NEA released the "Medium- and Long-Term Plan for the Development of Hydrogen Energy Industry (2021–2035)" to further promote the high-quality development of the hydrogen energy industry. In terms of local policy, more than 20 provinces, cities, and counties have introduced special policies for the hydrogen energy industry. In Jiangsu, Shandong, Zhejiang, Guangdong, and other economically developed coastal areas, a large number of relevant policies have been issued, among which county-level cities in the Jiangsu and Zhejiang regions have the highest enthusiasm for developing the hydrogen energy industry.

The industrial chain of hydrogen energy mainly includes: upstream hydrogen production; Midstream storage, transportation, and hydrogenation stations; and the downstream application of hydrogen energy. Spanning energy, chemical, and transportation sectors, it has a wide reach. Its rapid development will drive economic growth and accelerate the transition from fossil fuels to clean, low-carbon energy, reducing carbon emissions.

Industrial by-product hydrogen can be used as an important hydrogen source in the initial stage of hydrogen energy development. Hydrogen from water electrolysis has great potential for new energy utilization. In future development, "green hydrogen" will be produced from renewable energy to reduce the cost of hydrogen production.

In terms of hydrogen storage, our high-pressure gaseous hydrogen storage mainly depends on a 35 MPa bottle wrapped with carbon fiber. Low-temperature liquid hydrogen has been successfully used in aerospace engineering in China, but there is no relevant standard for civil use. Organic liquid hydrogen storage has been used in a fuel cell bus as a hydrogen storage demonstration case. Solid hydrogen storage has been commercially used in fuel cell submarines and demonstrated in distributed power generation and wind power hydrogen storage. Various hydrogen storage technologies will be described in detail in the following parts. In respect of hydrogen transportation, there are three main transportation modes in China, including gas hydrogen trailer transportation, gas hydrogen pipeline transportation, and liquid hydrogen tank truck transportation. Large-scale safe storage and transportation of hydrogen is the obstacle to the commercial application of hydrogen energy at present. High-pressure gaseous hydrogen storage is the future direction of development, so that it can be used in vehicle hydrogen storage tanks, transportation hydrogen storage tanks, and stationary hydrogen storage equipment. Large-scale transportation, such as liquid hydrogen tanker truck transportation and pipeline transportation, is also in urgent need of development. By 2050, as per the "White Paper on China's Hydrogen Energy and Fuel Cell Industry", hydrogen pipeline networks will spread across cities and villages, and vehicles will use higher-density, safer hydrogen storage technology.

In the downstream of the hydrogen energy industry, hydrogen energy can be widely used in traditional transportation fields, power generation, aerospace, chemical production, communications, construction, and so on. Take hydrogen fuel vehicles as an example. Proton exchange membrane fuel cell technology has been developed. Key components, such as membrane electrodes, bipolar plates, and proton exchange membranes, have the capability of localization. The small fuel cell stack industry is developing well, but the development of the auxiliary system key parts industry is relatively backward. The development of the system and vehicle industry is good, but the important parts depend on imports.

Hydrogen energy will become an important part of China's energy system. Hydrogen is expected to account for about 10% of China's energy system by 2050, with demand approaching 60 million tons and annual economic output exceeding 10 trillion yuan. The country will have more than 10,000 hydrogen refueling stations, and the transportation, industrial, and other fields will realize the widespread application of hydrogen energy. The output of fuel cell vehicles will reach 5.2 million units per year, the fixed power generation units will reach 20,000 units per year, and the fuel cell system capacity will reach 5.5 million units per year.

1.2 Hydrogen Storage and Transportation

The application of hydrogen energy needs to go through three steps: hydrogen energy production, hydrogen storage and transportation, and hydrogen utilization. Among them, the storage and transportation links are the key to the efficient utilization of hydrogen energy. However, the special physical and chemical properties of hydrogen make it difficult to store and transport, hampering the development of a hydrogen society.

Hydrogen is the lightest element in the periodic table and is the most abundant chemical element in the universe. Hydrogen has a very low density of 0.089 kg/m^3. Because of this, its storage usually requires high pressure or extremely low temperatures [3]. Moreover, the atomic radius of hydrogen is so small that it can pass through most pores that are invisible. Hydrogen can even penetrate thick steel plates at high temperatures and pressures. Besides, hydrogen is very active and is prone to combustion and explosion after leakage. Therefore, safety must be ensured in the process of storage and transportation. All these factors pose challenges to the storage and transportation technology of hydrogen.

This chapter reviews the hydrogen storage and transportation technology. How hydrogen is stored determines how it is transported. Therefore, this section focuses on hydrogen storage technologies, and the hydrogen transportation technology is briefly introduced in each part. The capture and release of hydrogen on materials may involve molecular adsorption, diffusion, and chemical bonding. Hydrogen can also be adsorbed in molecular/ionic form on suitable surfaces using pressure, temperature, and electrochemical potential to control its surface structure and bonding strength. So, hydrogen can be stored by many methods, including compressed hydrogen, liquid hydrogen, adsorbed hydrogen on materials, and so on [4] (Table 1.1)

Table 1.1 Hydrogen storage types and densities

	Kg H_2/kg	Kg H_2/m^3
Pressurized gas storage	0.01–0.05	2–16
Metal hydride	0.012–0.02	50–55
Liquid hydrogen	0.09–0.5	50–69

1.2.1 Compression and Cryogenics for Hydrogen Storage

1.2.1.1 Pressurized Hydrogen Gas

The most established hydrogen storage system is physical storage in the form of pressurized hydrogen gas. The density of hydrogen increases linearly with pressure near atmospheric pressure. But when it comes to higher pressure, a significant deviation from the ideal gas behavior occurs, as shown in Fig. 1.1. The hydrogen gas density increases slowly than the pressure, resulting in lower benefits from compressed gas at high pressure.

Depending on storage size and application, pressurized hydrogen gas storage technologies can be divided into three types [3]: (1) Large Underground Hydrogen Storage Systems; (2) Aboveground Pressurized Gas Storage Systems; (3) High-Pressure Hydrogen Tanks.

The first two types of pressurized gas storage systems are already used today in the natural gas business in various sizes and pressure ranges. And this application will be similar for hydrogen storage. But because of hydrogen's lower volumetric heating value, hydrogen storage systems of the same energy content will be more expensive than natural gas storage systems. The city of Kiel's public utility has been storing town gas with a hydrogen content of 60–65% in a gas cavern with a geometric volume of about 32,000 m^3 since 1971 [5]. Gaz de France (the French national gas company) has stored hydrogen-rich refinery by-product gases in an aquifer structure near Beynes, France. Imperial Chemical Industries of Great Britain stores hydrogen in the salt mine caverns near Teesside, United Kingdom [5]. Besides, the large salt cavity storage in Texas, USA, is also a famous example [6].

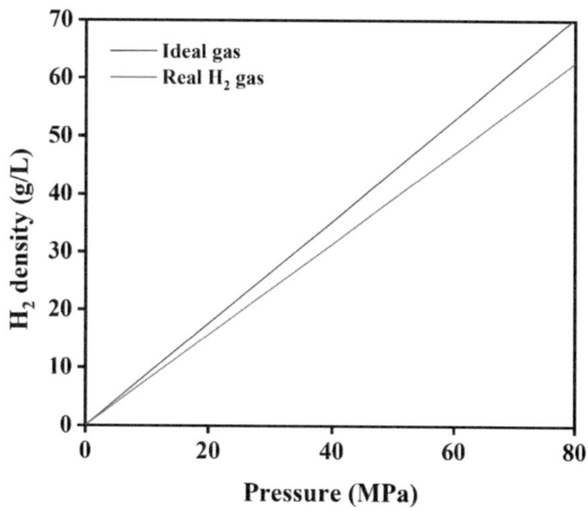

Fig. 1.1 Comparison of real hydrogen gas density to ideal gas as a function of pressure

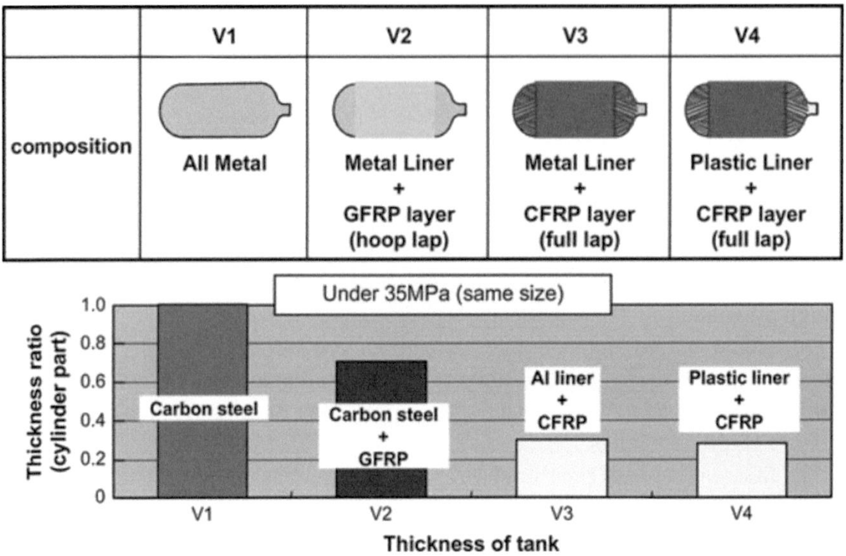

Fig. 1.2 Four types of high-pressure tank [7]

Now, when we talk about compressed hydrogen storage technology, we mostly refer to the last type, high-pressure hydrogen tanks. According to different materials, high-pressure hydrogen tanks can be classified into four types, as shown in Fig. 1.2.

Type V1 is the oldest type. It is an all-metal tank and hence the heaviest, typically employed for stationary use. Type V2 is made of a metallic liner with hoop reinforcement on the hoop part, with a glass fiber composite, and can withstand higher pressure than Type V1. It is also lighter than Type V1. However, both V1 and V2 tanks are unsuitable for portable applications due to their low hydrogen storage density and hydrogen embrittlement challenges. Type V3 is made of metal liner fully wrapped by carbon fibers in the 1980s. The metal liner is made of aluminum (Al), which solves the problem of embrittlement [8]. Type V3 offers a 25%–75% increase in capacity compared to Types V1 and V2. Type V4 is made of plastic liner fully wrapped by carbon fibers. Type V3 and Type V4 are widely accepted for portable applications, because of their light-weight properties compared to Types V1 and V2. However, they are much more expensive, which is one of the key issues for widespread applications. In 2010, Composites Technology Development Inc. developed an all-composite liner-less vessel, referred to as Type V5 [9]. Although it is significantly lighter than Type V4, the Type V5 cannot operate at pressures high enough for storing hydrogen and still needs a breakthrough [10].

All four types are commercially available, and the selection of storage is based on the final application, which requires a comparison between performance and cost. Most industrial hydrogen is stored in type V1 tanks (the pressure is from 150 to 300 bar), because they are the cheapest high-pressure vessels. When higher pressures are required for stationary applications, the type V2 tanks are preferred. Type V3 and type

V4 vessels are intended for portable applications because they have a higher energy density. Much progress has been achieved since they were first developed, such as weight loss, improved cycling performance, and other fibers than glass (Kevlar, carbon) [11].

Type V1 vessels can be manufactured from 3 different raw materials: plates, billets, and tubes. The main shape is formed by deep drawing, and the neck is formed by hot spinning. The liners of type V2 and V3 vessels can be manufactured in the same way. For type V4, its liner is polymeric. Hence, it is formed by cooling a polymer in the mold whose shape is the final designed liner shape. For all composite cylinder vessels, the composite can wrap the liner in three possible ways: hoop, polar, and helical. Figure 1.3 shows the possible wrapping ways. Generally, Type V2 is only hoop-wrapped, and a combination of hoop and polar wrapping is used for Types V3 and V4. The wrapping process can be performed by one filament winding machine. Then, the resin must be cured. After the wrapping process, a vessel for compressed hydrogen storage is made.

Many materials can be used for composite tanks wrapping, such as glass, aramids, and carbon fibers. Their mechanical properties, including tensile modulus, tensile strength, and elongation, are shown in Table 1.2. Considering the high pressure inside the hydrogen storage tank, carbon fiber is more suitable for wrapping. The epoxy resin is the most used material in the wrapping process for its good mechanical properties and good stability.

Considering the hydrogen embrittlement and permeation problems, the choice of the liner material requires much attention. The hydrogen embrittlement problem is related to metals. The hydrogen atoms can diffuse into some metal lattice and cause the loss of ductility and strength [12]. This increases the risk of hydrogen tanks bursting. After many accidents, a great deal of effort has been devoted to hydrogen embrittlement research, and quite a few mitigation measures have been developed.

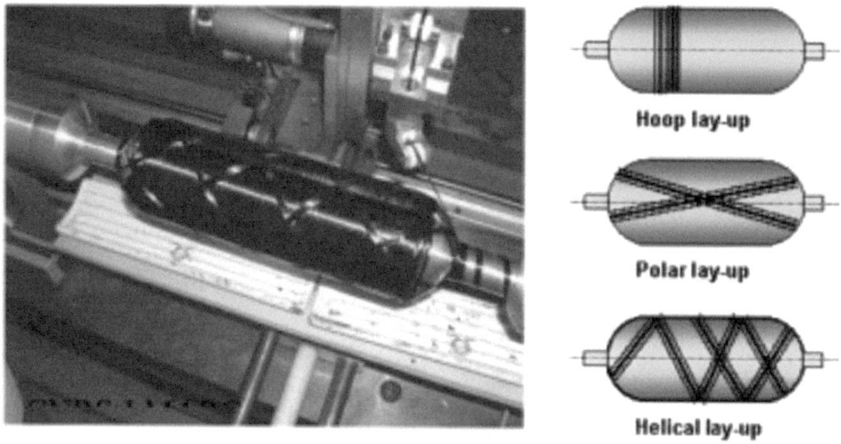

Fig. 1.3 Winding machine and the 3 winding possibilities [11]

Table 1.2 Range of fiber mechanical properties [11]

Fiber category	Tensile modulus (GPa)	Tensile strength (MPa)	Elongation (%)
Glass	~ 70–90	~ 3300–4800	~ 5
Aramid	~ 40–200	~ 3500	~ 1–9
Carbon	~ 230–600	~ 3500–6500	~ 0.7–2.2

Austenitic stainless steel is less susceptible to hydrogen embrittlement and is now widely used to produce hydrogen-related tanks. However, hydrogen embrittlement still occurs after a period of use. Zinc and nickel plating are found to be an effective way to create a barrier for hydrogen diffusion and can mitigate hydrogen embrittlement. Other prevention methods can be found in Ref. [13]. The permeation problem is related to polymers. Hence, it is a specific problem of type V4 vessels. It has been identified at the beginning and is an inherent problem for polymers in contact with gas. In addition, because hydrogen is a very small molecule, the permeation will be enhanced. Now, low-permeable polymers are still wanted for hydrogen storage.

Compared to other hydrogen storage methods, high-pressure vessels are simple to build and use. So, they become popular for hydrogen storage, especially for portable applications. Many manufacturers, including Ford, General Motors, Daimler-Chrysler, BMW, and Toyota, have long used high-pressure vessels in cars and buses [14]. In stationary applications, high-pressure hydrogen tanks also play an important role. The world's first 77 MPa multifunctional layered stationary hydrogen storage vessel with a volume of 2.5 m^3, shown in Fig. 1.4, has been put into operation in the demonstration hydrogen refueling station in China [15]. The inside diameter of the vessel is 700 mm, and the wall thickness is 200 mm, which makes the vessel more robust and damage-tolerant. Also, an online safety monitoring system is provided to ensure safety.

The high-pressure vessels are suitable for transporting hydrogen from the origin to end users or hydrogen refueling stations [16]. A long tube trailer can carry several high-pressure hydrogen tanks. In 2006, the pressure was usually between 16 and 20 MPa, and hydrogen delivered for a single trip was no more than 280 kg. Then, with the technological progress of high-pressure hydrogen storage vessels and the increase in pressure, hydrogen can be transported up to 800 kg in a single trip [16]. A large number of domestic companies are now entering this field to produce hydrogen storage vessels. It can be expected that the technological progress of hydrogen storage tanks will further reduce the cost of hydrogen energy storage and transportation, promoting the development of the hydrogen energy industry.

Except for the hydrogen transportation mentioned above, it is commercial practice in the chemical industries today to transport large quantities of gaseous hydrogen ($\leq$ 100 million scf/day) over long distances ($\geq$ 100 miles) at high pressures ($\leq$ 1500 psi) in pipelines specifically designed for hydrogen [17]. The cost of a pipeline designed for hydrogen is higher than that of natural gas lines because special attention would be needed. Besides, more power is needed to compress hydrogen than natural gas. But once the construction is complete, the subsequent cost of transporting large

Fig. 1.4 Multifunctional layered stationary hydrogen storage vessels [16]

quantities of hydrogen will be very low. Therefore, this is an attractive way for hydrogen transport in the future.

1.2.1.2 Liquid Hydrogen

The hydrogen density of compressed hydrogen storage largely depends on the hydrogen storage pressure. When the pressure increases to 69 MPa, the hydrogen density can increase to about 39 kg/m^3. Limited by the mechanical properties of the material, it is difficult to further increase the pressure that the high-pressure hydrogen storage tank can bear in a short time, which indicates that it is difficult to further increase the energy density of compressed hydrogen storage. Hence, it is necessary to try other strategies for hydrogen storage.

When hydrogen is cooled to − 253 °C, it is liquefied and can be stored in an adiabatic liquid hydrogen tank. In that condition, the hydrogen storage density can reach 70.6 kg/m^3, which is almost 848 times that of compressed hydrogen. So, it's more efficient than compressed hydrogen. In addition, liquid hydrogen tanks are usually not subjected to internal pressure and are only used to hold cryogenic liquids [18]. This means its energy density is not constrained by the development of high-strength materials. Table 1.3 compares several representative hydrogen storage technologies. Liquid hydrogen and ammonia are considered promising storage methods because of their high hydrogen storage density. And liquid hydrogen is more advantageous when high-purity hydrogen is required [19].

Table 1.3 Brief comparison of representative hydrogen storage technologies [20]

Properties	Compressed hydrogen	Metal hydride	Liquid hydrogen	Liquid ammonia
Density (kg/m^3)	39 (69 MPa, 25 °C)	1450	70.9 (1 atm, – 253 °C)	682 (1 atm, – 33.33 °C)
Boiling point (°C)	– 253	–	– 253	– 33.33 °C
Gravimetric hydrogen density (wt%)	100	7.1	100	17.8
Volumetric hydrogen density (kg-H_2/m^3)	42.2	–	70.9	120.3
Hydrogen release temp (°C)	–	250	– 253	350–900
Regeneration temp (°C)	–	–	–	400–600
Enthalpy change to release hydrogen (kJ/mol-H_2)	–	118	0.899	30.6

When talking about liquid hydrogen, the discussion relating to hydrogen isomers becomes very important. Hydrogen has two different spin isomers that can coexist, which were first observed in 1912 during experiments at low temperatures. Subsequently, they are proven to be ortho-hydrogen and para-hydrogen. Ortho-hydrogen is defined as a condition in which two protons in the nuclei of two hydrogen atoms rotate in the same direction. In contrast, para-hydrogen is defined as a condition in which the two protons in the nuclei of hydrogen atoms rotate in opposite directions. They have the same chemical properties and slightly different physical properties [21, 22]. At room temperature, normal hydrogen contains 75% ortho-hydrogen and 25% para-hydrogen [3]. As the temperature cools down, ortho-hydrogen is gradually converted to para-hydrogen. The equilibrium fraction of two forms is highly dependent on temperature and is shown in Fig. 1.5.

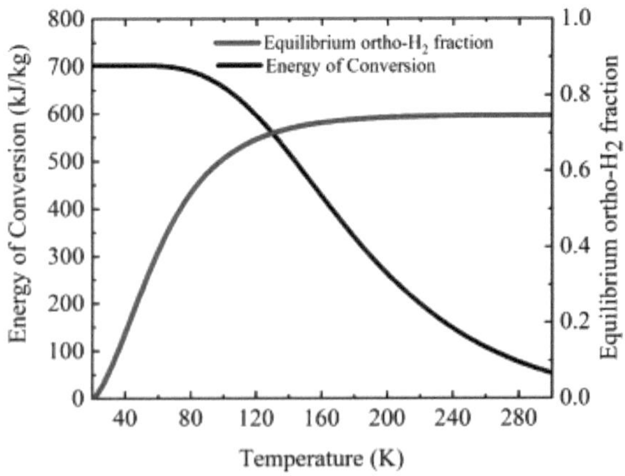

Fig. 1.5 Equilibrium fraction of ortho-hydrogen and energy of para-to-ortho hydrogen conversion [23]

Considering that para-hydrogen has a lower energy level than ortho-hydrogen, when gaseous hydrogen is liquefied at low temperature, the conversion of the two forms releases energy. The complete conversion of ortho-hydrogen yields 703 kJ/kg of heat. Based on this, it can be calculated that the conversion of normal hydrogen generates heat of 527 kJ/kg. However, the latent heat of vaporization of liquid hydrogen is only 446 kJ/kg, and the conversion of ortho-hydrogen is too slow. The conversion can happen weeks after liquefaction. Hence, boil-off may occur during storage, which leads to lower storage efficiency. This will be further described later.

A variety of hydrogen liquefaction technologies have been developed. Among them, the Linde–Hampson process is considered to be the oldest and simplest liquefaction process [24]. In this process, hydrogen gas is compressed and then cooled by heat exchange. Then, the isentropic Joule–Thomson expansion is carried out through the throttle valve. Since this process relies on the Joule–Thomson effect for liquefaction, high pressure is typically required. In addition, it is worth noticing that hydrogen gas heats up as it expands at room temperature. Hence, it must first be cooled to its conversion temperature (– 73 °C at 1 bar) or lower before expansion. Claude process and Brayton refrigeration process are two famous hydrogen liquefaction processes. They have similar steps and change some parts to improve the performance. The Claude process adds an expander to cool the feed, and the Brayton refrigeration process uses liquid nitrogen and helium to cool the hydrogen stream. A generic hydrogen liquefaction process based on the simple Claude process is shown in Fig. 1.6.

Theoretically, the minimum work required for hydrogen liquefaction is 2.7 kWh/kg [26]. However, the real energy requirement for hydrogen liquefaction is much larger than that. Liquefaction of hydrogen is a very energy-consuming process due to its very low boiling temperature (– 253 °C, 1 atm). In addition, since hydrogen cannot be cooled by throttling processes, including adiabatic and isenthalpic expansion, it requires pre-cooling during liquefaction, for example, by evaporation of liquid nitrogen. This precooling increases the energy consumed during liquefaction [27]. Further efforts are needed to reduce the energy consumption. In addition, the cost of liquefaction requires much attention in the future.

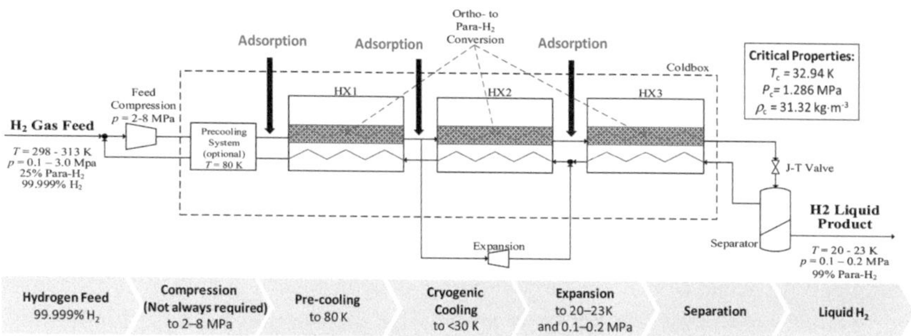

Fig. 1.6 Schematic diagram of the hydrogen liquefaction process based on the simple Claude cycle [25]

Storage of liquid hydrogen is also a great problem. Because of the extremely low temperatures, tanks in direct contact with liquid hydrogen must be designed and manufactured to withstand such low temperatures. Usually, liquid hydrogen containers are made of stainless steel and aluminum. To keep the tank cold, storage tanks for liquid hydrogen usually have two walls. There is a high vacuum between them to minimize heat transfer [28]. To ensure thermal insulation of storage tanks, technologies such as multi-layer [29], spray foam [30], and insulating glass microspheres [31] can also be used. However, the insulation can't be perfect, and liquid hydrogen stored in a warm environment will eventually vaporize. Heat transfer from the environment to the liquid increases the pressure inside the tank. Since tanks are not designed to maintain high pressure, hydrogen is allowed to escape through the safety valve, which is sometimes referred to as "boiling off" [32]. Large cryogenic hydrogen tanks tend to minimize the ratio of surface to volume, which is good for insulation [28]. And the geometry with the largest volume-to-surface area ratio is the sphere. Therefore, many liquid hydrogen storage tanks are spherical.

In addition to the traditional passive insulation technology, active insulation technology is also being applied to liquid hydrogen storage. Active thermal insulation technology is to actively realize heat transfer at the cost of energy consumption. The common method is to use the refrigerator to actively cool the object and balance the external heat leakage, to achieve a higher level of thermal insulation effect. This technology, first proposed by NASA at the end of the twentieth century, is for long-term in-orbit storage of cryogenic propellant for Mars exploration. In 2017, after years of work, NASA completed and tested zero-boiling-off storage (ZBO) technology of liquid hydrogen on the ground. The system is shown in Fig. 1.7. It uses cryogenic helium to cool the hydrogen. The Linde LR1620 closed reverse Brayton refrigeration cycle provides 390 W cooling at 20 K. The system has been tested successfully to achieve zero loss storage of liquid hydrogen.

Compared with passive adiabatic hydrogen storage, active adiabatic storage can achieve better adiabatic effect, lower liquid evaporation rate, and even ZBO storage, but it will increase energy consumption and generate heat. In addition, active adiabatic storage requires additional low-temperature equipment. So, it costs more, and the system is more complicated with low efficiency. Therefore, it is necessary to consider the need for active adiabatic storage technology from a cost point of view. The choice is different for different occasions (space or ground), different working media (liquid oxygen, nitrogen, or liquid hydrogen, helium), and different goals (low or zero evaporation).

For portable applications, passive adiabatic tanks are the main ones. According to the size of liquid hydrogen storage tanks, they can be roughly divided into large station storage tanks, medium transport storage tanks, and vehicle storage tanks. Many countries in the world have developed large liquid hydrogen storage tanks. For example, NASA had a 3200 m^3 large liquid hydrogen spherical tank in 1965. Liquid hydrogen storage tanks for transport are similar in structure to those for stations. China has been able to manufacture 300 m^3 portable liquid hydrogen storage tanks, which can store and transport more than 20 tons of hydrogen at a time. The vehicle's liquid hydrogen storage tank is different from the two mentioned above. It not only

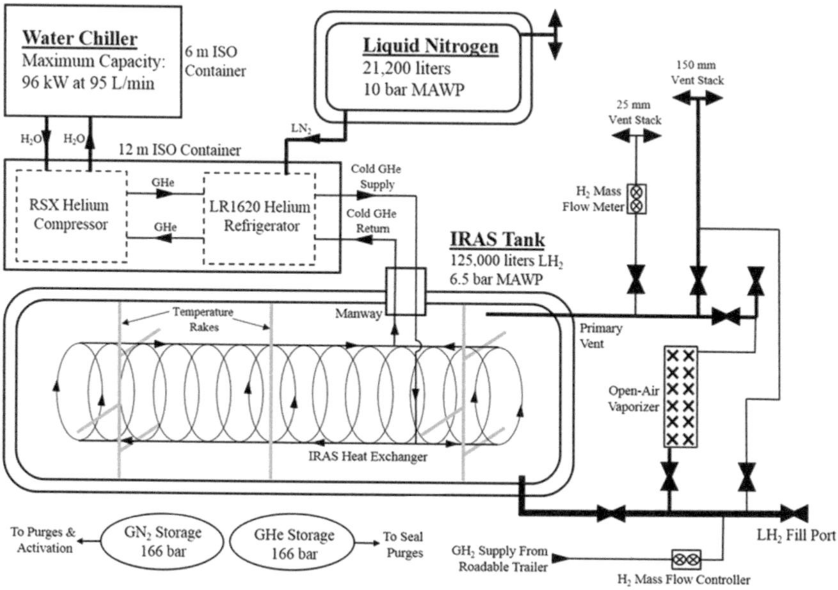

Fig. 1.7 Schematic of ZBO system [33]

has the function of storing liquid hydrogen, but also has the function of vaporizing liquid hydrogen to provide hydrogen to the vehicle. Therefore, the structure of this type of tank is much more complex than that simply storing liquid hydrogen. A typical automotive liquid hydrogen storage tank is shown in Fig. 1.8.

Liquid hydrogen, like compressed hydrogen, can be transported in two main ways. Among them, it is more common to use a truck, locomotive, or ship to transport the liquid hydrogen in a special cryogenic adiabatic tank. The long-distance transport of liquid hydrogen by low-temperature railway tanker is a fast and economical method

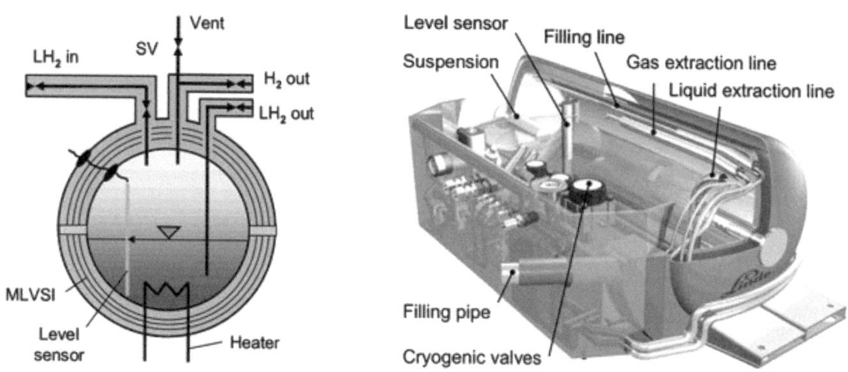

Fig. 1.8 Schematic of vehicle liquid hydrogen storage tank [34]

Fig. 1.9 Liquid hydrogen tanker truck [35]

of hydrogen transport. The cylindrical thermal insulated tank, which is commonly horizontally placed, has a storage capacity of up to 100 m^3 of liquid hydrogen. A special large capacity rail tank can even transport 120–200 m^3 of liquid hydrogen. Figure 1.9 shows the liquid hydrogen tanker truck. In the United States, NASA also builds large barges for transporting liquid hydrogen.

In addition, liquid hydrogen can also be transported through a special liquid hydrogen pipeline, because liquid hydrogen is a kind of low-temperature liquid, its storage containers and transfusion pipelines need a high adiabatic performance. Besides, it will have a certain amount of cold loss, so the adiabatic structure of pipeline containers is more complex. A liquid hydrogen pipeline is generally only suitable for short-distance transportation. At present, the liquid hydrogen pipeline is mainly used at the rocket launch site.

The cost of liquid hydrogen remains high due to the large investment in the liquid hydrogen plant and the high energy consumption in the liquefaction process. Liquid hydrogen is therefore suitable for situations where high energy density is required and evaporation is not a problem, such as commercial aircraft. However, even excluding the volume required for insulation, the volumetric energy density of liquid hydrogen is almost four times lower than that of kerosene [28]. This results in an unacceptably short range for commercial airliners.

1.2.1.3 Cryogenic Compressed Hydrogen

Although the above two hydrogen storage technologies have been developed for many years, there are still some deficiencies. Pressurized hydrogen gas needs a relatively large volume at ambient temperature. Liquid hydrogen continually absorbs heat from the environment and evaporates very easily. So, a new storage technology is needed to reduce the problems. Cold compression of hydrogen combines pressurized gas and liquid hydrogen to capture the advantages of both [36].

Figure 1.10 shows the relationship between temperature, pressure, and hydrogen storage density of the three hydrogen storage technologies. It can be seen that the temperature of cryogenic compressed hydrogen is between liquid hydrogen and compressed gaseous hydrogen, which has a wider range of storage parameters. And it can easily utilize the transport equipment of the above two storage technologies. Cryo-compressed hydrogen is a product of liquid hydrogen, which is compressed by a liquid hydrogen high-pressure pump to a high supercritical pressure. With a proper design of a liquid hydrogen high-pressure pump, the density of cryogenic compressed hydrogen at the outlet can exceed the density of liquid hydrogen storage at the inlet. In addition, both cryogenic compressed hydrogen and liquid hydrogen absorb heat from the surrounding environment when placed over a long period of time, causing the pressure in the tank to rise. The vessel for cryogenic compressed hydrogen can withstand higher pressure compared to the conventional liquid hydrogen vessel, which significantly extends the time before starting evaporation loss [37].

As shown in Fig. 1.11, a typical cryogenic pressure vessel consists of a high-pressure vessel and a vacuum-insulated outer layer so that the cryogenic pressure vessel can withstand internal pressure and have good insulation. Since hydrogen can be stored at low temperatures and high pressures at the same time, low-temperature compressed hydrogen storage can store more hydrogen in the same space as compressed hydrogen storage. And, compared to conventional liquid hydrogen storage, this technology has better thermal resistance. In addition, cryogenic pressure vessels also have some potential safety advantages, for example, the theoretical burst energy of cryogenic hydrogen is lower. Figure 1.12 shows the theoretical

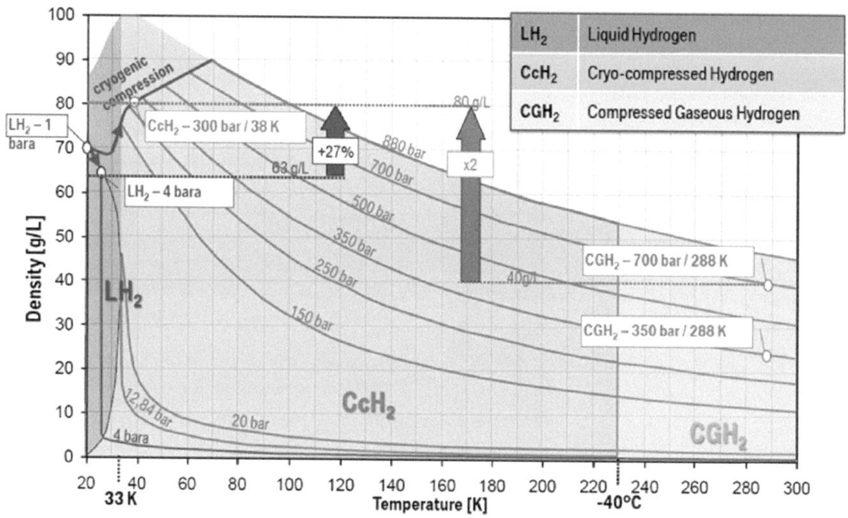

Fig. 1.10 Hydrogen density versus pressure and temperature from BMW [37]

maximum mechanical energy released by the sudden adiabatic expansion of high-pressure hydrogen at different temperatures to atmospheric pressure. Among them, the lower the temperature, the lower the theoretical maximum mechanical energy of hydrogen released [38]. Besides, a vacuum jacket around the high-pressure inner layer provides expansion volume that can mitigate the shock of hydrogen release. And the number and amplitude of pressure peaks in the cryogenic pressure vessel may be lower than those of compressed hydrogen.

The feasibility of cryogenic compressed hydrogen storage technology has been tested. Figure 1.13 shows a Generation 2 cryogenic pressure hydrogen storage vessel

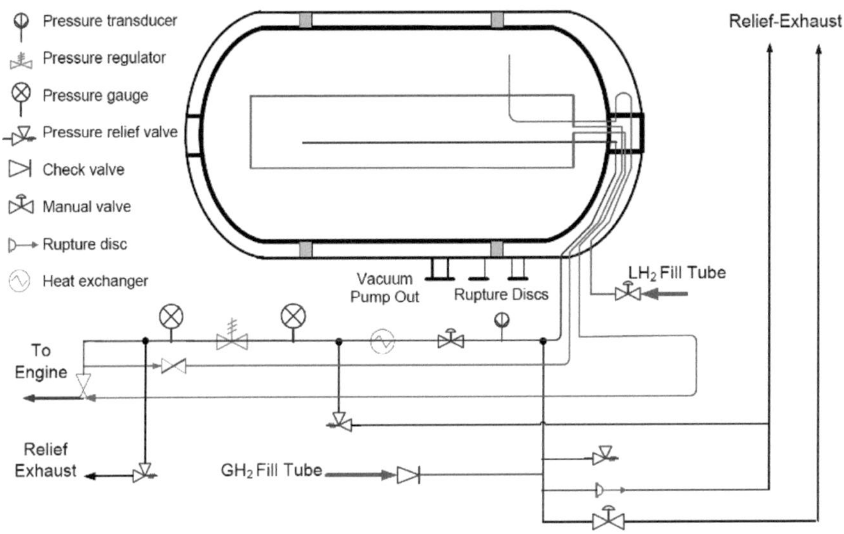

Fig. 1.11 A schematic of the Lawrence Livermore National Laboratory Gen-3 cryo-compressed H_2 storage tank system [28]

Fig. 1.12 Maximum mechanical energy (per kilogram of hydrogen) released upon instantaneous expansion of H_2 gas as a function of initial storage pressure at 80, 150, and 300 K [38]

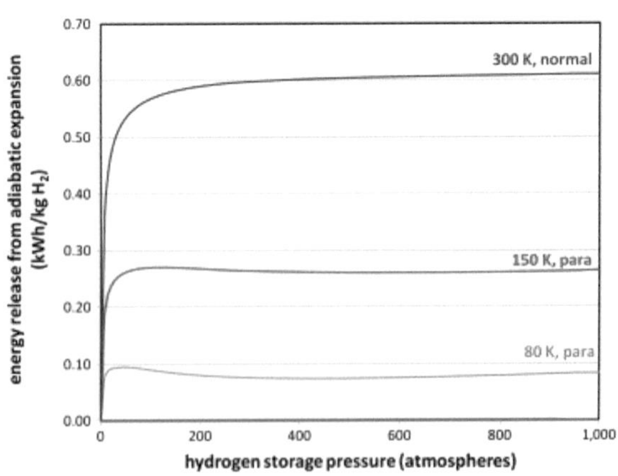

Fig. 1.13 Cryogenic pressure vessel installed onboard a hydrogen-fueled car [39]

designed by LLNL, installed onboard a Toyota Prius. The car then went on a 1050-km test drive. Under typical driving conditions, the cryogenic pressure vessel is expected to provide a driving range of about 800 km.

BMW is also working on cryogenic compressed hydrogen storage. In 2012, BMW unveiled a prototype cryogenic compressed hydrogen vessel with a weight and volume energy density of 5.4 wt% and 4.0 MJ/L, respectively. Its hydrogen evaporation rate is about 3–7 g/h [28]. After a long period of idleness, a large amount of hydrogen fuel remains in the tank [40]. This shows that cryogenic compressed hydrogen storage technology has less evaporation loss than liquid hydrogen storage technology, and it may be widely used in the future for hydrogen storage.

However, the current cost of cryogenic compressed hydrogen storage is still far above the US DOE's target [41], which may hinder its development. A major component of the cost is the composite pressure vessel wrapped by carbon fiber, which must meet the safety and working pressure requirements. Cost is probably the most challenging target for cryogenic compressed hydrogen, requiring future improvements in materials to reduce the cost of high-strength materials.

1.2.1.4 Slush Hydrogen

Slush hydrogen is defined as a cryogenic suspension of supercooled liquid and solid hydrogen mixed at the triple point (– 259.3 °C, 7.042 kPa). Compared with liquid hydrogen, slush hydrogen has a higher density and less fuel loss due to boiling off [42]. When 50% mass fraction of hydrogen is solid, the weight density and heat capacity of hydrogen increase by 15.5% and 18.3% respectively, compared with that of liquid hydrogen at boiling temperature. Besides, early studies on slush hydrogen showed that the loss rate was around 0.1% per day, which is significantly lower than the estimated modern losses of around 1% per day for liquid hydrogen [43]. The lower energy state of the hydrogen slush means that it requires more energy input to vaporize. At present, slush hydrogen is mainly used as space rocket fuel.

There are three main methods to produce slush hydrogen, including the nozzle expansion method, freeze/thaw method, and the auger method [42]. Nozzle expansion method depends on Joule-Thompson expansion to produce slush hydrogen. The liquid gas is transported into a nozzle and expands there [44]. The state transformation proceeds along an isenthalpic line. The freeze/thaw method cools the liquid to the triple-point temperature and produces a frozen layer on the surface of the liquid. Then the solid layer is controlled to melt slightly by adjusting the pressure. Repeating the freeze and thaw cycle numerous times, slush hydrogen is produced with about 50–65% solid [44]. Some researchers can even produce slush hydrogen with 65–85% solid. The Auger method produces slush hydrogen by scraping a frozen layer from a tube refrigerated by liquid helium. If designed appropriately, this method can produce slush hydrogen continuously. In addition, it can produce slush hydrogen at a higher pressure than triple-point pressure, resulting in less energy needed than the freeze–thaw method [42]. Table 1.4 compares some methods of slush hydrogen production.

Compared to liquid hydrogen storage, the low triple point of slush hydrogen becomes a safety problem because the low pressure draws air into the system through unexpected leaks. A mixture of liquid hydrogen and oxygen or air will explode even when shaken. Traction pressure can also bend the tank. The pressure above the triple point can be maintained by using gaseous helium or by layering the liquid layer of slush hydrogen. Although it may seem easy to use helium to maintain pressures above atmospheric pressure in slush storage vessels, the high cost of helium must be considered [42]. And the helium will dissolve into the hydrogen. At 14 K with a total pressure of 100 kPa, the solubility of He in liquid hydrogen is about 0.05%, which is higher than the allowable 0.002% "pollutant" in ultra-pure hydrogen for fuel cell applications [45]. This will increase the downstream utilization cost.

The literatures show that the fundamental understanding of slush hydrogen is well developed, and most of the current research and development focuses on engineering improvements on slush hydrogen systems. These improvements include the recovery of energy from the cryogenic liquid, the improvement of storage times, and efficiency for distributed applications. However, despite the advances in slush hydrogen and

Table 1.4 Summary of methods of slush hydrogen production [42]

Method	Size of particle (mm)	Maximum quality of slush (wt%)	Type of operation	Subcooling method
Freeze–thaw	0.5–0.7	6.5–8.5	Batchwise	Vacuum
Auger	0.1	65–95	Batchwise/continuous	Refrigeration by helium
Nozzle expansion	Less than 0.5	Not specified	Continuous	Joule–Thomson expansion
Helium injection	Not specified	55–70	Batchwise	Direct cooling

its applications in the space industry, there is no evidence of significant current industry interest in this technology, and it is rarely included in any techno-economic assessment of hydrogen transport technologies. This may be because slush hydrogen is too expensive, and other storage methods appear to be more achievable.

1.2.2 Liquid Carriers for Hydrogen Storage

1.2.2.1 Organic Hydrides

Compressed and cryogenic hydrogen storage requires a specially designed storage system to ensure safety, which increases the system cost. If hydrogen can be stored and transported cheaply, it will be more conducive to the promotion and application of hydrogen. Organic liquid hydrogen storage technology converts hydrogen into organic hydride for storage. As the converted organic hydride is safe, stable, and can easily utilize the existing storage and transportation system, the cost of the hydrogen storage and transportation system is greatly reduced. This advantage has attracted the attention of many researchers. The hydrogen storage process in liquid organic compounds is shown in Fig. 1.14. Hydrogen is produced and then used to hydrogenate a hydrogen-poor molecule, creating a hydrogen-rich molecule in which the hydrogen can be stored safely and stably. Then the organic hydride can be transported in tanks. The transport process can be done using existing chemical transport systems and does not need to be specifically designed. Last, at the destination, the organic hydride releases the hydrogen for use by dehydrogenation. This technology mainly uses some unsaturated organic matter to carry out reversible hydrogenation and dehydrogenation reactions with hydrogen to realize hydrogen storage.

Dehydrogenation reactions have been studied for a long time, but producing hydrogen was not the main goal. It wasn't until 1975 that a technical report from Exxon studies first proposed the idea of storing hydrogen in organic liquid molecules [47]. The idea is now widely accepted by the scientific community and has been extensively studied. Many organic hydrides have been recommended for hydrogen storage and transportation. Several common liquid organic hydrides will be briefly introduced below.

1. Formic Acid

Formic acid (FA, HCOOH) is a safe and convenient liquid hydrogen carrier with non-toxicity and excellent stability. It is suitable for transportation, handling, and storage, and is comparable to the diesel and gasoline used today. Formic acid contains 4.4% wt hydrogen and has a volume capacity of 53 g/L [48], which is higher than compressed hydrogen. It can be produced by hydrogenating carbon dioxide to reduce carbon dioxide concentrations in the environment. And it can also be broken down to produce hydrogen and carbon dioxide, as shown in Fig. 1.15. However, undesirable side effects may occur during formic acid formation and decomposition (Eq. 1.2), so

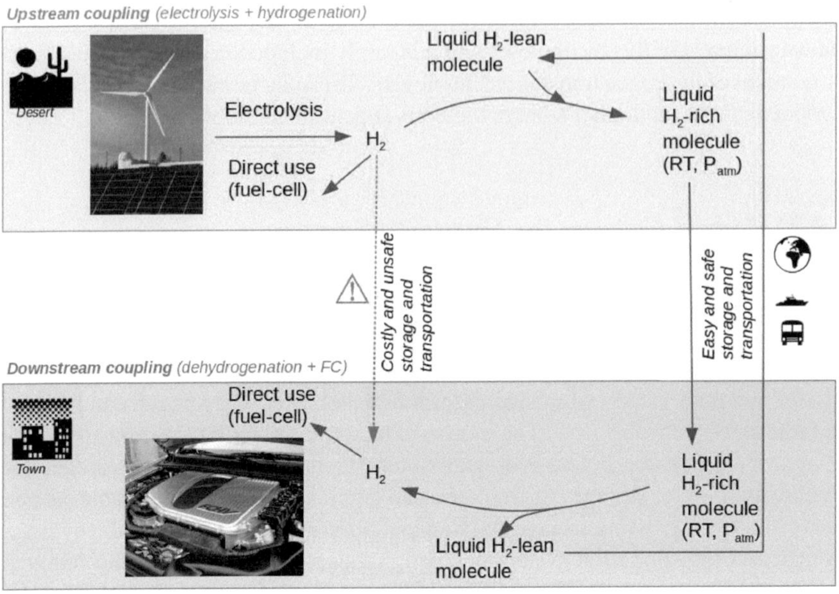

Fig. 1.14 General picture illustrating the hydrogen storage in liquid organic molecules [46]

the reactions that occur need to be strictly controlled. Hydrogen fixation and release should be supported by a suitable catalyst, which can reduce the activation energy and increase the reaction rate. On the other hand, catalysts can enhance the selectivity of the reaction and thus reduce the undesirable side reactions. Hence, the catalysts are the key to successful commercialization.

$$HCOOH \leftrightarrow H_2 + CO_2 \tag{1.1}$$

$$HCOOH \leftrightarrow H_2O + CO \tag{1.2}$$

Hydrogenation of carbon dioxide to formic acid has been extensively investigated, especially through the use of homogeneous catalysts. The first homogeneous approach was published in 1976, using a ruthenium phosphine complex as a catalyst [50]. Many transition metal complexes are then used in this section. The Ru, Rh, Ir, Pd, Ni, Fe, Ti, and Mo-based complexes show excellent activities. Some homogeneous catalyst reviews can be found in Refs. [51, 52]. Table 1.5 shows homogeneous catalysts for carbon dioxide reduction. To promote formic acid formation, a base is needed in the reaction system. Because it can shift the reaction balance to the side where formic acid is formed. Fortunately, it is not necessary to remove the base [48]. The adduct of formic acid and base allows the straightforward decomposition to hydrogen. Therefore, many catalysts are designed for use in alkaline aqueous solutions. For example, the rhodium–phosphane complex, $[RhCl(TPPTS)_3]$,

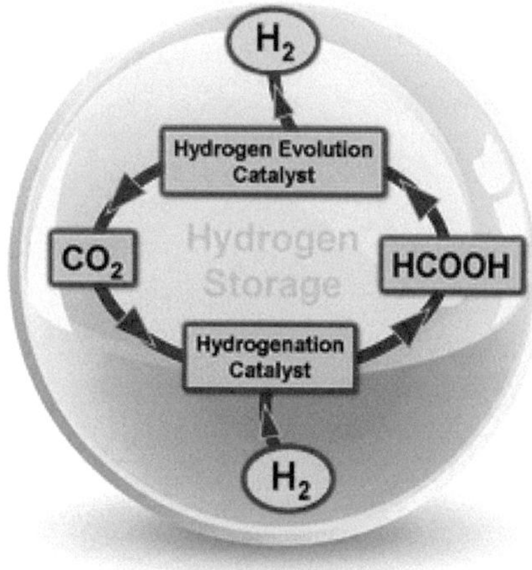

Fig. 1.15 Schematic of a cycle for hydrogen storage and utilization using formic acid [49]

is highly active for the hydrogenation of CO_2 in the water-amine mixture [53]. There is also the complex catalyst, $[RuCl_2(TPPMS)_2]_2$ (TPPMS = triphenylphosphine monosulfonate), used in sodium bicarbonate solution, where remarkable activity with a turnover frequency (TOF) of 9600 h^{-1} was observed [54]. However, the concentration of carbon dioxide and hydrogen in aqueous solution restricts the reaction rate. Therefore, applying supercritical conditions to improve the reaction outcome may be a good idea. Using a supercritical phase results in much higher reaction rates than those obtained under similar conditions in organic solvents. For example, $[RuCl(HCOO)(PMe_3)_4]$ achieves very high TOF values of 95,000 h^{-1} under supercritical conditions [55]. In addition to the homogeneous catalysts mentioned above, some catalysts capable of both hydrogenation and dehydrogenation have also received the attention of researchers. $[(Cp^*IrCl)_2(thbpym)]^{2+}$ can control the hydrogen consumption or production in the system as the PH value of the solution changes [48]. At ambient temperature and pressure, $[Ir(Cp^*)(pba)(H_2O)]^+$ can also achieve similar interconversion between hydrogen and formic acid for both directions, again depending on PH value [56]. In addition, some catalysts depend on temperature or pressure to switch the direction of the reaction [57]. The high catalytic activities of most catalysts mentioned above can only be achieved in an alkaline environment. However, the presence of a base complicates the separation of formic acid from the reaction mixture and reduces the hydrogen capacity for reversible hydrogen storage. Therefore, catalysts that can directly reduce CO_2 without a base will be more attractive and sustainable. The homogeneous catalyst $[RuCl_2(PTA)]$ (PTA = 1,3,5-Triaza-7-Phosphaadamantane) is a good choice. Using this catalyst, formic acid formation can be catalyzed in both water and DMSO solvents, and the catalyst can be reused many times without decreasing activity [58].

Compared with the extensive research on homogeneous catalytic hydrogenation of CO_2, the use of heterogeneous catalysts for CO_2 hydrogenation is still underdeveloped. Although the first attempt to synthesize formic acid from CO_2 using

Table 1.5 Some homogeneous catalysts for the reduction of CO_2 into formic acid [48]

Catalyst	Additives	Temp (°C)	TOF (h^{-1})
$[RhCl(TPPTS)_3]$	$NHMe_2$	81	7260
[Rh(hfacac)(dcpb)]	NEt_3	25	1335
$[RuCl_2(TPPMS)_2]_2$	$NaHCO_3$	80	9600
$[Cp^+Ir(dhbp)Cl]^+$	KOH	120	42,000
$[RuH_2(PMe_3)_4]$	NEt_3	50	1400
$[RuCl(OAc)(PMe_3)_4]$	NEt_3, C_6F_5OH	50	95,000
$[(Cp^+IrCl)_2(thbpym)]^{2+}$	HCOONa	80	53,800
$[(PNP)IrH_3]$	KOH	120	73,000
[(PNP)RuHCl(CO)]	DBU	120	$1.1 * 10^6$

heterogeneous catalysts was reported as early as 1935 [59]. The possible reasons for this are the unfavorable reaction conditions (high temperature and high pressure) compared to homogeneous catalysis and the low chemical selectivity. For example, in the presence of iron oxide and iron, the hydrogenation of carbon dioxide over nickel mainly forms methane [49]. Formic acid was observed only as an intermediate. The addition of base can make formic acid become the main product and methanol become the by-product. However, in both cases, catalytic activity was very low (TOFs < 1 h^{-1}). Besides, due to the chemical balance between carbonate and formic acid, the reaction could not proceed completely. The solubility of carbonate decreases with the increase of formate concentration. In addition, the products are mixed with the reaction, requiring additional separation steps. Inspired by the high activities of homogeneous catalysts, the immobilization of homogeneous catalysts on solid supports was also studied. At the same time, some scholars have used photochemical methods to decompose formic acid [60].

The catalysts for methanol decomposition can also be divided into homogeneous catalysis and heterogeneous catalysis. The activation of heterogeneous catalysts usually requires high temperature, while homogeneous catalysts can operate at lower temperatures, so homogeneous catalysts cause less CO pollution. This is important for fuel cell applications. Early studies on the homogeneous catalyst were focused on soluble Pt-, Rh-, Ir-, and Rh-based molecular catalysts [61]. But their catalytic activity and durability are low over a long period. By around 2008, research on homogeneous catalysts became hot as highly selective homogeneous catalysts under mild conditions were reported [62]. $[RuCl_2(PPh_3)_3]$ is an earlier-discovered catalyst that can operate at low temperatures and has excellent selectivity. The hydrogen it produces does not contain CO and can be used directly to generate electricity in fuel cells [63]. Another catalyst $[Ru(H_2O)_6]$ $(tos)_2$ (tos = toluene-4-sulfonate) has also been reported to achieve a conversion rate of over 90% at room temperature [64]. In 2014, Ru complexes with a PNP-pincer ligand were able to achieve TOF up to 257,000 h^{-1} and TON up to 706,500 without producing CO byproducts [62]. Besides

Table 1.6 Some homogeneous catalysts for the decomposition of formic acid [48]

Catalyst	Temp (°C)	TOF (h^{-1})
$RuBr_3 \cdot xH_2O/3PPh_3$	40	3630
$RuCl_3 \cdot xH_2O$/TPPTS	120	670
$Fe(BF_4)_2 \cdot 6H_2O/2PP_3$	80	5390
$[RuCl_2(PPh_3)_3]$	40	2688
$[(Cp^*IrCl)_2(thbpym)]^{2+}$	90	228,000

Ru-based catalysts, Ir-based catalysts were also found to have good catalytic performance. The complex [Cp*Ir(L)Cl] (HL = N-(quinolin-8-yl)benzenesulfonamide) is reported to have a high TOF of $2.97 \times 10^4\ h^{-1}$ and a good stability (> 100 h) at 60 °C [65].

The above-mentioned catalysts all use noble metal as the center metal for high activity. This increases the commercialization cost to some extent, which is not conducive to large-scale development. So, cheap homogeneous catalysts have gained attention. $Fe(BF_4)_2 \cdot 6H_2O$ is a non-noble metal homogeneous catalyst. It can convert all of the FA to generate H_2 and CO_2 without the CO byproduct (< 1 ppm) at 40 °C. A high TOF (9425 h^{-1}) was obtained at 80 °C, with a low catalyst loading (0.005 mol%) [66]. Nickel can also be used as a catalyst for formic acid decomposition. $Ni(PMe_3)_4$ shows catalytic activity, and its TON reached 80. Recently, Mn has also begun to be used as a catalyst. The TOF of $[(tBuPNNOP)Mn(CO)_2]Br$ can reach 8500 h^{-1} at 80 °C. In addition to the above-mentioned substances, many other catalysts have been found to have good catalytic capacity for the decomposition of formic acid [67–69]. The catalytic mechanisms are also being investigated [70]. Some homogeneous catalysts are summarized in Table 1.6.

Although the homogeneous catalysts have made remarkable progress in the field of formic acid dehydrogenation, they still have many problems, such as high cost, limited stability, and difficulties in separation and recovery from the reaction mixture. These problems limit the large-scale application of homogeneous catalysts. Heterogeneous catalysts can overcome the defects of homogeneous catalysts, so they are considered to be commercially viable. Since the 1930s, the decomposition of formic acid in the presence of heterogeneous catalysts has been studied, which is earlier than homogeneous catalysts. Many heterogeneous systems have been studied in detail until now, but most reactions occur in the gas phase and require heating to more than 100 °C [71]. Therefore, it can't supply hydrogen for a fuel cell directly. It is necessary to develop a heterogeneous catalyst suitable for the decomposition of liquid formic acid. For heterogeneous catalysts, the geometric structure and local coordinating environment have a great influence on catalyst performance. So, three important factors should be noted in the study: (1) the geometric structure of the metal active site, (2) interactions between the active sites and the support material, and (3) synergistic catalytic effects. The related mechanism study of heterogeneous catalysts can be referred to in the literature [72]. Silver, gold, palladium, and platinum can be used for formic acid dehydrogenation. For example, Pd/N-MSC-30 can achieve a TOF of 8415 h^{-1} at 55 °C. Au/ZrO_2 can also achieve a TOF of 1590 h^{-1}

Table 1.7 Some heterogeneous catalysts for the decomposition of aqueous formic acid

Catalyst	Temp (°C)	TOF (h^{-1})
Pd/C	25	64
PdAu/C–CeO_2	92	227
PdAuEu/C	92	387
AgPd/C	50	382
Pd/MSC-30	50	2623
AgPd/rGO	25	105

at 50 °C [73]. Until recently, most heterogeneous catalysts were composed of noble metals. Some attempts have also been made to incorporate non-noble metals into catalysts to reduce costs. The tri-metallic catalyst $Co_{0.30}Au_{0.35}Pd_{0.35}/C$ achieves 91% conversion at room temperature [74], which reduces the consumption of noble metals and improves the commercialization potential. In addition, as mentioned above, the supported material also has an important impact on the performance of the catalyst. For example, amine-functionalized Au@SiO_2 shows high activity and 100% selectivity for the reaction. However, unsupported or silica-supported Au NPs have no catalytic activity [75]. Some heterogeneous catalysts are summarized in Table 1.7.

Formic acid is a very potential hydrogen storage carrier. However, the current catalyst is still not enough to meet the commercial needs. More research on heterogeneous catalysts under mild conditions is needed in the future.

2. Methanol

Methanol (CH_3OH, MeOH) is the simplest alcohol and can be produced from biomass or by the reduction of CO_2 in the presence of appropriate catalysts. It contains 12.6 wt% hydrogen, and its volumetric hydrogen capacity is higher than that of cryogenic hydrogen, which makes it a convenient material to store hydrogen. To date, many reactions for methanol generation have been proposed. The most common method is produced from syngas in fossil fuels reforming, resulting in a product mixed with CO, H_2, and a small amount of CO_2 [76]. Cu/ZnO/Al_2O_3 catalysts are found to be highly active for this reaction, which can operate in a range of 50–100 atm and has a lifetime up to four years. In addition, much progress has been made in the direct conversion of methane into methanol. Methane, CO_2, and steam could be converted exclusively into a mixture of CO and H_2 at temperatures of 800–1000 °C and pressures of 5–40 bar over a nickel-based catalyst [76]. This method has also been utilized in practice. Recently, formic acid hydrogenation by a homogeneous catalyst has been reported as a novel methanol production method. But the methanol yield with the use of $[Cp{*}Ir(bpy)(H_2O)]^{2+}$ (Ir-18) in water was only up to 1.9% [77]. This approach is still at the laboratory stage, although there have been several related studies.

Among all the reactions, the direct hydrogenation of CO_2 to methanol is the most straightforward and desirable process (Eq. 1.6). The key to direct hydrogenation of CO_2 is to develop efficient catalysts. Fortunately, this method has been well developed because the CO_2 hydrogenation process is not radically different from the process based on natural gas. The similar Cu/ZnO-based catalysts can be used in both processes with high catalytic performance. Hence, many high-performance catalysts have already been developed and make this method a commercialized technology [78]. In 2011, the first plant operated by this reaction was put into operation in Iceland, and a production of 1600 tonnes per year has been achieved using renewable energy. Recently, a new class of NiGa catalysts was reported to have similar or better activity than conventional Cu/ZnO-based catalysts. They can work at low pressure, which is important for small-scale production. By the continuous development of catalysts, the direct hydrogenation of CO_2 will be more efficient and practical.

$$CO + 2H_2 \rightarrow CH_3OH \tag{1.3}$$

$$CH_4 + \frac{1}{2}O_2 \rightarrow CH_3OH \tag{1.4}$$

$$3CH_4 + CO_2 + 2H_2O \rightarrow 4CO + 8H_2 + 4CH_3OH \tag{1.5}$$

$$CO_2 + 3H_2 \rightarrow CH_3OH + H_2O \tag{1.6}$$

$$HCOOH + H_2 \rightarrow CH_3OH + H_2O \tag{1.7}$$

Converting methanol into hydrogen is another key point for a "methanol–hydrogen economy". This can be realized in many ways: stream reforming, partial oxidation, and methanol thermolysis [6]. Among them, methanol steam reforming [Eq. (1.2)] is the most attractive decomposition method, which has achieved steady progress in recent years [79–83]. Because this reaction causes the release of three moles of hydrogen per mole of methanol, higher than other reactions. Cu/ZnO/Al_2O_3 can also be applied in this reaction as a high-performance catalyst. External heating is needed because the reaction and evaporation of water and methanol are endothermic. Therefore, it requires high temperatures (200–300 °C) and high pressures (25–50 bar) to promote the reaction. The ideal way to use methanol as a hydrogen carrier is under mild conditions with high selectivity. Fortunately, an aqueous-phase methanol reforming process at temperatures below 100 °C was published in 2013, and several important advances have been achieved since that [84]. $[Fe(H)(X)(CO)(HN(C_2H_4PiPr_2)_2)]$ ($X = HBH_3$ or Br) can promote the aqueous-phase methanol reforming at 91 °C, with a TOF up to 702 h^{-1} and TON up to 10,000 [48]. Nevertheless, the catalytic performance is still low, and alkaline conditions are needed.

$$CH_3OH + H_2O \rightarrow CO_2 + 3H_2, \Delta H = -131\ \text{kJ/mol} \tag{1.8}$$

In summary, the catalytic activities and stabilities still need to be further improved for practical applications. Highly active catalysts under mild conditions will bring us closer to a “methanol–hydrogen economy”.

3. Cycloalkanes

Cycloalkanes are different from the above hydrogen storage materials because of their liquid state in all reactions. Their dehydrogenation and hydrogenation processes are reversible [85] and result in only gaseous hydrogen. Hence, there is no need to capture or store CO_2, and the produced hydrogen is relatively pure. These special reaction processes are mainly achieved by saturation and unsaturation of carbon–carbon bonds. These bonds should be stable during repeated hydrogenation and dehydrogenation. Ideal materials for hydrogen storage should also have low toxicity and cost. After these years development, cyclohexane, [86] methylcyclohexane, [87] decalin, [88] and bicyclohexyl, [89] are found to be potential candidates containing 6–8 wt% hydrogen and volumetric capacities of 60–62 g L^{-1}. In addition, because of their high boiling points, cycloalkanes are convenient for storing and transporting.

The hydrogenation process for cycloalkanes is well-established and has already been used in the petroleum industry. Many works have been made on the hydrogenation at ambient temperatures and pressures [90, 91]. Noble metals, such as Pt and Pd, display excellent catalytic performance for the hydrogenation reactions [92]. And nickel-based catalysts have also been studied to reduce costs [93].

For the dehydrogenation step, noble metals, particularly Pt metallic catalysts, also show high catalytic activities [87]. The addition of a second metal, such as W, Ir, Rh, Re, can further promote the reactions due to the electronic effect and the synergistic effect [48]. In addition, the nickel-based catalysts are also studied for their comparable performance and low cost [94]. However, the hydrogen release step requires a large enthalpy, and high temperatures are needed for the thermodynamics of the dehydrogenation reactions [85]. To maintain the hydrogen release temperatures in practical applications, a fraction of the hydrogen needs to be burned, thus reducing the real hydrogen storage capacities. Moreover, cycloalkanes may be gaseous at reaction conditions, and a separation step is required to yield a pure hydrogen stream. Besides, the harsh condition is a problem for the long-term stability of the catalysts. Coking, dealkylation, and catalyst deactivation must be taken into account in catalyst development.

Much progress has been achieved in using cycloalkanes for liquid hydrogen storage. However, the widespread use of cycloalkanes is still impeded by the lack of stable and high-performance catalysts. Some methods, such as the modification of the metal surface and the reduction of the reaction temperature, are adopted to prevent the formation of coke. Further development of efficient catalysts for the dehydrogenation process is still needed.

4. Nitrogen-substituted heterocycles

An organic heterocyclic hydrogen storage method was first reported in 2006 [48]. The related studies suggested that incorporation of N atoms into the rings could reduce the dehydrogenation enthalpy of the compound, which results in a lower temperature for hydrogen release [95]. Therefore, compared to cycloalkanes, nitrogen-substituted heterocycles consume less energy, making them superior candidates for hydrogen storage [96]. N-ethylcarbazole and dodecahydro-N-ethylcarbazole are found to be more effective than many other heterocycles for reversible hydrogen storage [97].

Dodecahydro-N-ethylcarbazole is liquid at room temperature with a boiling point higher than 200 °C. Theoretical gravimetric capacity and volumetric capacity are 5.8 wt% and 54 g L^{-1}, respectively. During repeated hydrogenation and dehydrogenation reactions, dodecahydro-N-ethylcarbazole is found to be stable as a hydrogen carrier. Its hydrogenation and dehydrogenation steps can be catalyzed by noble metal catalysts, such as Pd and Ir. However, the hydrogenation gives not only the completely dehydrogenated product of N-ethylcarbazole but also by-products because of low selectivity. Side products of octa- and tetrahydro-N-ethylcarbazole can also be found after the dehydrogenation reaction. Therefore, the amount of H_2 recovered from the dodecahydro-N-ethylcarbazole was lower than the theoretical yield.

N-ethylcarbazole is another important material for the future implementation of reversible hydrogen storage. It can also be catalyzed by noble metals, such as Pd, Ru, and Rh, and the catalytic performance is greatly influenced by the choice of both the metal and support. A commercial Ru/Al_2O_3 catalyst shows a selectivity of 97% for the hydrogenation reaction. Among noble metals, Ru was found to be the most active metal based on experimental and theoretical studies. However, it may suffer from a relatively low selectivity due to the weak adsorption and poor surface diffusion. In contrast, Rh catalysts with a moderate activity may show a higher selectivity under similar conditions (Fig. 1.16).

Compared to cycloalkanes, nitrogen-substituted heterocycles have a lower enthalpy of the dehydrogenation reactions and are more favorable as reversible hydrogen carriers for hydrogen storage. These materials can store hydrogen under moderate conditions. However, there are still some problems within this system. First is that the melting point of N-ethylcarbazole is 68 °C, which means it is solid at ambient temperature, increasing the hydrogen storage system complexity. Second is the requirement for catalysts with high catalytic performance, low cost, and high selectivity. With the continuous progress of related research, we believe that nitrogen-substituted heterocycles will be closer and closer to the day of practical use.

1.2.2.2 Inorganic Carriers

1. Ammonia

Ammonia is considered to be an attractive hydrogen storage medium because of its high hydrogen storage density (17.7 wt%), low storage pressure, and high stability for long-term storage. Furthermore, it currently plays an important role in agricultural fertilizer, refrigerant gas, and other chemicals. Therefore, its synthesis, handling, and

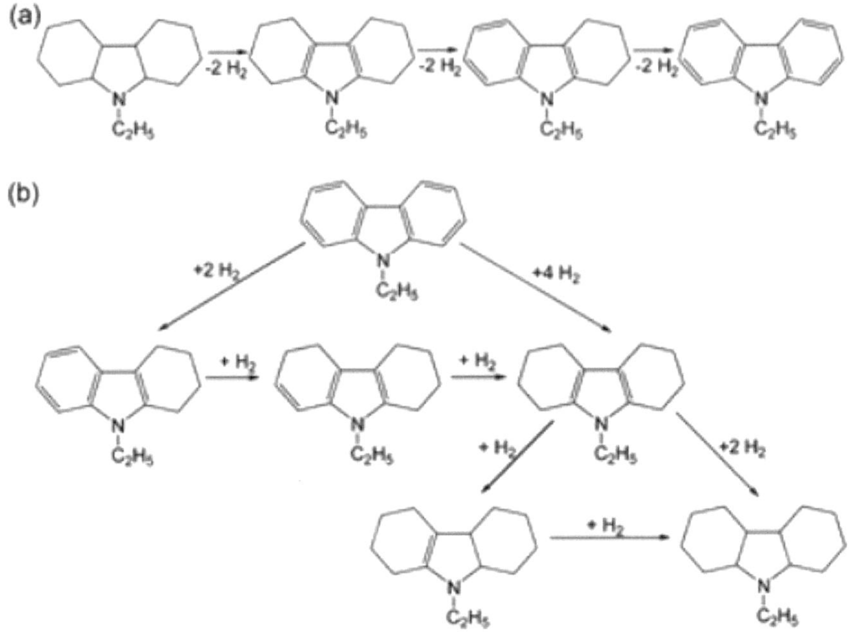

Fig. 1.16 Schematic of dehydrogenation and hydrogenation reaction pathways for dodecahydro-N-ethylcarbazole [48]

transportation is very mature, which reduces the cost for large-scale applications. However, there are still some challenges for ammonia to be a hydrogen carrier.

Table 1.8 shows some physical properties of ammonia. Its boiling temperature is only − 33.4 °C at atmospheric pressure. Therefore, it is a gas under standard conditions, with a relatively strong odor. Its combustion heat is higher than liquid hydrogen, and it can be used as fuel directly. The challenges faced by ammonia combustion include its narrow flammability range, and it is usually considered to be non-flammable during its storage and transportation. In addition, considering nitrogen in ammonia, the combustion of ammonia potentially leads to NO_x formation. Therefore, its combustion management is crucial. Moreover, as ammonia is categorized as a toxic chemical, it is crucial to reduce its negative impact on the environment [98].

Due to its chemical properties, ammonia contains a high volume of hydrogen. Both in the form of gas or liquid, ammonia shows a higher density than hydrogen. Liquid ammonia can store even more hydrogen in volumes than liquid hydrogen (121 kg/m^3 higher than 70.8 kg/m^3), and its density can reach 600 kg/m^3, resulting in easier transportation as a hydrogen carrier. Therefore, ammonia is a promising hydrogen carrier. In the following part, we will pay attention to its conversion technology.

Table 1.8 Physical characteristics of ammonia [98]

Properties	Unit	Value
Molar mass	g/mol	17.031
Density at STP	kg/m^3	0.769
Boiling point	°C	– 33.34
Flammability limit (equivalence ratio)	–	0.63–1.4
Heat of combustion	MJ/L	11.2

There are various methods for producing ammonia, including Haber–Bosch, electrochemical, thermochemical cycle, and plasma conversion technologies [99]. The dominating ammonia synthesis route, which is known as Haber–Bosch process, is invented by Fritz Haber and Carl Bosch about 100 years ago [100]. It accounts for approximately 85% of the total global production of ammonia [101]. The reaction can be expressed as:

$$3H_2 + N_2 \rightarrow 2NH_3, \Delta H_{27\,^\circ C} = -46.35\,kJ/mol \tag{1.9}$$

The reaction is exothermic, which means it can occur naturally at low temperatures. But in fact, the kinetics is too slow to be applicable for practical application at room temperature. To accelerate the reaction, high pressure and temperature are required. Reaction conditions of 300–550 °C and 200–350 bar are typical for effective ammonia synthesis from hydrogen and nitrogen. This condition is required due to the high dissociation energy of triple-bonded nitrogen over iron-based catalysts. About 30 MJ/kg-NH_3 of energy is required to maintain the high temperature and pressure [102], which increases the system costs. Therefore, many efforts have been carried out to reduce the extreme reaction conditions. Ru-based catalysts can catalyze ammonia synthesis at lower conditions (300–450 °C and 4–15 MPa) than iron-based catalysts. But they are too expensive to be used in industrial [103]. The addition of alkaline earth metal oxides and hydroxides can further improve the catalytic performance of Ru-based catalysts [104]. Some research reports that the temperature for ammonia synthesis can be lowered to 200 °C [105]. Transition metals can also enhance the performance of synthesis, due to the existence of scaling relations between the transition-state energy [106]. However, catalyzing the reaction under mild conditions remains to be further investigated to be applicable.

Apart from the methods mentioned above, ammonia can also be synthesized through an electrochemical process. This process is expected to realize higher energy performance, compared to Haber–Bosch process, as it consumes about 20% lower energy than the Haber–Bosch process [107]. Figure 1.17 shows the schematic of the electrochemical process for ammonia synthesis. The process is simple, potentially reducing system configuration and control complexity. Therefore, it may have lower investment costs compared to current ammonia synthesis systems.

The electrochemical reactions on both sides of the cells can be shown as:

$$N_2 + 6H^+ + 6e^- \leftrightarrow 2NH_3 \tag{1.10}$$

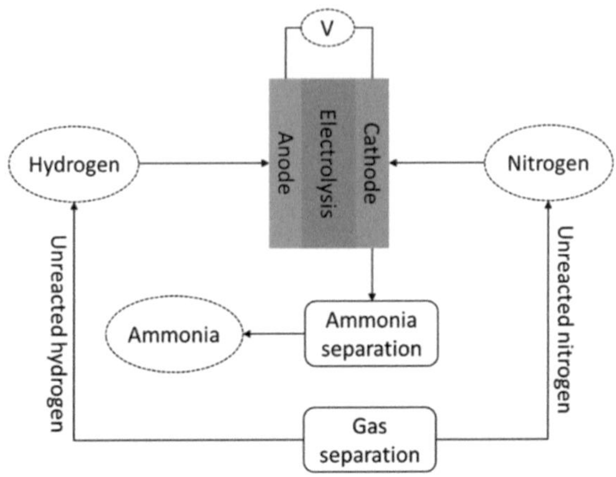

Fig. 1.17 Schematic diagram of electrochemical ammonia synthesis [98]

$$3H_2 \leftrightarrow 6H^+ + 6e^- \tag{1.11}$$

Like a fuel cell, the electrochemical system can be categorized based on the type of electrolyte used. Now, four types of electrolytes are available in ammonia synthesis: liquid electrolytes, molten salt, composite membranes, and solid-state electrolytes. Liquid electrolytes can operate under mild conditions (atmospheric temperature and pressure), but their system efficiency is lower than other types. $LiClO_4$ in H_2SO_4 and Li_2SO_4 in H_2SO_4 are typical liquid electrolytes used in lab experiments. This type of electrolyte is still far from application. Molten salt type electrolyte and composite electrolytes must operate at high temperature, as the salt must melt for ionic conducting. Their efficiency can reach 70%, which is higher than the system with liquid electrolytes. Among these types, solid electrolytes have a very wide operation temperature, depending on the specific materials used in the system. It ranges from room temperature to 800 °C. Please refer to the review for additional details on the electrochemical production of ammonia [108].

Thermochemical cycle, an alternative process for ammonia production, consists of two circulated processes: reduction and steam-hydrolysis. These reactions can be summarized as [109]:

$$Al_2O_3 + 3C + N_2 \rightarrow 2AlN + 3CO,\ \Delta H_{25\,^\circ C} = 708.1\ \text{kJ/mol} \tag{1.12}$$

$$2AlN + 3H_2O \rightarrow Al_2O_3 + 2NH_3,\ \Delta H_{25\,^\circ C} = -274.1\ \text{kJ/mol} \tag{1.13}$$

This thermochemical cycle can be conducted without a catalyst, under atmospheric pressure, and the process enables independent reaction control. However, the system does not require hydrogen for ammonia production, making it an unsuitable hydrogenation process for using ammonia as a hydrogen carrier. Besides, this system operates at a very high temperature, leading to limited materials. Therefore, no details will be given here.

After the ammonia is produced, it is also important to transport the ammonia to where it is needed for hydrogen production. Fortunately, the process of ammonia storage and transportation is relatively well established. As mentioned above, ammonia is relatively safe due to its higher auto-ignition temperature (650 °C) compared to hydrogen (520 °C). Considering its low density under atmospheric conditions, ammonia is generally compressed to a liquid for higher density. After that, the transportation can be performed via pipelines, tank-cars, and tanker vessels, like other chemicals. However, the boiling temperature for ammonia is only − 33 °C. To maintain the liquid form of ammonia during the transportation process, two methods are adopted. The first method is to increase its pressure at atmospheric temperature, and the second method is to decrease the temperature at atmospheric pressure. The methods are useful, and the infrastructure for propane can also be adopted to store liquid ammonia. Therefore, the transportation and storage of ammonia are well established.

Compared to the production, transportation, and storage processes of ammonia, the dehydrogenation process is likely to be the most challenging process of applying ammonia as a large-scale hydrogen carrier. The reverse reaction of ammonia synthesis (Haber–Bosch process) is the most common method of dehydrogenation. The decomposition of ammonia initiates at 200 °C, but complete conversion requires a higher temperature [110]. Ru is the most active catalyst for ammonia decomposition, but it is too expensive [111]. Alternative catalysts, such as Ni and Fe, are less active, which requires a higher temperature for complete conversion [112]. In addition, after ammonia decomposition, hydrogen will be mixed with nitrogen, which must be purified after separation for subsequent utilization. These all reduce the economic benefits of ammonia as a hydrogen carrier.

2. Ammonia borane

Ammonia borane (NH_3BH_3) is a stable solid at room temperature with high theoretical gravimetric and volumetric hydrogen, which are 19.6 wt% and 146 g L^{-1}, respectively [113]. It is an appropriate hydrogen carrier because of its high stability, nontoxicity, and high solubility in common solvents, such as water and methanol. Two methods can release hydrogen from ammonia borane: thermolysis and solvolysis [114]. However, the former method has some drawbacks. Thermal decomposition of ammonia borane occurs in three steps, with almost 6.5 wt% hydrogen released in each step. The complete release of hydrogen requires high temperature (> 500 °C) and a long induction time. Under acceptable conditions, only 13.0 wt% hydrogen can be achieved. Moreover, various by-products, such as ammonia, diborane, and borazine, can also be formed during the thermal decomposition. A lot of research has been done to lower the reaction temperature and reduce by-products, but the results are not good. In contrast, hydrolysis requires much simpler conditions and thus attracts more attention. The reaction can be expressed as:

$$NH_3BH_3 + (2 + x)H_2O \rightarrow NH_4BO_2 \cdot xH_2O + 3H_2 \tag{1.14}$$

As mentioned earlier, ammonia borane is stable in aqueous solution, and its solubility is high. Therefore, the hydrolysis of ammonia borane can occur at ambient temperature with an appreciable rate only in the presence of a suitable catalyst. Considerable effort has been devoted to the development of an efficient catalyst for the hydrolytic dehydrogenation. Noble metal-based catalysts, such as Pt, Ru, and Rh, are reported to achieve the best performances. Recent studies have focused on their particle size and supports. The metal nanoparticles tend to agglomerate during preparation, causing a significant loss in catalytic activity. Using surfactants or ligands in solution or using supporting materials with large surface area can mitigate the agglomeration [114]. But precious metals are still too costly for large-scale applications. Transition metals are more possible catalysts for ammonia borane dehydrogenation in industrial applications and Fe, Co, Ni, and Cu are the most studied. Among them, Co and Ni exhibit the highest activities for reactions. In addition, various alloys have been studied in this dehydrogenation process because of the synergistic effect. It is well known that the addition of a second metal may enhance the catalytic performance. The result show that some combinations of transition metals and noble metals have higher catalytic performance than monometallic noble catalysts. For example, Au@Co core–shell structured catalysts are reported to have excellent catalytic activity and long-term stability.

There is actually another way to release hydrogen. Methanolysis is a reaction similar to hydrolysis that can produce hydrogen at room temperature. It generates ammonium tetramethoxyborate ($NH_4B(OCH_3)_4$) as a byproduct and can only release 3.9 wt% hydrogen. As a result, there are fewer relevant studies compared to hydrolysis. But this method has a unique advantage, which is that it is easy to convert back to ammonia borane with NH_4Cl and lithium aluminium hydride. On the contrary, it is not possible to directly generate ammonia borane using H_2 for other hydrogen release methods. Regeneration of ammonia borane is difficult. Many reactions have been reported to produce ammonia borane, and they are expressed as [113]:

$$3THF\text{-}BH_3 + 3NH_3 \rightarrow [(NH_3)_2BH_2]^+[BH_4]^- + NH_3 - BH_3 + 3THF \tag{1.15}$$

$$LiBH_4 + NH_4Cl \rightarrow NH_3 - BH_3 + LiCl + H_2 \tag{1.16}$$

$$2LiBH_4 + (NH_4)_2SO_4 \rightarrow 2NH_3 - BH_3 + Li_2SO_4 + 2H_2 \tag{1.17}$$

$$NaBH_4 + (NH_4)_2CO_3 \rightarrow NH_3 - BH_3 + NaCO_3(NH_4) + H_2 \tag{1.18}$$

Considerable progress has been made in regenerating ammonia borane. However, their efficiency and cost still do not meet current requirements for large-scale hydrogen storage applications. Moreover, some ammonia may be released during the dehydrogenation process and should be removed before subsequent utilization, which increases the system costs. Significant improvements are still required for practical application.

3. Hydrazine

Anhydrous hydrazine (H_2NNH_2) is a liquid at ambient temperature, and it contains 12.5 wt% hydrogen. It can be decomposed in two ways: complete decomposition and incomplete decomposition (generates ammonia and hydrogen). The pathways depend on the catalyst and the reaction conditions. But in most of the reports, ammonia is formed due to the low selectivity. In addition, anhydrous hydrazine is explosive with a metal catalyst, making it dangerous to use. Hydrous hydrazine, such as hydrazine monohydrate ($H_2NNH_2{\cdot}H_2O$), is a much safer material for hydrogen storage. It contains 8.0 wt% hydrogen and is also liquid at room temperature. Moreover, it has the distinguished advantages of easy recharging, making it a promising hydrogen carrier. However, it also faces the problem of incomplete decomposition. Efficient and selective catalysts are urgently required to release hydrogen from hydrous hydrazine.

Many metal nanocatalysts (Fe, Co, Ni, Cu, Ru, Rh, Ir, Pt, and Pd) have been studied toward the decomposition of hydrous hydrazine. Among these metals, Rh shows the highest selectivity. Subsequent studies have been carried out on bimetallic alloys. 100% selectivity can be achieved by some alloy nanocatalysts due to the synergistic interactions. For example, the bimetallic Rh_4Ni nanocatalyst achieves a 100% selectivity for hydrogen generation at room temperature [115]. Some other results are shown in Table 1.9.

Reaction conditions, the distribution of the metals and the nature of the supports also play an important role in catalytic performance for hydrogen evolution from hydrous hydrazine. The presence of alkaline additives, such as NaOH, will promote the catalytic activity and hydrogen selectivity of the metal catalysts. This can be attributed to two reasons: promoting the rate-determining deprotonation step and inhibiting the formation of basic NH_3. Higher temperature and highly dispersed catalysts are also reported to enhance the catalytic performance. Wang et al. reported a graphene-supported RhNi catalyst, which exhibited 100% H_2 selectivity and a high catalytic activity. The possible reason is the synergistic effect of the graphene

Table 1.9 Catalytic activities of selected catalysts for the decomposition of hydrous hydrazine [48]

Catalyst	Temp (°C)	H_2 selectivity (%)	TOF (h^{-1})
Rh	25	43.8	2.5
$Ni_{0.93}Pt_{0.07}$	25	100	2.8
$Pt_{12}Ni_{48}$@G4-OH	70	100	240
$Ni_{60}Pt_{40}$-SF	25	100	120
$Pt_{0.8}Ni_{0.2}$@ZIF-8	50	100	69
NiFe	70	100	4.2
Ni_3Fe/C	20	100	556

support and the metals [116]. Due to the high cost of noble metals, many efforts have been dedicated to developing a catalyst that is completely free of them. A NiFe nanocatalyst with 100% hydrogen selectivity was then developed. And 9.86 wt% FeB/MWCNTs is reported to be a stable catalyst, achieving a very high TON of up to 114,480 in 30 h.

The regeneration of hydrazine is another important process for establishing a hydrazine-based hydrogen storage system. However, the direct hydrogenation of nitrogen is still impossible at present. The commercial production of hydrazine typically involves the use of various nitrogen-containing compounds, such as ammonia and urea. Therefore, the large-scale production of hydrazine needs to be combined with other chemical processes, such as the Haber–Bosch process.

Recent studies have shown that hydrous hydrazine is a promising candidate for hydrogen reservoirs. It has many advantages, including low cost, high hydrogen capacity, and the safety of liquid hydrazine hydrate. Remarkable progress has been achieved in the development of efficient catalysts for hydrogen generation from hydrous hydrazine, especially the development of non-precious metal catalysts with high performance. However, there are still some problems before practical use, and the biggest issue is its toxicity. Hydrazine is a potential carcinogenic agent, and people exposed to hydrazine are more prone to developing systemic diseases and cancer. Safety preparations must be made before using hydrazine on a large scale.

4. Hydrazine borane

Hydrazine borane ($N_2H_4BH_3$) is a stable and safe solid at room temperature. Because of its high hydrogen content (15.4 wt%), this substance has great potential to be a high-performance hydrogen carrier. Its synthesis process was first published in 1961 by mixing various reagents. The most effective synthesis process involves sodium borohydride and hydrazine hemisulfate at room temperature. The reactions can be expressed as:

$$2NaBH_4 + (N_2H_5)_2SO_4 \rightarrow 2N_2H_4BH_3 + Na_2SO_4 + 2H_2 \tag{1.19}$$

Several alternative synthesis processes for hydrazine borane have also been reported, typically involving a BH_3 source (mainly $NaBH_4$) and an N_2H_4 source ($N_2H_4 \cdot HCl$ and $MgCl_2 \cdot 4N_2H_4$). Details can be found in Ref. [48]. Overall, there are only limited reports on the production of hydrazine borane. In particular, there is a severe scarcity of direct pathways to synthesis via hydrogen, which leads to poor reversibility in the hydrogenation and hydrogenolysis reactions.

Similar to ammonia borane, the hydrogen in hydrazine borane can also be released via thermolysis or hydrolysis. Fortunately, hydrazine borane starts thermal decomposition at 65 °C, and higher temperature causes higher dehydrogenation rate and extent. About 6.5 wt% H_2 was reported to be released from hydrazine borane within 16 h at 140 °C [48]. On the contrary, the hydrolysis of hydrazine borane in solution requires suitable catalysts. The decomposition of the BH_3 group can be easily achieved, while the decomposition of the N_2H_4 moiety is much slower. Inspired by studies on the decomposition of hydrous hydrazine, some catalysts may also be useful for hydrazine borane. For example, a Rh_4Ni alloy can completely release hydrogen from hydrazine borane with a TOF value of 90 h^{-1}. However, the catalytic activity is still too low for practical applications. In addition, it faces similar problems to ammonia borane, including by-products, cost, and durability. Catalysts that exhibit high performance and selectivity for hydrazine decomposition are also urgently needed for hydrous hydrazine.

1.2.3 Solid Materials for Hydrogen Storage

Compressed hydrogen and liquid hydrogen are unlikely to meet the aggressive targets for on-board applications [117]. Therefore, solid-state hydrogen storage material, which offers improved volumetric hydrogen content, shows the potential advantage to be the best hydrogen carrier. A hydrogen fuel cell using a metal hydride (AlH_3) as a H_2 source yields surprisingly high energy densities, four times greater than even the best-performing conventional batteries (Li polymer) [117].

In solid-state storage systems, hydrogen is stored either by physisorption or chemisorption [118]. Chemisorption is the atomic hydrogen reacts with solids to form hydrides (metal, complex, and chemical hydrides). Metal hydrides have been recognized as one of the most feasible solutions to store hydrogen because of their high degree of safety, reversibility of hydrogenation/dehydrogenation, high volumetric energy densities of hydrogen, and low-pressure requirement [119].

In the physisorption method, molecular hydrogen is adsorbed on the surfaces of solids by van der Waals interactions, which are too weak, and a low temperature is needed to adsorb hydrogen. High surface area materials are more favorable for the adsorption process. Here are some typical materials: carbon nanotubes, fibers, fullerenes, activated carbon, zeolites, metal–organic frameworks (MOFs), covalent organic frameworks (COFs), and more recently, polymers of intrinsic microporosity (PIMs). The hydrogen can then be released whenever required by thermal stimulation or any other suitable technique.

1.2.3.1 Interstitial Hydrides

Intermetallic hydrides represent the hydrides of metals or alloys, in which hydrogen occupies interstitial sites of the metal sublattice and forms a metallic bond with metal atoms. This hydrogen absorption generally causes the sublattice expansion by 20–30%. The pure metals tend to form stable metal hydride, which needs high temperatures to release hydrogen. Hence, they are not suitable for large-scale hydrogen storage applications. On the contrary, alloy, which forms stable hydrides, shows a lower hydrogen release temperature, attracting more attention. Generally, the hydrogen-absorbing alloys consist of a hydride easily forming metal (Ti, V, Zr, Y) and a non-hydride-forming metal (Cr, Mn, Fe) [120]. Some exceptions to this exist, such as ZrV_2, which consists of two metals forming stable hydrides. Intermetallic hydrogen absorbing alloys are often categorized by their atomic ratio and crystal structure. Usually, hydrides easily formed by metals are named A, and non-forming elements are named B. There are several types of alloys for hydrogen storage, including AB_5, AB_3, A_2B_7, AB_2, AB, and A_2B. Solid solution alloys with a body-centered cubic (BCC) structure may also show high performance. Some typical interstitial hydrides are shown in Table 1.10.

AB_5-type alloys usually have a $CaCu_5$-type structure. The first AB_5-type alloy ($SmCo_5$) was found in 1969, via the investigation of magnetic degradation in hydrogen [121]. It absorbs hydrogen with an H/metal ratio of about 0.5. $LaNi_5$ is also a typical example of AB_5-type alloy, and it has more than twice the hydrogen capacity of $SmCo_5$. Its crystal structure is shown in Fig. 1.18. When it absorbs hydrogen, the hydride of $LaNi_5H_6$ is formed. The hydrogenation and dehydrogenation reactions can occur easily at room temperature without additional treatment [122]. However, the metal sublattice changes its position during the repeated hydrogen absorption and desorption cycles, making the $LaNi_5$ unsuitable for practical application. To solve the problem, the elements are both substituted. $MmNi_5$-based alloys, in which Ni can be replaced by Mn, Co, Al, etc., are reported to have better hydrogen storage properties [120]. Now, they have been successfully used as a negative material in commercial Nickel-Metal Hydride (Ni-MH) battery.

AB_2-type alloy usually has the Laves phase structure, forming various possible crystal structures, including hexagonal C14, cubic C15, and hexagonal C36. Figure 1.19 shows a C15 Laves phase structure. Ti and Zr are usually at the A

Table 1.10 Typical interstitial hydrides [120]

Type	Hydride	ρ_m (wt%)	ρ_v (kg/m^3)	ΔH (kJ/mol)
AB_5	$LaNi_5H_6$	1.4	92	– 31.2
AB_3	$LaMg_2Ni_9H_9$	1.3	–	– 24.0
AB_2	$TiCr_2H_4$	2.57	108	– 23
AB	$TiFeH_2$	1.89	119	– 28.1
BCC	$V_{22}Ti_{35}Cr_{43}$	2.5	156	– 39

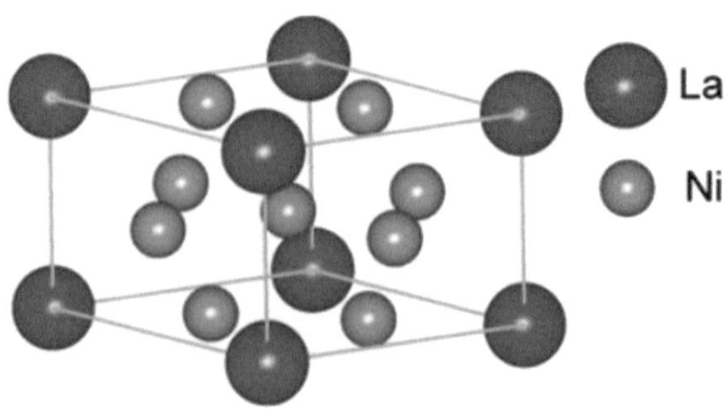

Fig. 1.18 Crystal structure of $LaNi_5$ [7]

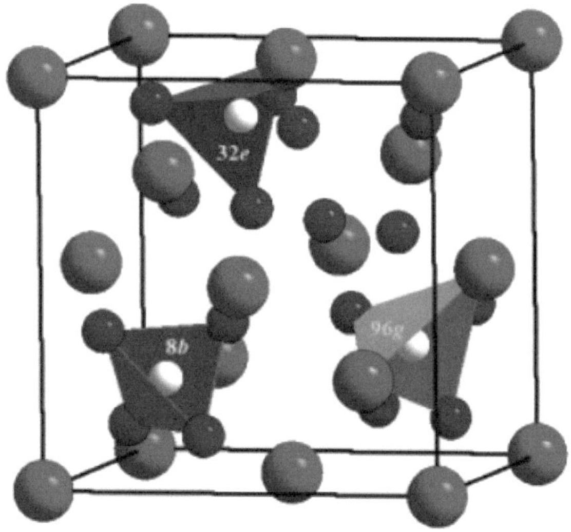

Fig. 1.19 The structural model of C15 Laves phase structure [124]

side, and Cr, Mn, and V can be found at the B side. Therefore, several typical alloys, such as $TiCr_2$, $ZrCr_2$, $ZrMn_2$, and ZrV_2, can be obtained. This type of alloy has the potential to absorb more hydrogen than AB_5 alloys, reaching the maximum hydrogen capacity of 1.9 wt% hydrogen in $TiMn_{1.5}$ [120]. In addition, AB_2 alloys have a wider temperature range of hydrogen absorption and desorption, suggesting the potential for applications [123].

The typical example of AB-type alloys is TiFe, which was found in 1974 [125]. Its hydrogenation and dehydrogenation can occur at room temperature, with a high hydrogen capacity similar to that of $TiMn_{1.5}$. However, its hydrogen absorption reaction suffers from the severe conditions for initial activation. Many works have been done to solve this problem [126]. The related studies show that its hydrogen absorption is a two-step reaction. The change of crystal structure in the hydrogenation process is shown in Fig. 1.20. Currently, many researchers are still attracted to extensively investigate TiFe alloy as a hydrogen carrier because of the low price of Ti and Fe.

The most well-known A_2B alloy is probably Mg_2Ni. It forms Mg_2NiH_4 after hydrogenation with a hydrogen capacity of 3.6 wt%. However, its hydrogenation and dehydrogenation occur at 300 °C, limiting its practical application. In addition, it is the only successful improvement in reaction temperature for Mg-based alloys as a hydrogen carrier. Various attempts are still needed to lower desorption temperatures.

AB_3-type alloys were first reported in 1997 [127]. Its crystal structure is a stacking of an AB_5 subunit and an AB_2 subunit. La–Mg–Ni-based alloys are typical AB_3-type alloys, such as La_2MgNi_9, $La_5Mg_2Ni_{23}$, and La_3MgNi_{14}. Their maximum H/metal ratio can reach 1.1 at 333 K in the pressure range of 10^{-3}–1.0 MPa. By modifying the ratio of AB_2 subunit to AB_5 subunit, A_2B_7-type and A_5B_{19}-type alloys can also be formed. Among these types of La–Mg–Ni-based alloys, AB_3-type alloys have the

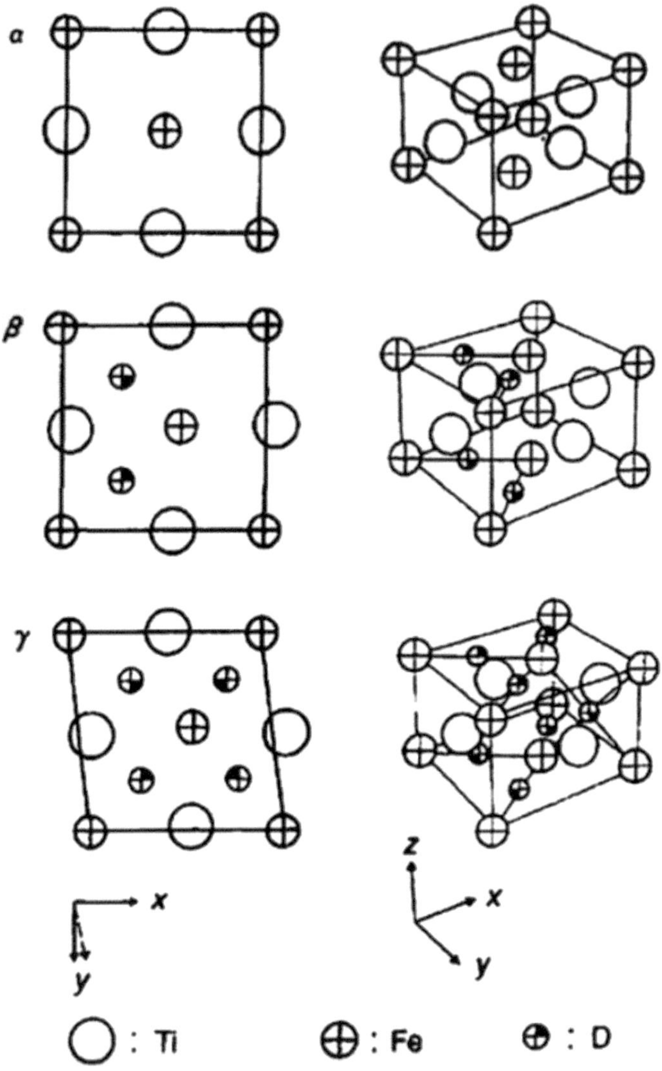

Fig. 1.20 Structure change of TiFe with hydrogenation [7]

highest hydrogen storage capacity because of their higher A_2B_4 subunit ratio [128]. However, their poor durability in hydrogen absorption and desorption cycles limits their practical applications.

Solid solution alloys with BCC structures are different from the above-mentioned alloys, which have a large hydrogen capacity reaching 3 wt% at room temperature. Much effort has been devoted to this type of alloy, finding Ti–Cr–Mo, Ti–Mn–V, and Ti–Cr–V alloys [129]. The BCC alloys absorb and desorb hydrogen via two-step reactions, and the hydrogen absorbed in the first plateau is hard to release. Moreover, it also faces many problems, such as difficult activation, sluggish kinetics, sloped plateau, and degraded cycling capacity [130].

1.2.3.2 Non-interstitial Hydrides

In the non-interstitial hydrides, hydrogen forms a covalent or ionic chemical bond with the neighboring atom. Typical non-interstitial hydrides include alanates ($M(AlH_4)_n$), amides ($M(NH_2)_n$), and borohydrides ($M(BH_4)_n$), and so on. Ammonia borane can also be categorized as this type, but has already been covered in the previous section. So, it won't be repeated here. Most of the non-interstitial hydrides have a larger hydrogen capacity than that of the interstitial hydrides, and some even reach 20 wt%. Therefore, it is expected as promising high-density hydrogen storage materials. Some typical non-interstitial hydrides are summarized in Table 1.11.

Alanates are generally expressed as $M(AlH_4)_n$, where H binds with Al via a covalent bond. $NaAlH_4$ is a typical alanate, which can release totally 7.8 wt% of hydrogen through three steps. The reactions can be expressed as:

$$3NaAlH_4 \leftrightarrow Na_3AlH_6 + 2Al + 3H_2 \tag{1.20}$$

$$Na_3AlH_6 \leftrightarrow 3NaH + Al + 1.5H_2 \tag{1.21}$$

$$3NaH \leftrightarrow 3Na + 1.5H_2\ (1.8\ \text{wt\%}) \tag{1.22}$$

However, the hydrogen release from NaH only occurs above 400 °C, and the effective hydrogen capacity of $NaAlH_4$ is believed to be 5.5 wt% [131]. Moreover, these reactions were initially thought to be poorly reversible. The addition of Ti-based catalyst is crucial for the realization of reversible hydrogen storage [131]. Despite the slow reaction rate, the discovery of reversibility was a remarkable progress in solid-state hydrogen storage. Now, $NaAlH_4$ has become one of the most well-studied reversible non-interstitial hydrides. Srinivasan et al. demonstrate a measured hydrogen capacity of nearly 4 wt% for $NaAlH_4$ at 160 °C in 100 reversible hydrogen cycles over Ti-based catalysts [132]. In addition to $NaAlH_4$, there are several other alanates, such as Na_2LiAlH_6, K_3AlH_6, Li_3AlH_6, and $Ti(AlH_4)_4$. However, some of these materials

Table 1.11 Typical non-interstitial hydrides [120]

Type	Hydride	ρ_m (wt%)	ρ_v (kg/m^3)
Alanate	$3NaAlH_4$	3.7	48
Amide	$LiNH_2$	605	68
Borohydride	$Mg(BH_4)_2$	11.1	110
Mg-based hydride	MgH_2	7.6	109

are poorly reversible, and others contain little hydrogen and are too stable to react. Therefore, $NaAlH_4$ is still the most promising alanate to store hydrogen.

Amides can be generally expressed as $M(NH_2)_n$, where H forms a covalent bond with N. Their thermal decomposition is prone to release ammonia rather than hydrogen. Hence, they are not considered to be a hydrogen carrier at first until the amide-hydride composite was discovered, realizing reversible hydrogen absorption and desorption [133]. The electrostatic interaction between amide and hydride may promote the formation of hydrogen. So, the metal amides used for hydrogen storage are the amide-hydride composites. $LiNH_2$-LiH is the first amide-hydride composite for hydrogen storage, which can release a total of 10.4 wt% of hydrogen. The reaction takes two steps at different temperatures:

$$LiNH_2 + 2LiH \rightarrow Li_2NH + LiH + H_2 \tag{1.23}$$

$$Li_2NH + LiH \rightarrow Li_3N + H_2 \tag{1.24}$$

Similar to that of $NaAlH_4$, an intermediate phase is formed in the release process. Therefore, high temperature is required for complete dehydrogenation. To reduce the reaction temperature and enhance the kinetics, the chemical composition is changed [134]. For example, $Mg(NH_2)_2$–2LiH starts hydrogen desorption below 90 °C and achieves a high hydrogen capacity of 5.6 wt% at around 150 °C. In addition, adequate mixing and introducing catalytic additives can also improve the dehydrogenation performance [135]. Further research is still necessary before application, and attention should be paid to the inhibition of ammonia production.

Borohydrides can be expressed as $M(BH_4)_n$, which consists of a metal cation and a polyatomic anion (BH_4^-). H binds with B via a covalent bond. In the previous section, we mentioned $NaBH_4$, which belongs to this type. It can be used to synthesize ammonia borane, but not suitable for direct use as a hydrogen carrier. In general, the alkali borohydrides require high temperatures for dehydrogenation because of their large decomposition enthalpy. $LiBH_4$, $Mg(BH_4)_2$, and $Ca(BH_4)_2$ are potential candidates for hydrogen storage [136], with a hydrogen capacity higher than 10 wt%. Take $LiBH_4$ as an example, its decomposition occurs above 380 °C and the reaction can be simply described as [117]:

$$2LiBH_4 \rightarrow 2LiH + 2B + 3H_2 \tag{1.25}$$

Its hydrogen capacity can be up to 13.5 wt%. But the temperature is still too high for hydrogen storage. Moreover, some unwanted intermediate compounds may form in decomposition reactions [137], and they are prone to accumulate in the repeated cycles, which degrades the reversibility of borohydrides. For example, $MgB_{12}H_{12}$ will be generated in the dehydrogenation reaction of $Mg(BH_4)_2$, which accounts for the degradation. Reversibility can be achieved only in extreme conditions, but it is impractical for portable applications. The addition of hydrides is reported to be a good strategy to address these problems, which can improve the reaction kinetics and reversibility by altering the dehydrogenation process. In addition, choosing metal with large electronegativity can lower the reaction temperature, which suppresses the formation of stable intermediate compounds.

Mg can be directly used to absorb hydrogen, forming MgH_2, with a hydrogen capacity of 7.6 wt%. Besides, it is the eighth abundant element on Earth. Hence, it has been extensively investigated as a potential candidate for hydrogen carrier [138]. In MgH_2, H binds with B via a covalent and ionic bond. Its hydrogenation reaction is thermodynamically favorable, but the diffusion of hydrogen in MgH_2 is slow. The diffusivity of hydrogen is highly dependent on temperature. Therefore, high temperature and high pressure are required for acceptable kinetics [139]. High temperature is also necessary for the hydrogen release process because of the high enthalpy (75 kJ mol^{-1} H_2). Forming an alloy [140], reducing the crystal grain size [141], and changing the reaction pathway [142] have been heavily investigated to reduce reaction temperatures. Furthermore, greater efforts are needed to enhance the kinetics of Mg.

1.2.3.3 High Surface Area Materials

High specific surface area materials offer an alternative way to store hydrogen. They can adsorb hydrogen onto their surface via weak van der Waals interactions, with the enthalpy between 4 and 10 kJ/mol. The interactions are too weak for low temperatures, are necessary for a large amount of hydrogen to be adsorbed. Again, due to the weak interactions, this adsorption is highly reversible, reducing the difficulty of hydrogen release and achieving very fast kinetics. Therefore, the system can be simplified, and the efficiencies of adsorption and desorption can be high. Furthermore, adsorption is a surface process, so materials with a high specific surface area can adsorb more hydrogen in the same volume. Porous materials have received much attention in this field, and some of them will be reviewed here.

Carbons have many advantages, including low cost, abundance, stability, and low density. Therefore, it is considered a promising choice for be sorption material. Many unique carbon structures with high surface area have been investigated, which possess a low volumetric density and a high surface area. Activated carbon, consisting of graphite crystallites and amorphous carbon, has a high specific surface area of 3000 m^2 g^{-1}. One of the most attractive aspects of this material is the wide range of material sources. It can be produced from carbonaceous material, such as litchi-trunk, coal, chitosan, seeds, corncob, and so on, through dry distillation. This reduces the price of activated carbon. Moreover, the material employed to form activated carbon affects the final hydrogen storage capacity, ranging between 0.5 and 5 wt% [143]. Some hydrogen capacities of activated carbon reported in the articles are summarized in Table 1.12. And the results show that the capacity can be improved through chemical treatment and doping. Even so, they still haven't met the demand.

Graphite is another popular carbon material. It is a naturally existing carbon material that can also be prepared artificially at high temperatures. It has a structure

Table 1.12 Experimentally measured hydrogen storage capacity of activated carbon [144]

Adsorbent	Hydrogen storage (wt%)	Conditions temp, pressure
AC	0.67	303 K, 10 MPa
AC	1.4	77 K, 0.1 MPa
AC	5.7	77 K, 3 MPa
AC (KOH-treated)	6.6	77 K, 4 MPa
AC (Ni-doped)	1.8	77 K, 0.1 MPa
AC (Pd-doped)	5.5	298 K, 8 MPa

that is planar and layered. Atoms in each layer are bonded to form a honeycomb-like network. Figure 1.21 shows the structure of graphite.

The distance between planes is only 3.354 Å, too small for the intercalation of hydrogen [145]. Therefore, graphite possesses low hydrogen capacity. Expanding the interlayer distance becomes a good strategy to enhance graphite's hydrogen storage capacity. Several methods have been investigated. For example, by bonding alkali to organic ligands, the distance between the layers can be increased [146]. The hydrogen capacity is reported to increase, but the larger distance may cause exfoliation at higher temperatures [147]. Some hydrogen storage capacities of various graphite samples are summarized in Table 1.13.

Fullerenes are a spherical structure, consisting of pentagonal or hexagonal rings. Generally, twelve pentagonal rings and two or even more hexagonal rings are formed by carbon atoms. Popular and well-known fullerenes include C70, C76, C84, C240, and C540. The hydrogen capacity of fullerene is low because of its stable structure and chemical inertness. The adsorption of metals onto the surface can enhance the

Fig. 1.21 Graphite structure [144]

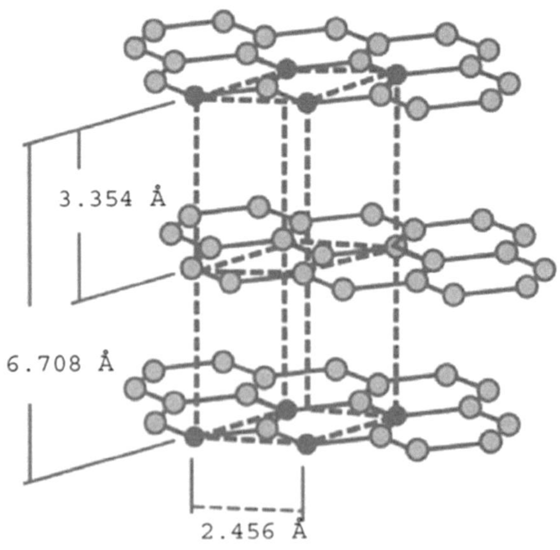

Table 1.13 Hydrogen storage capacities of graphite [144]

Adsorbent	Hydrogen storage (wt%)	Conditions temp, pressure
G (Ni-doped)	0.92	–
G (Cu-doped)	0.97	–
G (Fedoped)	3.9	–
G (Al-doped)	3.48	300 K, 100 MPa
G (Li-doped)	6.5	298 K, 2 MPa

binding strength of molecular hydrogen, which may be a method to increase the hydrogen capacity of fullerenes.

Metal organic frameworks (MOFs) are porous materials built from metal ions/clusters and organic linkers. MOFs have high surface areas, good stability, adjustable pore size, high void volume, and controllable properties, making them possible hydrogen storage materials [148]. Research on MOFs as hydrogen carriers has been growing since 2003. Modification of the organic linker, reaction temperature, and metal/ligand ratio can afford different MOFs, which may have high hydrogen capacity [149]. MOF-177 was reported to have a surface area of 5640 m^2/g and a hydrogen adsorption capacity of 7.6 wt% [148]. But the capacity is measured at −176 °C, and how to store large quantities of hydrogen at room temperature is still a problem. Moreover, their moisture instability and scale-up difficulties further hinder their commercial application.

1.3 Hydrogen Utilization

1.3.1 Industrial Production

1.3.1.1 Petroleum Industry

One of the most famous applications of hydrogen in the petroleum industry is hydrocracking [150]. Because low value distillate feedstocks contain various impurities such as asphaltenes, sulfur, and heavy metals, they cannot be used by public directly. Only after conversion into refined fuels with smaller hydrocarbons, such as gasoline and diesel, they can be easily utilized. A schematic of a hydrocracking system is shown in Fig. 1.22. Various high-value products can be achieved after conversation. The remaining hydrogen gas from the fractionators can be recycled.

Hydrocracking is considered to be the most suitable technology for conversion. It has two steps: cracking and hydrogenation [152]. Cracking is an endothermic reaction in which one or more C–C bonds are broken and free valences are thereby created. While hydrogenation reactions are exothermic, in which the olefins and aromatics are saturated by hydrogen [153]. Therefore, catalysts used for this reaction must have

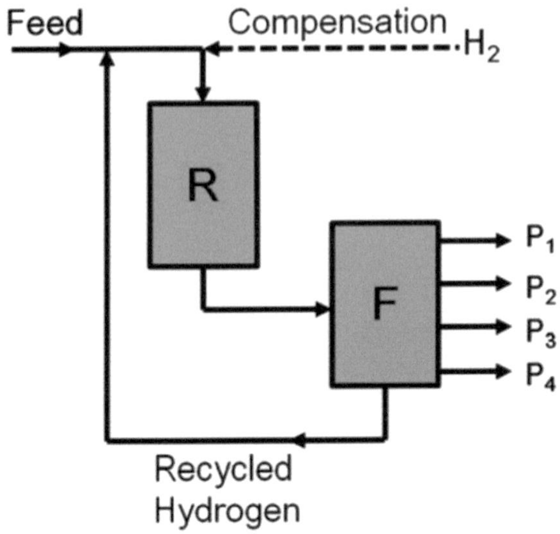

Fig. 1.22 Schematic diagram of a typical single-stage hydrocracking system. R, F, and P are reactor, fractionators, and product, respectively [151]

dual functionality. The properties of the final products after hydrocracking are highly dependent on the relative amounts and properties of the two functions of the catalysts [154]. For example, lower boiling products, such as gasoline and lighter materials, will be produced in the presence of catalysts with stronger cracking functions. While for catalysts with stronger hydrogenation functions, higher boiling products, such as middle distillates, will be produced. Depending on the specific catalyst employed, there are four possible reaction mechanisms for hydrocracking [153]. Noble metals, such as Pt and Pd, are reported to show excellent performance in hydrocracking. Non-noble metals, such as nickel, cobalt, molybdenum, and tungsten, are also common catalysts used in the industry. It is worth noticing that the constituents of heavy oil are harmful to catalyst activity, and there is a need to develop durable catalysts for hydroprocessing [155].

Various hydrocracking reactors, including fixed-bed, ebullated-bed, moving-bed, or slurry-phase reactors, have been developed. In general, the reaction conditions for each technology are different, and therefore, they require different feed and catalysts. The middle distillates can be used in fixed-bed reactors, while more complex feeds can be used in moving-bed reactors. Although some systems have been developed and commercialized, there are still some problems to be solved. For example, fixed-bed reactors cannot handle high metal impurities, and ebullated-bed reactors face high cost and low reactor efficiency [152]. The integrated systems with high conversion, high selectivity, high durability, and low cost will also be needed in the next few years.

Hydrotreatment is another important application of hydrogen in the petroleum industry. Petroleum products usually contain sulfur and nitrogen compounds, which lower the quality of the fuel and cause NO_x and SO_x pollution during their combustion. Hydrotreatment is the process in which hydrogen is used to hydrogenate sulfur and nitrogen compounds, and finally remove them in the form of H_2S and NH_3 gases [156].

1.3.1.2 Chemical Industry

In the chemical industry, hydrogen is used as a reactant to produce ammonia for fertilizer. The application of fertilizers has increased global food production and fed a larger population. As ammonia is the main raw material to produce fertilizers, ammonia synthesis is of great importance. The well-established method for the production of ammonia using nitrogen and hydrogen is known as the Haber–Bosch process. It was first developed by Fitz Haber in 1908 and then industrialized by Carl Bosch in 1911. It consumes about 50% of the hydrogen produced worldwide [150]. Fused-iron catalysts with a small amount of promoter are the most commonly used catalysts in ammonia synthesis plants [157]. High pressure and temperature are required to achieve the desired conversion.

It is worth noting that although hydrogen is used to synthesize ammonia, the plants do not use hydrogen as a feedstock. A modern ammonia manufacturing process is highly integrated, and fossil fuels, such as natural gas, oil, and coal, are used as feedstock to produce hydrogen [158]. Figure 1.23 shows a schematic of the methane-fed Haber–Bosch process. This conventional Haber–Bosch process leads to approximately 1.2% of global CO_2 emissions and is a significant polluter of the environment. Fortunately, fossil fuels are not necessary for the Haber–Bosch process. With the development of hydrogen generation technology, a cleaner and more efficient ammonia production process is expected.

1.3.1.3 Metallurgical Industry

Most metals exist in the form of compounds, so after mining, they must be reduced to obtain the metal. Carbon, carbon monoxide, and hydrogen are commonly used in metal reduction processes as reduction agents. However, the extensive use of carbon can lead to serious environmental pollution, so hydrogen has received more attention from researchers in recent years. Figure 1.24 shows the metal production using hydrogen.

The commercial application of hydrogen in non-ferrous metal industries is very limited. Only W, Mo, Ni, and Co use hydrogen in commercial production. Among them, Mo and W oxides can be directly reduced in a hydrogen atmosphere. While Ni and Co should form intermediate products before hydrogen reduction. Carbon is still dominating in current global metal reduction processes as a reduction agent. It is a stronger reducing agent compared to hydrogen. Fortunately, hydrogen has the potential to reduce many metal oxides based on thermodynamic knowledge, and quite a few have been successfully studied in the laboratory. Further large-scale application of hydrogen in the metallurgical industry is promising, and more effort needs to be devoted to this field.

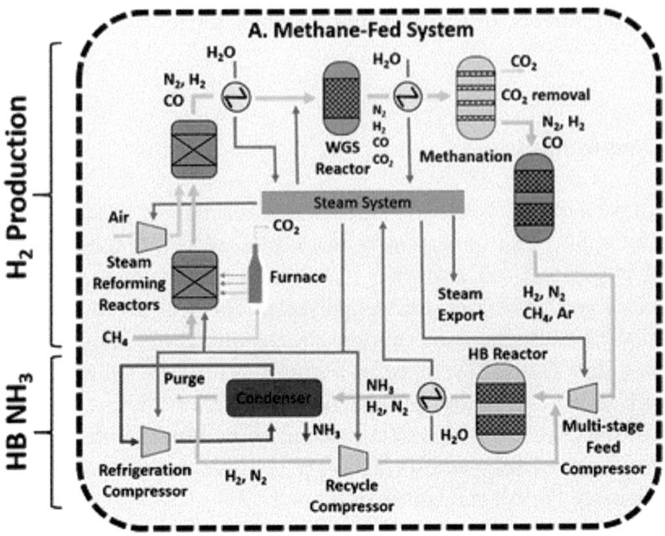

Fig. 1.23 A schematic of typical methane-fed Haber Bosch process [158]

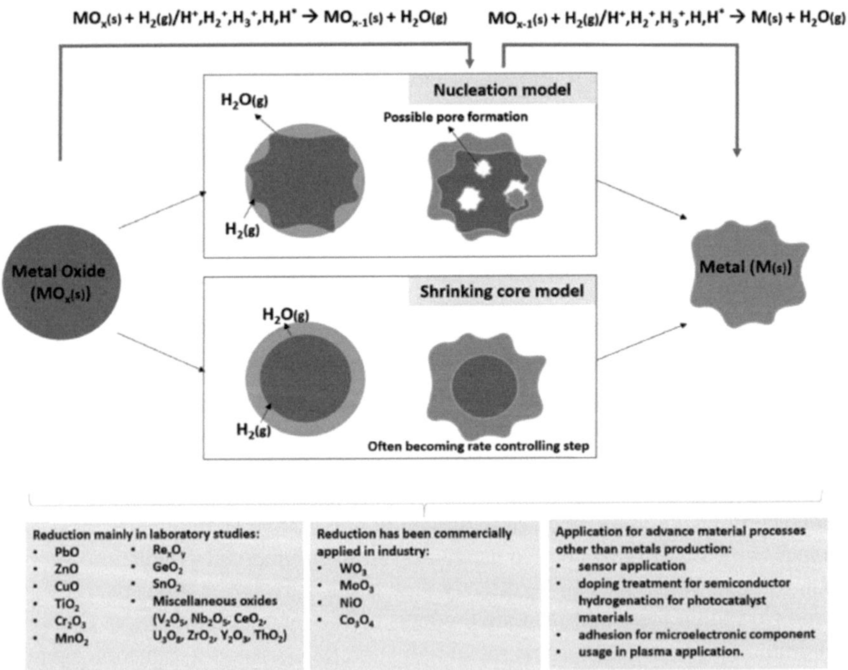

Fig. 1.24 Hydrogen in metal production [159]

On the contrary, hydrogen has been applied in iron production for a long time. Many reduction processes have been developed, including the blast furnace production process, direct reduction iron process, and smelting reduction iron process [160]. To reduce CO_2 emissions in the steel industry, replacing carbon with hydrogen is inevitable. Some suggestions and overlooks can be found in Ref. [160].

1.3.2 Combustion

1.3.2.1 Internal Combustion Engine

Compared to currently used fossil fuels, hydrogen contains no carbon, achieving the goal of zero carbon emissions. And it has a high specific energy density, almost three times the energy density of other fossil fuels, which makes hydrogen combustion very promising [161]. However, its minimum ignition energy in air is very low, which means it can be easily ignited [162]. This may cause the pre-ignition of fuel and further lead to a loss of control and knocking [163] in combustion, eventually resulting in engine failure. Moreover, the quenching distance of hydrogen is shorter than that of conventional hydrocarbon fuels. As a result, an increase in combustion heat losses may occur because of a higher temperature gradient near the wall of the combustion chamber. Therefore, hydrogen-fueled internal combustion engines need to be redesigned. For example, modifying engine geometry can reduce combustion heat losses. A non-platinum cold-rated spark plug can avoid pre-ignition and backfiring in the engine [164].

Based on the methods of mixture formation, the combustion can be categorized as premixed combustion (PC) and diffusive combustion (DC). In a PC, a homogenous mixture of fuel and air will fill the whole combustion chamber before ignition. While in DC, the pressurized fuel is directly injected into a closed combustion chamber just before ignition. As for the ignition methods, they can be divided into spark ignition (SI) and compression ignition (CI). In the SI engine, ignition is caused by sparks generated by spark plugs. While in the CI engine, the fuel is self-ignited due to the high gas temperature after the compression. Thus, four typical combustion types can be obtained. Hydrogen can be applied to all these types. Among them, SIPC and SIDC used to be major application targets because of their higher auto-ignition point. However, their low engine compression ratio limits the thermodynamic efficiency. Now, the combustion mode, known as dual-fuel hydrogen-diesel direct injection, attracts more attention for its ability to alleviate the power and compression ratio limitations. More details of the practical hydrogen internal combustion engine can be found in Ref. [161].

1.3.2.2 Hydrogen-Fueled Gas Turbines

Gas turbines are important in the supply of electrical energy because of their flexible and dispatchable power generation. However, conventional natural gas turbines cannot meet the demand for emission reductions. Further improvements in the thermal efficiency will have less influence on emissions. Hence, the use of hydrogen in gas turbines is regarded as a promising method to mitigate issues.

Numerous studies have previously explored the combustion of hydrogen in gas turbines. The results show that the hydrogen changes the combustion chemistry and affects the stability of the flame. For example, hydrogen flames have a very short ignition delay time and require higher flow rates to avoid flashback phenomena. Conventional gas turbines can only use a low share of hydrogen in the gas. Currently, the successful use of gas with higher hydrogen content has been reported in studies, but using pure hydrogen is still a problem for gas turbines. Specialized design is required to achieve stable combustion. In addition, pure hydrogen supply leads to higher NO_x emissions than natural gas firing. This is reported to be induced by the greater presence of H, O, and OH radicals in the flame. How to reduce NO_x emissions from hydrogen-fired gas turbines is also an issue that must be addressed before practical application.

1.3.3 Fuel Cells

Another important utilization method of hydrogen is fuel cells. Fuel cells are electrochemical devices that directly convert chemical energy into electrical energy. Because the conversion process does not require thermal work. Thus, it is more efficient. In addition, fuel cells operate more quietly and stably than conventional power plants because they have no moving parts. When hydrogen is used as fuel, the reaction product is only water, which does not pollute the environment. Therefore, fuel cells are also considered as one of the core components for building a hydrogen society in the future. Many automakers have already introduced cars that use fuel cells. However, there are still some problems that limit the further spread of fuel cells, such as the high cost of catalysts and the short operating life of fuel cells. To facilitate the realization of a hydrogen society, the following chapters will describe fuel cells and related research in detail.

1.3.4 Medicine

The application of hydrogen in medicine has gained significant attention in recent years due to its potential for therapeutic and diagnostic purposes. Hydrogen gas has been shown to have antioxidant and anti-inflammatory properties, which can be beneficial in the treatment of various medical conditions [165, 166]. An acute rat model was investigated [167] in which oxidative stress damage was induced in the brain by focal ischemia and reperfusion. The inhalation of H_2 gas markedly suppressed brain injury by buffering the effects of oxidative stress. Thus, H_2 can be used as an effective antioxidant therapy. One of the most promising areas of research is the use of hydrogen gas as a therapeutic agent for acute and chronic diseases. Studies have shown that inhalation of hydrogen gas can reduce oxidative stress and inflammation in the body, which may be beneficial in the treatment of conditions such as cardiovascular disease, stroke, and diabetes. In a clinical study in rodents, Motoaki Sano [168] found that H_2 gas had cardioprotective effects, prevented ischemia–reperfusion injury in rats, and enhanced the recovery of left ventricular function after hypoxia-reoxygenation.

Additionally, hydrogen gas has been explored for its potential as a diagnostic tool. Researchers have developed hydrogen breath tests, which can be used to detect various medical conditions such as lactose intolerance, small intestinal bacterial overgrowth, and Helicobacter pylori infection. For example, hydroxyl-phenyl-fluorescein (HPF) fluorescence signaling was used for the selective detection of highly reactive oxygen species (hROS), such as hydroxyl radicals (·OH) and peroxidase-activated intermediates [169]. These tests involve measuring the amount of hydrogen gas in a patient's breath after ingestion of a substrate that is metabolized by gut bacteria. Elevated levels of hydrogen gas in the breath can indicate the presence of a medical condition. Research has found that hydrogen has great potential to improve oxidative stress-related diseases by inhaling H_2 gas, injecting saline with dissolved H_2, or drinking water with dissolved H_2 (H_2-water) [167, 170]. At the same time, the right amount of H_2-water is still effective in improving obesity. In addition, results also showed a difference in hydrogen concentration between arterial and venous blood that could be used for medical diagnosis [171].

Moreover, the application of hydrogen energy in medicine extends beyond the use of hydrogen gas as a therapeutic and diagnostic tool. Hydrogen fuel cells have been developed as an alternative power source for medical devices and equipment [172, 173]. These fuel cells have the potential to provide a reliable and efficient source of power, without the need for traditional batteries or external power sources. This can be particularly useful in remote and underdeveloped areas, where access to reliable electricity is limited.

In conclusion, the application of hydrogen energy in medicine has the potential to revolutionize the way we diagnose and treat various medical conditions. The therapeutic and diagnostic properties of hydrogen gas, as well as the development of hydrogen fuel cells, represent a significant step forward in the field of medicine. As research in this area continues to progress, it is likely to see more widespread applications of hydrogen energy in various medical settings.

1.3.5 Agriculture

Hydrogen has emerged as a promising candidate for the agricultural sector, with multiple applications that can help increase crop yields, reduce pollution and greenhouse gas emissions, and enhance soil health [174–176]. One of the most significant benefits of hydrogen in agriculture is its potential as a clean and renewable energy source for agricultural machinery and equipment [177]. Hydrogen fuel cells can be used to power tractors, harvesters, and other farm machinery, helping to reduce dependence on fossil fuels and decrease the carbon footprint of agricultural operations. Moreover, using hydrogen fuel cells in agriculture can improve energy efficiency and reduce operational costs. Another important application of hydrogen in agriculture is its potential to improve soil health and crop yields. Hydrogen-rich water can help increase oxygen levels in soil [178], promoting the growth of beneficial soil microbes and enhancing nutrient uptake by plants [179]. Over time, this can lead to improved crop yields and soil quality. Additionally, hydrogen can be used to produce fertilizers that are free from harmful chemicals, promoting sustainable agricultural practices. Furthermore, hydrogen has the potential to help mitigate the impact of climate change on agriculture. For example, hydrogen can be used to power heating and ventilation systems in greenhouses, reducing the carbon footprint of greenhouse agriculture. It can also be used to produce biofuels from agricultural waste, such as corn straw and manure, which can help reduce greenhouse gas emissions from livestock operations. Finally, hydrogen can be used as a disinfectant in food processing and storage facilities, promoting the safety and quality of agricultural products [177, 180]. In summary, hydrogen has the potential to significantly improve the sustainability and efficiency of agriculture while reducing its environmental impact. As the world faces the challenges of climate change and food security, the integration of hydrogen into agriculture may prove to be a crucial step in building a more sustainable and resilient food system for the future.

In addition to these applications, hydrogen can also play a crucial role in improving the quality and safety of agricultural products. For instance, hydrogen can be used to sterilize agricultural equipment and tools, which can help reduce the risk of crop diseases and pest infestations [181].

Overall, the applications of hydrogen in agriculture are vast and varied, with the potential to significantly improve the sustainability and efficiency of farming operations while also reducing the environmental impact of agriculture. As the world continues to grapple with the challenges of climate change and food security, the integration of hydrogen into agriculture may prove to be a crucial step in building a more sustainable and resilient food system for the future.

1.4 Summary and Outlook

This chapter reviews the storage and utilization of hydrogen and shows the potential of hydrogen for energy storage and transfer. With the increasingly serious environmental and energy problems, hydrogen energy will become the most reliable and important energy source in the future. Hydrogen will play an important role in reducing carbon emissions and alleviating the energy shortage. Progress in science and technology has accelerated the process of turning this potential into reality. Figure 1.25 shows what a hydrogen society looks like in the future.

The large amount of electricity generated through renewable energy can be directly supplied to society on the one hand. On the other hand, the production of hydrogen using renewable energy is critical to decarbonizing the energy landscape [182]. Fossil fuels can be used as backup energy for society, directly for electricity generation, or for hydrogen production to meet energy needs. With the advantages of hydrogen storage and transportation, the hydrogen society will greatly improve the utilization rate of renewable energy, realize the emission reduction of the whole society, and greatly alleviate the social pollution problem.

In a hydrogen society, the transformation process of electric energy and hydrogen energy becomes the key point of society. As a key power generation device that can directly convert chemical energy from hydrogen into electric energy, the fuel cell is bound to be paid more attention by researchers. Meanwhile, advancements in fuel cell technologies, such as the development of high-performance gas-fed direct ammonia hydroxide exchange membrane fuel cells, offer promising solutions for efficient and clean energy conversion, further expanding the scope of hydrogen applications [183].

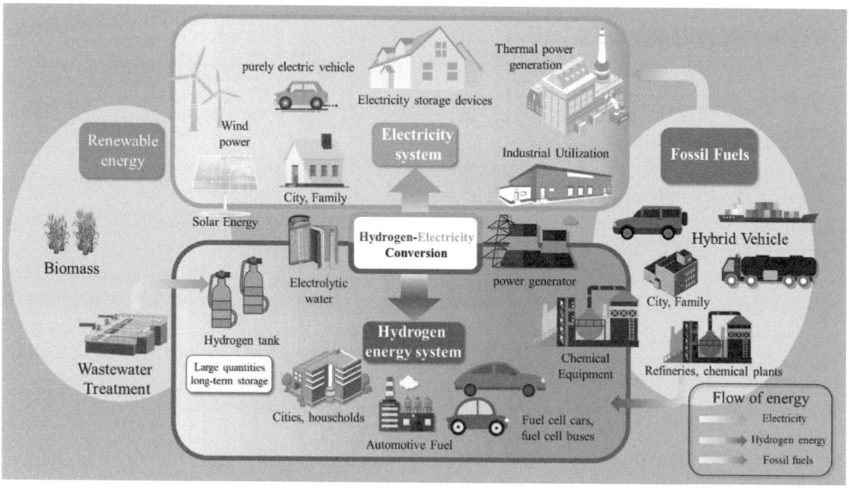

Fig. 1.25 Schematic of the hydrogen society

The following part of this book will start from the basic knowledge of fuel cells and then introduce the relevant research progress of several typical fuel cells in detail. The purpose of this part is to provide a reference for researchers in the field of fuel cells, and to help build a hydrogen society in the world.

References

1. Richards D, Yabar H. Promoting energy and resource recovery from livestock waste: Case study Yuge Farm, Japan[J]. Case Studies in Chemical and Environmental Engineering, 2023, 7: 100299.
2. EU strategies on hydrogen and energy system integration [EB/OL]. [2024-01-28]. https://www.europarl.europa.eu/thinktank/en/document/EPRS_ATA(2021)690599.
3. S.A. Sheriff, G.D. Yogi, E. Stefanakos, et al. Handbook of hydrogen energy[M]. CRC Press,2014.
4. Niaz S, Manzoor T, Pandith A H. Hydrogen storage: Materials, methods and perspectives[J]. Renewable and Sustainable Energy Reviews, 2015, 50: 457–469.
5. Yürüm Y. Hydrogen Energy System: Production and Utilization of Hydrogen and Future Aspects[M]. Springer Science & Business Media, 1995.
6. Andersson J, Grönkvist S. Large-scale storage of hydrogen[J]. International journal of hydrogen energy, 2019, 44(23): 11901–11919.
7. Sasaki K, Li H W, Hayashi A, et al. Hydrogen Energy Engineering: A Japanese Perspective[M]. Tokyo: Springer Japan, 2016.
8. Langmi H W, Engelbrecht N, Modisha P M, et al. Electrochemical Power Sources: Fundamentals, Systems, and Applications[M]. Elsevier, 2022: 455–486.
9. The first commercial Type V composite pressure vessel[EB/OL]. [2022-08-25]. https://www.compositesworld.com/articles/next-generation-pressure-vessels.
10. Balasooriya W, Clute C, Schrittesser B, et al. A review on applicability, limitations, and improvements of polymeric materials in high-pressure hydrogen gas atmospheres[J]. Polymer reviews, 2022, 62(1): 175–209.
11. Barthélémy H. Hydrogen storage–Industrial prospectives[J]. International journal of hydrogen energy, 2012, 37(22): 17364–17372.
12. Barthélémy H. Compatibility of metallic materials with hydrogen review of the present knowledge[J]. 2007.
13. Dwivedi S K, Vishwakarma M. Hydrogen embrittlement in different materials: A review[J]. International Journal of Hydrogen Energy, 2018, 43(46): 21603–21616.
14. Ni M. An overview of hydrogen storage technologies[J]. Energy exploration & exploitation, 2006, 24(3): 197–209.
15. Zheng J, Liu X, Xu P, et al. Stationary flat steel ribbon wound vessels for storage of high pressure hydrogen[C]. Pressure Vessels and Piping Conference. American Society of Mechanical Engineers, 2012, 55003: 565–569.
16. Zheng J, Liu X, Xu P, et al. Development of high pressure gaseous hydrogen storage technologies[J]. International journal of hydrogen energy, 2012, 37(1): 1048–1057.
17. Ogden J M. Prospects for building a hydrogen energy infrastructure[J]. Annual review of energy and the environment, 1999, 24(1): 227–279.
18. Xu W, Li Q, Huang M. Design and analysis of liquid hydrogen storage tank for high-altitude long-endurance remotely-operated aircraft[J]. International Journal of Hydrogen Energy, 2015, 40(46): 16578–16586.
19. Wijayanta A T, Oda T, Purnomo C W, et al. Liquid hydrogen, methylcyclohexane, and ammonia as potential hydrogen storage: Comparison review[J]. International Journal of Hydrogen Energy, 2019, 44(29): 15026–15044.

20. Aziz M. Liquid hydrogen: A review on liquefaction, storage, transportation, and safety[J]. Energies, 2021, 14(18): 5917.
21. Giauque W F, Johnston H L. Symmetrical and antisymmetrical hydrogen and the third law of thermodynamics. Thermal equilibrium and the triple point pressure[J]. Journal of the American Chemical Society, 1928, 50(12): 3221–3228.
22. Brickwedde F G, Scott R B, Taylor H S. The difference in vapor pressures of ortho and para deuterium[J]. The Journal of Chemical Physics, 1935, 3(11): 653–660.
23. Chen L, Xiao R, Cheng C, et al. Thermodynamic analysis of the para-to-ortho hydrogen conversion in cryo-compressed hydrogen vessels for automotive applications[J]. international journal of hydrogen energy, 2020, 45(46): 24928–24937.
24. Godula-Jopek A, Jehle W, Wellnitz J. Hydrogen storage technologies: new materials, transport, and infrastructure[M]. John Wiley & Sons, 2012.
25. Al Ghafri S Z S, Munro S, Cardella U, et al. Hydrogen liquefaction: a review of the fundamental physics, engineering practice and future opportunities[J]. Energy & environmental science, 2022, 15(7): 2690–2731.
26. Wilhelmsen Ø, Berstad D, Aasen A, et al. Reducing the exergy destruction in the cryogenic heat exchangers of hydrogen liquefaction processes[J]. International Journal of Hydrogen Energy, 2018, 43(10): 5033–5047.
27. Valenti G. Hydrogen liquefaction and liquid hydrogen storage[M]. Compendium of hydrogen energy. Woodhead Publishing, 2016: 27–51.
28. Rivard E, Trudeau M, Zaghib K. Hydrogen storage for mobility: A review[J]. Materials, 2019, 12(12): 1973.
29. Hedayat A, Hastings L J, Brown T. Analytical modeling of variable density multilayer insulation for cryogenic storage[C]. AIP conference proceedings. American Institute of Physics, 2002, 613(1): 1557–1564.
30. Xiaodai X, Sixian W, Chen C, et al. Thermal performance study for hybrid SOFI and MLI system used in space[C]. AIP Conference Proceedings. American Institute of Physics, 2014, 1573(1): 716–719.
31. Wang P, Ji L, Yuan J, et al. Modeling and optimization of composite thermal insulation system with HGMs and VDMLI for liquid hydrogen on orbit storage[J]. International Journal of Hydrogen Energy, 2020, 45(11): 7088–7097.
32. Petitpas G. Simulation of boil-off losses during transfer at a LH2 based hydrogen refueling station[J]. International Journal of Hydrogen Energy, 2018, 43(46): 21451–21463.
33. Notardonato W U, Swanger A M, Fesmire J E, et al. Zero boil-off methods for large-scale liquid hydrogen tanks using integrated refrigeration and storage[C]. IOP conference series: materials science and engineering. IOP Publishing, 2017, 278(1): 012012.
34. Durbin D J, Malardier-Jugroot C. Review of hydrogen storage techniques for on board vehicle applications[J]. International journal of hydrogen energy, 2013, 38(34): 14595–14617.
35. Zohuri B, Zohuri B. Cryogenics and liquid hydrogen storage[J]. Hydrogen Energy: Challenges and Solutions for a Cleaner Future, 2019: 121–139.
36. Aceves S M, Berry G D, Rambach G D, Insulated pressure vessels for hydrogen storage on vehicles[J]. International journal of hydrogen energy, 1998, 23(7): 583–591.
37. Barthélémy H, Weber M, Barbier F. Hydrogen storage: Recent improvements and industrial perspectives[J]. International Journal of Hydrogen Energy, 2017, 42(11): 7254–7262.
38. Aceves S M, Espinosa-Loza F, Ledesma-Orozco E, et al. High-density automotive hydrogen storage with cryogenic capable pressure vessels[J]. International Journal of Hydrogen Energy, 2010, 35(3): 1219–1226.
39. Aceves S M, Petitpas G, Espinosa-Loza F, et al. Safe, long range, inexpensive and rapidly refuelable hydrogen vehicles with cryogenic pressure vessels[J]. international journal of hydrogen energy, 2013, 38(5): 2480–2489.
40. Gupta R B, Basile A, Veziroglu T N. Compendium of hydrogen energy: Hydrogen storage, distribution and infrastructure[M]. Woodhead Publishing, 2016.
41. Lu X, Xie Z, Wu X, et al. Hydrogen storage metal-organic framework classification models based on crystal graph convolutional neural networks[J]. Chemical Engineering Science, 2022, 259: 117813.

42. Park Y M. Literature research on the production, loading, flow, and heat transfer of slush hydrogen[J]. International journal of hydrogen energy, 2010, 35(23): 12993–13003.
43. Atkinson K, Roth S, Hirscher M, et al. Carbon nanostructures: An efficient hydrogen storage medium for fuel cells[J]. Fuel Cells Bulletin, 2001, 4(38): 9–12.
44. Carney R R. "Slush hydrogen" production and handling as a fuel for space projects[C]. Advances in Cryogenic Engineering: Proceedings of the 1963 Cryogenic Engineering Conference University of Colorado College of Engineering and National Bureau of Standards Boulder Laboratories Boulder, Colorado August 19–21, 1963. Boston, MA: Springer US, 1964: 529–536.
45. Lamb K E, Webb C J. A quantitative review of slurries for hydrogen storage–Slush hydrogen, and metal and chemical hydrides in carrier liquids[J]. Journal of Alloys and Compounds, 2022, 906: 164235.
46. Meille V, Pitault I. Liquid Organic Hydrogen Carriers or Organic Liquid Hydrides: 40 Years of History[J]. Reactions, 2021, 2(2): 94–101.
47. Sultan O, Shaw H. Study of automotive storage of hydrogen using recyclable liquid chemical carriers[J]. NASA STI/Recon Technical Report N, 1975, 76: 33642.
48. Zhu Q L, Xu Q. Liquid organic and inorganic chemical hydrides for high-capacity hydrogen storage[J]. Energy & Environmental Science, 2015, 8(2): 478–512.
49. Enthaler S, Langermann J von, Schmidt T. Carbon dioxide and formic acid—the couple for environmental-friendly hydrogen storage?[J]. Energy & Environmental Science, 2010, 3(9): 1207–1217.
50. Inoue Y, Izumida H, Sasaki Y, et al. Catalytic fixation of carbon dioxide to formic acid by transition-metal complexes under mild conditions[J]. Chemistry Letters, 1976, 5(8): 863–864.
51. Jessop P G, Ikariya T, Noyori R. Homogeneous hydrogenation of carbon dioxide[J]. Chemical Reviews, 1995, 95(2): 259–272.
52. Jessop P G, Joó F, Tai C C. Recent advances in the homogeneous hydrogenation of carbon dioxide[J]. Coordination Chemistry Reviews, 2004, 248(21–24): 2425–2442.
53. Gassner F, Leitner W. Hydrogenation of carbon dioxide to formic acid using water-soluble rhodium catalysts [J]. Journal of the Chemical Society, Chemical Communications, 1993 (19): 1465–1466.
54. Elek J, Nádasdi L, Papp G, et al. Homogeneous hydrogenation of carbon dioxide and bicarbonate in aqueous solution catalyzed by water-soluble ruthenium (II) phosphine complexes[J]. Applied Catalysis A: General, 2003, 255(1): 59–67.
55. Munshi P, Main A D, Linehan J C, et al. Hydrogenation of Carbon Dioxide Catalyzed by Ruthenium Trimethylphosphine Complexes:The Accelerating Effect of Certain Alcohols and Amines[J]. Journal of the American Chemical Society, 2002, 124(27): 7963–7971.
56. Hull J F, Himeda Y, Wang W H, et al. Reversible hydrogen storage using CO_2 and a proton-switchable iridium catalyst in aqueous media under mild temperatures and pressures[J]. Nature Chemistry, 2012, 4(5): 383–388.
57. Boddien A, Gärtner F, Federsel C, et al. CO2-"Neutral" Hydrogen Storage Based on Bicarbonates and Formates[J]. Angewandte Chemie International Edition, 2011, 50(28): 6411–6414.
58. Moret S, Dyson P J, Laurenczy G. Direct synthesis of formic acid from carbon dioxide by hydrogenation in acidic media[J]. Nature Communications, 2014, 5(1): 4017.
59. Farlow M W, Adkins H. The Hydrogenation of Carbon Dioxide and a Correction of the Reported Synthesis of Urethans[J]. Journal of the American Chemical Society, 1935, 57(11): 2222–2223.
60. Liu K, Li Y. Photothermal Induced Dual-Interface: Accelerating Sustainable Hydrogen Evolution from Formic Acid[J]. ACS Applied Materials & Interfaces, 2024, 16(45): 62033–62042.
61. Coffey R S. The decomposition of formic acid catalysed by soluble metal complexes[J]. Chemical Communications (London), 1967(18): 923b–9924.
62. Onishi N, Kanega R, Kawanami H, et al. Recent Progress in Homogeneous Catalytic Dehydrogenation of Formic Acid[J]. Molecules, 2022, 27(2): 455.

63. Loges B, Boddien A, Junge H, et al. Controlled Generation of Hydrogen from Formic Acid Amine Adducts at Room Temperature and Application in H_2/O_2 Fuel Cells[J]. Angewandte Chemie International Edition, 2008, 47(21): 3962–3965.
64. Fellay C, Dyson P J, Laurenczy G. A viable hydrogen-storage system based on selective formic acid decomposition with a ruthenium catalyst[J]. Angewandte Chemie International Edition, 2008, 47(21): 3966–3968.
65. Wang X, Meng Q, Gao L, et al. Recent progress in hydrogen production from formic acid decomposition[J]. International Journal of Hydrogen Energy, 2018, 43(14): 7055–7071.
66. Boddien A, Mellmann D, Gärtner F, et al. Efficient dehydrogenation of formic acid using an iron catalyst[J]. Science, 2011, 333(6050): 1733–1736.
67. Mellone I, Peruzzini M, Rosi L, et al. Formic acid dehydrogenation catalysed by ruthenium complexes bearing the tripodal ligands triphos and NP 3[J]. Dalton Transactions, 2013, 42(7): 2495–2501.
68. Morris D J, Clarkson G J, Wills M. Insights into hydrogen generation from formic acid using ruthenium complexes[J]. Organometallics, 2009, 28(14): 4133–4140.
69. Manca G, Mellone I, Bertini F, et al. Inner-versus outer-sphere Ru-catalyzed formic acid dehydrogenation: a computational study[J]. Organometallics, 2013, 32(23): 7053–7064.
70. Gao Y, Kuncheria J, Puddephatt R J, et al. An efficient binuclear catalyst for decomposition of formic acid[J]. Chemical Communications, 1998(21): 2365–2366. https://doi.org/10.1039/A805789C.
71. Solymosi F, Koós Á, Liliom N, et al. Production of CO-free H2 from formic acid. A comparative study of the catalytic behavior of Pt metals on a carbon support[J]. Journal of Catalysis, 2011, 279(1): 213–219.
72. Liu M, Xu Y, Meng Y, et al. Heterogeneous Catalysis for Carbon Dioxide Mediated Hydrogen Storage Technology Based on Formic Acid[J]. Advanced Energy Materials, 2022, 12(31): 2200817.
73. Bi Q Y, Du X L, Liu Y M, et al. Efficient Subnanometric Gold-Catalyzed Hydrogen Generation via Formic Acid Decomposition under Ambient Conditions[J]. Journal of the American Chemical Society, 2012, 134(21): 8926-8933.
74. Wang Z L, Yan J M, Ping Y, et al. An Efficient CoAuPd/C Catalyst for Hydrogen Generation from Formic Acid at Room Temperature[J]. Angewandte Chemie International Edition, 2013, 52(16): 4406-4409.
75. Yadav M, Akita T, Tsumori N, et al. Strong metal–molecular support interaction (SMMSI): Amine-functionalized gold nanoparticles encapsulated in silica nanospheres highly active for catalytic decomposition of formic acid[J]. Journal of Materials Chemistry, 2012, 22(25): 12582-12586.
76. Olah G A, Goeppert A, Czaun M, et al. Bi-reforming of methane from any source with steam and carbon dioxide exclusively to metgas (CO–2H2) for methanol and hydrocarbon synthesis[J]. Journal of the American Chemical Society, 2013, 135(2): 648-650.
77. Onishi N, Laurenczy G, Beller M, et al. Recent progress for reversible homogeneous catalytic hydrogen storage in formic acid and in methanol[J]. Coordination Chemistry Reviews, 2018, 373: 317-332.
78. Wu J, Saito M, Takeuchi M, et al. The stability of Cu/ZnO-based catalysts in methanol synthesis from a CO2-rich feed and from a CO-rich feed[J]. Applied Catalysis A: General, 2001, 218(1-2): 235-240.
79. Cortright R D, Davda R R, Dumesic J A. Hydrogen from catalytic reforming of biomass-derived hydrocarbons in liquid water[J]. Nature, 2002, 418(6901): 964-967.
80. Palo D R, Dagle R A, Holladay J D. Methanol steam reforming for hydrogen production[J]. Chemical reviews, 2007, 107(10): 3992-4021.
81. Barelli L, Bidini G, Gallorini F, et al. Hydrogen production through sorption-enhanced steam methane reforming and membrane technology: A review[J]. Energy, 2008, 33(4): 554-570.
82. Sá S, Silva H, Brandão L, et al. Catalysts for methanol steam reforming—A review[J]. Applied Catalysis B: Environmental, 2010, 99(1-2): 43-57.

83. Kusche M, Enzenberger F, Bajus S, et al. Enhanced activity and selectivity in catalytic methanol steam reforming by basic alkali metal salt coatings[J]. Angewandte Chemie International Edition, 2013, 52(19): 5028-5032.
84. Rodríguez-Lugo R E, Trincado M, Vogt M, et al. A homogeneous transition metal complex for clean hydrogen production from methanol–water mixtures[J]. Nature Chemistry, 2013, 5(4): 342-347.
85. Biniwale R B, Rayalu S, Devotta S, et al. Chemical hydrides: a solution to high capacity hydrogen storage and supply[J]. International Journal of Hydrogen Energy, 2008, 33(1): 360-365.
86. Kariya N, Fukuoka A, Utagawa T, et al. Efficient hydrogen production using cyclohexane and decalin by pulse-spray mode reactor with Pt catalysts[J]. Applied Catalysis A: General, 2003, 247(2): 247-259.
87. Shukla A A, Gosavi P V, Pande J V, et al. Efficient hydrogen supply through catalytic dehydrogenation of methylcyclohexane over Pt/metal oxide catalysts[J]. International Journal of Hydrogen Energy, 2010, 35(9): 4020–4026.
88. Lazaro M P, Garcia-Bordeje E, Sebastian D, et al. In situ hydrogen generation from cycloalkanes using a Pt/CNF catalyst[J]. Catalysis Today, 2008, 138(3–4): 203–209.
89. Kustov L M, Tarasov A L, Tarasov B P. Intermetallide catalysts for hydrogen storage on the basis of reversible aromatics hydrogenation/dehydrogenation reactions[J]. International Journal of Hydrogen Energy, 2013, 38(14): 5713–5716.
90. Rossi L M, Machado G. Ruthenium nanoparticles prepared from ruthenium dioxide precursor: Highly active catalyst for hydrogenation of arenes under mild conditions[J]. Journal of Molecular Catalysis A: Chemical, 2009, 298(1): 69–73.
91. Lylykangas M S, Rautanen P A, Krause A O I. Liquid-Phase Hydrogenation Kinetics of Multicomponent Aromatic Mixtures on Ni/Al2O3[J]. Industrial & Engineering Chemistry Research, 2002, 41(23): 5632–5639.
92. Wang J, Chen H, Li Q. Influence of the Brönsted and Lewis acid sites in platinum/solid acid catalysts on the hydrogenation of benzene and toluene[J]. Reaction Kinetics and Catalysis Letters, 2000, 69: 277–284.
93. Murzin D, Salmi T, Smeds S, et al. Toluene and methylcyclohexane adsorption on nickel catalysts[J]. Reaction Kinetics and Catalysis Letters, 1997, 61(2): 227–236.
94. Yolcular S, Olgun Ö. Ni/Al2O3 catalysts and their activity in dehydrogenation of methylcyclohexane for hydrogen production[J]. Catalysis Today, 2008, 138(3): 198–202.
95. Moores A, Poyatos M, Luo Y, et al. Catalysed low temperature H2 release from nitrogen heterocycles[J]. New Journal of Chemistry, 2006, 30(11): 1675–1678.
96. Clot E, Eisenstein O, Crabtree R H. Computational structure–activity relationships in H2 storage: how placement of N atoms affects release temperatures in organic liquid storage materials[J]. Chemical Communications, 2007(22): 2231–2233.
97. Sotoodeh F, Huber B J M, Smith K J. The effect of the N atom on the dehydrogenation of heterocycles used for hydrogen storage[J]. Applied Catalysis A: General, 2012, 419–420: 67–72.
98. Aziz M, Wijayanta A T, Nandiyanto A B D. Ammonia as Effective Hydrogen Storage: A Review on Production, Storage and Utilization[J]. Energies, 2020, 13(12): 3062.
99. Li C, Wang T, Gong J. Alternative strategies toward sustainable ammonia synthesis[J]. Transactions of Tianjin University, 2020, 26: 67–91.
100. Kobayashi H, Hayakawa A, Somarathne K D K A, et al. Science and technology of ammonia combustion[J]. Proceedings of the combustion institute, 2019, 37(1): 109–133.
101. Rafiqul I, Weber C, Lehmann B, et al. Energy efficiency improvements in ammonia production—perspectives and uncertainties[J]. Energy, 2005, 30(13): 2487–2504.
102. Wang L, Xia M, Wang H, et al. Greening ammonia toward the solar ammonia refinery[J]. Joule, 2018, 2(6): 1055–1074.
103. Siporin S E, Davis RJ. Use of kinetic models to explore the role of base promoters on Ru/MgO ammonia synthesis catalysts[J]. journal of Catalysis, 2004, 225(2): 359–368.

104. Bielawa H, Hinrichsen O, Birkner A, et al. The ammonia-synthesis catalyst of the next generation: barium-promoted oxide-supported ruthenium[J]. Angewandte Chemie International Edition, 2001, 40(6): 1061–1063.
105. Kitano M, Inoue Y, Ishikawa H, et al. Essential role of hydride ion in ruthenium-based ammonia synthesis catalysts[J]. Chemical science, 2016, 7(7): 4036–4043.
106. Medford A J, Vojvodic A, Hummelshøj J S, et al. From the Sabatier principle to a predictive theory of transition-metal heterogeneous catalysis[J]. Journal of Catalysis, 2015, 328: 36–42.
107. Kyriakou V, Garagounis I, Vasileiou E, et al. Progress in the electrochemical synthesis of ammonia[J]. Catalysis Today, 2017, 286: 2–13.
108. Juangsa F B, Irhamna A R, Aziz M. Production of ammonia as potential hydrogen carrier: Review on thermochemical and electrochemical processes[J]. International Journal of Hydrogen Energy, 2021, 46(27): 14455–14477.
109. Gálvez M E, Halmann M, Steinfeld A. Ammonia Production via a Two-Step Al2O3/AlN Thermochemical Cycle. 1. Thermodynamic, Environmental, and Economic Analyses[J]. Industrial & Engineering Chemistry Research, 2007, 46(7): 2042–2046.
110. Minic D. Hydrogen Energy: Challenges and Perspectives[M]. InTech, 2012.
111. Mukherjee S, Devaguptapu S V, Sviripa A, et al. Low-temperature ammonia decomposition catalysts for hydrogen generation[J]. Applied Catalysis B: Environmental, 2018, 226: 162–181.
112. Giddey S, Badwal S P S, Munnings C, et al. Ammonia as a renewable energy transportation media[J]. ACS Sustainable Chemistry & Engineering, 2017, 5(11): 10231–10239.
113. Kantürk F A, Pişkin M B, Coşkuner B, et al. Synthesis, structural characterization, and hydrolysis of Ammonia Borane (NH3BH3) as a hydrogen storage carrier[J]. International Journal of Hydrogen Energy, 2013, 38(36): 16215–16228.
114. Akbayrak S, Özkar S. Ammonia borane as hydrogen storage materials[J]. International Journal of Hydrogen Energy, 2018, 43(40): 18592–18606.
115. Singh S K, Xu Q. Complete conversion of hydrous hydrazine to hydrogen at room temperature for chemical hydrogen storage[J]. Journal of the American Chemical Society, 2009, 131(50): 18032–18033.
116. Wang J, Zhang X B, Wang Z L, et al. Rhodium–nickel nanoparticles grown on graphene as highly efficient catalyst for complete decomposition of hydrous hydrazine at room temperature for chemical hydrogen storage[J]. Energy & Environmental Science, 2012, 5(5): 6885–6888.
117. Graetz J. New approaches to hydrogen storage[J]. Chemical Society Reviews, 2009, 38(1): 73–82.
118. Prabhukhot P R, Wagh M M, Gangal A C. A review on solid state hydrogen storage material[J]. Adv. Energy Power, 2016, 4(11).
119. Rusman N A A, Dahari M. A review on the current progress of metal hydrides material for solid-state hydrogen storage applications[J]. International Journal of Hydrogen Energy, 2016, 41(28): 12108–12126.
120. Chen Z, Ma Z, Zheng J, et al. Perspectives and challenges of hydrogen storage in solid-state hydrides[J]. Chinese Journal of Chemical Engineering, 2021, 29: 1–12.
121. Zijlstra H, Westendorp F F. Influence of hydrogen on the magnetic properties of SmCo5[J]. Solid State Communications, 1969, 7(12): 857–859.
122. Schlapbach L, Brundle C R. XPS study of the chemisorption induced surface segregation in LaNi5 and ThNi5[J]. Journal de Physique, 1981, 42(7): 1025–1028.
123. Johnson J R, Reilly J J. Reaction of hydrogen with the low-temperature form (C15) of titanium-chromium (TiCr2)[J]. Inorganic Chemistry, 1978, 17(11): 3103–3108.
124. Shen H, Zhang J, Paul-Boncour V, et al. AB2-type rare earth-based compounds with C-15 structure: Looking for reversible hydrogen storage materials[J]. Journal of Rare Earths, 2023.
125. Reilly J J, Wiswall R H. Formation and properties of iron titanium hydride[J]. Inorganic Chemistry, 1974, 13(1): 218–222.
126. Sujan G K, Pan Z, Li H, et al. An overview on TiFe intermetallic for solid-state hydrogen storage: microstructure, hydrogenation and fabrication processes[J]. Critical Reviews in Solid State and Materials Sciences, 2020, 45(5): 410–427.

127. Kadir K, Sakai T, Uehara I. Synthesis and structure determination of a new series of hydrogen storage alloys; RMg2Ni9 (R= La, Ce, Pr, Nd, Sm and Gd) built from MgNi2 Laves-type layers alternating with AB5 layers[J]. Journal of Alloys and Compounds, 1997, 257(1–2): 115–121.
128. Liu J, Qin C, Chen X, et al. Prolonging cycling life of AB3-type superlattice alloys by adjusting hydrogen absorption/desorption behaviors of [A2B4] and [AB5] subunits[J]. International Journal of Hydrogen Energy, 2024, 53: 946–957.
129. Sakaki K, Kim H, Asano K, et al. Hydrogen storage properties of Nb-based solid solution alloys with a BCC structure[J]. Journal of Alloys and Compounds, 2020, 820: 153399.
130. Akiba E, Okada M. Metallic hydrides III: body-centered-cubic solid-solution alloys[J]. MRS bulletin, 2002, 27(9): 699–703.
131. Bogdanović B, Schwickardi M. Ti-doped alkali metal aluminium hydrides as potential novel reversible hydrogen storage materials[J]. Journal of alloys and compounds, 1997, 253: 1–9.
132. Srinivasan S S, Brinks H W, Hauback B C, et al. Long-term cycling behavior of titanium doped NaAlH4 prepared through solvent-mediated milling of NaH and Al with titanium dopant precursors[J]. Journal of Alloys and Compounds, 2004, 377(1): 283–289.
133. Chen P, Xiong Z, Luo J, et al. Interaction of hydrogen with metal nitrides and imides[J]. Nature, 2002, 420(6913): 302–304.
134. Wang J, Li H W, Chen P. Amides and borohydrides for high-capacity solid-state hydrogen storage—materials design and kinetic improvements[J]. Mrs Bulletin, 2013, 38(6): 480–487.
135. Li C, Liu Y, Ma R, et al. Superior dehydrogenation/hydrogenation kinetics and long-term cycling performance of K and Rb Cocatalyzed Mg (NH2) 2-2LiH system[J]. ACS applied materials & interfaces, 2014, 6(19): 17024–17033.
136. Li H W, Yan Y, Orimo S, et al. Recent progress in metal borohydrides for hydrogen storage[J]. Energies, 2011, 4(1): 185–214.
137. Chong M, Karkamkar A, Autrey T, et al. Reversible dehydrogenation of magnesium borohydride to magnesium triborane in the solid state under moderate conditions[J]. Chemical Communications, 2011, 47(4): 1330–1332.
138. Crivello J C, Dam B, Denys R V, et al. Review of magnesium hydride-based materials: development and optimisation[J]. Applied Physics A, 2016, 122: 1–20.
139. Zhang J, Li Z, Wu Y, et al. Recent advances on the thermal destabilization of Mg-based hydrogen storage materials[J]. RSC Advances, 2019, 9(1): 408–428.
140. Zhu M, Lu Y, Ouyang L, et al. Thermodynamic tuning of Mg-based hydrogen storage alloys: a review[J]. Materials, 2013, 6(10): 4654–4674.
141. Wagemans R W P, Van Lenthe J H, De Jongh P E, et al. Hydrogen Storage in Magnesium Clusters: Quantum Chemical Study[J]. Journal of the American Chemical Society, 2005, 127(47): 16675–16680.
142. Vajo J J, Mertens F, Ahn C C, et al. Altering hydrogen storage properties by hydride destabilization through alloy formation: LiH and MgH2 destabilized with Si[J]. The Journal of Physical Chemistry B, 2004, 108(37): 13977–13983.
143. Ströbel R, Garche J, Moseley P T, et al. Hydrogen storage by carbon materials[J]. Journal of Power Sources, 2006, 159(2): 781–801.
144. Mohan M, Sharma V K, Kumar E A, et al. Hydrogen storage in carbon materials—A review[J]. Energy Storage, 2019, 1(2): e35.
145. Pinkerton F E, Wicke B G, Olk C H, et al. Thermogravimetric Measurement of Hydrogen Absorption in Alkali-Modified Carbon Materials[J]. The Journal of Physical Chemistry B, 2000, 104(40): 9460–9467.
146. Rüdorff W. Graphite intercalation compounds[M]. Advances in Inorganic Chemistry and Radiochemistry. Academic Press, 1959, 1: 223–266.
147. Parker G J. Guided-Wave Optical Communications: Materials[J]. Encyclopedia of Materials: Science and Technology, 2001: 3703–3707.
148. Gangu K K, Maddila S, Mukkamala S B, et al. Characteristics of MOF, MWCNT and graphene containing materials for hydrogen storage: A review[J]. Journal of Energy Chemistry, 2019, 30: 132–144.

149. Bétard A, Fischer R A. Metal–Organic Framework Thin Films: From Fundamentals to Applications.[J]. Chemical Reviews, 2012, 112(2): 1055–1083.
150. Abdalla A M, Hossain S, Nisfindy O B, et al. Hydrogen production, storage, transportation and key challenges with applications: A review[J]. Energy Conversion and Management, 2018, 165: 602–627.
151. Zhang J Z, Li J, Li Y, et al. Hydrogen generation, storage and utilization[M]. John Wiley & Sons, 2014.
152. Sahu R, Song B J, Im J S, et al. A review of recent advances in catalytic hydrocracking of heavy residues[J]. Journal of Industrial and Engineering Chemistry, 2015, 27: 12–24.
153. Weitkamp J. Catalytic Hydrocracking—Mechanisms and Versatility of the Process[J]. ChemCatChem, 2012, 4(3): 292–306.
154. Ward J W. Hydrocracking processes and catalysts[J]. Fuel Processing Technology, 1993, 35(1): 55–85.
155. Speight J G. New approaches to hydroprocessing[J]. Catalysis Today, 2004, 98(1): 55–60.
156. Ramachandran R, Menon R K. An overview of industrial uses of hydrogen[J]. International Journal of Hydrogen Energy, 1998, 23(7): 593–598.
157. Humphreys J, Lan R, Tao S. Development and Recent Progress on Ammonia Synthesis Catalysts for Haber–Bosch Process[J]. Advanced Energy and Sustainability Research, 2021, 2(1): 2000043.
158. Smith C, K. Hill A, Torrente-Murciano L. Current and future role of Haber–Bosch ammonia in a carbon-free energy landscape[J]. Energy & Environmental Science, 2020, 13(2): 331–344.
159. Rukini A, Rhamdhani M A, Brooks G A, et al. Metals Production and Metal Oxides Reduction Using Hydrogen: A Review[J]. Journal of Sustainable Metallurgy, 2022, 8(1): 1–24.
160. Liu W, Zuo H, Wang J, et al. The production and application of hydrogen in steel industry[J]. International Journal of Hydrogen Energy, 2021, 46(17): 10548–10569.
161. Yip H L, Srna A, Yuen A C Y, et al. A Review of Hydrogen Direct Injection for Internal Combustion Engines: Towards Carbon-Free Combustion[J]. Applied Sciences, 2019, 9(22): 4842.
162. Aleiferis P G, Rosati M F. Controlled autoignition of hydrogen in a direct-injection optical engine[J]. Combustion and Flame, 2012, 159(7): 2500–2515.
163. White C M, Steeper R R, Lutz A E. The hydrogen-fueled internal combustion engine: a technical review[J]. International Journal of Hydrogen Energy, 2006, 31(10): 1292–1305.
164. Huyskens P, Van Oost S, Goemaere P J, et al. The technical implementation of a retrofit hydrogen PFI system on a passenger car[R]. SAE Technical Paper, 2011.
165. Guo J, Yin C K, Zhong D L, et al. Formic Acid as a Potential On-Board Hydrogen Storage Method: Development of Homogeneous Noble Metal Catalysts for Dehydrogenation Reactions [J]. ChemSusChem, 2021, 14(13): 2655–81.
166. Li H J, Yan Y F, Feng S, et al. Ammonia Borane and Its Applications in the Advanced Energy Technology [J]. Journal of Energy Resources Technology-Transactions of the Asme, 2021, 143(11).
167. Kamimura N, Nishimaki K, Ohsawa I, et al. Molecular Hydrogen Improves Obesity and Diabetes by Inducing Hepatic FGF21 and Stimulating Energy Metabolism in db/db Mice [J]. Obesity, 2011, 19(7): 1396–403.
168. Hayashida K, Sano M, Ohsawa I, et al. Inhalation of hydrogen gas reduces infarct size in the rat model of myocardial ischemia–reperfusion injury [J]. Biochemical and Biophysical Research Communications, 2008, 373(1): 30–5.
169. Setsukinai K-i, Urano Y, Kakinuma K, et al. Development of Novel Fluorescence Probes That Can Reliably Detect Reactive Oxygen Species and Distinguish Specific Species*210 [J]. Journal of Biological Chemistry, 2003, 278(5): 3170–5.
170. Ohsawa I, Ishikawa M, Takahashi K, et al. Hydrogen acts as a therapeutic antioxidant by selectively reducing cytotoxic oxygen radicals [J]. Nature Medicine, 2007, 13(6): 688–94.
171. Ono H, Nishijima Y, Adachi N, et al. A basic study on molecular hydrogen (H2) inhalation in acute cerebral ischemia patients for safety check with physiological parameters and measurement of blood H2 level [J]. Medical gas research, 2012, 2(1): 21.

172. Fan L, Tu Z, Chan S H. Recent development of hydrogen and fuel cell technologies: A review [J]. Energy Reports, 2021, 7: 8421–46.
173. Singh R, Kaur N, Singh M. Bio-compatible bio-fuel cells for medical devices [J]. Materials Today: Proceedings, 2021, 44: 242–9.
174. Jin Q, Zhu K, Cui W, et al. Hydrogen-Modulated Stomatal Sensitivity to Abscisic Acid and Drought Tolerance Via the Regulation of Apoplastic pH in Medicago sativa [J]. Journal of Plant Growth Regulation, 2016, 35(2): 565–73.
175. Zhu Y, Liao W. A positive role for hydrogen gas in adventitious root development [J]. Plant signaling & behavior, 2016, 11(6): e1187359.
176. Xu S, Zhu S, Jiang Y, et al. Hydrogen-rich water alleviates salt stress in rice during seed germination [J]. Plant and Soil, 2013, 370(1): 47–57.
177. Jin Q, Zhu K, Cui W, et al. Hydrogen gas acts as a novel bioactive molecule in enhancing plant tolerance to paraquat-induced oxidative stress via the modulation of heme oxygenase-1 signalling system [J]. Plant, Cell & Environment, 2013, 36(5): 956–69.
178. Hu H, Li P, Wang Y, et al. Hydrogen-rich water delays postharvest ripening and senescence of kiwifruit [J]. Food Chemistry, 2014, 156: 100–9.
179. Xie Y, Mao Y, Zhang W, et al. Reactive Oxygen Species-Dependent Nitric Oxide Production Contributes to Hydrogen-Promoted Stomatal Closure in Arabidopsis [J]. Plant Physiology, 2014, 165(2): 759–73.
180. Chen M, Cui W, Zhu K, et al. Hydrogen-rich water alleviates aluminum-induced inhibition of root elongation in alfalfa via decreasing nitric oxide production [J]. Journal of Hazardous Materials, 2014, 267: 40–7.
181. Hu H, Zhao S, Li P, et al. Hydrogen gas prolongs the shelf life of kiwifruit by decreasing ethylene biosynthesis [J]. Postharvest Biology and Technology, 2018, 135: 123–30.
182. Boettcher S W. Introduction to Green Hydrogen[J]. Chemical Reviews, 2024, 124(23): 13095–13098.
183. Zhao Y, Wang T, Setzler B P, et al. A high-performance gas-fed direct ammonia hydroxide exchange membrane fuel cell[J]. ACS Energy Letters, 2021, 6(5): 1996–2002.

Chapter 2
Fuel Cells

2.1 Introduction

The initial design concepts for fuel cells can be traced back to 1839, when the British chemist William Robert Grove was inspired by William Nicholson and Anthony Carlisle, who electrolyzed water to hydrogen and oxygen, and designed a device that could recombine hydrogen and oxygen into water to generate electricity. In February 1839, Grove publicly reported and illustrated the device using platinum plates as electrodes and dilute sulfuric acid as electrolyte, which has been regarded as the prototype of the fuel cell [1]. Fifty years later, the term 'fuel cell' was firstly proposed by British chemist Ludwig Mond and Charles Langer in 1889. They used platinum black electrodes with a large specific surface area and the porous ceramic separator impregnated with dilute sulfuric acid, which improved the cell performance to an impressive level (3.5 mA cm^{-2} @ 0.73 V). In 1894, the German chemist Wilhelm Ostwald demonstrated the unparalleled thermodynamic efficiency and environmental friendliness of fuel cells, marking the establishment of the basic theoretical framework of fuel cell thermodynamics.

The demonstration application of fuel cells can be traced back to 1932, when the British engineer Francis Thomas Bacon constructed a 5 kW alkaline fuel cell stack and successfully applied it in forklifts and welding machines [2]. The Bacon cell attracted great attention and inspired the research boom of fuel cell technology. The first practical application of fuel cells occurred in the aerospace field, and NASA made outstanding contributions to the development of fuel cells during the 1960s. The proton exchange membrane fuel cells (PEMFCs) were first used as the power source for the Gemini spacecraft. In the subsequent Apollo moon-landing project, the alkaline fuel cells with lower corrosiveness were selected [3]. Due to the successful application of fuel cells in spacecraft and the emergence of the global energy crisis in the 1970s, many countries began to explore the feasibility of fuel cell application in ground energy sectors. During this period, small power plants with hundreds of

Y. Li, *Fuel Cell Fundamentals and Applications*,
https://doi.org/10.1007/978-981-96-9845-5_2

kilowatt phosphoric acid fuel cells (PAFCs), solid oxide fuel cells (SOFCs), and molten carbonate fuel cells (MCFCs) were successively constructed [4].

DuPont in the United States successfully developed a perfluorosulfonic acid polymer membrane (Nafion) for fuel cells, which brought people's research focus back to the PEMFCs. In 1993, Ballard Corporation of Canada announced the manufacture of the first PEMFC-powered bus [5]. An air-independent propulsion (AIP) submarine driven by PEMFCs was built and put into service by Germany in 2005 [6]. The fuel cell vehicles of "Mirai", "Clarity", and "Nexo" have been launched in Japan since 2014, indicating the official entry of fuel cell technology into the stage of commercial application [7]. Despite the significant progress of fuel cells in their long history, there is still a long way to go to realize the large-scale commercialization of fuel cells. Nowadays, fuel cells are developing towards higher power density, longer durability, and lower cost, which will play an important role in the future energy.

As one of the representative types of energy conversion devices, fuel cells follow a similar power generation principle to conventional batteries, which can directly convert chemical energy into electrical energy through electrochemical reactions [8]. The core difference between a fuel cell and a battery is that the chemical energy of the battery is stored inside, and once depleted, it can only be discarded or requires a long period of charging to continue to generate electricity. However, the chemical energy of fuel cells is obtained through an external supply. Theoretically, as long as the supply of fuels and oxidants is maintained, fuel cells can continuously produce electricity [9]. Previous studies have found that when the driving distance of vehicles exceeds 160 km, the vehicles driven by hydrogen-powered fuel cells have comprehensively outperformed lithium-ion batteries in terms of operating cost, refueling time, energy efficiency, and service lifetime, which demonstrates the potential of fuel cells in long-distance transportation [10]. Obviously, the fuel supply pattern of fuel cells is similar to traditional thermal power generation devices, but the essential difference between them is that fuel cells directly convert chemical energy into electricity, avoiding the multi-step and inefficient "chemical–thermal–mechanical–electrical" energy conversion processes. Therefore, fuel cells combine many advantages of traditional batteries and thermal power generation devices:

(1) High energy conversion efficiency. As an energy conversion technology without being limited by the Carnot cycle, fuel cells can directly convert chemical energy in reactants into electricity. For example, the theoretical thermodynamic efficiency of hydrogen–oxygen fuel cells under standard conditions is 83%, significantly higher than that of conventional internal combustion engines and thermal power-generation plants [11].
(2) Low environmental pollution. Water is the only product of hydrogen–oxygen fuel cells, achieving zero pollution during the energy conversion processes. Although fuel cells based on hydrocarbon fuels (CH_3OH, CH_3CH_2OH, etc.) emit a certain amount of carbon dioxide, they avoid the emission of atmospheric pollutants such as nitrogen oxides, sulfur oxides, and dust [12].
(3) High reliability. The fuel cell stack is composed of repetitive single cells without moving components, which means the operation of the fuel cells is very quiet

and suitable for military applications. Moreover, the all-solid-state structure of fuel cells indicates the potential for higher reliability and longer lifespan of fuel cell systems [13].

(4) Decoupled of capacity and power. The capacity and power of traditional batteries (typically, lithium-ion batteries) are interrelated, so the capacity and power cannot be arbitrarily scaled. However, the capacity of a fuel cell is determined by the size of the fuel storage tank, and the power is mainly determined by the size of the fuel cell stack. Therefore, the capacity and power of fuel cells are decoupled and can be made into different specifications from microwatts to megawatts, depending on the power requirements [14].
(5) Improved energy security. The hydrogen produced from the renewable energy can be used to power fuel cells, reducing the consumption of non-renewable energy while reducing the reliance on traditional petroleum energy and lowering the risk of energy supply.

Although fuel cells present many attractive advantages, they also have some representative shortcomings. The most frequently mentioned bottlenecks in fuel cell applications are the cost and performance. Firstly, the efficient operation of fuel cells commonly relies on the expensive electrocatalysts, resulting in fuel cell technology being competitive in only a few special application areas, such as the aerospace field [15]. Then, even though the power density of fuel cells has significantly improved in the past few decades, further improvement should be achieved if fuel cells are to be competitive in the field of portable devices and automobiles [16]. Apart from the aforementioned issues, there are still many intractable problems limiting the large-scale application of fuel cell technology, represented by hydrogen–oxygen fuel cell technology:

(1) High costs in hydrogen production. Hydrogen is not as ubiquitous as oxygen. The current mainstream hydrogen production pathway through fossil fuel reforming is unsustainable and accompanied by a large amount of carbon dioxide emissions. Integrating water electrolysis with renewable energy (typically, solar and wind energy) can avoid greenhouse gas emissions, but the electricity consumption of water electrolysis is enormous, resulting in the high cost of producing so-called green hydrogen. In addition, the huge investment in the construction of infrastructure and supporting equipment for hydrogen production and refueling further limits the applications [17].
(2) Difficulties in hydrogen storage and transportation. Three different hydrogen storage and transportation, including high-pressure gaseous, low-temperature liquid, and solid-state adsorption hydrogen storage and transportation, have their technical challenges. Among them, the costs for the first two types of hydrogen storage and transportation remain high, mainly due to the huge energy consumption and strict requirements for the related equipment and materials during the hydrogen storage and release process. In contrast, the hydrogen storage and release rates of materials (such as metal hydrides, carbon nanotubes, etc.) are relatively slow and are vulnerable to temperature and humidity, limiting the development of solid-state hydrogen storage and transportation [18].

Table 2.1 The gravimetric and volumetric energy density of hydrogen, methanol, and ethanol

Fuels	Storage T (°C)/P (MPa)	Gravimetric energy density (kWh/kg)	Volumetric energy density (kWh/L)
H_2 (g)	25/0.1	32.94	2.71×10^{-3}
H_2 (g)	25/70	32.94	1.29
H_2 (l)	– 253/0.1	32.94	2.33
CH_3OH (l)	25/0.1	6.10	4.50
CH_3CH_2OH (l)	25/0.1	7.99	6.28

(3) Low energy density. Although the gravimetric energy density of hydrogen is as high as 32.94 kWh kg^{-1}, the volumetric energy density of hydrogen (Table 2.1), whether in the gaseous or liquid state, is extremely low, which is adverse to its application in portable devices [19].
(4) Problems in hydrogen security. Hydrogen is a highly flammable and explosive gas, which raises concerns about the security of its storage, transportation, and utilization. The range of concentrations of hydrogen that can ignite and cause an explosion is remarkably broad, ranging from 4 to 75% in the air. The odorless and invisible inherent characteristics of hydrogen make it difficult to be detected, let alone the formidable challenge of hydrogen storage and transportation caused by the hydrogen embrittlement [20].
(5) Complex power generation system. The performance of the fuel cell system is sensitive to the operating temperature, humidity, and environmental toxicity. For example, the improved mass and charge transport resistance in low-humidity environments and the membrane swelling in high-humidity environments mean the demand for additional water and thermal management systems, which complicates and further increases the cost of the fuel cell power generation systems [21].

2.2 Fuel Cell Types

Fuel cells, with their wide array of types and applications, can be classified through various methodologies that primarily focus on key factors such as the ion-conductive electrolyte, the type of fuel and fuel supply method, the operating temperature, and the specific application domain. Among these, the classification based on the ion-conductive electrolyte is the most widely adopted, as it directly impacts the performance, efficiency, and operational characteristics of different fuel cells. For example, Proton Exchange Membrane Fuel Cells (PEMFCs) use a solid polymer membrane as the electrolyte and are favored for low-temperature applications like transportation and portable power devices. Alkaline Fuel Cells (AFCs), which use an aqueous alkaline solution, are more tolerant to fuel impurities, making them suitable for specialized uses such as space missions. Solid Oxide Fuel Cells (SOFCs) employ a ceramic electrolyte and operate at high temperatures, making them ideal for large-scale, stationary power generation. Phosphoric Acid Fuel Cells (PAFCs), with their liquid phosphoric acid electrolyte, operate at moderate temperatures and offer a balance of efficiency and durability, commonly applied in stationary power systems. Molten Carbonate Fuel Cells (MCFCs), which use a molten carbonate salt as the electrolyte, also operate at high temperatures and are typically used in industrial applications due to their high efficiency. In addition to electrolyte type, fuel cells can also be categorized based on the type of fuel they use. Hydrogen Fuel Cells (HFCs), considered the cleanest option, directly convert hydrogen into electricity, with water and heat as byproducts, while Direct Methanol Fuel Cells (DMFCs) and Direct Ethanol Fuel Cells (DEFCs) use methanol and ethanol, respectively, as fuel sources, providing alternative energy options for portable devices. Fuel cells can also be distinguished by their fuel supply approaches, with active fuel cells requiring mechanisms like pumps or compressors to deliver fuel, and passive fuel cells relying on natural diffusion. Furthermore, fuel cells are classified by their operating temperature ranges: low-temperature fuel cells like PEMFCs, which operate below 100 °C, medium-temperature fuel cells in the 100–400 °C range, and high-temperature fuel cells such as SOFCs and MCFCs, which operate above 400 °C. Lastly, fuel cells are categorized based on their application domains, with stationary fuel cells used for power generation in fixed installations and portable fuel cells designed for mobile applications. While each classification methodology provides valuable insights, the most prevalent system remains the one based on the ion-conductive electrolyte, as it most directly correlates with the operational features and application suitability of the different fuel cell types. The following sections will further explore the differences among the five primary fuel cell types (Fig. 2.1), shedding light on their unique characteristics, advantages, and challenges.

2.2.1 Alkaline Fuel Cells

Among the five types of fuel cells, the practical application of alkaline fuel cells was the earliest, which can be traced back to the Apollo program in the 1960s. Basically, the alkaline fuel cell is mainly composed of a porous matrix (typically asbestos film) sandwiched between the anode and cathode electrodes. The transmitted ions in alkaline fuel cells are OH^-, and the alkaline solution inside the porous matrix is usually KOH with a concentration of 85 wt% at an operating temperature of 250 °C and a concentration of 35–50 wt% at an operating temperature lower than 120 °C. The advantages of AFCs include: (i) low system cost due to the utilization of non-precious catalysts, (ii) quick electro-kinetics in alkaline media, and (iii) weak corrosion towards electrode materials compared to acidic environment [22]. Despite the above merits, the problem of poor CO_2 tolerance in AFCs is extremely striking. The carbon dioxide in the air reacts with the alkaline solution to generate carbonate ions. On one hand, the aggregation of carbonate ions hinders the transmission of OH^- ions, which increases the internal resistance and thus weakens the output power of

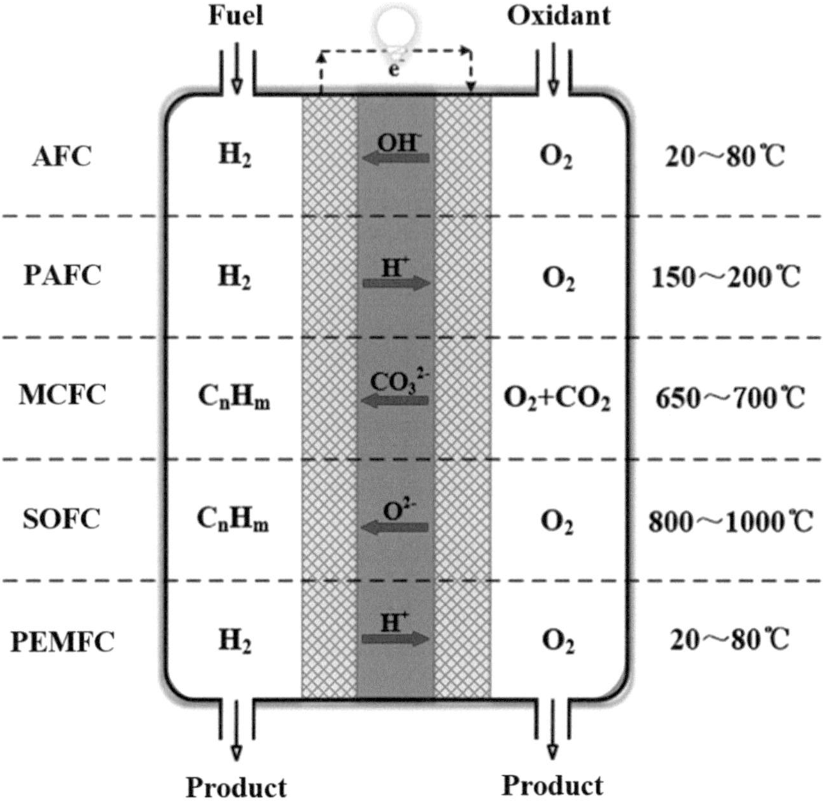

Fig. 2.1 The schematic diagram of fuel cell classification and working temperature, and the principle of different types of fuel cells

AFCs. On the other hand, the generated carbonate ions are prone to deposition inside the porous electrodes, leading to higher mass transfer resistance and degrading the lifespan of AFCs. Hence, the early application of AFCs mainly focused on space environments with scarce CO_2. However, the invention and development of anion-exchange membranes (AEMs) have refocused people's attention on AFCs. Because when replacing the liquid alkaline solution with solid AEMs, the problem of poor CO_2 tolerance in AFCs is effectively alleviated, which not only improves the energy density and performance of AFCs, but also prolongs their lifespans.

2.2.2 Proton Exchange Membrane Fuel Cells

The electrolyte of PEMFCs is a polymer-electrolyte membrane that conducts H^+ ions from the anode to the cathode. The most commonly used solid polymer-electrolyte membrane in PEMFCs is the perfluorosulfonic acid membrane (Nafion) produced by DuPont in the United States. The Nafion membrane possesses the merits of excellent mechanical, chemical, and thermal stability, high proton conductivity, and good resistance to hydrogen and oxygen crossover. However, the proton conductivity of the Nafion membrane is highly dependent on the liquid water in the membrane, which limits the operating temperature of PEMFCs to below 100 °C. Thus, the PEMFCs belong to the category of low-temperature fuel cells, which have the typical characteristics of high-power density, quick startup, and response. Due to the low operating temperature, the redox reactions in both anode and cathode electrodes of PEMFCs rely on the presence of precious metal electrocatalysts (typically Pt/C), resulting in the high cost of the PEMFCs system [23]. Therefore, the cost and performance are still predominant bottlenecks limiting the large-scale commercial application of PEMFCs. With the rapid development of PEMFCs technology in the past few years, PEMFCs have been one of the most important energy alternatives in the fields of transportation, small-scale power plants, and portable electronic devices.

2.2.3 *Phosphoric Acid Fuel Cells*

Phosphoric Acid Fuel Cells utilize a liquid phosphoric acid (100 wt%) electrolyte, which is impregnated into a silicon carbide matrix, offering a robust medium for ion conduction. In this system, hydrogen ions (H^+) are transferred from the anode to the cathode under the influence of the electrochemical potential difference, with platinum-based electrocatalysts such as Pt/C commonly employed to facilitate the reactions at both electrodes. The electrolyte, consisting of highly concentrated H_3PO_4, is highly stable, contributing to the overall durability of the fuel cell and allowing it to operate at elevated temperatures in the range of 150–220 °C, classifying it as a medium-temperature fuel cell. One of the key advantages of PAFCs lies in their relatively high-power generation efficiency, typically around 40%, but when coupled with a cogeneration system, the overall efficiency can soar to as high as 80% [24]. This makes PAFCs particularly attractive for applications where both electricity and heat are required, such as in industrial or residential settings. However, despite their potential, PAFCs still face several challenges that need to be addressed for broader application. These include issues such as the volatilization of the phosphoric acid electrolyte at high temperatures, which can reduce the long-term stability and performance of the cell, as well as CO poisoning of the electrocatalysts, which can degrade the efficiency of the electrochemical reactions. Additionally, PAFCs suffer from a relatively long warm-up period, which can be a disadvantage in situations requiring rapid deployment or start-up. Despite these challenges, PAFCs have reached a relatively mature level of technological development compared to other types of fuel cells, and they are frequently used in distributed power generation for small-scale, stationary applications such as in hospitals, schools, and remote areas. Their ability to provide reliable, efficient power in such contexts underscores their continuing role in the transition to cleaner energy sources.

2.2.4 Molten Carbonate Fuel Cells

The electrolyte of MCFCs is usually the mixed carbonate of alkali metals (Li, Na, K) immersed in a porous ceramic matrix. To maintain the molten state of alkali metals and ensure the high conductivity of carbonate ions, the operating temperature of this type of fuel cell should be set between 600 and 700 °C, which can be classified into the category of high-temperature fuel cells [25]. The high operating temperature makes it possible to use non-precious metal catalysts in both the anode and cathode of MCFCs, typically, nickel aluminum or nickel chromium alloys for the anode, and nickel oxide for the cathode, which directly reduces the cost of the MCFCs system. In addition, various types of fuels, including hydrogen, natural gas, methane, and coal gas, can be supplied to the MCFCs, making their fuel applicability broader. The MCFCs are suitable for fixed applications, such as small and medium-sized power plants or commercial electricity consumption. The initial investment in their power generation system is high, but the system's operating cost is low. It should be specifically pointed out that the energy efficiency of molten carbonate fuel cells is as high as 60%. If combined with gas or steam turbines for cogeneration, their energy utilization efficiency can be further improved to 80%.

2.2.5 Solid Oxygen Fuel Cells

Among the various types of fuel cells, Solid Oxide Fuel Cells operate at the highest temperatures, typically ranging from 800 to 1000 °C, and are categorized as high-temperature fuel cells. The key feature of SOFCs is their use of a solid electrolyte, most commonly yttria-stabilized zirconia (YSZ), which is a highly conductive material for oxygen ions (O^{2-}). Under the extreme conditions within the fuel cell, O^{2-} ions are transferred from the cathode to the anode, enabling the electrochemical reactions that generate electricity. The high operating temperature of SOFCs offers several advantages. It accelerates the electrochemical reaction rate at the electrodes, improving the overall efficiency of the system. Additionally, the elevated temperature reduces the reliance on expensive precious metal electrocatalysts [26], such as platinum, which are commonly used in lower-temperature fuel cells. This characteristic not only lowers operational costs but also enhances the fuel adaptability of SOFCs, enabling them to efficiently operate on a wide range of fuels, including natural gas, biogas, and even coal-derived syngas. This fuel flexibility gives SOFCs a significant edge in terms of their potential for widespread application, especially in large-scale and industrial settings. SOFCs are well-suited for use in large fixed power plants, where they can provide efficient, high-capacity electricity generation. They are also increasingly used in small and medium-sized distributed power supply systems, which are ideal for locations requiring reliable and independent power sources, such as remote areas or critical infrastructure. Moreover, SOFCs are being explored for small household cogeneration systems, where they can simultaneously generate electricity and useful heat for residential use, contributing to energy efficiency and sustainability. With their high efficiency, fuel flexibility, and broad range of applications, SOFCs represent a promising technology in the future of clean energy.

2.3 Thermodynamics of Fuel Cells

2.3.1 *Gibbs Free Energy*

As is known, the fuel cell is an energy conversion device that can directly convert chemical energy stored in fuels and oxidants into electrical energy. Thermodynamics is the science that investigates energy and energy conversion, therefore, fuel cell thermodynamics is crucial for understanding various theoretical boundary parameters in the fuel cells [27]. The thermodynamics of fuel cells can be mainly used to (1) predict whether a certain chemical reaction in a fuel cell can occur spontaneously, (2) calculate the reversible open-circuit voltage (OCV) of the fuel cell, and (3) calculate the theoretical thermodynamic efficiency of the fuel cell. In thermodynamic systems, the internal energy (U) is the energy possessed in the irregular motion and interaction of microscopic particles. The expression for enthalpy is $H = U + pV$, which can reflect the maximum heat energy extracted from the fuel under atmospheric pressure conditions. The Gibbs free energy (G) represents the maximum theoretical value that a reaction system can convert chemical energy into electrical energy. Under the constant temperature and pressure conditions, the maximum electrical energy that a fuel cell can output is equal to the negative value of the Gibbs free energy change (ΔG) during the reaction process, which can be expressed as:

$$W_{\text{elec}} = -\Delta G \tag{2.1}$$

In the equation, the electrical energy (W) is achieved by moving a certain amount of charge (Q) under the potential difference (E):

$$W_{\text{elec}} = -\Delta G = QE = nFE \tag{2.2}$$

where n represents the number of transferred electrons, which are 2 and 4 for the hydrogen oxidation reaction and the oxygen reduction reaction, respectively. F represents the Faraday constant, and it is 96,485 C mol^{-1}.

2.3.2 *Reversible Voltage*

When discussing the thermodynamics of fuel cells, the concept of reversibility is often mentioned, and the reversible voltage of the fuel cell refers to the voltage at which the fuel cell is in thermodynamic equilibrium. The reversible voltage of fuel cells is closely related to their thermodynamics. The reversible voltage under standard conditions can be expressed as:

$$E_0 = -\frac{\Delta G_0}{nF} \tag{2.3}$$

where E_0 and ΔG_0 are the reversible voltage and Gibbs free energy change of fuel cells under the standard conditions. It should be noted that the equations regarding reversible voltage are only applicable to the fuel cells in thermodynamic equilibrium. Once the balance is disrupted, such as when the fuel cell begins to produce electricity, the equation for reversible voltage will no longer apply. Generally, the reversible voltage is used to estimate the maximum voltage of the fuel cell; for example, the maximum voltage under standard conditions can be obtained from Eq. 2.3. However, fuel cells often operate under non-standard conditions, and it is crucial to predict the reversible voltage of fuel cells under non-standard conditions.

Firstly, the impact of operating temperature on the reversible voltage is investigated as follows. According to the definition of fuel cell thermodynamics, the differential equation for Gibbs free energy can be expressed as:

$$\mathrm{d}G = -S\mathrm{d}T + V\mathrm{d}p \tag{2.4}$$

assuming that the fuel cells operate at ordinary pressure, the above equation can be transformed into:

$$\left(\frac{\mathrm{d}G}{\mathrm{d}T}\right)_p = -S \tag{2.5}$$

by substituting Eq. 2.3, the following equation is obtained:

$$\left(\frac{\mathrm{d}E}{\mathrm{d}T}\right)_p = \frac{S}{nF} \tag{2.6}$$

the effect of operating temperature on the reversible voltage of fuel cells can be obtained by integrating Eq. 2.6:

$$E_T = E_0 + \frac{\Delta S}{nF}(T - T_0) \tag{2.7}$$

where T represents the operating temperature of the fuel cell. E_T is the reversible voltage of the fuel cell at operating temperature T. ΔS is the entropy change of the fuel cell's overall reaction. The entropy change is negative for the vast majority of fuel cells, which leads to a decrease in the reversible voltage as the operating temperature increases. It is estimated that for every 100 °C increase in operating temperature of hydrogen–oxygen fuel cells, their reversible voltage decreases by

23 mV [28]. Although raising the operating temperature has adverse effects on the fuel cell thermodynamics, the actual cell performance significantly improves with the increase of operating temperature. This can be mainly attributed to the influence on electrochemical kinetics, which will be further discussed in the future.

The above equations assume that fuel cells operate at ordinary pressure, but in reality, fuel cells can operate at any pressure. For an isothermal process, the change of Gibbs free energy can be expressed as:

$$\mathrm{d}G = V\mathrm{d}p \tag{2.8}$$

for the ideal gases such as hydrogen and oxygen, they all satisfy the ideal gas state equation:

$$pV = RT \tag{2.9}$$

where V is the molar volume ($\mathrm{m}^3\ \mathrm{mol}^{-1}$). Combining formulas 2.8 and 2.9, and then integrating to obtain the pressure effect on the Gibbs free energy change:

$$\Delta G = \Delta G_0 + RT\left(\frac{p}{p_0}\right) \tag{2.10}$$

where p represents the operating pressure of the fuel cell. p_0 is the standard atmospheric pressure.

For any fuel cell chemical reaction:

$$aA + bB \rightarrow mC + nD \tag{2.11}$$

according to the thermodynamics of fuel cells, the change in Gibbs free energy of the reaction can be expressed by the activity of reactants and products as follows:

$$\Delta G = \Delta G_0 + RT\left(\frac{a_M^m \cdot a_N^n}{a_A^a \cdot a_B^b}\right) \tag{2.12}$$

by combining the relationship between Gibbs free energy change and reversible voltage, the reversible voltage of fuel cells can be obtained as a function of activity:

$$E = E_0 - \frac{RT}{nF}\left(\frac{a_M^m \cdot a_N^n}{a_A^a \cdot a_B^b}\right) \tag{2.13}$$

this is the famous Nernst equation, through which the reversible voltage of the fuel cells can be calculated under any operating conditions [29].

2.4 Kinetics of Electrochemical Reaction

2.4.1 Activation Energy

As mentioned earlier, the Gibbs free energy change of fuel cells is negative, meaning that from an energy perspective, the transition from reactants to products is spontaneous. However, the spontaneity does not guarantee that a reaction will occur, nor does it give the reaction rate at which the reaction occurs. This indicates that the electrochemical reaction is not only influenced by the fuel cell thermodynamics, but also limited by the electrochemical kinetics, which directly determines the rate of electrochemical reaction. To achieve the transition from reactants to products, the Gibbs free energy of the reaction must cross a barrier at a certain energy peak, which is called the activation energy. The magnitude of the activation energy directly affects the probability of the reactant being in an activated state, which in turn determines the rate of electrochemical reaction. The relationship between the probability of reactants being in an activated state and the activation energy can be expressed as:

$$P = \mathrm{e}^{-\Delta G^{*}/RT} \tag{2.14}$$

where P is the probability that the reactant is in an activated state, ΔG^{*} is the activation energy. According to the transition state theory, the rate of electrochemical reaction (v) is determined by three parameters: (1) the concentration of the reactant, (2) the frequency of the activated reactant decay into the product, and (3) the probability of the reactant being in the activated state:

$$v = cfP = c\frac{kT}{h}\mathrm{e}^{-\Delta G^{*}/RT} \tag{2.15}$$

where c is the concentration of the reactant, f is the frequency of the activated reactant decay into the product, k represents the Boltzmann constant, and h is the Planck constant. After understanding the relationship between electrochemical reaction rate and activation energy, let's focus our attention again on the fuel cells. Compared to the rate of electrochemical reactions, current density is more important in fuel cells. According to Faraday's law, the current density generated by the electrochemical reaction is a direct reflection of the electrochemical reaction rate. For any reversible reaction, there exist the forward current density (j_1) and the backward current density (j_2), which can be respectively represented as:

$$j_1 = nFv_1 = nFc_1f_1\mathrm{e}^{-\Delta G_1^{*}/RT} \tag{2.16}$$

$$j_2 = nFv_2 = nFc_2f_2\mathrm{e}^{-\Delta G_2^{*}/RT} \tag{2.17}$$

where the subscripts 1 and 2 represent the corresponding parameters in the forward and backward reactions, respectively. For example, v_1 and v_2 are the forward and backward reaction rates. It is worth emphasizing that when a fuel cell is in thermodynamic equilibrium, the forward reaction current density is equal to the backward reaction current density, and there is no net current output from the fuel cell:

$$j_0 = j_1 = j_2 \tag{2.18}$$

where j_0 is the exchange current density, which is a very important parameter in the fuel cells. It represents that even if the fuel cell does not produce a net current, the forward and backward reactions of the fuel cell are still occurring at a reaction rate of j_0.

2.4.2 Butler–Volmer Equation

According to the fuel cell thermodynamics, we know that when a fuel cell is in equilibrium, the reversible voltage precisely offsets the difference in Gibbs free energy of reactants and products, and there is no net current output from the fuel cell. Therefore, to produce electricity, it is necessary to change the cell voltage and break the balance state of the fuel cell. In the electrochemical reaction process, changing the voltage can affect the Gibbs free energy of charged species, which leads to a variation in the activation energy of the electrochemical reaction and ultimately affects the reaction rate of both forward and backward reactions. Assuming that the reversible potential of an electrochemical reaction generates an overpotential of η, which is equivalent to an increase in anodic potential of η or a decrease in cathodic potential of η. The appearance of anode overpotential makes the anodic potential more positive and prone to oxidation reactions [30]. Similarly, the appearance of cathode overpotential makes the cathodic potential more negative and prone to occur reduction reactions. In short, whether for the anode or cathode, the overpotential causes the electrochemical reaction to proceed in a forward direction. The occurrence of overpotential reduces the activation energy of the forward reaction and increases the activation energy of the backward reaction. At this time, the activation energies of the forward and backward reactions are:

$$\Delta G_1^{**} = \Delta G_1^{*} - \alpha n F \eta \tag{2.19}$$

$$\Delta G_2^{**} = \Delta G_2^{*} + (1 - \alpha) n F \eta \tag{2.20}$$

where ΔG_1^{**} and ΔG_2^{**} are respectively the activation energies of the forward and backward reactions with the overpotential of η. ΔG_1^{*} and ΔG_2^{*} are respectively the activation energies of the forward and backward reactions in the equilibrium state. α is the charge transfer coefficient, which depends on the symmetry of the activation energy in the electrochemical reaction. By substituting the corresponding formulas 2.16 and 2.17, the forward and backward current densities are transformed into:

$$j_1 = nFc_1f_1\mathrm{e}^{-(\Delta G_1{}^* - \alpha nF\eta)/RT} = j_0\mathrm{e}^{\alpha nF\eta/RT} \tag{2.21}$$

$$j_2 = nFc_2f_2\mathrm{e}^{-[\Delta G_2^* + (1-\alpha)nF\eta]/RT} = j_0\mathrm{e}^{-(1-\alpha)nF\eta/RT} \tag{2.22}$$

subtracting the two equations to obtain the net current density:

$$j = j_0\left(\mathrm{e}^{\alpha nF\eta/RT} - \mathrm{e}^{-(1-\alpha)nF\eta/RT}\right) \tag{2.23}$$

the above equation is the famous Butler–Volmer (BV) equation, which is considered as the cornerstone of electrochemical kinetics [31].

2.5 Mass and Charge Transport

2.5.1 Charge Transport

Through the in-depth exploration of electrochemical kinetics, we disrupted the equilibrium of the electrochemical reactions by sacrificing a certain degree of reversible voltage, achieving the net current output. To utilize the generated current, a circuit must be formed, which involves the transfer of charges (including ions and electrons). Taking the PEMFCs as an example, the electrons generated by the hydrogen oxidation reaction are transferred from the anode to the cathode through an external circuit. At the same time, the protons are transferred from the anode to the cathode through the PEM, forming a closed loop to compensate for the system charge difference. Three main driving forces in fuel cells drive the charge transport: (1) the electrochemical driving force generated by the potential difference, (2) the chemical driving force due to the concentration difference, and (3) the mechanical driving force caused by the pressure difference. Among them, the electrochemical driving force plays a dominant role in the charge transfer process of the fuel cell [32]. The transport characteristics and mechanisms of ions and electrons in fuel cells are substantially different. In most cases, ion transport is much more difficult than electron transport, so people pay more attention to the investigation of ion transport [33].

To explore the characteristics and mechanisms of charge transport, it is necessary to understand and be familiar with the microstructure of perfluorosulfonic acid (PFSA) membrane. The PFSA resin materials are composed of the main chain of polytetrafluoroethylene (PTFE) backbone and the side chain of sulfonic acid group. Among them, the PTFE main chain is hydrophobic, while the sulfonic acid group is hydrophilic. Therefore, PFSA is a unique material with amphiphilicity. The hydrophilic phase can provide channels for proton transport after absorbing water, while the hydrophobic phase ensures the mechanical size and morphology of the membrane. Based on the unique hydrophilic and hydrophobic phase separation structure of PFSA materials, two widely recognized proton conduction mechanisms have been developed in the academic community: the vehicle mechanism and the Grotthuss mechanism [34]. The Grotthuss mechanism is vividly described as proton hopping, which refers to the protons hopping along the hydrogen-bond network on the surface of adjacent hydronium ions, usually occurring in the solid polymer electrolyte with high hydration. While in the case of the solid polymer electrolyte with low hydration or an incomplete hydrogen-bond network, protons will migrate with water molecules and diffuse in the form of hydronium ions, which is called the vehicle mechanism. It should be emphasized that the charge transfer within the membrane and the catalyst layer (CL) is influenced by numerous factors such as environment, materials, and electrode structure. Only by mastering their changing patterns and reaction mechanisms can fuel cells' performance be optimized in a targeted manner.

2.5.2 *Diffusive Mass Transport*

To maintain the electricity generation by fuel cells, it is necessary to continuously supply fuel and oxidant to the anode and cathode of the fuel cells, while ensuring the effective exclusion of the products. The process of reactant supply and product exhaust is referred to the mass transport in fuel cells. There are two types of mass transport within the fuel cell: the diffusive mass transport occurring in the electrodes with micrometer and nanoscale pores, and the convective mass transport occurring in the flow channels with centimeter or millimeter structures [35]. In this section, the focus will first be on the diffusive mass transport, which is the mass transfer induced by the concentration gradients. In the stable operation of fuel cells, the diffusion flow rate of diffusive mass transport caused by the concentration gradient is exactly equal to the consumption rate of reactants in the catalytic layer:

$$j = nFJ_{\text{diff}} \tag{2.24}$$

where j is the current density of the fuel cell in the stable state, which is a reflection of the electrochemical reaction rate. J_{diff} is the diffusion flow rate from the flow channel to the catalytic layer, which can also be expressed by the diffusion equation:

$$J_{\text{diff}} = -D^{\text{eff}} \frac{c_R^* - c_R^0}{\delta} \tag{2.25}$$

where D_{diff} is the effective diffusion coefficient of reactants within the catalytic layer. c_R^* and c_R^0 respectively represent the concentration of reactants in the catalytic layer and flow channel. δ is the thickness of the catalytic layer. It can be observed from formula 2.25 that there is a limit on the diffusion flow rate, that is, when the concentration of reactants in the catalytic layer decreases to 0, the diffusive mass transport reaches its maximum value. When the above situation is applied to the current density of fuel cells, the concept of limiting current density will appear:

$$j_L = nFD^{\text{eff}} \frac{c_R^0}{\delta} \tag{2.26}$$

where, j_L is the limiting current density of the fuel cell [36]. The key to improving the limiting current density of fuel cells lies in the enhancement of the diffusive mass transport.

2.5.3 *Convective Mass Transport*

The convective mass transport refers to the mass transfer due to the motion of the fluid containing both liquid and gas under the action of a mechanical force. In the stable operation of fuel cells, the convective mass transport is equal to the diffusion mass transport. The quality of convective mass transport in fuel cells is mainly determined by the convective mass transfer coefficient, which is influenced by the flow channel design and geometric dimensions. The existing research indicates that increasing the flow velocity, reducing channel and gas diffusion layer (GDL) thickness can enhance the convective mass transport in fuel cells. In addition, improper flow channel and size design in fuel cells not only worsens the convective mass transport, leading to a decrease in fuel cell performance, but in severe cases, it can also cause the reverse polarity of the fuel cell, resulting in irreversible damage to the fuel cell materials. At present, there are various designs for fuel cell flow channels, but the most commonly used designs can be divided into the following three categories: (1) the parallel flow channels. The advantage of parallel flow channels is that the pressure drop of the fluid from the inlet to the outlet is small, but the distribution of reactants in this channel may be uneven, and the reaction products are prone to aggregation, which damages the convective mass transport. (2) Interdigitated flow channels, which effectively promote the forced convection of reactants in the, but this flow channel results in a high pressure drop loss. (3) Serpentine flow channels or parallel-serpentine flow channels, which have a good compromise between convective mass transport and pressure drop loss, so it is widely used in practical fuel cells [37].

2.6 Summary and Outlook

The large-scale deployment of fuel cells, known for their low-carbon emissions, environmental sustainability, and high efficiency, plays a crucial role in achieving global carbon neutrality goals. Among the various types of fuel cells, PEMFCs, which are powered by hydrogen, have attracted considerable attention in the commercial market over the past few decades. This is particularly evident in the transportation sector, where PEMFCs have shown significant promise due to their exceptional power density and the simplicity of their reaction products, primarily water. These advantages make PEMFCs an attractive option for clean energy solutions, especially in areas that demand high efficiency and low emissions. However, despite the growing number of PEMFCs applications, several challenges remain in realizing the large-scale deployment of hydrogen-powered fuel cells. These challenges primarily stem from issues related to fuel cell performance, which continue to be a critical barrier to widespread adoption. Specifically, the performance of fuel cells is heavily influenced by various types of voltage losses, including the voltage loss caused by fuel crossover, activation loss, Ohmic loss, and concentration loss. When fuel crossover and internal current occur in a fuel cell, it is equivalent to generating current inside the fuel cell, even if the external circuit current density of the fuel cell is zero. The extremely low current density inside a fuel cell can lead to higher voltage losses, so the actual open-circuit voltage of the fuel cell is significantly lower than the theoretical reversible voltage. Activation loss occurs due to the energy required to initiate electrochemical reactions at the electrodes, particularly at lower temperatures or under high current densities. Ohmic losses are associated with the resistance of the fuel cell components, such as the electrolyte and the electrodes, which hinder the efficient flow of ions and electrons. Concentration loss arises from the depletion of reactants near the electrode surface, especially when the fuel cell is operating at higher power levels. In a word, in the actual operation of fuel cells, the polarization losses mentioned above coexist and jointly affect the actual operating voltage of fuel cells. The schematic of the impact of the polarization losses on the actual operating voltage of fuel cells is shown in Fig. 2.2.

The continued development of PEMFCs will need to address these key issues. Innovations in materials, electrode design, and fuel cell architecture could help mitigate these losses and improve efficiency. Furthermore, advances in hydrogen production, storage, and distribution systems are crucial for the practical integration of PEMFCs into various applications, including vehicles, stationary power generation,

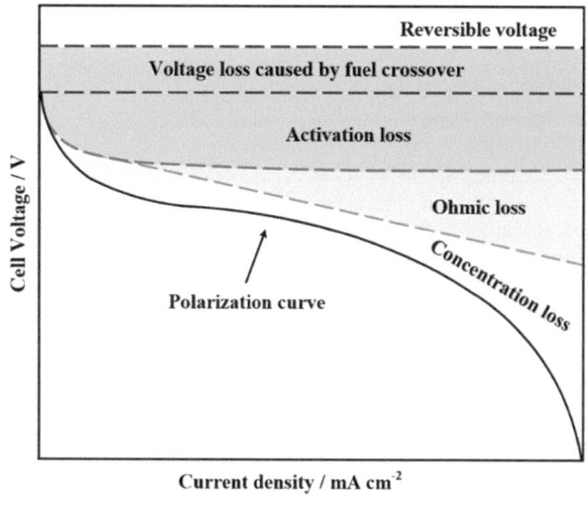

Fig. 2.2 The schematic of polarization losses in the fuel cell

and portable devices. The future of hydrogen-powered fuel cells depends not only on improving the technology itself but also on creating an integrated infrastructure that supports their widespread adoption. In conclusion, while PEMFCs hold great promise for a sustainable, low-carbon future, the road to large-scale commercialization is still lined with technical challenges. Overcoming these obstacles will require continuous research, innovation, and investment in both fuel cell technology and the hydrogen economy. As these advancements unfold, hydrogen-powered fuel cells could become a cornerstone of global efforts to reduce greenhouse gas emissions and combat climate change.

References

1. Brusso B C. Electrochemistry and the development of the hydrogen fuel cell [J]. IEEE Industry Applications Magazine, 2021, 27(5): 8–13.
2. Appleby AJ, Foulkes FR. Fuel cell handbook [M]. US: Van Nostrand Reinhold, 1988.
3. Gülzow E. Alkaline fuel cells [J]. Fuel cells, 2004, 4(4): 251–255.
4. Brenscheidt T, Janowitz K, Salge HJ, et al. Performance of ONSI PC25 PAFC cogeneration plant [J]. International journal of hydrogen energy, 1998, 23(1): 53–56.
5. Prater KB. Solid polymer fuel cell developments at Ballard [J]. Journal of Power Sources, 1992, 37(1–2): 181–188.
6. Psoma A, Sattler G. Fuel cell systems for submarines: from the first idea to serial production [J]. Journal of Power Sources, 2002, 106(1–2): 381–383.
7. Wang Y, Diaz DFR, Chen KS, et al. Materials, technological status, and fundamentals of PEM fuel cells–a review [J]. Materials today, 2020, 32: 178–203.
8. Carrette L, Friedrich K A, Stimming U. Fuel cells: principles, types, fuels, and applications [J]. ChemPhysChem, 2000, 1(4): 162–193.
9. Stambouli A B. Fuel cells: The expectations for an environmental-friendly and sustainable source of energy [J]. Renewable and Sustainable Energy Reviews, 2011, 15(9): 4507–4520.
10. Li S, Djilali N, Rosen M A, et al. Transition of heavy-duty trucks from diesel to hydrogen fuel cells: opportunities, challenges, and recommendations [J]. International Journal of Energy Research, 2022, 46(9): 11718–11729.
11. Du H, Zhao C X, Lin J, et al. Carbon nanomaterials in direct liquid fuel cells [J]. The Chemical Record, 2018, 18(9): 1365–1372.
12. An L, Zhao T S, Li Y S. Carbon-neutral sustainable energy technology: Direct ethanol fuel cells [J]. Renewable and Sustainable Energy Reviews, 2015, 50: 1462–1468.
13. Wang S, Jiang S P. Prospects of fuel cell technologies [J]. National Science Review, 2017, 4(2): 163–166.
14. Mekhilef S, Saidur R, Safari A. Comparative study of different fuel cell technologies [J]. Renewable and Sustainable Energy Reviews, 2012, 16(1): 981–989.
15. Baroutaji A, Wilberforce T, Ramadan M, et al. Comprehensive investigation on hydrogen and fuel cell technology in the aviation and aerospace sectors [J]. Renewable and sustainable energy reviews, 2019, 106: 31–40.
16. Pollet BG, Staffell I, Shang JL. Current status of hybrid, battery and fuel cell electric vehicles: From electrochemistry to market prospects [J]. Electrochimica Acta, 2012, 84: 235–249.
17. Prater DN, Rusek JJ. Energy density of a methanol/hydrogen-peroxide fuel cell [J]. Applied Energy, 2003, 74(1–2): 135–140.
18. Chalk SG, Miller JF. Key challenges and recent progress in batteries, fuel cells, and hydrogen storage for clean energy systems [J]. Journal of Power Sources, 2006, 159(1): 73–80.
19. Ning F, He X, Shen Y, et al. Flexible and lightweight fuel cell with high specific power density [J]. ACS nano, 2017, 11(6): 5982–5991.

20. Fan L, Tu Z, Chan S H. Recent development of hydrogen and fuel cell technologies: A review [J]. Energy Reports, 2021, 7: 8421–8446.
21. Rashidi S, Karimi N, Sunden B, et al. Progress and challenges on the thermal management of electrochemical energy conversion and storage technologies: Fuel cells, electrolysers, and supercapacitors [J]. Progress in Energy and Combustion Science, 2022, 88: 100966.
22. Ferriday T B, Middleton P H. Alkaline fuel cell technology-A review [J]. International journal of hydrogen energy, 2021, 46(35): 18489–18510.
23. Lü X, Qu Y, Wang Y, et al. A comprehensive review on hybrid power system for PEMFC-HEV: Issues and strategies [J]. Energy Conversion and Management, 2018, 171: 1273–1291.
24. Sammes N, Bove R, Stahl K. Phosphoric acid fuel cells: Fundamentals and applications [J]. Current opinion in solid state and materials science, 2004, 8(5): 372–378.
25. Cui C, Li S, Gong J, et al. Review of molten carbonate-based direct carbon fuel cells [J]. Materials for Renewable and Sustainable Energy, 2021, 10: 1–24.
26. Singh M, Zappa D, Comini E. Solid oxide fuel cell: Decade of progress, future perspectives and challenges [J]. International Journal of Hydrogen Energy, 2021, 46(54): 27643–27674.
27. Taner T. The novel and innovative design with using H2 fuel of PEM fuel cell: Efficiency of thermodynamic analyze [J]. Fuel, 2021, 302: 121109.
28. Jiang SP, Li Q, Jiang SP, et al. Fuel Cell Electrochemistry [J]. Introduction to Fuel Cells: Electrochemistry and Materials, 2022: 69–122.
29. Sahu IP, Krishna G, Biswas M, et al. Performance study of PEM fuel cell under different loading conditions [J]. Energy Procedia, 2014, 54: 468–478.
30. Mann RF, Amphlett JC, Peppley BA, et al. Application of Butler–Volmer equations in the modelling of activation polarization for PEM fuel cells [J]. Journal of Power Sources, 2006, 161(2): 775–781.
31. Li X, Tan H, Ni Z, et al. Select sensitivity parameters for proton exchange membrane fuel cell model: An identification method from analytical Butler-Volmer equation [J]. Journal of Power Sources, 2024, 608: 234330.
32. Zhang J, Zhu W, Huang T, et al. Recent insights on catalyst layers for anion exchange membrane fuel cells [J]. Advanced Science, 2021, 8(15): 2100284.
33. Zhao Z, Zhang M, Du W, et al. Strong and flexible high-performance anion exchange membranes with long-distance interconnected ion transport channels for alkaline fuel cells [J]. ACS Applied Materials & Interfaces, 2022, 14(33): 38132–38143.
34. Annapragada R, Vandavasi K R, Kanuparthy P R. Metal-organic framework membranes for proton exchange membrane fuel cells: A mini-review [J]. Inorganica Chimica Acta, 2023, 546: 121304.
35. Mojica F, Rahman M A, Sarker M, et al. Study of converging-diverging channel induced convective mass transport in a proton exchange membrane fuel cell [J]. Energy Conversion and Management, 2021, 237: 114095.
36. Zhong G, Xu S, Liu L, et al. Effect of experimental operations on the limiting current density of oxygen reduction reaction evaluated by rotating-disk electrode [J]. ChemElectroChem, 2020, 7(5): 1107–1114.
37. Wang J. Theory and practice of flow field designs for fuel cell scaling-up: A critical review [J]. Applied Energy, 2015, 157: 640–663.

Chapter 3
Fuel Cell Diagnostic Tools

3.1 Introduction

A fuel cell is an electrochemical device consisting of several components. Its power output is affected not only by the components but also by the interfaces between different components. Therefore, the performance of the fuel cell system cannot be simply obtained by summing up that of its independent internal components. In addition, the complex physical and chemical processes inside the cell make it difficult to analyze the performance [1]. For example, in a hydrogen–oxygen fuel cell, the electrochemical reaction will generate water. An increase in water content can cause water flooding, hindering the transmission of reactants. Then, the decrease in reactants will slow down the reaction rate, leading to a reduction in water generation rate. This is the complex interaction between water transport and reactant transport [2]. Therefore, to understand the complex physical and chemical processes inside the cell and better optimize the design of fuel cells, it is very important to quantitatively analyze various aspects of the cell performance. Characterization techniques for fuel cells can be used to quantitatively compare fuel cell systems. Some characterization techniques can distinguish various sources of internal losses in a fuel cell, thereby providing necessary guidance for cell design [3]. Moreover, the experimental data obtained from testing can provide a foundation and reference for cell simulation and analysis models [4]. Therefore, characterization methods play an irreplaceable role in the development of fuel cells.

The most commonly used indicators for evaluating cell performance are polarization characteristics [5]. They are often tested using an electrochemical workstation and plotted as polarization curves for comparison. Because polarization characteristics can characterize the macroscopic output characteristics of the cell, they are also commonly used as the ultimate indicator for optimizing cell design. However, they are strongly influenced by operating conditions. To ensure fairness in comparing cell performance, the tested cell is generally run under the same conditions. However, a

Y. Li, *Fuel Cell Fundamentals and Applications*,
https://doi.org/10.1007/978-981-96-9845-5_3

single polarization curve can only reflect the total effects of all characteristic parameters, including design structure, material properties, and working conditions. It cannot meet the requirements for analyzing the complex processes inside the cell. Therefore, in experiments, other electrochemical characterization techniques are often combined to further analyze the cell. For example, Electrochemical Impedance Spectroscopy (EIS) [6] and Cyclic voltammetry (CV) [4] can provide a more in-depth understanding of the activation loss, ohmic loss, and mass transport loss inside the cell, but the experimental results may be difficult to interpret. The following chapters will provide some explanations of electrochemical testing methods.

In fuel cells, the membrane electrode is the core component and has a significant impact on all aspects of cell performance [7]. Therefore, offline characterization of cell materials will help to analyze the mechanisms that affect cell performance and deepen the understanding of internal transport mechanisms and durability. In this regard, since the scale of electrode materials is generally in the micrometer or even nanometer range, the resolution of ordinary optical microscopes cannot meet their observation needs. At this point, new testing equipment such as scanning electron microscopes, atomic force microscopes, and transmission electron microscopes can more accurately observe the internal structure of the electrode, which is beneficial for analyzing the core components of the cell [8].

In addition to the electrochemical and offline material characterization techniques mentioned above, the distribution of gas concentration [9], current density [10], and temperature [11] during cell operation is often not uniform. Therefore, it is important to use distribution characterization testing techniques to analyze the distribution characteristics inside the cell. Fuel cell durability [12], cold start [13], and intramembrane transport properties[14] are also important indicators that have a significant impact on the commercial prospects of the cell. It is necessary to have appropriate characterization techniques to test the macroscopic properties of the cell.

At the same time, electrocatalysts are still one of the key materials restricting the further commercialization of fuel cells. Various in situ characterization techniques are widely used in the evolution of catalysts' morphology and electronic structure, as well as in the identification of reactants and intermediates during catalysis. Therefore, in situ characterization has become an indispensable characterization method in fuel cells catalytic reactions, which not only helps us to explore the catalytic mechanism more deeply, but also makes outstanding contributions to the further design of various effective catalysts. Most of the top journals published in the past years contain in situ characterization, and the more commonly used in situ characterization includes in situ infrared spectroscopy, in situ Raman spectroscopy, in situ XPS, and in situ XRD.

This chapter introduces and discusses the most commonly used and effective fuel cell characterization techniques. We will focus on in situ electrochemical characterization techniques, as these techniques provide various useful information when the fuel cell is operating. In addition, the micro-materials of the cell play a decisive role in the cell performance, so offline material characterization techniques will also be introduced. The macroscopic properties of the cell are of great significance for the commercialization of the fuel cell. In order to better understand the lifetime, cold start, and other characteristics, this content will be introduced at the end of this chapter. In addition, we only give a brief description of in situ characterization in the summary and outlook section. Readers can refer to relevant materials if they want to know more about the mechanism of various in situ characterization methods.

3.2 Electrochemical Characterization Methods

3.2.1 Cyclic Voltammetry

Cyclic voltammetry (CV) is a useful characterization method that can be used to study the properties, mechanisms, and kinetic parameters of electrode reactions. It can also be used to quantitatively determine reactant concentrations, electrode surface coverage of adsorbates, and electrode active area [15]. During the testing process, the scan is performed at a certain scan rate between two voltages, and the resulting current–voltage curve is called the cyclic voltammogram. A typical CV curve is shown in Fig. 3.1.

In fuel cells, CV cycles are commonly used to characterize the activity of catalyst materials[16]. However, due to the poor reversibility of electrochemical reactions, it is almost impossible to obtain the results shown in Fig. 3.1b. Taking Pt as the catalyst in a hydrogen–oxygen fuel cell as an example, during CV measurements, hydrogen at 100% relative humidity at room temperature is generally introduced into the anode as a reference electrode and a counter electrode, which acts as a dynamic hydrogen electrode. An inert gas (nitrogen or argon) is introduced into the cathode as a working electrode and an induction electrode. In the experiment, it is necessary to pay attention to the selection of the potential scan range and potential scan rate. The scan voltage range is generally 0.05–1.2 V. When the applied working voltage exceeds 0.8 V, the carbon corrosion current will increase sharply, which may cause

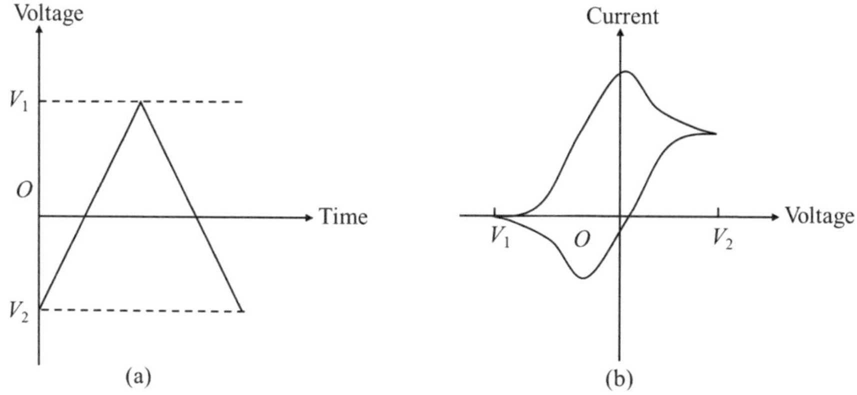

Fig. 3.1 Typical CV curve: **a** the variation of voltage with time, **b** the variation of current with voltage

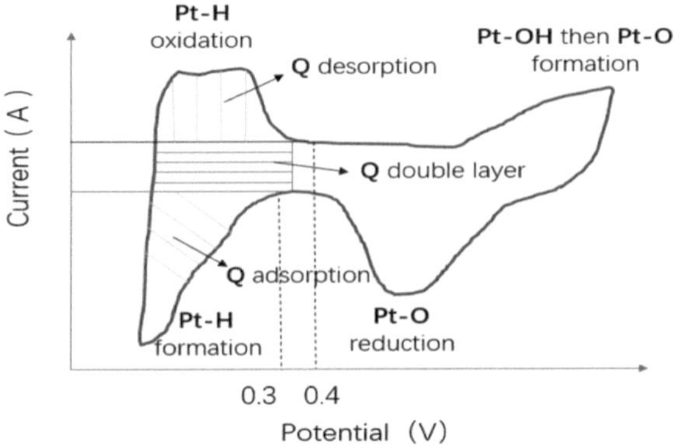

Fig. 3.2 A typical CV curve of Pt/C catalyst

a deterioration in the performance of the catalyst layer. Therefore, some researchers set the scan voltage range to 0.05–0.8 V. The selection range of the scan rate is 1–1000 mV s^{-1}, but to minimize the activation loss of porous electrodes, a smaller scan frequency (10 mV s^{-1}) is generally preferred. The scan rate is also a very important parameter in CV testing, which will be further discussed in later sections. Common CV curves for Pt/C are shown in Fig. 3.2.

From the figure, it can be seen that the voltage increases from 0 V, and the current increases accordingly. The total measured current has two components, one of which is the capacitive charging current caused by the linear change in voltage, also known as non-Faradaic current. As the potential change rate is constant, the current magnitude is constant. The second current response is non-linear, corresponding to the hydrogen adsorption reaction on the surface of the anode catalyst of electrochemical activity. When the voltage continues to increase, the second current gradually increases to a peak, and then decreases as the entire catalyst surface reaches hydrogen saturation. When the voltage increases above 0.4 V, the second current almost disappears. From the figure, it can be seen that there is a plateau region in the current change, in which only the non-Faradaic current exists, so the current size will not change with the change in potential. As the voltage further increases, the catalyst surface will undergo an oxidation reaction, and the total current will continue to increase. With the reversal of the potential scan direction, the direction of the tested current also changes. However, in this region, there is only non-Faradaic current, so there is a current plateau region corresponding to the forward scan. When the potential is reduced to below 0.3 V, a significant negative current can be observed in the figure, indicating hydrogen adsorption and even hydrogen evolution[17]. The reactions that occur in the electrode material, mainly in the 0–0.3 V range, can be expressed by Eq. (3.1):

$$Pt{-}H_{ads} \leftrightarrow Pt + H^{+} + e^{-} \tag{3.1}$$

In theory, the amount of hydrogen adsorption and desorption on the surface of Pt in the catalytic material should be equal. The amount of adsorbed hydrogen on the Pt surface can be inferred by detecting the current generated during hydrogen desorption, which in turn yields the effective reaction area. In the calculation process, particular attention needs to be paid to the influence of double-layer charging and discharging. Because the double-layer charging and discharging are relatively stable throughout the entire CV process, a baseline can be selected from the region where the current is relatively flat between 0.4–0.6 V, and an area is enclosed by the baseline and the desorption peak, which is then integrated to obtain the total charge. This total charge corresponds to the area under the desorption peak of the hydrogen gas, subtracting the double-layer current from the CV. After obtaining the total charge through integration, Electrochemical Active Surface Area (ECSA) can be calculated using the following formula:

$$ECSA = \frac{q}{\Gamma L} \tag{3.2}$$

where Γ is the specific charge required to oxidize or reduce a monolayer of the adsorbed species, which is usually taken as 210 $\mu C/cm^2$ for Pt; L represents the loading size of Pt in the catalyst layer, with units of gPt/cm^2 electrode.

To calculate the effective reaction area of the cell using the hydrogen adsorption current, the impact of the double-layer charge–discharge must be considered. Therefore, a double-layer baseline must be selected to determine the integration region. However, the current plateau obtained in the experiment may not be as smooth as predicted, and different baseline selections can affect the ECSA calculation. This is especially true for alloy catalysts, as their surface reactions are more complex, making it difficult to determine the current baseline.

In addition to the above method, the ECSA in Proton exchange membrane fuel cells (PEMFCs) can also be calculated by the current generated by CO oxidation [18]. For PEMFCs that use alloy catalysts, the CO-stripping method is more effective. CO-stripping method measures the electrochemical response of the removal of CO to calculate ECSA. Before measurement, hydrogen gas is fed into the anode, and a mixture of CO and inert gas (such as nitrogen) is fed into the cathode for a certain period, allowing CO to be adsorbed onto the electrode. The adsorption process of CO is very rapid, and the platinum surface will be completely covered by CO. During this process, the cell potential should be kept as low as possible (lower than the CO dissolution voltage). After the catalyst surface is fully covered, the residual CO gas inside the cell is blown out with pure inert gas for a few minutes. The cell is then subjected to cyclic scanning between two voltages at a certain scanning rate, but the maximum voltage must be higher than 0.9 V (the CO dissolution voltage). CO adsorbs tightly on the platinum catalyst surface in the form of a coordination bond due to the presence of lone electron pairs. When the voltage is higher than 0.9 V, CO will be quickly and completely oxidized. The reaction can be expressed as:

$$CO + H_2O \rightarrow CO_2 + 2H^+ + 2e^- \tag{3.3}$$

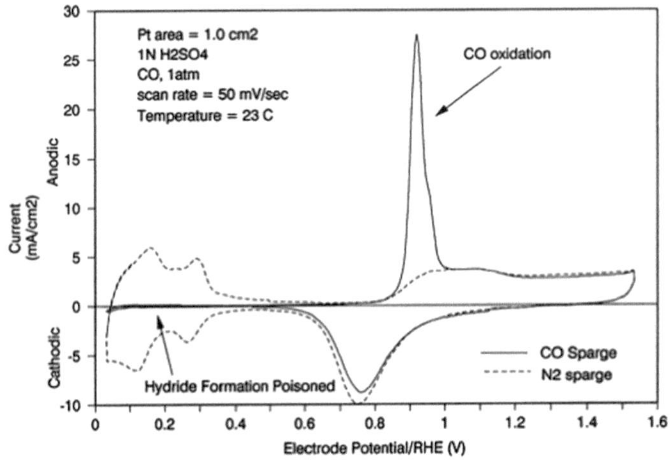

Fig. 3.3 CO stripping voltammetry curve [19]

Carbon monoxide is ultimately oxidized to carbon dioxide and flows out of the catalyst layer. At this point, testing will continue, and the second cyclic voltammetry characteristic curve will be scanned, as shown in Fig. 3.3.

As can be seen from the figure, there is a significant difference between the curves obtained from the two CO stripping voltammetry tests. In the first forward scan, the oxidation reaction occurred due to CO adsorption on the Pt surface, resulting in a sudden increase in current. By the second forward scan, all of the CO adsorbed on the Pt surface had been oxidized, so no increase in current due to CO desorption was observed when scanning the same voltage range. The area enclosed by the two current curves obtained from the two scans indicates the amount of charge required for CO adsorption oxidation. Since the two scans were performed under the same test conditions, the double-layer charge–discharge effect of the cell should be the same, so this effect is canceled out. There is no need to artificially select a baseline when calculating the charge amount, making the calculation more objective and more accurate. The equation used to calculate ECSA with CO stripping voltammetry is the same as Eq. (3.2), but with a different value of Γ, which is 420 $\mu C/cm^2$ [20].

3.2.2 *Electrochemical Impedance Spectrum*

Impedance is an important parameter in the process of electrochemical energy conversion and transmission and can reflect the reaction kinetics and transport processes at the interface of the electrode and electrolyte. Therefore, by measuring the impedance spectrum of the cell system, the dynamic characteristics of the electrochemical reaction process and the performance of the cell can be deeply investigated. The electrochemical reaction process inside a cell generally includes electrolyte resistance, adsorption of electroactive species, charge transfer at the electrode surface, and mass transport from the bulk solution to the electrode surface. Each process can be seen as an electronic component or a simple circuit. The entire reaction process can be represented by a circuit consisting of resistors, capacitors, or constant-phase elements connected in parallel or series [21]. The EIS method applies a small AC voltage to the fuel cell system and measures the corresponding current response, which can determine the impedance of the system and analyze the various losses of the cell [22].

During Electrochemical Impedance Spectrum (EIS) testing, the fuel cell system is connected to an electrochemical impedance analyzer, which applies small AC voltage perturbations at different frequencies and measures the resulting current response of the system. Throughout the testing process, the cell needs to maintain transient stability, which means that the perturbation signal must be as small as possible relative to the stable voltage of the test system. The entire system remains in a linear state during the entire test and can fully recover to its initial state after the test is completed. Since the cell is not a simple pure resistance to current, there must be a phase angle deviation between the response signal output by the cell and the perturbation signal. Figure 3.4 shows the relationship between voltage perturbation and current response.

As can be seen from the figure, the current response of the cell output has the same frequency as the sine voltage signal input, but there is a certain phase shift. By

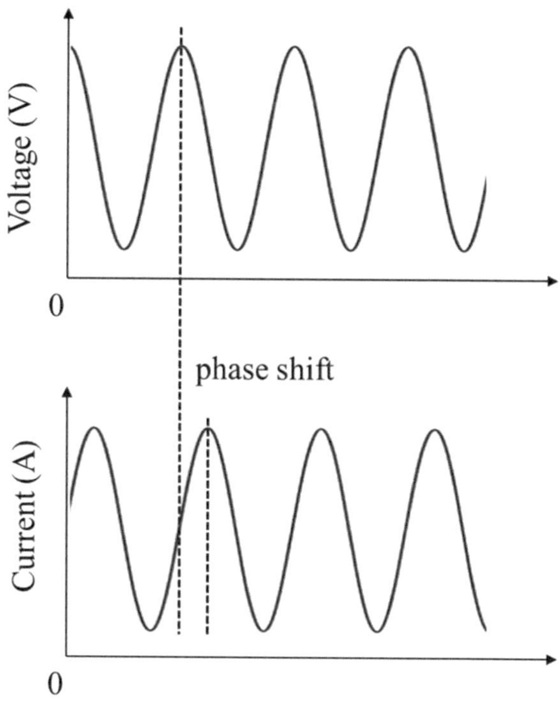

Fig. 3.4 Diagram of voltage disturbance and current response

analyzing the difference in amplitude and phase angle between the two at different frequencies, the impedance characteristics of the system can be obtained.

The relationship between impedance and current, and voltage can be expressed by the following formula:

$$V(t) = V_0 \cos(\omega t) \tag{3.4}$$

$$i(t) = I_0 \cos(\omega t - \phi) \tag{3.5}$$

$$Z = \frac{V(t)}{i(t)} = Z_0(\cos\phi + j\sin\phi) \tag{3.6}$$

where V(t) represents the change of voltage over time, V_0 is the amplitude of voltage, $i(t)$ represents the change of current over time, I_0 is the current amplitude, Z represents impedance, ω is the angular frequency, and ϕ is the phase difference. Each impedance can be represented by an impedance number and a phase angle. They can also be represented by a real part and an imaginary part on the complex plane. When plotting impedance data, the real part of impedance is generally used as the horizontal axis and the imaginary part as the vertical axis. The impedance data graph plotted in this way is called a Nyquist plot [23]. In addition to this method, impedance data can also be plotted in the form of a Bode plot [24]. Each has its advantages and disadvantages, where a Bode plot can give the change curve of impedance amplitude and phase angle frequency, but cannot directly give the real and imaginary parts of impedance at a specific frequency. In contrast, a Nyquist plot can give the real and imaginary parts of cell impedance at different frequencies, but cannot directly obtain the test frequency of data points. For better analysis of the sources of losses within the cell, we mainly introduce the Nyquist plot here. By combining some knowledge of impedance analysis, we can analyze the ohmic loss, activation loss, and other losses of the cell from the Nyquist plot. Figure 3.5 shows a hypothetical fuel cell Nyquist curve with three marked impedance regions.

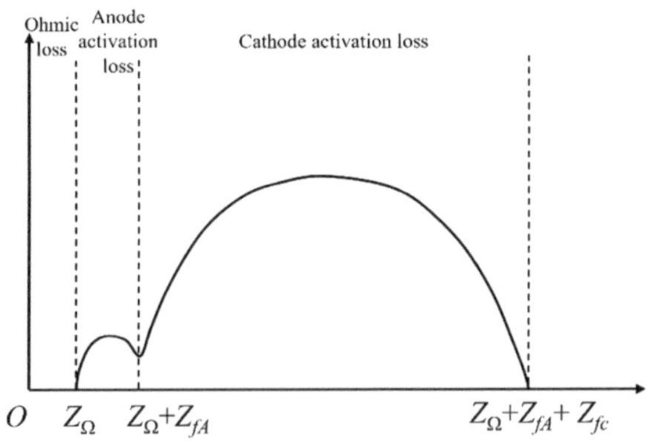

Fig. 3.5 Hypothetical fuel cell Nyquist curve

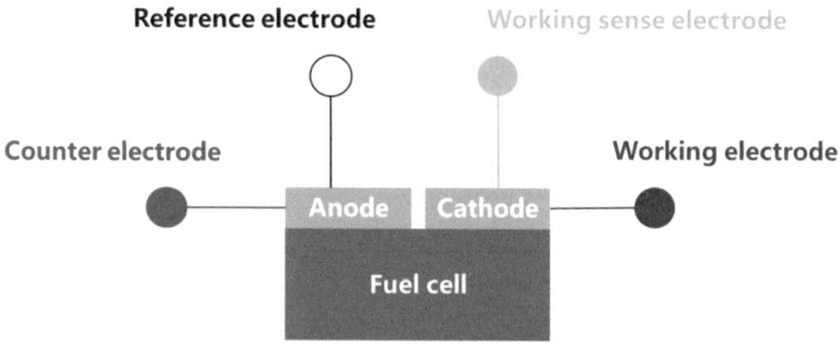

Fig. 3.6 Four-electrode test system for fuel cell

The curve in Fig. 3.5 represents a series of data points, with each point representing the impedance magnitude of the fuel cell at a specific sine frequency. As the frequency of the input voltage decreases, the impedance points move from left to right. A typical fuel cell curve is made up of two smaller arcs, representing the anode activation loss and cathode activation loss, respectively. The intersection of the small arc with the real axis represents the size of the cell resistance. To draw a Nyquist plot like Fig. 3.5, dozens or even hundreds of impedance data points at different sine voltage frequencies are needed. The disturbance frequency range selected for fuel cell testing is generally between 100 kHz and 100 mHz, and the voltage signal is controlled within 20 mV to ensure that there is an approximate linear relationship between the output response and the input signal. In addition, to ensure accuracy in testing, four electrodes are commonly used to measure the impedance of fuel cells, as shown in Fig. 3.6.

The working electrode and the reference electrode of the testing device are connected to the cathode of the fuel cell, while the counter electrode and the sensing electrode are connected to the anode. The Nyquist plot of the fuel cell is recorded and plotted through the fuel cell testing system. To better understand the reasons for the impedance changes inside the cell and analyze the various losses of the cell, it is necessary to have a certain understanding of the impedance characteristics of the basic components. The basic principles of electrochemical impedance are briefly introduced below. For a more in-depth understanding, reading relevant books [22] is recommended.

All kinds of charge transport processes that occur inside a fuel cell can be represented by basic circuit components. Common electrical components include resistance, capacitance, constant phase element, and Warburg impedance. Resistance and capacitance are basic electrical components that have been studied in high school. The input voltage and output current of a resistor always satisfy Ohm's law, and its impedance is a constant independent of frequency, with a zero imaginary part. Therefore, a resistance is represented as a point with a value of R on the real axis of the Nyquist plot, as shown in Fig. 3.7a. Similarly, a capacitor can be expressed as a straight line coinciding with the imaginary axis on the Nyquist plot, as shown in Fig. 3.7b.

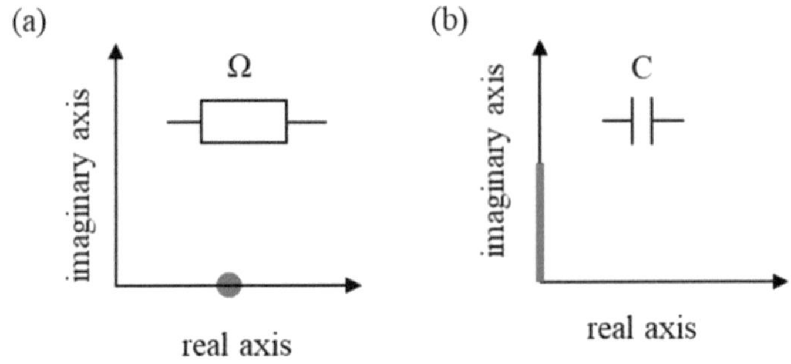

Fig. 3.7 Nyquist diagram of resistance and capacitance

A constant phase element is an imaginary component that does not exist. It is a concept created by researchers who found that the electrical components in real life do not perfectly exhibit the impedance in a fuel cell. Its formula is:

$$Z = \frac{1}{Y_0}(j\omega)^{-\alpha} \tag{3.7}$$

where α is the dispersion coefficient, Y_0 is the deviation of the double layer from the pure capacitance equivalent component. When $\alpha = 0$, the constant phase element exhibits pure resistance characteristics; when $\alpha = 1$, the constant phase element exhibits pure capacitance characteristics; and when $\alpha = -1$, the constant phase element exhibits pure inductance characteristics. As shown in Fig. 3.8, different values of α correspond to different Nyquist plots. The α value of the component is usually between 0 and 1.

When the α value of the constant phase element is 0.5, the constant phase element exhibits semi-infinite diffusion Warburg impedance characteristics. Warburg impedance is an electrical component used to characterize mass transfer processes. For a thick diffusion layer under constant conditions, material diffusion can be approximately considered as semi-infinite diffusion. At this time, the impedance is called semi-infinite diffusion Warburg impedance and can be expressed as:

$$Z_w = \frac{\sigma}{\sqrt{\omega}}(1 - j) \tag{3.8}$$

where σ is the Warburg coefficient of the material, which characterizes the effectiveness of the material reaching or leaving the effective reaction interface. It is defined as:

$$\sigma = \frac{RT}{(nF)^2 A\sqrt{2}}\left(\frac{1}{C\sqrt{D}}\right) \tag{3.9}$$

where n is the number of transferred electrons in the reaction, F is the Faraday constant, A is the electrode area, C is the total concentration of the substance, and

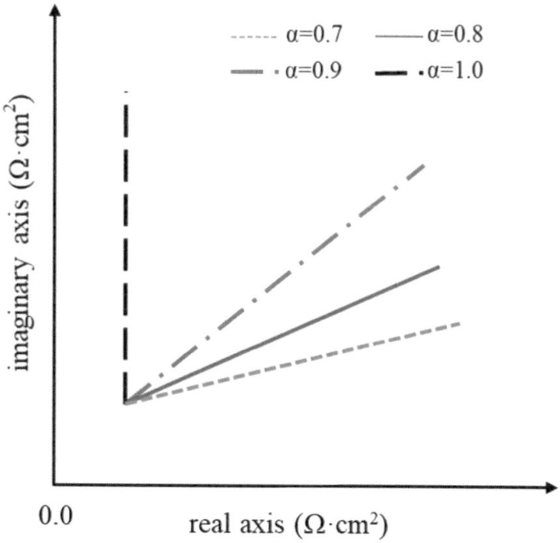

Fig. 3.8 Nyquist diagram of constant phase element under different α

D is the diffusion coefficient of the substance. It should be noted that Warburg impedance is related to frequency, substance concentration, and diffusion. The higher the substance concentration and diffusion coefficient, the smaller the Warburg coefficient, which means the Warburg impedance is smaller and the mass transport resistance can be ignored. The higher the frequency, the faster the reaction changes, and the substance does not need to be transported over long distances. Therefore, the Warburg impedance is also smaller. Figure 3.9 shows the Warburg impedance of semi-infinite diffusion.

The Warburg impedance formula indicates that the Warburg impedance increases linearly with decreasing frequency, and the real and imaginary parts of the impedance formula are identical. Therefore, it appears as a line with a slope of 1 on the Nyquist

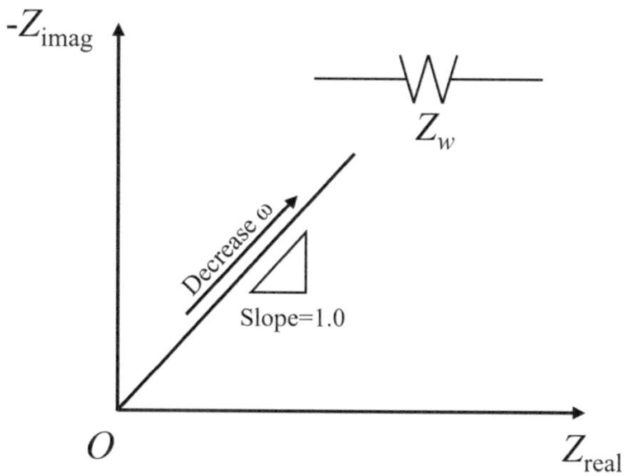

Fig. 3.9 The Warburg impedance of semi-infinite diffusion

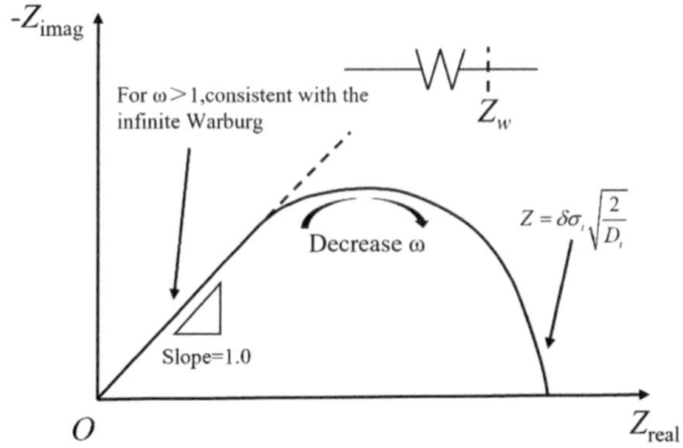

Fig. 3.10 Porous bounded Warburg impedance

plot. However, this impedance formula only exists under the semi-infinite diffusion condition, which does not apply to the diffusion layer in a fuel cell. In experiments, it is rare to see curves resembling semi-infinite diffusion impedance, and more often than not, the mass transfer loss curve eventually returns to the real axis. In such cases, a porous bounded Warburg impedance is used to characterize the diffusion process in the finite thickness diffusion layer. The impedance expression is:

$$Z_0 = \frac{\sigma}{\sqrt{\omega}}(1 - j)\tanh\left(\delta\sqrt{\frac{j\omega}{D}}\right) \tag{3.10}$$

where δ is the thickness of the diffusion layer. The corresponding Nyquist plot is shown in Fig. 3.10.

From the graph, it can be seen that in the high-frequency region, the porous bounded Warburg impedance exhibits a similar trend to the semi-infinite diffusion Warburg impedance due to the rapid changes. As the frequency decreases, the diffusion layer thickness for mass transfer exceeds the electrode thickness, causing the porous bounded Warburg impedance to eventually return to the real axis. The intersection of the low-frequency real-axis impedance reflects the thickness information of the catalyst layer.

To now, we have a basic understanding of several commonly used electrical components in EIS. Then, we need to use the above electrical components to simulate the electrochemical reaction process inside the fuel cell. This circuit-based representation of fuel cell behavior is also known as an equivalent circuit model [25]. Generally, the reaction process inside a fuel cell consists of three steps: Ohmic conduction, electrochemical reactions, and mass transport [26]. Here, we will provide a detailed explanation of the circuit models used to characterize these processes.

Ohmic conduction: The equivalent circuit for Ohmic conduction inside a fuel cell is very simple, as it is just a constant resistance that is independent of frequency. Therefore, the Nyquist plot for Ohmic losses is depicted as in Fig. 3.7a. However, during testing, the fuel cell may show inductance characteristics in the high-frequency

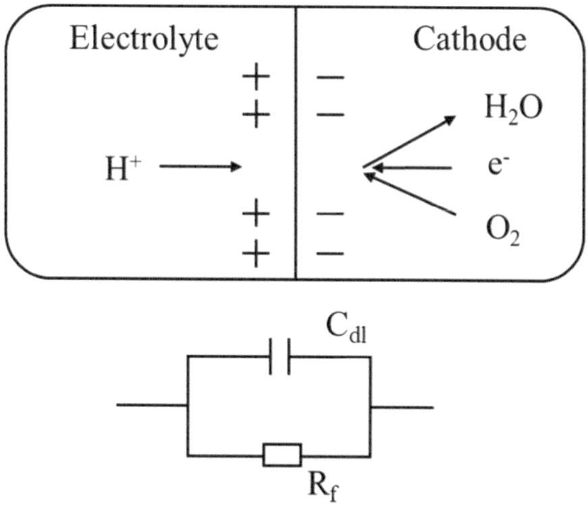

Fig. 3.11 Physical representation and equivalent circuit modeling of electrochemical reaction interfaces

region, which is caused by the inductive characteristics of the testing leads and should not be included in the Ohmic conduction inside the fuel cell.

Electrochemical reactions: The process of electrochemical reactions is relatively complex and is commonly represented by two parallel resistors and a capacitor. Figure 3.11 provides a physical representation of the electrochemical reaction interface and its equivalent circuit model.

In the diagram, R_f represents the Faraday resistance, which reflects the kinetic characteristics of the electrochemical reaction, while C represents the double-layer capacitance, which reflects the capacitance properties of the interface. The kinetic processes at the electrochemical interface can be described using the Tafel equation, and therefore, the Faraday resistance can also be quantitatively analyzed in this equation. Taking the differential of the Tafel equation yields the following formula:

$$R_f = \frac{d\eta_{act}}{di} = \frac{RT}{\alpha nF}\frac{1}{i} = \frac{RT}{\alpha nF}\frac{1}{i_0 e^{\alpha nF\eta_{act}/(RT)}} \tag{3.11}$$

where R_f represents the Faraday resistance, i_0 represents the exchange current. As can be seen from the formula, the Faraday resistance is independent of frequency, so the kinetic characteristics of the cell can be represented by a pure resistance. As the current density of the electrode material increases, or as the activation overpotential increases, the reaction becomes more favorable, leading to a decrease in the Faraday resistance at the electrochemical interface.

In addition to the electrochemical reaction process described above, ions and electrons will be adsorbed on both sides of the electrochemical reaction interface due to the potential difference between the electrode and the electrolyte. This adsorption process causes the interface to act as a capacitor C, which can affect the output of the cell during discharge. In fact, in Sect. 3.2.1, where CV is introduced, we have already mentioned the appearance of a plateau region during CV testing, which is because only the charging and discharging of the capacitor leads to a stable current in this potential range. Considering both of the above effects, to model the electrochemical reaction interface, both resistance and capacitance need to be considered. And there are two ways to connect these two components: in series or in parallel. Considering that electrochemical reactions and capacitance charging and discharging can occur synchronously at the interface, using an RC parallel unit to model the electrochemical reaction interface is a more appropriate choice. The impedance of the RC parallel unit can be expressed as:

$$\frac{1}{Z} = \frac{1}{R_f} + j\omega C \tag{3.12}$$

$$Z = \frac{R_f}{1 + j\omega R_f C} = \frac{R_f}{1 + (\omega R_f C)^2} - j\frac{\omega R_f^2 C}{1 + (\omega R_f C)^2} \tag{3.13}$$

The Nyquist plot of the RC parallel circuit is shown in Fig. 3.12, where a clear semicircle is observed.

The left intersection point of the semicircle and the real axis at 0 indicates that at high frequencies, there is almost no resistance at the interface. As the frequency decreases, the point on the semicircle moves to the right. The right intersection point of the semicircle and the real axis has a horizontal coordinate of R, indicating that at low frequencies, the current is almost entirely provided by electrochemical reactions, exhibiting significant resistance characteristics. The frequency at the vertex of the semicircle is determined by the RC time constant of the interface, as given by the formula:

$$\omega = 1/(R_f C) \tag{3.14}$$

Using this formula, the equivalent capacitance of the electrochemical reaction interface can be determined. However, the above equivalent circuit is generally only

Fig. 3.12 Nyquist plot of the RC parallel circuit

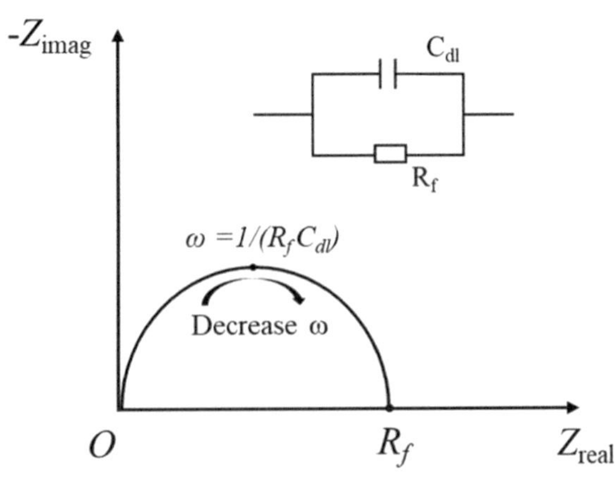

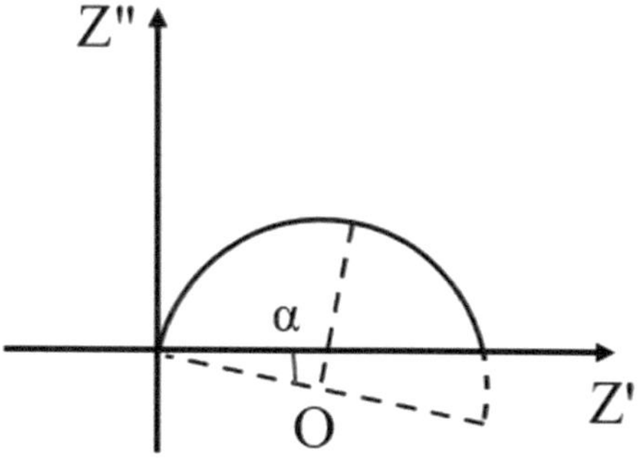

Fig. 3.13 Nyquist plot of the RQ circuit

applicable to the electrochemical interface composed of electrodes and solutions. For electrochemical reactions occurring in hydrogen–oxygen fuel cells, the characteristics of the double-layer capacitance are not the same as those of pure capacitance. This phenomenon is known as the "diffusion effect" in the literature, so an electrical component that deviates from pure capacitance is needed to simulate electrochemical reactions. The electrical component introduced here is the constant phase element. The value of α for the component is generally between 0 and 1, exhibiting non-pure capacitance characteristics. The impedance of the parallel connection of the constant phase element and the resistor is shown as follows:

$$Z = \frac{1}{R_f^{-1} + Y_0\omega^\alpha\left(\cos\left(\frac{\pi\alpha}{2}\right) + j\sin\left(\frac{\pi\alpha}{2}\right)\right)} \tag{3.15}$$

where Y_0 is the equivalent element that the double layer deviates from the pure capacitance, α is the dispersion coefficient. The Nyquist plot of the RQ circuit is shown in Fig. 3.13.

As can be seen from the figure, the RQ parallel circuit is not a semicircle like the RC parallel circuit, but only an arc on the semicircle. Different values of the dispersion coefficient will affect the degree of curve deviation from a semicircle.

Mass transportation: For hydrogen fuel cells, significant mass transport loss exists at the cathode due to the gas–liquid two-phase flow process [27]. This mass transport loss can be modeled by Warburg circuit component, which can be characterized by semi-infinite diffusion impedance or finite layer diffusion impedance. These two impedances have been described in detail in the previous article. The Nyquist diagrams of them are shown in Figs. 3.9 and 3.10. These two impedances have a great relationship with electrode structure. And the model is generally chosen according to the measured impedance spectra.

Now that we have a preliminary understanding of the various reaction processes in the fuel cell, we can use all the circuit components mentioned above to assemble a complete fuel cell equivalent circuit model. Typical fuel cell polarization contains mass transport loss, activation loss, and ohmic loss. To simplify the model, the semi-infinite diffusion impedance is used here to characterize the mass transport loss in the cell. Activation loss is divided into cathode activation loss and anode activation loss, which need to be characterized by two RQ circuits connected in series. Ohm loss is the simplest and is still represented by a constant resistance. The Nyquist diagram of the obtained ideal fuel cell is shown in Fig. 3.14.

The fuel cell equivalent circuit in the figure consists of two parallel RQ circuits, a Warburg component, and an ohmic resistor. The internal mass transfer loss of the fuel cell is significant at low frequency, but not at high frequency [28], so the Warburg component is in parallel with the cathode constant phase element. In the Nyquist plot, the curve intersects the real axis at a point R and gradually moves to the right as the test frequency decreases. The intersection of the image and the real axis represents the magnitude of the ohm resistance, the first arc represents the anode activation impedance, the second arc represents the cathode activation impedance, and the last diagonal line represents the cathode mass transport loss [29]. It can be seen that the arc diameter of the cathode is larger than that of the anode, and the arc symbolizes the electrochemical kinetic impedance, which indicates that the reaction of the cathode is not as fast as that of the anode. By establishing the equivalent circuit model to

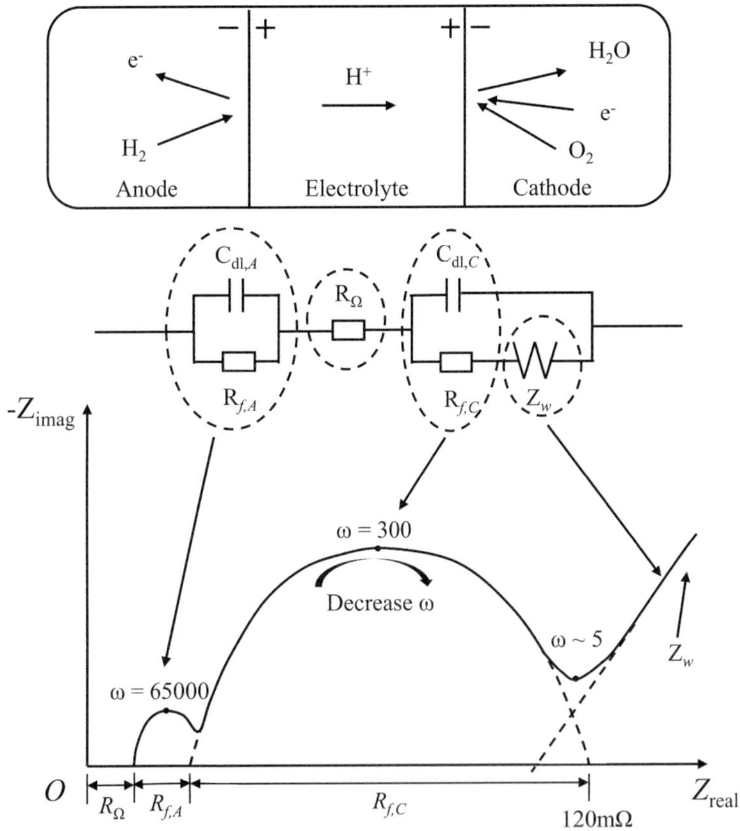

Fig. 3.14 Physical, circuit, and Nyquist diagrams for a simple fuel cell impedance model

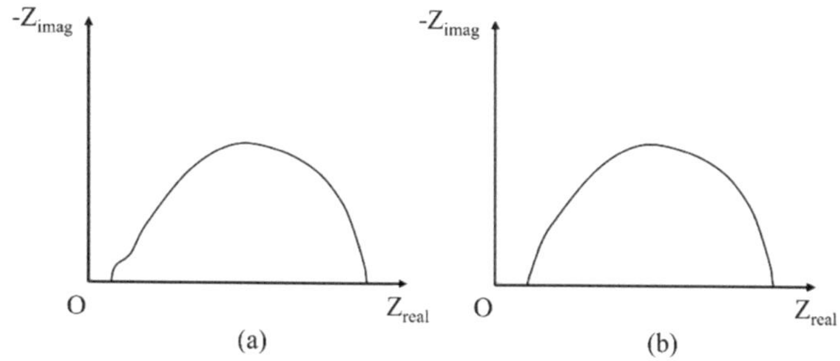

Fig. 3.15 Examples of experimentally obtained Nyquist plots

fit the experimental curve, we can obtain various coefficients of the fuel cell, such as transmission characteristics, electric double layer capacitor, and electrochemical reaction coefficient. However, the EIS method itself also has some defects, which restrict its further application. For example, it may be difficult to clearly distinguish each impedance from the experimental curve, as shown in Fig. 3.15.

The cathode impedance in the figure covers part of the anode impedance, making it impossible to obtain an accurate arc radius size. The calculated value can only be used as an auxiliary reference. In addition, in the high-frequency region, the test wire may show inductance characteristics. While in the low frequency region, the irreversible reaction inside the cell may also show certain inductance characteristics, which will affect numerical fitting. However, under different current densities, the various impedances of the cell are not constant. In experiments, it is necessary to choose a suitable working current according to research requirements. Finally, even if the above-mentioned situation is weakened by selecting a suitable interval, there are often several different fitting circuits for the obtained curve, and the electrochemical parameters obtained through different circuits are also different. It is necessary to have a good understanding of the electrochemical reaction process to obtain more accurate results. Overall, the EIS method is a very powerful electrochemical method that can obtain a large number of electrochemical parameters from testing, but this technology is relatively complex, and some experimental data cannot be explained. Therefore, caution should be taken when using it.

3.2.3 Linear Sweep Voltammetry

Linear sweep voltammetry (LSV) is a simple electrochemical technique. This method is similar to cyclic voltammetry, but linear sweep voltammetry involves only a unidirectional linear scan from the initial potential to the final potential [30]. This is particularly useful for irreversible systems where a reverse scan would not reveal more information. The voltage scan applied in LSV testing is shown in Fig. 3.16.

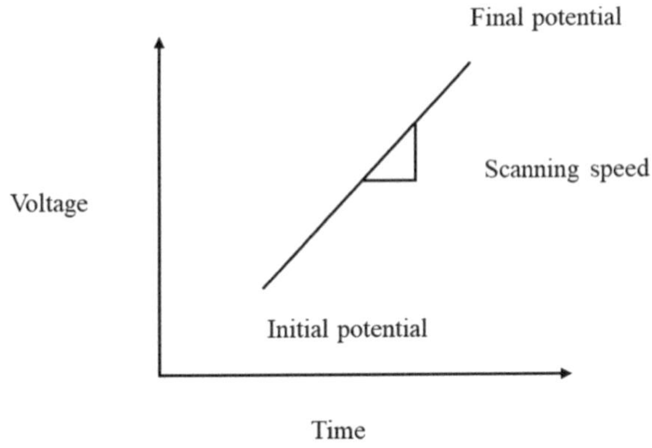

Fig. 3.16 Variation of applied voltage in typical LSV tests

Like cyclic voltammetry, linear sweep voltammetry can also be used to test with a three-electrode system. It consists of a working electrode, a counter electrode, and a reference electrode. The three electrodes are connected to a potentiostat and placed in an electrochemical cell with the solution. The potentiostat controls the potential between the working electrode and reference electrode and measures the current at the counter electrode to plot a graph of potential versus current. Figure 3.17 shows typical results obtained from LSV testing.

As shown in the figure, when scanning began, no electrochemical reaction occurred, so the measured current was 0. As the voltage is further scanned to the right, overpotential drives the reaction. In Fig. 3.17a, the electrochemical reaction rate and mass transfer rate reached equilibrium, so the current reached its limit and no longer changed. In Fig. 3.17b, the electrochemical reaction is faster than the mass transfer rate. So, after reaching peak current, the current density decreased due to a lack of reactants on the electrode surface. Therefore, the results obtained by LSV testing are highly dependent on testing conditions. For example, when the potential change rate used by LSV slows down or reactant transfer faster, the current drop after peak current will gradually become gentler and eventually level off, resulting in results similar to those shown in Fig. 3.17a. In fuel cell catalyst material testing,

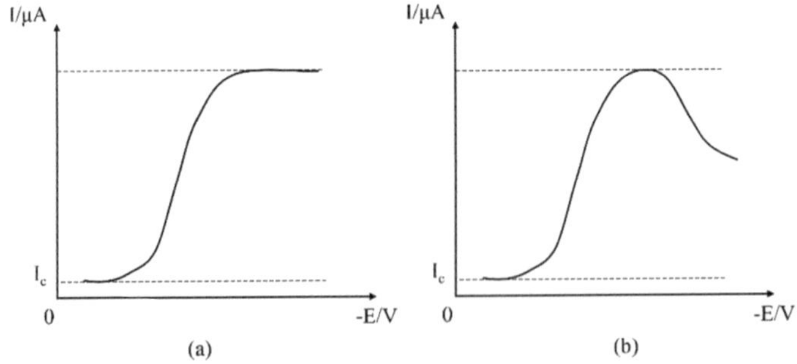

Fig. 3.17 Typical LSV test results

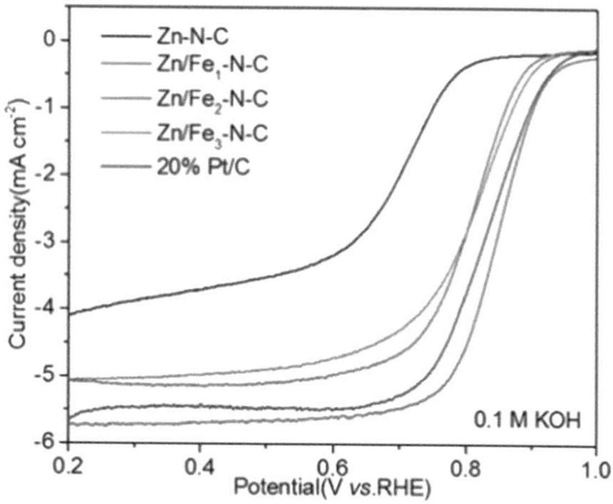

Fig. 3.18 LSV curves of various catalysts in O_2-saturated 0.1 M KOH with a rotation rate of 1600 rpm [31]

we often apply rotation to working electrodes and other methods to increase the mass transport rate on electrode surfaces. The true experimental results are shown in Fig. 3.18 [31].

In the LSV curves, we are interested in the onset potential, half-wave potential, and limiting current density. These parameters can be used to calculate electrochemical parameters for the catalyst material.

In addition to testing electrode materials, LSV can also be used to test entire fuel cells. It is commonly used to characterize gas permeation in fuel cells and can also be used to determine the lifetime of membrane electrodes. During fuel cell operation, hydrogen gas inevitably permeates through the membrane from the anode to the cathode and reacts with oxygen in the cathode catalyst layer, generating the hydrogen crossover current. Although this current is relatively small compared to normal operating currents, it can significantly reduce the open-circuit voltage of the fuel cell. Hydrogen permeation also inhibits catalyst agglomeration and growth. At high potentials, platinum will react to produce platinum ions, and hydrogen that permeates from the anode reacts with it to produce small platinum particles, which inhibits catalyst agglomeration and growth. This process may also occur inside ionomer membranes and affect their durability.

For example, in a hydrogen fuel cell, hydrogen gas is introduced into the anode and nitrogen gas into the cathode. A linear sweep voltage (0–0.8 V) is applied across both sides of the cell at a scan rate of 1–100 mV s^{-1} while recording voltage and response current to obtain corresponding linear voltammetry curves. Figure 3.19 shows a typical hydrogen crossover current by LSV [32].

In LSV measurements, hydrogen ions adsorbed on the surface of platinum in the range of 0.05–0.3 V are oxidized and result in a desorption peak. When the voltage exceeds 0.3 V, all hydrogen ions on the working electrode will be fully oxidized. At this point, due to the absence of active gas at the cathode of the cell, the measured current is generated solely from the anodic permeation of the hydrogen gas oxidation

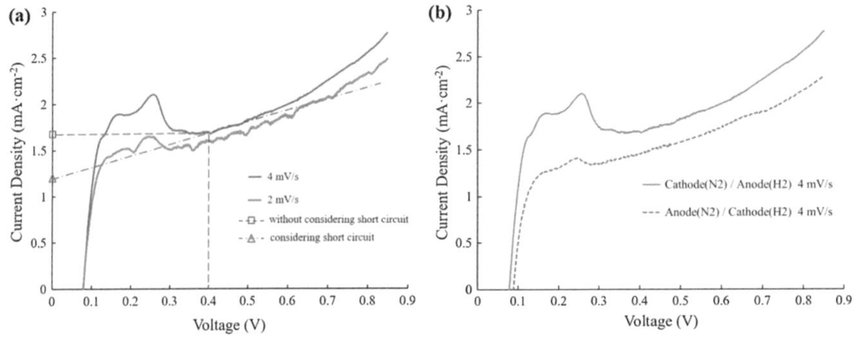

Fig. 3.19 Typical hydrogen crossover current by LSV [32]

reaction. Based on the current values in the range of 0.3–0.6 V, we can use Faraday's law to calculate the amount of permeated hydrogen gas:

$$J = \frac{i}{nF} \tag{3.16}$$

where J is the permeated hydrogen gas flow rate, i is the hydrogen permeation current, and n is the number of electrons transferred in the reaction. The lifespan of the membrane electrode is positively correlated with the amount of permeated hydrogen gas. When membrane degradation becomes more severe, the amount of permeated hydrogen gas also increases. Therefore, the amount of permeated hydrogen gas can be used as an indicator of membrane degradation.

3.2.4 Current Interrupt

EIS is widely used for the characterization of PEM fuel cells. However, when EIS is applied for on-site diagnosis and control of fuel cells, it has some practical limitations, such as the requirement for large equipment size, weight, and cost [33]. In comparison, the current interrupt method has the advantage of low hardware requirements and can be applied to high-power fuel cell systems. It is a fast way to distinguish between the ohmic and non-ohmic losses of the cell. Generally, it can also be performed synchronously with polarization curve testing [34]. Therefore, the current interrupt method is widely accepted in fuel cell research, especially for the characterization of large fuel cells.

Figure 3.20 shows the applied current change for the current interrupt method. Assuming the cell was operating at a certain current, and then the current passing through the cell was suddenly interrupted (dropping to 0 A). At this time, according to Ohm's law, the cell's ohmic losses will immediately disappear, and the voltage will rise instantly. However, in addition to the ohmic losses, there are also non-ohmic losses inside the cell, including activation losses and mass transport losses [35]. Since these two losses are related to electrochemical reactions and mass transport processes, they cannot change abruptly with the current but change slowly over time. The typical result obtained from the test is shown in Fig. 3.21.

The voltage difference between the solid line and the first dashed line in the graph represents the Ohmic loss, which undergoes an abrupt change with the variation of the current. The Ohmic loss at this current can be calculated from this segment using the following formula [36]:

$$R = \frac{\Delta V}{I} \tag{3.17}$$

where ΔV is the instantaneous voltage recovery size in the voltage graph, I is the current size before interruption. The voltage difference between the two dashed lines represents the sum of activation loss and mass transfer loss, which cannot undergo an abrupt change with the current variation, and therefore shows a slow increase in

Fig. 3.20 Variation of applied current by current interruption method

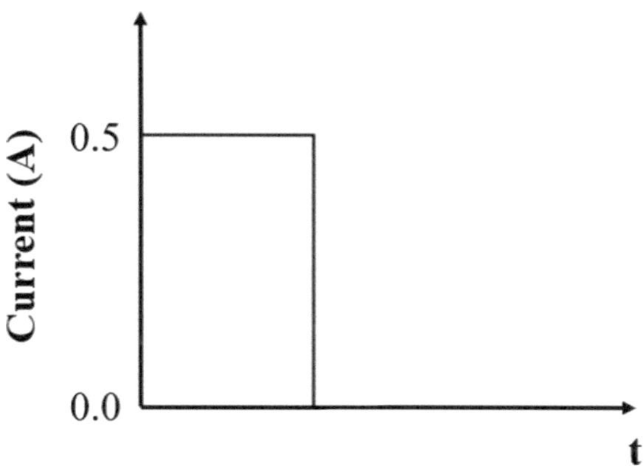

Fig. 3.21 Ideal voltage response

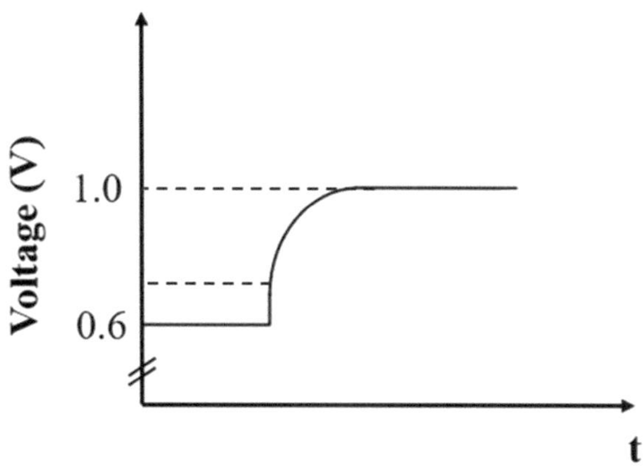

the graph. By combining with the polarization curve, the activation loss and mass transport loss can be further distinguished. This calculation process involves fitting with the Tafel formula, which will be introduced in the polarization curve section. It is worth noting that the current interruption method requires high accuracy in capturing voltage step changes. Rapid acquisition of instantaneous voltage data (in milliseconds) is crucial for separating the cell's Ohmic and non-Ohmic losses.

3.2.5 Polarization Curve

The most common way to evaluate fuel cells is to measure their polarization curves. The polarization curve is an image of current versus voltage, which reflects the comprehensive performance of the fuel cell under certain operating conditions and is also the most important indicator for evaluating the performance of the cell under normal conditions [37]. Through the polarization curve, we can obtain some basic information about the cell, such as open circuit voltage, peak power density, and limiting current density [38]. There are two commonly used measurement methods for polarization curves: (1) constant potential measurement method, which controls the output voltage of the cell and records the current response at the corresponding voltage; (2) constant current measurement method, which controls the output current of the cell system and records the voltage response at the corresponding current. In theory, when the cell is in a steady state, the polarization curves obtained by these two test methods should be the same. In addition to these two methods, LSV can obtain a similar curve with low scanning speeds. The lower the scan speed used in the test, the more similar the current–voltage image obtained to a polarization curve. A typical polarization curve of a proton exchange membrane fuel cell is shown in Fig. 3.22.

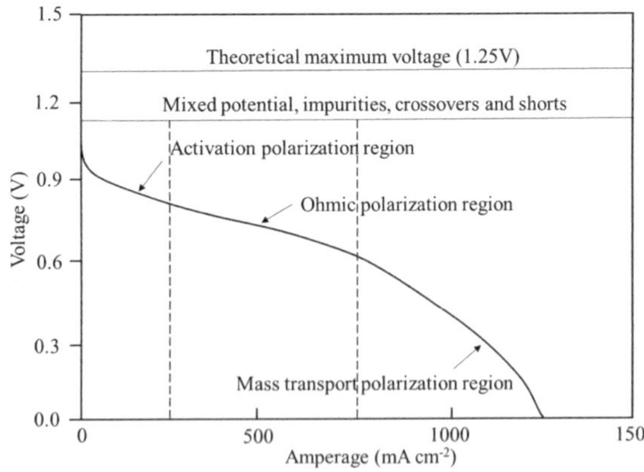

Fig. 3.22 Typical proton exchange membrane fuel cell polarization curves

The theoretical potential of the cell and the typical polarization curve are marked in the figure, and the polarization curve is divided into three sections according to current density [39]. Taking hydrogen fuel cells as an example, the theoretical potential is calculated according to the Nernst equation. The theoretical potential equation for hydrogen fuel cells is as follows:

$$E = E^0 - \frac{RT}{2F} Ln\left(\frac{a_{H_2O}}{P_{H_2}\sqrt{P_{O_2}}}\right) \tag{3.18}$$

where E^0 represents the discharge voltage of the cell under standard conditions, R is the gas constant, F is Faraday's constant, a is activity, and P is pressure. It can be seen from the figure that the potential of the cell decreases as the current increases. When the current tends to zero, the electrode surface reaches electrochemical equilibrium, and various losses, such as ohmic loss, do not need to be considered at this time. Then the output voltage of the cell reaches maximum. This maximum voltage is called the open-circuit voltage. In practical applications, the measured open circuit voltage is always lower than the theoretical voltage, as shown in Fig. 3.22. This difference is mainly due to the crossover of fuel from anode to cathode [39]. In addition, current leakage will also reduce the measured open-circuit voltage.

When the current increases slightly from zero, the output voltage of the cell drops sharply. The voltage rapid drop in this section is generally known as activation polarization, which is due to a slow electrochemical reaction. This can be seen from the Bolter-Volmer equation commonly used in electrochemical reactions.

$$i = i_0\left[\exp\left(\frac{\alpha nF\eta_{act}}{RT}\right) + \exp\left(-\frac{\beta nF\eta_{act}}{RT}\right)\right] \tag{3.19}$$

where, i_0 is the exchange current density, α and β are transfer coefficients, and η is the activation overpotential. With the cell output current increasing, the activation overpotential needs to increase simultaneously, as can be seen from the formula. The activation overpotential describes the voltage deviation from the equilibrium potential. This means that the larger the cell current, the more severe the degree of voltage deviation.

With the continued increase of current, the influence of Ohm's resistance on voltage gradually becomes apparent. At this time, the voltage drop is mainly caused by internal resistance, including contact resistance between fuel cell components and proton-conducting membrane proton resistance. The resulting voltage drop follows Ohm's law, so it often shows linear characteristics in the figure. In PEMFCs, proton resistance of the polymer electrolyte membrane contributes most to total ohmic resistance [40]. Targeted modification of the electrolyte membrane to reduce cell internal resistance can effectively improve cell performance.

In the high current density range, due to the slower transfer rate of reactants and products than the reaction rate, the cell potential drops rapidly, which is called concentration polarization. The transfer of reactants and products in fuel cells is called mass transport, which consists of convection and diffusion. The concentration polarization of fuel cells is closely related to mass transfer. Convection is volume movement of liquid under mechanical force and is the main mass transfer mode in a flow channel. Diffusion dominates in the gas diffusion layer, which is generated by concentration gradients. Both transport processes will affect cell performance. Reaction can only take place after reaching the catalytic site, so multiple mass transfers must be carried out before reaction [41]. This leads to the reactant concentration in the catalyst layer being much lower than that in the flow channel, which can lead to a situation where the cell output current cannot be increased with decreasing voltage. Many tests have found the existence of this limiting current and can infer mass transfer resistance inside the cell from it [42].

Combining the effects of the above three types of losses, the polarization curve of the fuel cell first decreases, then levels off, and finally decreases rapidly again. When testing an actual fuel cell system, it is necessary to ensure the stability of the tested system at each data point collected. Taking a constant current test as an example, at each current, constant current discharge should be maintained for as long as possible so that the cell reaches a stable state before recording voltage data. In addition, as many data points as possible should be measured to draw a more accurate polarization curve. Before testing, the cell should be purged with reactants for a period of time to ensure the exclusion of air inside the cell. In addition, a cell needs to be activated so that its performance reaches an optimal state. It should be noted that appropriate assembly pressure is also critical for obtaining optimal performance of the cell [43]. If assembly pressure is too low, contact resistance between internal components of cell will increase and even cause gas leakage; if assembly pressure is too high, gas diffusion layer, catalyst layer or membrane components may be excessively squeezed and damaged during assembly process, thus increasing mass transport resistance and hydrogen permeation [44]. In short, the polarization curve of a cell reflects comprehensive performance under the cooperation of various components inside the cell. To obtain accurate cell performance, every detail should be done well in testing.

Fuel cells contain three types of loss, which can be identified and separated by polarization curves combined with the interrupted current method to obtain their values. The specific separation steps are as follows: (1) Calculate the cell resistance under various discharge conditions through the interrupted current method; (2) Calculate the cell voltage after compensating for Ohm loss through Ohm's law; (3) Draw a curve graph with the logarithm of current density as the horizontal axis; (4) Select 3–4 points for linear fitting in the low current density area of the curve and calculate the regression equation; (5) Bring the current density into the regression equation to calculate the voltage y, then the difference between the theoretical voltage and the y value is the activation polarization; (6) Concentration polarization can be calculated by the difference between the IR-free voltage and the y value on the regression equation. The separated impedances are shown in Fig. 3.23.

As shown in the figure, the voltage plotted has been compensated for Ohm loss and is higher than the cell voltage measured in the experiment. The size of activation polarization and concentration polarization is plotted in the figure. The size of the Ohm loss can be calculated by Ohm's law. Activation polarization depends on the

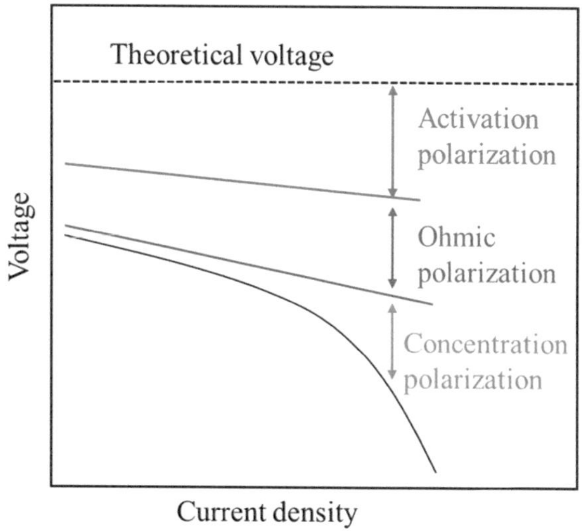

Fig. 3.23 Schematic diagram of the separation of three types of impedance

Tafel formula. When the overpotential value is large, the electrochemical reaction kinetics formula can be simplified to obtain the following formula:

$$\eta_{act} = a + b \log i \tag{3.20}$$

where a = − (2.303RT/αnF)logi_0 b = 2.303RT/αnF. At this time, activation polarization and the logarithm of current density are linearly related. Therefore, in Fig. 3.23, the logarithm of current is used as the horizontal axis. In addition, based on previous learning, we know that in a low current density region, voltage loss mainly comes from activation polarization. Therefore, by fitting the curve in the low current density region, we can approximately obtain a linear relationship between activation polarization and the logarithm of current density. Then we can distinguish the values of the three losses inside the cell with some convenience.

3.3 Ex-Situ Characterization of Materials

3.3.1 Porous Structure and Permeability Characterization

Substance transport has a significant impact on fuel cell performance [45]. The higher the concentration of reactants within the Catalyst Layer (CL), the higher the power density the cell can reach. Figure 3.24 plots the oxygen concentration at various locations in the cell [46].

As can be seen from the figure, the oxygen concentration decreases continuously from the flow channel to the CL. Compared with the flow channel, the oxygen

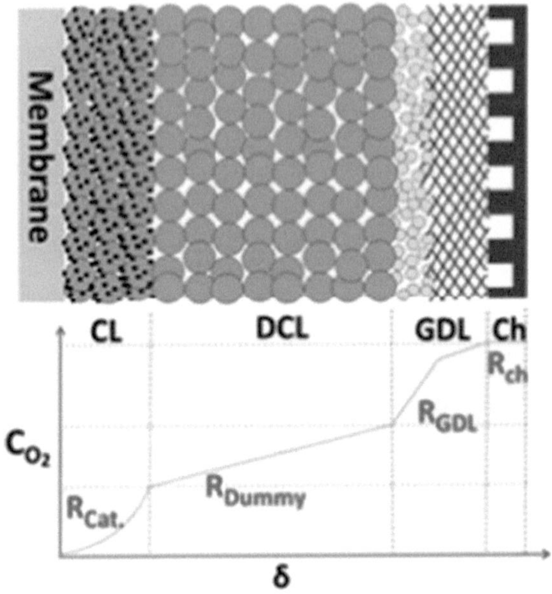

Fig. 3.24 Schematic of oxygen concentration at different positions in the fuel cell [46]

concentration in the porous material decreases rapidly, which indicates that the porous materials cause a large mass transport resistance [47]. So, the porous material is the main component leading to the difficulty of mass transport in the cell. The porous structure and permeability are important indicators for judging the mass transport ability of porous materials. Therefore, it is necessary to obtain their values. Generally, the material porosity can be inferred from the mass and volume of the material. But the electrode materials applied in fuel cells will be prepared in multiple steps, so it is difficult to accurately calculate their porosity. In addition, other structural parameters such as pore size and pore distribution can also affect mass transport. Hence, it is necessary to characterize the pore properties of materials by experimental methods. The measured results can also be used for modeling studies. The two most commonly used characterization methods in the fuel cell are mercury intrusion porosimetry and the gas adsorption method.

3.3.1.1 Mercury Intrusion Porosimetry

Mercury intrusion porosimetry (MIP) is a method for measuring the total connected porosity, volume of pores, pore size distribution, and surface area of porous materials [48]. This method takes advantage of the fact that mercury has a high surface tension and cannot enter the pores of the material easily. So, it can only flow into the pores when an external pressure is applied. The higher the external pressure, the smaller the pore radius through which mercury can enter. By measuring the amount of mercury that enters the pores at different pressures, the pore volume of the corresponding pore size can be obtained. The ratio of the total pore volume to the material volume can be used to calculate the porosity of the material [49]. By combining the Washburn equation, the relationship between the applied pressure and the capillary pore diameter can be established, and more detailed information on the pore size distribution can be obtained. The following equation is the Washburn formula.

$$p = \frac{2\sigma}{r}\cos\theta \tag{3.21}$$

where σ is the surface tension of mercury, θ is the mercury contact angle, r is the intrusion radius of cylindrical pores, and p is the external pressure. It is worth noting that there may be some small pores that mercury cannot enter in real porous materials, due to the complex pore structure. Besides, excessive pressure may lead to microstructural collapse, so MIP cannot accurately measure the volume of small pores. During the test, the material is placed in a vacuum sample chamber, and mercury is injected into the porous sample through a capillary. The mercury in the capillary and the metal material outside the tube are equivalent to the metal plates at both ends of the capacitor, and the capillary is equivalent to the insulating plate. The entire device constitutes a capacitor. As the experiment progresses, the length of the mercury column in the capillary changes, causing a change in the capacitance of the capacitor. The sensor collects changes in electrical signals, which can be quickly converted into changes in mercury volume through data processing, and then calculates data such as porosity. The test results are usually expressed in three types of graphs: pore diameter, cumulative volume, and differential volume or log differential volume. Figure 3.25 shows MIP results of the Pt/V_A catalyst powder, electrodes with different I/C ratios, and the blank ethylene tetrafluoroethylene substrates in the plot of pore size versus log differential volume.

The orange line in the figure is the measurement of a blank substrate. Pore sizes larger than 5 μm are due to the gap between the electrodes and the walls of the MIP sample compartment, which need to be excluded from the data. After some corrective calculations, the porosity calculated by the MIP method can be derived.

Table 3.1 shows that the porosity obtained from MIP is much lower than the theoretically calculated porosity, and the difference between the measured and calculated values increases with the increase of I/C ratio. This indicates that the ionomer may block the small pores in the electrode during the membrane electrode assembly (MEA) preparation process. As the I/C ratio increases, the extent of pore blocking intensifies, resulting in the theoretically calculated porosity not being used in the simulation.

3.3.1.2 Gas Adsorption

Gas adsorption is a method using adsorption phenomena to characterize the porous structure properties of materials. The measurement principle of this method relies

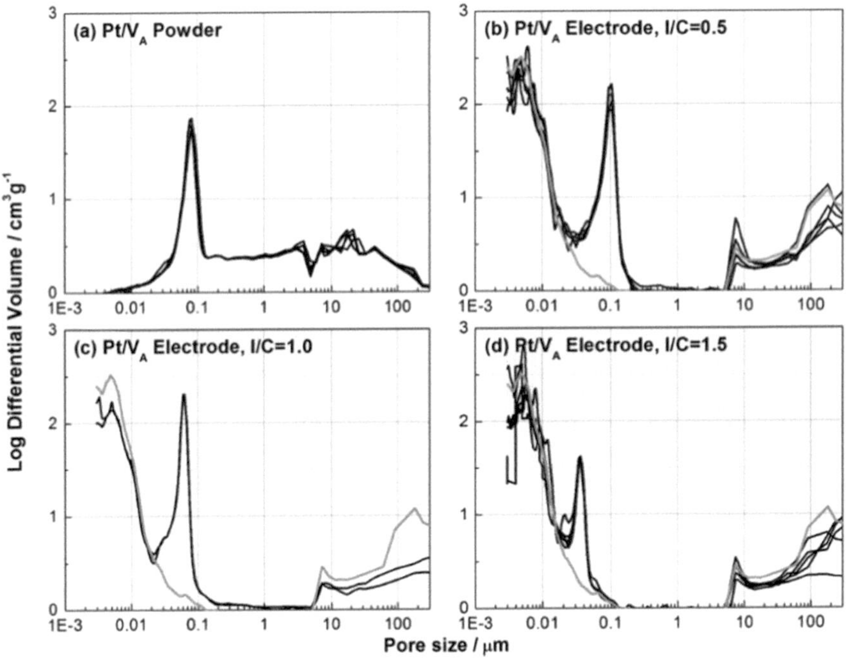

Fig. 3.25 MIP results of Pt/VA catalyst powder and electrodes [50]

Table 3.1 Porosity results [50]

	Total V_{Hg} ($cm^3\ g^{-1}$)	V_{Hg} (> 25 nm)	ε from MIP (%)	ε from densities (%)
Pt/V_A powder	0.726	0.695	83.0	
Pt/V_A, I/C = 0.5	2.255	0.870	49.8	68.1
Pt/V_A, I/C = 1.0	2.075	0.760	41.0	58.1
Pt/V_A, I/C = 1.5	1.867	0.468	17.7	48.1

on the reversible physical adsorption of gas molecules (adsorbate) on the surface of the measured sample (adsorbent) at ultra-low temperatures, and this adsorption has a certain equilibrium adsorption amount. By measuring the equilibrium adsorption amount, the specific surface area, pore volume, and pore size distribution of porous materials can be obtained using theoretical models [51]. The pore volume and surface area of mesoporous and microporous parts can also be calculated through complete adsorption–desorption curves. Nitrogen, chlorine, and oxygen are usually used as the adsorbate in experiments. In the actual adsorption process, the material surface may adsorb multiple layers of gas molecules. If the gas adsorption amount is directly used for calculation, it will overestimate the surface area of porous materials. The Brunauer Emmett Teller (BET) surface area determination method is one of the most commonly used techniques in gas adsorption for porous material characterization [52]. This method establishes the relationship between single-layer saturated adsorption amount V_m and multi-layer adsorption amount V. It is possible to calculate the monolayer adsorption V_m from V, thus obtaining a more accurate specific surface area [53].

$$\frac{p}{V(p_0 - p)} = \frac{1}{V_m C} + \frac{C-1}{V_m C}\frac{p}{p_0} \tag{3.22}$$

where, P is the equilibrium pressure of the adsorbed gas at the adsorption temperature, P_0 is the saturated vapor pressure, V is the total volume of the adsorbed gas when the equilibrium pressure is P, V_m is the volume of gas required to cover the surface of the material with one layer of gas, C is a constant related to adsorption. Unlike the electrode surface measured by cyclic voltammetry, the surface area obtained by the BET method is independent of electrochemical activity. As a distinction, the surface area calculated by cyclic voltammetry is also called the electrochemically active surface area. During the measurement process, it is necessary to first extract the gas from the pre-treated sample and then cool it to liquid ammonia temperature (77 K). Then gradually increase the pressure inside the sample tube. Nitrogen will undergo physical adsorption on the sample surface. The true specific surface area of the sample can be calculated according to the adsorption isotherm equation. The maximum pressure that nitrogen adsorption can measure is atmospheric pressure. Large pores in porous materials require higher pressures to adsorb adsorbates, so measuring large pores will cause larger errors. Figure 3.26 shows the experimentally measured nitrogen adsorption–desorption isotherm and pore size distribution.

The adsorption–desorption isotherms shown in Fig. 3.26a can be classified as type IV, implying the appearance of the mesoporous structure. The calculated pore size distribution (Fig. 3.26b) further confirms the existence of a mesoporous structure. According to the calculated values, the specific surface area of the porous C–Ni rises to 6.2 $m^2\ g^{-1}$, much higher than that of pristine porous Ni (0.15 $m^2\ g^{-1}$), which means the modified porous C–Ni is more suitable as an electrode substrate for extending the triple-phase boundary.

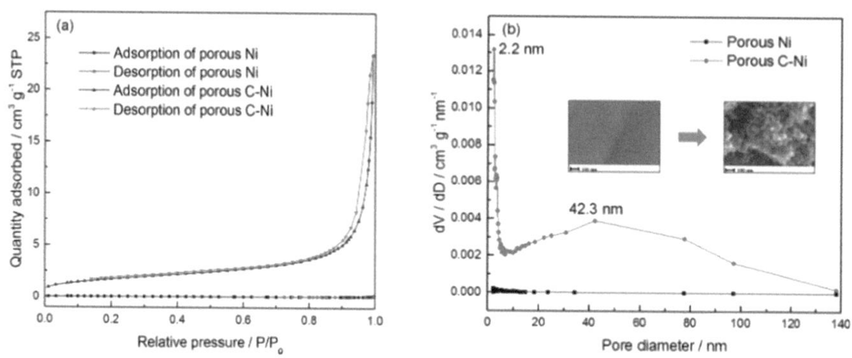

Fig. 3.26 Nitrogen adsorption–desorption isotherms (**a**) and pore size distribution (**b**) [54]

Among the two methods of characterizing porous structures, MIP is inaccurate for measuring nanoscale pores, while the gas adsorption method cannot measure the adsorption amount of large pores. Therefore, different methods should be chosen to measure the specific surface area and porosity of different samples according to the pore size. Microporous and mesoporous samples are recommended to use the gas adsorption method, and a large pore distribution is recommended to use MIP. In addition, the advantage of MIP is not only measuring the pore size distribution, but also measuring the particle size, compressibility, permeability, and other material parameters.

In addition to maintaining high porosity and surface area, permeability is also extremely important for porous materials. If most of the pores are not directly connected or blocked, effective gas transport cannot be guaranteed [55]. This will cause a significant increase in mass transport loss of the cell, greatly reducing the performance of the cell. The permeability of the sample can be calculated from Darcy's law.

$$Q = -\frac{kA}{\mu}\frac{\Delta P}{L} \tag{3.23}$$

where Q is the volumetric flow rate of the fluid, k presents the permeability of the porous material, A is cross section area of the porous material, μ is the viscosity of the fluid, ΔP is the differential pressure between the porous material, and L is the thickness of the porous material. Figure 3.27 demonstrates a laboratory-fabricated water permeation device.

The key in the measurement is to keep the differential pressure constant while being able to measure the flow rate through the porous material. The permeability of the material can then be calculated using Darcy's law.

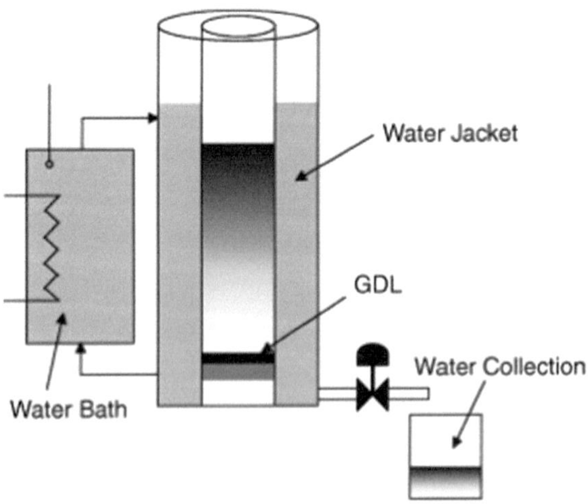

Fig. 3.27 Schematic of the water flow measurement across a GDL [56]

3.3.2 *Morphology Characterization*

The morphology of porous materials has an important influence on the performance and surface mass transfer ability of the porous electrode [57]. For example, under the same intrinsic activity, the catalyst loaded on the high surface area support will have better performance. Some special surface morphologies can also delay the decay of the Pt catalyst and improve the durability of the cell. Besides, understanding the morphology of materials is the first step to understanding their properties. Therefore, in this part, three common morphology characterization techniques in the fuel cell field are briefly introduced.

3.3.2.1 Scanning Electron Microscopy (SEM)

The scanning electron microscopy stands for scanning electron microscope, which is a microscope that uses electrons to form images. The electron is generated by an electron gun at the top of the microscope and then focused on the sample by electromagnetic fields and lenses. Once the high-energy electron beam hits the surface of the sample material, secondary electrons, backscattered electrons, characteristic X-rays, and other signals will be generated. The receiver collects these signals and converts them into images after amplification. Among them, secondary electrons are the main signals used in scanning electron microscopy. They are excited and detached from the outermost electrons (5 –10 nm) of the sample by atomic beams and are very sensitive to the surface morphology of the sample, which can accurately reflect the surface morphology characteristics of the sample. In addition, because the electron beam is very narrow, SEM micrographs have a large depth of field and can produce characteristic three-dimensional appearances. Figure 3.28 shows a simple schematic of the observation principle of SEM.

SEM generates images in a sequence similar to knitting. The scanning generator guides the electron beam to every corner of the sample and pauses there for a while. For example, if the scanner starts from the upper left corner of the sample, the corresponding image will appear in the upper left corner of the monitor. The electron beam moves to the right and stops again quickly, repeating this process until the first row is completed. Then, the scanner moves down to repeat this process until it reaches another corner of the sample. The brightness of the displayed image is determined by the number of electrons collected by the secondary electron detector. The more electrons at that position, the brighter it is in the image. The magnification range of SEM is from about 10 times to over 500,000 times, which is about 250 times the best optical microscope magnification limit. So, it is often used in cell research to characterize the morphology of the catalyst layers, microporous layers, and diffusion layers, which helps to optimize the structure of each component and study the degradation mechanism. Figure 3.29 shows the catalyst layer structure taken by SEM.

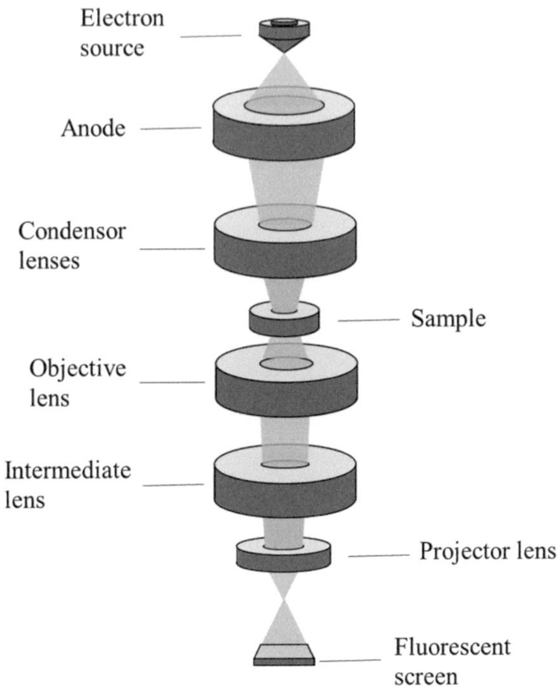

Fig. 3.28 Schematic of the SEM principle

Figure 3.29 presents the morphology of catalyst layers with different A3 contents. From the figure, we can directly observe the changes in the catalyst layer surface. When the A3 content is low, the catalyst layer surface is relatively dense, and only small pores can be seen. However, as the A3 content increases, larger holes appear on the catalyst layer, and the surface particles become more agglomerated. This means that as the A3 content increases, the catalyst layer will have fewer exposed active sites, which will weaken the cell performance. But on the other hand, when the A3 content is low, the catalyst layer surface is relatively dense, and only small pores can be observed, which means that such a catalyst layer may not be suitable for mass transport. Therefore, the optimal catalyst layer performance should be achieved at a relatively moderate A3 content [58].

3.3.2.2 Atomic Force Microscope (AFM)

The atomic force microscope is an analytical method that can investigate the structure of almost any solid surface. It is based on the effect of interatomic van der Waals forces to study the surface structure and properties of substances, which overcomes the limitation of SEM that can only image conductive surfaces. AFM uses a probe to approach the sample surface. One end of the probe is fixed to the end of a cantilever that is extremely sensitive to weak forces. As the probe approaches the surface of

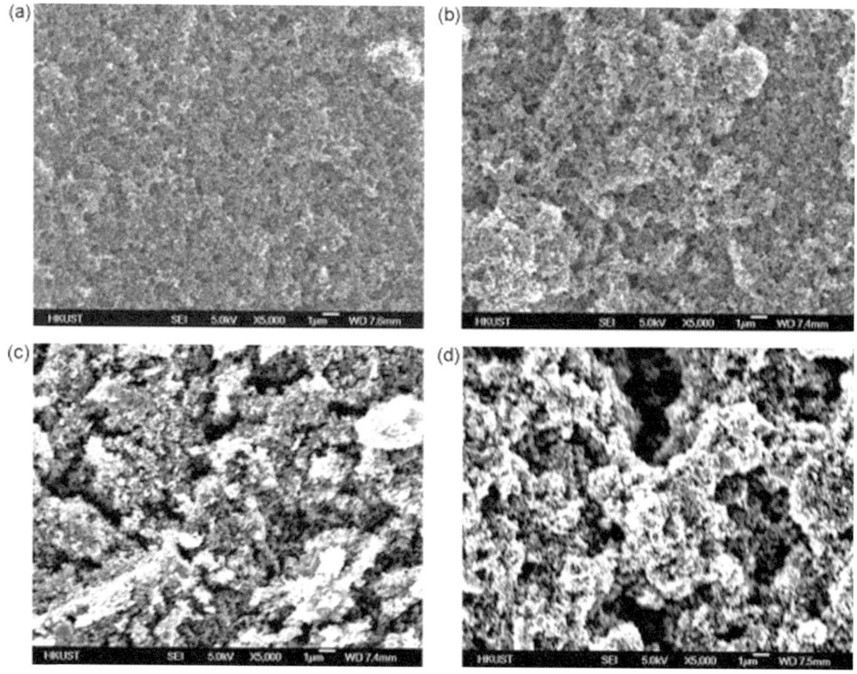

Fig. 3.29 SEM images of catalyst layers with different A3 contents **a** 0 wt%, **b** 5 wt%, **c** 10 wt%, **d** 30 wt% [58]

the sample, Van der Waals forces are generated between the tip of the needle and the sample, and the cantilever will deform or change its state of motion as a result. During the scanning process, the probe will undulate in the direction perpendicular to the surface of the sample, and by detecting these movements with the use of an optoelectronic detection system, information such as the morphological structure and roughness of the sample surface can be obtained with a resolution of up to the nanometer scale [59]. Figure 3.30 shows a schematic of the AFM principle.

Unlike SEM, which can only provide two-dimensional images, AFM can perform measurements in the vertical direction and provide high-resolution three-dimensional images of the sample surface. In addition, AFM is not constrained by the high vacuum conditions which is necessary for SEM. And it can be made at atmospheric pressure without any special treatment of the sample's surface. Besides, it does not damage the sample. However, the imaging of AFM relies on the movement of the probe on the sample surface, so the imaging range is small, slow, and complicated. In addition, due to the limitations of the probes, AFM is unable to accurately detect steep surface changes and is highly susceptible to being damaged due to collisions. Figure 3.31 shows the picture obtained by probing the synthesized catalyst sample with AFM.

It can be visualized from the figure that $CuNiTiO_3$ has a rougher surface than $NiTiO_3$. Based on the data obtained from AFM, it can be calculated that the surface

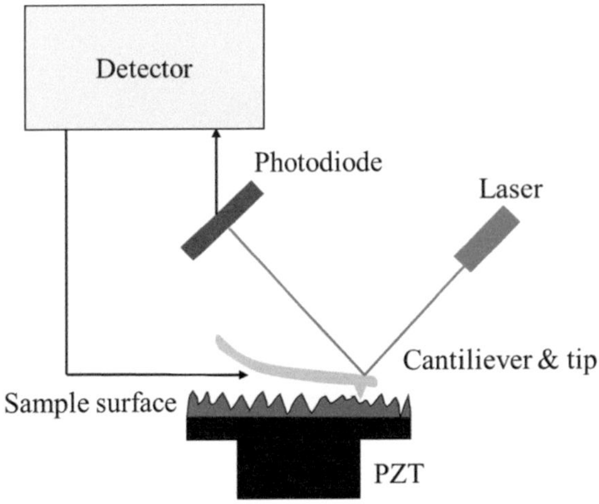

Fig. 3.30 Schematic of AFM principle

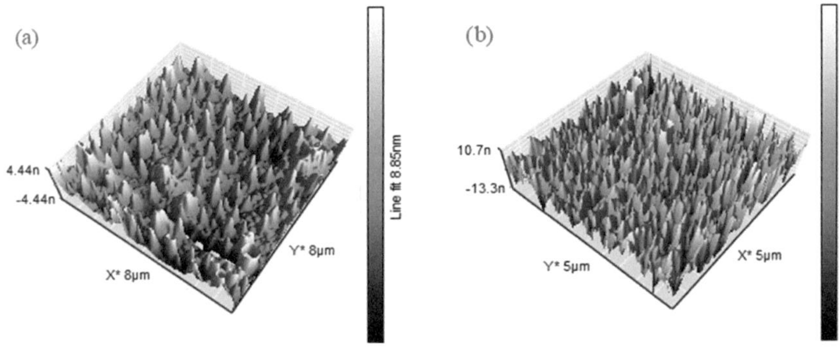

Fig. 3.31 AFM images of **a** $NiTiO_3$ and **b** $CuNiTiO_3$ [60]

roughness of the catalyst increases by a factor of 1.7 with the addition of copper as a dopant.

3.3.2.3 Transmission Electron Microscopy (TEM)

The transmission electron microscopy (TEM) is another electron microscopy technique, but its technical principle is different from that of SEM described above, and is closer to that of optical microscopy [61]. The optical microscope observes the light waves through the sample, but due to the wavelength limitation, it is unable to see the structure at the nanometer level. TEM utilizes a high-energy electron beam with a shorter wavelength as a light source and uses an electromagnetic field as a lens. The electron beam scatters or diffracts from the sample. TEM detects the electron beam

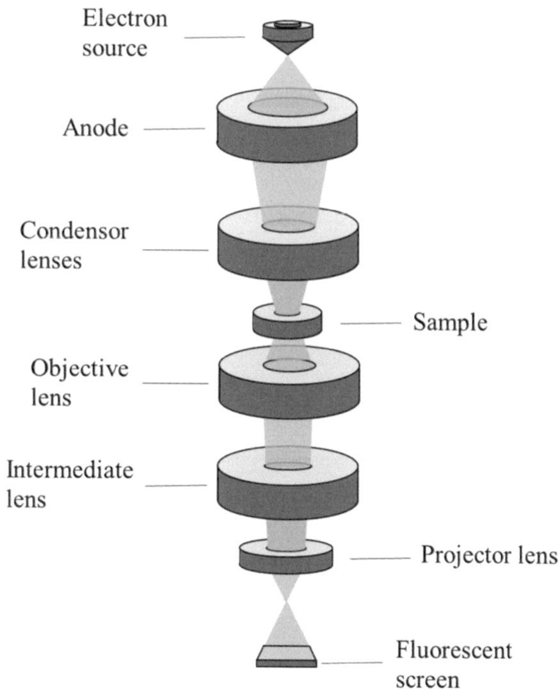

Fig. 3.32 Schematic of TEM

passing through the lower surface of the sample with a resolution of up to 0.2 nm. Figure 3.32 illustrates the principle of TEM.

Since TEM requires the electron beam to pass through the sample, and the electrons are easily scattered or absorbed by the sample with low penetration, the samples are usually prepared as ultrathin slices of less than 150 nm. Compared with SEM, TEM can provide higher resolution images, and can directly observe the particle size, distribution, and morphology. In the field of PEMFC, TEM is commonly used to observe the morphology and size of catalysts. The particle size and its distribution can be calculated in combination with statistics. Figure 3.33 shows the morphology picture of the Pd/C catalyst obtained by TEM.

The figure shows the morphology pictures of Pd/C catalysts with different running times under the same conditions. It can be seen that at the beginning, the Pd particles were spherical and well dispersed on the carbon carriers, and no obvious agglomeration occurred. With the increase of running time, the Pd particles gradually agglomerated and grew. From the TEM data, it can be analyzed that the average particle size of the Pd particles at the beginning is 2.25 nm, while the average particle size of the Pd particles has grown to 3.45 nm at 300 h. The agglomeration and growth of the Pd particles means that there is a decrease in ECSA, which also indicates that the performance of the cell will decrease with the increase in the running time.

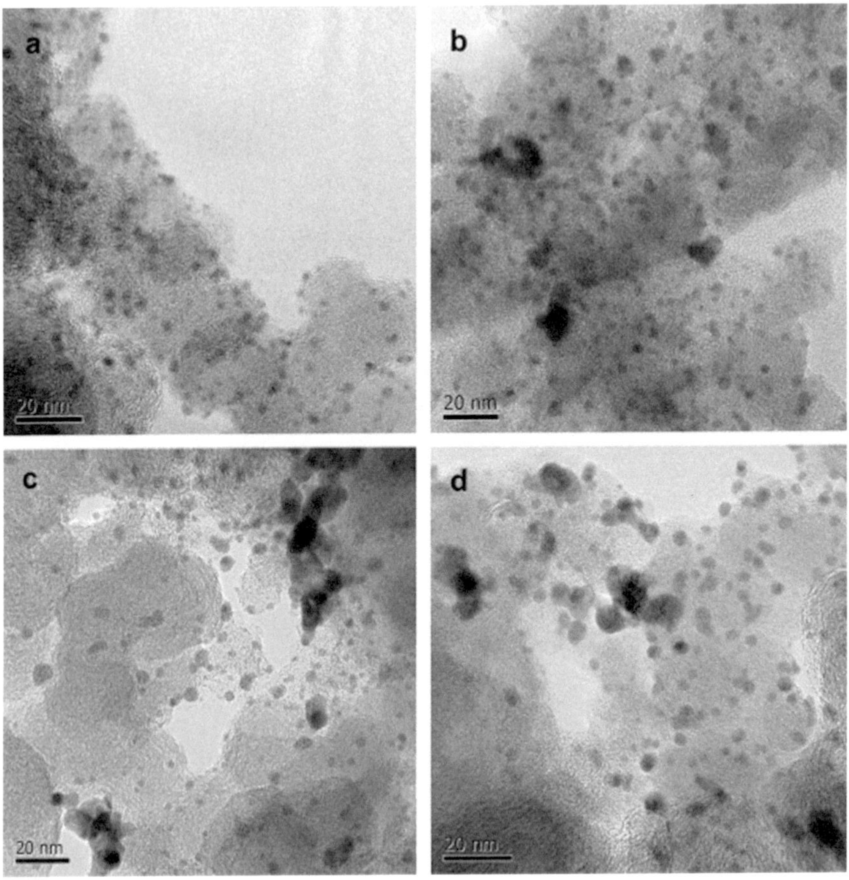

Fig. 3.33 TEM images of the Pd/C at 0 h **a**, 100 h **b**, 300 h **c** and 500 h **d** [62]

3.3.2.4 Grading Analysis

In fuel cells, the catalyst particle size has an important impact on the cell performance. Taking Pt/C catalyst as an example, the smaller the catalyst particle size, the more active sites can be exposed under the same loading capacity, and the better the membrane electrode performance [63]. However, too small a catalyst particle size will make Pt easy to decay, resulting in rapid weakening of the MEA, which makes it difficult to meet the lifetime requirements of fuel cells. For this reason, commercialized Pt/C catalysts generally control the Pt at about 3 nm. For other catalytic materials, there are similar constraints between performance and lifetime, so there must be an optimal size for the catalyst particle. To better analyze and judge the particle size distribution of catalytic materials, this subsection will give a brief introduction to the X-ray diffraction (XRD) method. In addition to this method, the TEM

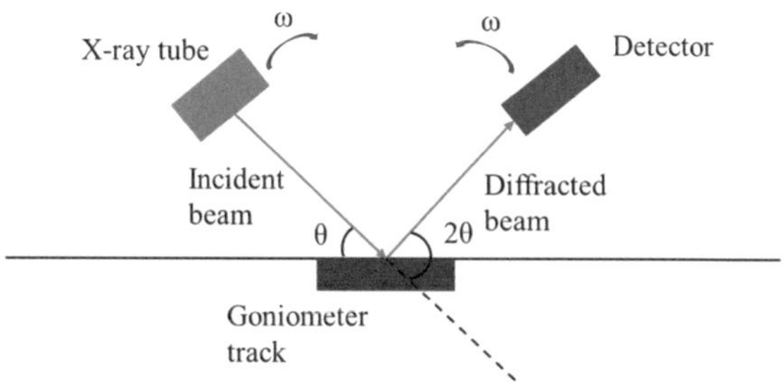

Fig. 3.34 Schematic of X-ray diffractometer

in the previous section, due to the image resolution of 0.1 nm, can also be used to measure and statistically count the particle size in the image.

X-ray diffraction (XRD) is a simple method to determine the average particle size of nanocrystals in powders, single crystals, or polycrystals bulk materials by using the diffraction phenomenon of X-rays in crystals to obtain the characteristics of the post-diffraction X-ray signals [64].

For crystalline materials, the XRD diffraction pattern is characterized by intense peaks (X-ray diffraction intensification results) only at specific positions because the atomic arrangement is long-range ordered in three-dimensional space. The characterization of a diffraction pattern can be considered to consist of two aspects: (a) The pattern of distribution of diffraction lines in space, which is determined by the size, shape and orientation of the crystal cell; (b) The intensity of the diffraction lines, which depends on the type of atoms and their positions in the crystal cell. Thus, these diffraction patterns act like fingerprints of the crystal [64]. The schematic diagram of the XRD measurement is briefly illustrated in Fig. 3.34.

Each crystal has a uniform arrangement of atoms and therefore a unique X-ray diffraction pattern, which is an important basis for phase analysis using XRD. XRD can yield specific quantitative structural information, such as crystal structure, orientation, and compounds. In the field of PEMFC, this information is important for the development of new electrodes, catalysts, or electrolyte materials.

The Bragg equation is the cornerstone of XRD theory and reflects the relationship between diffraction and crystal structure. When X-rays are irradiated into a crystal, the difference in the optical path between the two neighboring crystal planes is 2dsinθ. If the optical path difference is equal to n times the wavelength of the X-rays, the diffraction intensities of the X-rays will be enhanced by each other, and conversely, the diffraction intensities will remain unchanged or diminish in other cases. The following equation is the Bragg equation.

$$2d \sin \theta = n\lambda \tag{3.24}$$

where θ is the angle of incidence, d is the crystal plane distance, n is a diffraction level, and λ is the wavelength of incoming rays. Obviously, the Bragg equation allows one to use the known wavelength of X-rays to solve the crystal facet distance, d, and thus obtain information about the crystal structure.

The Scherrer formula is the theoretical basis for the measurement of grain size. It describes the relationship between the grain size and the half-peak width of the diffraction peaks. The wider the half-peak width, the finer the grain size. The formula is as follows.

$$L = \frac{k\lambda}{(B - b)\cos\theta} \tag{3.25}$$

where, L is the average thickness of the grain in the direction vertical to the grain surface, k is Scherrer's constant, generally taken as 0.89, λ is the wavelength of X-rays; θ is the angle of Bragg diffraction (Brag angle), $B - b$ is the half width of the diffraction peak of the measured sample, which needs to be converted to radians in the calculation, b is the instrument correction factor. Since the grain sizes in the real material are not the same, the calculated result of this method is the average size of the different-sized grains. Figure 3.35 illustrates the XRD image of the Pd/Ni foam metal.

From the figure, it can be seen that several peaks exist at different 2θ angles for the prepared foam electrode. Depending on the location where the peaks appear, the elements and crystallographic orientation contained in the electrode can be distinguished. The elements and crystallographic directions indicated by each peak have been labeled in the figure. Figure 3.36 illustrates the XRD results in another literature.

The XRD method can also be used to identify the size of catalyst particles. Based on the XRD data in Fig. 3.36 and the Scherrer equation, average crystallite sizes in various catalytic materials can be calculated. For example, they are calculated to be 4.4 nm in Pd/CF and 4.3 nm in Pd_1Ni_1/CF.

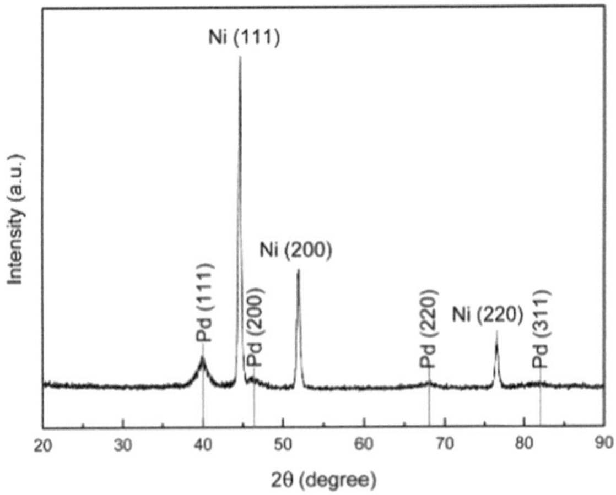

Fig. 3.35 XRD pattern of the layer-reduced Pd/Ni foam electrode [65]

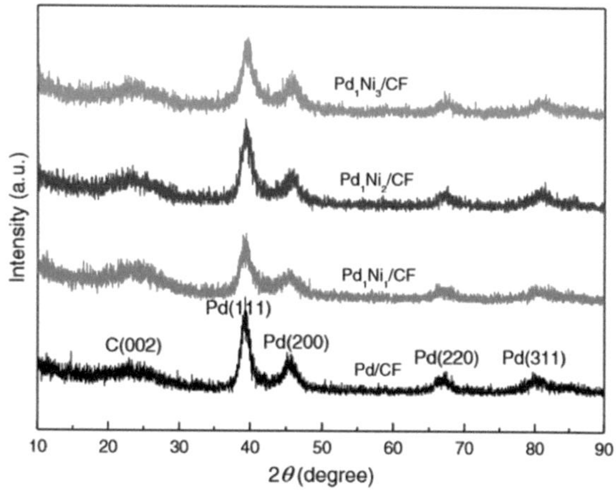

Fig. 3.36 XRD patterns of the Pd/CF and PdNi/CF [66]

3.3.3 *Elemental Analysis*

Elemental analysis techniques can show the distribution and quantity of elements in the material, which can be used to verify whether the material preparation is successful or not, and play an important role in the research of catalytic materials for cells. Here are some typical elemental analysis methods.

X-ray energy dispersive spectrometer (EDS) is a technique often used for the analysis of elements, where an electron beam is bombarded on the surface of a sample in a vacuum chamber and the material is excited to emit characteristic X-rays. The characteristic wavelength of the X-rays is determined by the characteristic energy ΔE released during the energy level transitions and is only related to the constituent elements of the sample. In the previous section, we learned that SEM and TEM also use high-energy electron beams to bombard the sample. Therefore, the technique can be used to obtain data from both SEM and TEM while observing the microscopic morphology. Figure 3.37 shows the SEM and EDS images obtained from the experiment.

The images in Fig. 3.37 were obtained on the same SEM equipment, so that it is possible to analyze the distribution of elements at the same location immediately after obtaining the micro-morphological results of the material. From (3.37 h, i), it can be observed that the elements are uniformly dispersed on the carbon substrate. In addition to the characteristic X-rays in the SEM, back-reflected electrons, generated by the interaction of the high-energy electron beam with the surface material of the sample, can also be used to reflect the compositional distribution of the sample. The back-emission electrons are part of the incident electrons reflected from the nuclei in the sample, and the reflection coefficient is related to the inclination angle of the sample surface and the atomic number of the sample, which can be used to study the surface morphology and compositional distribution of the sample. However, due to

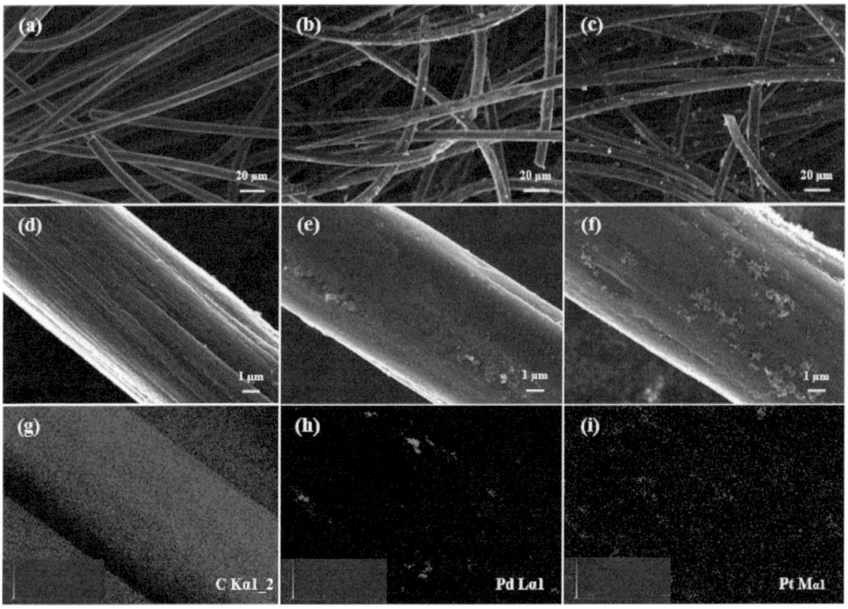

Fig. 3.37 SEM images and elemental distribution [67]

the low resolution, it is generally not used to analyze the surface morphology, but to observe the distribution of components on the sample surface.

X-ray photoelectron spectroscopy (XPS) is a technique that uses X-rays to analyze samples. The measurement principle is the photoelectric effect. The photoelectric effect occurs when X-rays are used to irradiate the surface area of a sample, and the inner electrons or valence electrons of the sample are excited and emitted [68]. These electrons, which are excited by photons, are called photoelectrons, and XPS measures the kinetic energy of the photoelectrons as they leave the surface of the material. Theoretically, the kinetic energy equals the difference between the energy of the incident photon and the binding energy of the electrons. Since the energy of the incident X-rays is known, the binding energy of the electron can be calculated. The photoelectron binding energy depends on the atomic properties and orbitals, from which the composition of the sample can be analyzed. This enables qualitative analysis of the surface elements, especially chemical analysis, such as the chemical composition of the material surface and the chemical binding state of each element. Figure 3.38 shows a schematic of the XPS test. The XPS test system is installed in an ultra-vacuum environment and consists of an X-ray source, sample stage, extraction lenses, analyzer, and detector.

In the field of fuel cells, X-ray photoelectron spectroscopy is commonly used for the development of new materials for electrodes. Figure 3.39 illustrates the XPS results of the Pt_4Co@NC-900 sample from the literature [69].

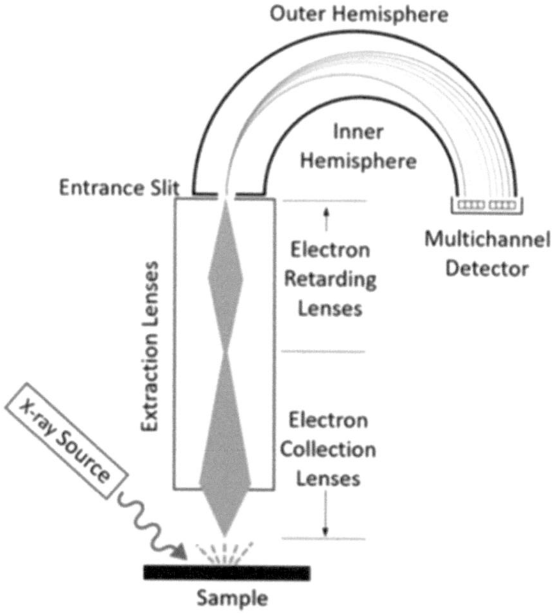

Fig. 3.38 Diagram of X-ray photoelectron spectroscopy test [68]

From the figure, not only can the elements contained in the sample be analyzed, but also the chemical bonds formed between the elements can be identified. For example, in Fig. 3.39b, the presence of five types of bonds, C–C, C–N, C–O, C=O, and $\pi - \pi^*$, can be recognized, and C–C has the highest percentage of all chemical bonds at 56.53%. Using this technique, XPS, the actual bonding within the catalyst can be analyzed, which is important for the construction of cell catalysts.

3.4 Fuel Cell System Characterization

3.4.1 Current Mapping

In fuel cell operation, due to the electrochemical reactions along the flow field, there is a non-uniform distribution of reactant concentration in the cell, which further results in a non-uniform current density and temperature distribution. This heterogeneity will have an important impact on the performance and durability of the cell. For example, uneven temperature distribution may cause thermal stress, which in turn may cause deformation or even damage to the membrane electrode [70]. The ability to measure the heterogeneity helps optimize the fuel cell design and extend the lifetime.

Among them, the current density distribution can directly reflect the inhomogeneity of the cell, so a brief introduction is provided here. The widely used current

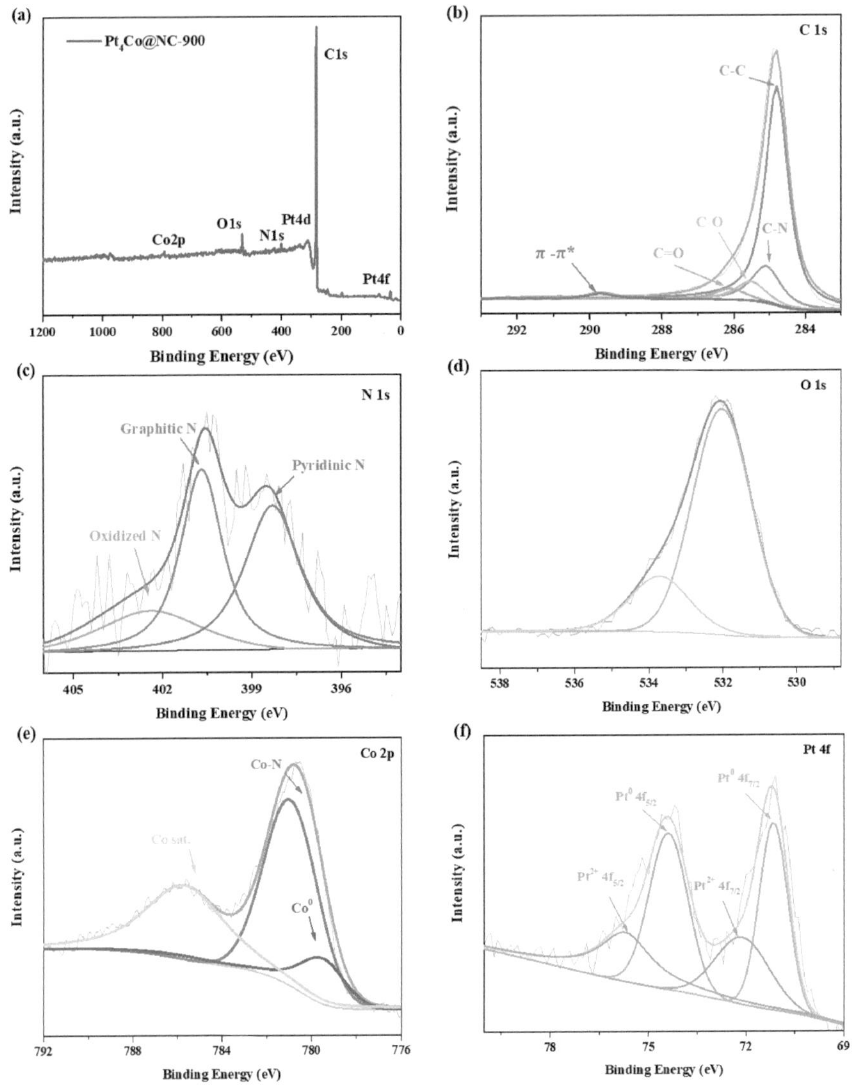

Fig. 3.39 **a** Full XPS spectra of Pt_4Co@NC-900 and XPS scan of **b** C 1s, **c** N 1s, **d** O 1s, **e** Co 2p, and **f** Pt 4f [69]

mapping methods can be divided into four types according to the test method: sub-cell method, printed circuit board method, resistor network method, and Hall effect measurement method. These four methods will be introduced one by one below.

(1) Sub-cell method. A large cell is divided into multiple sub-cells, and each sub-cell is controlled by a separate load. Then, by measuring the current density of each sub-cell, the current density distribution of the entire large cell can be

obtained. This method is easier to think of, but it is difficult to manufacture and destroys the original structure of the cell. As a result, this approach was proposed relatively early but has been applied less in recent years.

(2) Printed circuit board method. Segmentation of all the components of the cell is difficult to realize, thus limiting the application of the sub-cell method. Unlike the previous method, the printed circuit board method requires only the replacement of the collector plate and has therefore received more attention recently. The printed circuit board used in this method has been divided into several insulated areas. The current in each area is measured to obtain the distribution of current density in the cell. It is worth noting that such a method is affected by the in-plane current in the electrode, which does not accurately reveal the heterogeneity of the current distribution in the catalyst layer in actual operation. Fortunately, it has been demonstrated that in-plane currents within porous electrodes have less effect.

(3) Network resistance method. This method is similar to the printed circuit board method in principle. The difference is that it directly detects the voltage on both sides of the resistor, which is then converted into current distribution data by Ohm's law. Resistors are embedded in a non-conductive thin plate and placed between the flow field plate and the collector plate. Therefore, this method has the same advantages of low cost and ease of use. However, this method is equally susceptible to in-plane currents. Relatively low in-plane currents are a prerequisite for accurate measurements.

(4) Hall effect measurement method. The most important feature of Hall effect measurements is the difference in the detection principle. The Hall effect is a process in which a transverse electric field is developed in a solid material when an electric current flows through a magnetic field region in this material. The magnitude of the induced voltage is determined by the current flowing through the magnetic field. Hence, the Hall sensor can detect current density. This measurement method also requires splitting the collector plate. In addition, the test requires an external electric field and several Hall sensors on the end plate. However, magnetic fields in the environment may affect the test results. It also has problems caused by splitting the collector plate, like other methods, which will not be repeated here.

Each of the above measurement methods has its advantages and disadvantages. Among them, the printed circuit board method and the Hall effect measurement method are more popular due to their accuracy and ease of operation. In recent years, based on these research methods, some new types of measurement methods have been improved, but all of them also face the problems of difficult manufacturing and low resolution. Developing more convenient, efficient, and accurate measurement methods for current density distribution is urgent.

3.4.2 Temperature Mapping

Local overheating can cause cell performance degradation and durability reduction [71]. Therefore, it is important to detect the temperature distribution inside the fuel cell. There are many types of temperature distribution measurement techniques, and the commonly used techniques include the contact measurement method and the infrared detection method.

(1) Contact measurement method. As the name suggests, this type of method takes temperature measurements by directly touching the location where the temperature needs to be measured. The common temperature measurement points include the proton exchange membrane, the diffusion layer surface, the middle of the flow channel, or the back of the flow field plate. Temperature sensors can be categorized into thermal resistance detectors and thermocouple detectors based on the principle of measuring temperature. Thermal resistance detectors rely on a linear relationship between metal resistance with temperature growth. It directly measures the value of the resistance of the probe and relies on a pre-supplied formula to convert to the temperature of the probe. These sensors are usually made of platinum, copper, or nickel metal materials. Thermocouple measurements are based on the Seebeck effect. Two conductors of different materials are connected at both ends to form a circuit, and when the temperature of the joints is different, an electric potential is generated in the circuit. From this potential, the temperature can be calculated. These sensor probes can be made very small and are therefore commonly used for temperature measurement in fuel cells. It is worth noting that contact measurement can only measure the temperature of one point at a time. Therefore, a large number of probes need to be placed in the cell to get the temperature distribution over the whole surface.

(2) Infrared detection method. The most important feature of infrared detection compared to the above methods is that the temperature can be measured directly on the whole surface [72]. According to the theory of thermal radiation, an object is emitting electromagnetic waves at all times, and the wavelength of the radiation is related to the temperature of the object. An infrared camera can image such electromagnetic waves at infrared wavelengths and therefore visualize the temperature distribution of an object. However, this radiation resembles natural light and can be so easily blocked as to be unobservable. So, infrared cameras can only observe the temperature distribution on the surface of an object. And there is no point in observing the temperature of the fuel cell surface, because the heterogeneity of temperature exists within the cell. To observe the internal temperature, the fuel cell often needs to be made transparent. In addition, this measurement does not reveal the temperature distribution in the through-plane direction.

In recent years, with the development of temperature measurement technology, more techniques that can realize high-precision temperature field measurement have been proposed [73]. However, all temperature measurement techniques require different degrees of changes to the cell structure, and it is difficult to measure the temperature deep inside the membrane electrode. The temperature distribution of the flow field and end plate measured by the above methods does not directly represent the temperature distribution of the membrane electrode. It is the temperature distribution inside the membrane electrode that affects the life of the electrode the most, which is most worth measuring. The non-destructive acquisition of the temperature distribution of the membrane electrode in normal operation still needs further research.

3.4.3 Water Distribution Visualization

For PEMFCs, the internal gas–liquid two-phase flow has a great impact on the cell performance and lifetime [74]. Liquid water icing at low temperatures also directly affects the cold-start performance of the cell. So, it is especially important to obtain the water distribution in the flow channel diffusion layer and MEA in situ without any damage. Measurement methods of liquid water distribution inside the PEMFC can be divided into direct and indirect methods. The direct method is a test technique based on visualization instruments to directly observe the liquid water flow and phase transition process inside the cell. Typical visualization techniques include X-ray imaging, neutron imaging, nuclear magnetic resonance, and transparent cells. The indirect method is to provide feedback on the internal water distribution of the cell through external parameters, such as pressure drop, high-frequency impedance, performance fluctuation, and so on. Compared with indirect methods, visual characterization techniques have the advantages of higher accuracy, temporal, and spatial resolution. Therefore, we will select two common visual characterization techniques to introduce here.

3.4.3.1 Transparent Cell

Transparent cells are the most obvious means of visualization. This method allows the researchers to directly observe the distribution and transport process of liquid water in the flow channel during the operation of the cell by replacing a portion of the cell components with transparent material [75]. In a fuel cell, the component that prevents direct observation of liquid water in the flow channel is the flow field plate, which is therefore replaced with a transparent plate. Considering that the flow field plate also has the function of collecting and distributing material transport, it is necessary to add a flow field sheet that can fulfill the above functions to keep the cell in normal operation. Figure 3.40 shows a schematic diagram of a typical transparent cell structure.

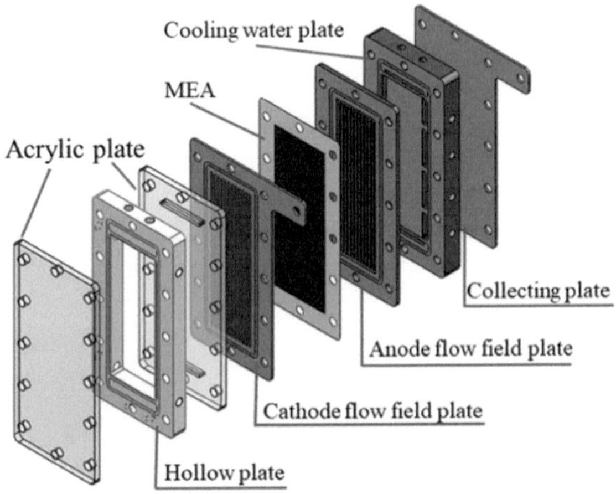

Fig. 3.40 Schematic of a transparent cell [76]

The equipment needed for the transparent cell experiment is relatively simple. Only a digital camera needs to be added in addition to the conventional cell performance test system, so this method is favored due to its lower cost and simplicity of testing. Typical images obtained from the experiment are shown in Fig. 3.41.

The liquid water distribution in the flow channel of an alkaline direct ethanol fuel cell under different current densities is shown in Fig. 3.41. It can be seen that there is an obvious gas–liquid interface in the flow channel. And it can be visualized that the liquid water content in the flow channel experienced an increase and then a decrease with the increase of current density. This implies a change in the main water transport mechanism of the cell at different current densities. At low current densities, the water at the cathode mainly originates from the transmembrane transport of anodic water. As the reaction proceeds, the rate of anodic water production increases, leading to an increasing water concentration at the interface of the anodic catalyst layer, which increases the transmembrane flux of water, and thus, cathodic flooding becomes more and more serious. However, when the current density exceeds 200 mA cm^{-2}, the role of electro-osmotic drag in water transport becomes more and more important. Since ions flow from cathode to anode inside the membrane, water molecules will also move with the ions from cathode to anode, leading to a decrease in cathode water volume, and thus, the cathode flow channel liquid water is significantly reduced at 300 mA cm^{-2}. The photographs taken by the transparent cell show the possibility of cathode water flooding in alkaline direct ethanol fuel cells and make an important contribution to the understanding of the flooding mechanism.

Except for water content, optically transparent cells can be used to study the distribution of temperature, oxygen, carbon dioxide, water vapor, and other concentrations inside the cell based on other dyes. For example, the oxygen concentration change from the inlet to the outlet in the flow channel is probed by laser irradiation of the oxygen-sensitive dye complex. With the improvement of camera resolution and acquisition rate, the optical transparent cell visualization technique is expected

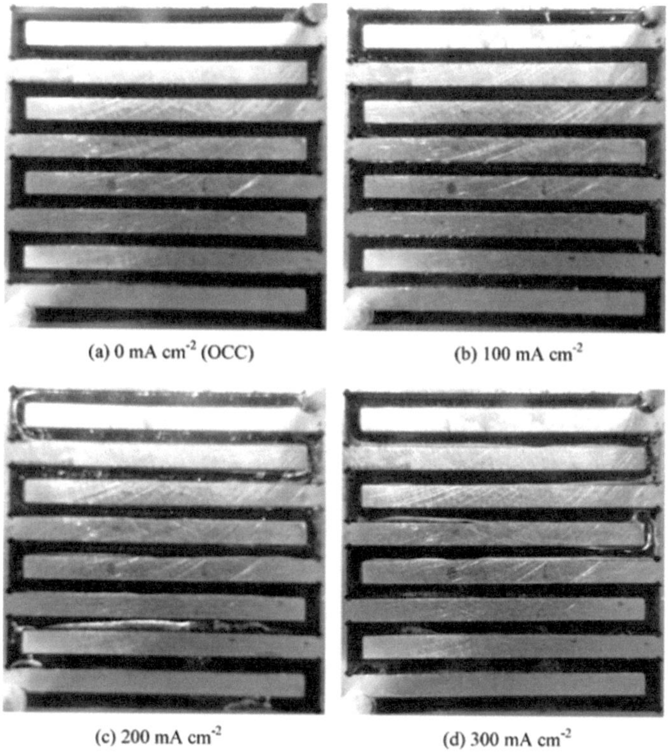

Fig. 3.41 Images of liquid water flow behavior at different current densities [77]

to play an increasingly important role in revealing the water transport mechanism inside the cell.

However, optically transparent cells have some drawbacks that limit their application. The first is that it is difficult to achieve precise control of the cell temperature. Because of the switch to transparent materials, the thermal conductivity of the end plates decreases and making it difficult to install heating pads. For small cells, the temperature field can be approximated as uniform, but for cells with a larger reaction area, the assumption of a uniform temperature field cannot be met, which may affect the performance of the cell and research analysis. Secondly, visualized transparent cells have the possibility of the visualization window being obscured during the reaction. For example, water vapor condenses on the surface of the window, resulting in the impossibility of clearly observing the gas–liquid two-phase distribution inside. Finally, optically transparent cells cannot analyze the water distribution inside the electrode. The gas diffusion and catalyst layers inside the cell play an important role in the cell reaction, and transparent materials are unable to achieve the same performance. Therefore, they cannot be visualized by this method now, we need other technology to investigate the water in MEA.

Fig. 3.42 Schematic of a neutron imaging system [78]

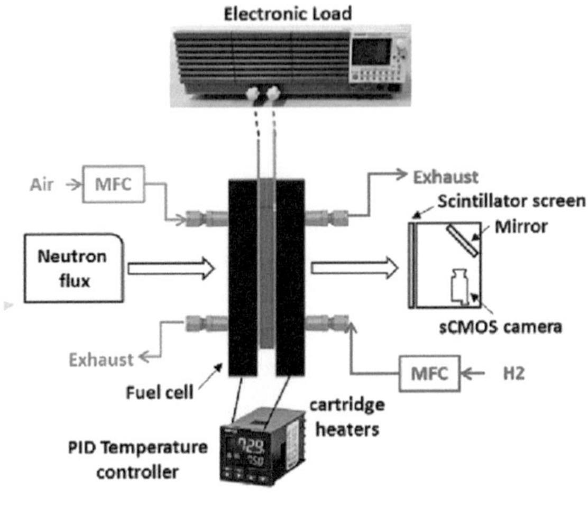

3.4.3.2 Neutron Imaging

Neutron imaging (NI) is an imaging technique that uses a neutron beam as a source of rays to photograph the shadows of an object. A simple imaging system is shown in Fig. 3.42. The neutron beam is irradiated onto and through the sample. Some of these neutrons scatter and react with the nuclei in the sample, resulting in a decrease in the number of neutrons transmitted. The neutron camera captures the neutrons projected and generates a picture based on the number of neutrons to obtain information about the material inside the sample and its distribution.

Based on the above description, we can see that neutron imaging is based on the interaction between neutrons and nuclei, independent of electrons. Therefore, compared with X-rays, neutrons are easily absorbed by light materials (like hydrogen) but can penetrate heavy materials (like lead). In fuel cells, materials such as metal, graphite, and water have different absorption properties for neutrons, and it happens that neutron rays are more sensitive to hydrogen in water, so neutron imaging is often used to qualitatively or even quantitatively analyze the water content inside the cell. However, neutron generation requires large-scale equipment such as nuclear reactors, neutron generators, or accelerators, which are expensive and complicated to operate, and currently, only a few national laboratories can utilize neutron imaging. Nevertheless, neutron imaging has great advantages in visualizing the water distribution and transport mechanism of proton exchange membrane fuel cells due to its high sensitivity to hydrogen and good spatial resolution. Figure 3.43 shows the water distribution inside the cell obtained by neutron imaging.

In Fig. 3.43, the channel inlet is in the upper right corner, the outlet is located in the lower left corner, and the dark area in the center of the fuel cell is the cell heater. A large amount of water is visible in the picture near the fuel cell outlet (lower left corner).

Fig. 3.43 NI images of fuel cell [79]

3.5 Summary and Outlook

The fuel cell is a complex system that involves multiple material transfer processes and electrochemical reaction processes. It is of great significance to use appropriate characterization tools to optimize cell performance. At the same time, the results of characterization tests are also an important basis for subsequent cell simulation analysis. Therefore, this chapter summarizes some common test methods in the field of fuel cells, including electrochemical characterization, material characterization, and fuel cell characterization.

For electrochemical characterization methods, this chapter focuses on five methods: CV, EIS, LSV, CI, and PC. Through these electrochemical characterization methods, various polarization losses in the cell can be quantitatively analyzed, which is of great significance for further understanding the reaction mechanism inside the cell. CV can provide rich information for electrode performance characterization and design optimization; EIS can effectively distinguish different losses in cells, and, combined with equivalent circuit models, can derive more electrochemical parameters; LSV can characterize electrochemical activity and measure hydrogen permeation; CI method can quickly distinguish between ohmic impedance and non-ohmic impedance; PC can measure the comprehensive performance of the cell under specific working conditions and is the most basic index for cell performance. In addition to these electrochemical characterization methods, there are still some other methods that have not been introduced, such as chronoamperometry, which can be used to test cell stability. Each electrochemical characterization has its strengths and weaknesses in application. Researchers need to learn more about the specific details of electrochemical characterization to comprehensively apply various electrochemical characterization methods to better reveal the electrochemical reaction process inside the cell.

In the Ex-site material characterization section, this chapter divides some commonly used methods into four parts: structure characterization, morphology characterization, grading analysis, and element analysis. The pore structure and permeability of porous electrodes are of great significance for material transfer inside the cell; the morphology of the catalyst layer will have an important impact on MEA performance and durability; element analysis and grading analysis are extremely important in catalyst development. Through these Ex-site characterization methods, researchers can deepen their understanding of electrochemical processes in the catalyst layer.

In the fuel cell characterization section, this chapter introduces two methods: temperature and current mapping and cell visualization technology. The two characterization methods focus on distribution. Since internal current density, component concentration, and temperature distribution are not uniform during cell operation, studying their distribution will help improve cell performance and durability. Currently used visualization techniques mainly include transparent cell and neutron imaging. But these technologies have their advantages and limitations. In-site acquisition of water distribution without damage is still a future development direction.

In addition, many other in situ characterization methods have been applied to the characterization of fuel cells. For example, the fuel cell cathode catalysts within the MEA will be deactivated and degraded due to long-term exposure to acidic conditions [80]. To better understand the degradation mechanism of cathode catalysts, three-dimensional (3D) imaging techniques, especially computed tomography (CT), have been widely utilized. CT provides an efficient method for reconstructing 3D images of samples from a series of transmitted images, particularly suitable for measurements using hard X-ray detectors [81]. This method facilitates the accurate determination of the spatial distribution of Pt catalysts within PEMFC under conditions of high water and high gas content. At the same time, X-ray-based 3D imaging techniques, such as probes, have been developed to visualize heterogeneous structures in real materials. The researchers used X-ray computed tomography (XCT) technology to monitor the in-situ degradation of the cathode catalyst throughout its lifetime [82]. The XCT technology provides a complete and lossless 3D visualization of the MEA, revealing the impact of carbon corrosion on its thickness and the formation of cracks.

With the continuous development of research equipment, it is believed that more characterization methods and technologies will be introduced into cell research to bring more possibilities for improving cell performance and durability. In addition, developing more comprehensive characterization methods for cells by combining various characterization techniques will also become an important development direction in the future.

References

1. Wang X R, Ma Y, Gao J, et al. Review on water management methods for proton exchange membrane fuel cells[J]. International Journal of Hydrogen Energy, 2021, 46(22): 12206–12229.
2. Can E M, Mufundirwa A, Wang P, et al. Superhydrophobic fluorinated carbon powders for improved water management in hydrogen fuel cells[J]. Journal of Power Sources, 2022, 548: 232098.
3. Meyer Q, Zeng Y, Zhao C. In Situ and Operando Characterization of Proton Exchange Membrane Fuel Cells[J]. Advanced Materials, 2019, 31(40): 1901900.
4. Van Der Linden F, Pahon E, Morando S, et al. A review on the Proton-Exchange Membrane Fuel Cell break-in physical principles, activation procedures, and characterization methods[J]. Journal of Power Sources, 2023, 575: 233168.
5. Liu L, Liu T, Ding F, et al. Exploration of the Polarization Curve for Proton-Exchange Membrane Fuel Cells[J]. ACS Applied Materials & Interfaces, 2021, 13(49): 58838–58847.
6. Ren G, Qu Z, Wang X, et al. Electrospun fabrication and experimental characterization of highly porous microporous layers for PEM fuel cells[J]. International Journal of Hydrogen Energy, 2024, 55: 455–463.
7. Shangguan Z, Li B, Ming P, et al. Understanding the functions and modifications of interfaces in membrane electrode assemblies of proton exchange membrane fuel cells[J]. Journal of Materials Chemistry A, 2021, 9(27): 15111–15139.
8. Xing Y, Liu L, Li Z, et al. Performance of an ePTFE-Reinforced Membrane Electrode Assembly for Proton-Exchange Membrane Fuel Cells[J]. Energy & Fuels, 2022, 36(18): 11177–11185.
9. Pan M, Meng X, Li C, et al. Impact of nonuniform reactant flow rate on the performance of proton exchange membrane fuel cell stacks[J]. International Journal of Green Energy, 2020, 17(11): 603–616.
10. Peng L, Shao H, Qiu D, et al. Investigation of the non-uniform distribution of current density in commercial-size proton exchange membrane fuel cells[J]. Journal of Power Sources, 2020, 453: 227836.
11. Yu W, Sun L, Zhu W, et al. Investigation of non-uniform temperature distribution in air-cooled proton exchange membrane fuel cells with multizone temperature measurement[J]. International Journal of Green Energy, 2023: 1–11.
12. Cha H-C, Chen C-Y, Shiu J-Y. Investigation on the durability of direct methanol fuel cells[J]. Journal of Power Sources, 2009, 192(2): 451–456.
13. Luo Y, Jiao K. Cold start of proton exchange membrane fuel cell[J]. Progress in Energy and Combustion Science, 2018, 64: 29–61.
14. Li S, Wei R, Shen Q, et al. Numerical Study on the Water Transport Through the Membrane of Proton Exchange Membrane Fuel Cells[J]. International Journal of Electrochemical Science, 2020, 15(8): 7152–7162.
15. Rafiee M, Abrams D J, Cardinale L, et al. Cyclic voltammetry and chronoamperometry: mechanistic tools for organic electrosynthesis[J]. Chemical Society Reviews, 2024: 10.1039.D2CS00706A.
16. Carter R N, Kocha S S, Wagner F, et al. Artifacts in Measuring Electrode Catalyst Area of Fuel Cells through Cyclic Voltammetry[J]. ECS Transactions, 2007, 11(1): 403–410.
17. Do C L, San Pham T, Nguyen N P, et al. Properties of Pt/C nanoparticle catalysts synthesized by electroless deposition for proton exchange membrane fuel cell[J]. Advances in Natural Sciences: Nanoscience and Nanotechnology, 2013, 4(3): 035011.
18. Prass S, St-Pierre J, Klingele M, et al. Hydrogen Oxidation Artifact During Platinum Oxide Reduction in Cyclic Voltammetry Analysis of Low-Loaded PEMFC Electrodes[J]. Electrocatalysis, 2021, 12(1): 45–55.
19. Bellows R J, Marucchi-Soos E P, Buckley D T. Analysis of reaction kinetics for carbon monoxide and carbon dioxide on polycrystalline platinum relative to fuel cell operation. Ind Eng Chem Res, 1996, 35: 1235–1242
20. Brightman E, Hinds G, O'Malley R. In situ measurement of active catalyst surface area in fuel cell stacks[J]. Journal of Power Sources, 2013, 242: 244–254.

21. Dotelli G, Ferrero R, Stampino P G, et al. Analysis and Compensation of PEM Fuel Cell Instabilities in Low-Frequency EIS Measurements[J]. IEEE Transactions on Instrumentation and Measurement, 2014, 63(7): 1693–1700.
22. Yuan X-Z, Song C, Wang H, et al. Electrochemical Impedance Spectroscopy in PEM Fuel Cells[M]. London: Springer London, 2010.
23. Baricci A, Casalegno A. Experimental Analysis of Catalyst Layer Operation in a High-Temperature Proton Exchange Membrane Fuel Cell by Electrochemical Impedance Spectroscopy[J]. Energies, 2023, 16(12): 4671.
24. Iranzo A, Muñoz M, Pino Fco J, et al. Non-dimensional analysis of PEM fuel cell phenomena by means of AC impedance measurements[J]. Journal of Power Sources, 2011, 196(9): 4264–4269.
25. Zhao L, Dai H, Pei F, et al. A Comparative Study of Equivalent Circuit Models for Electro-Chemical Impedance Spectroscopy Analysis of Proton Exchange Membrane Fuel Cells[J]. Energies, 2022, 15(1): 386.
26. Zhao F, Virkar A. Dependence of polarization in anode-supported solid oxide fuel cells on various cell parameters[J]. Journal of Power Sources, 2005, 141(1): 79–95.
27. Zhuge W, Zhang Y, Ming P, et al. Numerical simulation of three-dimensional gas/liquid two-phase flow in a proton exchange membrane fuel cell[J]. Frontiers of Energy and Power Engineering in China, 2007, 1(3): 305–310.
28. Aït-Idir W, Wu P, Sgarbi R, et al. Oxygen diffusion impedance in proton exchange membrane fuel cells – insights into electrochemical impedance spectra and equivalent electrical circuit modeling[J]. Electrochimica Acta, 2023, 472: 143430.
29. Rezaei Niya S M, Hoorfar M. Study of proton exchange membrane fuel cells using electrochemical impedance spectroscopy technique – A review[J]. Journal of Power Sources, 2013, 240: 281–293.
30. González-Velasco J. The linear sweep voltametric method: An application to the study of reversible and irreversible processes[J]. Electroanalysis, 1994, 6(9): 711–724.
31. Xue J, Li Y, Hu J. Nanoporous bimetallic Zn/Fe–N–C for efficient oxygen reduction in acidic and alkaline media[J]. Journal of Materials Chemistry A, 2020, 8(15): 7145–7157.
32. Pei P, Wu Z, Li Y, et al. Improved methods to measure hydrogen crossover current in proton exchange membrane fuel cell[J]. Applied Energy, 2018, 215: 338-347.
33. Rubio M A, Urquia A, Dormido S. Diagnosis of PEM fuel cells through current interruption[J]. Journal of Power Sources, 2007, 171(2): 670–677.
34. Hosseini M G, Ahadzadeh I. Application and comparison of current interruption and electrochemical impedance spectroscopy methods to study a microbial fuel cell[J]. Instrumentation science & technology, 2013, 41(1): 72-81.
35. Tashima D, Sakaguchi Y, Hidaka H, et al. Estimating the optimal number of membrane electrode assembly catalyst layers for proton exchange membrane fuel cell by considering open circuit voltage and polarization[J]. Materials Chemistry and Physics, 2010, 122(2–3): 544–547.
36. Cooper K R, Smith M. Electrical test methods for on-line fuel cell ohmic resistance measurement[J]. Journal of Power Sources, 2006, 160(2): 1088–1095.
37. Mohsin M, Raza R, Mohsin-ul-Mulk M, et al. Electrochemical characterization of polymer electrolyte membrane fuel cells and polarization curve analysis[J]. International Journal of Hydrogen Energy, 2020, 45(45): 24093–24107.
38. Salva J A, Iranzo A, Rosa F, et al. Optimization of a PEM fuel cell operating conditions: Obtaining the maximum performance polarization curve[J]. International Journal of Hydrogen Energy, 2016, 41(43): 19713–19723.
39. Wang H. PEM Fuel Cell Diagnostic Tools[J].
40. Tong G, Xu X, Yuan Q, et al. Simulation study of proton exchange membrane fuel cell cross-convection self-humidifying flow channel[J]. International Journal of Energy Research, 2021, 45(3): 4036–4047.
41. Li X, Wang X, He J, et al. Construction of Homogeneous Catalyst Layers at Proton Exchange Membrane Fuel Cell Cathodes[J]. Journal of The Electrochemical Society, 2023, 170(4): 044511.

42. Baker D R, Caulk D A, Neyerlin K C, et al. Measurement of Oxygen Transport Resistance in PEM Fuel Cells by Limiting Current Methods[J]. Journal of The Electrochemical Society, IOP Publishing, 2009, 156(9): B991.
43. Atyabi S A, Afshari E, Wongwises S, et al. Effects of assembly pressure on PEM fuel cell performance by taking into accounts electrical and thermal contact resistances[J]. Energy, 2019, 179: 490–501.
44. Taymaz I, Benli M. Numerical study of assembly pressure effect on the performance of proton exchange membrane fuel cell[J]. Energy, 2010, 35(5): 2134–2140.
45. Song H, Liu Y-T, Zhang W-S, et al. Rational design of carbon network structure in microporous layer toward enhanced mass transport of proton exchange membrane fuel cell[J]. Journal of Power Sources, 2022, 539: 231623.
46. Wang C, Cheng X, Lu J, et al. The Experimental Measurement of Local and Bulk Oxygen Transport Resistances in the Catalyst Layer of Proton Exchange Membrane Fuel Cells[J]. The Journal of Physical Chemistry Letters, American Chemical Society, 2017, 8(23): 5848–5852.
47. Huang T, Yi D, Ren X, et al. Optimization of gas diffusion layer thickness for high-temperature proton exchange membrane fuel cells[J]. Ionics, 2024.
48. Froboese L, Titscher P, Westphal B, et al. Mercury intrusion for ion- and conversion-based battery electrodes – Structure and diffusion coefficient determination[J]. Materials Characterization, 2017, 133: 102–111.
49. Webb P A. An Introduction To The Physical Characterization of Materials by Mercury Intrusion Porosimetry with Emphasis On Reduction And Presentation of Experimental Data[J].
50. Yu Z, Carter R N, Zhang J. Measurements of Pore Size Distribution, Porosity, Effective Oxygen Diffusivity, and Tortuosity of PEM Fuel Cell Electrodes[J]. Fuel Cells, 2012, 12(4): 557–565.
51. Matsukawa Y, Linsenmann F, Plass M A, et al. Gas sorption porosimetry for the evaluation of hard carbons as anodes for Li- and Na-ion batteries[J]. Beilstein Journal of Nanotechnology, 2020, 11: 1217–1229.
52. Sinha P, Datar A, Jeong C, et al. Surface Area Determination of Porous Materials Using the Brunauer–Emmett–Teller (BET) Method: Limitations and Improvements[J]. The Journal of Physical Chemistry C, 2019, 123(33): 20195–20209.
53. Barrett E P, Joyner L G, Halenda P P. The Determination of Pore Volume and Area Distributions in Porous Substances. I. Computations from Nitrogen Isotherms[EB/OL]. ACS Publications. American Chemical Society 2002–05–01/2023–04–02. https://doi.org/10.1021/ja01145a126.
54. Sun X, Li Y, Li M-J. Highly Dispersed Palladium Nanoparticles on Carbon-Decorated Porous Nickel Electrode: An Effective Strategy to Boost Direct Ethanol Fuel Cell up to 202 mW cm –2[J]. ACS Sustainable Chemistry & Engineering, 2019, 7(13): 11186–11193.
55. Brandon N P, Brett D J. Engineering porous materials for fuel cell applications[J]. Philosophical Transactions of the Royal Society A: Mathematical, Physical and Engineering Sciences, 2006, 364(1838): 147–159.
56. Park S, Popov B N. Effect of a GDL based on carbon paper or carbon cloth on PEM fuel cell performance[J]. Fuel, 2011, 90(1): 436–440.
57. Li B, Liu Y, Guo Y, et al. Controlling the microscopic morphology and permeability of catalyst layers in proton exchange membrane fuel cells by adjusting catalyst ink agglomerates[J]. International Journal of Hydrogen Energy, 2021, 46(63): 32215–32225.
58. Li Y S, Zhao T S, Liang Z X. Effect of polymer binders in anode catalyst layer on performance of alkaline direct ethanol fuel cells[J]. Journal of Power Sources, 2009, 190(2): 223–229.
59. Patel A N, Kranz C. (Multi)functional Atomic Force Microscopy Imaging[J]. Annual Review of Analytical Chemistry, 2018, 11(1): 329–350.
60. Rezaei A, Aber S, Roberts D J, et al. Synthesis and study of CuNiTiO3 as an ORR electrocatalyst to enhance microbial fuel cell efficiency[J]. Chemosphere, 2022, 307: 135709.
61. Sun C, Müller E, Meffert M, et al. On the Progress of Scanning Transmission Electron Microscopy (STEM) Imaging in a Scanning Electron Microscope[J]. Microscopy and Microanalysis, 2018, 24(2): 99–106.
62. Li Y S, Zhao T S. Understanding the performance degradation of anion-exchange membrane direct ethanol fuel cells[J]. International Journal of Hydrogen Energy, 2012, 37(5): 4413–4421.

63. Deng X, Zhang J, Fan Z, et al. Understanding and Engineering of Multiphase Transport Processes in Membrane Electrode Assembly of Proton-Exchange Membrane Fuel Cells with a Focus on the Cathode Catalyst Layer: A Review[J]. Energy & Fuels, 2020, 34(8): 9175–9188.
64. Ameh E S. A review of basic crystallography and X-ray diffraction applications[J]. The International Journal of Advanced Manufacturing Technology, 2019, 105(7–8): 3289–3302.
65. Li Y, He Y. Layer reduction method for fabricating Pd-coated Ni foams as high-performance ethanol electrode for anion-exchange membrane fuel cells[J]. RSC Advances, 2014, 4(32): 16879.
66. Li Y, Lv J, He Y. A Monolithic Carbon Foam-Supported Pd-Based Catalyst towards Ethanol Electro-Oxidation in Alkaline Media[J]. Journal of The Electrochemical Society, 2016, 163(5): F424–F427.
67. Sun X, Lou Y, He Y L, et al. A Na-ion direct formate fuel cell converting solar fuel to electricity and hydrogen[J]. Journal of Power Sources, 2021, 499: 229960.
68. Stevie F A, Donley C L. Introduction to X-ray photoelectron spectroscopy[J]. Journal of Vacuum Science & Technology A: Vacuum, Surfaces, and Films, 2020, 38(6): 063204.
69. Zhou N, Li Y. PtCo incorporated porous carbon nanofiber as a promising oxygen reduction electrocatalyst[J]. International Journal of Hydrogen Energy, 2021, 46(76): 37884–37894.
70. Raileanu Ilie V A, Martemianov S, Thomas A. Investigation of the local temperature and overheat inside the membrane electrode assembly of PEM fuel cell[J]. International Journal of Hydrogen Energy, 2016, 41(34): 15528–15537.
71. Rahgoshay S M, Ranjbar A A, Ramiar A, et al. Thermal investigation of a PEM fuel cell with cooling flow field[J]. Energy, 2017, 134: 61–73.
72. Ulsh M, Porter J M, Bittinat D C, et al. Defect Detection in Fuel Cell Gas Diffusion Electrodes Using Infrared Thermography[J]. Fuel Cells, 2016, 16(2): 170–178.
73. Moller S, König J, Resagk C, et al. Influence of the illumination spectrum and observation angle on temperature measurements using thermochromic liquid crystals[J]. Measurement Science and Technology, 2019, 30(8): 084006.
74. Anderson R, Zhang L, Ding Y, et al. A critical review of two-phase flow in gas flow channels of proton exchange membrane fuel cells[J]. Journal of Power Sources, 2010, 195(15): 4531–4553.
75. Tüber K, Pócza D, Hebling C. Visualization of water buildup in the cathode of a transparent PEM fuel cell[J]. Journal of Power Sources, 2003, 124(2): 403–414.
76. Pei H, Xiao C, Tu Z. Experimental study on liquid water formation characteristics in a novel transparent proton exchange membrane fuel cell[J]. Applied Energy, 2022, 321: 119349.
77. Li Y S, Zhao T S, Chen R. Cathode flooding behaviour in alkaline direct ethanol fuel cells[J]. Journal of Power Sources, 2011, 196(1): 133–139.
78. Wu Y, Xu L, Zhou S, et al. Water management and mass transport of a fractal metal foam flow-field based polymer electrolyte fuel cell using operando neutron imaging[J]. Applied Energy, 2024, 364: 123204.
79. Pekula N, Heller K, Chuang P A, et al. Study of water distribution and transport in a polymer electrolyte fuel cell using neutron imaging[J]. Nuclear Instruments and Methods in Physics Research Section A: Accelerators, Spectrometers, Detectors and Associated Equipment, 2005, 542(1): 134–141.
80. Garcia-Sanchez, Daniel, et al. "Local impact of load cycling on degradation in polymer electrolyte fuel cells." Applied Energy 259 (2020): 114210.
81. James, J. P.; Choi, H. W.; Pharoah, J. G. X-ray computed tomography reconstruction and analysis of polymer electrolyte membrane fuel cell porous transport layers. Int. J. Hydrogen Energy 2012, 37 (23), 18216−18230.
82. White, R. T.; Wu, A.; Najm, M.; Orfino, F. P.; Dutta, M.; Kjeang, E. 4D in situ visualization of electrode morphology changes during accelerated degradation in fuel cells by X-ray computed tomography. J. Power Sources 2017, 350, 94−102.

Chapter 4
Proton Exchange Membrane Fuel Cells: Fundamentals

4.1 Introduction

Under the dual pressures of global energy transition and environmental protection, finding clean, efficient, and sustainable energy solutions has become a common challenge for governments, enterprises, and research institutions [1, 2]. With the depletion of traditional fossil fuels and their negative impact on the environment, the development of new energy technologies has become particularly urgent [3, 4]. In this context, the Proton Exchange Membrane Fuel Cell (PEMFC), as an innovative energy conversion device, stands out with its unique advantages and brings new possibilities for the future energy structure. The working principle of the PEMFC is based on an electrochemical reaction, which skillfully combines hydrogen and oxygen to produce water and release electricity at the same time. This process not only realizes efficient conversion of energy but also almost completely avoids the emission of harmful substances, such as carbon dioxide, nitrogen oxides, and other particulate matter. Therefore, PEMFC is regarded as one of the representatives of "green energy", and is an important tool for realizing the goals of a low-carbon economy and sustainable development.

In 1800, British scientists Sir Anthony Carlisle and William Nicholson first discovered the decomposition of water into hydrogen and oxygen using electricity, known as hydrogen production by water electrolysis [5]. On this basis, Sir William Robert Grove proposed the concept of the fuel cell in 1839, from which he is widely regarded as the father of the fuel cell. In addition, Grove argued that the process of water electrolysis could be reversed and that hydrogen could be combined with oxygen to produce an electric current, which eventually became the initial principle of the fuel cell [6]. Later in 1842, Sir William Robert Grove designed a gaseous voltaic battery consisting of 50 pairs of tubes filled with oxygen and hydrogen, respectively, and platinized platina foil, which first realized gas-generated electric current [7]. In 1889, Ludwig Mond and Car Langer attempted to improve upon Grove's electrodes by making it porous. In this work, each porous electrode consists of a diaphragm made

Y. Li, *Fuel Cell Fundamentals and Applications*,
https://doi.org/10.1007/978-981-96-9845-5_4

of porous non-conducting material (plaster of Paris, earthenware, asbestos, pasteboard) and a porous platinum or gold foil, which acts as a collector in contact with the active surface of platinum black. In this way, the hydroxide fuel cell was able to produce a current density of 3.5 mA cm^{-2} at an output voltage of 0.73 V. The structure of this cell was essentially similar to that of modern low-temperature fuel cells, and the concept of the modern fuel cell was born [8]. In 1893, Friedrich Wilhelm Ostwald experimentally determined the interconnection of the various components of a fuel cell: electrodes, electrolyte, oxidizing and reducing agents, anions, and cations. This laid the foundation for further research in the field of fuel cells [5]. Thomas Francis Bacon built on this foundation to develop the first fuel cell consisting of hydrogen and oxygen, which was put into practical use in 1933. He first experimented with an alkaline fuel cell using potassium hydroxide electrolyte instead of acids known since the early experiments of Grove. And to prevent the electrolyte from flooding the pores of the electrodes, in 1939, he constructed a cell with nickel electrodes operating at high pressure (200 atm) [5].

The advent of Teflon (polytetrafluoroethylene or PTFE) materials in 1950 set off the development of PEMFC. It was first used in fuel cells with platinum electrodes and acidic electrolytes, as well as in fuel cells with carbon electrodes and alkaline electrolytes [8]. In the early 1960s, General Electric developed the first polymer membrane fuel cell for use in the Gemini program, and in the mid-1960s, General Motors experimented with a fuel cell-powered van (these were developed by the Union Carbide Corporation). In the 1970s, DuPont developed a perfluoro sulfonate membrane with high electrical conductivity to further extend fuel cell life. In 1989, Perry Technologies, in cooperation with Ballard, successfully demonstrated a polymer electrolyte membrane (PEM) fuel cell-powered submarine. And in 1993, Ballard Power Systems demonstrated a fuel cell-powered bus. By the end of the twentieth century, almost all the automotive companies supported by the U.S. Department of Energy had built and exhibited fuel cell-powered vehicles, signaling the birth of a new industry [9].

Since it was first proposed in the 1960s, PEMFCs have experienced a long journey from laboratory research to commercialization. Early research focused on materials science and engineering design, as scientists continued to explore more efficient catalysts, more durable proton exchange membranes [10], and optimized cell structures. After decades of efforts, the technological level of PEMFCs has been significantly improved, the cost has been significantly reduced, the performance is more stable and reliable, and the conditions for large-scale commercialization have gradually been met.

Nowadays, PEMFCs have shown great application potential in many fields. In the field of transportation, they are used as the core power source of electric vehicles, promoting the electrification process of the automobile industry; in the distributed power generation system, they provide clean and reliable power supply for families, commercial buildings and industrial facilities; and in the market of portable electronic devices, the miniaturized PEMFCs have become a new type of mobile power supply option. In addition, PEMFCs have also found a place in aerospace, military, and other special fields, providing these industries with quiet, efficient, and environmentally friendly power solutions.

4.2 Principles

PEMFCs are energy conversion devices that generate electricity directly from fuels through electrochemical reactions. The electrolyte of PEMFCs is the proton-conducting polymer electrolyte membrane, which distinguishes it from other types of fuel cells. Therefore, PEMFCs are also known as polymer electrolyte membrane fuel cells. The fundamental of the PEMFCs is shown in Fig. 4.1. At a temperature of 60–80 °C and a pressure of 1–2 atm [4], the hydrogen is transported to the anode and oxidized to produce electrons and protons in a process known as hydrogen oxidation reaction (HOR). The electrons produced flow from the anode to the cathode through an external circuit, while the protons produced flow from the anode to the cathode through the membrane. Oxygen is transported to the cathode and combines with the arriving protons and electrons to form water in a process known as the oxygen reduction reaction (ORR). The electrochemical reaction converts the chemical energy stored in the fuel into electricity and releases heat. The anodic half-reaction, cathodic half-reaction, and the overall reaction in the PEMFCs are shown in Eqs. (4.1–4.3), respectively:

$$\text{Anode: } H_2 \rightarrow 2H^+ + 2e^- \tag{4.1}$$

$$\text{Cathode: } \frac{1}{2}O_2 + 2H^+ + 2e^- \rightarrow H_2O \tag{4.2}$$

$$\text{Overall: } H_2 + \frac{1}{2}O_2 \rightarrow H_2O \tag{4.3}$$

Thermodynamics is the key to understanding the conversion of chemical energy to electrical energy in fuel cells. According to the second law of thermodynamics, the maximum electrical work W output from the PEMFCs is:

$$W = \Delta G = \Delta H - T\Delta S = -nFE \tag{4.4}$$

where ΔG is the Gibbs free energy change, ΔH, T, and ΔS are the enthalpy change, temperature, and entropy change, respectively. n and E represent the number of electrons transferred by the reaction and the theoretical reversible potential difference, and F is the Faraday constant.

Considering that the PEMFC operates in a non-standardized state, the Gibbs free energy change of the reaction is expressed by the activity of the reactants and products:

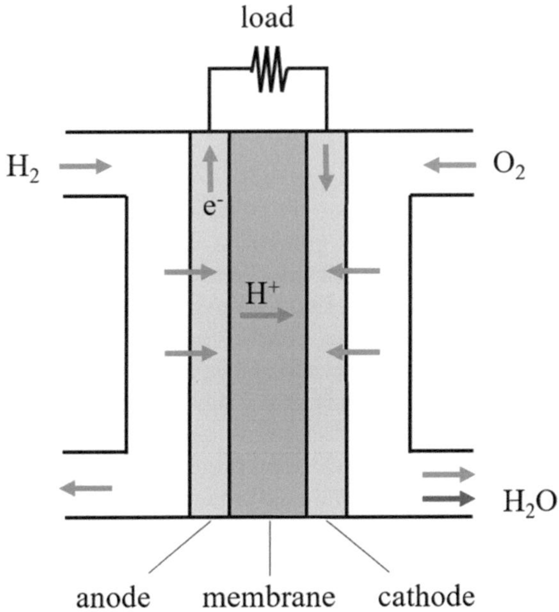

Fig. 4.1 Schematic of PEMFC working principle

$$\Delta G = \Delta G_0 + RT \ln\left(\frac{\alpha_{H_2O}}{\alpha_{H_2}\alpha_{O_2}^{0.5}}\right) \tag{4.5}$$

where ΔG_0 is the Gibbs free energy change in the standard state, and R, α are the ideal gas constant and the activity of the reactants and products, respectively. Equations (4.5) illustrates how the Gibbs free energy of the system changes as a function of reactant and product activity (e.g., concentration or gas pressure).

Further combined with Eqs. (4.4) and (4.5), the theoretical reversible potential difference E of the PEMFCs can be expressed as a function of chemical activity:

$$E = E_r - \frac{RT}{nF} \ln\left(\frac{\alpha_{H_2O}}{\alpha_{H_2}\alpha_{O_2}^{0.5}}\right) \tag{4.6}$$

where E_r is the theoretical reversible potential difference in the standard state. $\Delta G_0 = -237.3$ kJ mol^{-1}, $E_r = 1.23$ V. This important result is the well-known Nernst Equation, which sketches the relationship between reversible electrochemical cell voltages and substance concentrations, gas pressures, and so on. The Nernst Equation is also at the core of fuel cell thermodynamics.

In electrochemical reactions, the free energy of a charged substance is sensitive to the potential, which will further affect the magnitude of the reaction activation energy barrier. Thus, this provides an effective method to manipulate the magnitude of the activation energy barriers of electrochemical reactions by varying the cell voltage. Under equilibrium conditions, the current densities of the forward and reverse reactions are equal and both are j_0. Once far from equilibrium, the forward current density j_1 and reverse current density j_2 are:

$$j_1 = j_0 e^{anF\eta/(RT)} \tag{4.7}$$

$$j_2 = j_0 e^{-(1-a)nF\eta/(RT)} \tag{4.8}$$

The net current ($j_1 - j_2$) is:

$$j = j_1 - j_2 = j_0\left(e^{anF\eta/(RT)} - e^{-(1-a)nF\eta/(RT)}\right) \tag{4.9}$$

where η and n represent the voltage lost to overcome the activation energy of the electrochemical reaction and the number of electrons transferred, respectively; j_0 represents the exchange current density. a is the transfer coefficient, which depends on the symmetry of the activation energy barriers and indicates how the change in the potential at the reaction interface alters the magnitude of the activation energy barrier in the forward and reverse directions. Equation (4.9) is the well-known Butler-Volmer equation, which illustrates that the current produced by an electrochemical reaction increases exponentially with the activation overpotential. It is considered the cornerstone of electrochemical kinetics. The desire to obtain a higher current from the fuel cell must come at the expense of a loss of voltage.

Due to irreversible losses in a PEMFC, its output voltage is lower than the theoretical reversible potential difference. A typical PEMFC polarization curve is shown in Fig. 4.2. The irreversible losses in PEMFC consist of four components:

(1) Voltage loss due to fuel permeation and internal short-circuit current. An ideal proton exchange membrane can completely isolate the anode hydrogen and cathode oxygen, in fact, a very small amount of hydrogen can penetrate the

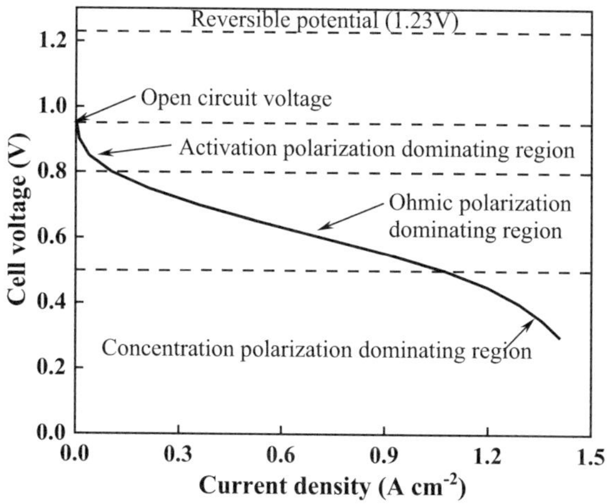

Fig. 4.2 The typical polarization curve of PEMFC

proton exchange membrane to reach the cathode and directly react to produce current. Ideally, the proton exchange membrane is a poor conductor of electrons, but there are still a very small number of electrons that do not pass through the external circuit but rather through the proton exchange membrane directly to the cathode. The effect of both amounts to a small reverse current, which causes a large activation loss and makes the open-circuit voltage of the PEMFCs significantly lower than the theoretical reversible potential difference. The development of a selection of better permeable proton exchange membranes can reduce the voltage losses due to fuel permeation and internal short circuit currents.

(2) Activation loss. Activation loss is caused by slow kinetic reactions on the electrode surface and is highly nonlinear. The activation loss is related to the exchange current density at the electrode surface, where a higher exchange current density indicates a faster kinetic reaction rate and a smaller activation loss. The kinetic rate of the anodic hydrogen oxidation reaction is significantly higher than that of the cathodic oxygen reduction reaction, so the activation loss of PEMFCs is mainly generated by the cathode. Activation losses can be reduced by increasing the reaction temperature, using more active catalysts, increasing the roughness of the electrodes, increasing the concentration of reactants, and increasing the operating pressure. In the low current density region (or high voltage region), activation losses dominate due to sufficient reactants and low current density.

(3) Ohmic losses. The ohmic loss consists of two parts: one is the electron transport loss due to the bulk resistance of each component and the contact resistance between components, and the other is the proton transport loss due to the bulk resistance of the proton exchange membrane and ionomer. The ohmic loss is proportional to the current density and varies linearly. Due to the high conductivity of the components, ohmic losses are usually caused by proton transport. The bulk phase resistance can be reduced by using components with higher conductivity, the contact resistance can be reduced by developing integrated component materials or using tighter assemblies, and the ionic resistance can be reduced by improving membrane hydration, using thinner or higher proton conductivity proton exchange membranes. In the medium current density region (or medium voltage region), the voltage varies linearly with the current, and ohmic losses dominate.

(4) Concentration loss. Concentration loss occurs because the rate of reactant consumption on the catalyst surface is greater than the rate of reactant supply, which makes the concentration of reactants on the catalyst surface smaller than the concentration of reactants in the catalyst layer. Concentration loss can be reduced by rational flow field design, optimizing the electrode structure, and increasing the reactant supply concentration. In the high current density region (or low voltage region), the reactants are consumed rapidly, the reactant concentration on the electrode surface decreases rapidly, and the concentration loss dominates.

In PEMFCs, the above four losses exist simultaneously and jointly affect the output voltage. Due to the low voltage of a single cell, the output voltage is usually increased by connecting multiple single cells in series to form a fuel cell stack, which meets the needs of different application scenarios.

For any energy conversion device, efficiency is essential. The efficiency η_{cell} of a PEMFC is defined as the electrical work W divided by the enthalpy change ΔH. The Gibbs free energy ($\Delta G = \Delta H - T\Delta S$) in a fuel cell determines the maximum limit for the conversion of chemical energy into electrical work. The maximum limiting efficiency of a PEMFC is further expressed as:

$$\eta_{cell} = \frac{W}{\Delta H} = 1 - T\frac{\Delta S}{\Delta H} \tag{4.10}$$

According to Eqs. (4.4) and (4.10), the efficiency of the PEMFC is less than 1 and depends on the temperature. At room temperature and pressure, for the hydrogen–oxygen fuel cell, there are $\Delta H_0 = -\ 285.8\ \text{kJ mol}^{-1}$ and $\Delta G_0 = -\ 237.3\ \text{kJ mol}^{-1}$, and the efficiency of the PEMFC is 83% in the standard state.

Complex multi-physical coupling phenomena occur within the PEMFC, including substance transport, electrochemical reactions, charge transport, hydrothermal transport, and gas–liquid two-phase flow. The efficiency of these transport processes and electrochemical reactions profoundly affects the performance of the PEMFC. The workflow of the PEMFC includes:

(1) The humidified reactant gas is pumped into the cathode and anode flow field, transported and distributed over the entire active area of the electrodes by convection through the flow field, and then in the form of diffusion through the diffusion layer and the microporous layer to the three-phase interface in the catalyst layer. Proton exchange membranes and ionomers adsorb water vapor into dissolved water to act as proton transport carriers. The high efficiency of the reactant gas transport and the uniformity of its distribution over the active area of the electrode affect the electrochemical reaction rate and its uniformity, and if the reactant gas is not uniformly distributed, it will also reduce the efficiency of the catalyst.

(2) Hydrogen and oxygen are catalyzed by the catalyst at the three-phase interface of the catalyst layer of the anode and cathode, respectively, where the electrochemical reaction occurs. The concentration of reactants at the three-phase interface affects the electrochemical reaction rate, and a good supply of reactant gas can reduce the concentration loss in PEMFC. The hydroxide reaction forms protons and electrons, and the oxygen reduction reaction consumes protons and electrons and produces water. At room temperature, water in the PEMFC exists in three forms, i.e., dissolved water, liquid water, and water vapor. Dissolved water and water vapor are converted to each other in the form of adsorption/desorption, and water vapor and liquid water are converted to each other in the form of condensation/evaporation, and the three kinds of water are kept in dynamic equilibrium. Among them, the dissolved water content reflects the

degree of hydration of the proton exchange membrane and ionomer, which determines the proton conductivity and affects the ohmic loss of the cell.

(3) Electrons are generated in the anode catalyst layer, conducted to the anode collector plate through the microporous layer, diffusion layer, bipolar plate, and then conducted to the cathode collector plate through the external circuit, and arrive at the cathode three-phase interface through the bipolar plate, diffusion layer, microporous layer, and catalyst layer of the carbon carrier, and the directional migration of electrons forms the electric current and does the work externally. Protons generated in the anode catalyst layer, through the anode catalyst layer ionomer, proton exchange membrane, cathode catalyst layer ionomer, in the form of hydrated hydrogen ions, are transported to the cathode three-phase interface, where electrons, protons, and oxygen in the three-phase interface undergo electrochemical reactions.

(4) Liquid water is transported from the pores of the catalyst layer, microporous layer, and diffusion layer to the flow field under capillary pressure and discharged under the airflow. The accumulation of liquid water will block the pores of the catalyst layer and diffusion layer, which increases the reactant gas transport resistance in the pores on one hand, and increases the reactant local transport resistance on the catalyst surface on the other hand. Overall, the accumulated liquid water affects the material transport of reactants and increases the concentration loss of the cell.

(5) Heat sources in the cell include latent heat caused by the phase transition of water, Joule heat generated by electron and proton transport, and irreversible heat generated by electrochemical reactions. The heat generated in the cell is conducted to the bipolar plate through the catalyst layer, the microporous layer, and the diffusion layer, and then discharged through the cooling flow field. Temperature significantly affects the reaction kinetics, substance transport and water adsorption and desorption, and phase transition. In addition, the uniformity of temperature distribution will affect the uniformity of electrochemical reaction rate.

4.3 Components

The basic components of the PEMFC are shown in Fig. 4.3, including proton exchange membrane (PEM), catalyst layer, gas diffusion layer, and bipolar plate (BP). The bipolar plate is also known as the flow field plate, and the cathode and anode flow fields are arranged on either side of it. The cathode and anode gas diffusion layer, the catalyst layer, and the proton exchange membrane are also referred to as the membrane electrode assembly (MEA). The auxiliary assemblies of the PEMFC include the collector plate, the end plate, the insulating plate, the gaskets, and the fastening bolts.

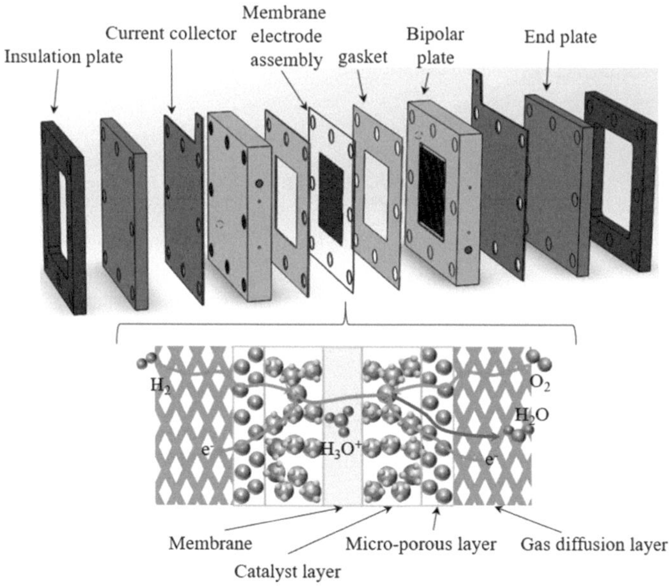

Fig. 4.3 The components of PEMFC and the working procedure of PEMFC

4.3.1 Proton Exchange Membrane

The main functions of the PEM include: first, transferring protons from the anode to the cathode to participate in the oxygen reduction reaction; second, the PEM is used to isolate the cathode and anode reaction gases as well as to prevent electron transport within the membrane. The transfer rate of protons, which decides the reaction rate of the cell, is affected by the thickness, material, and preparation process of the PEM. To improve the proton transfer efficiency of the PEM and to ensure the stability of the membrane, the design and preparation of the PEM should usually meet the following requirements:

(1) Ensure the denseness of the membrane to reduce the transmittance of H_2. In the process of membrane preparation, when the preparation process and material selection are the same, the thicker membrane theoretically reduces the transmittance of H_2. However, the thicker membrane will affect the transfer rate of the proton, which will increase the cell's ohmic impedance. Therefore, the optimization of the PEM thickness is one of the main focuses of the current PEM research.
(2) The use of materials that isolate electron transport and facilitate proton transfer to prevent the direct transmission of electrons through the internal channels of the PEM. The current provided by the PEMFC for external loads is derived from the electron transport in the external circuit of the cell. If electrons are directly transported from the anode CL to the cathode CL through the PEM, it will lead

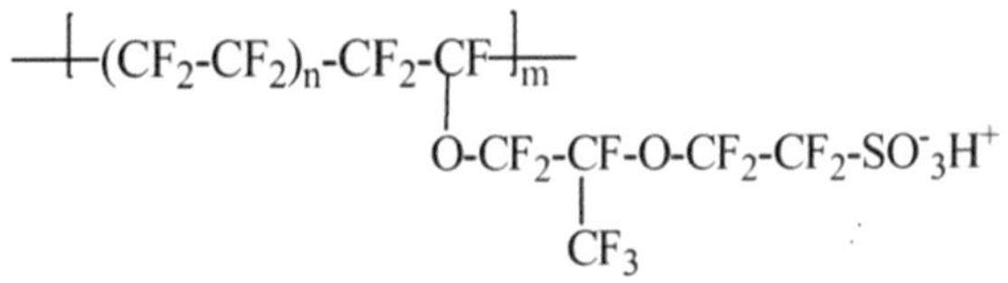

Fig. 4.4 Chemical structure of perfluorinated sulfonic-acid membrane [11]

to a reduction in the external current. Thus, the selection of the materials used in the preparation of the PEM is an important factor in the isolation of the electron transport and the facilitation of the proton transfer.

(3) The thickness of PEM is generally a few micrometers to tens of micrometers, and its thickness is relatively thin. Thus, good mechanical properties and mechanical strength in the design of PEM is a focus of research.

(4) Reducing the production cost of PEMs while meeting the cell performance is an important factor in accelerating the commercialization of PEMFCs.

At present, the main manufacturers of the PEM include DuPont (USA, which produces PEM with perfluorinated sulfonic acid (PFSA) as the main material), Gore (USA, which produces PEM with PTFE and PFSA as the main materials), and Ballard (Canada, which produces PEM with trifluorinated styrenesulfonic acid (TFSS) as the main material), among which the Nafion® series of proton exchange membranes produced by DuPont is the most widely applied in PEMFC. The PFSA membranes are the most commonly used type of PEM in commercial PEMFC, and their structure is shown in Fig. 4.4. The membrane consists of a hydrophobic polytetrafluoroethylene backbone (PTFE) and hydrophilic sulfonic acid groups ($SO_3^-H^+$) as side chains. The PTFE backbone provides mechanical strength, and the sulfonic acid groups provide proton transport channels. In the presence of water, protons (H^+) in the pores combine with water to form hydrated ions (H_3O^+) and detach from the sulfonic acid group side chains. Once there is enough water in the pore space, hydrated hydrogen ions can be transported in the aqueous phase, forming a channel for proton conduction.

PFSA materials have good chemical and thermal stability, and their service life in fuel cells has reached more than 60,000 h, facilitating their commercialization. Due to the large number of sulfonic acid groups in the polymer side chain modification, the PFSA membrane represented by Nafion has a low ionic group equivalent value (1100–1200 g mol^{-1}) and excellent proton conductivity. However, Nafion still has many disadvantages in the process of use, for example:

(1) Nafion has a low glass transition temperature, thus limiting its use to temperatures that are not too high (should be below 80 °C). From a kinetic point of view, lower temperatures have an inhibitory effect on the redox reaction rate, reducing the cell efficiency.

(2) The proton transfer of Nafion has to take place in a hydrated state. When the humidity in the reactant gas is low, the proton conductivity of Nafion decreases, which seriously affects its performance output.

(3) Insufficient barrier efficiency to reactants. For example, in DMFC, Nafion's good affinity for methanol leads to easier penetration of methanol molecules from the

anode to the cathode of the cell, resulting in loss of open-circuit voltage (OCV) as well as poisoning of the catalyst.

(4) High cost.
(5) Complex and harsh synthesis process, resulting in low yield of Nafion.

Since the synthesis of Nafion requires multi-step reactions with harsh reaction conditions, and the F-containing substrate will also pollute the environment after degradation, based on PFSA, Ballard® has produced the BAM-3G series of proton exchange membranes with polyvinylidene fluoride as the main chain and non-fluoroaromatic groups as the side chains [12]. This type of PEM retains the original advantages of perfluorosulfonic acid membranes, but also has a relatively simple synthesis process. The reduction of fluorine content in BAM-3G series PEMs is also in line with the trend of green energy development. However, the high IEC leads to the problem of water absorption and swelling, which restricts its further development.

Non-fluorinated proton exchange membranes are another important component of PEMs. Non-fluorinated proton exchange membranes mainly include sulfonated polyaryl ether (SPAE), sulfonated polyethersulfone (SPE), sulfonated polyaryl ether sulfone (SPAE), polybenzimidazole (PBI), sulfonated polyetherimide (SPEI), polystyrene (PS), and their derivatives. The chemical structural formulas of several common non-fluorinated proton exchange membranes are shown in Fig. 4.5. Sulfonated polyarylene ether materials are polymers with benzene rings and ether bonds in their molecular main chains and modifiable side chains. Compared with PFSA polyelectrolytes, non-fluorinated proton exchange membranes have a broad development prospect due to their excellent processability, good thermal stability, mechanical properties, and low cost.

Due to the low strength and complicated preparation process of PFSA-type polymer electrolyte materials, the main chain and side chains contain a large number

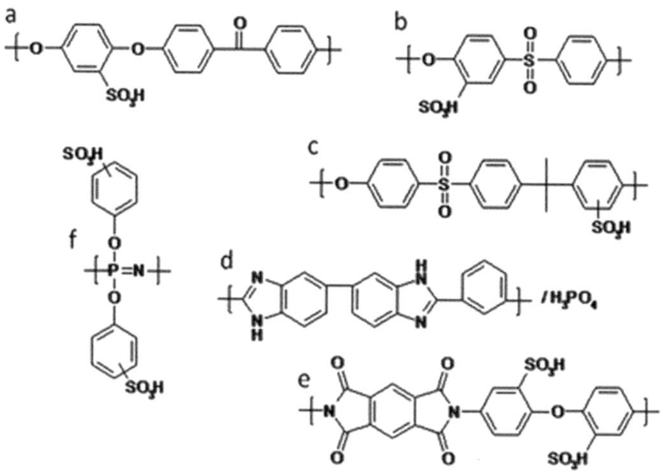

Fig. 4.5 Chemical structural formulae of several non-fluorinated proton exchange membranes [13]

of C–F bonds with high surface energy, which limits their further in situ post-modification treatment. For non-fluorinated proton exchange membranes, although regulating the length of polymer chain segments and designing new structural monomers can effectively improve the performance of PEMs, there are some drawbacks to these methods. For example, increasing the content of sulfonic acid groups too much can improve the proton conductivity of PEMs, but it will also reduce the dimensional stability of the material, and even cause it to dissolve in hot water with excessive water absorption. The excessive swelling will cause the PEM to peel off from the catalyst during the operation of the fuel cell, which increases the interfacial resistance of the cell and leads to the degradation of the cell performance. Therefore, doping and blending methods are also utilized in the literature to modify Nafion membranes. Such composite membranes obtained by doping and blending of two-phase or multi-phase materials can synergize the advantages of individual materials and jointly enhance the comprehensive performance of the materials. Moreover, the modification methods are more targeted and can solve the various problems of PEM simply and efficiently. Modification methods mainly include:

(1) Organic-organic blend modification. Organic-organic blend modification usually refers to the use of polymers containing specific structures or exhibiting special properties as modifiers, combined with the polymer matrix to be modified, to improve the performance of composite membranes. This modification method is commonly used to solve the problems of low strength and poor alcohol barrier properties of Nafion. Meanwhile, some researchers have further improved the proton conductivity of Nafion and its output power and durability in fuel cells by modifying the micro-phase separation structure of Nafion.
(2) Organic-inorganic doping modification. Organic-inorganic doping modification refers to the introduction of inorganic nanoparticles into organic polymer electrolytes by physical doping, in situ synthetic grafting, or sol–gel to achieve the purpose of optimizing the performance of PEM. The inorganic particles, due to their strong, rigid structure, have a rate of dimensional change that hardly varies with temperature, which allows them to improve the dimensional stability of the composite film to a certain extent. The introduction of inorganic particles can also make the PEM have a denser microstructure, reduce the fuel permeability of the PEM, and improve the cell performance of the PEM. At present, commonly used inorganic nanoparticle modifiers are mainly divided into: cerium oxide and other metal oxides to improve the durability of the proton exchange membrane; silica, titanium dioxide and other inorganic nanoparticles to improve the water retention capacity of the proton exchange membrane; heteropolyacids, ionic liquids, and other inorganic fillers to improve the proton conductivity of the PEM in low humidity; carbon materials such as functionalized zero-dimensional carbon quantum dots, one-dimensional carbon nanowires, two-dimensional graphene, and three-dimensional carbon nanofiber mats; and metal–organic skeleton (MOF) materials that carry a large number of modifiable functional groups.

(3) Electrostatic spinning modification. Electrostatic spinning technology can change the morphology of the material, which is conducive to solving the problem of phase separation between the modifier and the PEM matrix when organic-organic blending modification. At the same time, mixing inorganic particles into the spinning solution for spinning, or in situ growth on the fiber surface, is conducive to the controlled distribution of inorganic modifiers in the PEM matrix. Electrostatic spinning refers to the process of polymer curing into fibers under the action of a high-voltage electric field, successively through jetting, stretching, and phase separation. The specific process of electrostatic spinning is as follows: Firstly, a high-voltage DC power supply is used to build a high-voltage electrostatic field between the nozzle of the spinning tube and the receiving device; Then the polymer solution in the spinning tube is charged under the action of the electrostatic field, and when the polymer charge exceeds the critical value, which means the electric field force is greater than the surface tension, the polymer solution will be sprayed in the form of jets to the receiving device; In the process, the jet will gradually decrease in diameter due to traction, splitting and solvent evaporation; Finally, the polymer will be attached to the receiver plate in the form of fibers to obtain the electrostatically spun fiber mats. The morphology and diameter of the nanofibers are also affected by the molecular weight of the polymer, the concentration of the spinning solution, the distance between the nozzle and the receiver, the experimental temperature, and humidity.

4.3.2 Catalyst Layer

The catalyst layer (CL) is where the electrochemical reaction occurs in the PEMFC, and the structural design of the catalyst layer will directly affect the overall performance, cost, and durability of the fuel cell. As shown in Fig. 4.6, the catalyst layer consists of a carbon substrate, catalyst, ionomer, and voids. And the most commonly used catalyst in PEMFCs is platinum-on-carbon (Pt/C), which is a stretch-loaded catalyst with nanoscale platinum particles dispersed and loaded on carbon powder. To allow for the timely export of protons generated during the reaction, the catalyst is also encapsulated with a proton-conducting ionomer, which also acts as a binder in the catalyst layer. During the electrochemical reaction, the gaseous fuel reaches the surface of the catalyst by diffusion, and an electrochemical reaction occurs. The electrons generated during the electrochemical reaction are conducted by the carbon carrying platinum, and the protons are conducted by the ionomer, which forms the three-phase interface (electrocatalyst/ionomer/gas fuel). The density of the three-phase interface within the catalyst layer will directly affect the catalyst utilization efficiency and discharge capacity of the cell. Therefore, a well-designed catalyst layer requires not only a highly active catalyst but also good fuel transport properties, electron conduction, and proton conduction.

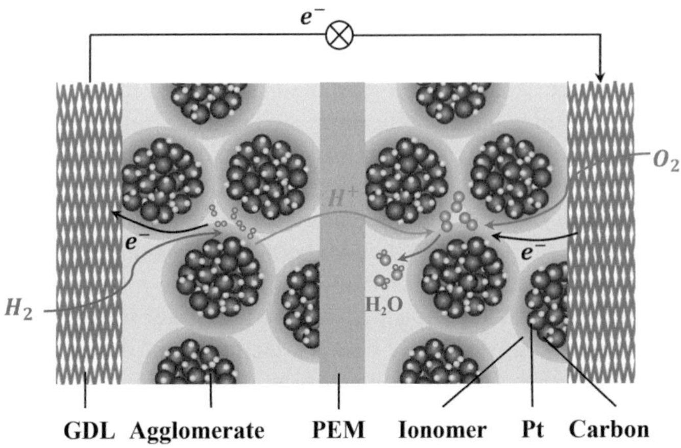

Fig. 4.6 Schematic of the three-phase interface within the catalyst layer

The electrocatalysts with high activity also require excellent resistance to poisoning. Currently, platinum–carbon or platinum-based alloy catalysts are used as electrocatalysts due to their high activity, however, their limited source and high price are not conducive to the reduction of cell cost and hinder the commercialization of PEMFCs. Therefore, the electrocatalysts suitable for PEMFC need to have high electrocatalytic activity, large specific surface area, good electrical conductivity and chemical stability, and strong corrosion resistance. At present, the commonly used commercial catalysts are Pt/C catalysts with Vulcan XC-72 as the carbon carrier, and the Pt loading is usually in the range of 10–60%.

Carbon, especially carbon black and graphitized carbon, is the most commonly used carrier in PEMFC. Carbon carriers are commonly used to load fuel cell catalysts due to their low cost, high specific surface area (101–103 m^2/g), and chemical stability and affinity for Pt (or its alloys) nanoparticles. The single carbon black particles consist of a pseudo-spherical arrangement of several thick layers of quasi-graphite microcrystals with a width of 20–30 Å, which generates a complex porous material with a high specific surface area. Primary carbon black particles tend to aggregate to form carbon particle agglomerates that exhibit a bimodal pore size distribution. These pores include mesopores (2–20 nm, secondary pores) formed between primary particles and macropores (> 20 nm) caused by coalescence, forming agglomerated particles with a size of ~ 200 nm. In addition, there are pore structures < 2 nm within the primary carbon particles, termed micropores. Ketjen Black and Vulcan XC-72 are two carbon carriers commonly used in PEMFC, and they have significantly different effective surface areas, pore volumes, and proportions of micropores. Ketjen Black has a large specific surface area (~ 890 m^2/g), primarily due to its large proportion of micropores (< 2 nm). In contrast, Vulcan XC-72 has a smaller specific surface area (~ 220 m^2/g). The porous structure of the carbon carriers has an important effect on the ionomer and Pt particle distribution, water absorption, and ORR activity. For example, Ketjen Black has a larger pore volume in the secondary pore size range (2–20 nm) than low specific surface area carbon at the same ionomer loading.

The ionomer in the catalyst layer is generally a perfluorosulfonic acid resin of the same composition as the membrane. Usually, the ionomer solution and the loaded catalyst can be mixed to form a catalyst ink dispersion and deposited on the PEM or GDL by spraying or other means. In addition to playing an important role in the proton exchange between the catalyst layer and the membrane, the ionomer needs to have non-electronic conductivity, gas/water permeability, and can be used as a physical binder for the catalyst/carrier particles. In addition, it must have a good physical compatibility with the membranes used, thus reducing the proton resistance. Although the ionomers preferentially adsorb onto the pre-formed Pt/C particle agglomerates to form an ultrathin coating, there is still more evidence that the distribution of the ionomers in the catalyst layer is not completely homogeneous, and that zones of ionomer agglomeration may exist. Such agglomeration of the ionomers may help the adhesion between the catalyst particles on the one hand. But on the other hand, the non-uniform distribution of the ionomers may affect the transport of gas, water, and protons.

The structure of the catalyst layer is important for the rate of energy conversion and transfer in fuel cell electrodes. The low voltages and power densities observed in fuel cell polarization curves are due to limitations in electrochemical kinetics and poor mass transport in the membrane electrodes. In addition, the ohmic losses observed at moderate current density states are also commonly attributed to the proton resistance of the ionomers in the membrane and catalyst layers. All these losses are related to the catalyst layer structure. Control over the structure of the catalyst layer usually relies on the preparation of solution-based catalyst inks, since all active materials are dispersed in the ink. Catalyst inks contain a catalyst (e.g., Pt/C) loaded on a carrier, an ionomer, and a dispersing solvent. The non-homogeneous microstructure of the catalyst ink is of great significance for the structure and properties of the formed catalyst layer. The dispersing solvent plays a crucial role in the micromorphology of the catalyst and ionomers in the catalyst ink and in the microstructure of the catalyst layer, such as the size of the aggregates of the catalyst/ionomer particles, the solution viscosity, the rate of curing, and the basic physical properties and mass transfer properties of the catalyst layer.

4.3.3 Gas Diffusion Layer

Gas diffusion layer (GDL) is one of the important components of the fuel cell, and its main functions include supporting the catalyst, supporting the electrode structure, conducting electrons and heat, as well as transporting reactive gases and water. The GDL is in direct contact with the catalyst layer, and the pore network in the GDL is responsible for transferring reactive gases in the flow field to the catalyst layer as well as transferring the water produced in the catalyst layer to the flow field. Therefore, it plays a direct role and influence on the MEA water balance and water management. Physical processes occurring in the GDL include diffusive transport of gases, heat transfer, aqueous phase change, two-phase flow, electron transport, and droplet dynamics on the surface of the GDL. GDL needs to have good gas permeability, high electrical conductivity, a flat surface, adequate mechanical strength, and chemical and thermal stability.

There are many types of backing layers for GDLs reported in the literature, such as carbon fiber paper, carbon cloth, carbon black paper, nonwoven fabric, and porous metal. Woven carbon cloth and nonwoven carbon paper are two typical GDL materials because they (i) are stable in acidic environments, (ii) provide high gas permeability and good electrical conductivity, and (iii) are elastic when compressed. Carbon paper is widely used in commercial fuel cell applications due to its low manufacturing cost. Carbon cloth is made by spinning and weaving carbon yarns, which are then carbonized or graphitized. Carbon paper consists of carbon fibers that are graphitized at high temperatures (> 2000 °C) to improve electrical conductivity and mechanical strength, and then bonded together with an adhesive. Carbon fibers can be made from polyacrylonitrile (PAN), recombinant fractions of oil or coal, cellulose, or other substances with excellent stiffness, strength, electrical conductivity, and chemical resistance. In the manufacturing process, the virgin fibers are first carbonized in an inert thermal environment to reduce other elements (e.g., oxygen and hydrogen), resulting in carbon fibers with high carbon content and mechanical strength. The carbon fibers are cut short and dispersed in water with a binder, typically 5–15% of the weight of the carbon paper. The fiber solution is then used to produce carbon paper rolls by filtration or other methods. The final carbon paper has an average pore size of 10 μm and a porosity of about 0.7–0.8.

Figure 4.7 illustrates the basic structure of a GDL prepared from carbon fiber paper. The GDL consists of a backing layer (BL) and a microporous layer (MPL). The BL is made of carbon fibers and adhesives that are in contact with the flow field. The MPL is made of carbon powder and a hydrophobic agent that is in contact with a catalyst layer. The BL primarily supports the overall structure of the GDL, and the MPL is used to distribute the delivery of water and gas transport. The GDL's ability to manage water is largely dependent on the appropriate internal porosity and effective pore size distribution. Higher porosity is beneficial for reducing mass transfer resistance, improving material transfer efficiency, discharging product water in time, reducing liquid saturation, and preventing deactivation of three-phase reaction sites due to flooding by water. However, too high porosity will lead to lower electronic conductivity through the surface, and the thermal conductivity and mechanical properties of the GDL will be reduced. The water management capability of the GDL affects the cell performance at high current densities. A well-designed GDL can drain quickly, whereas an improperly designed GDL reduces the cell's water management capability and can lead to electrode flooding in severe cases. After flooding occurs, the transport of reactive gases through the GDL is blocked, which leads to rapid degradation of cell performance. Current research on GDLs includes two main aspects: (1) optimizing the pore size, hydrophilicity, and thickness of BLs to construct gas–liquid transport channels; and (2) optimizing the structure and materials of MPLs to improve the water management capability.

Fig. 4.7 The sketch of the bilayer structure of GDL by SGL [14]

The liquid water inside the fuel cell needs to continue to move towards the outer layers instead of being trapped in the pores of the GDL or CL. To enhance the water transport process, the electrode must be sufficiently hydrophobic. To make the carbon paper or cloth hydrophobic, the carbon paper or cloth is treated with a hydrophobic agent. Hydrophobic agents used include polytetrafluoroethylene (PTFE), polyvinylidene fluoride (PVDF), and fluorinated ethylene propylene (FEP). The GDL hydrophobic treatment methods include impregnation, spraying, and brushing. A representative hydrophobic treatment method is to dip the GDL into a suspension containing hydrophobic polymers, then dry and heat above 350 °C to remove the surfactant and uniformly distribute the hydrophobic polymers. The amount of hydrophobic agent in the whole GDL was controlled by the impregnation time and the concentration of the suspension. The GDL preparation process and SEM images of a typical GDL sample are shown in Fig. 4.8.

Commonly used methods for preparing MPLs on backing layers include spraying, coating, and electrostatic spinning. Spraying and coating technologies have become more mature after years of research, and GDLs currently used commercially are usually prepared by spraying or coating. In the GDLs prepared by spraying or coating, the MPL needs carbon paper as the backing layer and cannot exist alone. In contrast, MPLs prepared using electrostatic spinning technology can be independent of the substrate layer. The types of toner often used in the preparation of MPL are shown in Table 4.1.

The selection of GDL in the fuel cell stack design process has a significant impact on the fuel cell performance, which is usually a comprehensive trade-off between thickness, specific gravity, compression resilience, thickness, porosity, PTFE content, electrical conductivity characteristics, thermal conductivity characteristics, and gas diffusion characteristics. At present, GDL has mature scale production technology, and fuel cell manufacturers mostly use carbon paper products from Toray of Japan, Ballard of Canada, and SGL of Germany, among which Toray occupies a larger market share. Table 4.2 summarizes several commonly used GDL brands and their main characteristics.

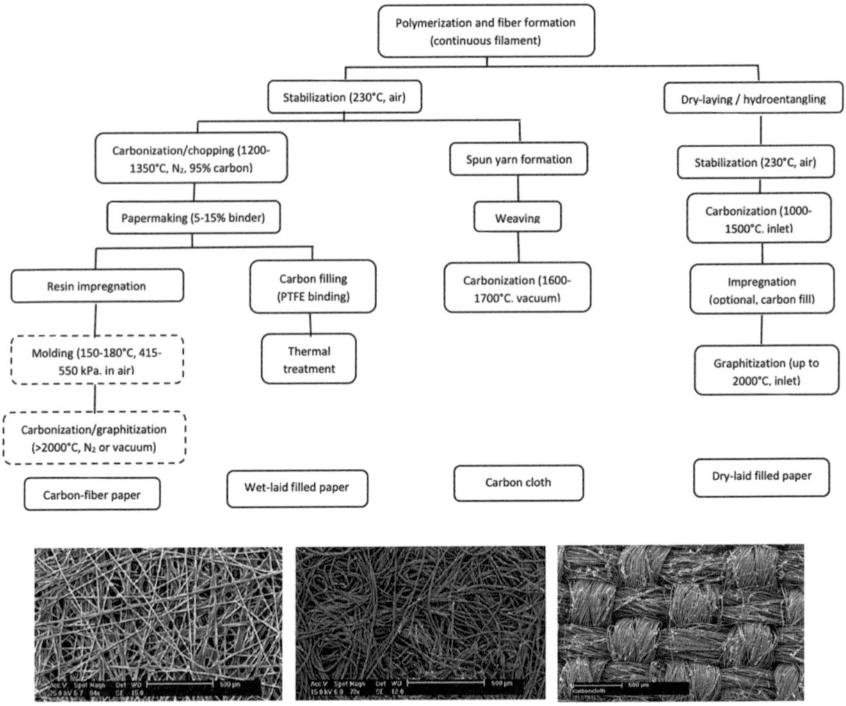

Fig. 4.8 Various processing routes for GDL material preparation, and three typical GDL materials (left to right): Carbon paper Toray H-060, Carbon paper Freudenberg C2, and Carbon cloth [15]

Table 4.1 Typical carbon materials of MPL in GDL

Carbon type	V ($cm^3\ g^{-1}$)	APR (mm)	VP ($cm^3\ g^{-1}$)	VS ($cm^3\ g^{-1}$)	APR_P (mm)	APR_S (mm)
Asbury 850	0.346	3.5	0.212	0.134	0.29	8.6
Mogul L	0.276	6.0	0.157	0.119	0.20	13.6
Vulcan XC-72	0.489	1.8	0.319	0.170	0.24	4.9
Shawinigan Acetylene Black	0.594	1.7	0.368	0.226	0.27	4.3

V is pore volume; APR is average pore radius; Subscripts P and S mean primary and secondary, respectively

4.3.4 Membrane Electrode Assembly

The membrane electrode assembly (MEA), including the GDL, PEM, and CL, is the core component of the PEMFC. As shown in Fig. 4.9, the gas fuel reaches the catalytic

Table 4.2 Commercial GDL cloth and papers [15]

GDL brand	ELAT	Toray	Sigracet	Freudenberg	JNT
Type	Cloth	Paper	Paper	Paper	Paper
Porosity	~ 0.31 to 0.8	~ 0.78 to 0.8	~ 0.75 to 0.9	~ 0.46 to 0.61	< 0.9
Permeability (10^{-12} m^2)	~ 2.156 to 10	~ 1	~ 0.35 to 170	~ 1 to 90	1–10[a]
Electrical resistance (mΩ cm^2, through plane)	~ 2.4 to 5.5[b]	~ 2 to 5	~ 1 to 3	~ 4 to 10	< 20 @1 MPa
Areal weight (g m^{-2})	~ 130 to 250	~ 44 to 167[c]	~ 54 to 140	~ 65 to 150	~ 70 to 110
Thickness (μm)	~ 406 to 490	~ 110 to 370	~ 190 to 325	~ 150 to 290	~ 200 to 350

[a] Unit: [cc/cm^2 s]; [b] GDL-graphite bipolar plate resistance; [c] Estimate based on bulk density and paper thickness

site through diffusion in the membrane electrode, the electrochemical reaction occurs, the electrons generated are conducted by the carbon carrier or platinum, and the protons are conducted by the ionomer, forming a three-phase interface. Therefore, the density of the three-phase interface within the membrane electrode will directly affect the catalyst utilization efficiency and discharge capacity of the PEMFC, which further determines the performance and lifetime of the cell. A well-designed MEA requires not only a highly active catalyst, but also good fuel transport properties, electron conduction, and proton conduction.

The membrane electrode assembly provides channels for protons, electrons, reactive gases, and water for the electrochemical reaction in PEMFC, respectively:

(1) Material transport channel, where the reactive gases and reaction products can be smoothly transported in the porous catalytic layer and gas diffusion layer.

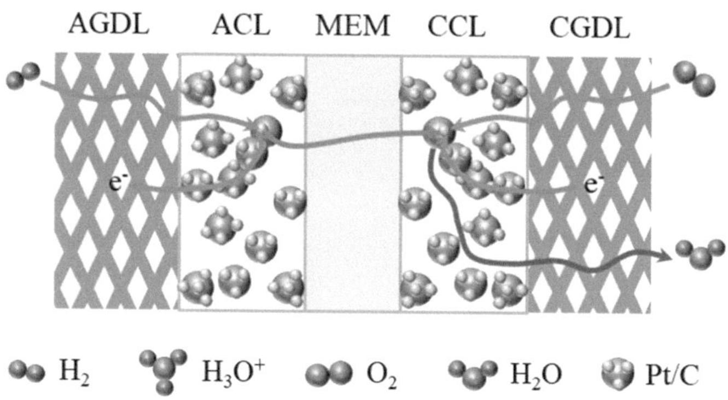

Fig. 4.9 Schematic of the MEA and the three-phase interface

(2) Proton transport channel, which conducts along the Nafion polymer (proton conductor) and passes from the anodic catalytic layer to the cathodic catalytic layer.
(3) Electron transport channel, where electrons are transported in the catalytic layer relying on the electronic conductivity of the catalyst or catalyst carrier, and then pass through the gas diffusion layer to the load side of the external circuit.

As the site for electrochemical reactions to occur, an ideal MEA should have the following characteristics: (1) good adhesion between catalyst and membrane; (2) abundant three-phase interface, with good contact between the reactants, electrolyte, and conductive catalyst, and matching each other; (3) low electrical resistivity between the catalyst layer and the membrane, and between the catalyst layer and the gas diffusion layer; (4) Water and gas can easily enter and exit the catalytic active region; (5) good electrical conductivity of the membrane, catalyst layer, and gas diffusion layer; and (6) short seepage paths allow the generated water to be discharged efficiently, avoiding flooding.

The preparation method, assembly process, physical and chemical properties, materials used, and operating conditions of MEA can have important effects on the performance of PEMFC. There are usually two ways to prepare a membrane electrode assembly (as shown in Fig. 4.10), which are the gas diffusion electrode (GDE) method and the catalyst-coated membrane (CCM) method. The GDE method prepares the catalyst slurry onto the pre-treated diffusion layer to produce a porous gas diffusion electrode, and further, the produced porous gas diffusion electrode and the treated membrane are formed into an MEA by hot pressing, which has a thicker catalytic layer and a lower catalyst utilization rate. The CCM method prepares the catalytic layer on both sides of the proton exchange membrane by the transfer decal method or direct spraying method to form a three-in-one membrane electrode assembly, which has a relatively thin catalytic layer with a thickness of about 10 μm.

Compared with the CCM method, the GDE method can reduce the electrical contact resistance between the catalyst layer and gas diffusion layer, but the ionic

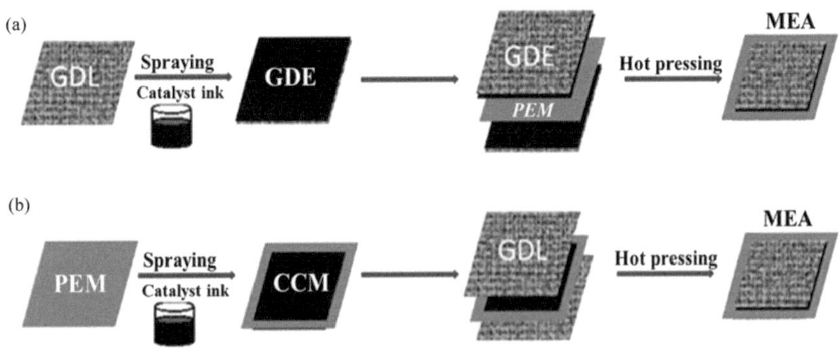

Fig. 4.10 Schematic diagram of the preparation process of MEA by **a** the GDE method and **b** the CCM method [16]

contact between the membrane and the catalyst is poor. There is also a reduction in catalyst utilization due to catalyst penetration into the GDL. Using the CCM method, the interfacial contact between the membrane and the catalyst layer can be enhanced, and the catalyst loading can be reduced. It further promotes the charge transfer kinetics and reduces the activation energy of water electrolysis, which is more advantageous than the membrane electrodes prepared by the GDE method. Therefore, the CCM method is the mainstream method for MEA preparation. The CCM method can enhance the interfacial contact between the membrane and the catalyst layer and reduce the catalyst loading, promote the charge transfer kinetics, and reduce the activation energy of water electrolysis, which is more advantageous than the membrane electrodes prepared with GDE, and therefore it is the mainstream method of MEA preparation nowadays.

Whether it is the GDE method, the CCM method, or the gradient design of the membrane electrode assembly, the catalyst nanoparticles, Nafion proton-conductors, and pore structures inside the MEA are randomly distributed in a disordered manner. This leads to the proton transport channel, gas transport channel, and electron transport channel between the catalyst Pt/C are highly susceptible to being interrupted, and a large number of reaction dead zones exist. It further results in a small reactive area and a few three-phase reaction interfaces for electrochemical reactions. In addition, there exists a large mass transfer resistance in the curved and variable-size gas mass transfer channels, which will lead to a mismatch between the electrochemical reaction rate and the mass transfer flux, causing a serious concentration polarization phenomenon, and largely reducing the utilization of the catalyst.

To solve the above problems, the nano-array structure and porous ordered structure are introduced into the membrane electrode preparation process. By designing and optimizing the structures of the catalyst/ carbon carrier interface, the gas/liquid mass transfer interface, and the contact interface of the proton exchange membrane/ catalyst/gas diffusion layer in MEA, the ordering of the electron, proton transfer and material transport channels and the maximization of the three-phase reaction interface of the membrane electrode assembly are realized, and meanwhile, the precious metal loading is effectively reduced to improve the stability of the electrode structure.

Therefore, a well-designed MEA structure needs to focus on the reaction mechanism at the three-phase interface to improve the catalyst activity, and at the same time, good fuel transport properties, electron conduction, and proton conduction are also needed to further realize the perfect matching of the above three processes.

4.3.5 Bipolar Plates

As the core component of PEMFC, bipolar plates account for approximately 60% of the total weight and 20% of the total cost of the fuel cell stack. The functions of the bipolar plates and the flow field include: (i) isolation of the cathode and anode to avoid infiltration of the cathode and anode reactant gases, (ii) support of the membrane electrode assembly to maintain the stability of the cell structure, (iii) collection and conduction of current, (iv) thermal conductivity and dissipation of heat, and maintenance of the uniformity and stability of the cell temperature, (v) transport and distribution of the reactant gases, and (vi) timely discharge of product water. Runners and ribs are the basic components of the flow field. Gas and liquid transport through the runners. Electrons and heat transport through the ribs. To ensure the efficient transmission of fluid, electricity, and heat, the opening ratio of the flow field is generally 45–75% [17].

Bipolar plates should have high electrical and thermal conductivity, good corrosion resistance, and inertness to reactant gases. Common materials for bipolar plates include graphite, composites, and metals [18], and each of the three materials has its advantages and disadvantages. Graphite materials have high electrical conductivity and good corrosion resistance, but high gas permeability and brittleness; thus, graphite bipolar plates are usually thicker. Composite materials generally use graphite as filler and polymer resin as binder, which are characterized by high corrosion resistance and strength, but the production cost is high. Metallic materials have good ductility, electrical and thermal conductivity, but are prone to corrosion in acidic environments and require anti-corrosion treatment.

To obtain high output power, PEMFCs need to operate at high current densities, leading to an increase in the heat production of PEMFCs, an increase in the concentration loss, and an exacerbation of the risk of electrode water flooding. Researchers have developed different forms of flow fields from the perspectives of heat and mass transport enhancement and water transport regulation. According to the presence or absence of channels and ribs, flow fields can be categorized into channel-rib flow fields and porous flow fields.

Channel-rib flow fields include parallel flow field, serpentine flow field, interdigitated flow field, bio-inspired flow field, and dot matrix flow field. The typical channel-rib flow field configuration is shown in Fig. 4.11.

A parallel flow field is characterized by the presence of several parallel and continuous flow paths in the flow field without bending, and the inlet and outlet sides of these parallel flow paths converge into the same main flow path. The parallel flow field has a shorter flow and lower pressure drop. Parallel flow field and its improved types are still the mainstream designs for current commercial applications due to their simple structure and high stability, such as the partially narrow flow channel used in Toyota's second-generation Mirai [19], and the zigzag flow field used in Honda's Clarity [20].

The serpentine flow field is characterized by a continuous flow path from inlet to outlet, and the flow path has continuous multiple bends in the flow field. As the active area of the PEMFC electrodes increases, the single flow channel leads to a sharp increase in the along-stream flow resistance and pump power loss. Therefore, the serpentine flow field is further categorized into single-channel serpentine flow field and multi-channel serpentine flow field, which are distinguished from each other by the difference in the number of flow channels.

The finger-inserted flow field is characterized by the discontinuity between the inlet and outlet flow channels, and the reactant gas in the inlet flow channel reaches the outlet flow channel through forced convection from the under-rib region and then

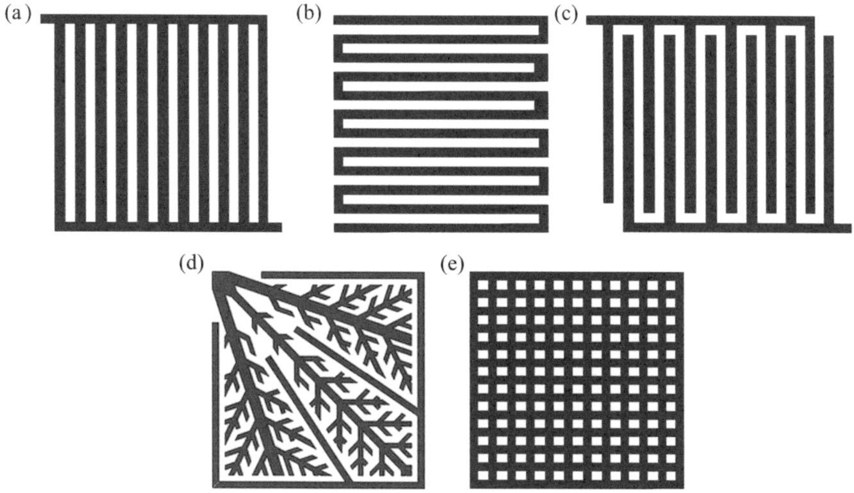

Fig. 4.11 The typical channel-rib flow field configurations: **a** parallel flow field, **b** serpentine flow field, **c** interdigitated flow field, **d** bio-inspired flow field, and **e** dot matrix flow field

flows out. The finger-type flow field has a higher mass transport capacity, but the pressure drop is larger.

The flow channel distribution of the bionic flow field is similar to the distribution of leaf veins in a leaf or the distribution of bronchial tubes in a lung, with the distinctive features of the main flow channel and the sub-flow channel (secondary flow channel). At the runner bifurcation, the cross-sectional area of the sub-runners is determined according to Murray's law [21]. The bionic flow field is more difficult to process, and scaling up is not easily achieved.

The punctate flow field is also known as the Pin-type flow field. The point-type flow field is characterized by the absence of continuous ribs, the short ribs are distributed in arrays, and there are no guided flow paths between the inlet and outlet of the flow field, so that the reactant gas can flow freely. The pressure drop in the fractional flow field is small, but the gas flows along the path of least resistance between the inlet and outlet, resulting in a weak flow area or flow dead zone, leading to uneven distribution of reactants and liquid water accumulation. In addition, the short-ribbed leeward zone may form a relatively stable return zone, resulting in an insufficient supply of reactants within the return zone [22].

The porous flow field has no flow channels or ribs, and the reactant gases are transported through a continuous pore space. The typical porous flow field structure is shown in Fig. 4.12. The 3D fine-mesh flow field (3D fine-mesh flow field) developed by Toyota [23] can be considered as an ordered porous flow field, which consists of a series of small fish-scaled cells arranged in a regular sequence. It both guides the reactant gas transport into the electrode and effectively promotes the liquid water discharge from the electrode, but its disadvantage is the complexity of processing. Most of the remaining porous flow fields reported in the literature are constructed

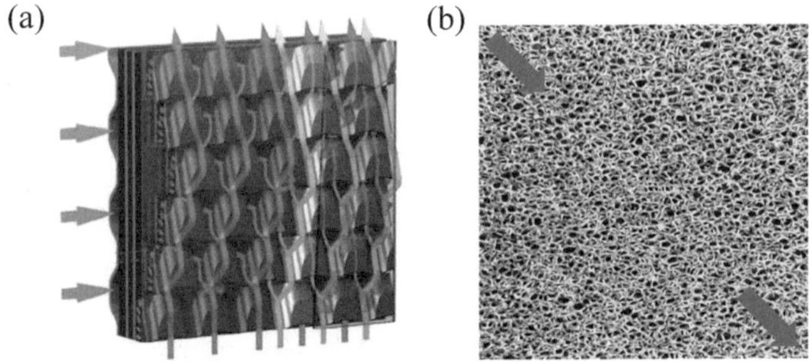

Fig. 4.12 The typical porous flow field configurations: **a** 3-D fine mesh flow field [22] and **b** metal foam flow field [24]

using porous materials, including metal foam [24], carbon foam [25], and so on. Due to the absence of ribs, porous flow fields can promote uniform distribution of reactant gases, but the pressure drop is also larger [26]. The disordered porous flow field can promote the self-humidification of PEMFC, but the absence of guided flow channels makes it face the problem of liquid water retention under high humidification conditions [27].

4.4 Proton Exchange Membrane Fuel Cells Systems

The PEMFC systems can be divided into five subsystems, including the fuel cell stack subsystem, the hydrogen supply subsystem, the air supply subsystem, the thermal management subsystem, and the control subsystem. The fuel cell stack consists of a series of interconnected single fuel cells to build up the voltage, which is the key component of the whole fuel cell system. The fuel supply subsystem and the air supply subsystem, respectively, deliver fuel and air into the anode and cathode of the fuel cell to ensure the chemical reaction, while the thermal management subsystem delivers cooling liquid into the liquid channels to remove the waste heat to maintain the proper operational temperature of the stack.

4.4.1 Fuel Cell Stack Subsystems

The PEMFC stack subsystem is the key component of the PEMFC systems, in which the electrochemical reaction occurs. The fuel cell stack is composed of multiple groups of single cells connected in series, anode and cathode current collector plates, and end plates. Figure 4.13 shows the fuel cell stack used in the Toyota Mirai vehicle. Each single cell is composed of an MEA, two adjacent gaskets, and two plates (cathode and anode). The bipolar plate serves as the anode in one cell and the cathode

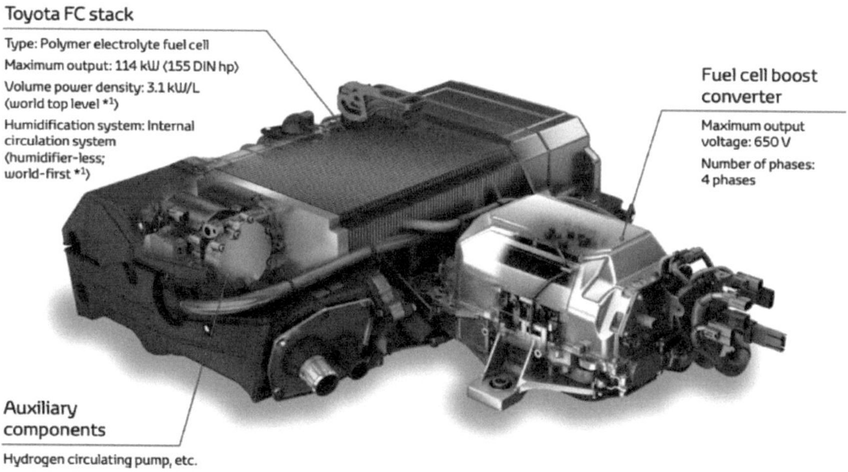

Fig. 4.13 Toyota Mirai fuel cell stack [28]

in the next cell. The current collector plates are used to collect the electron current and conduct it to the external load, thus forming the electrical circuit. The stack is formed by stacking parts including end plates, current collector plates, gaskets, bipolar plates and MEAs in sequence and position, and compressing them with screws or hydraulic machinery at a certain compression speed. The whole stack is kept together by tie rods, bolts, a shroud, or some other arrangement.

During the operation of the PEMFC stack, the inflow and outflow of fuel, air, and coolant are simultaneously occurring in the stack. Therefore, it is necessary to pay attention to the distribution and the sealing of fluids inside the stack in the stack structure design process. In addition, as an output power source, the stack usually has high current and high voltage, so the electrical insulation design of the stack is also one of its basic functional requirements.

Uniform distribution of reactants and coolant to each single cell is important. It is necessary to uniformly distribute fuel and oxidant into the anode and cathode, respectively, and efficiently remove the produced water and unreacted reactants outside the stack. In addition, it is also important to distribute the coolant into the cooling flow field to remove the heat and maintain the optimal temperature of the stack. Because fuel cell performance is sensitive to the flow rate of the reactants, uneven flow distribution will result in uneven performance between the cells. The poor consistency of single-cell performance between the cells leads to poor stability of the PEMFC stack. The good consistency of cell performance strongly relies on the rational design of the stack. The fluid distribution function of the stack is realized by the stack-level distribution and the cell-level distribution. The stack-level distribution is realized by the manifold of the stack, through which the fuel, oxidant, and coolant are respectively guided and distributed to each single cell, while the cell-level distribution is realized by the channel and flow fields on the bipolar plates.

The stack sealing refers to the effective isolation between the fuel, the oxidant, and the coolant through the stack sealing structure to prevent mutual leakage and external leakage among the three. Even a small amount of leakage of fuel and oxidant can lead to the direct chemical reaction between them, resulting in a parasitic current and performance degradation. And a serious leakage may even cause the electrodes and the stack to burn. As for the coolant, its leakage into the anode or cathode flow fields will lead to the pollution of the electrode, which increases the mass transport resistance of reactants. What's worse, its adsorption on the catalyst surface will reduce the effective reaction area of the cell, resulting in cell performance degradation and causing irreversible damage to the electrode in severe cases. The leakage of fuel, oxidant, and coolant caused by poor sealing will reduce the reliability of the stack and system, and may cause safety accidents. The sealing of the stack is usually achieved by the cooperation of the bipolar plate, the sealing frame, and the sealing element.

The performance consistency of the single cells in the stack refers to the consistency of the performance of each single cell during the working process of the stack. The higher the consistency, the better the working stability of the stack. The extreme situation of poor consistency of the single cells in the stack will lead to the voltage reversal phenomenon, which is a unique phenomenon in which the output voltage of the single cell changes from positive to negative during the process of stack power generation. During the operation of the stack, when the fuel and oxidant are sufficiently and effectively guided into each single cell, the hydrogen oxidation reaction at the anode and the oxygen reduction reaction at the cathode will continuously take place. However, when anode fuel or cathode oxidant starvation occurs, the current of the stack will turn to drive the reverse reaction in the anode or cathode, causing the voltage of the single cell to reverse to negative. When the hydrogen starvation occurs due to the insufficient supply of hydrogen, the electric potential generated by the working stack will promote the water electrolysis in the anode to provide the protons and electrons to maintain the current conduction in the hydrogen-starvation single cell, while the oxygen reduction reaction in the cathode is normal. In this situation, oxygen is generated at the anode. When oxygen starvation occurs due to the insufficient supply of oxygen, the anodic hydrogen oxidation reaction is normal, while the protons and electrons that respectively reach the cathode through the membrane and external circuit will combine to generate hydrogen. In this situation, hydrogen is generated at the cathode. The voltage reversal will cause the combustion reaction of oxygen and hydrogen within the cathode or anode, which will damage the MEA. In addition, since the catalyst carrier, the gas diffusion layer, and the graphite bipolar plate are carbon materials, the voltage reversal will also lead to carbon corrosion.

4.4.2 Hydrogen Supply Subsystems

The hydrogen supply subsystem provides wet high-pressure hydrogen for the stack, which usually consists of a hydrogen storage tank, a pressure reducing valve, a proportional valve, a circulating device, a vapor separator, an exhaust valve, and so on. The high-pressure hydrogen in the storage tank is reduced by a pressure-reducing valve, then, with the flow rate adjusted by a proportional valve, finally flows into the stack to participate in the reaction. The hydrogen is excessively supplied to the PEMFC stack from the high-pressure tanks to prevent reactant starvation and avoid water flooding. There are four modes for the hydrogen supply subsystem [25, 26], as shown in Fig. 4.14.

(1) The first is the flow-through mode. In this mode, the excessive hydrogen flows through the anode flow channels continuously, whereas only a portion of the hydrogen in the fuel stream is consumed in the anode compartment. The excess hydrogen has the potential to minimize the liquid water and nitrogen accumulation in the anode channel. However, it leads to fuel waste and potential dangers.
(2) The second is the dead-end mode. In this mode, the purge valve is closed, and the hydrogen is exactly supplied and consumed inside the anode catalyst layer.

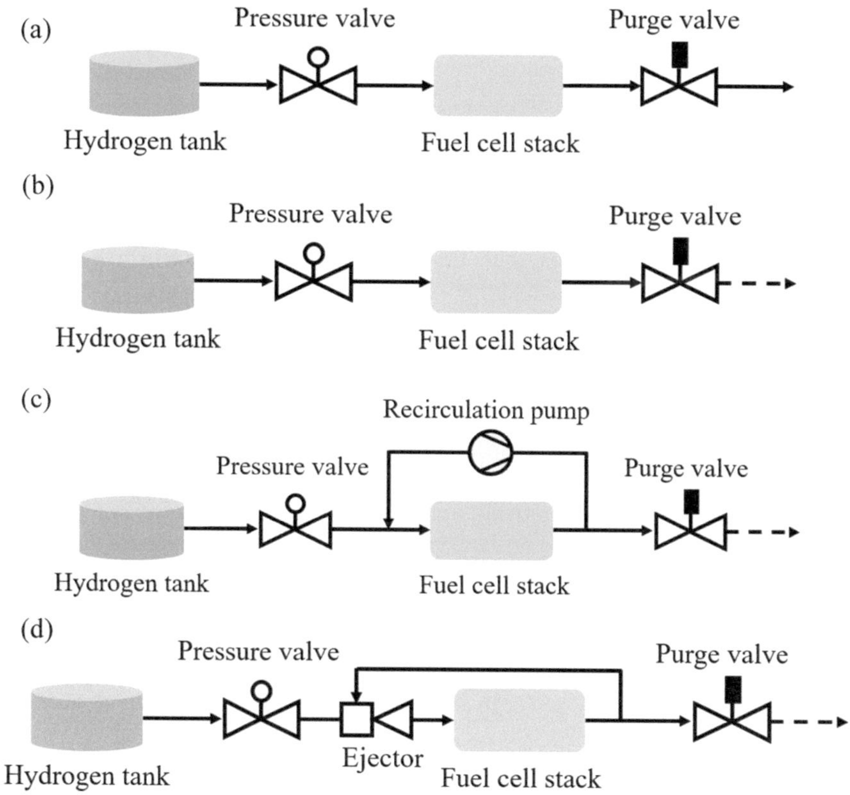

Fig. 4.14 Hydrogen supply subsystems: **a** flow-through mode, **b** dead-end mode, **c** active recirculation mode, **d** passive recirculation mode

However, it leads to nitrogen gas contaminants and liquid water accumulation in the anode, thus reducing the reaction efficiency and electrical load support capability. Therefore, in this mode, the purge valve needs to be opened intermittently according to a certain control strategy.

(3) The third is the active recirculation mode. In this mode, the compressor is used to circulate the unreacted hydrogen gas to the anode inlet of the fuel cell stack. Therefore, the utilization rate of hydrogen is improved, and the inlet hydrogen can be humidified to some extent. However, the application of a compressor also increases the system complexity and the extra energy consumption.
(4) The fourth is the passive recirculation mode. In this mode, the ejector is used to circulate the unreacted hydrogen gas to the anode inlet of the fuel cell stack.

Because the active and passive recirculation modes can increase the hydrogen utilization rate, avoid the hydrogen waste, maintain the pressure difference between anode and cathode, and maintain membrane humidity, they are usually adopted in the hydrogen supply subsystem. However, water and nitrogen can permeate into the anode through the MEA due to pressure and concentration gradients and gradually accumulate in the anode, which leads to performance degradation, hydrogen concentration reduction, and even carbon corrosion. Therefore, it is necessary to decrease nitrogen concentration and remove liquid water by the purging process. There are two kinds of purge strategies: discontinuous purge and continuous purge. The discontinuous method is based on discontinuous purge and uses the measured hydrogen content at the outlet of fuel cell stacks as a purge trigger. For the continuous method, the opening of the purge valve is controlled to purge continuously by using the parameter of hydrogen content at the stack outlet.

In the hydrogen supply subsystem, the key equipment includes a hydrogen circulating pump and an ejector. The hydrogen circulating pump, as a hydrogen circulating power supply component, needs to provide a sufficient flow rate and sufficiently high-pressure hydrogen under high sealing requirements. Claw circulation pumps have the advantages of being oil-free, having good reliability, high efficiency, high output pressure, etc., but the processing requirements are high. There is another type of circulating pump, the scroll vacuum pump, as shown in Fig. 4.15. The pump produces a fine force balance through the floating scroll compression technology, which minimizes the friction loss between the moving and fixed scrolls, improves the compression efficiency, stability, reliability, and service life of the product, and reduces the cost of use. Meanwhile, the topless sealing structure of the moving and fixed scrolls avoids contamination problems. The circulating pump is an active control component with the advantages of precise control and individual adjustment, which was once widely used. However, it has been gradually replaced in recent years due to the disadvantages of high hardware cost and high auxiliary power consumption, which makes it prone to failure.

The ejector is a passive component that does not require auxiliary power consumption under the same reflux conditions, which is conducive to reducing hydrogen consumption and improving system efficiency. The ejector is a device that uses jets to mix different pressure fluids to transfer energy and mass. The schematic of the

Fig. 4.15 The scroll vacuum pump [29]

ejector is shown in Fig. 4.16. The operating principle of ejectors is that the pressure energy of the primary hydrogen from pressure valves is converted to velocity energy by adiabatic expansion in the nozzle. The pressure drop of the motive fluid creates a low-pressure zone before the mixing region. Due to the low-pressure zone, the secondary hydrogen (recirculation hydrogen) moves toward it and mixes with the primary hydrogen in the mixing region. Mixed hydrogen enters the diverging portion of the ejector, where its velocity energy is converted into pressure energy. The advantages of the ejector are no moving parts, no extra power consumption, simple structure, reliable operation, easy installation and maintenance, and good sealing. However, the ejector in the current application has poor adaptability to the wide range of variable load conditions of the fuel cell, especially in the small flow operating conditions, where the ejection coefficient is low or cannot even eject the reflux hydrogen. To enhance the adaptability of the ejector, there have been solutions such as parallel connection of multiple ejectors, adaptive adjustment of ejectors, series connection of ejectors and pulse control valves, and parallel connection of ejectors and hydrogen circulating pumps etc. Hydrogen circulation systems will be developed with the goal of high compactness, high efficiency, low energy consumption, and no pollution, so ejectors are the way of the future.

4.4.3 Air Supply Subsystems

The air supply system is a system that filters, humidifies, and regulates the pressure of the air entering the fuel cell, including an air filter, air compressor, heat exchanger, humidifier, and back pressure valves [30]. In fuel cell systems, the hydrogen supply

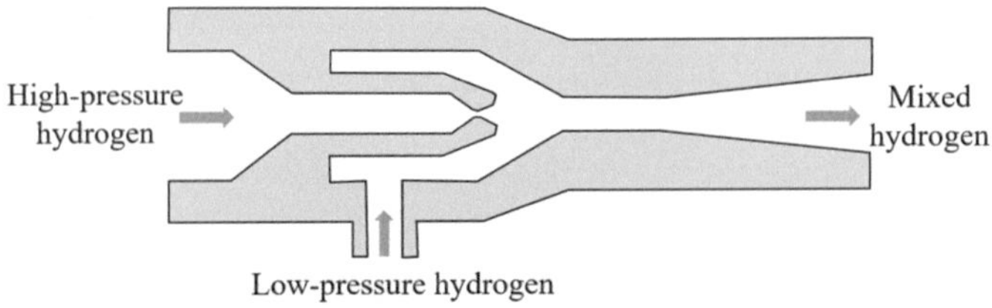

Fig. 4.16 Schematic of ejector

tends to be sufficient, and the response speed of the hydrogen supply system tends to be greater than that of the air supply system due to its smaller size and relatively simple actuating devices. Therefore, the dynamic output performance of the fuel cell system is often determined by the air supply subsystem. The air supply subsystem is responsible for providing clean and moist oxygen to the cathode of the fuel cell stack. As shown in Fig. 4.17, the typical working process of the air supply subsystem is as follows: (1) The air first flows through an air filter to remove particles (solid or oil), and then enters the cathode manifold after being properly pressurized by an air compressor; (2) Since the temperature of the air passing through the air compressor is very high, the air is cooled down by a heat exchanger, humidified by a humidifier, and then fed into the cathode flow field of the fuel cell stack; (3) The residual air after the electrochemical reaction enters into the humidifier for moisture heat recovery; (4) Finally, the air is discharged from the system through a pressure regulating valve to assist in regulating the air pressure in the system.

(1) Air filter

The function of the air filter includes filtering particles in the air to prevent particles from clogging the fuel cell electrodes. In addition, the air filter needs to remove harmful impurity gases (SO_2, NH_3, NO_x, etc.) to avoid poisoning the fuel cell catalyst. The key component of the air filter is the filter element, which contains at least two layers, one for physical filtration and the other for chemical adsorption, as shown in Fig. 4.18. Some air filters add a specialized muffler structure to reduce noise. Some air filters are designed to intelligently control whether the air passes through the chemisorption layer or not, so that it does not pass through the chemisorption

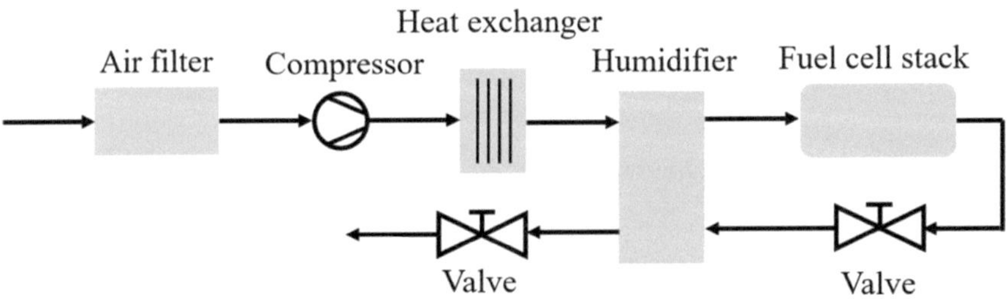

Fig. 4.17 Schematic of air supply systems

layer when the concentration of harmful gases is low. Such a design can save filter element materials, which is conducive to prolonging the life of the filter element, but the structure is complex. Air filters use physical filtration to remove particles in the air, the principle includes direct interception, inertial impact, and diffusion interception. Direct interception mainly removes particles with a diameter larger than the pore size of the filter material, and diffusion interception removes particles with a diameter smaller than the pore size of the filter material. Air filter chemical adsorption works by selecting materials with large surface area, low density and high porosity (such as activated carbon, activated carbon fiber, alumina) as a carrier, supplemented by surface modification technology or loaded catalysts, not only to achieve high efficiency adsorption effect, but also to improve the air flow pore, reducing the air transmission resistance. The design parameters of air filters are generally air flow, inlet resistance, ash capacity, and filtration efficiency. After the flow rate is determined, the parameter that has the greatest impact on the performance of the fuel cell is the air intake resistance. To reduce the air intake resistance of the air filter, the design should consider reducing the thickness of the filter element and increasing the area of the filter element.

(2) Air compressor

The air compressor can increase the power density and efficiency of the fuel cell as well as reduce the size of the fuel cell system by increasing the cathode inlet pressure. However, the parasitic power consumption of the air compressor is large, accounting for about 80% of the auxiliary power consumption of the fuel cell, and its performance directly affects the efficiency, compactness, and water balance characteristics of the fuel cell system. Requirements for the air compressor in the PEMFC system include oil-free compressed air, low power consumption, high efficiency, smaller size, lighter weight, low noise, and good dynamic response capability. The design and selection of the air compressor should take into account all aspects of the system, such as noise, pressure ratio, flow rate, mass, and efficiency, to optimize the performance of the fuel cell system.

According to the different working principles, the commonly used compressors are vane type, screw type, centrifugal type, scroll type and roots type. (1) Vane compressors form a compression cavity by engaging two sets of sliding vane chains at 90° to each other and compressing air through cross-rotation; (2) Screw-type air

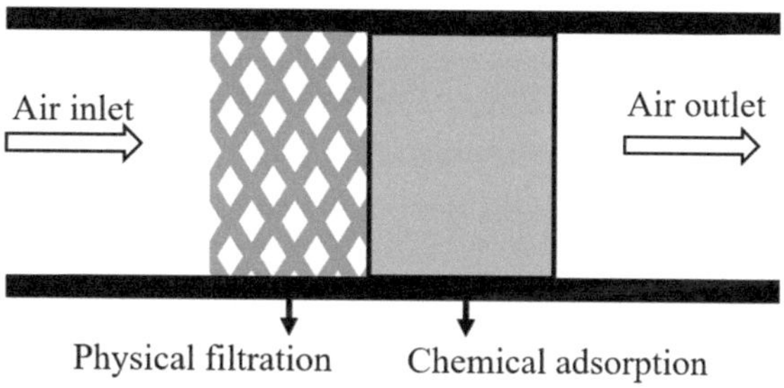

Fig. 4.18 Schematics of the filter

compressor utilizes the air groove formed between the screws to compress air. It has a simple and reliable structure, a wide flow rate range, good pressure/flow rate characteristics, and is an ideal air compressor for fuel cells. However, the noise problem of screw air compressor should not be ignored, and the measures taken to minimize the noise further increase the cost, quality and complexity of the system; (3) Centrifugal air compressor belongs to vane air compressor, which is characterized by compact structure, fast response, long life and high efficiency. Centrifugal air compressors convert mechanical energy into pressure energy of the gas by rotating the impeller in the impeller and diffuser flow channel, utilizing the centrifugal pressurizing and narrowing diffuser effect. However, a centrifugal air compressor in a low flow rate will experience a wheezing phenomenon, which will seriously affect the performance of the system and the service life of the air compressor; (4) Scroll air compressor has high efficiency, low noise, simple structure, light weight, and high reliability; (5) The working speed of Roots-type air compressors is low, so it is unnecessary to use air bearings with a complicated structure. Roots-type air compressor has a wider high-efficiency operation area, which can improve the fuel economy of the whole working condition. Table 4.3 shows the types of air compressors used in the currently developed fuel cell systems.

Scroll, screw and centrifugal air compressors have better overall performance. However, scroll and screw air compressors have mutual friction between the blades, higher noise and mass, and it is difficult to work with the turbine matching. It is impossible to recover the exhaust energy, which can only be realized by matching the turbine with the centrifugal compressor at present. Centrifugal compressors are one of the most promising ways of pressurizing air, as they offer the best comprehensive effect in terms of density, efficiency, and noise reduction. The typical structure of a centrifugal compressor is shown in Fig. 4.19.

(3) Humidifier

Gases humidification is important for PEMFCs to guarantee the membrane hydration for achieving its high ion conductivity. In contrast, low humidification or non-humidification operation may accelerate the membrane degradation process due to the radical formation and membrane dehydration [33]. The humidification

Table 4.3 Air compressor type used in commercial fuel cell systems [31]

Company	Toyota	Honda	Daimler	General motors	Hyundai
Vehicle	Mirai	FCX Clarity	FCV A-class B-class	Chevrolet Equinox	FCV ix35
Air compressor type	6 leaves roots compressor	2-stage centrifugal compressor	Centrifugal compressor	Centrifugal compressor	Centrifugal compressor

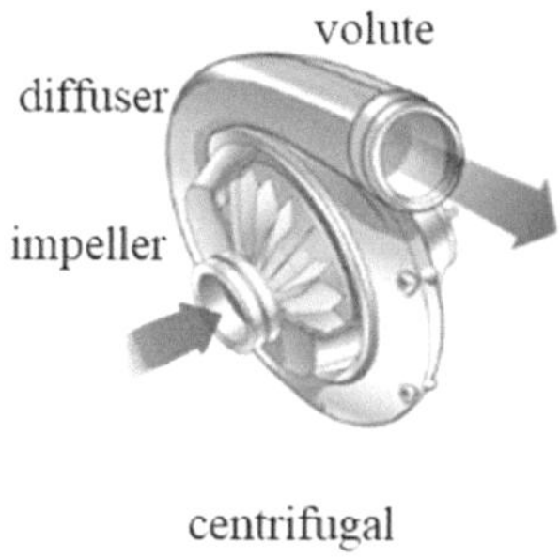

Fig. 4.19 Structure diagrams of centrifugal compressor [32]

methods include gas bubbling humidification, direct water (liquid/vapor) injection, enthalpy wheel humidifiers, membrane humidification, and exhaust gas recirculation. In Fig. 4.20, the membrane humidifier and enthalpy wheel humidifier are the two most prospective candidates for air humidification in a large power fuel cell system, as heat and water from the cathode exhaust can be recycled to humidify intake air [34].

The enthalpy wheel humidifier uses a hygroscopic core with porous honeycomb-shaped columns that can absorb the hot moisture from the fuel cell exhaust gas and deliver it to the dry inlet gas through rotation. The dry gas is humidified and heated by the humidifier and then flows into the fuel cell. The diameter and thickness of the enthalpy wheel are the structural factors that affect the humidification level. Larger diameter and thickness lead to higher humidification levels. In practice, the size needs to be determined according to the power of the fuel cell. The motor rotation speed and the gas flow rate are the operational factors that can be used to adjust the humidification level. The advantages of the enthalpy wheel humidifier include low pressure drop, lightweight, and low cost. The disadvantage of the enthalpy wheel humidifier is that it may cause gas leakage and cross-flow through the rotation. The gas leakage mainly consists of two components. One is the gas leaking between the interface of the seal and the core, the other is the gas carried over by the rotating core.

Membrane humidifier can be divided into water-to-gas and gas-to-gas humidifier, where the liquid water and moist gas are respectively used to humidify the other side's dry air. Furthermore, the gas-to-gas humidifier is widely used. The membrane

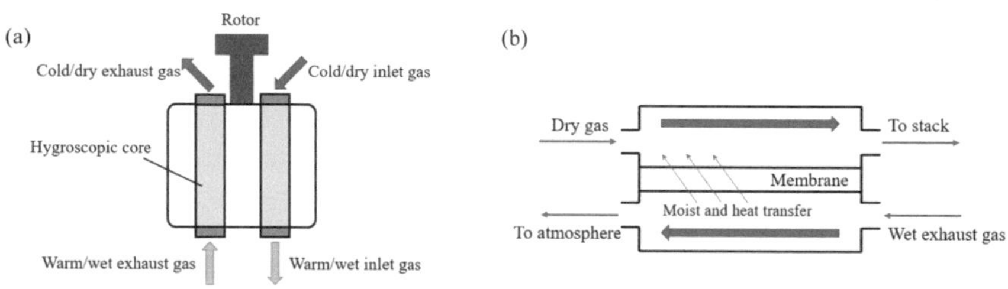

Fig. 4.20 Schematics of **a** enthalpy wheel humidifier and **b** planar membrane humidifier

humidifier adopts a membrane (usually a Nafion membrane) to separate the wet side and dry side. Moisture and heat are transferred across the membrane to the dry side, while the gas is not allowed to transfer through the membrane, neglecting the unavoidable small amount of crossover. The advantages of the membrane humidifier include no moving parts and easy scaling up. However, it is difficult to realize the exact control over the humidification level. Besides, the membrane may be blocked by the impurities from liquid water for the liquid-to-gas membrane humidifier in long-term operation. According to the different shapes, membrane humidifiers can be divided into planar humidifiers and tubular humidifiers. The former is usually used for gas-to-gas humidifiers, and the latter for liquid-to-gas humidifiers. The structure of the planar membrane humidifier is similar to the PEMFC membrane electrode assembly, while the tubular membrane humidifier mainly consists of a large number of hollow fiber membrane tubes bundled in a cylindrical cavity. The main structure of the planar membrane humidifier is composed of several thin plates with channels, and the materials are mostly graphite plates. Therefore, the design of the channels can be largely borrowed from the design of the fuel cell bipolar plate channels. In addition, the design of the membrane humidifier's channels needs to consider the pressure drop to reduce the system energy consumption.

4.4.4 Thermal Management Subsystems

Electricity generation in the fuel cell stack is accompanied by heat generation. The stack can overheat if the heat generation rate is too high. However, temperature is a key factor that affects electrochemical reaction, reactants transport, water balance, and ion transport, which further affect the cell performance and durability. A higher temperature that exceeds the recommended operating temperature range of the stack can decrease activation barriers for reactants, while it also leads to membrane dehydration. However, a lower temperature not only decreases the cell performance but also increases the risk of water flooding in the cell. If stack cooling is not uniform, the thermal gradient within the stack can harm cell performance by causing cells to operate at different voltages and enhancing degradation mechanisms. Therefore, thermal management is vital for the PEMFC stack to operate within its optimal temperature range and to avoid thermal gradients. As shown in Fig. 4.21, the thermal management subsystem scheme includes passive and active cooling of the fuel cell stack, the selection of which depends on the stack size, operating environment, and operating strategy. The passive cooling method involves: (1) cooling via natural convection of air against the external surface area of the fuel cell stack; (2) cooling via the free or forced convection of reactant gases through the fuel cell stack. The active thermal management system is generally composed of pumps, radiators, pipes, and extra bypasses, with coolant heaters being required for cold start circumstances. The liquid cooling instead of gas cooling is often used in an active thermal management system for high power density PEMFC stacks, because the volumetric heat capacities of liquids are much greater than the volumetric heat capacities of gases.

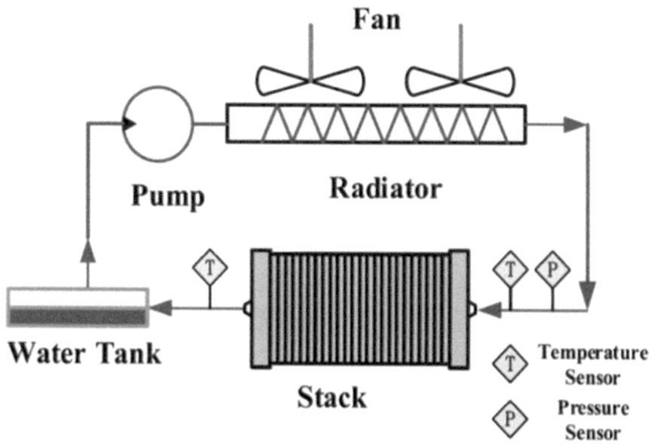

Fig. 4.21 Layout of a fuel cell thermal management system [35]

Most automotive fuel cell stacks (in the range of 50–90 kW) are liquid cooled using either deionized water or a water–glycol mixture.

To resolve the local overheating and non-uniform temperature distribution within the stack, cooling flow field design and coolant selection are important [36]. In Fig. 4.22, the cooling flow field that is sandwiched between the anode and cathode flow fields is etched on the bipolar plate. Most of the heat is generated and accumulated in the MEA cathode side, and then released via heat conduction and removed via forced heat convection between the cooling channel walls and cooling medium. Due to the relatively high thermal conductivity of BPs materials (commonly made of graphite, metallic, or composite plates), the overall temperature control performance by active liquid cooling is primarily determined by this convective heat transfer process, in which the cooling flow field executes a dominant role. Currently, four main cooling channel configurations are widely adopted, i.e., parallel, serpentine, wavy, and 3D fine mesh flow fields. In commercial applications, bipolar plates with wavy channels and their refined versions have been used in the PEMFC stack from Honda Clarity FCEVs since 2006 [37].

Liquid-cooling PEMFCs integrated into FCEVs generally utilize a mixture of 50% deionized water and 50% ethylene glycol (EG) as their cooling medium. Due to ion contamination from the bipolar plates or ionic production from oxidation of the glycol, water gets contaminated and becomes more electrically conductive with many ions absorbed, which interrupts the flow of electricity throughout the BPs, resulting in the accelerated degradation of the PEMFC stack. Therefore, to effectively remove these ions, a real-time deionized configuration (generally equipped with a deionization filter) is essential to assist in the cooling loop scheme. Nanofluids are another alternative coolant due to their features of high thermal conductivity but low electrical conductivity, which could improve the performance of temperature control for the PEMFC stack. In this regard, oxide nanoparticles like CuO, Al_2O_3, SiO_2, ZnO, and TiO_2, dispersed in water or EG (as the base fluid) as a promising heat transfer medium, have been investigated in numerous studies.

Fig. 4.22 Schematic of Honda's bipolar plates [38]

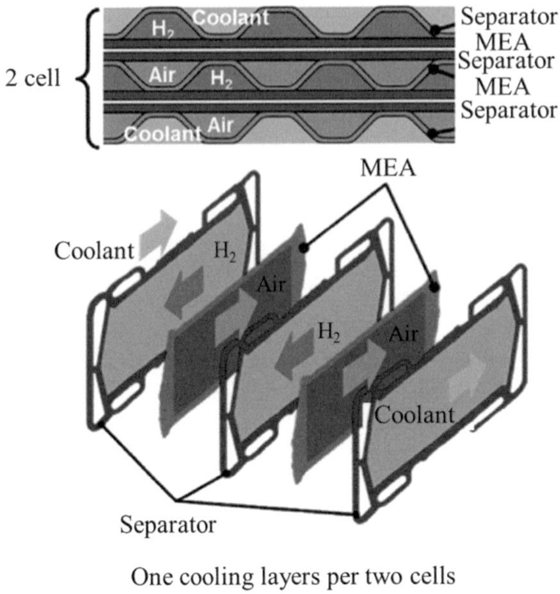

4.4.5 *Control Subsystems*

As shown in Fig. 4.23, control systems for fuel cell stacks mainly consist of PEMFC stacks, controllers, hydrogen supply subsystems, air supply subsystems, humidification subsystems, heat management subsystems, and power converters. Here, controllers regulate the flow and pressure of hydrogen and air, humidity, stack temperature, and output power to satisfy the operational requirements of PEMFCs under different operation conditions.

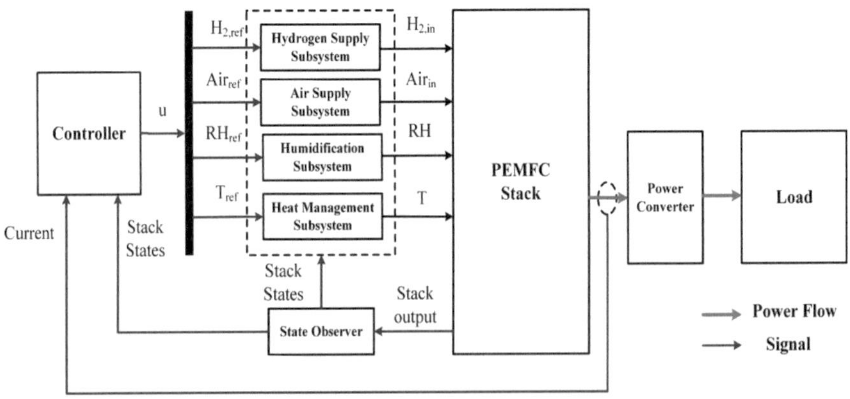

Fig. 4.23 Control structure of a PEMFC stack [35]

Hydrogen supply control systems mainly consist of valves, recirculation pumps, controllers, and hydrogen flow pressure sensors in PEMFC systems, in which controllers can regulate hydrogen flow and pressure by controlling the opening of valves and recirculation pumps (Fig. 4.24). The main control objective of hydrogen supply systems is to regulate hydrogen flow and pressure through system valves to meet the operational requirements of PEMFC systems, in which the dynamic and static control performance of hydrogen supply systems can determine output power and operational performance.

Air supply control systems mainly consist of air compressors, controllers, and air flow pressure sensors in PEMFC systems, in which controllers regulate air flow and pressure by controlling the compressor (Fig. 4.25). The main control objective of air supply systems is to supply adequate air flow to PEMFC stacks during operation, in which oxygen is supplied through a compressor driven by an electric motor and a valve.

Humidification control systems in PEMFCs mainly consist of controllers, humidifiers, humidity sensors, and heaters, in which controllers can regulate PEMFC stack

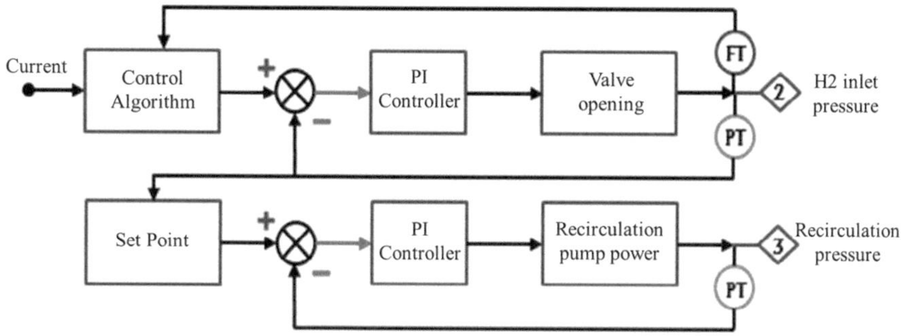

Fig. 4.24 Control structure block of a hydrogen supply system in a PEMFC system [39]

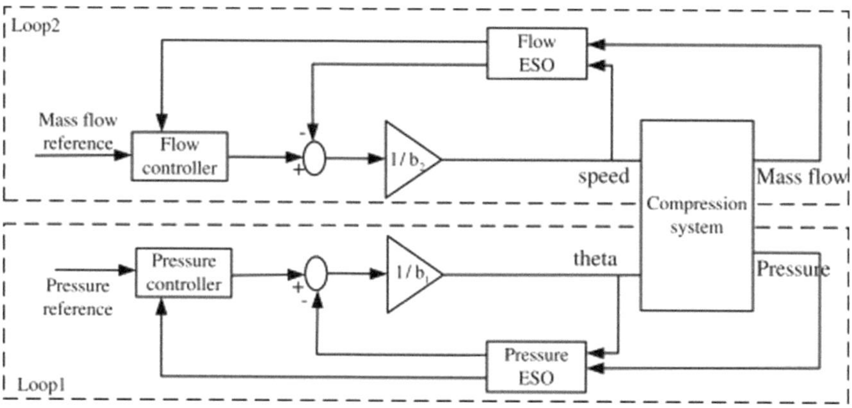

Fig. 4.25 Control structure block of an air supply system in a PEMFC system [40]

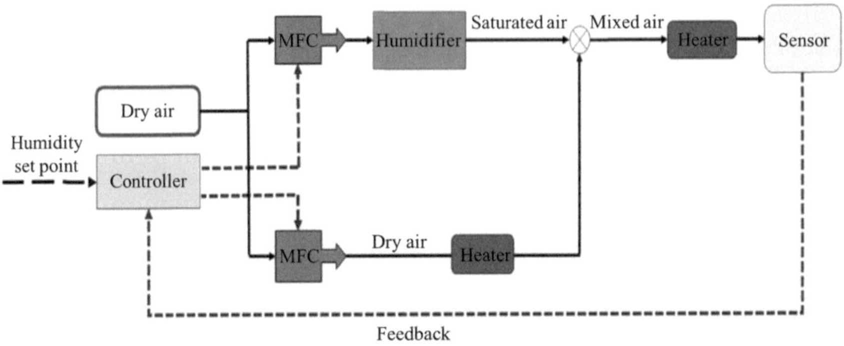

Fig. 4.26 Control structure block of a humidification system in a PEMFC system [41]

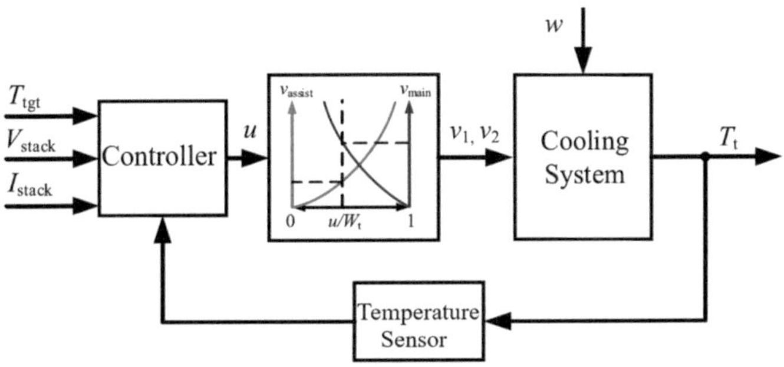

Fig. 4.27 Control structure block of a thermal management system in a PEMFC system [42]

humidity by controlling humidifiers and heaters (Fig. 4.26). The main objectives of PEMFC humidification control systems are to maintain internal water balance in membranes and external water balance in gas diffusion layers and flow channels [36], in which cathode or anode humidifiers are always used to prevent membrane dehydration and flooding in cathodes and anodes.

PEMFC thermal management control systems mainly consist of controllers, coolant heaters, or radiator fan-based cooling systems, and temperature sensors, in which controllers can regulate PEMFC stack temperatures by controlling either the coolant heater or the radiator fan or both (Fig. 4.27). The main control objective of thermal management systems is to regulate stack temperature to enhance the performance and durability of PEMFCs.

4.5 Applications

PEMFCs are applied in vehicles, marine, aviation, and stationary applications. The information is summarized as follows.

4.5.1 Vehicle Applications

Fuel cell electric vehicles (FCEVs) use electricity to power an electric motor. However, in contrast to other electric vehicles, FCEVs use a hydrogen fuel cell to generate electricity, rather than just drawing it from a battery. The amount of energy stored in the vehicle is determined by the size of the hydrogen fuel tank. As defined by the U.S. Department of Energy (DOE), the major components of an FCEV include:

(1) Battery (auxiliary): in an electrically driven vehicle, the low-voltage auxiliary battery provides power to start the vehicle before the traction battery is engaged; it also provides power for vehicle accessories.
(2) Battery Pack: this high-voltage battery stores the energy generated by regenerative braking and provides supplemental power to the electric traction motor.
(3) DC/DC converter: this device converts the high-voltage DC power from the traction battery pack into the low-voltage DC power needed to operate the vehicle accessories and to charge the auxiliary battery.
(4) Electric traction motor (FCEV): utilizes power from the fuel cell and traction battery pack to drive the vehicle's wheels. Some vehicles use a motor-generator with both drive and regenerative functions.
(5) Fuel Cell Stack: an assembly of individual membrane electrodes that use hydrogen and oxygen to generate electricity.
(6) Fuel Filler: a nozzle of a fuel dispenser attached to a container on a vehicle used to fill a fuel tank.
(7) Fuel Tank (Hydrogen): storage of hydrogen in the vehicle until it is needed by the fuel cell.
(8) Power electronic controller (FCEV): this unit manages the flow of electrical energy delivered by the fuel cell and the traction battery, and controls the speed of the electric traction motor and the torque it produces.
(9) Thermal System (Cooling)—(FCEV): this system maintains the proper operating temperature range for the fuel cell, electric motor, power electronics, and other components.
(10) Transmission (Electric): the transmission transmits mechanical power from the electric traction motor to drive the wheels.

Fuel cell electric vehicles can be further categorized into pure fuel cell electric vehicles (PFCEVs) and fuel cell hybrid electric vehicles (FCHEVs). In a PFCEV, the power source of the system is only the fuel cell stack, with no other power generation units. The structure of the system is relatively simple and includes only a fuel tank, an FC stack, a DC–DC power converter, an inverter, and an electric motor, as shown in Fig. 4.28a. The FCHEV consists of a fuel cell stack and a battery pack, with the FC connected to the DC bus using a unidirectional DC–DC converter (UDC) and the battery connected to a bidirectional DC–DC converter, as shown in Fig. 4.28b. During the operation of the FCHEV, the battery is used to provide power

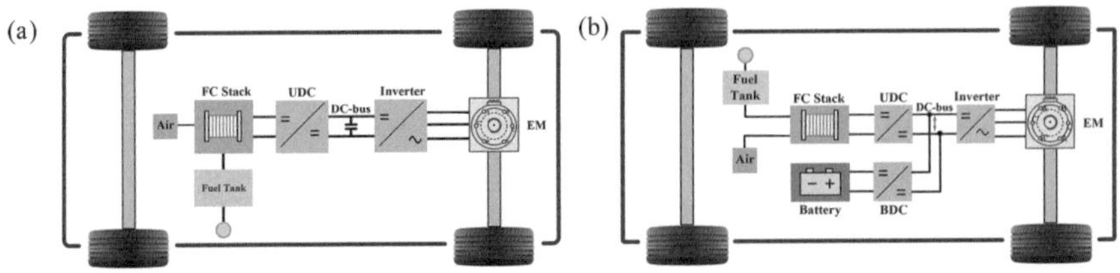

Fig. 4.28 The powertrain of FCEV and FCHEV [43]

to complete the initial startup of the vehicle to prevent the fuel cell from operating in the inefficient zone. Therefore, the battery is required to provide a high current to start the electric motor. After the vehicle is started, the fuel cell is activated to provide power to keep the electric motor running. At this point, the same power from the fuel cell recharges the battery. The FCHEV has four operating modes, including fuel cell mode, auxiliary mode, charging mode, and energy recovery mode.

(1) Fuel cell mode: The fuel cell directly supplies power to the motor, and all power is supplied by the fuel cell.
(2) Auxiliary mode: The fuel cell and the battery work together to output power to the motor.
(3) Charging mode: The power from the fuel cell is used to charge the battery, similar to the role of a range extender.
(4) Energy recovery mode: When decelerating, the motor converts the recovered energy into electricity to charge the battery.

Commercially available fuel cell vehicles include passenger cars, light commercial vehicles, buses, and trucks. Some automakers produce both fuel cell stacks and vehicles, such as General Motors, Toyota, and Honda. Some automakers, such as Ford, Mazda, DaimlerChrysler, Hyundai, Fiat, and Volkswagen, have completed the development of fuel cell vehicles by purchasing fuel cell stacks. Some commercially available fuel cell passenger cars are shown in Table 4.4. The main difference between different fuel cell vehicles produced by different automobile manufacturers lies in the arrangement of the fuel cell stack, hydrogen cylinder, and power cell in the vehicle, and the different arrangement positions also affect the shape of these components. The main arrangement has the following four categories: (a) hydrogen cylinders are placed longitudinally along the vehicle's central axis as well as horizontally at the rear of the vehicle, while the battery is rear-mounted and the fuel cell stack is front-mounted; (b) hydrogen cylinder is placed transversely at the rear, the battery is placed in the middle like a pure electric car, and the fuel cell stack is front-mounted; (c) hydrogen cylinder is placed transversely at the rear, the battery is rear-mounted and the fuel cell stack is middle-mounted; (d) hydrogen cylinder is placed transversely at the rear, the battery is rear-mounted, and the fuel cell stack is front-mounted. In general, in the latest generation, most manufacturers have adopted the rear-mounted power cell and front-mounted fuel cell stack (system), while hydrogen cylinders can be mounted both longitudinally and horizontally.

Table 4.4 The commercial FCEVs produced by different automobile manufacturers [43]

Appearance	Name	Year (interval)	Plug (Y/N)	Range (km)	Top speed (km/h)	FC power (kW)	Country
	Toyota Mirai FCEV	2015–Now	No	480	160	114	Japan
	Honda Clarity	2019–Now	Yes	590	178	103	Japan
	Mercedes-Benz GLC FCELL	2020–Now	Yes	478	160	100	Germany
	Hyundai Nexo	2020–Now	Yes	570	179	95	South Korea

A vehicle with a maximum loaded weight of more than 26,000 lbs is classified as a heavy-duty vehicle. Heavy-duty vehicles (HDV) contribute more to greenhouse gas emissions because they are more powerful. In America, medium and heavy-duty vehicles account for approximately 23% of the greenhouse gas emissions, and heavy-duty vehicles also account for approximately 40–60% of the NO_x and PM emissions. To reduce the emission pollution issues, alternative powertrain technologies for heavy-duty vehicles are investigated, such as battery-electric and hydrogen fuel cell electric heavy-duty vehicles. The driving range of a fuel cell HDV depends not only on the payload but also on the amount of hydrogen stored in the vehicle. A typical fuel cell HDV carrying two hydrogen tanks (40–60 kg of hydrogen at 350 bar) has a range of about 500–1000 miles. The typical hydrogen refueling rate for a heavy-duty truck is about 120 g/s, or a refueling time of about 16 min. It is typically categorized as end-of-life when the voltage drops 10% relative to the start-of-life voltage. The goal for fuel cell HDVs is to achieve a service life of 25,000 h, or 500,000 miles, or 12 years. The current average service life of light-duty vehicles is around 2000–3000 h, and the 5000 h goal has not yet been achieved. However, buses with more than 25,000 h of operation have been reported in the open literature.

Table 4.5 compares the cost and efficiency metrics of diesel HDVs, pure electric HDVs, and fuel cell HDVs [44]. It can be found that the cost of fuel cell HDV is still higher than that of diesel HDV, but it is lower than that of purely electric HDV. In addition, fuel cell HDV has a larger nominal driving range. If the price of fuel cells drops to around $100/kWh, the total cost of fuel cell HDVs can compete with diesel HDVs. With technological advances in fuel cell systems, this low price is likely to be realized within the next 10–20 years. Since fuel cell HDVs have the highest specific energy, fuel cell HDVs can drive similar mileage to low-capacity diesel, but weigh less.

Table 4.5 Comparison between three heavy-duty vehicle powertrains [44]

Criteria	Diesel	Battery	Hydrogen fuel cell
Tailpipe emissions	Yes	No	No
Total vehicle cost	119,000–134,000 USD	164,641–585,000 USD	135,503–249,900 USD
Well-to-tank efficiency	~ 86%	~ 55.3%	~ 76%
Tank-to-wheel efficiency	~ 23%	~ 68%	~ 45%
Fuel consumption	6.5 miles/gallon	0.5 miles/kWh	5.5–9.2 miles/kg H_2
Range	975–1950 miles	62–500 miles	660–1104 miles
Refueling time	6–12 min	2.85–20 h	16.67 min
Specific Energy	42.9 MJ/kg	0.432–0.792 MJ/kg	118 MJ/kg

Table 4.6 further compares the advantages and disadvantages of diesel HDVs, pure electric HDVs, and fuel cell HDVs [44]. It can be found that the fuel cell HDV has lower emissions and greater specific energy. In addition, fuel cell HDVs are more efficient than diesel HDVs and have shorter refueling times than battery HDVs. However, the disadvantages are that fuel cell HDVs are currently the most costly, have the most infrastructure requirements, and are experiencing the slowest technology development. As infrastructure increases, large-scale commercialization of fuel cell HDVs will be enhanced.

Table 4.6 Advantages and disadvantages of the three powertrains [44]

Technology	Advantages	Disadvantages
Diesel	Lowest vehicle cost No infrastructure investment required Long range and high payload Faster refueling time than BEVs Large market with widely available parts and vehicles	High greenhouse gas emissions Source of local air pollution (high tailpipe emissions) High refueling and maintenance costs Low energy efficiency
Battery	Reduces greenhouse gas emissions Reduces local air pollution (no tailpipe emissions) Lower refueling and maintenance costs than ICE vehicles Higher energy efficiency than ICE Less infrastructure investment is required for FCEV	Infrastructure investment is required Higher vehicle cost than diesel Long recharging time Limited range Limited cargo weight and size due to the large battery
Fuel cell	Reduces greenhouse gas emissions Reduces local air pollution (no tailpipe emissions) Higher energy efficiency than ICE Faster refueling time than BEVs High specific energy	High initial hydrogen fuel cost Heavy infrastructure development is required The highest vehicle cost compared to diesel or battery Slow FCEV development

Fig. 4.29 Applications of fuel cell buses [45]

Fig. 4.30 Schematics of a fuel cell bus [45]

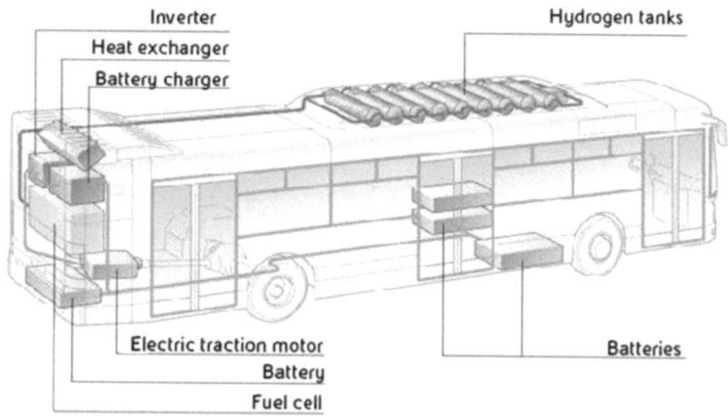

City buses are another scenario for the application of fuel cell systems, which have been successfully operated in Japan, Germany, and the UK, as shown in Fig. 4.29. The main benefits of buses over passenger cars are: (1) they are huge amount of hydrogen can be easily stored on board; (2) the driving routes are fixed, which is conducive to the arrangement of hydrogen refueling stations as well as the improvement of the utilization rate of hydrogen refueling stations. The schematics of a fuel cell bus are shown in Fig. 4.30.

4.5.2 Marine Applications

At present, the application of PEMFCs in vehicles has been perfected and taken shape, and the application of PEMFCs in vessels can partly draw on its experience. However, the operating conditions of vessels are very different from those of vehicles. For example, the power of marine fuel cells (about 500–1000 kW) is higher than that of vehicle fuel cells. In this case, marine fuel cells cannot simply stack multiple single fuel cells, and in-depth research is needed on the configuration between single cells, water and thermal management systems, explosion-proof design, and so on.

In addition, the vessels are in a humid marine environment, and it is necessary to focus on the impact of high humidity and salt spray marine environment on the performance of fuel cells. Furthermore, the multi-dimensional operation scenarios of vessels also pose greater challenges to the safety of fuel cell systems.

The application of fuel cells on vessels can be traced back to their application on submarines. This is because PEMFC as a submarine AIP (Air Independent Propulsion) power source has the following advantages: (1) high energy conversion efficiency; (2) low operating temperature, less heat radiation to the sea water; (3) the product is only water, without the need to discharge waste to the outside of the boat; (4) ultra-quiet operation, a very low acoustic signal characteristic. Therefore, the stealth effect of the submarine using PEMFC is excellent, and the survivability and combat effectiveness of the submarine have been greatly improved. In the early 1980s, German company Siemens began to investigate the PEMFC as a submarine's propulsion power. In 2003, Germany test-flown the world's first modernized AIP PEMFC-powered submarine—Class 212A U31 submarine. The submarine consists of 9 groups of fuel cell modules with a power of 34 kW respectively, and the total output power of the system is 306 kW, which can make the submarine dive 1250 nautical miles at a speed of 4.5 knots for 15–17 days, which is 5 times of the diesel-electric submarine's dive time. The submarine also carries two liquid oxygen storage tanks (14 tons each). The problem to be solved by fuel cells in submarines is the storage of hydrogen. Because hydrogen is flammable and explosive, there is a threat of explosion once the concentration of hydrogen is more than 4%. Therefore, the hydrogen content in the cabin space needs to be strictly controlled. Conventional storage of liquid hydrogen in tanks, although low-cost, is not suitable for submarines. Because when attacked by torpedoes, the liquid hydrogen stored in tanks is prone to secondary detonation under the shock vibration generated after the explosion, leading to more serious secondary accidents.

With the gradual development of fuel cell technology, fuel cells have been gradually emphasized in the field of civil ships, mainly focused on small ships. The ship emissions of diesel propulsion systems account for 2.6% of CO_2 emissions, 15% of NO_x emissions, and 13% of SO_x in total global emissions, which are hundreds of times higher than automobiles. Facing strict maritime emission regulations, FCs with low emissions and high energy efficiency are developing rapidly in the field of ships. The fuel cell ship's propulsion system mainly consists of an FC system, a DC/DC converter, an energy management strategy, a battery, and a propulsion motor. Likewise, the FC provides the main power required by the propulsion system, and the battery supplies the peak power and recovers the surplus and braking energy.

In 2003, MTU Friedrichshafen GmbH of Germany developed the fuel cell excursion boat "Sailing Boat", which is 12 m long. The ship is equipped with 4 sets of 1.5 kW hydrogen fuel cells and 9 sets of storage batteries, which can produce a maximum power output of 20 kW. The boat has a maximum range of 225 km at a sailing speed of 14.8 km/h. In 2008, the German company Alster-Touristik GmbH developed the world's first fuel-cell-powered passenger vessel, FCS Alsterwasser, which can carry more than 100 passengers. The power system reaches the displacement of 72 tons and a top speed of 8 knots with two 48 kW PEMFC systems and a lead-gel battery system. 12 hydrogen fuel storage tanks at 35 MPa were used, which can carry 50 kg of hydrogen fuel at a time, and can run continuously for 2 to 3 days. The successful operation of the vessel proves the reliability of hydrogen fuel cells. In 2010, the French classification society (BV) developed the fuel cell passenger ship "NemoH_2" for navigation in the Amsterdam Canal, which adopts a hybrid powertrain of 70 kW fuel cells and 55 kW lead-acid batteries. In 2016, South Korea developed a 20-m-long fuel cell excursion ship equipped with a 50 kW fuel cell. With an output power of 86 kW, the ship has a speed of 7.5 knots (13.9 km/h) and a maximum range of 300 km. In 2021, China's first fuel-cell yacht, the "Li Hu", led by the Dalian Maritime University, passed a test voyage, which utilized a hybrid power system consisting of a 70 kW fuel cell and a lithium-ion battery. In 2022, All American Marine (AAM) and shipowner Switch Maritime officially announced the launch of the Sea Change, the world's first 100% hydrogen fuel cell-powered commercial vessel. The 70-foot-long, 75-passenger, zero-emission, hydrogen fuel cell-powered electric ferry completed its first hydrogen-fueled trial run in late 2021. The vessel utilizes a 360 kW fuel cell and a 246 kg hexagonal hydrogen storage tank. In 2023, the inland PEMFC-powered vessel "Sanxia Qingzhou I" was launched, which was developed and constructed by China Three Gorges Group in Zhongshan, Guangdong Province. The ship is equipped with a 500 kW PEMFC and lithium battery power system, with an overall length of 49.9 m, a depth of 3.2 m, a design draft of 1.75 m, a maximum speed of 28 km/h, and a range of up to 200 km.

4.5.3 Aviation Applications

Fuel cell application scenarios in aviation include manned aircraft and unmanned aerial vehicles (UAVs). In the application of manned aircraft, the major research institutions and projects include Boeing, Airbus, German Aerospace Center, and the ENFICA-FC consortium.

In 2007, the Boeing Company first published research results on a manned airplane powered by PEMFC. They developed a small manned two-seater prototype airplane powered primarily by a PEMFC stack, with auxiliary power provided by a lithium-ion battery [46]. It was the first time in aviation history that a straight level manned flight was made with fuel cells as the only power source. During the takeoff and climb phases, the aircraft has a power requirement of 36 kW, which is supplied by both fuel cells and lithium-ion batteries. During the level flight phase, the aircraft has a power requirement of 15 kW, supplied primarily by the fuel cell. The benefit of this hybrid power source design is that it allows for the maximization of energy utilization throughout the system, allowing for the minimization of the size and weight of the system. The PEMFC stack is fueled by compressed hydrogen and pressurized air as the oxidizer. The output voltage of the fuel cell stack is 200 V, and the maximum output power of the system is 20 kW net or 24 kW gross.

In 2012, the German Aerospace Center developed a fuel cell system-powered aircraft, Antares DLR-H2 [47], which is equipped with two payload nacelles. The pod consists of a hydrogen storage tank pod (installed on the right-hand side) and a fuel cell pod (installed on the left-hand side). The tank pod contains a 350 bar tank system, which is capable of storing up to 4.9 kg of compressed gaseous hydrogen. In September 2012 long distance flight testing was performed to study the performance of the system and to identify potential for future optimization. Within this flight campaign, six single flights and four cross-country flights have been accomplished.

In 2013, the European Commission funded the ENFICA-FC project developed a light sport aircraft RAPID200 powered by a fuel cell for inner city flight purposes [48], as shown in Fig. 4.31. A feasibility study was carried out to provide a preliminary definition of new forms of Inter-City aircraft power systems that could be provided by fuel cell technologies. A hybrid battery/fuel cell system was chosen, where the fuel cell system's power was designed to be 20 kW. The hydrogen storage system consists of two Dynatek L026 tanks with a capacity of 26 L and a working pressure of 350 bar. A maximum speed of 135 km/h was recorded during runs 6–7.

Compared with manned aviation, PEMFC has been more widely used in the field of UAVs. Fuel cell-powered UAVs have the characteristics of long endurance, zero pollution, and low noise. In addition, the thermal infrared characteristic feature of fuel cell UAVs is not obvious, which is advantageous in long-endurance reconnaissance missions. There are four common types of UAVs: single-rotor UAVs, multi-rotor UAVs, fixed-wing UAVs, and hybrid fixed-wing vertical takeoff and landing (VTOL) UAVs. A typical UAV mission consists of five main work phases: (1) the takeoff phase, in which the UAV is accelerated from the takeoff site; (2) the climb phase, in which the UAV's high power output arrives at a specified flight altitude; (3) the cruise phase, in which the UAV performs uniform fixed-altitude flights; (4) the descent phase, in which the UAV descends to a lower altitude; and (5) the landing phase, in which the UAV comes to land at a standstill. Among them, the power demand of the UAV is larger in the takeoff phase and climb phase, but the flight time is shorter. The power demand in the UAV cruise phase is smaller, but the flight time is longer, and the energy used in the cruise process accounts for more than 90% of the total energy of the flight process. Based on the above reasons, fuel cell UAVs usually use fuel cells as the main energy source and lithium batteries as the auxiliary power source, constituting a hybrid power system. Thus, a PEMFC-powered UAV consists

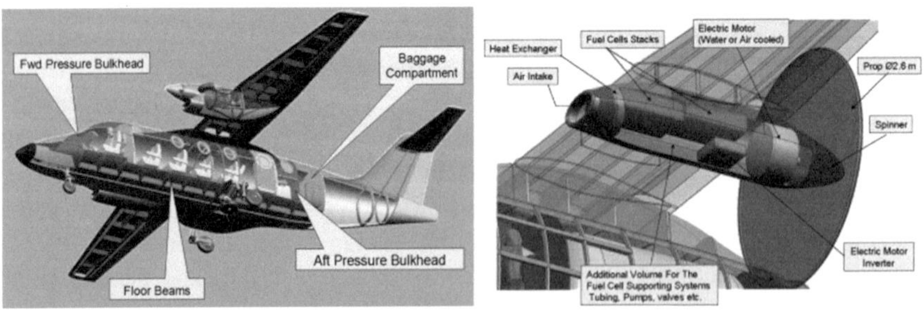

Fig. 4.31 All-electric intercity airplane powered by a fuel cell fueled by hydrogen [48]

of conventional electric motors and propellers, a lithium-ion battery, and a fuel cell stack. Some fuel cell UAVs are summarized in Table 4.7.

The first attempt to develop a fuel cell-based UAV power system was made in the U.S. In 2003, Boeing and the U.S. Department of Defense jointly developed a fuel cell-based UAV propulsion system and completed an operational demonstration. In 2005, Georgia Tech designed a powertrain architecture suitable for all-electric aircraft and proposed a fuel cell powertrain design, and conducted a flight demonstration of an unmanned aerial vehicle with a 500W fuel cell and customized water cooling in 2007. The Spider Lion UAV developed by the U.S. Naval Research Laboratory (NRL) completed a 3-h, 19-min flight using a 95 W proton exchange membrane fuel cell. In 2009, the U.S. NRL developed the Ion Tiger, an unmanned aerial vehicle equipped with a 550 W fuel cell, weighing 15.9 kg, with a wingspan of 5.2m, and realized a 26-h flight demonstration with it. And in 2013, an improved version of the Ion Tiger achieved a continuous flight time of up to 48 h. The Korea Aerospace Research Institute (KARI) has developed a small drone that combines fuel cell and solar technology. In 2012, China's Tongji University developed the country's first pure fuel cell drone, the Feiyue I, which has a range of two hours. The Feiyue I uses a 1 kW proton exchange membrane fuel cell to power the UAV's cruise, with a payload of 1 kg. Also in 2012, the Thunderbird, a fuel cell UAV developed by China's Liaoning General Aviation Research Institute, made its maiden flight successfully, using a 10 kW proton exchange membrane fuel cell.

Take a fixed-wing UAV for example, Boeing Research & Technology Europe has designed, developed, and subsequently bench and flight tested, in a wide range of different operational conditions, an electric Unmanned Air Vehicle (UAV) powered by a hybrid energy source, as shown in Fig. 4.32. The energy source features a 200 We Polymer Electrolyte Membrane (PEM) fuel cell system fed by a chemical hydride hydrogen generator that produces highly pure hydrogen at the fuel cell operating

Table 4.7 Fuel cell-powered UAVs [48]

Organization (date)	Reactant storage type	Endurance (est.)
AeroVironment (2003)	H_2 sodium borohydride	0.2 h
AeroVironment (2005)	H_2 cryogenic	24 h
FH Wiesbaden (2005)	H_2 gaseous	90 s
Naval Research Lab (2006)	H_2 gaseous	3.3 h
Georgia Inst. of Tech. (2006)	H_2 gaseous	0.75 h
CSU Los Angeles (2006)	H_2 gaseous	0.75 h
DLR/HyFish (2006)	H_2 gaseous	0.25 h
CSULA/OSU (2007)	H_2 gaseous	12 h
KAIST (2007)	H_2 sodium borohydride	10 h
AeroVironment (2007)	H_2 sodium borohydride	9 h
Naval Research Laboratory—Ion Tiger UAV (2009)	H_2 gaseous	26 h
Naval Research Laboratory—Ion Tiger UAV (2013)	H_2 cryogenic	48 h

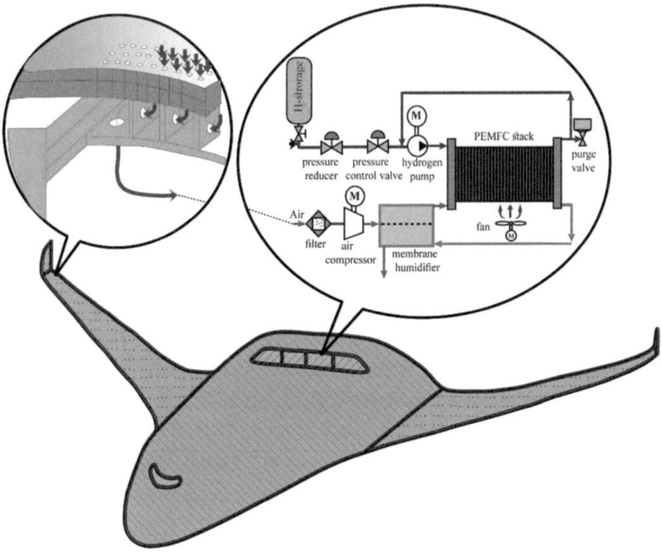

Fig. 4.32 Unmanned air vehicle (UAV) [50]

pressure from the controlled hydrolysis of Sodium Borohydride ($NaBH_4$), resulting in 900 Wh of energy from 1 L of chemical solution. Equipped also with high-specific-energy Lithium Polymer batteries, this fuel cell-powered UAV can achieve flight durations close to 4 h [49].

4.5.4 *Train Applications*

Currently, the power demand for hydrogen rail trains is up to 750 kW, which is difficult to achieve only by increasing the number of single cells. Therefore, corresponding energy storage systems are needed to provide sufficient energy for the traction motors as an auxiliary power source when the train is in operation. Due to the slow dynamic response of the fuel cell, it is not possible to do emergency stops and fast starts, so it is necessary to respond faster to the lithium battery or supercapacitor as an auxiliary power source. Therefore, hydrogen fuel cell trains usually use a hybrid power system combining hydrogen fuel cells, power batteries and supercapacitors, where the hydrogen fuel cells are directly connected to the bus to provide continuous and stable power during the driving process, and the excess energy is stored in the power batteries or supercapacitors and other energy storage systems.

With the increasing attention to environmental issues worldwide, several countries have established the strategic goal of realizing zero-emission railroads in the future. And hydrogen energy, as an important alternative energy source, has received extensive attention in the railroad field. Due to the obvious advantages of fuel cells in heavy-duty applications, hydrogen-powered trains have great potential for development worldwide. As the cost of the whole industrial chain of hydrogen energy decreases rapidly, it will play an increasingly important role in the railroad sector. Since 2006, when Japan developed the world's first E995-series hybrid train equipped with hydrogen fuel cells and storage batteries, countries such as France, Germany, and the United Kingdom have actively carried out research and development and testing of hydrogen fuel cell trains.

In the United States, in 2019, Stadler US and the San Bernardino County Transportation Authority (SBCTA) in California introduced zero-emission passenger rail technology to the United States. With this, the first commercial hydrogen-powered passenger train (Flirt H_2) program in the United States was launched. As part of the Redlands Passenger Rail Project, the Flirt H_2, with 108 passenger seats and ample standing room, will have a top speed of 130 km/h and will connect Southern California Railway's Redlands and San Bernardino stations over a distance of nine miles, with service expanding to approximately 60 miles in the future. Delivery is expected in 2023, with operations in 2024.

In Germany, in 2016, European rail vehicle manufacturer Alstom demonstrated the iLint fuel cell train in Berlin. In 2017, the hydrogen fuel cell train was successfully tested in a walk-around test on the Salzgitter test line in Germany at a speed of 80 km/h. The iLint fuel cell train is based on the Coradia Lint 54 internal combustion train set and retains the interior of the prototype. The fuel cell and hydrogen storage tanks are mounted on the roof, while the lithium-ion battery and power conversion unit are mounted under the floor. The front bogie of each vehicle is a power bogie. The vehicle has a top speed of 140 km/h, an electric motor with an output of 320 kW, and a continuous travel distance of 600–800 km. And the fuel cell has an output of 200 kW. 24 hydrogen tanks with a storage pressure of 35 MPa and a total of 94 kg of hydrogen are mounted on each vehicle. iLint Fuel Cell Train was tested in the Czech Republic on the Velim Loop Line at a top speed of 140 km/h in the Czech Republic before entering commercial operation in Germany in 2018. By 2021, the iLint train had traveled more than 200,000 km, and orders from Germany, France, and Italy had totaled 59 trains.

In the United Kingdom, the University of Birmingham's Center for Railway Research and Education (BCRRE) and Porterbrook Railway Passenger Transport in the United Kingdom developed the first hydrogen-powered train in the United Kingdom, the HydroFLEX, over a period of 2 years. The train is a conversion of a Type 319 internal combustion train, with an integrated hydrogen drive system consisting of hydrogen tanks, fuel cells, lithium-ion batteries, electric motors, and a control system. Four high-pressure hydrogen tanks (with a storage pressure of 350 bar) hold 44 pounds of liquid hydrogen, which is regulated by a pressure-drop regulator and held at a constant pressure of 8.5 bar. The fuel cell system can generate up to 100 kW of electricity. A single hydrogen refueling can provide a range of up to 800 km with existing hydrogen equipment configurations, and a maximum operating speed of 120 km/h. The train can seamlessly switch to the UK's rail grid, drawing power from either the 750 V DC third rail or the 25 kV overhead power grid. This gives the train the flexibility to run on most routes. In May 2019, Vivarail UK and

ArcolaEnergy announced a partnership agreement to develop a hydrogen fuel cell train set. The prototype modular hydrogen-fueled train will be a modification of the type-230 train, consisting of 2 vehicles in a formation: one with 2 battery modules; the other with hydrogen fuel cells and hydrogen storage tanks, with the equipment mounted under the floor of the train. The concept train will be used to demonstrate the capability and test the performance of the hydrogen drive system, while the final hydrogen train will consist of four vehicles, including two battery-controlled vehicles and two intermediate vehicles equipped with fuel cells and hydrogen storage tanks.

In China, in 2013, Southwest Jiaotong University rolled out China's first fuel cell electric vehicle "Blue Sky", which uses Ballard's HD6-150K PEMFC stack (150 kW) and two 120 kW permanent magnet synchronous motors to form a hybrid power system, with a range of 24 h at light load and a maximum speed of 65 km/h. In 2015, the world's first hydrogen-energy modern train, developed by CRRC Qingdao Sifang Rolling Stock Co., Ltd., officially rolled off the production line, applying the world's leading technology of permanent magnet synchronous motor direct drive system and articulated bogie. The power system consists of a Ballard FCveloCity-HD6 V2 PEMFC module, supercapacitor, and lithium-titanium-acid batteries, where the PEMFC rated power is 150 kW. There is a carbon fiber high-pressure hydrogen storage tank on top of the body, which can withstand a high pressure of 100 MPa, and can be filled with 11.7 kg of hydrogen (35 MPa) in 3–5 min. The whole vehicle is designed using modularization, with a maximum speed of 80 km/h. In 2016, the world's first commercially available fuel-cell-super-capacitor hybrid power tram, which was developed by Southwest Jiaotong University and CSR Tangshan, rolled off the assembly line, with a maximum speed of 70 km/h. The hydrogen refueling takes only 15 min, and can travel more than 40 km.

In Japan, in 2019, the East Japan Railway (JR) announced that it would invest 4 billion yen in the development of an FV-E991 series train equipped with fuel cells, lithium-ion batteries, and high-pressure hydrogen tanks. The train consists of a bullet train and a trailer, with the hydrogen technology provided by Toyota. The train travels at a maximum speed of 100 km/h on a 1067 mm gauge track. JR began testing the train in 2021 and plans to use the technology for commercial applications by 2024. The power conversion unit of the FV-E991 series train consists of a VVVF inverter unit (which converts the electrical energy generated by the fuel cell unit into three-phase alternating current to drive the traction motor) and an auxiliary power supply unit (which supplies power to the air conditioner, etc.). The hydrogen storage tank is a common type of composite container with a plastic liner used in automobiles. This type of hydrogen storage tank has a capacity of 51 L/tank, with 5 tanks as 1 unit and a total of 4 units mounted on the roof of the vehicle. During traction, hydrogen is supplied to the fuel cell stack from the hydrogen storage unit, and electricity is generated by the electrochemical reaction between hydrogen and oxygen. The generated electricity is boosted by a booster chopper and supplied to the VVVF inverter unit. The main circuit battery, which is another source of energy, also supplies power to the VVVF inverter, which converts the supplied power into three-phase alternating current to drive the traction motor. During braking, the regenerative power generated by the traction motor charges the main circuit battery. The auxiliary power supply unit is also powered by the fuel cell and the main circuit battery.

4.5.5 Combined Heat and Power Systems

Fuel cells will also generate heat during the power generation process, so fuel cells can be applied in a combined heat and power (CHP) system, thus improving the energy utilization efficiency. By combining electrical and thermal efficiencies, fuel cell-based CHP systems can achieve a total efficiency of nearly 90%, which is higher than conventional CHP systems. In addition, since fuel cells do not produce greenhouse gases or harmful pollutants during operation, this gives them a significant advantage over conventional CHP systems. Fuel cell CHP systems can be used in industrial or residential buildings. Since PEMFCs can be categorized into low-temperature PEMFCs (LT-PEMFCs) and high-temperature PEMFCs (HT-PEMFCs) according to their operating temperatures, there are 2 types of PEMFC-based CHP systems, as shown in Fig. 4.33. Compared with the low-temperature PEMFC-CHP system, the high-temperature PEMFC-CHP system has a simpler fuel pretreatment structure. However, the high-temperature PEMFC cannot realize a cold start like the low-temperature proton exchange membrane fuel cell. Meanwhile, the HT-PEMFC system generates a large amount of low-grade energy, making it suitable for adsorption-cooling-based combined cooling, heating, and power generation (CCHP) systems. CHP can be categorized based on the amount of power into small-CHP and micro-CHP. The small-CHP refers to CCHP systems that generate less than 100 kW of power, while micro-CHP refers to systems that generate less than 15 kW of power. Small and micro systems are suitable for different residential, tertiary, and industrial applications, such as residential buildings, hospitals, supermarkets, and sports centers, where the heat demand is closely related to the electricity demand of the users. Figure 4.34 shows a typical domestic CHP system configuration. The fuel cell generates electricity that can be used directly in the home or can be fed into or output from the grid. The heat generated during power generation can be used for purposes such as space heating or domestic hot water. Compared to the PEMFC system, a fuel treatment system is added to the CHP system. The fuel treatment system is mainly used for fuel pre-treatment to convert the fuel into usable hydrogen. Currently, hydrogen in CHP systems is mainly derived from local methane reforming.

The key components of a fuel cell CHP system include:

(1) Fuel cell pack: Hydrogen and oxygen combine to produce electricity, heat, and water.
(2) Fuel processor: Converts hydrocarbon fuels such as natural gas into hydrogen and carbon dioxide.
(3) Inverters, Grid Connectors, and Power Electronics: Convert the DC output of the fuel cell bank into AC power to meet the building's energy needs or to feed back into the grid.

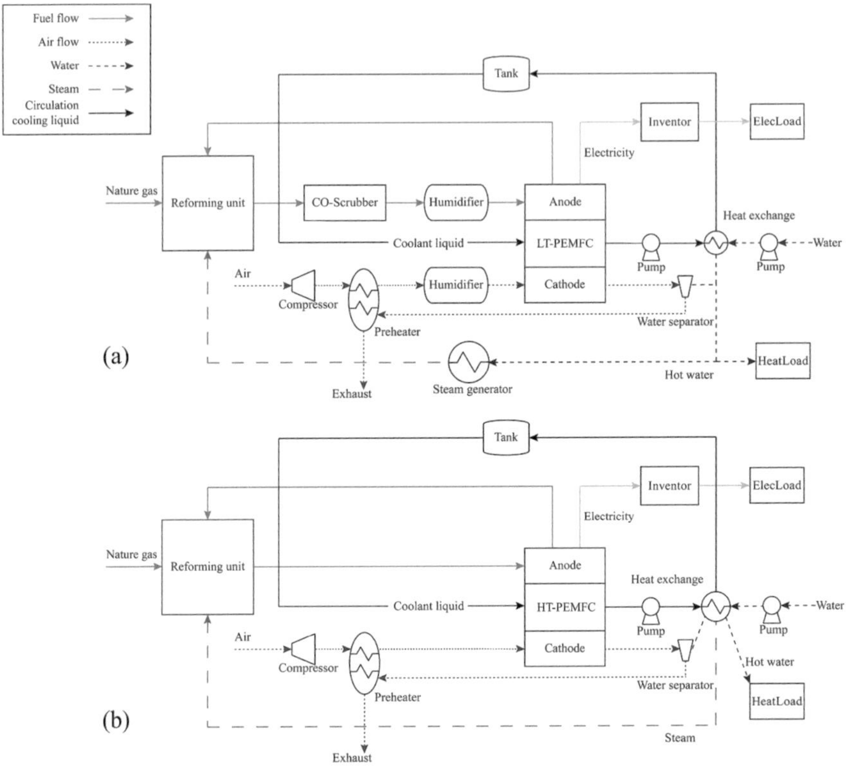

Fig. 4.33 Typical layout of **a** a LT-PEMFC-based CHP system and **b** a HT-PEMFC-based CHP system [51]

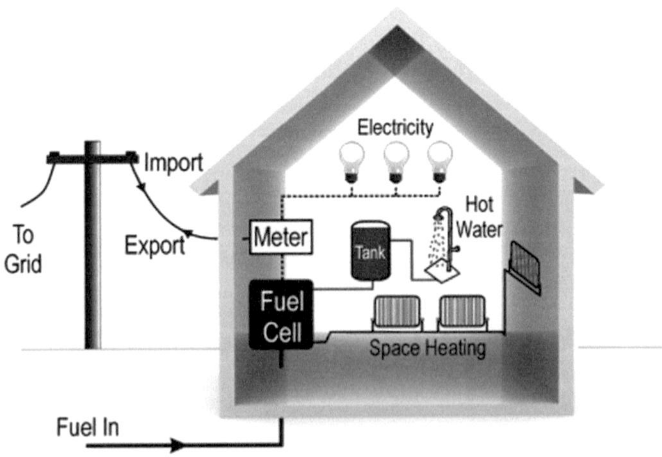

Fig. 4.34 Fuel cell micro CHP in a domestic home [52]

(4) Heat recovery system: Used to recover heat from the fuel cell to make it a CHP system.

In addition to the aforementioned key components, the CHP system requires some auxiliary components, including:

(1) Boiler: To provide the peak heat load in conjunction with the fuel cell.
(2) Hot water tank: To store the heat output from the fuel cell.
(3) Smart meters: To measure and record energy production and consumption.
(4) Internet connection: To facilitate remote monitoring and data collection.

Japan is a world leader in the development and deployment of fuel cell-based CHP systems. In 2009, Panasonic launched the world's first home fuel cell cogeneration system, ENE-FARM, which generates electricity using hydrogen extracted from natural gas. In 2013, Japan's Tokyo Gas and Panasonic developed the next-generation ENE-FARM residential system. The system is expected to have a system efficiency of 95% (based on a low calorific value), while the fuel cell stack has a lifetime of 60,000 h and a minimum power output of 0.2 kW. Panasonic plans to have a production capacity of at least 15,000 units per year, while Tokyo Gas is aiming to sell 12,000 units per year, with a total volume of 300,000 units to be sold by 2020. Japan aims to install ENE-FARM systems in 10% of households by 2030. In 2021, Panasonic launched the H2 KIBOU, a 5 kW pure hydrogen fuel cell generator.

In Europe, German manufacturer Bosch Thermotechnik installed 70 CHP systems in homes in Germany, the UK, the Netherlands and France. The system is supplied by Japan's Aisin Seiki (a subsidiary of the Toyota Group), which has a net electrical efficiency of 45% and rated electrical and thermal outputs of 0.7 kW and 0.7 kW, respectively. In 2014, Viessmann (a manufacturer of heating systems) launched a mass-production program of the Vitovalor 300-P system for the European market. The system consists of a 0.75 kW PEMFC stack unit (developed by Panasonic), a peak load boiler, and a hot water tank with a heat output of up to 19 kW. In England, 2G Energy installed the UK's first 100% hydrogen-based CHP system at Kirkwall Airport, which works in tandem with the airport's existing heating system to meet the heat and power needs of the main airport buildings. In America, Duke Energy operates a cogeneration system that provides steam and 2.8 MW of electricity to Clemson University, as well as 15 MW of electricity to the public grid. The system can operate in island mode in the event of a grid outage.

Table 4.8 summarizes some typical fuel cell CHP products [53].

4.5.6 Uninterruptible Power Supply Systems

Data centers, banks, hospitals, and other institutions require backup power systems. PEMFC systems can be used as an uninterruptible power supply, but commercial applications are less common at this time. The advantage of PEMFCs as an uninterruptible power supply is that they can provide more reliable and stable operation.

Table 4.8 Fuel cell micro CHP products [53]

Manufacturer (country of origin)	Electrical capacity (kW)	Electrical efficiency (%)	Thermal output (kW)	Auxiliary heater included	Commercial availability	Partners/projects	Comments
Baxi (UK)	1	32	1.7	20 kW	2015	Ballard/Callux	Requires external heater
Toshiba (JPN)	0.7	35	1	–	Japan—2009 EU—2015	EneFarm	80,000 operational hours expected
Viessmann (GER)	0.75	37	1.3	19 kW	Germany—2014 Europe—2015	Panasonic	Uses Japanese stack
Elcore (GER)	0.3	33	0.6	–	Enefield—2013	Enefield	Low electrical/heat output means fuel cell runs continuously
Dantherm Power (DEN)	1.7, 2.5 and 5	–	–		Danish field trials	Ballard	Only short duration tests thus far
Panasonic (JPN)	0.7	40	0.9	Yes	Japan—2011 Europe—2014	EneFarm	European R&D stared in 2012
JX Eneos (JPN)	0.7	40	–	–	Japan–2011	EneFarm	Now pursuing SOFC technology
Vaillant (GER)	5	–	25–50	–	–	PlugPower	Aimed at multi-family homes
Plug Power (USA)	0.3–3	30	1.65	Yes (7–25 kW)	–	–	Operates on natural gas

UK United Kingdom, *JPN* Japan, *GER* Germany, *DEN* Denmark, *AUS* Australia, *SUI* Switzerland, *USA* United States of America

In addition, PEMFCs can withstand harsher environments than batteries and do not require the addition of extreme cooling equipment. Some applications are listed as follows. In 2019, Denmark-based Ballard Power Systems Europe signed Framework Agreements for providing 30 PEMFC backup power systems with 5 kW FCgen®-H2PM units with Danish fibre-optic broadband network operators Eniig and Fibia [49]. In 2020, Proton Motor Power Systems signed a framework agreement with APEX Energy Teterow GmbH in Germany to supply a further 10 container-integrated 100 kW fuel cell systems. PM has also received an order from real estate company Vonovia to provide a 36 kW system for a housing project in Bochum [54]. In 2020, AdKor and SFC company provided 2.9 kW PEMFC stacks for German radio tower sites as the backup power [55].

4.6 Summary and Outlook

Proton exchange membrane fuel cells are electrochemical devices that directly convert hydrogen into electricity. They are compact, environmentally friendly, and safe. Consequently, they are expected to play a significant role in the promotion of hydrogen use on a large scale. The key components of PEMFC include proton exchange membrane, catalyst layer, gas diffusion layer, and bipolar plate, which should have a lifetime of over 40,000 h [56]. The development of MW-scale PEMFC stacks is required for heavy-duty transport decarbonization, hence, the development of flow fields over 1000 cm is imminent [57]. The PEMFC system includes a fuel cell stack subsystem, a hydrogen supply subsystem, an oxygen supply subsystem, a thermal management subsystem, and a control subsystem. PEMFCs have been preliminarily applied in vehicle, marine, aviation, and stationary scenarios. To promote large-scale commercialization of fuel cells, it is necessary to continually develop low-cost and stable alternative materials, including reducing the amount of catalyst, developing better proton exchange membranes, and designing more efficient flow fields [58]. The two-phase fluid transport in a multi-cell stack increases the complexity of reactant diffusion to the porous electrodes compared to a single-cell stack, as higher current generation naturally increases water and heat production. Ensuring optimal hydration and uniform heat distribution is critical to maintain the durability and overall electrochemical performance of the MEA [59]. On the other hand, the selling price and the construction of infrastructure facilities related to hydrogen refueling stations are the most critical factors affecting and determining the commercialization and popularization of PEMFCs [60]. With the progress of science and technology, the government's attention and financial investment, the problems of cell safety, lifetime, hydrogen fuel storage, high system cost, and the lag in the construction of hydrogen refueling station supporting facilities will be gradually resolved.

References

1. Usman M, Balsalobre-lorente D. Environmental concern in the era of industrialization: Can financial development, renewable energy and natural resources alleviate some load? [J]. Energy Policy, 2022, 162: 112780.
2. Raihan A, Muhtasim D A, Farhana S, et al. Nexus between carbon emissions, economic growth, renewable energy use, urbanization, industrialization, technological innovation, and forest area towards achieving environmental sustainability in Bangladesh [J]. Energy and Climate Change, 2022, 3: 100080.
3. Achakulwisut P, Erickson P, Guivarch C, et al. Global fossil fuel reduction pathways under different climate mitigation strategies and ambitions [J]. Nature Communications, 2023, 14(1): 5425.
4. Zhang G, Kandlikar SG. A critical review of cooling techniques in proton exchange membrane fuel cell stacks. Int J Hydrogen Energy 2012;37:2412–29. https://doi.org/10.1016/j.ijhydene.2011.11.010.
5. Andújar JM, Segura F. Fuel cells: History and updating. A walk along two centuries. Renew Sustain Energy Rev 2009;13:2309–22. https://doi.org/10.1016/j.rser.2009.03.015.
6. Chaurasia PBL, Ando Y, Tanaka T. Regenerative fuel cell with chemical reactions. Energy Convers Manag 2003;44:611–28. https://doi.org/10.1016/S0196-8904(02)00066-3.
7. Grove WR. LXXII. On a gaseous voltaic battery. London, Edinburgh, Dublin Philos Mag J Sci 1842;21:417–20. https://doi.org/10.1080/14786444208621600.
8. Appleby AJ. From Sir William Grove to today: fuel cells and the future. J Power Sources 1990;29:3–11. https://doi.org/10.1016/0378-7753(90)80002-U.
9. Barbir F. PEM Fuel Cells. 2013. https://doi.org/10.1016/C2011-0-06706-6.
10. Hu X, Yang B, Ke S, et al. Review and Perspectives of Carbon-Supported Platinum-Based Catalysts for Proton Exchange Membrane Fuel Cells [J]. Energy & Fuels, 2023, 37(16): 11532–66.
11. Choi J S, Sohn J Y, Shin J. A comparative study on EB-radiation deterioration of nafion membrane in water and isopropanol solvents [J]. Energies, 2015, 8(6): 5370–5380.
12. Kraytsberg A, Ein-eli Y. Review of Advanced Materials for Proton Exchange Membrane Fuel Cells [J]. Energy & Fuels, 2014, 28(12): 7303–30.
13. Peron J, Shi Z, Holdcroft S. Hydrocarbon proton conducting polymers for fuel cell catalyst layers [J]. Energy & Environmental Science, 2011, 4(5): 1575–91.
14. Lin R, Chen L, Zheng T, et al. Interfacial water management of gradient microporous layer for self-humidifying proton exchange membrane fuel cells [J]. International Journal of Heat and Mass Transfer, 2021, 175: 121340.
15. Chen Q, Niu Z, Li H, Jiao K, Wang Y. Recent progress of gas diffusion layer in proton exchange membrane fuel cell: Two-phase flow and material properties. Int J Hydrogen Energy 2021;46:8640–71. https://doi.org/10.1016/j.ijhydene.2020.12.076.
16. Gong X, Ruan M, Song P, et al. A cost-effective and highly efficient dissymmetric membrane electrode assembly designed for fuel cells [J]. Journal of Power Sources, 2021, 489.
17. Qiu D, Peng L, Yi P, Lai X, Lehnert W. Flow channel design for metallic bipolar plates in proton exchange membrane fuel cells: Experiments. Energy Convers Manag 2018;174:814–23. https://doi.org/10.1016/j.enconman.2018.08.070.
18. Liu Q, Lan F, Zeng C, Chen J, Wang J. A review of proton exchange membrane fuel cell's bipolar plate design and fabrication process. J Power Sources 2022;538:231543. https://doi.org/10.1016/j.jpowsour.2022.231543.
19. Hu K, Zhao P, Wang S, Wang Y. Three-dimensional multiphase simulation of a partially narrowed flow field configuration for a high-performance polymer electrolyte membrane fuel cell. Appl Therm Eng 2023;223:119986. https://doi.org/10.1016/j.applthermaleng.2023.119986.
20. Liao Z, Wei L, Dafalla AM, Guo J, Jiang F. Analysis of the impact of flow field arrangement on the performance of PEMFC with zigzag-shaped channels. Int J Heat Mass Transf 2021;181:121900. https://doi.org/10.1016/j.ijheatmasstransfer.2021.121900.

21. Dang DK, Zhou B. Liquid water transport in PEMFC cathode with symmetrical biomimetic flow field design based on Murray's law. Int J Hydrogen Energy 2021;46:21059–74. https://doi.org/10.1016/j.ijhydene.2021.03.161.
22. Li X, Sabir I. Review of bipolar plates in PEM fuel cells: Flow-field designs. Int J Hydrogen Energy 2005;30:359–71. https://doi.org/10.1016/j.ijhydene.2004.09.019.
23. Nonobe Y. Development of the fuel cell vehicle mirai. IEEJ Trans Electr Electron Eng 2017;12:5–9. https://doi.org/10.1002/tee.22328.
24. Chen X, Yang C, Sun Y, Liu Q, Wan Z, Kong X, et al. Water management and structure optimization study of nickel metal foam as flow distributors in proton exchange membrane fuel cell. Appl Energy 2022;309. https://doi.org/10.1016/j.apenergy.2021.118448.
25. Kim J, Cunningham N. Development of porous carbon foam polymer electrolyte membrane fuel cell. J Power Sources 2010;195:2291–300. https://doi.org/10.1016/j.jpowsour.2009.10.053.
26. Wang Y, Pang Y, Xu H, Martinez A, Chen KS. PEM Fuel cell and electrolysis cell technologies and hydrogen infrastructure development - a review. Energy Environ Sci 2022;15:2288–328. https://doi.org/10.1039/d2ee00790h.
27. Bao Z, Niu Z, Jiao K. Gas distribution and droplet removal of metal foam flow field for proton exchange membrane fuel cells. Appl Energy 2020;280. https://doi.org/10.1016/j.apenergy.2020.116011.
28. Aminudin MA, Kamarudin SK, Lim BH, Majilan EH, Masdar MS, Shaari N. An overview: Current progress on hydrogen fuel cell vehicles. Int J Hydrogen Energy 2023;48:4371–88. https://doi.org/10.1016/j.ijhydene.2022.10.156.
29. Nicholas N, Shaffer B. All-Metal Scroll Vacuum Pumps for Tritium Processing Systems [J]. Fusion Science and Technology, 2020, 76(3): 366–72.
30. Shen KY, Park S, Kim YB. Hydrogen utilization enhancement of proton exchange membrane fuel cell with anode recirculation system through a purge strategy. Int J Hydrogen Energy 2020;45:16773–86. https://doi.org/10.1016/j.ijhydene.2020.04.147.
31. Wu Y, Bao H, Fu J, Wang X, Liu J. Review of recent developments in fuel cell centrifugal air compressor: Comprehensive performance and testing techniques. Int J Hydrogen Energy 2023. https://doi.org/10.1016/j.ijhydene.2023.04.262.
32. Li Y, Pei P, Ma Z, Ren P, Huang H. Analysis of air compression, progress of compressor and control for optimal energy efficiency in proton exchange membrane fuel cell. Renew Sustain Energy Rev 2020;133. https://doi.org/10.1016/j.rser.2020.110304.
33. Chang Y, Qin Y, Yin Y, Zhang J, Li X. Humidification strategy for polymer electrolyte membrane fuel cells – A review. Appl Energy 2018;230:643–62. https://doi.org/10.1016/j.apenergy.2018.08.125.
34. Xingsheng L, Shihu M, Juntao Z, Hong Z, Weilin Z, Yangjun Z. Comparative study of membrane humidifier and enthalpy wheel humidifier for large power fuel cell system. J Fuel Cell Sci Technol 2009;6:0145011–3. https://doi.org/10.1115/1.2971175.
35. Wu D, Peng C, Yin C, et al. Review of System Integration and Control of Proton Exchange Membrane Fuel Cells [J]. Electrochemical Energy Reviews, 2020, 3(3): 466–505.
36. Yu Y, Chen M, Zaman S, Xing S, Wang M, Wang H. Thermal management system for liquid-cooling PEMFC stack: From primary configuration to system control strategy. ETransportation 2022;12:100165. https://doi.org/10.1016/j.etran.2022.100165.
37. Tanaka S, Nagumo K, Yamamoto M, Chiba H, Yoshida K, Okano R. Fuel cell system for Honda CLARITY fuel cell. ETransportation 2020;3:100046. https://doi.org/10.1016/j.etran.2020.100046.
38. Tanaka S, Nagumo K, Yamamoto M, et al. Fuel cell system for Honda CLARITY fuel cell [J]. eTransportation, 2020, 3: 100046.
39. Rabbani A, Rokni M. Effect of nitrogen crossover on purging strategy in PEM fuel cell systems [J]. Applied Energy, 2013, 111: 1061–70.
40. Zhao D, Zheng Q, Gao F, et al. Disturbance decoupling control of an ultra-high speed centrifugal compressor for the air management of fuel cell systems [J]. International Journal of Hydrogen Energy, 2014, 39(4): 1788–98.

41. Raman S, Swaminathan S, Bhardwaj S, et al. Rapid humidity regulation by mixing of dry and humid gases with feedback control for PEM fuel cells [J]. International Journal of Hydrogen Energy, 2019, 44(1): 389–407.
42. Fang C, Xu L, Cheng S, et al. Sliding-mode-based temperature regulation of a proton exchange membrane fuel cell test bench [J]. International Journal of Hydrogen Energy, 2017, 42(16): 11745–57.
43. İnci M, Büyük M, Demir M H, et al. A review and research on fuel cell electric vehicles: Topologies, power electronic converters, energy management methods, technical challenges, marketing and future aspects [J]. Renewable and Sustainable Energy Reviews, 2021, 137: 110648.
44. Cunanan C, Tran MK, Lee Y, Kwok S, Leung V, Fowler M. A Review of Heavy-Duty Vehicle Powertrain Technologies: Diesel Engine Vehicles, Battery Electric Vehicles, and Hydrogen Fuel Cell Electric Vehicles. Clean Technol 2021;3:474–89. https://doi.org/10.3390/cleantechnol3020028.
45. Pramuanjaroenkij A, Kakaç S. The fuel cell electric vehicles: The highlight review. Int J Hydrogen Energy 2023;48:9401–25. https://doi.org/10.1016/j.ijhydene.2022.11.103.
46. Lapeña-Rey N, Mosquera J, Bataller E, Ortí F, Dudfield C, Orsillo A. Environmentally friendly power sources for aerospace applications. J Power Sources 2008;181:353–62. https://doi.org/10.1016/j.jpowsour.2007.11.045.
47. Dyantyi N, Parsons A, Sita C, Pasupathi S. PEMFC for aeronautic applications: A review on the durability aspects. Open Eng 2017;7:287–302. https://doi.org/10.1515/eng-2017-0035.
48. Romeo G, Borello F, Correa G, Cestino E. ENFICA-FC: Design of transport aircraft powered by fuel cell & flight test of zero emission 2-seater aircraft powered by fuel cells fueled by hydrogen. Int J Hydrogen Energy 2013;38:469–79. https://doi.org/10.1016/j.ijhydene.2012.09.064.
49. Baroutaji A, Wilberforce T, Ramadan M, Olabi AG. Comprehensive investigation on hydrogen and fuel cell technology in the aviation and aerospace sectors. Renew Sustain Energy Rev 2019;106:31–40. https://doi.org/10.1016/j.rser.2019.02.022.
50. Ballard backup power for Danish broadband. Fuel Cells Bull 2019;2019:1–1. https://doi.org/10.1016/s1464-2859(19)30040-9.
51. Yu S, Fan Y, Shi Z, Li J, Zhao X, Zhang T, et al. Hydrogen-based combined heat and power systems: A review of technologies and challenges. Int J Hydrogen Energy 2023. https://doi.org/10.1016/j.ijhydene.2023.05.187.
52. Elmer T, Worall M, Wu S, Riffat SB. Fuel cell technology for domestic built environment applications: State of-the-art review. Renew Sustain Energy Rev 2015;42:913–31. https://doi.org/10.1016/j.rser.2014.10.080.
53. Nielsen ER, Prag CB, Bachmann TM, Carnicelli F, Boyd E, Walker I, et al. Status on Demonstration of Fuel Cell Based Micro-CHP Units in Europe. Fuel Cells 2019;19:340–5. https://doi.org/10.1002/fuce.201800189.
54. Proton Motor signs framework deal with APEX, Vonovia order. Fuel Cells Bull 2020;2020:8–8. https://doi.org/10.1016/s1464-2859(20)30153-x.
55. SFC, adKor order 500 Ballard stacks for backup power systems at German radio towers. Fuel Cells Bull 2020;2020:6–7. https://doi.org/10.1016/s1464-2859(20)30059-6.
56. Mensharapov R M, Ivanova N A, Spasov D D, et al. PEMFC performance at nonstandard operating conditions: A review [J]. International Journal of Hydrogen Energy, 2024, 96: 664-79.
57. Rocha C, Knöri T, Ribeirinha P, et al. A review on flow field design for proton exchange membrane fuel cells: Challenges to increase the active area for MW applications [J]. Renewable and Sustainable Energy Reviews, 2024, 192: 114198.
58. Sun D, Zheng J, Zhang J, et al. Recent Advance in Surface Modification for Proton Exchange Membrance Fuel Cell Metal Bipolar Plate [J]. Materials Protection, 2024, 57(7): 1–13.

59. Sarjuni C A, Lim B H, Majlan E H, et al. A review: Fluid dynamic and mass transport behaviour in a proton exchange membrane fuel cell stack [J]. Renewable and Sustainable Energy Reviews, 2024, 193: 114292.
60. Hong J, Yang J, Weng Z, et al. Review on proton exchange membrane fuel cells: Safety analysis and fault diagnosis [J]. Journal of Power Sources, 2024, 617: 235118.

Chapter 5
Proton Exchange Membrane Fuel Cells: Electrocatalysts

5.1 Introduction

Global warming has promoted research and development of alternative energy technologies. Developing environmentally clean alternative power generation has become a worldwide problem. As the ultimate technology for clean power generation, proton exchange membrane fuel cells (PEMFCs) have attracted much attention because they can efficiently convert chemical energy directly into electrical energy.

The previous chapter introduced that the membrane electrode assembly (MEA) is an important part of the PEMFC. The MEA includes a proton exchange membrane (PEM), a gas diffusion layer (GDL), and a catalyst layer (CL). Among them, the electrocatalyst is one of the key materials of fuel cells. Due to the complexity of the cathode and anode reactions of fuel cells and the influence of reaction potential energy barriers, catalysts are critical to improving the performance of fuel cells. An electrocatalyst is a catalyst that speeds up chemical reactions by increasing the rate of chemical reactions or reducing overpotential to provide another reaction pathway with lower activation energy and promotes electrochemical reactions such as oxidation and reduction, without consuming before or after the reaction. Electrocatalysts have properties that help transfer electrons between electrodes and reactants, thereby facilitating electrochemical conversions.

The electrocatalyst layer is usually prepared from an electrocatalyst and a proton exchange resin solution. It has a thin-layer porous structure and has electrocatalytic activity for hydrogen oxidation or oxygen reduction. The thickness of the catalyst layer is generally 5–10 μm. In PEMFC, an electrochemical reaction occurs between gas (taking hydrogen and oxygen as examples), protons, and electrons, and the reaction is shown in formula (5.1). Therefore, energy conversion from chemical to electrical can only occur when all three access and interact with each other on the surface of the electrocatalyst.

$$O_2 + 2H_2(4H^+ + 4e^-) \rightarrow \text{electrochemical energy} + 2H_2O \tag{5.1}$$

Y. Li, *Fuel Cell Fundamentals and Applications*,
https://doi.org/10.1007/978-981-96-9845-5_5

At present, the PEMFC catalyst layer has a high Pt loading. The amount of Pt required for fuel cell cycling is about 50 g per car and 100 g per bus. While taking into account the performance and cost of the fuel cell, it is a huge challenge to improve the catalyst activity and reduce the dosage of Pt precious metal.

For PEMFC under acidic conditions, the overpotential of the hydrogen oxidation reaction (HOR) of the anode is very small, and it can work under extremely low platinum loading (0.05 mg/cm^{-2}) without causing obvious energy and performance degradation. The cathodic exchange current density of the oxygen reduction reaction (ORR) is low, and the reaction involves a complex four-electron transfer step, which is the control step of the total reaction of the fuel cell. The cathode ORR reaction process is complex, with many intermediate products, and the reaction rate is much lower than the anode fuel oxidation reaction. The complex ORR process causes a serious current efficiency loss in low-temperature fuel cells, and the resulting decrease in battery efficiency accounts for up to 80% of the total cell efficiency loss. Therefore, the study of electrocatalysts with high activity and high stability is of great significance in promoting the large-scale commercialization of PEMFC.

In recent years, research on PEMFC catalysts has been advancing. The electrocatalyst must be electrically conductive to allow the electron transfer. It must be porous to enable reactant gases to travel, as they only travel through voids. The electrocatalyst must be in sufficiently close contact with the ionomer to allow protons to travel, but not so close that the catalyst is exposed to ionomer poisoning. Nanoscience research shows that as the particle size decreases at the nanometer scale, the specific surface area increases sharply. Due to the insufficient coordination number of surface atoms, high activity occurs, and it is easy to adsorb other molecules and cause chemical reactions. The progress of nanotechnology directly promotes the progress of electrocatalysts. The structure of nanoparticles, especially their surface composition, strongly affects the catalytic effect. Electrocatalytic efficiency is the comprehensive result of individualization and intensification of nanoparticles, because the adsorption of reaction molecules on the surface of the catalyst is a prerequisite for catalysis. The excellent molecular adsorption and dissociation behavior of Pt makes Pt-based materials always the focus of catalyst research. However, Pt is expensive, and the resource shortage problem is becoming increasingly prominent. Research on reducing the Pt loading and exploring non-platinum catalysts is still in depth.

This chapter will introduce the development process and reaction mechanism of catalysts for different reactions at the cathode and anode of PEMFC.

5.2 Electrocatalyst Oxygen Reduction Reaction

Fuel cells based on PEMs or alkaline anion exchange membranes (AEMs) possess extremely slow cathode and anode reaction rates, which make them inefficient [1–3]. To speed up the cathode reactions, it is necessary to use electrode catalysts in those situations. The efficient catalysts are thus needed both on the anode and cathode of the fuel cell to reduce its overpotential reaction. To address this, umpteen research institutions and researchers have conducted a large amount of research and development and found that Pt-based materials are the ideal catalyst for the acceleration of cathode and anode reaction rates [4–8]. Unfortunately, Pt has a very limited quantity of reserves, and as a result, its price is very high, which would increase manufacturing costs. Additionally, cathode overpotential for ORR is significantly higher than anode overpotential for oxygen oxidation reactions, meaning that catalysts from Pt are mostly found in the cathode of the fuel cell [9–12]. For reducing the production cost of fuel cell systems, it is necessary to utilize low Pt catalysts or catalysts without precious metals on the cathode side.

5.2.1 *Pt Electrocatalyst*

5.2.1.1 ORR Mechanism of Pt Surface

The oxygen reduction process is a complicated reaction that involves multi-electron metastasis and several intermediate intermediates. To better grasp the reaction process, the researchers do a huge variety of experimental characterization and simulation computations. ORR mechanism and dynamical rate are highly dependent on the electrode catalyst, electrode voltage, and electrolyte environment. Therefore, revealing the mechanism and process of ORR is extremely challenging. Most researchers report the ORR process of platinum surface to be divided into two categories (Fig. 5.1): oxidation of oxygen molecule to water or OH^- by reduction with 4 electrons [1, 2].

Alkaline medium:

$$2H_2O + O_2 + 4e^- \rightarrow 4HO^- \tag{5.2}$$

Acidic medium:

$$4H^+ + O_2 + 4e^- \rightarrow 2H_2O \tag{5.3}$$

And another is a slow two-electron ($2e^-$) two-step process in which the oxygen molecule acquires $2e^-$ to generate H_2O_2.

Alkaline medium:

$$O_2 + H_2O + 2e^- \rightarrow HO_2^- + OH^- \tag{5.4}$$

$$HO_2^- + H_2O + 2e^- \rightarrow 3OH^- \tag{5.5}$$

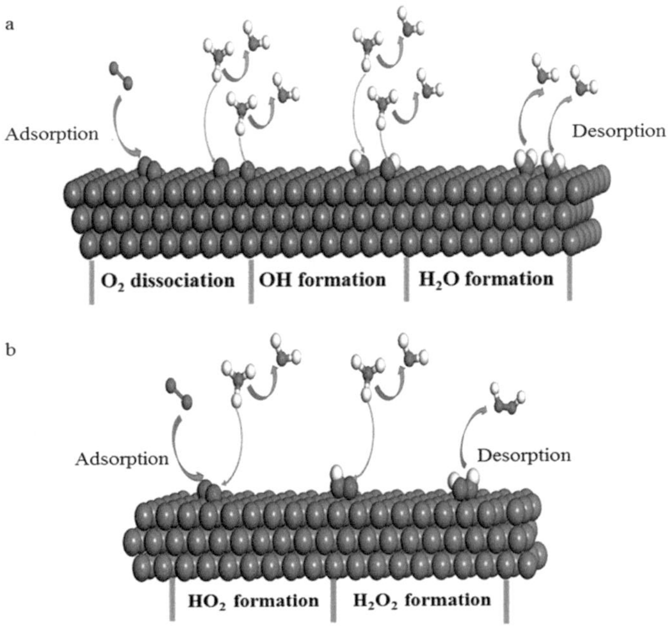

Fig. 5.1 Two ORR pathways: **a** full oxygen reduction (4-electron pathway); **b** imperfect oxygen reduction (2-electron pathway) [14]

Acidic medium:

$$O_2 + 2H^+ + 2e^- \rightarrow H_2O_2 \tag{5.6}$$

$$H_2O_2 + 2H^+ + 2e^- \rightarrow 2H_2O \tag{5.7}$$

The direct $4e^-$ process is projected to have more activity than the undesired $2e^-$ process, and it is required for highly effective electrocatalysts [13, 14].

Following a vast number of experiments, the researchers think that the oxygen-reducing activity of the catalyst is dependent on the intermediate species (e.g., O and OH, etc.) and the catalyst surface. This binding cannot be too weak, or the catalyst will not be able to effectively adsorb and dissociate O_2, and it cannot be too strong, or the catalyst will not be able to cause the OH* and O* and the like from the reaction site to inhibit future reactions [15–18].

Figures 5.2 and 5.3 depicts a "volcanic" relationship diagram of various metal catalyst oxygen reduction activity and metal to oxygen by Nørskov et al. [19]. Based on this graph, it is feasible to estimate that the Pt has the best oxygen reduction activity in pure metals, since the binding energy (ΔE_O) of the Pt surface to oxygen is the most "suitable" metal material. However, it can be shown from Fig. 5.2 that Pt may be linked to oxygen more than the theoretical ideal oxygen binding energy, and

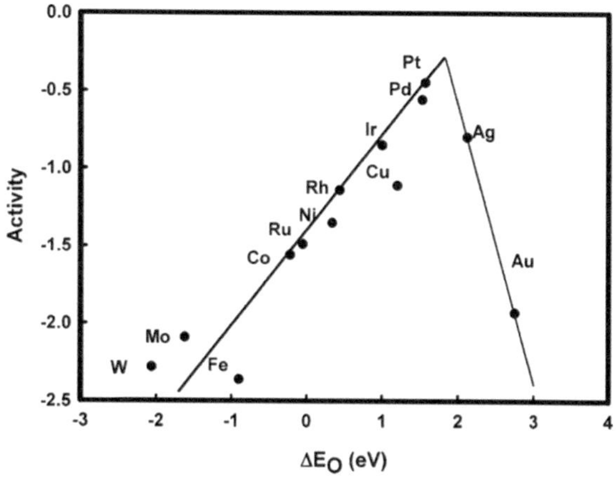

Fig. 5.2 Trends in oxygen reduction activity were plotted as a function of the oxygen-binding energy [20]

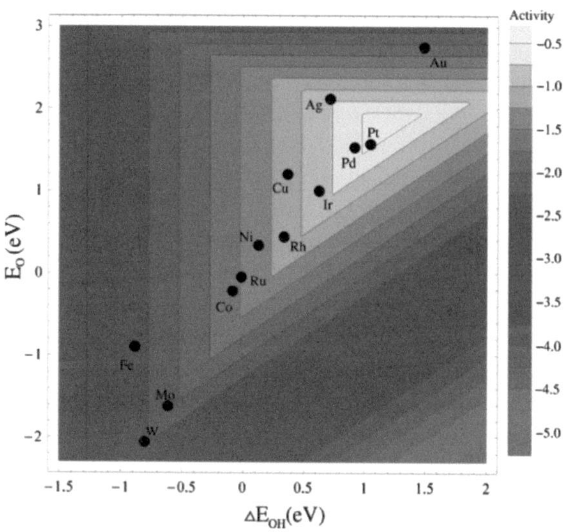

Fig. 5.3 Trends in oxygen reduction activity were plotted as a function of both the O and the OH binding energy [21]

Pt activity does not reach the top of the volcano map, implying that the electronic and structural state of the Pt surface can be altered. The catalyst can be optimized to obtain unequaled oxygen binding energy and oxygen reduction catalytic activity.

5.2.1.2 Pt/C Catalyst

As the catalyst is energized during the operation of the fuel cell. To improve the utilization of platinum, reduce the amount of platinum, and disperse platinum in the form of higher nanoscale particles on the support that bears the load to make it conductive and corrosion resistant [22–24]. The current commercial support is

Vulcan XC-72 carbon black, which has an average particle size of about 30 nm and a specific surface area of 250 m^2/g [25].

There are two main methods for loading platinum onto a carrier with high dispersion: chemical method and physical method, which specifically include: (1) the colloidal platinum sol method, (2) the ion exchange method, (3) the H_2PtCl_6 direct reduction method, and (4) the vacuum sputtering method. Among them, the vacuum sputtering method is a mature physics method suitable for the mass production of platinum electrodes [25, 26]. It takes a sputtering metal (such as platinum) as the sputtering source, as the cathode, and a sputtering object (such as carbon paper in the diffusion layer of the electrode) as the anode, with a high voltage between the electrodes, the platinum particles on the sputtering source can be sputtered onto the carbon paper as nanoscale particles. To improve the dispersion of the platinum particles sputtered onto the carbon paper and to increase the thickness of the electrode to accommodate the movement of the electrode reaction interface during operation, ion etching can also be used to prepare a thin layer of nanoscale carbon on the surface of the carbon paper, followed by sputtering of nanoscale platinum [27–30]. The preparation of platinum electrodes for nanoscale electrocatalysts by the vacuum sputtering method, suitable for mass production, has great potential for development, which can greatly reduce the amount of platinum for PEMFC.

Pt/C catalyst is the only commercial fuel cell catalyst due to its outstanding catalytic activity. Platinum is dispersed in a conductive carrier and loaded in nanometer form with a catalyst. The platinum catalyst is the current commercial fuel cell catalyst, but its disadvantages are quite apparent in the ORR. The platinum is loaded in a nanometer shape and dispersed in a conductive, corrosion-resistant carrier. In reality, aggregation or mobility of Platinum particles on the Pt/C catalyst carbon carriers, or dissolution of Pt particles in the electrolyte, as well as carbon carrier degradation, might result in diminished Pt/C catalytic activity. The lowering of the catalyst active surface area, in particular, is critical to the decrease of Pt/C catalyst fuel cell life [31, 32].

The current mainstream Pt/C catalyst has a low mass activity (MA, 0.21 A $mg^{-1}{}_{Pt}$) at 0.9 V (vs. RHE) and specific activity (SA, 0.30 mA cm^{-1}). Its performance and the target of the US Department of Energy (DOE) still have a large gap: mass activity reaches 0.44 A $mg^{-1}{}_{Pt,}$ and the specific activity reaches 0.72 mA cm^{-1} [32, 33]. (2) Pt belongs to precious metals, with limited supplies on the planet and high costs; the high consumption of fuel cell catalysts raises the catalyst cost. Reducing Pt content in the fuel cell catalyst, increasing Pt efficiency, and improving the activity of the Pt base catalyst will create a scientific challenge that researchers will have to solve.

Due to the scarcity and high price of Pt, enhancing the utilization of Pt while achieving good catalytic performance and long-term stability has become a common concern. By controlling the morphology and structure of Pt, including core–shell structure, porous structure, polyhedral structure, and nanostructure, the size of Pt particles can be controlled within a certain range so that the ORR will not be hindered by the specific activity due to the small size and will not reduce the surface area due to the large size. In addition, Pt alloys can effectively improve the activity of catalysts.

Also, changing the morphology and structure of the catalyst can increase the exposure of its active sites and enhance its activity and stability.

The current carbon carrier materials [26], such as carbon black (CB), carbon nanotubes (CNTs), carbon nanofibers (CNFs), carbon nanohorns, and graphene (Table 5.1), can increase the utilization of Pt while improving the performance of electrocatalysts. Carbon substrates should have a brute high surface area, suitable porosity, high corrosion resistance, and excellent electrical conductivity to favor high dispersion of PtNPs and accelerated Carbon substrates electron transfer at the electrode–electrolyte interface [34, 35]. Oxidizing materials with higher corrosion resistance than carbon materials are also sufficient to support Pt NPs for electrocatalysts.

It has been investigated that there is a significant difference in the durability of CB loaded Pt and multi-walled carbon nanotubes loaded Pt (Pt/CNTs) catalysts [43, 44]. During the test, Pt/C exhibited a significant decrease in electrochemical surface area, while Pt/CNTs showed less loss of electrochemical surface area, mainly because CNTs have a stronger resistance to electrochemical oxidation than carbon black and thus exhibit better stability [45].

The surface chemistry and morphology of the carriers have important effects on the dispersion, particle size and activity of Pt catalysts [46]. CNFs have received much attention as electrocatalytic carriers due to their unique structural, mechanical and electrical properties. Due to the morphological characteristics of CNFs, CNFs can be used as electrocatalyst carriers without any pretreatment [47, 48]. Owing to the graphite planar edges which can serve as a suitable location for stabilizing small Pt particles, the porous structure of CNFs and the better interaction between Pt and the carrier through the oxygen groups on the surface of the carrier improve the stability of the catalyst under acidic conditions.

Graphene has the advantages of high electrical conductivity, specific surface area, mechanical strength, and excellent stability, which are necessary for electrocatalyst carriers [49–51]. The surface of graphene oxide (GO) has different oxygen-containing functional groups (OH, COOH), and the surface functional groups are beneficial for Pt dispersion and prevent Pt agglomeration, thus anchoring Pt nanoparticles on the carrier surface [52, 53]. The durability of the carbon carrier ORR was successfully solved, and Pt/RGO has higher stability and electrocatalytic performance compared with the conventional Pt/C-formed composites [51].

Table 5.1 Several carbon support materials and their typical characteristic

Carbon support	Surface area ($m^2\ g^{-1}$)	Particle size (nm)	Conductivity (S cm^{-1})	References
Carbon black	254	30 nm	2.77	[25]
Carbon nanotubes	50–1000	1–5 μm	0.3–400	[36–38]
Carbon nanofibers	50–1000	30–100 μm	200–900	[37, 39, 40]
Carbon nanohorns	300–400	40–50 nm	4.95–7.07	[41]
Graphene	Graphene	10–20 nm	10^3 to 10^4	[40, 42]

5.2.1.3 Pt-Based Binary Alloys

The relatively inexpensive transition metal M (Fe, Co, Ni, Au, Ag, Pd, Cr, Cu, Mo, Mn, etc.) and Pt formation alloy can reduce the use of Pt, but also improve the activity of catalyst [54–68]. Figures 5.4 and 5.5 are a “volcano” type plot of Pt base alloy (Pt_3Ni, Pt_3Co, Pt_3Fe, Pt_3Sc, Pt_3Ti, Pt_3Y, Pt/Pd), and the model of Pt activity and oxygen absorbent energy [69–72]. As seen, the surface oxygen-absorbing energy of the Pt alloy is higher than the Pt and the activity is higher, which can guide to better modify the Pt through alloying effect, and then improve the ORR activity of the catalyst.

One of the major approaches to maximize the combination of the Pt surface with oxygen is through Pt and other metal alloying (Table 5.2). After the Pt is created

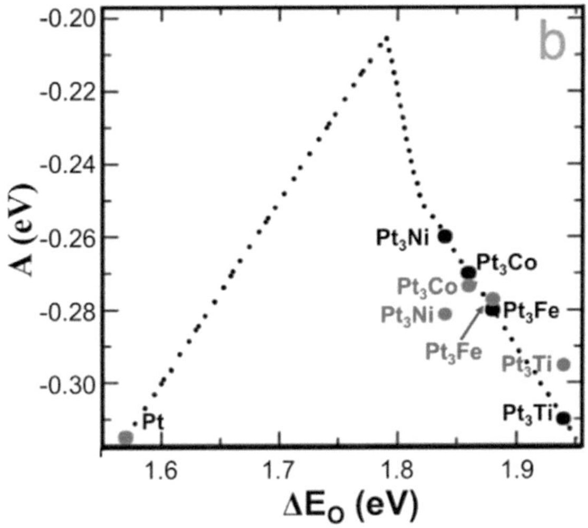

Fig. 5.4 Model of the activity (A) as a function of the O adsorption energy (OAE) for bimetallic systems [69]

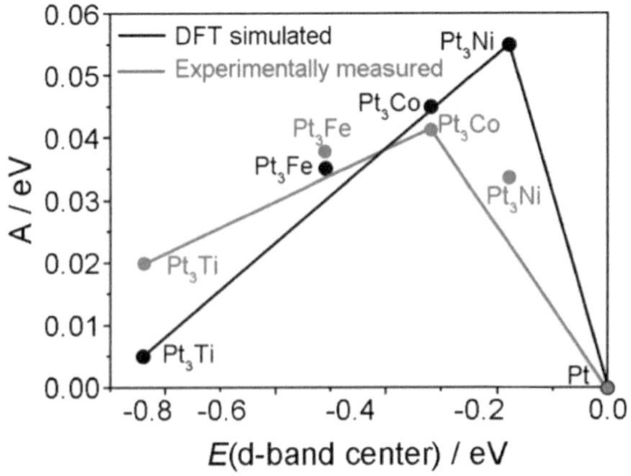

Fig. 5.5 Oxygen reduction activity versus the experimentally measured d-band center relative to platinum [71]

by the alloy, the Pt can be altered to the bonding strength of the Pt and oxygen reduction in the form of oxygen reduction, thereby increasing its oxygen reduction activity. Furthermore, this alloy technique can minimize the overall quantity of Pt in the catalyst as well as the fuel cell's production cost. People are now researching binary alloys and ternary alloys [73–76]. According to research, the 3D transition metal in the alloy supplementary transition metal, can inhibit the combination of the Pt surface with OH*.

A bimetallic Pt-based nanocatalyst, for example, may greatly improve D-band center catalytic activity by effectively sieving [77, 78]. As a result, new techniques for creating improved Pt-based nanostructures for prospective applications in electrochemical and heterogeneous processes must be developed and implemented. In addition to the electronic structural modification caused by alloys, the shape of the catalyst is an effective technique for electrocatalytic reactions because electrochemical and non-uniform reactions occur [79–81]. In addition to the electronic structural modification created by alloys, the geometry of the catalyst is a useful technique for electrocatalytic reactions because electrochemical and non-uniform reactions often occur on the surface of the catalyst. Thus, the ideal catalyst should possess a high surface-active area to offer an adequate activity center and suitable adsorption energy to improve the adsorption strength of the reaction intermediate. The past study [82] presented a facile CVD method for producing monodisperse Pt–Co bimetallic alloy nanoparticles, which shows that carbon-supported Pt–Co bimetallic nanoparticles possess superior uniformity and dispersion after the long-term cycles. Electrochemical experiments indicate that the Pt_3Co catalyst exhibits exceptional performance, with superior mass activity, specific activity, and long-term stability. Moreover, one-dimensional (1D) Pt-Ni bundle nanospheres (BNSs) were prepared by a one-pot solvothermal method to reduce different ratios of Pt and Ni precursors in crude amine [10]. 1D Pt-Ni (Pt-Ni) nanoalloys were used for the ORR with high mass and specific activities, nearly 17 and 14 times higher than commercial Pt-C (Pt/C) catalysts. The catalysts have high stability with minimal activity decay after cycling. Both experimental results and theoretical calculations reveal the presence of fewer strongly bonded Pt-Oxygen (Pt-O) sites due to strain and ligand effects. In addition, the Pt-Ni BNC electrocatalyst showed better performance compared to commercial Pt/C in both hydrogen-air PEMFC and aspirated PEMFC setups at room temperature and ambient pressure (Fig. 5.6). The Structural characterization and theoretical calculations indicate that Pt-Ni alloy nanostructures have the best oxygen intermediate adsorption energy for Pt.

The Pt alloying with other metals can optimize the catalyst structure and promote higher ORR activity compared to commercial Pt/C catalysts, mainly due to the synergistic effect of ligands and geometric effects to enhance Pt utilization efficiency and optimized adsorption strength. The Ga precursors were initially adsorbed on the surface of Pt NW seeds by using and then reduced to Ga adsorption atoms by an autocatalytic surface reduction process [83]. These reduced Ga-treated atoms were then diffused into Pt NWs to form PtGa alloy NWs at high reaction temperatures over a long period, based on an interatomic diffusion process that is essential for the synthesis of intermetallic nanocrystals.

Table 5.2 ORR performance of recently carbon-supported Pt catalysts in 0.1 M $HClO_4$ solution reported

Catalyst	Half-wave potential E1/2 (V)	Mass activity (A $mgPt^{-1}$) at 0.9 V	ECSA ($m^2\ g^{-1}$)	References
PtNi-BNCs/C	NA	3.52	68.2	Science, 2019, 36, 850–856
PtNi-BNSs/C	NA	1.89	43.5	Science, 2019, 36,850–856
PtCo@NGNS	0.95	1.29	76	Angew. 2022, 134, e202115835
Pt/40Co-NC-900	0.92	1.15	45	Nano Lett. 2018, 18, 4163
PtCu-NFs	0.92	0.82	NA	Appl. Catal. B. 2021, 1, 282
Int-PtNiN/KB	0.93	1.83	85.9	ACS Catal. 2020, 10, 10637
7 wt%-FePt/rGO	0.94	1.96	NA	J. Am. Chem. Soc. 2020, 142, 14190–14200
Pt–Co ND-NF	0.95	0.939	35.6	Appl. Catal. B. 2021, 281, 119460
Pt@Ni ZIF-NC	0.902	0.38	74	J. Energy. Chem. 2022, 65, 48–54
PtBi/C	0.886	0.14	71.34	ACS Catal. 2018, 8, 5581–5590
PtFe/Pt-V NW	0.959 V	3.65	124.5	J. Mater. Chem. A. 2021, 9, 12845
PtCu NSs/C	0.893	0.42	73.9	J. Catal. 2019, 375, 164–170
Pt_3Co/C	NA	0.217	52	Top. Catal. 2008, 49, 241
PtFe-Fe_2C	0.91	1.50	42.5	Chem. 2018, 4 1153–1
$PtCo_3$/HSAC	NA	51	0.36	ECS Transactions. 2010, 33, 333
PtCo NW	0.92	0.200	NA	Nano Energy. 2014, 10, 135
$PtCo_3$-H600	0.903	0.74	NA	Adv. Funct. Mater. 2019, 29, 1–9

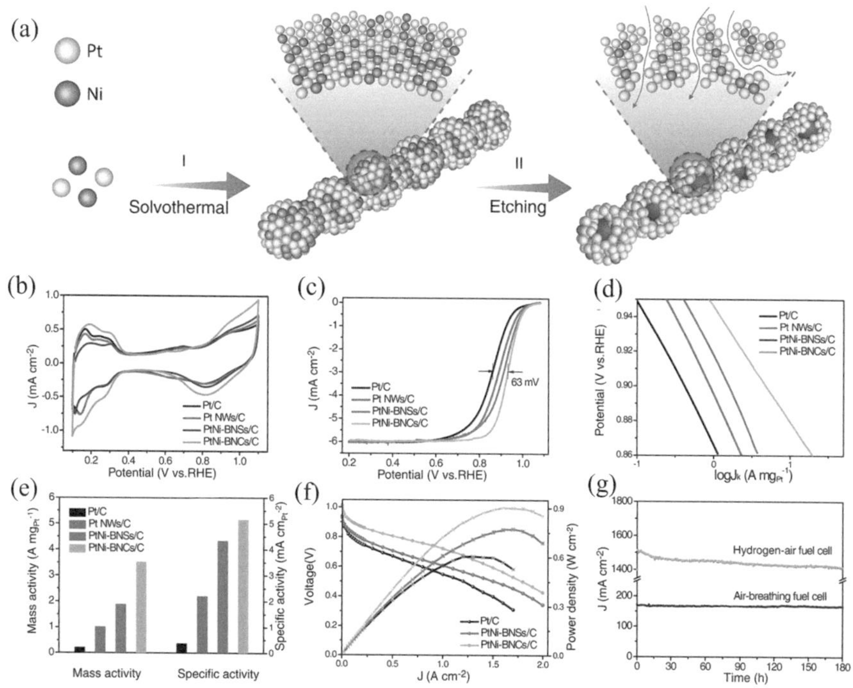

Fig. 5.6 **a** Schematic illustration of the preparation of PtNi-BNCs. **b** CVs, **c** LSVs, **d** corresponding Tafel plots, and **e** mass and specific activities. **f** H_2-air fuel cell polarization plots. **g** Stability test of H_2-air and air-breathing fuel cells at 0.6 V [10]

Noble metals extremely alloyed nanoclusters have been widely used as excellent heterogeneous catalysts for many chemical and electrochemical reactions, facilitating the study of metal particle size, composition, ligands, and supports on the catalytic properties of supported metal catalysts [84–93]. It was found that colloidal solutions of PtCu alloy NCs produced by a modified base-glycol method can stabilize PtCu alloy nanoclusters with tunable Pt/Cu ratios and similar average diameters of PtCu alloy NCs (Fig. 5.7) [94]. The prepared colloidal PtCu NCs were immobilized on Melem-modified carbon support (MMC) to prepare robust electrocatalysts (PtCu/MMC). $Pt_{75}Cu_{25}$/MMC exhibited good durability and optimal electrochemical mass activity over most of the reported PtCu alloy catalysts.

In ORR catalysts, Pt–Pd is widely used in the research of ORR electrocatalysis. The Pt and Pd atoms were alloyed to form nanoparticle alloy catalysts. Feng et al. [95] synthesized porous Pt–Pd nanocrystals by a simple wet chemical method using PVP as a co-stabilizer and urea as a co-structural guide, and the Pt–Pd nanocrystals prepared by this method were easy to scale up. Under the same test conditions, Pt–Pd NDs have higher electrochemical activity than Pt/C and Pd/C. The synergistic effect between Pt–Pd NDs with porous structure and Pt and Pd is the main reason for the good electrocatalytic activity of Pt–Pd NDs.

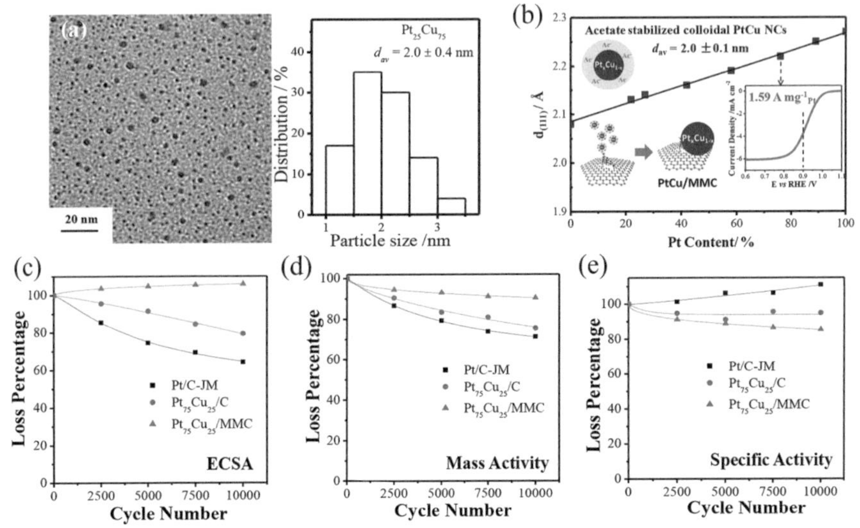

Fig. 5.7 **a** TEM images and size distributions of colloidal $Pt_{25}Cu_{75}$. **b** Size distributions of the prepared PtCu NCs with different Pt/Cu ratios. **c** ECSA, **d** mass activity, and **e** specific activity as a function of potential cycles during the durability tests (10,000 potential cycles between 0.6 and 1.0 V) [94]

In general, Pt and transition metals can form alloyed nanocatalysts, which often leads to lattice strain (lattice mismatch) between Pt and the transition metal. Lattice strain, either stretching or shrinking, can modulate the surface electronic structure of alloyed catalysts by adjusting the distance between surface atoms and thus their catalytic activity [96, 97]. For Pt, the contraction distance of Pt–Pt on the catalyst surface atoms can change the d-band structure and modulate the adsorption energy of intermediates (O*, OH*) to improve the ORR activity [98–100]. PtAuNWs with controlled composition and tunable facets were synthesized to improve the electrocatalytic activity and stability of ORR. Meanwhile, Au has high catalytic activity for many reactions (e.g., CO oxidation) at the nanoscale [101]. PtAu nanocatalysts exhibit synergistic catalytic effects, which change the structure of Pt electronic bands by altering the surface adsorption. The synergistic catalytic effect significantly improved the electrocatalytic activity and stability of Pt catalysts, and improved the CO tolerance of Pt catalysts. The electrocatalysts exhibited higher catalytic activity and stability, indicating that the addition of Au could greatly improve the catalytic performance [102].

5.2.1.4 Multi-composited Pt-Based Alloys

The dependence of the development of Pt alloys on improving the ORR new material with high activity and stability of the catalyst is desperately needed. The aim focuses on how to improve the catalytic activity and durability of the catalyst. The activity of the catalyst for oxygen reduction activity of the catalyst strength is one of the catalysts are excellent judging criteria, has now been confirmed that the catalytic activity of the alloy catalyst is superior to pure Pt catalyst. Based on the binary ordered alloy to join the third metal, and the preparation of the ternary alloy, through further optimization of Pt electronic structure, implementation of Pt catalysts to further improve electrochemical activity (Table 5.3). To improve the durability of the catalyst, using particle migration in the process of growth and transition metal dissolution will directly affect the stability of the catalyst itself. How to prevent these problems is a great problem to solve and improve the stability of the catalyst [103, 104]. Pt-based multi-metal satisfactorily attracted researchers due to their high metal utilization and favorable catalytic activity. Improvement in catalyst activity is mainly achieved through the regulation of metal addition. These anomalous not only govern the electronic structure of Pt in Pt to boost the catalytic process, but also expose more surfaces. Provides an active region with a high concentration of surface-active sites to improve catalytic performance. Nanoparticles encapsulated can effectively prevent the migration of nanoparticles in the process of high-temperature annealing, reunion, and avoid the loss caused by high-temperature sintering activity. Given the transition metal dissolution, the preparation of the ternary alloy catalyst system, because of its atomic arrangement in an orderly way, can effectively reduce the problem of metal dissolution and realize the stability of Pt catalysts [105–108].

Core–shell structured Pt-based nanomaterials are one of the main research directions for multi-composited Pt-based alloys [109–115]. However, the structural instability of the ultra-thin shell layer and the leaching of core components result in the materials stability that can not meet the requirements for smooth battery operation. Given this, a PtPdAu ternary alloy oxygen reduction catalyst with high activity, high stability, and low Pt loading is reported [116]. Based on the different diffusion rates and electronegativities of Au, Pd, and Pt atoms, the researchers used electrochemical cyclic voltammetry to treat the surface structure of the sub-nanometer Pt–Pd–Au

Table 5.3 Summary of the ORR performance of Pt-based alloys catalysts reported in recent years

Catalyst	MA (A $mgPt^{-1}$) at 0.9 V versus RHE	SA (mA cm^{-2})	References
Pd/PtFe	2.71	4.32	Science Bulletin 2021, 66, 44–51
Au@Co@PtCoAu	0.69	1.73	Adv. Funct. Mater. 2020, 30, 2001575
$Pt_2CuW_{0.25}/C$	0.75	1.43	Adv. Funct. Mater. 2020, 30, 1908230
L10–Pt–Ni–Co	2.28	4.38	Nano Lett. 2018, 18, 4163
PtCuNi nanoframe	0.86	1.65	Angew. Chem. Int. Ed. 2019, 58, 15471–15477
$Fe_3Pt/Ti_{0.5}Cr_{0.5}N$	0.64	1.28	Adv. Energy Mater. 2019, 9, 1803040

shell layer and finally obtained a PtPdAu ternary alloy shell layer oxygen reduction catalyst with high efficiency and high stability (Fig. 5.8). The mass activity of the catalyst is 15 times higher than that of commercial Pt/C, and the stability test remains unchanged after 70,000 revolutions. DFT calculations show that the surface atomic composition of the final Pt and Pd reduces the free energy change of the catalyst surface from *OH to water, thus improving the ORR catalytic activity.

PtAu/TiO_2 NWs were prepared by electrostatic spinning and subsequent thermal annealing [117]. To elucidate and optimize the effect of 1D nanostructures, a series of 1D PtAu/TiO_2 NWs catalysts with different Pt, Au, and Ti atomic ratios were prepared. Among them, $Pt_1Au_1/(TiO_2)_1$ NWs were the best, with higher mass activity (MA) and specific activity (SA) than commercial Pt/C. The enhanced activity observed in $Pt_1Au_1/(TiO_2)_1$ NWs was attributed to the electron transfer from TiO_2 to Pt NPs and affected the unfilled d-band vacancies of Pt (Fig. 5.9).

Pt and Pt alloy nanoparticles (NPS) below 2 nm are not suitable for ORR. However, whether the same trend applies to the nano line (NW_S) and nanosheets of the Pt base still needs to be resolved, because there is no standardized method to prepare these Pt nanostructures [118–123]. Jiang et al. provide a comprehensive method for fabricating nanoscale Pt alloy NWs with diameters about 4 atomic layer thickness, spanning from monometallic PtNWs to bimetallic PtNi and PtCo NWs and trimetallic PtNiCo NWs [124]. In stark comparison to Pt alloy NPs, submicron Pt alloy NWs exhibit extraordinary mass and specific activities that are 32.3 and 26.9 times greater than those of standard Pt/C, respectively (Fig. 5.10). The improved ORR performances are attributable to the catalytically active sites in the subnanometer Pt alloy NWs, according to calculation results. The electrocatalytic activity is also extremely stable under ORR conditions, with low activity decline.

Alloying pure Pt with secondary metals reduces the use of scarce Pt metals while improving the performance of mass activity compared to pure Pt [125], which has led to the development of active and durable Pt-based electrocatalysts with a wide range of compositions [126, 127]. However, although studies to date have led to a significant increase in ORR activity, there is room for further improvement in the stability of catalysts with Pt_3Ni nanostructures. The elemental Ni in these nanostructures gradually leaches out under harmful and corrosive ORR conditions, leading to a rapid loss of performance [128, 129]. Therefore, the synthesis of Pt-based nanostructures with both high catalytic activity and durability remains an important open challenge [130].

Since the surface and near-surface characteristics of catalysts have a great impact on their catalytic performance, some studies have employed surface engineering strategies to further explore and improve the performance of Pt_3Ni-based nanocatalysts on Pt due to the volume extension Pt_3Ni (111) surface is one of the most efficient catalytic surfaces for ORR [131]. ORR catalysts with high activity and stability were developed based on controlled doping of various transition metals onto the dispersed and octahedral Pt surface (called M–Pt_3Ni/C, where M could be V, Cr, Mn, Fe, Co, Mo, W, or Re). The Mo–Pt_3Ni/C catalyst demonstrated the highest ORR performance, showing 81 and 73-fold improvements over Pt/C catalysts. The addition of Mo enhances the catalytic activity and stability of the Pt_3Ni catalyst.

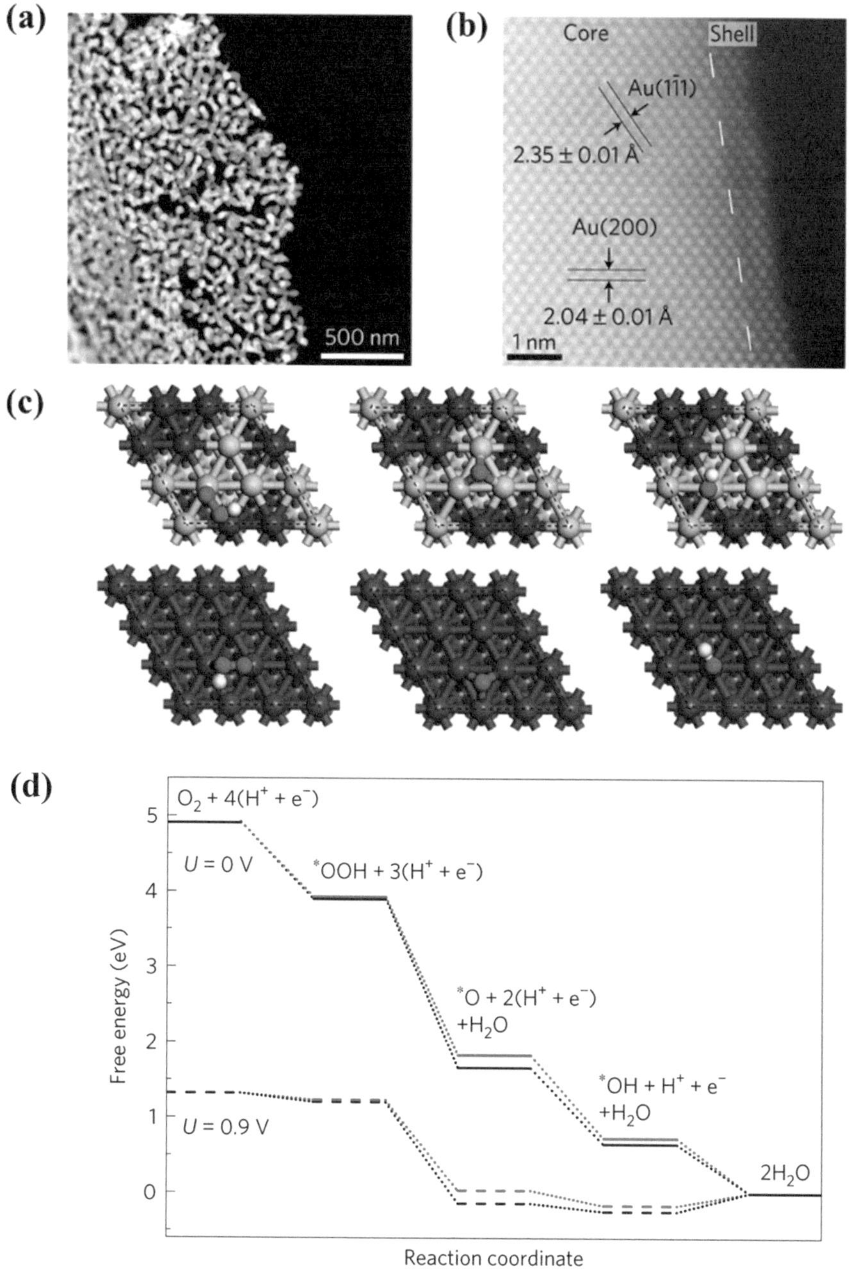

Fig. 5.8 **a** Low-magnification HAADF image and **b** High-resolution HAADF image of a surface region of the electrocatalyst Pt–Pd–Au. **c** and **d** Calculated adsorption configurations of the intermediate species of ORR on the surfaces of the Pt–Pd–Au (111) and the pure Pt (111) models and the calculated free energy profiles of the ORR steps [116]

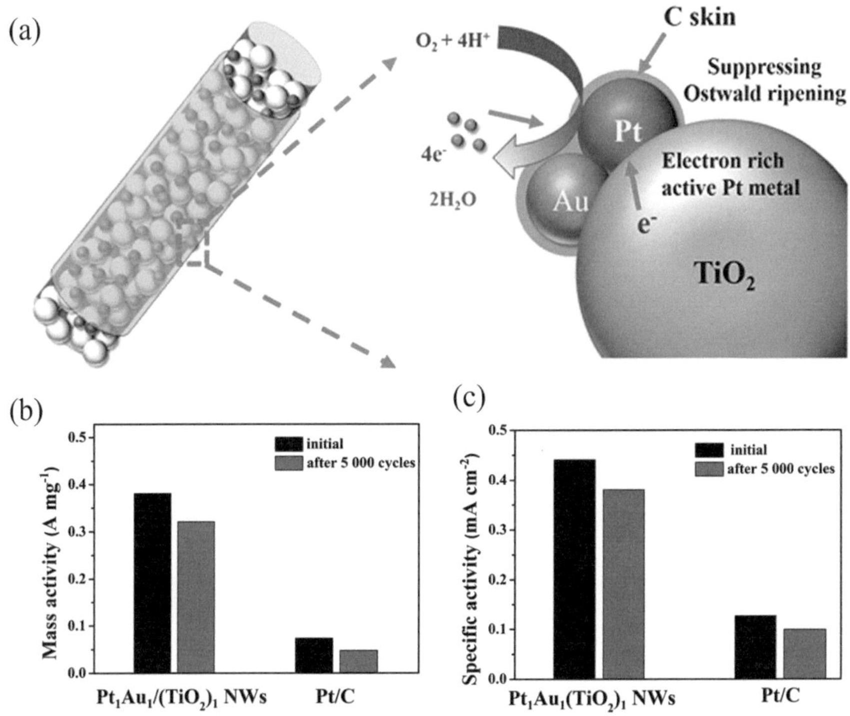

Fig. 5.9 **a** Structural illustration of PtAu/(TiO_2) NWs. **b** and **c** Mass and specific activities of 20% Pt/C and $Pt_1Au_1/(TiO_2)_1$ NW catalysts before and after the durability test [117]

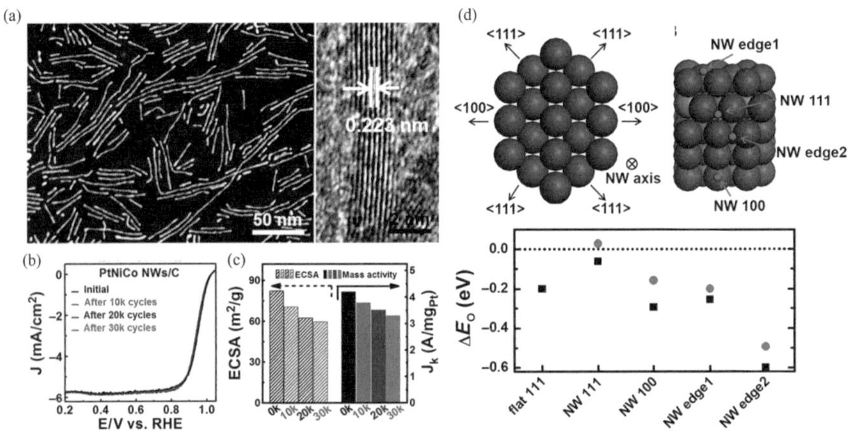

Fig. 5.10 **a** Representative STEM and HRTEM images of PtNiCo NWs. **b** The ORR polarization curves of PtNiCo NWs before and after different potential cycles in an O_2-saturated 0.1 M $HClO_4$ aqueous solution. **c** The changes in ECSA and mass activities of PtNiCo NWs before and after different potential cycles. **d** DFT calculations of E_O [124]

5.2.1.5 Pt Single Atom Catalyst

Carbon material-loaded catalysts with atomically dispersed metal sites, also known as single-atom catalysts (SACs), have attracted much attention from researchers in the field of electrocatalytic ORR in recent years due to their theoretical maximum atomic utilization efficiency and excellent catalytic activity [132, 133]. To further improve the electrocatalytic activity of SACs, there are two strategies: one is to enhance the intrinsic activity of SACs by adjusting the electronic structure of the active sites, and the other is to add more monodisperse active sites by increasing the metal loading density.

SACs have two advantages: one is the reduction of material cost due to the high abundance of non-precious metals, and the other is the substantial increase in atomic utilization because each metal atom can become an active site [134, 135]. A typical representative of these is carbon material-loaded SACs with atomically dispersed metal centers coordinated by nitrogen and/or carbon atoms [136], usually denoted as $M\text{-}N_x(C_y)$, which exhibit performance in ORR catalysis even comparable to that of commercial Pt/C catalysts.

At present, the methods for preparing SACs include solution chemistry, atomic layer deposition, and so on. Among them, the conventional solution chemistry method is one of the mainstream methods for the preparation of precious metal monatomic materials, but the diffusion migration of metal ions and atoms in the precursor solution leads to the formation of many nanocrystals. Therefore, the catalysts obtained by this method have a very serious agglomeration of precious metal atoms [137–140]. The catalysts obtained by atomic layer deposition can also only maintain atomic-level dispersion at very low loadings (typically less than 0.4 wt%), which severely limits the electrocatalytic performance of noble metal platinum. Therefore, high Pt loading and uniform single-atom dispersion are key factors for the development of Pt-based SACs.

Because of this, a photochemical solid-phase reduction method has been reported to prepare atomic platinum catalysts. The method simply uses nitrogen-doped carbon material as a carrier to adsorb noble metal acid ions or salt ions. Then, the obtained precursor powder is irradiated with a certain intensity of ultraviolet light for a while to acquire a high loading amount of noble metal atomic catalysts. The loading of platinum reached 3.8 wt% and was completely dispersed in a monoatomic state. The catalysts exhibit excellent ORR electrocatalytic performance in acidic and alkaline media. The method is remarkable for its simplicity and ease of batch preparation and is also suitable for the preparation of monoatomic catalysts of other noble metals such as Au, Ru, and Ir. The method effectively avoids the agglomeration of metal atoms in the aqueous phase preparation, washing, and drying processes of noble metal monoatoms, and enables the preparation of batches. The monatomic Pt is present in a $Pt\text{-}N_4$ coordination structure (Fig. 5.11). The pyridine-type $Pt\text{-}N_4$ structure is more stable and is the most likely active site. Monatomic Pt catalysts have positive half-wave potential, high specific mass activity and specific activity, low Tafel slope, low hydrogen peroxide yield, and high stability, which is better than 20 wt% commercial Pt carbon catalysts [141]. The unique chemical and electronic structure around the

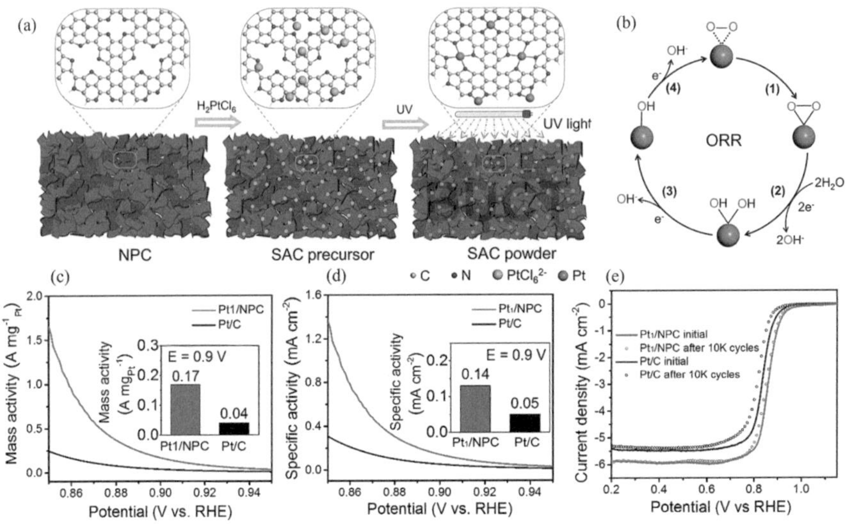

Fig. 5.11 **a** Schematic illustration of the formation of Pt_1/NPC catalyst. **b** Proposed mechanism for ORR catalyzed by Pt_1/NPC. **c** ORR mass activity at different potentials of Pt_1/NPC and Pt/C; inset shows the mass activity at 0.9 V versus RHE. **d** Specific activity; inset shows the specific activity at 0.9 V versus RHE. **f** Polarization curves were recorded in O_2-saturated 0.1 M KOH before and after 10 K CV cycles [141]

platinum, due to the strong electron transfer between the Pt and the neighboring nitrogen atoms, may be responsible for the excellent electrocatalytic performance of Pt-based SACs.

To reduce the catalyst cost for PEMFCs, the development of efficient Pt-based SACs (Pt_1) for ORR is an urgent problem. The first preparation of Pt loading of about 5 wt% by hydrothermal ethanol reduction using nitrogen-doped activated carbon (Black Pearl 2000, NBP) as a carrier [142]. Pt-based electrocatalysts (Pt_1/NBP) with a Pt loading of about 5 wt% were prepared by hydrothermal ethanol reduction. The coordination structure of isolated Pt atoms was reconstructed by further high-temperature pyrolysis of Pt_1/NBP to produce unique nitrogen-anchored Pt single atoms (Pt_1@Pt/NBP), which retained 47.8% of isolated Pt atoms and well-dispersed Pt nanoparticles on the carbon carrier. Electrochemical test results showed that the Pt_1@Pt/NBP catalyst exhibited excellent ORR performance and stability in the high potential region, as well as fast ORR kinetics. As a cathode catalyst for PEMFCs, the mass activity at a cell voltage of Pt_1@Pt/NBP (0.9 V) reached 8.7 times that of commercially available Pt/C. Physical characterization and density flooding theory (DFT) calculations indicate that the doped N heteroatom in the carbon carrier can effectively promote the atomic diffusion of Pt, while high-temperature pyrolysis can reconfigure the coordination structure of isolated Pt atoms adjacent to the nitrogen atom and generate unique pyridine nitrogen-anchored Pt single-atom active sites, resulting in an efficient four-electron ORR pathway.

In recent years, nitrogen-ligated transition metal SACs have been considered as one of the best alternatives to conventional Pt catalysts. A Pt single-atom oxygen reduction catalyst with high activity, high stability, high conversion efficiency (TOF), and only 0.026 wt% of Pt has been simulated and experimentally realized using DFT calculations by assembling a highly catalytically active Pt into an M-N-C structure for the ORR. The isolated Pt single atom is unable to complete the oxygen–oxygen bond breakage by lateral adsorption, making the oxygen reduction process tend towards the 2-electron product, peroxide, rather than the 4-electron product required for fuel cells. The catalytic performance of the Pt single-atom site is enhanced by the design of the coordination structure, which can improve the selectivity of the reaction and achieve the 4-electron oxygen reduction of Pt single atoms. Phosphorus with more valence electrons was selected to participate in the construction of Pt atomic active sites. A variety of different coordination configurations of Pt atoms were constructed and predicted by DFT theoretical simulations (Fig. 5.12), and it was found that when the phosphorus atom replaced the α-carbon in the Pt-N3 site, the d-band center of Pt became deeper, which was favorable for the oxygen reduction [143]. Meanwhile, this catalyst was successfully synthesized using experiments. The catalyst was tested to contain only about 0.026 wt% Pt, but exhibited oxygen reduction activity comparable to that of commercial Pt/C (20 wt%) with a TOF of 6.80 s^{-1} (at 0.9 VRHE), which is nearly 170 times higher than that of commercial Pt/C.

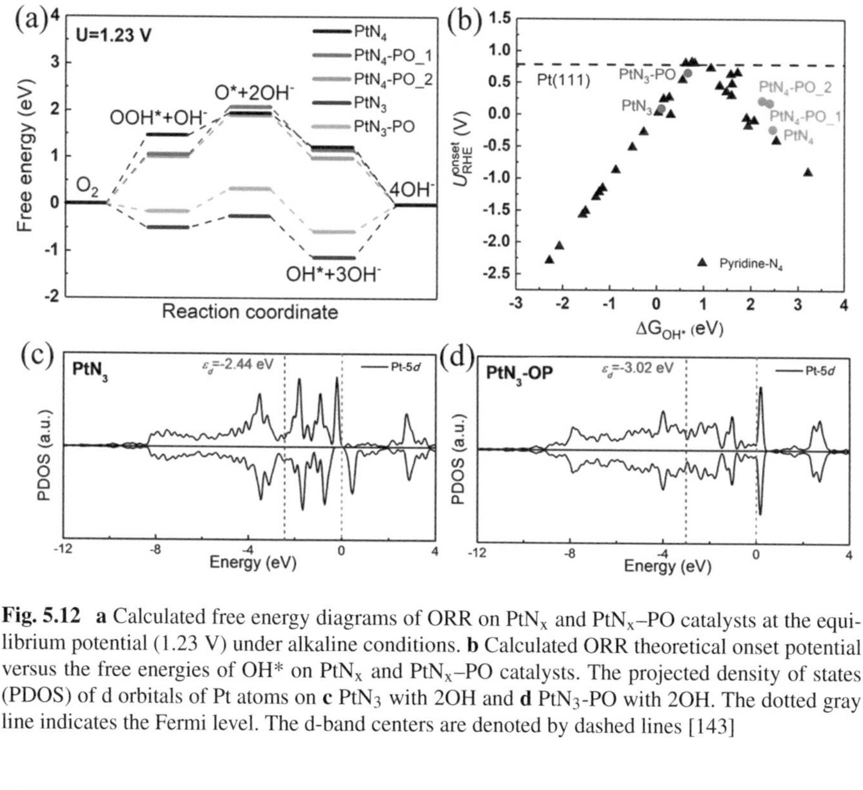

Fig. 5.12 **a** Calculated free energy diagrams of ORR on PtN_x and PtN_x–PO catalysts at the equilibrium potential (1.23 V) under alkaline conditions. **b** Calculated ORR theoretical onset potential versus the free energies of OH* on PtN_x and PtN_x–PO catalysts. The projected density of states (PDOS) of d orbitals of Pt atoms on **c** PtN_3 with 2OH and **d** PtN_3-PO with 2OH. The dotted gray line indicates the Fermi level. The d-band centers are denoted by dashed lines [143]

To date, platinum (or platinum-based alloy) nanoparticles (NPs) loaded on carbon carriers remain the most promising ORR catalysts due to their inherent high activity. Current schemes for the synthesis of monodisperse Pt nanoparticles typically use hydrocarbon molecules as surface-covered ligands; however, utilizing Pt nanoparticles as catalysts for ORR necessitates the removal of these ligands to expose the surface sites. It was found that highly durable ORR catalysts can be achieved without removing the ligands, and instead, the natural ligands are converted into ultra-thin conformal graphite layers by simple thermal annealing, and that the annealing temperature is a key factor in determining the ORR performance of the Pt catalyst [143, 144]. Pt nanoparticles treated at 500 °C exhibited very poor ORR activity, while Pt nanoparticles treated at 700 °C showed high activity and extraordinary stability [145]. In-depth characterization showed that the heat treatment from 500 to 700 °C induced graphitization, thus gradually opening the pores in the carbon shell. Importantly, this graphite-shell encapsulated Pt catalyst has excellent ORR stability and remains largely active after 20,000 cycles in the membrane electrode assembly. Furthermore, this ligand-based carbonization strategy can be extended to modify commercial Pt/carbon catalysts to give them significantly improved stability. The work demonstrates the feasibility of improving the ORR performance of common Pt nanoparticles by using natural surface ligands. A simple and effective method (using natural ligands on the surface of colloidal Pt nanoparticles) was developed to prepare stable graphite-shell-coated Pt catalysts. Simple thermal annealing of the Pt nanoparticles allows the conversion of the hydrocarbon ligands into a bilayer, conformal graphitized shell, which is strong enough to protect the Pt nanoparticles from migration, agglomeration, and dissolution while maintaining the activity of the surface sites to a large extent. The ORR activity of the obtained carbon-coated Pt catalysts depends strongly on the thermal annealing temperature, which determines the microstructure of the graphite layer. In particular, the long-term stability of the graphite-shell-coated Pt catalysts after thermal treatment at 700 °C was superior to that of the commercial Pt/carbon catalysts. Importantly, this in situ mating-based carbonization strategy is versatile and can also be applied to improve the stability of industrial Pt/C catalysts, thus opening a new avenue for the development of highly durable ORR catalysts for PEMFCs (Table 5.4).

5.2.2 Nonprecious Metal Catalysts

The issues of high cost-effectiveness of Pt-based catalysts for commercialization and their toxic stability have not been solved, however, even though Pt-based catalysts have shown good performance in PEMFCs. The kinetic reactions of the cathode ORR are 6 orders of magnitude slower than those of the anode HOR, so the researches on the catalysts for the proton exchange membrane fuel cells are mainly focused on the cathode aspect. Given the cost structure of fuel cell stack systems, catalysts account for the highest percentage of the stack cost, and as the number of cells used in the system increases, the proportion of catalyst cost increases, and fuel cells

Table 5.4 Summary of the ORR performance of Pt-based single atom catalyts reported in recent years

Catalyst	ECSA (m^2/g Pt)	Mass activity (A mg_{Pt}^{-1}) at 0.9 V versus RHE	Specific activity (mA cm^{-2})	References
J-PtNWs/C	118	13.6	11.5	Science, 2016, 35, 1414–1419
PtNW/SG	24.5 (0.1 M KOH)	0.167	0.675	Adv. Mater. 2015, 27, 1229–1234
Pt_1@Pt/NBP	34.5 (0.1 M KOH)	0.214	0.62	ACS Catal. 2021, 11, 1, 466–475
Pt_1/NPC	NA (0.1 M KOH)	0.17	0.14	ACS Catal. 2018, 8, 8450–8458
PtNW/S-CNT	15.7 (0.1 M $HClO_4$)	0.272	1.61	ACS Sustainable Chem. Eng. 2018, 6, 93–98
Pt-NfnPPy	45.9 (0.1 M KOH)	0.136	0.296	Sci. Rep. 2015, 5, 16100
Pt-PANI-GDL	12 (0.1 M $HClO_4$)	0.147	0.108	J. Mater. Chem. A, 2017, 5, 15260–15265
Pt/VC-PANI	75.5 (0.5 M $H2SO_4$)	0.64	1.28	Fuel Cells 12,2012, No. 6, 939–945

require high catalyst performance. The high cost and low durability of Pt-based catalysts are the main limitations for proton exchange membrane fuel cells. Excellent catalytic performance of catalysts is necessary for high-performance batteries. The real commercialization of non-precious metal catalysts still needs to solve the issues of high activity and stability. Therefore, the development of highly active, low-cost, and strongly stable non-precious metal catalysts to replace Pt-based catalysts is more than a missing part of the PEMFC commercialization applications. Currently, the research on non-precious metal catalysts is mainly focused on three directions, namely, composite metal oxides, non-precious metal, and N co-doped carbon-based (M-N-C) and Chevrel phase materials. It has been identified for several decades that M-N-C has been found to have excellent ORR activity and stability, and has certain resistance to methanol oxidation, thus receiving focused attention.

5.2.2.1 M-N-C Catalysts

The beginnings of M-N-C materials as oxygen reduction catalysts can be traced back to the research on nitrogen-containing metal macrocyclic compounds, which was first discovered by Jasinski [146] in 1964 when cobalt phthalocyanine macrocyclic compounds as precursors showed good catalytic activity for oxygen reduction in alkaline environments. A series of oxygen reduction catalysts with porphyrins, phthalocyanines, and other non-precious metal-N_4 macrocyclic compounds as precursors were subsequently prepared. It was found that the catalyst oxygen reduction performance was mainly influenced by the pyrolysis temperature, non-precious metal, and carbon matrix, among which, the ORR activity and stability of catalysts were significantly improved after high-temperature heat treatment. However, owing to the relatively high cost-effectiveness of transition metal macrocycles, it was found that, furthermore, economical M-N-C catalysts could be prepared by heat treatment of a mixture of nitrogen-rich organic matter and transition metal salts in the form of a carbon support. Moreover, the metal type plays a decisive role in the catalytic performance, and the order of oxygen reduction activity is Fe > Co > Mn > Cu > Ni [147] in non-precious metal carbon matrix materials with different metal centers. The ORR catalysts fabricated from transition metals, nitrogen-rich substances, and carbon-based materials as precursors have been studied with considerable achievements. Although the oxygen reduction activity of M-N-C catalysts still suffers from a certain gap with commercial Pt/C, such catalysts have emerged as the most promising alternative to ORR catalysts that could potentially replace Pt-based materials [148–151]. Ongoing technological innovations using simple synthesis methods and inexpensive feedstocks have laid the foundation for large-scale production.

To systematically design and synthesize higher efficient and robust ORR catalysts, it is necessary to investigate the active sites of the catalysts obtained after pyrolysis. However, the active site properties of M-N-C catalysts have been controversial, and the understanding of their reaction mechanism is unclear. One view is that the M-N_x structure formed by atomically dispersed metal ligated with nitrogen is the active center; another view is that the N-doped carbon-coated metal, metal nitride, metal carbide, and metal oxide nanoparticles (M@NC) are the main active center [152]. Meanwhile, the co-existence of M-Nx groups and M@NC in some catalysts may play an important role in the high onset reduction potential, which may be explained by a two-site mechanism requiring two adjacent active sites, M-N_x groups and M@NC particles, to effectively synergize to promote the 4-electron ($4e^-$) process. The diversity of catalytically active sites is also a reasonable explanation for the proliferation of various catalytic materials.

Today, M-N-C-catalysts obtained from the pyrolysis of elemental precursors and where the metal is either Fe and/or Co, are the most promising non-platinum-group-metal (non-PGM) catalysts for the ORR in PEMFC [153, 154]. However, to replace platinum-based catalysts, non-PGM catalysts for ORR must still meet the following three criteria: a volumetric activity $\geq$ 1/10 that of Pt-based catalysts, excellent mass-transport properties, and high durability. The first two criteria are essential for achieving high power density. Although poor volumetric activity had been a major challenge for M-N-C-catalysts since their discovery, recent breakthroughs have increased their activity to about 1/10 that of Pt-based catalysts [155–158]. This level is sufficient for practical applications [159, 160], because M/N/C-catalyst loadings of 1–5 mg cm^{-2} are technically and economically feasible, whereas typical platinum loadings at the cathode are only 0.1–0.4 mg cm^{-2} to restrain cost [161]. However, to date, these thicker M-N-C-catalyst cathodes have not yet shown sufficient power density because of poor mass-transport properties intrinsic to the catalyst. To achieve high performance of Fe-N-C catalysts in PEMFC, there are two strategies: (1) increasing the intrinsic activity of Fe-N-C or (2) increasing the number of active sites on a given electrode. Besides, the activity of electrochemical catalysts can also be enhanced by the improvement in conductivity for fast electron transport.

Here, the volumetric activity target at 0.8 $V_{iR\text{-}free}$ has been achieved by a couple of highly active Fe-N-C catalysts. In 2016, U.S. DOE changed the activity target to an area current density of 0.044 A cm^{-2} @0.9 $V_{iR\text{-}free}$. This activity target is rather challenging because most PGM-free catalysts could not even deliver a considerable current at this very high voltage. Thus, little literature reported relevant activities referring to this new target. Even for the catalysts that have achieved the 2020 DOE volumetric activity target, they are still far from satisfying the new target of 0.044 A cm^{-2}@0.9 $V_{iR\text{-}free}$. To date, Zelenay and co-workers reported the highest activity of 0.016 A cm^{-2} at 0.9 $V_{iR\text{-}free}$ under DOE testing protocol, using a hierarchical porous Fe–N-C catalyst ((CM + PANI)-Fe-C). As shown in Fig. 5.6a, the voltage corresponding to 0.044 A cm^{-2} in the iR-free polarization curve is 0.87 V, which is 30 mV lower than the 2020 activity target and 10 mV lower than the 2018 activity target. For the application of Fe-N-C catalysts in PEMFC, the intrinsic activity is not the only important factor [162]. The mass transfer in the catalyst layer also plays an important role in determining the fuel cell power performance in the practical voltage range, usually around 0.6–0.7 V [159, 163]. In this fuel cell voltage range, we make a balance between energy conversion efficiency and power density of PEMFCs. To improve the mass transfer ability, the structure of Fe-N-C catalysts should be intentionally designed [164–166]. For evaluating power density, a fuel cell voltage of 0.6 V is meaningful because it allows for an acceptable balance between energy efficiency and power density for well-performing catalysts. The highest power densities at 0.6 V reported before this work in H_2-O_2 PEMFC comprising M-N-C-based cathodes have been 0.38 W cm^{-2} [158], 0.37 W cm^{-2} [167], 0.32 W cm^{-2} [155] and 0.27 W cm^{-2} [157] with 1.8, 1, 1.5 and 2 bar back-pressures, respectively. This is significantly lower than the typical value of ca. 1–1.2 W cm^{-2} attained by a Pt-based cathode with a loading of 0.3 mg_{Pt} cm^{-2} tested under identical operating conditions. The challenge of achieving a comparable power density with M-N-C-based cathodes at 0.6 V has been especially difficult because, for M-N-C-catalysts, high microporosity is required to obtain high catalytic activity [167, 168], whereas high mesoporosity or macroporosity has been commonly viewed as a requirement for efficient transport of ORR-related species (O_2, protons and H_2O) in the catalyst layer [169, 170], As discussed above, ORR volumetric performance is related to catalytic activity because mass transport limitations are related to electrode thicknesses, in which a thick MEA will lower mass transport and decrease power density [155]. Therefore, it is important to attain the same kinetic current density on NPMC cathodes as Pt-based cathodes without exceeding the thickness; however, the volumetric current densities of state-of-the-art Fe-N-C catalysts are still far from the DOE target. This can be explained by the facts that Fe-N-C catalysts are carbonized at high temperatures (> 700 °C) and are more hydrophobic and that although the porous structure of Fe-N-C catalysts can contribute to higher surface areas for active site dispersion and mass transport enhancement, it also lightens bulk density and cause these catalysts to be much thicker than Pt/C at same mass loading. Therefore, the fabrication of thin Fe-N-C catalyst layers with fast mass transfer is a key issue in future studies.

In addition, although Fe-N-C has the optimal ORR catalytic activity under acidic conditions, due to the easy dissolution of iron to form iron ions under acidic conditions, and the by-product of the ORR reaction, H_2O, forms Fenton reagent (Fenton) with strong oxidizing properties, which has an extremely destructive effect on the proton exchange membrane and catalyst. Therefore, exploring non-iron-based non-Platinum catalysts is the development direction of non-Platinum catalysts, and thus, Co-N-C materials have become another research hotspot. The substitution of Fe-based catalysts with Co-based ones in PEMFC can alleviate the catalytic reaction to H_2O_2. Nevertheless, to date, Co-N-C catalysts have exhibited inferior ORR activity and higher yields of the product of the $2e^-$ reduction of oxygen, H_2O_2, than Fe-N-C catalysts. Based on the latest research results, a reasonable Co-N-C catalyst should have high active site density, reasonable microstructure and pore size distribution, and good corrosion resistance. The recent improvement of catalytic activity is an urgent problem, and a lot of work is still needed for the reasonable design and preparation of membrane electrodes suitable for non-Pt-based catalysts.

Although M-N-C SACs show great prospects in catalysis, their demerits are also obvious. On one hand, being subjected to only one kind of specific active site in most of the M-N-C SACs, it is tough for M-N-C SACs to break the linear scaling relations between intermediates when it comes to complicated reactions involving multiple intermediates [171], except for a small number of reports [172]. As a result, M-N-C SACs are usually hard to improve the catalytic activity of reactions involving multiple intermediates to an optimal level. Additionally, some reactions need adjacent active sites to catalyze the reaction. For example, single Pt atom catalysts are usually inclined to catalyze ORR toward H_2O_2 instead of H_2O owing to a lack of continuous active sites [173–175]. On the other hand, to maintain the single-atom configuration, the loading amount of metal is often limited to avoid aggregation. Hence, the overall activity of M-N-C SACs is not very pleasant despite high activity per atom [176].

Inspired by bimetallic nanocatalysts, M-N-C-based dual-atomic-site catalysts (M-N-C DASCs) have aroused great interest as an extension of M-N-C SACs to solve the above problems. Just like bimetallic nanocrystal catalysts, M-N-C DASCs have more chances to adjust the d-band center by the interaction of electronic orbits compared with M-N-C SACs, thus optimizing the adsorption energy of intermediates on the active sites. Thereby, M-N-C DASCs have great potential to elevate the intrinsic activity of catalysts. In addition, DASCs are more likely to be multi-functional catalysts, or to break the linear scaling relationship and optimize the adsorption energy of multiple intermediates due to more than one kind of adsorption site. Moreover, M-N-C DASCs possess more accessible active sites, and the loading amount of metal atoms in M-N-C DASCs can be promoted further, thus enhancing the overall catalytic performance and promoting the development of M-N-C SACs for practical application in the future. Though still at the early stages, the study of M-N-C DASCs has drawn much attention (Fig. 5.13).

A uniformly dispersed Zn/Co-NC*x*-C diatomic high-efficiency oxygen reduction catalyst has been prepared based on the competitive coordination strategy [178]. Owing to the active site of Zn-Co diatomic pair with high activity, the catalyst showed remarkable catalytic activity and stability under both alkaline ($E_{\text{onset}} = 1.004$ V, $E_{1/2} = 0.861$ V) and acidic conditions ($E_{\text{onset}} = 0.97$ V, $E_{1/2} = 0.796$ V). The catalyst was applied to a power density of 705 mW cm^{-2} (about 78% of commercial Pt/C) in H_2/O_2 fuel cells (Fig. 5.14) and exhibited no degradation of activity after 8 h of stability testing. The above results indicate that the catalyst has great practical application potential in fuel cells. In this research, M-N-C DACs have been successfully applied as the cathode catalyst of fuel cells, greatly reducing the cost of the catalytic layer,

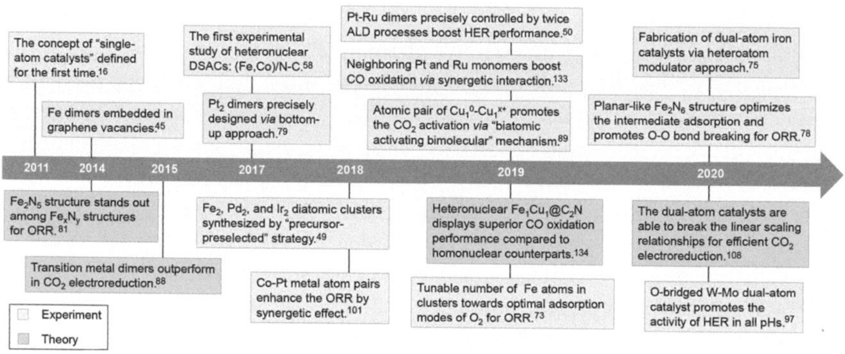

Fig. 5.13 A timeline of major developments of M-N-C DASCs [177]

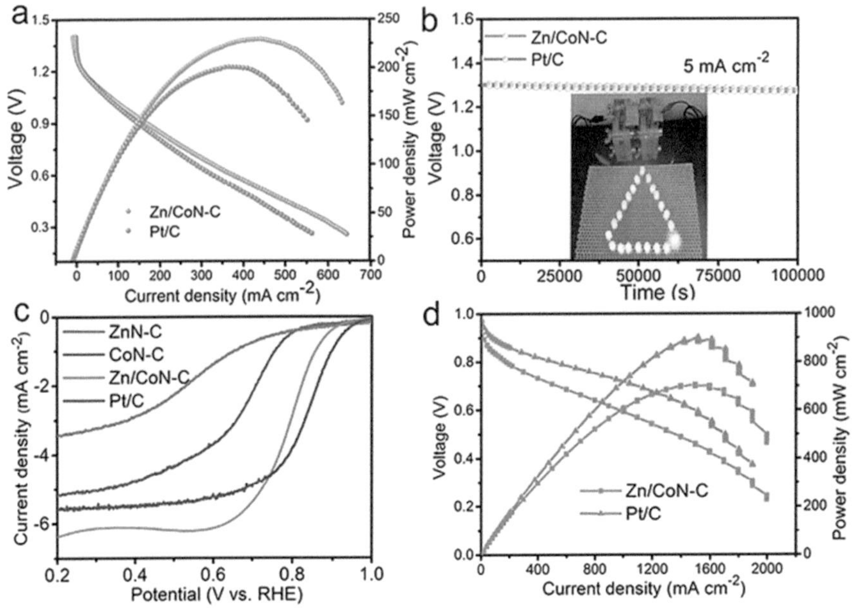

Fig. 5.14 **a** Polarization and power density curves of primary Zn-air batteries using Zn/CoN-C and Pt/C as ORR catalyst in 6 M KOH electrolyte. **b** Galvanostatic discharge of the fabricated Zn-air batteries, and the inset is the photograph of a series of LED powered by two liquid Zn-air batteries in series. **c** LSV curves of ORR for different catalysts in 0.1 M $HClO_4$ solution. **d** H_2/O_2 fuel cell polarization plots [178]

which is a breakthrough in the field of fuel cells. This increases the potential for commercial development of fuel cells and is expected to facilitate the development of fuel cell vehicles in the future.

5.2.2.2 Transition Metal Oxides/Sulfides

Transition metals are seen as an effective alternative to precious metal catalysts due to their abundant reserves, low cost, ease of preparation, and chemical activity in alkaline electrolytes. In particular, transition metal oxides such as Fe, Co, and Mn exhibit excellent oxygen reduction activity in alkaline electrolytes and are expected to replace precious metal Pt catalysts in PEMFC. There have been extensive studies on the oxygen reduction activity of transition metal oxides, but the biggest drawback of transition metal oxides is their poor electrical conductivity (10^{-6} to 10^{-5} S cm^{-1}) [179], so in practical applications, it is often necessary to composite transition metal oxides with graphene or carbon nanotubes with high electrical conductivity. Therefore, in practical applications, it is often necessary to compound transition metal oxides with graphene or carbon nanotubes, which have high electrical conductivity [180, 181]. For instance, a nitrogen-doped graphene, CNT, and Co_3O_4 composite catalyst, NG/CNT/Co_3O_4 was prepared [182] using a solvothermal method by mixing graphene oxide, CNT and Co ions. the authors concluded that the synergistic effect of the three components in the catalyst resulted in excellent oxygen reduction activity and electrochemical stability in alkaline electrolytes.

Additionally, transition metals can yield a wide range of oxides with different crystal structures due to their diverse valence states. Many metal oxides, especially group IV and V metal oxides, are chemically stable in acidic electrolytes, but the low electrical conductivity and lack of oxygen adsorption sites on the metal oxide surface result in very low ORR activity. These problems can be solved by surface modification, doping, alloying and formation of highly dispersed nanoparticles. In addition, the electrocatalytic activity of transition metals depends mainly on the formation of redox active centers at the ORR potentials when switching between different valence states. The electrocatalytic activity is often improved by modulating their chemical composition, oxidation state, structure, morphology and crystal structure. ABO_3-type chalcogenide oxides, such as those with rare earths or alkali metals at the A-site and transition metals at the B-site, are widely used as catalysts for PEMFC. The performance of chalcogenide oxides can be tuned by replacing the cations at the A or B sites to obtain specific structures, oxygen contents and electrocatalytic properties. ORR catalytic performance can be improved by cation doping, surface optimization, nanostructures and composites.

For that, the Co_3O_4-N-rm GO catalyst [183] has high oxygen reduction activity in alkaline media. The high oxygen reduction activity of Co_3O_4-N-rm GO catalysts prepared by the group was found. It was further found that the oxygen reduction catalytic performance of the $MnCo_2O_4$-N-rm GO composite structure catalysts obtained when part of the Co atoms in Co_3O_4 were replaced by Mn atoms could be further enhanced. Stability tests showed that the composite catalyst has better electrochemical stability than commercial platinum carbon. The inversion of spinel structure could be achieved by modulating the octahedral site atomic species of Co_3O_4 spinel [184], thus regulating the catalytic activity of spinel (Fig. 5.15). The results showed that the {Co}[FeCo]O_4 spinels with the reversed structure exhibited higher oxygen reduction activity than the ortho-spinels {Co}[Co_2]O_4 and {Co}[Fe_2]O_4 with higher oxygen reduction activity, and the half-wave potential even exceeds that of commercial Pt/C 40 mV. DFT calculations show that the trans {Co}[FeCo]O_4 in spinel, the octa faceted position of Fe atoms and Co atoms have an electronegativity difference between them, which allows the charge of both atoms to be transferred and polarization, which facilitates the O–O bond breakage, thus enhancing its catalytic activity. This study reveals the conformational relationship between the spinel structure and oxygen reduction activity, and provides a theoretical basis for the design of spinel-based catalysts.

It has been shown that in addition to monometallic oxides, oxides containing two or more transition metals also have excellent ORR catalytic activity [185–188]. Crystal structure and morphology-controlled porous spinel Co controlled porous spinel $Co_xMn_{3-x}O_4$ is developed [189], where the cubic $Co_xMn_{3-x}O_4$ nanorods have excellent ORR catalytic activity, and their ORR starting potentials reach 0.9 V

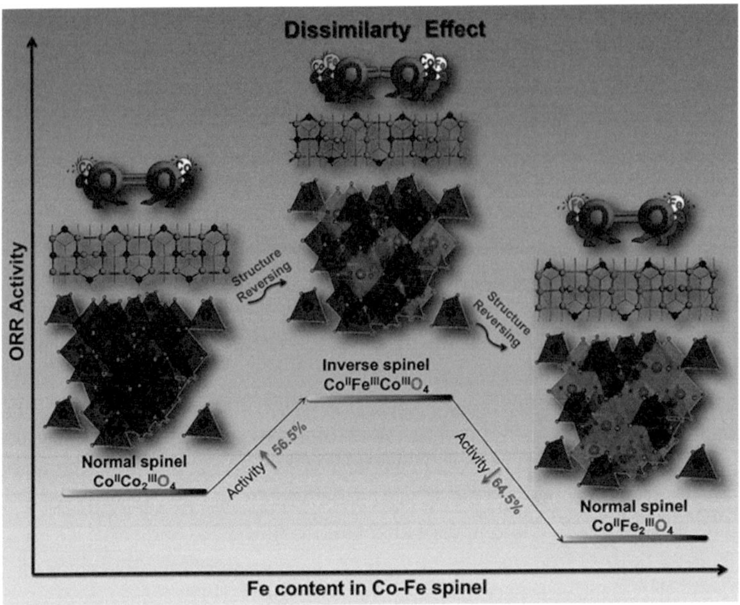

Fig. 5.15 The spinel structures of {Co}[Co_2]O_4, {Co}[FeCo]O_4, and {Co}[Fe_2]O_4 and the corresponding oxygen adsorptions and ORR activities (Fe, green; Co, blue; Absorbed O, magenta; Lattice O, red) [184] (color figure online)

and half-wave potentials reach 0.72 V, which is very close to the catalytic performance of commercial Pt/C catalysts. The catalytic performance of the commercial Pt/C catalyst is very close to that of the commercial Pt/C catalysts. The catalytic performance was very close to that of commercial Pt/C catalysts, and also had better stability than that of commercial Pt/C catalysts, and the catalytic performance did not change after 10,000 cycles. The catalytic performance did not deteriorate after 10,000 cycles. The catalytic performance of the tetragonal structure of $Co_xMn_{3-x}O_4$ is slightly lower than that of the cubic $CoMn_{3-x}O_4$ nano rods. This is due to the metal sites on the surface of the cubic structure on an O_2 adsorption energy of than the tetragonal structure, so the cubic structure has more active sites with the same surface area. The cubic structure has more active sites with the same surface area. Although transition metal oxide catalysts have low cost, high selectivity high selectivity and good catalytic performance, they are a promising class of non-precious metal catalysts. However, the preparation process of transition metal oxide catalysts is complex, and the oxides are prone to slow decomposition and nanoparticle and the current density is much lower than that of commercial Pt/C catalysts, which are the problems that need to be solved for transition metal oxide catalysts.

Since transition metal oxide oxygen reduction catalysts can only be used under alkaline conditions, limiting their applications, attention was turned to transition metal sulfides to develop catalysts that can be used under acidic conditions. DFT calculations have shown that the negatively charged sulfur atoms in Co_9S_8 can act as active sites to facilitate the adsorption of oxygen-containing species and promote O–O bond breaking [13], thus exhibiting oxygen reduction activity comparable to that of precious metal platinum–carbon catalysts in acidic electrolytes. Nevertheless, the catalytic activity of Co_9S_8 prepared in the present experiments cannot reach the level of Pt, probably due to the large size of the present Co_9S_8 particles and the weak synergy with the carbon group [190]. The composite catalyst Co_9S_8/NHCS with nitrogen-doped hollow carbon nanospheres loaded with monodisperse Co_9S_8 nanoparticles was prepared [191] by intercalating aminobenzene sulfonate into cobalt-aluminum hydrotalcite, followed by simple high-temperature carbonization and hydrochloric acid etching, which resulted in a strong interaction between Co_9S_8 and nitrogen-doped carbon groups are generated simultaneously during the high temperature process, thus providing a strong interaction between the two and allowing the catalyst to exhibit better oxygen reduction activity and stability in alkaline media than commercial Pt/C catalysts.

5.2.2.3 Other Oxygen Reduction Catalysts

Metal carbides with good electrical conductivity and special electronic structure have been studied more intensively in ORR and hydrogen precipitation reactions (HER). Tungsten carbide has similar electronic properties to Pt, and its ORR activity in acidic media is better than that of some transition metal carbides [192]. WC is directly applied to fuel cell cathodes, which is limited by its poor corrosion resistance in acidic media. The introduction of the second metal tantalum (Ta) in WC can form a W-Ta alloy in the material, which can improve the stability of the material in acidic media [193].

Perovskite oxides are a promising family of electrocatalysts for the ORR because of their unique physical and chemical properties, such as variable crystal structure and non-stoichiometric chemistry. Studies have shown that the catalytic properties of perovskite oxides in the ORR are largely related to oxygen vacancies, which alter their electronic and crystal structures and surface chemistry. As emerging ORR catalytic materials, ABO_3 perovskite oxides with unique physical and chemical properties have also aroused a great deal of research interest in recent years. The unique physical and chemical properties, such as variable crystal structure and non-stoichiometric chemistry, influence the electronic structure of the active sites and promote the ORR performance by improving the adsorption of O_2 and activating the O–O bonds [194, 195]. For perovskite oxide $ABO_{3-\delta}$, increasing the value of δ can gradually result in crystal structure distortion. Significant distortion would change the ideal cubic structure to an orthogonal or tripartite structure or even cause the collapse of the structure, accompanied by changes in the electronic structure and surface chemistry [196]. Therefore, the activity of the ORR is also affected. Table 5.1 summarizes the ORR performance of ABO_3 perovskite oxide electrocatalysts with different vacancies, along with important parameters of the perovskites, which represent different levels of vacancies. It is clear that too many oxygen vacancies do not benefit the

ORR. The best ORR performance is often obtained from those perovskite oxide catalysts with moderate contents of oxygen vacancies. A high oxygen vacancy content would weaken the electron holes and redox couples, leading to poor ORR activity. Furthermore, too many oxygen vacancies may result in structure collapse [197, 198]. A new method of using 5% H_2 in Ar to synthesize single-crystal $Ca_2Mn_2O_5$ with oxygen vacancies at an evaluated temperature was reported [199], and the resultant $Ca_2Mn_2O_5$ showed a higher ORR activity than that of $CaMnO_3$. However, excessive oxygen vacancies formed at high temperatures deteriorated the structural stability and electrical conductivity, and even resulted in structural collapse. Therefore, a proper content of oxygen vacancies is needed to regulate the electronic/crystal structures and surface chemistry of perovskite oxides for improved ORR performance.

By introducing indium metal into ZIF-67 to control the content of the derived nitrogen component to regulate the metal composition ratio, the intermetallic compound $Co_3InC_{0.7}N_{0.3}$ anti-chalcogenide structure nanomaterials were fabricated with high ORR performance [200]. The half-wave potential (0.854 V) in alkaline medium even exceeds that of most chalcogenide-structured materials. $(La_{0.8}Sr_{0.2})_{1+x}MnO_3$ ($x = 0$, 0.05, 0.1) (denoted LS, $LS_{1.05}$ M, $LS_{1.1}$ M, respectively) multiphase chalcogenide oxides were synthesized by the polymer-assisted chemical solution (PACS) method in alkaline media. Compared to pristine LSM, $LS_{1.05}$ M has a 21% higher diffusion-controlled ORR current density, 87% higher current density at 0.8 V (relative to Ag/AgCl 3.5 M), and an ultra-low electrochemical impedance [201]. By introducing F anions at the O-site of the double chalcogenide oxide ($Sr_2Fe_{1.5}Mo_{0.5}O_{6-\delta}$, SFM), Sun's team has designed a new type of oxide with excellent ORR activity of SOFC cathode with excellent ORR activity. The obtained doped F of SFM cathode ($Sr_2Fe_{1.5}Mo_{0.5}O_{6-x-\delta}F_x$, the SFMF, $x = 0$, 0.1, 0.1 0.2 and 0.3) shows improved crystal structure symmetry and enhanced electrocatalytic activity. At 800 °C, the F-doped SFM of the polarization resistance is significantly reduced to 0.072 Ω cm^{-2}, which may be due to its oxygen reduction capacity being enhanced. In addition, the F doping reduces the metal–oxygen bond energy, which accelerates the O^- ion transfer in the electrochemical process transport, thus significantly increasing the ORR activity [202]. David Portehault*'s team reported [203] the synthesis in molten salts of layered chalcogenide $La_{0.5}Sr_{1.5}MnO_4$ (l-LSMO) and pseudocubic chalcogenide $La_{0.7}Sr_{0.3}MnO_3$ (pc-LSMO) ligand-free nanoparticles. These particles are readily realized as ORR electrocatalysts under alkaline conditions. They are sensitive to the oxygen in water 4-electron reduction showed high ORR selectivity.

5.2.3 *Metal-Free Catalysts*

Carbon materials are considered candidates for metal-free catalysts due to their catalytic activity, stability, and cost-effectiveness. In addition, heteroatom (N, S, P, and B) doping or structural modulation can enhance the catalytic activity, and heteroatom-doped carbon materials are effective metal-free electrocatalysts. The doping of heteroatoms improves the surface properties of transition metals, the electronic structure of the electrocatalyst surface is optimized, and the activation area is increased, thus improving the electrocatalytic performance of transition metal materials. Due to the different doping atoms, the properties of transition metal hybrids show diversity, and the following is a specific analysis of the effect of carbon materials doped with different heteroatoms on their oxygen reduction properties.

(1) N-doped carbon materials. When the mobile π electrons are activated in the sp^2 hybridized carbon materials, the different electronegativity (2.55 for C and 3.04 for N) causes the off-domain effect, which makes the center of the electron cloud near the nucleus of the carbon atom start to move slowly toward the nitrogen atom and eventually makes the surrounding carbon atoms exhibit positive electrical properties. Gong's group found that the doping of nitrogen into carbon nanotube arrays can change the way oxygen is adsorbed [204]. Subsequently, the above changes are verified by density flooding theory simulations [205]. The functionalization of nitrogen not only alters the chemistry of the carbon surface but also changes the alkalinity, hydrophilicity/oleophilicity, and electronic structure of the material by providing a pair of lone pair electrons and distributing their movement into the undoped graphene lattice [206]. Strelko and his colleagues suggested that the addition of nitrogen atoms could reduce the band gap and increase the charge mobility of the band-gap graphite lattice, and showed by quantum chemical calculations that Pyrrole groups have the highest charge mobility at edge positions and those where quaternary N atoms are bound (quaternary N), and this change in carbon structure ultimately reduces the electron work function at the carbon-liquid or gas interface compared to a pure carbon surface [207]. Depending on the way of C bonding, they can be classified into four types: graphitic N, pyrrole N, pyridine N, and oxidized N. Graphitic N refers to the substitution of N atoms for C atoms in the graphite plane. Graphitic N atoms have been reported experimentally and theoretically as ORR catalytic sites due to their reduced adsorption energy. The introduction of graphitic N in the edge structure of graphene enhances the catalytic activity, and such a structure not only increases the first electron transfer rate but also favors a reduction pathway that exhibits 4 electrons [208]. The substitution of pyridine N is similar to that of graphitic N in that the N atom replaces the C atom in the graphitic plane, except that the pyridine N is often positioned at the edge of the carbon material. Experiments found that the amount of graphitic N contained in the nitrogen-doped graphene catalyst determined the ultimate current density, and pyridine N facilitated the conversion of the reaction from the 2-electron regime to the 4-electron regime and increased the starting potential of the reaction [206]. In contrast, pyrrolic N is a structure that forms a five-membered ring with five carbon atoms. Oxidized N refers to two carbon atoms and one oxygen atom, respectively, and its position is often at the edge of the material [209], as shown in Fig. 5.16 [210].

(2) S, P-doped carbon materials. S and P have the same electronegativity as C (S is 2.58 and P is 2.19), so they will have less effect on the charge density of

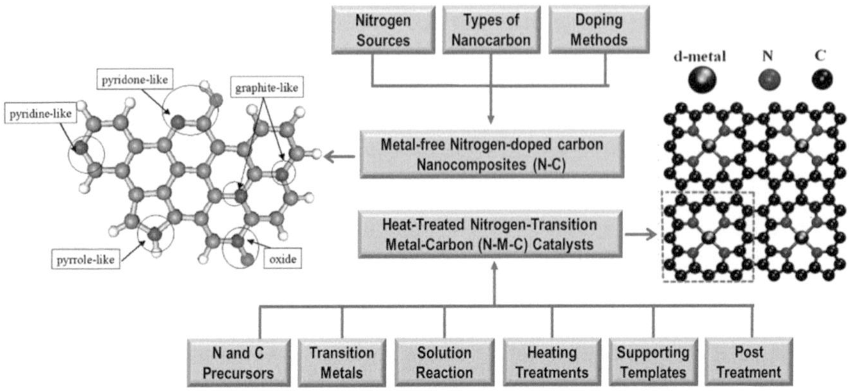

Fig. 5.16 Effect of synthetic parameters on performance of metal-free nitrogen-doped (N-C) and transition metal-derived carbon (M-N-C) catalysts [63]

the material. But we know that S or P are larger than the atomic radius of C, and therefore the atomic radius of S atoms is in the upper part of the graphene plane, so the catalytic effect is mainly on the S atoms on the edge of the carbon material. In P element-doped carbon materials, the bonding mode is mainly P–C and P–O bonds, and their positions are also present in the C on the edges of the material. Related studies have also demonstrated that combining S or P doping with N-doped carbon materials is a good way to improve ORR properties. Choi et al. prepared a P- and N-doped carbon material electrocatalyst by pyrolysis of DCDA (C-N source) and phosphoric acid (P source), and experimentally demonstrated that the addition of P elements could improve the catalytic activity of NDC. The optimized PNDC of ORR mass activity was higher than that of NDC, improved 4 times or more [211]. Five nitrogen-sulfur co-doped carbon materials are prepared by selecting five different amino acids as raw materials by pyrolysis, and their electrocatalytic activities are compared for fuel cell ORR. Results showed that the starting potentials of all five synthesized nitrogen-sulfur double-doped carbon catalyst materials were significantly increased [212].

(3) B, F-doped carbon materials. Although B, F (B is 2.04 and F is 3.98) are electronegative with C which differs from that of which leads to a large difference in the electronegativity of C atom's spin density will be weaker, but the same can indirectly affect the ORR performance. For example, in 2012, Woo et al. synthesized N, B, and P ternary-doped carbon ORR catalysts, demonstrating that the addition of B and P heteroatoms can change the properties of carbon and improve the ORR activity of N-doped carbon catalysts. It was pointed out in the literature that the proportion of pyridine-type N could be increased by B doping to achieve the expansion of the number of sp^2-carbon structures.

In summary, the reasons for the enhancement of the oxygen reduction activity of carbon materials caused by single heteroatom doping can all be attributed to the following two factors: (1) breaking the electroneutrality of pure carbon materials, causing an asymmetry in charge density or spin density; (2) activating the electrons on the C atom near the heteroatom, making it easy to transfer to the O atom. Considering the single doped heteroatom brings less active sites, or the active sites are too strong or too weak for the adsorption of oxygen species resulting in insufficient activity of the active sites, etc., the use of two or more elements co-doped to make full use of the synergistic effect between heteroatoms can improve the adsorption ability of carbon materials to oxygen species and further improve the heteroatom-doped carbon The activity and stability of heteroatom-doped carbon materials can be further improved. For heteroatom-doped carbon materials, improving and exploring synthetic methods to achieve carbon materials with high specific surface and reasonable multi-stage pore structure, combining experimental characterization and theoretical simulations to clarify the reactive sites in different carbon materials, regulating the relative content of different nitrogen-doped forms in nitrogen-doped carbon materials and creating independent electron transport channels, exploring and optimizing the conditions and types of heteroatom co-doping are the keys to develop efficient and stable carbon materials catalysts and realize large-scale production in the future.

5.3 Electrocatalysts Hydrogen Oxidation Reaction

As a new class of fuel cells, the development of anion exchange membrane fuel cells (AEMFCs) is still in its infancy, and one of the important issues is the extremely slow hydrogen oxidation reaction (HOR) kinetics [213–221]. The HOR kinetics in acidic media is fast, and only a very low Pt loading ($< 0.05\ mg_{pt}/cm^2$) is required for operation [222–224]. However, in an alkaline medium, the HOR rate is 2–3 orders of magnitude lower than in the acid, even at the Pt electrode surface, and the reaction mechanism is not yet clear [225, 226]. This implies that the required Pt loading at the anodic end is significantly higher than $0.05\ mg_{pt}/cm^2$, which inevitably poses a cost problem for AEMFCs [227]. Therefore, the study of HOR in alkaline media and the development of efficient and low-cost HOR catalysts are urgently needed. The electrode reactions are as follows.

Anode (H_2 oxidation reaction):

$$2H_2 + 4OH^- \rightarrow 4H_2O + 4e^- \tag{5.8}$$

Cathode (O_2 reduction reaction):

$$O_2 + 2H_2O + 4e^- \rightarrow 4OH^- \tag{5.9}$$

Total reaction:

$$O_2 + 2H_2 \rightarrow 2H_2O \tag{5.10}$$

The slow alkaline HOR reaction rate weakens the advantage of AEMFCs in terms of low cost and limits the development of AEMFCs. Due to this limitation, the design, preparation, and evaluation of high-performance, low-cost alkaline HOR electrocatalysts around the alkaline HOR mechanism have been carried out worldwide [228, 229].

5.3.1 Overview of Hydrogen Oxidation Mechanism

(1) HOR in acidic media

The design of efficient catalysts is inseparable from the study of the catalytic mechanism. HOR in acidic media is relatively simple, and the process is as follows:

$$H_2 + M \rightarrow 2M\text{ - - - }2H_{ads}\ \text{(Tafel)} \tag{5.11}$$

$$H_2 + M \rightarrow M\text{ - - - }H_{ads} + H^+ + e^-\ \text{(Heyrovsky)} \tag{5.12}$$

$$M\text{ - - - }H_{ads} \rightarrow H^+ + e^-\ \text{(Volmer)} \tag{5.13}$$

It is generally accepted that the HOR process on Pt surfaces in acidic media follows the Tafel-Volmer procedure. H_2 is first adsorbed on the Pt surface by homolytic adsorption to form M-H_{ads}, followed by the oxidation reaction. where Tafel is the rate-determining step (RDS). Therefore, the catalytic material needs to have sufficient chemisorption capacity for H_2 with sufficient chemisorption capacity. This is important for the design of non-precious metal basic HOR catalysts is also very important. Typically, the dissociation of adsorbed H_2 molecules generally requires at least two atomic sites or vacancies that are adjacent and not adsorbed on the surface, and H_{ads} is adsorbed on the metal surface in the atomic state. Experiments [230] showed that H_2 preferentially adsorbs on higher coordination sites. In contrast, the strong chemisorption of H atoms requires 3–7 neighboring metal atoms, also known as the group effect [230, 231]. Studies [232] confirmed that satisfying the H_2 dissociative adsorption sites must include the formation of independent vacancies and spaces for subsequent diffusion and polymerization, and these coupling effects determine the surface activity of the catalyst. Scanning tunneling microscopy observations indicate that H_2 molecules dissociate and adsorb on the active sites of Pd (111) crystalline surface, requiring at least three or more vacancies at the active sites. At higher H_{ads} coverage, the H_{ads} at higher H_{ads} coverage, there is an indirect repulsion between Had, resulting in a significant decrease in the metal-hydrogen binding energy (HBE). and the production of weakly adsorbed and active H_{ads} species. Therefore, controlling the number of adjacent active sites is expected to be used for HBE modulation of catalysts to enhance the catalytic activity.

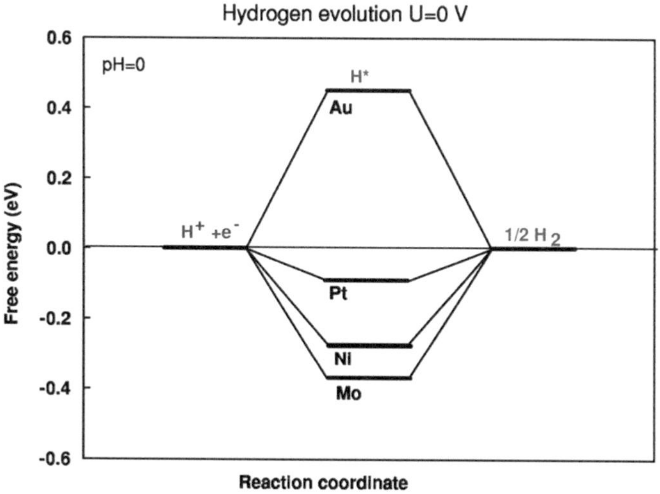

Fig. 5.17 Volcano plot for HOR/HER on bulk monometallic surfaces in acidic electrolytes [236]

Since Tafel is a tachyonic step in acidic media, the HOR kinetics is mainly controlled by the adsorption energy of the metal and the reaction intermediate H_{ads}. As shown in Fig. 5.17, the exchange current density of the hydrogen electrode reaction and the HBE on different metal surfaces showed a volcano-type relationship, and the catalytic activity was higher due to the moderate HBE on the Pt surface, which was distributed near the top of the volcano curve [233–236].

(2) HOR in alkaline media

(2.1) The chemical formula of HOR:

$$H_2 + 2OH^- \rightarrow 2H_2O + 2e^- \tag{5.14}$$

It is generally accepted that the basic HOR reaction would involve the following three reaction steps (H_{ads} represent H atoms in the adsorbed state, and M represents the electrocatalytic surface active site) (Fig. 5.18).

$$H_2 + M \rightarrow 2M\text{ - - - }2H_{ads}\ \text{(Tafel)} \tag{5.15}$$

$$H_2 + 2OH^- + M \rightarrow M\text{ - - - }H_2O + e^-\ \text{(Heyrovsky)} \tag{5.16}$$

$$M\text{ - - - }H_{ads} + OH^- \rightarrow M + H_2O + e^-\ \text{(Volmer)} \tag{5.17}$$

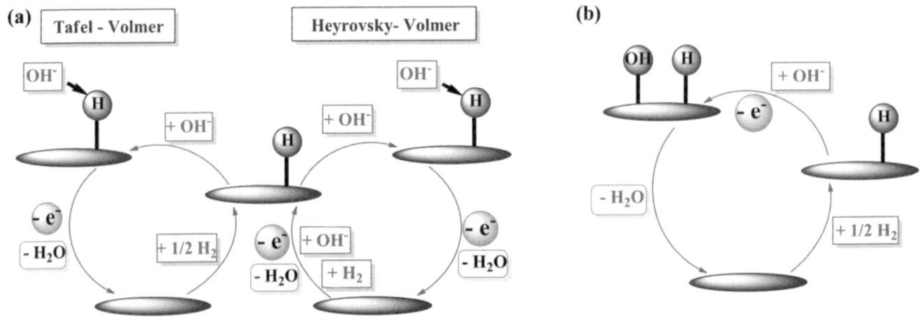

Fig. 5.18 Possible alkaline HOR pathways with the involvement of OH^- in an electrolyte (**a**) or OH_{ads} on the surface of an electrocatalyst (**b**) [218]

(2.2) The Tafel step is a speed control step

Preparing a series of Ru_xPt_y alloy catalysts and found that the optimized RuPt catalysts had a Tafel slope of ~ 30 mV dec^{-1} by calculating a Tafel slope of ~ 30 mV dec^{-1}. Usually, a Tafel slope of ~ 30 mV dec^{-1} means that the Tafel is a rapid control step. For a slope of ~ 118 mV dec^{-1}, Heyrovsky or Volmer is the rapid control step. Therefore, St. John concluded that the quick control step of HOR in this system is the Tafel step.

(2.3) Heyrovsky steps are speed control steps

Investigating [225] the HOR kinetics of Pt in acid/alkaline media, and found that the fitted curve at the Pt electrode surface HOR only proceeds through the Heyrovsky-Volmer step, and when Heyrovsky is the tachyon step fits better with the Butler-Volmer equation. The activation energy barrier of the Volmer step was found to be smaller than that of the Heyrovsky step by DFT theoretical calculations, and the transfer coefficient α further confirmed that the Heyrovsky step is the rate-determining step.

(2.4) The Volmer step is a speed control routine

Studying [226] the HOR of Pt/C, Pd/C, and Ir/C in acid/alkaline media, and determined that the tachyon step for HOR on Pt/C surfaces is the Volmer step by the charge transfer impedance of underpotential deposited hydrogen (H_{upd}). Rheinlander [238] found the transfer coefficient $\alpha_a + \alpha_c = 1$ on polycrystalline Pt surfaces by calculation. Therefore, the HOR in alkaline media proceeds through the Tafel-Volmer step, and the Volmer step is the rate-determining step. Studying [239] the HOR of Pt/C, Ir/C, Pd/C, and Rh/C in the pH = 0–13 range, and based on the fitted Butler-Volmer equation with the transfer coefficients $\alpha_a + \alpha_c = 1$, it was determined that the HOR is through the Tafel-Volmer step. Elbert [240–242] constructed Ru@Pt core–shell structures to study HOR in alkaline media, and theoretically calculated that the standard activation energies of the Tafel and Volmer steps are smaller than those of the Heyrovsky step. Therefore, they concluded that the alkaline HOR is carried out through the Tafel-Volmer step, and the Volmer (2.5) step is the rapid control step.

In conclusion, the calculated Tafel slope, the fit of the Butler-Volmer equation, the transfer coefficient, and the theoretical calculation of the reaction energy barrier are the main methods currently used to determine the reaction process and to determine the tacho-control step. It should be noted that the Tafel slope needs to be calculated with extra care, and the Butler-Volmer equation is usually used to find the Tafel slope only for strongly polarized regions with overpotentials above 120 mV. However, for most of the better-performing basic HOR catalysts, the potential range above 120 mV is not suitable for most of the better-performing alkaline HOR catalysts because they reach the ultimate current density in the ultimate diffusion region too early. The potential interval is too low to meet the approximation of the Tafel equation. Therefore, the value of the Tafel slope needs to be analyzed in the context of the actual catalytic system before the determination of the rapid control step. Currently, there is a strong consensus on the basic HOR that the Tafel-Volmer step is the most widely accepted radical reaction step for the alkaline HOR process, and the Volmer step is the rapid control step. Therefore, accelerating the adsorbed Had oxidative desorption on the catalyst surface is currently the main way to facilitate the reaction.

(3) Alkaline HOR electrocatalyst activity index

There are currently two views on the course of alkaline HOR, and accordingly, two views exist to evaluate the performance of electrocatalysts. (1) The strength of hydrogen bonding energy is the only criterion to evaluate the activity of basic HOR. (2) The hydrogen bonding energy and the strength of oxophilicity jointly affect the performance of basic HOR.

The strength of HBE is the only criterion.

As shown in Fig. 5.19, designing three Pt-based electrocatalysts: Pt/C, PtNi/C, and acid-treated PtNi/C. PtNi/C and acid-treated PtNi/C electrocatalysts [243] have similar hydrogen bonding energy and are weaker than that of Pt/C; PtNi/C is more oxygenophilic than Pt/C and acid-treated PtNi/C. The order of activity of the three electrocatalysts from highest to lowest is. Acid-treated PtNi/C = PtNi/C > Pt/C. The trend of electrocatalyst activity is consistent with the trend of hydrogen bonding energy strength, indicating that hydrogen bonding energy is an important indicator of activity. In addition, the activity of the acid-treated IrNi/C electrocatalyst was similar to that of IrNi alloy/C, again suggesting that OH_{ads} are not an important indicator affecting activity [244].

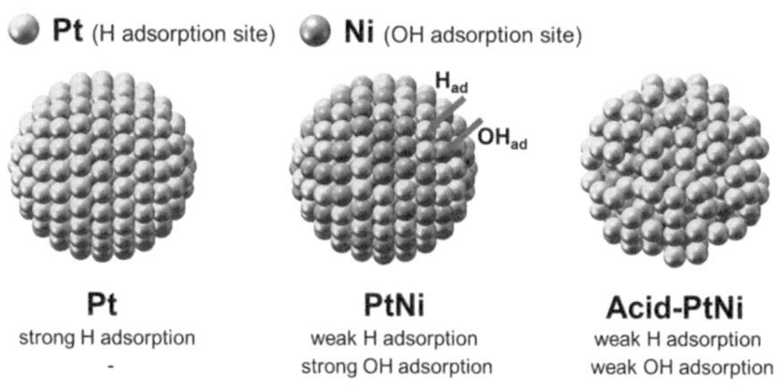

Fig. 5.19 Scheme of the Pt, PtNi, and Acid-PtNi nanoparticle model catalysts [243]

Founding [226] by CV that the peak position of the HUPD desorption peak of Pt/C gradually shifted to higher potential with increasing pH, and HUPD was difficult to be desorbed, and Durst attributed the slow alkaline HOR reaction rate to the strong Pt–H bond energy. /C, Ir/C and Rh/C HUPD desorption peaks were gradually shifted to higher potentials with increasing pH (Fig. 5.20), and the metal-H bonding energy was enhanced. However, the CO dissolution onset potential gradually shifts to a lower potential (Fig. 5.20e), which is more likely to form the adsorbed OH. If the adsorbed OH is involved in the basic HOR reaction course, the basic HOR performance should be enhanced, however, the opposite is true. Therefore, Zheng et al. ruled out that the strength of the electrophilic properties affects the alkaline HOR activity and concluded that the strength of the hydrogen bonding energy is an important criterion for evaluating the alkaline HOR activity. Zhuang et al. [239, 245, 246] examined the alkaline HOR activity of commercial Pt/C and commercial PtRu/C and found that the doping of Ru can significantly enhance the alkaline HOR performance. The CO dissolution curves indicate that in the alkaline medium, the CO onset potential of PtRu/C in the alkaline medium is higher than that of Pt/C, and the pro-oxidant performance is weaker. However, CV and DFT calculations indicate that the hydrogen bonding energy of PtRu/C is moderate. Therefore, it is proposed that the increased rate of PtRu/C-catalyzed HOR reaction does not originate from the pro-oxidant effect, but the electronic effect of the alloying element Ru weakens the Pt–H bond energy, thus accelerating the oxidative desorption of adsorbed state H, a rate control step of HOR.

(3.1) Combined effect of hydrogen bonding energy and oxophilicity

The study [241] found that the introduction of $Ni(OH)_2$ species on the surface of the Pt (111) working electrode significantly increased the HOR/HER reaction rate compared to that of the Pt (111) working electrode. The adsorbed H is bonded to the adsorbed OH to form adsorbed H_2O, and the adsorbed H_2O is further desorbed from the surface of the electrocatalyst to complete the alkaline HOR reaction. The experiments of Strmcnik demonstrated that the hydrogen bonding energy and oxophilicity together affect the performance of alkaline HOR electrocatalysts.

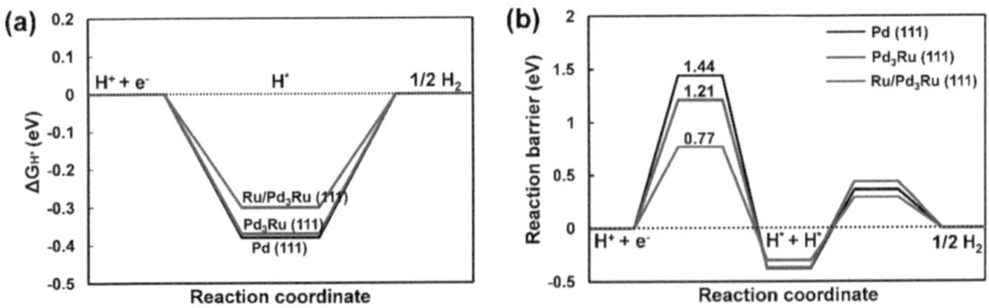

Fig. 5.20 **a** Gibbs free energy profile (ΔG_{H*}) on different catalyst surfaces; **b** reaction barriers for Volmer-Tafel pathway in an alkaline solution. Only the reaction barriers for the rate-determining step (Volmer reaction) are marked in the graph [248]

When $Pd/CeO_2/C$ was used as the anode electrocatalyst for HEMFCs [247], the peak power density was four times higher than that of Pd/C. Miller et al. suggested that the interaction between Pd and CeO_2/C carriers weakened the Pd–H bond energy and the introduced CeO_2 facilitated the adsorption of OH species, which together enhanced the alkaline HOR kinetics Qin et al. [248] systematically investigated the activity of pure Pd/C Ru@Pd core–shell/C and Ru cluster modified PdRu alloy/C electrocatalysts (Fig. 5.20) and found that the Ru cluster-modified PdRu alloy/C electrocatalysts had the best performance, The moderate hydrogen bonding energy and the DFT calculation showed that $Ru/Pd_3Ru/C$ electrocatalyst could form OH^* at − 0.26 eV, representing its strong oxygen affinity (Fig. 5.21). The moderate hydrogen bonding energy and strong oxophilicity were used to explain its high activity. Obtaining [249] Ni/NiO/C electrocatalysts with an excellent performance by direct pyrolysis of Ni-metal organic frameworks, and DFT calculations showed that the hydrogen bonding energies of Pt (111) Ni/NiO NiO (111), and Ni (111) were − 0.40 eV, − 0.20 eV, − 1.27 eV and − 0.54 eV, and the hydroxide binding energies were 0.87 eV, 0.72 eV and − 0.07 eV, respectively. The hydrogen and hydroxide bonding energies of Ni/NiO are closer to those of Pt/C, contributing to its better performance.

Apparently, different research groups have different views on the alkaline HOR course, and Song et al. suggested that the contradiction lies in the simplicity of the conventional description of the alkaline HOR course, which should include the adsorption and dissociation of H_2, diffusion of OH^-, formation of OH_{ads}, bonding

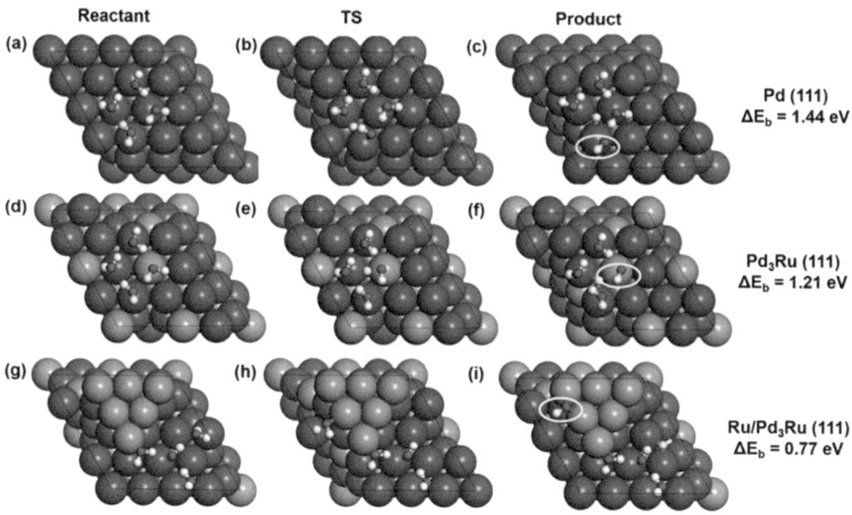

Fig. 5.21 Top view of reactants, TSs, and products of Volmer reactions in alkaline media for Pd(111) (**a**–**c**), $Pd_3Ru(111)$ (**d**–**f**), and $Ru/Pd_3Ru(111)$ (**g**–**i**). Corresponding reaction barriers (ΔE_b) are shown on the right of the figures. Color code, Pd: dark blue; Ru: pink; O: red; H: white. Yellow-colored atoms represent the protons involved in the reactions. Yellow circled OH are the produced OH* species in Volmer reactions [248] (color figure online)

of HO_{ads}-H_{ads}/OH-H_{ads}, and desorption of H_2O_{ads} (Fig. 5.22), The rate control steps should be different for different electrocatalysts, corresponding to different activity evaluation indexes; whether OH is involved in the basic HOR course in ionic or adsorbed form can be characterized by more in situ techniques, such as in situ Raman in situ infrared in situ electron microscopy, which provides a solution to the heated debate on this issue.

5.3.2 *Precious Metal Catalysts*

Currently, most of the studies on alkaline HOR catalysts are focused on Pt-based noble metals. This is because Pt-based noble metals can provide a large amount of fingerprint information in experimental exploration, which is beneficial for the establishment of conformational relationships and analysis of catalytic mechanisms, and guides the design of subsequent catalysts.

(1) The dimensional effects of catalysts

As shown in Fig. 5.23 [250], the performance of Ru/C with different particle sizes in alkaline HOR. The results showed that the mass to activity and area to the activity of Ru/C showed a volcano-shaped curve with particle size. The results showed that Ru/C of ~ 3 nm had a rougher surface structure and higher activity of ligated unsaturated Ru atoms, and thus had the best activity, even higher than that of commercial Pt/C [251]. Subsequently, Ir/C nanoparticles of 3–12 nm were fabricated in the same way [252]. Similar to Pd/C, the area-specific activity of Ir/C increased with the increase of particle size, but the mass-specific activity decreased with the increase of particle size, which was very unfavorable for cost reduction. Rh/C catalysts with different particle sizes were prepared by the impregnation reduction method, and the highest catalytic activity was obtained for Rh/C of ~ 2 nm [253].

(2) Effect of composition on performance

Alloying is a commonly used catalyst modification strategy, and the electronic structure modulation of the alloy can effectively optimize the HBE on the catalyst surface and also enhance the hydroxide binding energy of the catalyst surface, which can significantly improve the catalytic performance. Studies [254] showed that the catalytic activity of PtRu/C alloy catalyst in 0.1 M NaOH electrolyte was 3–4 times higher than that of Pt/C, which means that the use of PtRu alloy can effectively reduce the amount of Pt used in the anode of AEMFCs, and it is almost the same as that of PEMFCs. In studies [245, 255], the peak power densities of single cells assembled with PtRu/C and Pt/C as anode catalysts reached ~ 2 W cm^{-2}, respectively, which again emphasized the high catalytic activity of PtRu alloys and the promising development of AEMFCs. A PtNb/NbOx-C catalyst was designed [256] by co-precipitation with a catalytic activity of 0.8 mA cm^{-2} Pt, which is better than that of commercial Pt/C. A PtRh aerogel alloy catalyst was prepared

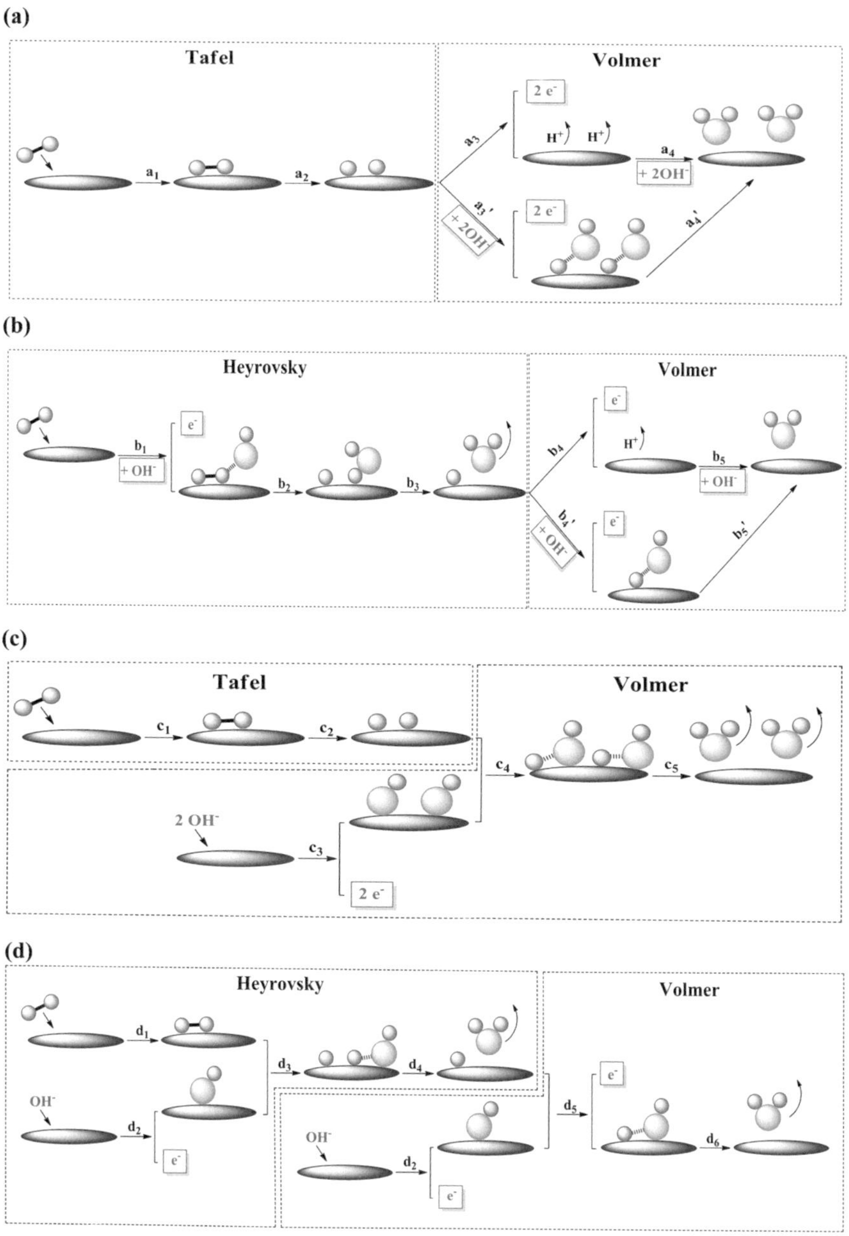

Fig. 5.22 Possible alkaline HOR pathways with the involvement of OH^- in an electrolyte (**a**, **b**) or adsorbed OH_{ads} on the surface of an electrocatalyst (**c**, **d**) in detail; a_1, c_1, d_1, H_2 adsorption; a_2, c_2, dissociation of H_2; a_3, b_4, H_{ads} desorption; a_3', a_4, b_4', b_5, c_4, d_5, HO–H bond formation; b_1, H_2 adsorption with the effect of OH^-, b_2, d_3, dissociation of H_2 and HO–H bond formation; a_4', b_3, b_5', c_5, d_4, d_6, H_2O desorption/diffusion; c_3, d_2, OH_{ads} formation; represents H atom; represents O atom [218]

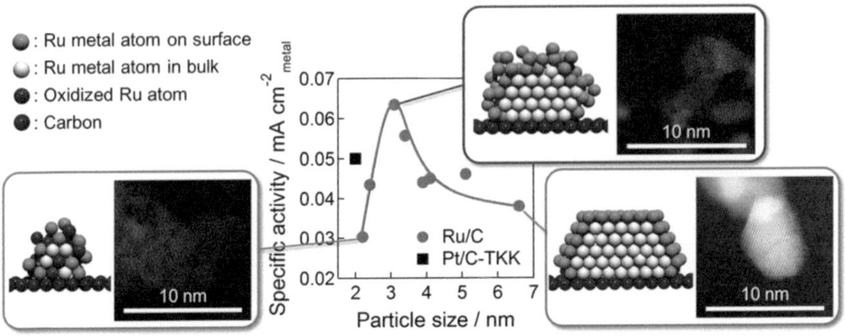

Fig. 5.23 Alkaline HOR specific activity of Ru/C at different particle sizes with different structures [250]

[257] with a unique lamellar structure, hierarchical pore arrangement and a large number of low-coordination active sites, which resulted in a double optimization of the electronic structure and hydroxide performance (Fig. 5.24). The unique lamellar structure, hierarchical pore arrangement, and a large number of low coordination active sites resulted in a doubly optimized electronic structure and hydroxide performance with catalytic activity up to 1.25 mA cm^{-2} Pt + Rh, which is 5.5 times higher than that of commercial Pt/C. Another research [237] found that by comparing the alkaline HOR kinetics of Pt-rich $Pt_{0.8}Ru_{0.2}$ and Ru-rich $Ru_{0.8}Pt_{0.2}$ alloys. Although the Hupd peak of the PtRu alloy is only negatively shifted by ~ 5 mV compared to Pt, the HOR activity is significantly enhanced with a threefold increase in the exchange current density compared to Pt. Therefore, the ligand effect of Ru cannot be neglected and the calculated Tafel slope indicates that the ligand effect of Ru alters the tacho-control step of HOR, significantly reducing the overpotential required for the reaction.

(3) Effect of specific morphology on performance

The performance of electrocatalysts depends not only on the particle size but also on the surface structure. The same composition of electrocatalysts may have different basic HOR activities due to different surface structures. Pt nanowires were prepared using a phase-transfer-$NaBH_4$ reduction method [258], and the basic HOR performance was superior to that of Pt nanoparticles (NPs), attributed to the anisotropic morphology of PtNWs with a one-dimensional structure, higher surface area, and fewer harmful defective active sites. The alkaline HOR performance of Ru@Pt core–shell electrocatalysts, RuPt alloys, and pure Pt electrocatalysts was systematically investigated [242], and the activity order was from high to low: Ru_1@Pt_1 > Ru_1@$Pt_{0.5}$ > $Ru_{0.6}Pt_{0.4}$ > Pt (Fig. 5.25). Compared with the $Ru_{0.6}Pt_{0.4}$ alloy electrocatalyst, the Ru@Pt core–shell electrocatalyst has better performance due to the superior hydrogen bonding energy of Pt in the shell layer.

The construction of bifunctional interfaces between metal nanoparticles and oxides to weaken the hydrogen bonding energy or enhance the adsorption of

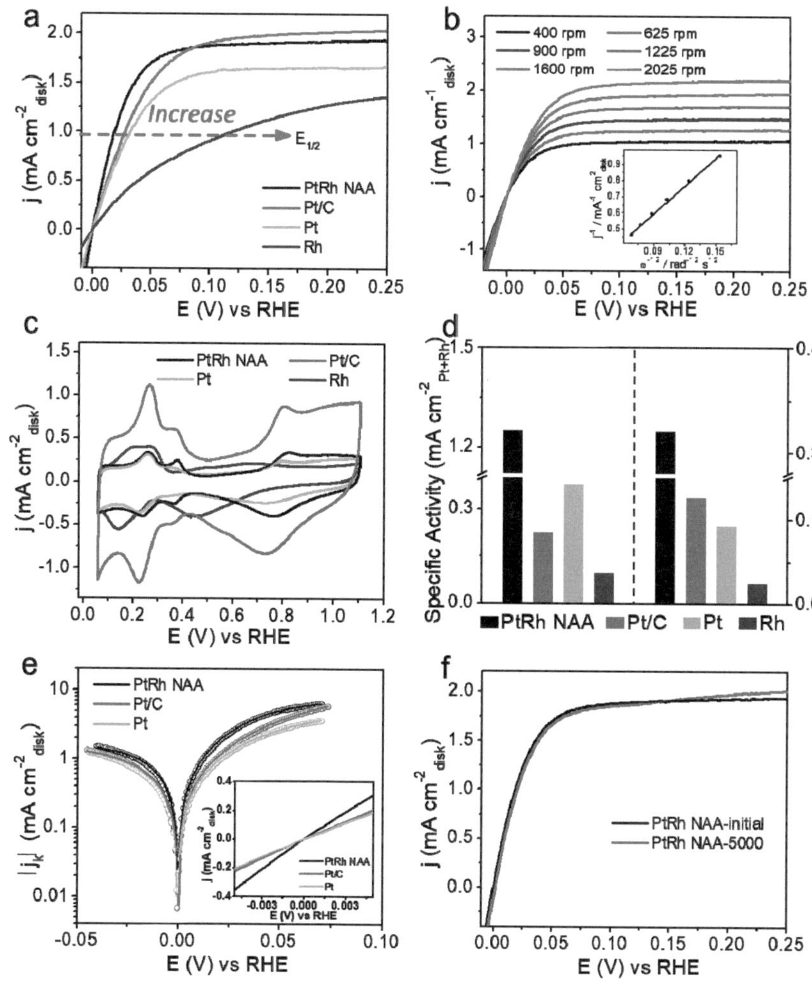

Fig. 5.24 Evaluation of HOR performance of as-prepared PtRh NAA electrocatalyst. **a** LSV polarization curves, **b** HOR polarization curves, **c** cyclic voltammograms, **d** specific activities (SA) and mass activities (MA), **e** Tafel plots, the inset shows micropolarization region (− 5 to 5 mV), **f** LSV curves are recorded before and after 5000 potential cycles of different electrocatalysts [257]

hydroxide may be an effective way to improve the alkaline HOR activity of electrocatalysts. In addition, the strong interaction between metal nanoparticles and carriers may also be beneficial to improve the durability of electrocatalysts. Bhowmik et al. [259] have found that the alkaline HOR activity and durability of Pd/CNX are superior to those of Pd/C, and the authors attributed the improved activity and durability to the synergistic interaction of Pd NPs with the carrier g-C_3N_4 and the strong interaction.

The preparation of Ir/CeO_2/C heterostructure electrocatalysts by constructing Ir and CeO_2 interfaces [260]. The authors attributed the enhanced Ir/CeO_2/C activity

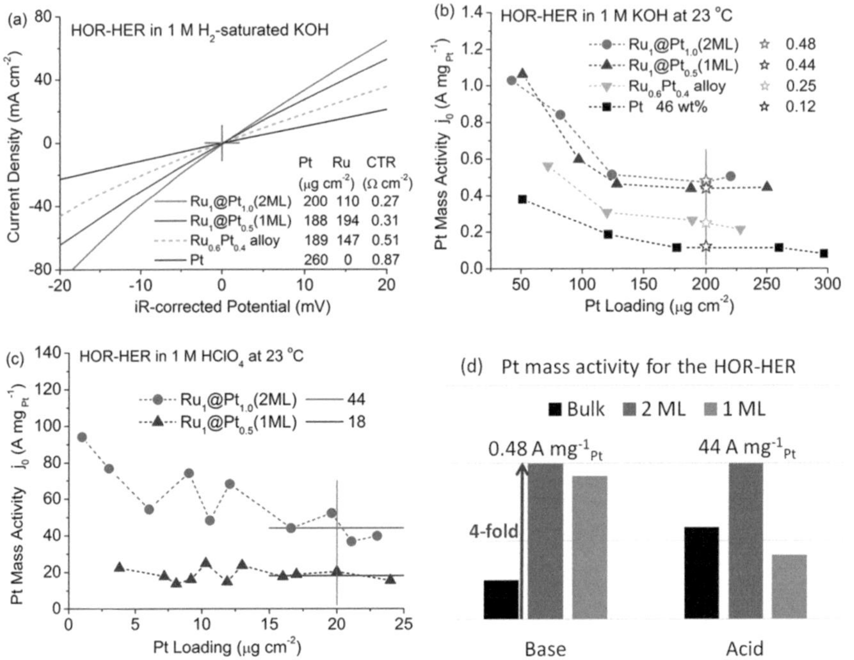

Fig. 5.25 **a** Typical HOR-HER polarization curves for carbon-supported nanocatalysts on 1 cm^2 GDEs in 1 M KOH. **b** and **c** Pt mass-normalized exchange currents as a function of Pt loading for the HOR-HER in base and acid, respectively. **d** Comparison of the Pt mass activities of the Pt (black), $Ru_1@Pt_1$(2 ML) (red), and $Ru_1@Pt_{0.5}$ (1 ML) (blue) nanocatalysts. The heights of the bars are normalized to the maximal value of 0.48 A (mg of Pt)$^{-1}$ in base and that of 44 A (mg of Pt)$^{-1}$ in acid [242] (color figure online)

to CeO_2 facilitating the adsorption of OH species, and the excellent durability of Ir/CeO_2/C to CeO_2 effectively inhibiting Ostwald ripening of Ir particles. Enhanced alkaline HOR activity of Pd-based electrocatalysts was achieved by constructing a Pd-CeO_2 interface [247]. The power density was 5 times higher than that of Pd/C The authors concluded that the enhanced Pd/CeO_2/C alkaline HOR activity was mainly due to the weakening of the Pd-H bond energy by CeO_2 [261] and the facilitated adsorption of OH species by CeO_2 [262]. In addition, the increased stability of both electrocatalysts was attributed to the interaction of Pd with CeO_2.

5.3.3 Nonprecious Metal Catalysts

The fundamental way to reduce catalyst costs is to develop inexpensive non-precious metal catalysts. In general, relatively little attention has been paid to the development of alkaline HOR electrocatalysts [263–266]. In decades of research, only nickel and its metal alloys have been active HOR PGM-free catalysts. In addition to moderate activity, Ni has low stability at potentials above ~ 0.1 V concerning the reversible hydrogen electrode (RHE) due to its relatively strong binding affinity for oxygen species [267–269]. In contrast, to obtain a considerable power output, the HOR catalyst should function at potentials up to 0.3 V relative to the RHE [213]. Unfortunately, so far, Ni has not been able to be used as a satisfactory alkaline HOR electrocatalyst in batteries. Therefore, many efforts have been made to enhance the alkaline HOR activity of Ni-based electrocatalysts in place of noble metals.

The earliest Ni-based anode HOR material used in alkaline fuel cells was a fine-grained nickel-aluminum alloy with a porous structure, Raney Ni [270]. In addition, Raney Ni alloys and NiB have been used as HOR electrocatalysts in alkaline fuel cells (AFCs), but only for space shuttles, submarines, etc., regardless of cost [259, 271, 272]. Modification of the electronic properties of the Ni surface by inhibiting the d-electron orbital reactivity of Ni, e.g., appropriate transition metal modification of the Ni surface seems to be useful in making the Ni surface more resistant to oxidation Cr and W modified Ni NP have been used for HOR can act as a stable catalyst for HOR in alkaline media [266]. Despite the low cell performance, it demonstrates the potential of using PGM-free electrocatalysts in the anode.

The alkali metal HOR activity of the $Co_{0.12}Ni_{5.10}Mo$ electrocatalyst made by electrodeposition at Sheng W C was better than that of Ni, but still much lower than that of Pt. The catalyst exhibited a relatively high HOR activity (20 times higher than Ni) among non-precious metal catalysts. Under sufficient loading, CoNiMo catalysts are expected to outperform Pt, thus providing a promising low-cost route for alkaline or basic membrane fuel cells. DFT calculations show that CoNiMo has a hydrogen binding energy (HBE) similar to Pt, but much lower than Ni. The formation of polymetallic bonds changes the HBE of Ni and may be an important factor in the enhanced HOR activity.

Recently, synthesizing Ni_{1-x} Cux/C with different Ni: Cu atomic ratios by impregnation followed by freeze-drying [273]. The activity of $Ni_{0.95}Cu_{0.05}$/C (14 μAcm^{-2}) was similar to that of CoNiMo (15 μAcm^{-2}), but two times lower than that of Ni/N-CNT (28 μAcm^{-2}) [274, 275]. The addition of a small amount of Cu significantly improved the electrocatalytic activity of Ni in the hydroxide reaction, while increasing the fraction of Cu above 5% led to its decay. The enhanced alkaline HOR activity of NiCu/C compared to Ni/C may be caused by the electron effect introduced by Cu, which leads to a decrease in the adsorption energy of the hydrogen intermediates. Ni NP loaded on N-doped carbon nanotubes (Ni/N-CNT) by a wet chemical method is to further improve the alkaline HOR activity of Ni-based electrocatalysts [275]. The exchange current density of the catalyst was still one order of magnitude lower than that of commercial Pt/C, but it increased 21-fold concerning pure Ni NPs (Fig. 5.26), and the relatively high activity could be attributed to the synergistic effect of N-CNT with Ni.

Nickel nitride (Ni_3N) with platinum-like properties is of high research value and has received much attention from researchers, and has the potential to replace platinum group precious metal catalyst materials. The unique catalytic properties of nickel and nickel nitride nanostructures have been investigated in a variety of catalytic

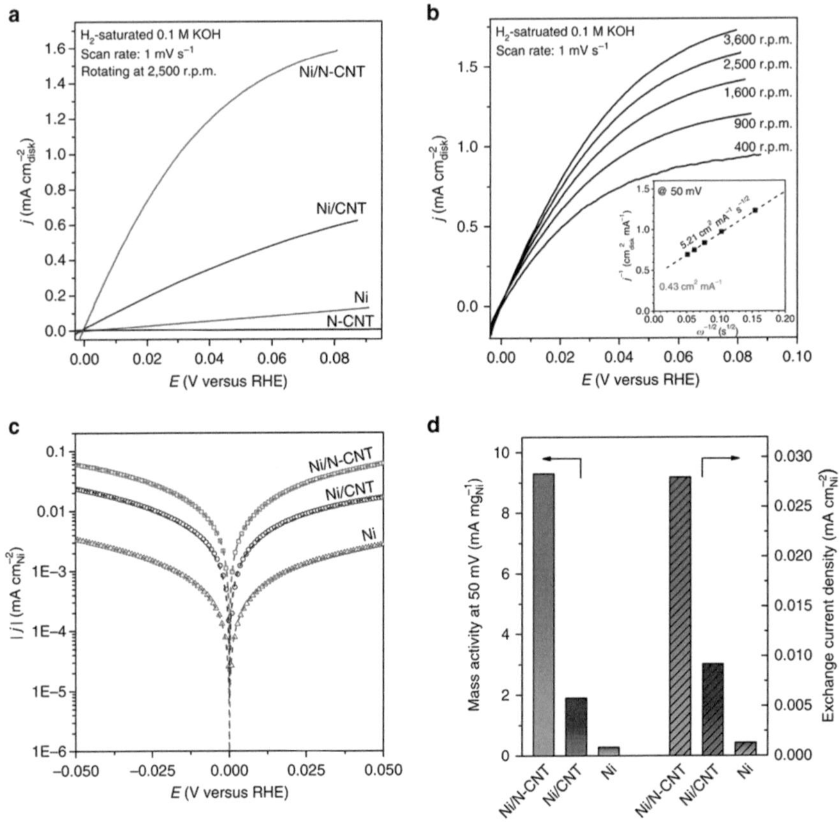

Fig. 5.26 **a** Polarization curves of Ni/N-CNT, Ni/CNT, Ni (all of the three catalysts with a loading of 0.25 mg_{Ni} cm^{-2}), and N-CNT (0.1 mg_C cm^{-2}) catalysts in H_2-saturated 0.1 M KOH at a scan rate of 1 mV s^{-1} and rotating speed of 2500 r.p.m. **b** Polarization curves of Ni/N-CNT in H_2-saturated 0.1 M KOH at a scan rate of 1 mV s^{-1} and various rotating speeds. Inset is the Koutecky–Levich plot at an overpotential of 50 mV. **c** HOR/HER Tafel plots of the specific current density on Ni/N-CNT, Ni/CNT, and Ni in H_2-saturated 0.1 M KOH. The dashed lines indicate the Butler–Volmer fitting. **d** Mass activity at 50 mV and exchange current density (patterned) of the Ni/N-CNT, Ni/CNT, and Ni, respectively [275]

and electrocatalytic fields. In recent years, several methods have emerged to synthesize nickel nitride. Physical methods include ion implantation, plasma nitriding, and sputtering techniques [276–279]. Chemical methods include solvothermal reactions, nitride complexation, and ammonolysis reactions of nickel and nickel salt precursors [280, 281]. Ni_3N intrinsically has metallic properties with high electron conductivity. Chen showed that when metal-nitrogen bonds are formed, the resulting metal nitrides can change the density of states in the unfilled d-band of Ni, causing the nitride surface to exhibit electron-feeding capabilities, similar to those of group VIII noble metals. Ni nitrides are expected to have higher catalytic activity than Ni [282].

Ni_3N to be an active HOR catalyst [283], and when loaded with carbon, Ni_3N achieved the highest mass-averaged activity of PGM-free catalysts to date, and the loaded catalyst Ni_3N/C was the highest mass-averaged activity of PGM-free catalysts to date. Specifically, Ni_3N nanoparticle samples were prepared by reducing suitable Ni precursors under NH_3 atmosphere, and NiO nanoparticle precursors were grown on carbon black (Vulcan-XC 72R) carriers by an oleylamine-assisted method to prevent Ni_3N nanoparticle aggregation. The reduction of the d-band of Ni and the interfacial charge transfer from Ni_3N to carbon carriers weakened the binding energy of hydrogen and oxygen species, thus the reduction of the d-band of Ni and the interfacial charge transfer from Ni_3N to the carbon carrier weaken the binding energy of hydrogen and oxygen species, resulting in significant HOR activity and stability.

5.3.4 *New Development on Hydrogen Oxidation Catalysts*

(1) Interstitial compound catalyst

Ni/NF on nickel foam (NF) surface by electrodeposition [284], followed by heat treatment under NH_3 atmosphere to obtain the final products $Ni_3N/Ni/NF$, $Ni_3N/Ni/NF$ with an exchange current density of 3.08 mA cm^{-2}, which is 1.4 and 17.6 times higher than that of Pt/NF and Ni/NF, respectively, and with stronger tolerance to CO. DFT calculations show that the unique electronic and geometric structure of Ni_3N/Ni makes the Gibbs free energy of H_{ads} close to 0 (Fig. 5.27). Moreover, the close contact between Ni_3N/Ni and nickel foam makes the electron transfer rate between the collector and the active site significantly faster, and the graded topology and higher pore distribution improve the utilization of the active site and accelerate the mass transfer rate. The graded topology and higher pore distribution improved the utilization of active sites, accelerated the mass transfer rate, and thus significantly enhanced the catalytic activity. Results [285] showed that the key to the formation of the Ni_3N phase is that the metallic Ni phase is difficult to be completely converted to the Ni_3N phase at nitriding temperatures below 400 °C, while above 450 °C, the N in the Ni_3N phase will escape from the lattice and be converted to metallic Ni phase. Electrochemical tests showed that the mass activity of Ni_3N nanoparticles at 50 mV overpotential was 29.75 A g^{-1}, and the exchange current density normalized to mass was 10.3 A g^{-1}. The DFT theoretical calculations confirmed that the interstitially doped N reduced the HBE of the catalyst and the energy barrier for water formation.

The interstitial compound Ni_3N is a hot research topic for Ni-based HOR catalysts. The nitriding conditions are mainly affected by the precursor type, particle size, and nitriding atmosphere, and there is no universal standard. However, it is still doubtful whether Ni_3N can exist stably at the working potential of HOR, and it is difficult to determine whether the active center is Ni_3N.

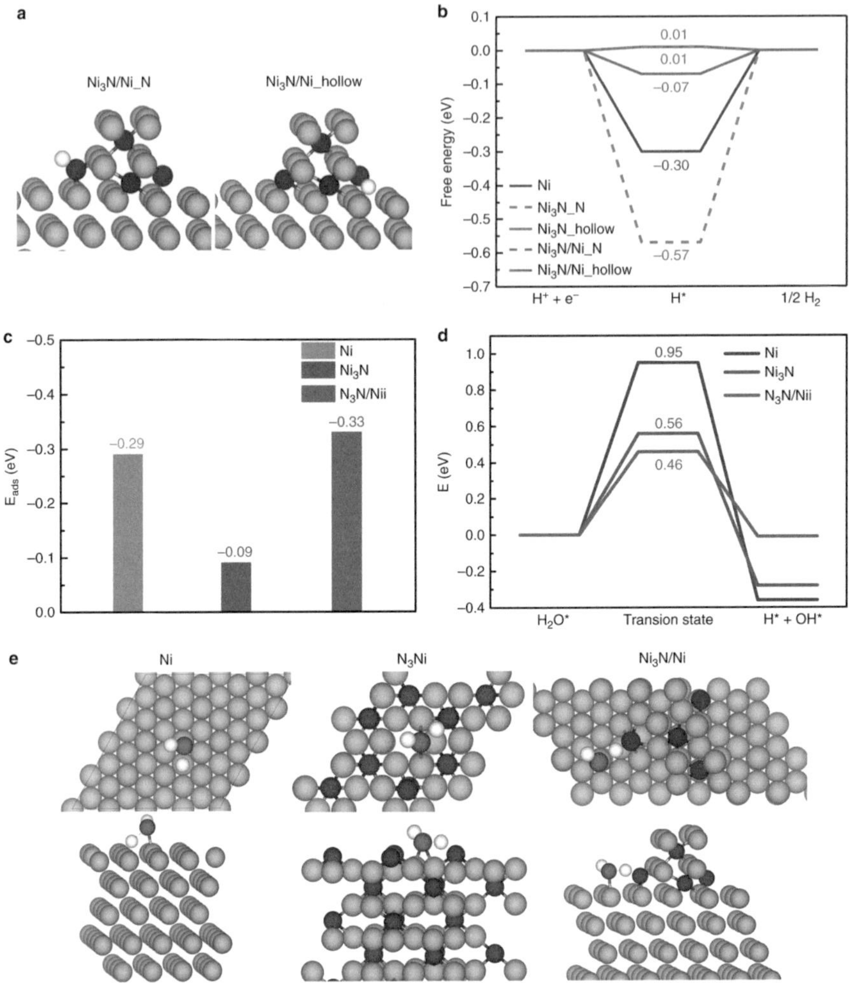

Fig. 5.27 DFT calculation results. **a** DFT-optimized structure of hydrogen adsorption at two interfacial sites of Ni_3N/Ni. b Hydrogen adsorption free energy (ΔGH*) on Ni, Ni_3N, and Ni_3N/Ni. **c** Adsorption energy of water on Ni, Ni_3N, and Ni_3N/Ni. Energy barrier for water dissociation on Ni, Ni_3N, and Ni_3N/Ni. **e** Transition state structures for water dissociation over Ni, Ni_3N, and Ni_3N/Ni, showing both the top view (top of **e**) and side view (bottom of **e**) of each structure. Color code: Ni: gray; N: blue; O: red; H: white [284] (color figure online)

(2) Heterogeneous structure catalyst

Preparing a series of Ni/NiO/C catalysts by pyrolysis of Ni-based MOFs precursors [249]. Among them, Ni/NiO/C-700 has the best catalytic activity with an exchange current density that is 14 times higher than that of pure Ni and has excellent CO tolerance and stability. DFT calculations confirmed a moderate HBE and OHBE at the Ni/NiO interface, resulting in higher activity [286]. A carbon-loaded small-size

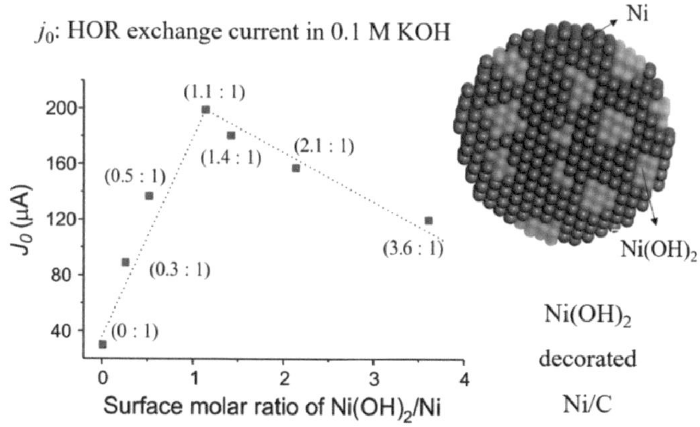

Fig. 5.28 A "volcano plot" was established between the HOR exchange current (j_0) and the surface molar ratio of $Ni(OH)_2/Ni$ in 0.1 M KOH [287]

Ni catalyst was developed using electrodeposition [287]. The catalytic activity of this Ni catalyst is very dependent on the post-treatment conditions, and it was treated by exposure to air and high-potential oxidation, respectively. The results show that the catalysts with high-potential oxidation have better catalytic activity, with a mass-to-mass activity of 22 $Ag_{Ni}{}^{-1}$. This is due to the increased Gibbs free energy of H_{ads} on the catalyst surface for high-potential oxidation, which allows HOR to occur at a lower potential. The potentiation oxidation method was used to partially oxidize the synthesized Ni/C catalysts [287]. Electrochemical tests showed that the surface $Ni(OH)_2/Ni$ ratio showed a volcano-type relationship with the catalytic activity. The $Ni(OH)_2/Ni$ ratio of 1.1:1 showed the highest catalytic activity (Fig. 5.28), again emphasizing the importance of heterogeneous structure formation on the catalyst surface for activity enhancement.

5.4 Summary and Outlook

PEMFCs have emerged as one of the most promising clean energy technologies, due to their high efficiency, low environmental impact, and versatile applications in both mobile and stationary power generation. However, the performance and cost-effectiveness of PEMFCs are heavily dependent on the efficiency of their catalysts, primarily the platinum (Pt)-based catalysts used in the anode and cathode. The catalyst plays a crucial role in facilitating the electrochemical reactions of hydrogen oxidation and oxygen reduction, which are key to the fuel cell's operation.

In recent years, research on catalysts for PEMFCs has focused on improving catalyst activity, durability, and reducing the usage of precious metals like platinum. Several strategies have been developed, such as the use of non-precious metal catalysts, platinum alloy catalysts, and the development of new supports or nanomaterials to enhance catalytic properties. For instance, platinum alloys with metals like palladium, nickel, or cobalt have shown promising improvements in catalytic performance by reducing the platinum loading while maintaining high efficiency. Furthermore, the use of carbon-based supports, such as graphene and carbon nanotubes, has enhanced the dispersion of catalyst particles, improving the overall activity and stability of the catalysts.

Another critical area of research is the improvement of the catalyst's resistance to degradation over time. Fuel cells are susceptible to catalyst poisoning and corrosion, which can reduce their efficiency and lifespan. Researchers are focusing on enhancing the structural stability of catalysts through novel synthesis methods and protective coatings. In addition, advances in understanding the fundamental reaction mechanisms at the atomic level are helping to design catalysts that are both more efficient and more robust in real-world applications.

The future of PEMFC catalyst research lies in the development of cost-effective, high-performance catalysts with minimal dependence on precious metals. One promising direction is the exploration of transition metal-based catalysts, which could provide a viable alternative to platinum. These catalysts, such as those based on iron, cobalt, and nickel, are abundant and inexpensive, yet show considerable catalytic activity under certain conditions. However, challenges remain in improving the activity and stability of these non-precious metal catalysts to match or exceed the performance of platinum-based ones.

In conclusion, while significant progress has been made in the development of PEMFC catalysts, continued innovation is needed to address the challenges of cost, performance, and durability. The ongoing research in alternative materials, nanotechnology, and advanced computational tools is expected to yield breakthroughs that will enable PEMFCs to become a mainstream technology in the global transition to a hydrogen-based economy.

References

1. H. Cruz-Martínez, M.M. Tellez-Cruz, O.X. Guerrero-Gutierrez, C.A. Ramírez-Herrera, M.G. Salinas-Juarez, A. Velazquez-Osorio, O. Solorza-Feria, Mexican contributions for the improvement of electrocatalytic properties for the oxygen reduction reaction in PEM fuel cells, Int. J. Hydrogen Energy, 2019, 44, 12477–12491.
2. M.K. Debe, Electrocatalyst approaches and challenges for automotive fuel cells, Nature, 2012, 486, 43–51.
3. T. Jacob. The mechanism of forming H_2O from H2 and O2 over a Pt catalyst via direct oxygen reduction, Fuel Cell, 2006, 6, 159–181.
4. X. Tian, X. Zhao, Y.-Q. Su, L. Wang, H. Wang, D. Dang, B. Chi, H. Liu, E. J. M. Hensen, X. W. Lou and B. Y. Xia, Science, 2019, 366, 850.
5. M. Luo, Z. Zhao, Y. Zhang, Y. Sun, Y. Xing, F. Lv, Y. Yang, X. Zhang, S. Hwang, Y. Qin, J.-Y. Ma, F. Lin, D. Su, G. Lu and S. Guo, Nature, 2019, 574, 81–85.
6. Z. Wang, Q. Lei, Z. Wang, H. Yuan, L. Cao, N. Qin, Z. Lu, J. Xiao and J. Liu, Chem. Eng. J., 2020, 395, 125180.
7. J. Liu, D. Zhu, T. Ling, A. Vasileff and S.Z. Qiao, Nano Energy, 2017, 40, 264–273.

8. J. Liu, H. Yuan, Z. Wang, J. Li, M. Yang, L. Cao, G. Liu, D. Qian and Z. Lu, Chem. Commun., 2019, 55, 10860–10863.
9. L. Chong, J. Wen, J. Kubal, F. G. Sen, J. Zou, J. Greeley, M. Chan, H. Barkholtz, W. Ding and D. J. Liu, Science, 2018, 362, 1276–1281.
10. Tian, X., Zhao, X., Su, Y. Q., Wang, L., Wang, H., Dang, D., Xia, B. Y. Engineering bunched Pt-Ni alloy nanocages for efficient oxygen reduction in practical fuel cells. Science, 2019, 366, 850–856.
11. J. Liu, Z. Wang, K. Su, D. Xv, D. Zhao, J. Li, H. Tong, D. Qian, C. Yang and Z. Lu, ACS Appl. Mater. Interfaces, 2019, 11, 25854–25862.
12. Shan, J., Ling, T., Davey, K., Zheng, Y., Qiao, S. Z. Transition-metal-doped RuIr bifunctional nanocrystals for overall water splitting in acidic environments. Advanced Materials, 2019, 3, 1900510.
13. S. Dou, et al., Etched and doped Co9S8/graphene hybrid for oxygen electrocatalysis, Energy Environ. Sci., 2016, 9(4), 1320–1326.
14. Stacy, et al., The recent progress and future of oxygen reduction reaction catalysis: a review, Renewable Sustainable Energy Rev., 2017, 69, 401–414.
15. Holton, O. T., & Stevenson, J. W. The role of platinum in proton exchange membrane fuel cells. Platinum Metals Review, 2013, 57(4), 259–271.
16. de Morais, R. F., Franco, A. A., Sautet, P., Loffreda, D. Coverage-dependent thermodynamic analysis of the formation of water and hydrogen peroxide on a platinum model catalyst. Physical Chemistry Chemical Physics, 2015, 17(17), 11392–11400.
17. Zhang, L., Xia, Z. Mechanisms of oxygen reduction reaction on nitrogen-doped graphene for fuel cells. The Journal of Physical Chemistry C, 2011, 115(22), 11170–11176.
18. Lim, D. H., Wilcox, J. Mechanisms of the oxygen reduction reaction on defective graphene-supported Pt nanoparticles from first-principles. The Journal of Physical Chemistry C, 2012, 116(5), 3653–3660.
19. Nørskov, J. K., Rossmeisl, J., Logadottir, A., Lindqvist, L. R. K. J., Kitchin, J. R., Bligaard, T., Jonsson, H. Origin of the overpotential for oxygen reduction at a fuel-cell cathode. The Journal of Physical Chemistry B, 2004, 108(46), 17886–17892.
20. J. K. Nørskov, J. Rossmeisl, A. Logadottir, L. Lindqvist, J. R. Kitchin, T. Bligaard and H. Jónsson, J. Phys. Chem. B, 2004, 108, 17886–17892
21. J. K. Norskov, et al., Origin of the overpotential for oxygen reduction at a fuel-cell cathode, J. Phys. Chem. B, 2004, 108(46), 17886–17892
22. E. Antolini, Composite materials: an emerging class of fuel cell catalyst supports, Appl. Catal., B, 2010, 100(3–4), 413–426.
23. H. Lv and S. Mu, Nano-ceramic support materials for low temperature fuel cell catalysts, Nanoscale, 2014, 6(10), 5063–5074.
24. Cho, K. Y., Yeom, Y. S., Seo, H. Y., Lee, A. S., Do, X. H., Hong, J. P., Yoon, H. G. Fine-sized Pt nanoparticles dispersed on PdPt bimetallic nanocrystals with non-covalently functionalized graphene toward synergistic effects on the oxygen reduction reaction. Electrochimica Acta, 2017, 257, 412–422.
25. Fedotov, A. A., Grigoriev, S. A., Lyutikova, E. K., Millet, P. Fateev, V. N. Characterization of carbon-supported platinum nano-particles synthesized using magnetron sputtering for application in PEM electrochemical systems. International journal of hydrogen energy, 2013, 38(1), 426–430.
26. Antolini, E. Carbon supports for low-temperature fuel cell catalysts. Applied Catalysis B: Environmental, 2009, 88(1–2), 1–24.
27. Figueiredo, J. L., Pereira, M. F. R., Serp, P., Kalck, P., Samant, P. V., & Fernandes, J. B. Development of carbon nanotube and carbon xerogel supported catalysts for the electro-oxidation of methanol in fuel cells. Carbon, 2006, 44(12), 2516–2522.
28. Lee, K., Zhang, J., Wang, H., Wilkinson, D. P. Progress in the synthesis of carbon nanotube- and nanofiber-supported Pt electrocatalysts for PEM fuel cell catalysis. Journal of applied electrochemistry, 2006, 36(5), 507–522.

29. C.A. Bessel, K. Laubernds, N.M. Rodriguez, R.T.K. Baker. Graphite nanofibers as an electrode for fuel cell applications. J Phys Chem B, 2001, 105, 1115–1118.
30. Z.R. Ismagilov, M.A. Kerzhentsev, N.V. Shikina, A.S. Lisitsyn, L.B. Okhlopkova, Ch.N. Barnakov, et al. Development of active catalysts for low Pt loading cathodes of PEMFC by surface tailoring of nanocarbon materials. Catal Today, 2005, 102–103.
31. Y. Y. Shao, G. P. Yin and Y. Z. Gao, Understanding and approaches for the durability issues of Pt-based catalysts for PEM fuel cell, J. Power Sources, 2007, 171(2), 558-566.
32. C. L. Li, et al., Emerging Pt-based electrocatalysts with highly open nanoarchitectures for boosting oxygen reduction reaction, Nano Today, 2018, 21, 91–105.
33. Liu, M., Zhao, Z., Duan, X., Huang, Y. Nanoscale structure design for high-performance Pt-based ORR catalysts. Advanced Materials, 2019, 31(6), 1802234.
34. W. H. Yang, Q. H. Zhang, H. H. Wang, Z. Y. Zhou and S. G. Sun, Preparation and utilization of a sub-5 nm PbO_2 colloid as an excellent co-catalyst for Pt-based catalysts toward ethanol electro-oxidation, New J. Chem., 2017, 41(20), 12123–12130.
35. R. M. Antoniassi, J. C. M. Silva, A. Oliveira Neto and E. V. Spinacé, Synthesis of Pt+SnO_2/C electrocatalysts containing Pt nanoparticles with preferential (100) orientation for direct ethanol fuel cell, Appl. Catal., B, 2017, 218, 91–100.
36. S. Tang, G. Sun, J. Qi, S. Sun, J. Guo, Q. Xin and G. M. Haarberg, Review of New Carbon Materials as Catalyst Supports in Direct Alcohol Fuel Cells, Chin. J. Catal., 2010, 31(1), 12–17.
37. A. Bharti and G. Cheruvally, Influence of various carbon nano-forms as supports for Pt catalyst on protonexchange membrane fuel cell performance, J. Power Sources, 2017, 360, 196–205.
38. S. L. Knupp, W. Li, O. Paschos, T. M. Murray, J. Snyder and P. Haldar. The effect of experimental parameters on the synthesis of carbon nanotube/nanofiber supported platinum by polyol processing techniques, Carbon, 2008, 46(10), 1276–1284.
39. B. Escobar, R. Barbosa, M. Miki Yoshida and Y. Verde Gomez, Carbon nanotubes as support of well dispersed platinum nanoparticles via colloidal synthesis, J. Power Sources, 2013, 243, 88–94.
40. T. Yamada, H. Yabutani, T. Saito and C. Y. Yang. Temperature dependence of carbon nanofiber resistance, Nanotechnology, 2010, 21(26), 265707.
41. J. Lee, K. Kim, W. I. Park, B. H. Kim, J. H. Park, T. H. Kim, S. Bong, C. H. Kim, G. Chae, M. Jun, Y. Hwang, Y. S. Jung and S. Jeon, Uniform graphene quantum dots patterned from self-assembled silica nanodots, Nano Lett., 2012, 12(12), 6078-6083.
42. C. Zhu, D. Liu, Z. Chen, L. Li and T. You, Superior catalytic activity of Pt/carbon nanohorns nanocomposites toward methanol and formic acid oxidation reactions, J. Colloid Interface Sci., 2018, 511, 77–83.
43. He D P, Mu S c, Pan M. Improved carbon nanotube supported Pt nanocatalyts with lyophilization method. International Journal of Hydrogen Energy, 2012, 37: 4699–4703.
44. Lv H F,Cheng N C, Mu S C, et al. Heat-treated mul-ti-walled carbon nanotubes as durable supports for PEM fuel cell catalysts. Electrochimica Acta, 2011,58(5):736–742.
45. Mardle, P., Ji, X., Wu, J., Guan, S., Dong, H., & Du, S. Thin film electrodes from Pt nanorods supported on aligned N-CNTs for proton exchange membrane fuel cells. Applied Catalysis B: Environmental, 2020, 260, 118031.
46. Sebastián, D., et al. "Influence of carbon nanofiber properties as electrocatalyst support on the electrochemical performance for PEM fuel cells." International Journal of Hydrogen Energy, 2010, 35, 9934–9942.
47. Calvillo, L., Gangeri, M., Perathoner, S., Centi, G., Moliner, R., & Lázaro, M. J. Effect of the support properties on the preparation and performance of platinum catalysts supported on carbon nanofibers. Journal of Power Sources, 2009, 192(1), 144–150.
48. Ruiz-Camacho, B., Palafox-Segoviano, J. A., Pérez-Díaz, P. J., & Medina-Ramírez, A. Synthesis of supported Pt nanoparticles by sonication for ORR: Effect of the graphene oxide-carbon composite. International Journal of Hydrogen Energy, 2021, 46(51), 26027–26039.
49. Y. Sun, Q. Wu and G. Shi, Graphene-based new energy materials, Energy Environ. Sci., 2011, 4(4), 1113–1132.

50. G. Saravanan and S. Mohan, Pt nanoparticles embedded on reduced graphite oxide with excellent electrocatalytic properties, Appl. Surf. Sci., 2016, 386, 96–102.
51. Sharma, S., Ganguly, A., Papakonstantinou, P., Miao, X., Li, M., Hutchison, J. L., Ukleja, S. Rapid microwave synthesis of CO tolerant reduced graphene oxide-supported platinum electrocatalysts for oxidation of methanol. The Journal of Physical Chemistry C, 2010, 114(45), 19459–19466.
52. Mu, X., Xu, Z., Ma, Y., Xie, Y., Mi, H., Ma, J. Graphene-carbon nanofiber hybrid supported Pt nanoparticles with enhanced catalytic performance for methanol oxidation and oxygen reduction. Electrochimica Acta, 2017, 253, 171–177.
53. Kaplan, B. Y., Haghmoradi, N., Biçer, E., Merino, C., Gürsel, S. A. High-performance electrocatalysts supported on graphene-based hybrids for polymer electrolyte membrane fuel cells. International journal of hydrogen energy, 2018, 43(52), 23221–23230.
54. X. X. Du, Y. He, X. X. Wang, J. N. Wang, Energy Environ. Sci. 2016, 9, 2623–2632;
55. H. Chen, D. Wang, Y. Yu, K. A. Newton, D. A. Muller, H. AbruÇa, F. J. DiSalvo, J. Am. Chem. Soc. 2012, 134, 18453–18459;
56. W. Xiao, M. A. L. Cordeiro, G. Gao, A. Zheng, J. Wang, W. Lei, M. Gong, R. Lin, E. Stavitski, H. L. Xin, D. Wang, Nano Energy, 2018, 50, 70–78.
57. B. C. Beard, P. N. Ross, J. Electrochem. Soc. 1990, 137, 3368-3374;
58. D. Wang, H. L. Xin, R. Hovden, H. Wang, Y. Yu, D. A. Muller, F. J. DiSalvo, H. D. AbruÇa, Nat. Mater. 2013, 12, 81;
59. X. X. Wang, S. Hwang, Y.-T. Pan, K. Chen, Y. He, S. G. Karakalos, H. Zhang, J. S. Spendelow, D. Su, G. Wu, Nano Lett. 2018, 18, 4163–4171.
60. L. Zou, J. Li, T. Yuan, Y. Zhou, X. Li, H. Yang, Nanoscale 2014, 6, 10686–10692;
61. Ji, X., Lee, K. T., Holden, R., Zhang, L., Zhang, J., Botton, G. A., Nazar, L. F. Nanocrystalline intermetallics on mesoporous carbon for direct formic acid fuel cell anodes. Nature Chemistry, 2010, 2(4), 286–293.
62. D. Xia, G. Chen, Z. Wang, J. Zhang, S. Hui, D. Ghosh, H. Wang, Chem. Mater. 2006, 18, 5746–5749.
63. T. Ghosh, B. M. Leonard, Q. Zhou, F. J. DiSalvo, Chem. Mater. 2010, 22, 2190–2202;
64. Z. Cui, H. Chen, M. Zhao, D. Marshall, Y. Yu, H. AbruÇa, F. J. DiSalvo, J. Am. Chem. Soc. 2014, 136, 10206–10209.
65. Z. Qi, C. Xiao, C. Liu, T. W. Goh, L. Zhou, R. Maligal-Ganesh, Y. Pei, X. Li, L. A. Curtiss, W. Huang, J. Am. Chem. Soc. 2017, 139, 4762–4768.
66. M. Bele, P. Jovanovic, A. Pavlisic, B. Jozinovic, M. Zorko, A. Recnik, E. Chernyshova, S. Hocevar, N. Hodnik, M. Gaberscek, Chem. Commun. 2014, 50, 13124–13126.
67. D. Y. DeSario, F. J. DiSalvo, Chem. Mater. 2014, 26, 2750–2757.
68. J. S. Kanady, P. Leidinger, A. Haas, S. Titlbach, S. Schunk, K. Schierle-Arndt, E. J. Crumlin, C. H. Wu, A. P. Alivisatos, J. Am. Chem. Soc. 2017, 139, 5672–5675.
69. V. Stamenkovic, B.S. Mun, K.J. Mayrhofer, P.N. Ross, N.M. Markovic, J. Rossmeisl, J.K. Nørskov, Changing the activity of electrocatalysts for oxygen reduction by tuning the surface electronic structure, Angew Chem Int (Ed.) 2006, 45, 2897–2901.
70. J. Greeley, I.E.L. Stephens, A.S. Bondarenko, T.P. Johansson, H.A. Hansen, T.F. Jaramillo, J. Rossmeisl, I. Chorkendorff, J.K. Nørskov, Alloys of platinum and early transition metals as oxygen reduction electrocatalysts, Nat. Chem., 2009, 552–556.
71. A. Kulkarni, S. Siahrostami, A. Patel, J.K. Nørskov, Understanding catalytic activity trends in the oxygen reduction reaction, Chem. Rev., 2018, 118 2302–2312.
72. H.A. Hansen, J. Rossmeisl, J.K. Nørskov, Surface Pourbaix diagrams and oxygen reduction activity of Pt, Ag and Ni (111) surfaces studied by DFT, Phys. Chem. Chem. Phys., 2008, 10, 3722–3730.
73. J. Greeley, I. E. L. Stephens, A. S. Bondarenko, T. P. Johansson, H. A. Hansen, T. F. Jaramillo, J. Rossmeisl, I. Chorkendorff and J. K. Nørskov, Nat. Chem., 2009, 1, 552–556
74. A. Chalgin, F. Shi, F. Li, Q. Xiang, W. Chen, C. Song, P. Tao, W. Shang, T. Deng and J. Wu, CrystEngComm, 2017, 19, 6964–6971.

75. V. Tripkovic, H. A. Hansen, J. Rossmeisl and T. Vegge, Phys. Chem. Chem. Phys., 2015, 17, 11647–11657.
76. S. H. Noh, B. Han and T. Ohsaka, Nano Res., 2015, 8, 3394–3403
77. R. Lin, et al., Stability of High-Performance Pt-Based Catalysts for Oxygen Reduction Reactions, Adv. Mater., 2018, 30(17), 8.
78. K. Bhunia, S. Khilari and D. Pradhan, Trimetallic PtAuNi alloy nanoparticles as an efficient electrocatalyst for the methanol electrooxidation reaction, Dalton Trans., 2017, 46(44), 15558-15566.
79. M. Sankar, et al., Designing bimetallic catalysts for a green and sustainable future, Chem. Soc. Rev., 2012, 41(24), 8099–8139.
80. V. Dal Santo, et al., Bimetallic heterogeneous catalysts for hydrogen production, Catal. Today, 2012, 197(1), 190–205.
81. I. Khalakhan, et al., Irreversible structural dynamics on the surface of bimetallic PtNi alloy catalyst under alternating oxidizing and reducing environments, Appl. Catal., B, 2020, 264, 118476.
82. Choi, D. S., Robertson, A. W., Warner, J. H., Kim, S. O., & Kim, H. Low-temperature chemical vapor deposition synthesis of Pt-Co alloyed nanoparticles with enhanced oxygen reduction reaction catalysis. Advanced materials, 2016, 28(33), 7115–7122.
83. Gao, L., Li, X., Yao, Z., Bai, H., Lu, Y., Ma, C., Huang, H. Unconventional p-d hybridization interaction in PtGa ultrathin nanowires boosts oxygen reduction electrocatalysis. Journal of the American Chemical Society, 2019, 141(45), 18083–18090.
84. W.Z. Li, C.H. Liang, W.J. Zhou, J.S. Qiu, Z.H. Zhou, G.Q. Sun, Q. Xin, Preparation and characterization of multiwalled carbon nanotube-supported platinum for cathode catalysts of direct methanol fuel cells, J. Phys. Chem. B, 2003, 107, 6292–6299.
85. J. Qi, L.H. Jiang, M.Y. Jing, Q.W. Tang, G.Q. Sun, Preparation of Pt/C via a polyol process-Investigation on carbon support adding sequence, Int. J. Hydrogen Energy, 2011, 36, 10490–10501.
86. Y.Y. Mu, H.P. Liang, J.S. Hu, L. Jiang, L.J. Wan, Controllable Pt nanoparticle deposition on carbon nanotubes as an anode catalyst for direct methanol fuel cells, J. Phys. Chem. B, 2005, 109, 22212–22216.
87. F.J. Li, Y. Chen, D.M. Tang, Z.L. Jian, C. Liu, D. Golberg, A. Yamada, H.S. Zhou, Performance-improved Li-O_2 battery with Ru nanoparticles supported on binder-free multi-walled carbon nanotube paper as cathode, Energy Environ. Sci., 2014, 7, 1648-1652.
88. L.W. Zhang, N. Zheng, A. Gao, C.M. Zhu, Z.Y. Wang, Y. Wang, Z.J. Shi, Y. Liu, A robust fuel cell cathode catalyst assembled with nitrogen-doped carbon nanohorn and platinum nanoclusters, J. Power Sources, 2012, 220 449–454.
89. J.N. Kuhn, C.-K. Tsung, W.Y. Huang, G.A. Somorjai, Effect of organic capping layers over monodisperse platinum nanoparticles upon activity for ethylene hydrogenation and carbon monoxide oxidation, J. Catal., 2009, 265209–215.
90. H.P. Yang, Y.F. Zhu, Glucose biosensor based on nano-SiO2 and "unprotected" Pt nanoclusters, Biosens. Bioelectron., 2007, 22, 2989–2993.
91. C. Lian, H.Q. Liu, C. Xiao, W. Yang, K. Zhang, Y. Liu, Y. Wang, Solvent-free selective hydrogenation of chloronitrobenzene to chloroaniline over a robust Pt/Fe3O4 catalyst, Chem. Commun., 2012, 48, 3124–3126.
92. C.Y. Chen, F. Chen, L. Zhang, S.X. Pan, C.Q. Bian, X.M. Zheng, X.J. Meng, F.S. Xiao, Importance of platinum particle size for complete oxidation of toluene over Pt/ZSM-5 catalysts, Chem. Commun., 2015, 51, 5936–5938.
93. P. Sonstrom, M. Baumer. Supported colloidal nanoparticles in heterogeneous gas phase catalysis: on the way to tailored catalysts, Phys. Chem. Chem. Phys., 2011, 13, 19270–19284.
94. Liu, Y., Chen, L., Cheng, T., Guo, H., Sun, B., & Wang, Y. Preparation and application in assembling high-performance fuel cell catalysts of colloidal PtCu alloy nanoclusters. Journal of Power Sources, 2018, 395, 66–76.
95. Lv, J. J., Zheng, J. N., Li, S. S., Chen, L. L., Wang, A. J., & Feng, J. J. Facile synthesis of Pt-Pd nanodendrites and their superior electrocatalytic activity. Journal of Materials Chemistry A, 2014, 2(12), 4384–4390.

96. Y. Feng, Q. Shao, F. Lv, L. Bu, J. Guo, S. Guo, X. Huang, Adv. Sci. 2020, 7, 1800178.
97. D. Zhang, F. Wu, M. Peng, X. Wang, D. Xia, G. Guo, J. Am. Chem. Soc. 2015, 137, 6263–6269.
98. H. Zhu, G. Gao, M. Du, J. Zhou, K. Wang, W. Wu, X. Chen, Y. Li, P. Ma, W. Dong, F. Duan, M. Chen, G. Wu, J. Wu, H. Yang, S. Guo, Adv. Mater. 2018, 30, 1707301.
99. M. Luo, Y. Sun, X. Zhang, Y. Qin, M. Li, Y. Li, C. Li, Y. Yang, L. Wang, P. Gao, G. Lu, S. Guo, Adv. Mater. 2018, 30, 1705515;
100. Q. Liu, L. Du, G. Fu, Z. Cui, Y. Li, D. Dang, X. Gao, Q. Zheng, J. Goodenough, Adv. Energy Mater. 2019, 9, 1803040.
101. T. Ishida and M. Haruta, Gold catalysts: towards sustainable chemistry, Angew. Chem., 2007, 46(38), 7154–7156.
102. Y. Y. Feng, G. H. Song, Q. Zhang, H.-S. Hu, M. Y. Feng, J.-Y. Wang and D.-S. Kong, Catalytic performance of nonalloyed bimetallic PtAu electrocatalysts for methanol oxidation reaction, Int. J. Hydrogen Energy, 2017, 42(51), 30109–30118.
103. J. Liang, N. Li, Z. Zhao, L. Ma, X. Wang, S. Li, X. Liu, T. Wang, Y. Du, G. Lu, J. Han, Y. Huang, D. Su and Q. Li, Angew. Chem., Int. Ed., 2019, 58, 2–9.
104. R. M. Aran-Ais, F. Dionigi, T. Merzdorf, M. Gocyla, M. Heggen, R. E. Dunin-Borkowski, M. Gliech, J. SollaGullon, E. Herrero, J. M. Feliu and P. Strasser, Nano Lett., 2015, 15, 7473–7480.
105. K. Li, X. Li, H. Huang, L. Luo, X. Li, X. Yan, C. Ma, R. Si, J. Yang and J. Zeng, J. Am. Chem. Soc., 2018, 140, 16159–16167.
106. S. Luo, W. Chen, Y. Cheng, X. Song, Q. Wu, L. Li, X. Wu, T. Wu, M. Li, Q. Yang, K. Deng and Z. Quan, Adv. Mater., 2019, 31, 1903683.
107. J. Liu and J. Zhang, Chem. Rev., 2020, 120, 2123–2170.
108. Y. Yao, D. S. He, Y. Lin, X. Feng, X. Wang, P. Yin, X. Hong, G. Zhou, Y. Wu and Y. Li, Angew. Chem., Int. Ed., 2016, 55, 5501–5505.
109. Wang, X. et al. Palladium-platinum core-shell icosahedra with substantially enhanced activity and durability towards oxygen reduction. Nat. Commun., 2015, 6, 7594.
110. Wang, X. et al. Pd@Pt core-shell concave decahedra: a class of catalysts for the oxygen reduction reaction with enhanced activity and durability. J. Am. Chem. Soc., 2015, 137, 15036–15042.
111. Gong, K. P., Su, D. & Adzic, R. R. Platinum-monolayer shell on AuNi0.5Fe nanoparticle core electrocatalyst with high activity and stability for the oxygen reduction reaction. J. Am. Chem. Soc., 2010, 132, 14364–14366.
112. Wang, C. et al. Multimetallic Au/FePt3 nanoparticles as highly durable electrocatalyst. Nano Lett., 2011, 11, 919–926.
113. Sasaki, K. et al. Highly stable Pt monolayer on PdAu nanoparticle electrocatalysts for the oxygen reduction reaction. Nat. Commun., 2012, 3, 1115.
114. Adzic, R. R. Platinum monolayer electrocatalysts: tunable activity, stability, and self-healing properties. Electrocatalysis, 2012, 3, 163–169.
115. Weissmüller, J. et al. Charge-induced reversible strain in a metal. Science, 2003, 300, 312–315.
116. Li, J., Yin, H. M., Li, X. B., Okunishi, E., Shen, Y. L., He, J., Ding, Y. Surface evolution of a Pt-Pd-Au electrocatalyst for stable oxygen reduction. Nature Energy, 2017, 2(8), 1–9.
117. X. Deng, S. Yin, X. Wu, M. Sun, Z. Xie and Q. Huang, Synthesis of PtAu/TiO2 nanowires with carbon skin as highly active and highly stable electrocatalyst for oxygen reduction reaction, Electrochim. Acta, 2018, 283, 987–996
118. M. Shao, A. Peles, K. Shoemaker, Electrocatalysis on platinum nanoparticles: Particle sizeeffect on oxygen reduction reaction activity.Nano Lett., 2011, 11, 3714–3719.
119. E. Antolini, Structural parameters of supported fuel cell catalysts: The effect of particlesize, inter-particle distance and metal loading on catalytic activity and fuel cell performance. Appl. Catal. B Environ., 2016, 181, 298-313.
120. F. J. Perez-Alonso, D. N. McCarthy, A. Nierhoff, P. Hernandez-Fernandez, C. Strebel, I. E. L. Stephens, J. H. Nielsen, I. Chorkendorff, The effect of size on the oxygenelectroreduction activity of mass-selected platinum nanoparticles.Angew. Chem. Int. Ed., 2012, 51, 4641–4643.

121. F. Calle-Vallejo, J. Tymoczko, V. Colic, Q. H. Vu, M. D. Pohl, K. Morgenstern, D. Loffreda, P. Sautet, W. Schuhmann, A. S. Bandarenka, Finding optimal surface sites onheterogeneous catalysts by counting nearest neighbors.Science, 2015, 350, 185–189.
122. M. Nesselberger, S. Ashton, J. C. Meier, I. Katsounaros, K. J. J. Mayrhofer, M. Arenz, Theparticle size effect on the oxygen reduction reaction activity of Pt catalysts: Influence ofelectrolyte and relation to single crystal models.J. Am. Chem. Soc., 2011, 133, 17428–17433.
123. Y. Sun, L. Zhuang, J. Lu, X. Hong, P. Liu, Collapse in crystalline structure and decline incatalytic activity of Pt nanoparticles on reducing particle size to 1 nm.J. Am. Chem. Soc., 2007, 129, 15465–15467.
124. Jiang, K., Zhao, D., Guo, S., Zhang, X., Zhu, X., Guo, J., Huang, X. Efficient oxygen reduction catalysis by subnanometer Pt alloy nanowires. Science advances, 2017, 3(2), e1601705.
125. A. S. Aricò, P. Bruce, B. Scrosati, J. M. Tarascon, W. van Schalkwijk, Nat. Mater., 2005, 4, 366–377.
126. S. Guo, S. Zhang, S. Sun, Angew. Chem. Int. Ed. Eng., 2013, l, 8526–8544.
127. J. Wu, A. Gross, H. Yang, Nano Lett.11, 798-802(2011).
128. J. Zhang, H. Yang, J. Fang, S. Zou, Nano Lett., 2010, 10, 638–644.
129. X. Huanget al., Adv. Mater., 2013, 25, 2974–2979.
130. H. Zhu, S. Zhang, S. Guo, D. Su, S. Sun,J. Am. Chem. Soc., 2013, 135,7130–7133.
131. Huang, X., Zhao, Z., Cao, L., Chen, Y., Zhu, E., Lin, Z., Huang, Y. High-performance transition metal-doped Pt_3Ni octahedra for oxygen reduction reaction. Science, 2015, 348(6240), 1230–1234.
132. X. F. Yang, A. Q. Wang, B. T. Qiao, J. Li, J. Y. Liu, T. Zhang, Acc. Chem. Res., 2013, 46, 1740.
133. J. Guo, J. Huo, Y. Liu, W. Wu, Y. Wang, M. Wu, H. Liu, G. Wang, Small Methods, 2019, 3, 1900159.
134. L. Zhang, K. Doyle-Davis, X. L. Sun, Energy Environ. Sci. 2019, 12, 492.
135. B. Zhang, G. Sun, S. Ding, H. Asakura, J. Zhang, P. Sautet, N. Yan, J. Am. Chem. Soc. 2019, 141, 8185.
136. M. Zhou, S. Bao, A. J. Bard, J. Am. Chem. Soc. 2019, 141, 10117.
137. Li, F., Li, Y., Zeng, X. C. Chen, Z. Exploration of high-performance single-atom catalysts on support M1/FeOx for CO oxidation via computational study. ACS Catal., 2015, 5, 544–552.
138. Cao, X., Ji, Y. & Luo, Y. Dehydrogenation of propane to propylene by a Pd/Cu Single-Atom catalyst: insight from first-principles calculations. J. Phys. Chem. C, 2015, 119, 1016–1023.
139. Liu, X., Sui, Y., Duan, T., Meng, C. & Han, Y. Monodisperse Pt atoms anchored on N-doped graphene as efficient catalysts for CO oxidation: a first-principles investigation. Catal. Sci. Technol., 2015, 5, 1658-1667.
140. Lin, J. et al. Remarkable performance of Ir1/FeOx single-atom catalyst in water gas shift reaction. J. Am. Chem. Soc., 2013, 135, 15314–15317.
141. Li, T., Liu, J., Song, Y., Wang, F. Photochemical solid-phase synthesis of platinum single atoms on nitrogen-doped carbon with high loading as bifunctional catalysts for hydrogen evolution and oxygen reduction reactions. ACS catalysis, 2018, 8(9), 8450–8458.
142. Liu, J., Bak, J., Roh, J., Lee, K. S., Cho, A., Han, J. W., Cho, E. Reconstructing the coordination environment of platinum single-atom active sites for boosting oxygen reduction reaction. ACS Catalysis, 2020, 11(1), 466–475.
143. J. J. Humphrey, S. Sadasivan, D. Plana, V. Celorrio, R. A. Tooze, D. J. Fermin, Chemistry, 2015, 21, 12694.
144. L. M. Rossi, J. L. Fiorio, M. A. S. Garcia, C. P. Ferraz, Dalton Trans., 2018, 47, 5889.
145. Zhu, X., Tan, X., Wu, K. H., Haw, S. C., Pao, C. W., Su, B. J., Lu, X. Intrinsic ORR Activity Enhancement of Pt Atomic Sites by Engineering the d-Band Center via Local Coordination Tuning. Angewandte Chemie International Edition, 2021, 60(40), 21911–21917.
146. Jasinski R. A new fuel cell cathode catalyst [J]. Nature, 1964, 201(4925): 1212–1213.
147. Alt H, Binder H, Sandstede G. Mechanism of the electrocatalytic reduction of oxygen on metal chelates [J]. Journal of Catalysis, 1973, 28(1): 8-19.

148. J. Tian, A. Morozan, M. T. Sougrati, M. LefHvre, R. Chenitz, J. P. Dodelet, D. Jones, F. Jaouen, Optimized synthesis of Fe/N/C cathode catalysts for PEM fuel cells: a matter of iron-ligand coordination strength [J]. Angew. Chem. Int. Ed. 2013, 52, 6867–6870.
149. Y. C. Wang, Y. J. Lai, L. Song, Z. Y. Zhou, J. G. Liu, Q. Wang, X. D. Yang, C. Chen, W. Shi, Y. P. Zheng, M. Rauf, S. G. Sun, S-Doping of an Fe/N/C ORR catalyst for polymer electrolyte membrane fuel cells with high power density [J]. Angew. Chem. Int. Ed. 2015, 54, 9907–9910.
150. J. Wang, Z. Huang, W. Liu, C. Chang, H. Tang, Z. Li, W. Chen, C. Jia, T. Yao, S. Wei, Y. Wu, Y. Li, Design of N-coordinated dual-Metal Sites: a stable and active Pt-free catalyst for acidic oxygen reduction reaction [J]. J. Am. Chem. Soc. 2017, 139, 17281–17284.
151. X. Wan, X. Liu, Y. Li, R. Yu, L. Zheng, W. Yan, H. Wang, M. Xu, J. Shui, Fe-N-C electrocatalyst with dense active sites and efficient mass transport for high-performance proton exchange membrane fuel cells [J]. Nat. Catal. 2019, 2, 259–268.
152. M. X. Chen, M. Z. Zhu, M. Zuo, S. Q. Chu, J. Zhang, Y. Wu, H. W. Liang, X. L Feng, Identification of catalytic sites for oxygen reduction in metal/nitrogen-doped carbons with encapsulated metal nanoparticles [J]. Angew. Chem. Int. Ed. 2020, 59, 1627–1633.
153. H. A. Gasteiger, N. M. Markovi, Just a dream-or future reality? [J]. Science, 2009, 324(5923), 48-49.
154. Frédéric, Jaouen, Eric, Proietti, Michel, & Lefèvre, et al. Recent advances in non-precious metal catalysis for oxygen-reduction reaction in polymer electrolyte fuel cells [J]. Energy Environ. Sci., 2011, 4, 114–130.
155. Lefèvre, Michel, Proietti, Eric, Jaouen, & Frédéric, et al. Iron-based catalysts with improved oxygen reduction activity in polymer electrolyte fuel cells [J]. Science, 2009, 324, 71–74.
156. H. T. Chung, C. M. Johnston, K. Artyushkova, M. Ferrandon, D. J. Myers, P. Zelenay, Cyanamide-derived non-precious metal catalyst for oxygen reduction [J]. Electrochemistry Communications, 2012, 12(12), 1792–1795.
157. L. B. Wu, Y. Nabae, S. Moriya, K, Matsubayashi, N. M. Islam, S. Kuroki, M. Kakimoto, J. Ozaki, S. Miyata, Retracted article: Pt-free cathode catalysts prepared via multi-step pyrolysis of Fe phthalocyanine and phenolic resin for fuel cells [J]. Chem. Commun., 2010, 46, 6377–6379.
158. G. Wu, K. L. More, C. M. Johnston, P. Zelenay, High-performance electrocatalysts for oxygen reduction derived from polyaniline, iron, and cobalt [J]. Science, 2011, 332(6028), 443–447.
159. H. A. Gasteiger, S. S. Kocha, B. Sompalli, F. T. Wagner. Activity benchmarks and requirements for Pt, Pt-alloy, and non-Pt oxygen reduction catalysts for PEMFCs [J]. Appl. Catal. B, 2005, 56, 9–35.
160. F. T. Wagner, B. Lakshmanan, M. F. Mathias, Electrochemistry and the future of the automobile. Journal of Physical Chemistry Letters [J], 2010, 1(14), 2204–2219.
161. Hydrogen and Fuel Cells Program Annual Merit Review. 2016. (https://www.hydrogen.energy.gov/pdfs/progress16/v_a_1_zelenay_2016.pdf).
162. A. Serov, A. D. Shum, X.H. Xiao, V. D. Andrade, K. Artyushkova, I.V. Zenyuk, P. Atanassov, Nano-structured platinum group metal-free catalysts and their integration in fuel cell electrode architectures [J]. Appl. Catal. B Environ. 2018, 237, 1139–1147.
163. D. Malko, A. Kucernak, T. Lopes, In situ electrochemical quantification of active sites in Fe-N/C non-precious metal catalysts [J]. Nat. Commun. 2016, 7, 13285–13291.
164. C. Zhang, Y. C. Wang, B. An, R. Y. Huang, C. Wang, Z. Y. Zhou, W. B. Lin, (2016). Networking pyrolyzed zeolitic imidazolate frameworks by carbon nanotubes improves conductivity and enhances oxygen-reduction performance in polymer-electrolyte-membrane fuel cells [J]. Adv. Mater. 2017, 29, 1604556–1604562.
165. S. Stariha, K. Artyushkova, A. Serov, P. Atanassov. Non-PGM membrane electrode assemblies: optimization for performance [J]. International Journal of Hydrogen Energy, 2015, 40(42), 14676–14682.
166. D. Banham, T. Kishimoto, T. Sato, Y. Kobayashi, K. Narizuka, J. I. Ozaki, Y. J. Zhou, E. Marquez, K. Bai, S. Y. Ye, New insights into non-precious metal catalyst layer designs for proton exchange membrane fuel cells: improving performance and stability [J]. J. Power Sources, 2017, 344(15), 39–45.

167. F. Jaouen, M. Lefèvre, J. P. Dodelet, M. Cai, Heat-treated Fe/N/C catalysts for O2 electroreduction: are active sites hosted in micropores? [J] J. Phys. Chem. B 110, 5553–5555.
168. Jaouen F, Herranz J,Lefevre M, et al. Cross-Laboratory Experimental Study of Non-Noble-Metal Electrocatalysts for the Oxygen Reduction Reaction [J]. ACS Applied Materials & Interfaces, 2009, 1(8):1623–1639.
169. A. Ignaszak, S. Ye, E. Gyenge, A study of the catalytic interface for O2 electroreduction on Pt: the interaction between carbon support meso/microstructure and ionomer (nafion) distribution. Journal of Physical Chemistry C, 2008, 113(1), 298-307.
170. E. Antolini, Carbon supports for low-temperature fuel cell catalysts [J]. Appl. Catal., B 2009, 88, 1–24.
171. J. Zhang, Q. A. Huang, J. Wang, J. Wang, Y. Zhao, Supported dual-atom catalysts: preparation, characterization, and potential applications [J]. Chinese Journal of Catalysis, 2020, 41(5), 783–798.
172. G. Sun, Z. J. Zhao, R. Mu, S. Zha, L. Li, S. Chen, K. Zang, J. Luo, Z. Li, S. C. Purdy, A. J. Kropf, J. T. Miller, L. Zeng, J. Gong, Breaking the scaling relationship via thermally stable Pt/Cu single atom alloys for catalytic dehydrogenation [J]. Nat. Commun. 2018, 9, 4454–4452.
173. S. Siahrostami, A. Verdaguer-Casadevall, M. Karamad, D. Deiana, P. Malacrida, B. Wickman, M. Escudero-Escribano, E. A. Paoli, R. Frydendal, T. W. Hansen, I. Chorkendorff, I. E. Stephens, J. Rossmeisl, Enabling direct H2O2 production through rational electrocatalyst design [J]. Nat. Mater. 2013, 12, 1137.
174. S. Yang, J. Kim, Y. J. Tak, A. Soon, H. Lee, Single-Atom Catalyst of Platinum Supported on Titanium Nitride for Selective Electrochemical Reactions [J]. Angew. Chem., Int. Ed.2016, 55, 2058.
175. C. H. Choi, M. Kim, H. C. Kwon, S. J. Cho, S. Yun, H. T. Kim, K. J. Mayrhofer, H. Kim, M. Choi, Tuning selectivity of electrochemical reactions by atomically dispersed platinum catalyst [J]. Nat. Commun. 2016, 7, 10922.
176. X. F. Yang, A. Wang, B. Qiao, J. Li, J. Liu, T. Zhang, Single-Atom Catalysts: A New Frontier in Heterogeneous Catalysis [J]. Acc. Chem. Res. 2013, 46, 1740.
177. Zhang W, Chao Y, Zhang W, et al. Emerging dual-atomic-site catalysts for efficient energy catalysis [J]. Advanced Materials, 2021, 33(36): 2102576.
178. Lu Z, Wang B, Hu Y, et al. An isolated zinc–cobalt atomic pair for highly active and durable oxygen reduction [J]. Angewandte Chemie, 2019, 131(9): 2648–2652.
179. Zhang Y, Hu Y, Li S, et al. Manganese dioxide-coated carbon nanotubes as an improved cathodic catalyst for oxygen reduction in a microbial fuel cell [J]. Journal of Power Sources, 2011, 196(22): 9284–9289.
180. K. Kumar, C. Canaff, J. Rousseau, et al. Effect of the Oxide-Carbon Heterointerface on the Activity of Co3O4/NRGO nanocomposites toward ORR and OER [J]. Journal of Physical Chemistry C, 2016, 120(15).
181. J. Xiao, C. Chen, J. Xi, et al. Core-shell Co@Co3O4 nanoparticle-embedded bamboo-like nitrogen-doped carbon nanotubes (BNCNTs) as a highly active electrocatalyst for the oxygen reduction reaction [J]. Nanoscale, 2015, 7(16):7056–64.
182. Shan-Shan, Li, Huai-Ping, et al. Flexible nitrogen-doped graphene/carbon nanotube/Co3O4 paper and its oxygen reduction activity [J]. Nanoscale, 2014.
183. Y. Liang, Y. Li, H.Wang, et al. Co3O4 nanocrystals on graphene as a synergistic catalyst for oxygen reduction reaction [J]. Nature materials, 2011, 10(10): 780–786.
184. G P. Wu, Jun, et al. A Strategy to Promote the Electrocatalytic Activity of Spinels for Oxygen Reduction by Structure Reversal [J]. Angew. Chem. Int. Ed., 2015.
185. I. Arindam, W. M. Prashanth, R. S. Nastaran, B. Arno, D. Chittaranjan, T. Massimo, et al. Unification of catalytic water oxidation and oxygen reduction reactions: amorphous beat crystalline cobalt iron oxides [J]. J. Am. Chem. Soc, 2014, 136(50): 17530–17536.
186. A. Zhao, J. Masa, W. Xia, A. Maljusch, M. G.Willinger, G. Clavel, , et al. Spinel mn–co oxide in n-doped carbon nanotubes as a bifunctional electrocatalyst synthesized by oxidative cutting [J]. J. Am. Chem. Soc. 2014, 136, 21, 7551–7554.

187. F. Cheng, J. Shen, B. Peng, et al. Rapid room-temperature synthesis of nanocrystalline spinels as oxygen reduction and evolution electrocatalysts [J]. Nature Chem, 2011, 3(1): 79–84.
188. H. Yang, F. Hu, Y. Zhang, L. Shi, Q. Wang. Controlled synthesis of porous spinel cobalt manganese oxides as efficient oxygen reduction reaction electrocatalysts [J]. Nano research, 2016, 9(1): 207–213.
189. R. A. Sidik, A. B. Anderson, Co9S8 as a catalyst for electroreduction of O2: quantum chemistry predictions [J]. The Journal of Physical Chemistry B, 2006, 110(2): 936–941.
190. J. Wang, L. Li, X. Chen, et al. Monodisperse cobalt sulfides embedded within nitrogen-doped carbon nanoflakes: an efficient and stable electrocatalyst for the oxygen reduction reaction [J]. J. Mater. Chem. A, 2016, 4,11342–11350.
191. S. Sheng, X. Wang, X. Zhou, et al. A comprehensive review of Pt electrocatalysts for the oxygen reduction reaction: Nanostructure, activity, mechanism and carbon support in PEM fuel cells [J]. J. Mater. Chem. A, 2017, 5,1808–1825.
192. D. B. Mahesh, L. Geunsik, S. L. Jae, Screening of Oxygen-Reduction-Reaction-Efficient Electrocatalysts Based on Ag-M (M = 3d, 4d, and 5d Transition Metals) Nanoalloys: A Density Functional Theory Study [J]. Energy Fuels, 2017, 31, 1874–1881.
193. D. Chen, C. Chen, Z. M. Baiyee, Z. Shao and F. Ciucci, Nonstoichiometric Oxides as Low-Cost and Highly-Efficient Oxygen Reduction/Evolution Catalysts for Low-Temperature Electrochemical Devices [J]. Chem. Rev., 2015, 115, 9869–9921.
194. D. Yan, Y. Li, J. Huo, R. Chen, L. Dai and S. Wang, Defect Chemistry of Nonprecious-Metal Electrocatalysts for Oxygen Reactions Adv. Mater., 2017, 29, 1606459–1606478.
195. T. Yamamoto, Y. Kobayashi, N. Hayashi, C. Tassel, T. Saito, S. Yamanaka, M. Takano, K. Ohoyama, Y. Shimakawa, K. Yoshimura and H. Kageyama, (Sr1-xBax)FeO2 ($0.4 \leq x \leq 1$): A New Oxygen-Deficient Perovskite Structure [J]. J. Am. Chem. Soc., 2012, 134, 11444–11454.
196. K. J. May, C. E. Carlton, K. A. Stoerzinger, M. Risch, J. Suntivich, Y. L. Lee, A. Grimaud and Y. Shao-Horn, Influence of Oxygen Evolution during Water Oxidation on the Surface of Perovskite Oxide Catalysts [J]. J. Phys. Chem. Lett., 2012, 3, 3264–3270.
197. M. Risch, A. Grimaud, K. J. May, K. A. Stoerzinger, T. J. Chen, A. N. Mansour and Y. Shao-Horn, Structural Changes of Cobalt-Based Perovskites upon Water Oxidation Investigated by EXAFS [J]. J. Phys. Chem. C, 2013, 117, 8628–8635.
198. J. Kim, X. Yin, K. C. Tsao, S. Fang and H. Yang, Ca2Mn2O5 as Oxygen-Deficient Perovskite Electrocatalyst for Oxygen Evolution Reaction [J]. J. Am. Chem.Soc., 2014, 136, 14646–14649.
199. H. Zhang, W. Xia, H. Shen, et al. Antiperovskite Intermetallic Nanoparticles for Enhanced Oxygen Reduction [J]. Angew. Chem. Int. Ed. 2020, 59(5): 1871–1877.
200. W. Xu, N. Apodaca, H. Wang, et al. A-site Excessive (La0.8Sr0.2)1+xMnO3 Perovskite Oxides for Bifunctional Oxygen Catalyst in Alkaline Media [J]. ACS Catalysis, 2019, 9(6): 5074–5083
201. L. Zhang, W. Sun, C. Xu, et al. Attenuating a metal-oxygen bond of a double perovskite oxide via anion doping to enhance its catalytic activity for the oxygen reduction reaction [J]. Journal of Materials Chemistry A, 2020, 8(28): 14091–14098
202. F. Gonell, C. M. Sánchez-Sánchez, V. Vivier, et al. Structure-Activity Relationship in Manganese Perovskite Oxide Nanocrystals from Molten Salts for Efficient Oxygen Reduction Reaction Electrocatalysis [J]. Chemistry of Materials, 2020, 32(10): 4241–4247
203. J. Greeley, I. Stephens, A. Bondarenko, et al. Alloys of platinum and early transition metals as oxygen reduction electrocatalysts [J]. Nature Chemistry, 2009, 1(7): 552–556.
204. W. He, Y. Wang, C. Jiang, et al. Structural effects of a carbon matrix in non-precious metal O2-reduction electrocatalysts [J]. Chemical Society Reviews, 2016, 45(9): 2396–2409.
205. R. Arrigo, M. Havecker, S. Wrabetz, et al. Tuning the acid/base properties of nanocarbons by functionalization via amination [J]. J. Am. Chem. Soc. 2010, 132, 28, 9616–9630.
206. V. Strelko, V. Kuts, P. Thrower, On the mechanism of possible influence of heteroatoms of nitrogen, boron and phosphorus in a carbon matrix on the catalytic activity of carbons in electron transfer reactions [J]. Carbon, 2000, 10(38): 1499–1503.

207. H. Kim, K. Lee, S. I. Woo, et al. On the mechanism of enhanced oxygen reduction reaction in nitrogen-doped graphene nanoribbons [J]. Physical Chemistry Chemical Physics, 2011, 13(39): 17505–17510.
208. L. Lai, J. R. Potts, D. Zhan, et al. Exploration of the active center structure of nitrogen-doped graphene-based catalysts for oxygen reduction reaction [J]. Energy Environ. Sci., 2012, 5(7): 7936–7942.
209. K. Gong, F. Du, Z. Xia, et al. Nitrogen-doped carbon nanotube arrays with high electrocatalytic activity for oxygen reduction [J]. Science, 2009, 323(5915): 760–764.
210. C. H. Choi, S. H. Park, S. I. Woo, Phosphorus-nitrogen dual doped carbon as an effective catalyst for oxygen reduction reaction in acidic media: Effects of the amount of P-doping on the physical and electrochemical properties of carbon [J]. J. Mater. Chem., 2012, 22, 12107–12115.
211. C. H. Choi, S. H. Park, S. I. Woo, Heteroatom doped carbons prepared by the pyrolysis of bio-derived amino acids as highly active catalysts for oxygen electro-reduction reactions [J]. Green Chemistry, 2011, 13(2): 406–412.
212. C. H. Choi, S. H. Park, S. I. Woo, Binary and ternary doping of nitrogen, boron, and phosphorus into carbon for enhancing electrochemical oxygen reduction activity [J]. ACS Nano, 2012, 6(8): 7084–7091.
213. Setzler, B. P., Zhuang, Z., Wittkopf, J. A., et al., Activity targets for nanostructured platinum-group-metal-free catalysts in hydroxide exchange membrane fuel cells [J]. Nat. Nanotechnol. 2016, 11: 1020–1025.
214. Firouzjaie, H. A., Mustain, W. E., Catalytic Advantages, Challenges, and Priorities in Alkaline Membrane Fuel Cells [J]. ACS Catal. 2019, 10: 225–234.
215. Omasta, T. J., Wang, L., Peng, X., et al., Importance of balancing membrane and electrode water in anion exchange membrane fuel cells [J]. J. Power Sources 2018, 375:205–213.
216. Ramaswamy, N., Mukerjee, S., Alkaline Anion-Exchange Membrane Fuel Cells: Challenges in Electrocatalysis and Interfacial Charge Transfer [J]. Chem. Rev. 2019, 119: 11945–11979.
217. DOE, US. Fuel Cell Multi-Year Research, Development and Demonstration Plan. Department of Energy Washington, DC: 2012, 1–34.
218. Cong, Y., Yi, B., Song, Y., Hydrogen oxidation reaction in alkaline media: From mechanism to recent electrocatalysts [J]. Nano Energy 2018, 44: 288–303.
219. Davydova, E. S., Mukerjee, S., Jaouen, F., et al., Electrocatalysts for Hydrogen Oxidation Reaction in Alkaline Electrolytes [J]. ACS Catal. 2018, 8: 6665–6690.
220. Jia, Q., Liu, E., Jiao, L., et al., Current understandings of the sluggish kinetics of the hydrogen evolution and oxidation reactions in base [J]. Curr. Opin. Electrochem. 2018, 12: 209–217.
221. Tian, X., Zhao, P., Sheng, W., Hydrogen Evolution and Oxidation: Mechanistic Studies and Material Advances [J]. Adv. Mater. 2019, 31: 1808066.
222. Maruyama, J., Inaba, M., Katakura, K., et al., Influence of Nafion® film on the kinetics of anodic hydrogen oxidation [J]. J. Electroanal. Chem. 1998, 447: 201–209.
223. Neyerlin, K. C., Gu, W., Jorne, J., et al., Study of the Exchange Current Density for the Hydrogen Oxidation and Evolution Reactions [J]. J. Electrochem. Soc. 2007, 154: B631.
224. Bagotzky, V., Osetrova, N., Investigations of hydrogen ionization on platinum with the help of micro-electrodes [J]. J. Electroanal. Chem. Interfacial Electrochem. 1973, 43: 233–249.
225. Sheng, W., Gasteiger, H. A., Shao-Horn, Y., Hydrogen Oxidation and Evolution Reaction Kinetics on Platinum: Acid vs Alkaline Electrolytes [J]. J. Electrochem. Soc. 2010, 157: B1529.
226. Durst, J., Siebel, A., Simon, C., et al., New insights into the electrochemical hydrogen oxidation and evolution reaction mechanism [J]. Energy Environ. Sci. 2014, 7: 2255–2260.
227. Schmidt, T., Ross Jr, P., Markovic, N., Temperature dependent surface electrochemistry on Pt single crystals in alkaline electrolytes: Part 2. The hydrogen evolution/oxidation reaction [J]. J. Electroanal. Chem. 2002, 524: 252–260.
228. W. E. Mustain, M. Chatenet, M. Page, Y. S. Kim., Durability Challenges of Anion Exchange Membrane Fuel Cells. Energy Environ [D]. Energy Environ. Sci. 2020, 13 (9), 2805–2838.

229. Y. Zhao, H. Yu, D. Yang, J. Li, Z. Shao, B. Yi, High-Performance Alkaline Fuel Cells Using Crosslinked Composite Anion Exchange Membrane [J]. Journal of Power Sources 2013, 221, 247–251.
230. Christmann, K., Interaction of hydrogen with solid surfaces [J]. Surf. Sci. Rep. 1988, 9: 1–163.
231. Paal, Z., Menon, P., Hydrogen effects in catalysis. Mar-cel Dekker, New York 1988.
232. Mitsui, T., Rose, M., Fomin, E., et al., Dissociative hydrogen adsorption on palladium requires aggregates of three or more vacancies [J]. Nature 2003, 422: 705–707.
233. Breiter, M., Reaction mechanisms of the H2 oxidation/evolution reaction. Handbook of fuel cells 2010.
234. Conway, B., Tilak, B., Interfacial processes involving electrocatalytic evolution and oxidation of H2, and the role of chemisorbed H [J]. Electrochim. Acta 2002, 47: 3571–3594.
235. Skúlason, E., Tripkovic, V., Björketun, M. E., et al., Modeling the electrochemical hydrogen oxidation and evolution reactions on the basis of densit functional theory calculations [J]. J. Phys. Chem. C 2010, 114: 18182–18197.
236. Nørskov, J. K., Bligaard, T., Logadottir, A., et al., Trends in the exchange current for hydrogen evolution [J]. J. Electrochem. Soc. 2005, 152: J23.
237. St. John, S., Atkinson, R. W., Unocic, R. R., et al., Ruthenium-Alloy Electrocatalysts with Tunable Hydrogen Oxidation Kinetics in Alkaline Electrolyte [J]. J. Phys. Chem. C 2015, 119: 13481–13487.
238. Rheinländer, P. J., Herranz, J., Durst, J., et al., Kinetics of the Hydrogen Oxidation/Evolution Reaction on Polycrystalline Platinum in Alkaline Electrolyte Reaction Order with Respect to Hydrogen Pressure [J]. J. Electrochem. Soc. 2014, 161: F1448–F1457.
239. Zheng, J., Sheng, W., Zhuang, Z., et al., Universal dependence of hydrogen oxidation and evolution reaction activity of platinum-group metals on pH and hydrogen binding energy [J]. Sci. Adv. 2016, 2: e1501602.
240. Wang, J. X., Springer, T. E., Adzic, R. R., Dual-Pathway Kinetic Equation for the Hydrogen Oxidation Reaction on Pt Electrodes [J]. J. Electrochem. Soc. 2006, 153: A1732.
241. Wang, J. X., Springer, T. E., Liu, P., et al., Hydrogen oxidation reaction on Pt in acidic media: Adsorption isotherm and activation free energies [J]. J. Phys. Chem. C 2007, 111: 12425–12433.
242. Elbert, K., Hu, J., Ma, Z., et al., Elucidating Hydrogen Oxidation/Evolution Kinetics in Base and Acid by Enhanced Activities at the Optimized Pt Shell Thickness on the Ru Core [J]. ACS Catal. 2015, 5: 6764–6772.
243. Lu S., Zhuang Z. Investigating the influences of the adsorbed species on catalytic activity for hydrogen oxidation reaction in alkaline electrolyte [J]. Journal of the American Chemical Society, 2017. 139(14): 5156–5163.
244. Liu D., Lu S., Xue Y., et al. One-pot synthesis of IrNi@Ir core-shell nanoparticles as highly active hydrogen oxidation reaction electrocatalyst in alkaline electrolyte [J]. Nano Energy, 2019, 59: 26–32.
245. Wang Y., Wang G., Li G., et al. Pt-Ru catalyzed hydrogen oxidation in alkaline media: Oxophilic effect or electronic effect? Energy & Environmental Science, 2015, 8(1): 177–181.
246. Li Q., Peng H., Wang Y., et al. The comparability of Pt to Pt-Ru in catalyzing the hydrogen oxidation reaction for alkaline polymer electrolyte fuel cells operated at 80 °C [J]. Angewandte Chemie International Edition, 2018, 131(5): 1456–1460.
247. Miller H. A., Lavacchi A., Vizza F., et al. Pd/C-CeO2 anode catalyst for high-performance platinum-free anion exchange membrane fuel cells [J]. Angewandte Chemie International Edition, 2016, 55(20): 6004–6007.
248. Qin X., Zhang L., Xu G.-L., et al. The role of Ru in improving the activity of Pd toward hydrogen evolution and oxidation reactions in alkaline solutions [J]. ACS Catalysis, 2019, 9(10): 9614–9621.
249. Yang Y. s Sun X., Han G., et al. Enhanced electrocatalytic hydrogen oxidation on Ni/NiO/C derived from a nickel-based metal-organic framework [J]. Angewandte Chemie International Edition, 2019, 58(31): 10644 -10649.

250. Ohyama, J., Sato, T., Yamamoto, Y., et al., Size specifically high activity of Ru nanoparticles for hydrogen oxidation reaction in alkaline electrolyte [J]. J. Am. Chem. Soc. 2013, 135: 8016–8021.
251. Zheng, J., Zhou, S., Gu, S., et al., Size-Dependent Hydrogen Oxidation and Evolution Activities on Supported Palladium Nanoparticles in Acid and Base [J]. J. Electrochem. Soc. 2016, 163: F499–F506.
252. Zheng, J., Zhuang, Z., Xu, B., et al., Correlating Hydrogen Oxidation/Evolution Reaction Activity with the Minority Weak Hydrogen-Binding Sites on Ir/C Catalysts [J]. ACS Catal. 2015, 5: 4449–4455.
253. Wang, H., Abruña, H. D., Rh and Rh Alloy Nanoparticles as Highly Active H2 Oxidation Catalysts for Alkaline Fuel Cells [J]. ACS Catal. 2019, 9, 5057–5062.
254. Ramaswamy, N., Ghoshal, S., Bates, M. K., et al., Hydrogen oxidation reaction in alkaline media: Relationship between electrocatalysis and electrochemical double-layer structure [J]. Nano Energy 2017, 41: 765–771.
255. Li, Q., Peng, H., Wang, Y., et al., The Comparability of Pt to Pt-Ru in Catalyzing the Hydrogen Oxidation Reaction for Alkaline Polymer Electrolyte Fuel Cells Operated at 80 degrees C [J]. Angew. Chem. Int. Ed. 2019, 58: 1442–1446.
256. Ghoshal, S., Jia, Q., Bates, M. K., et al., Tuning Nb Pt Interactions to Facilitate Fuel Cell Electrocatalysis [J]. ACS Catal. 2017, 7: 4936–4946.
257. Jin, Y., Chen, F., Wang, J., et al., Lamellar platinum rhodium aerogels with superior electrocatalytic performance for both hydrogen oxidation and evolution reaction in alkaline environment [J]. J. Power Sources 2019, 435: 226798.
258. Scofield M. E., Zhou Y., Yue S., et al. Role of chemical composition in the enhanced catalytic activity of Pt-based alloyed ultrathin nanowires for the hydrogen oxidation reaction under alkaline conditions [J]. ACS Catalysis, 2016,6(6): 3895–3908.
259. Bhowmik T., Kundu M. K., Baman S. Palladium nanoparticle-graphitic carbon nitride porous synergistic catalyst for hydrogen evolution/oxidation reactions over a broad range of pH and correlation of its catalytic activity with measured hydrogen binding energy [J]. ACS Catalysis, 2016, 6(3): 1929–1941.
260. Qin B., Yu H., Chi J., et al. A novel Ir/CeO2-C nanoparticle electrocatalyst for the hydrogen oxidation reaction of alkaline anion exchange membrane fuel cells [J]. RSC Advances, 2017, 7(50): 31574– 31581.
261. Yu H., Davydova E. S., Ash U., et al. Palladium-ceria nanocatalyst for hydrogen oxidation in alkaline media: Optimization of the Pd-CeO2 interface [J]. Nano Energy, 2019, 57: 820–826.
262. Miller H. A., Vizza F., Marelli M., et al. Highly active nanostructured palladium-ceria electrocatalysts for the hydrogen oxidation reaction in alkaline medium [J]. Nano Energy, 2017,33: 293–305.
263. Piana M, Boccia M, Filip A, et al. H2/air alkaline membrane fuel cell performance and durability, using novel ionomer and non-platinum group metal cathode catalyst [J]. Journal of Power Sources, 2010, 195(18): 5875–5881.
264. Lu S, Jing P, Huang A, et al. Alkaline polymer electrolyte fuel cells completely free from noble metal catalysts [J]. China Basic Science, 2009, 105(52): 20611–20614.
265. Gao L, Wang Y, Li H, et al. Nickel nanocatalyst within h-BN shell for enhanced hydrogen oxidation reactions [J]. Chemical Science, 2017, 8(8): 5728–5734.
266. Hu Q, Li G, Pan J, et al. Alkaline polymer electrolyte fuel cell with Ni-based anode and Co-based cathode [J]. International Journal of Hydrogen Energy, 2013, 38(36): 16264– 16268.
267. Kenjo T. Chromium-doped Raney-nickel catalyst for hydrogen electrodes in alkaline fuel-cells [J]. Journal of the Electrochemical Society, 1985, 132(2): 383–386.
268. Xiro S Y, SChWinZ S. Long-term hydrogen oxidation catalysts in alkaline fuel cells [J]. Journal of Power Sources, 2000, 87(1/2): 101–105.
269. Mund K, Richter G, Sturm F V. Titanium-containing Raney nickel catalyst for hydrogen electrodes in alkaline fuel cell systems [J]. Journal of the Electrochemical Society, 1977, 124(1): 1–6.

270. Ewe H, Justi E, Schmitt A. Untersuchungen zur Konservierung von Raney-Nickel-Misc h-Katalysatoren [J]. Energy Conversion, 1975, 14(2): 35–41.
271. Kenjo T. Doping effects of transition metals on the polarization characteristics in Raney nickel hydrogen electrodes [J]. Electrochimica Acta, 1988, 33(1): 41–46.
272. Shim J-P, Park Y-S, Lee H-K, et al. Hydrogen oxidation characteristics of Raney nickel electrodes with carbon black in an alkaline fuel cell [J]. Journal of Power Sources, 1998, 74(1): 151–154.
273. Sheng W C, Bivens A P, Myint M, et al. Non-precious metal electrocatalysts with high activity for hydrogen oxidation reaction in alkaline electrolytes [J]. Energy & Environmental Science, 2014, 7(5): 1719–1724.
274. Cherstiouk O V, Simonov P A, Oshchepkov A G, et al. Electrocatalysis of the hydrogen oxidation reaction on carbon-supported bimetallic NiCu particles prepared by an improved wet chemical synthesis [J]. Journal of Electroanalytical Chemistry, 2016, 783: 146–151.
275. Zhuang Z, Giles S A, Zheng J, et al. Nickel supported on nitrogen-doped carbon nanotubes as hydrogen oxidation reaction catalyst in alkaline electrolyte [J]. Nature Communications, 2016, 7: 10141–10141.
276. Dhunna R, Lal C, Avasthi D K, et al. Irradiation induced effects on Ni3N/Si bilayer system [J]. Vacuum, 2009, 83(12): 1448–1453.
277. Vempaire D, Miraglia S, Sulpice A, et al. Structure and magnetic properties of nickel nitride thin film synthesized by plasma-based ion implantation [J]. Journal of Magnetism & Magnetic Materials, 2004, 272: 843–844.
278. Kahraman F, Karadeniz S. Characterization and Wear Behavior of Plasma Nitrided Nickel Based Dental Aloy [J]. Plasma Chemistry & Plasma Processing, 2011, 31: 595–604.
279. Ningthoujam R S, Gajbhiye N S, Sharma S. Reduction mechanism of $Ni2^+$ into Ni nanoparticles prepared from different precursors: Magnetic studies [J]. Pramana, 2009, 72(3): 577–586.
280. Choi J, Gillan E G. Solvothermal Metal Azide Decomposition Routes to Nanocrystalline Metastable Nickel, Iron, and Manganese Nitrides [J]. Inorganic Chemistry, 2009, 48(10): 4470–4477.
281. Desmoulins-Krawiec S, Aymonier C, Loppinet-Serani A, et al. Synthesis of nanostructured materials in supercritical ammonia: nitrides, metals and oxides [J]. Journal of Materials Chemistry, 2004, 14(2): 228–232.
282. Zhang Y, Ouyang B, Xu J, et al. 3D Porous Hierarchical Nickel-Molybdenum Nitrides Synthesized by RF Plasma as Highly Active and Stable Hydrogen-Evolution-Reaction Electrocatalysts [J]. Advanced Energy Materials, 2016, 6(11): 1600221–1600221.
283. Ni W, Krammer A, Hsu C-S, et al. Ni3N as an Active Hydrogen Oxidation Reaction Catalyst in Alkaline Medium [J]. Angewandte Chemie International Edition, 2019, 58(22): 7445–7449.
284. Song, F., Li, W., Yang, J., et al., Interfacing nickel nitride and nickel boosts both electrocatalytic hydrogen evolution and oxidation reactions [J]. Nat. Commun. 2018, 9: 4531.
285. Wang, T., Wang, M., Yang, H., et al., Weakening hydrogen adsorption on nickel via interstitial nitrogen doping promotes bifunctional hydrogen electrocatalysis in alkaline solution [J]. Energy Environ. Sci. 2019, 12: 3522–3529.
286. Oshchepkov, A. G., Bonnefont, A., Pronkin, S. N., et al., Nanostructured nickel nanoparticles supported on Vulcan carbon as a highly active catalyst for the hydrogen oxidation reaction in alkaline media [J]. J. Power Sources 2018, 402: 447–452.
287. Pan, Y., Hu, G., Lu, J., et al., $Ni(OH)_2$-Ni/C for hydrogen oxidation reaction in alkaline media [J]. J. Energy Chem. 2019, 29: 111–115.

Chapter 6
Proton Exchange Membrane Fuel Cells: Macro-model

6.1 Introduction

The basic structure of a Proton Exchange Membrane Fuel Cell (PEMFC) single cell is intricate and sophisticated, comprised of multiple vital components meticulously stacked together. These include the cathode flow field plate, cathode diffusion layer, cathode catalyst layer, the core proton exchange membrane, anode catalyst layer, anode diffusion layer, and anode flow field plate. Each layer serves a unique and crucial role, collectively facilitating the efficient electrochemical reaction and effective conversion of hydrogen and oxygen into electricity and water.

In addressing the complexity of PEMFCs, modeling studies employ a range of spatial dimension strategies, spanning from zero-dimensional (0D) to three-dimensional (3D), tailored to different research requirements and depths. The 0D model simplifies the entire fuel cell into an idealized "black box," focusing on the overall system inputs (such as fuel and oxidant supply) and outputs (current, voltage, heat, etc.), disregarding internal intricacies. This approach is suited for rapid assessments and macroscopic system integration analysis, particularly favored in the study of large-scale, highly integrated fuel cell systems due to its simplicity and high computational efficiency.

As the model dimensions increase, so does the precision and complexity of the analysis. Early one-dimensional (1D) models, while not fully encompassing parameter variations within the plane, effectively explore along the thickness direction, capturing the material transport and electrochemical reaction characteristics within the diffusion layers, catalyst layers, and the proton exchange membrane. These models concentrate on processes perpendicular to the flow field plate surface, providing a preliminary framework for understanding the physical and chemical mechanisms inside the cell.

Two-dimensional (2D) models expand the research scope further, considering dynamics along the flow direction as well as cross-sectional analysis perpendicular to the flow, which helps elucidate the impact of channel designs, gas diffusion,

Y. Li, *Fuel Cell Fundamentals and Applications*,
https://doi.org/10.1007/978-981-96-9845-5_6

and the layout of electrochemically active areas on performance. This is pivotal for optimizing flow field designs and enhancing reaction efficiency.

The most exhaustive, three-dimensional (3D) models provide comprehensive analysis and prediction of physical parameter distributions within the PEMFC in real spatial dimensions, such as concentration gradients of hydrogen, oxygen, and water, intricate fluid flow structures, uneven current density distributions, and complex temperature field variations. 3D models reveal local effects and non-uniformities, playing a decisive role in deeply understanding the intricate interaction mechanisms within the cell and precisely guiding the optimization of operational strategies and refined structural design adjustments.

With the continuous advancement of computer technology, the application of big data analytics and Machine Learning (ML) techniques has become increasingly prevalent across various domains, particularly in monitoring systems where ML enables the implementation of adaptive and self-learning threshold settings, revolutionizing the monitoring paradigm. However, the prerequisite for deploying ML-based monitoring strategies lies in the monitored entity exhibiting predictable behavioral patterns. Leveraging this technology not only facilitates accurate prediction of fuel cell degradation rates but also enables real-time monitoring of their operational status, thereby providing robust data support and intelligent decision-making groundwork for optimizing fuel cell performance and prolonging their service life.

In the ongoing evolution and innovation of PEMFC technology, modeling research plays an irreplaceable pivotal role. This domain not only necessitates the combination of highly accurate theoretical predictions with experimental validations but also requires in-depth exploration of the intricate physicochemical mechanisms within the cell. Through meticulously constructed and continually optimized mathematical and physical models, researchers can unveil the mysteries of the PEMFC's interior, gaining insight into the subtle interplay across various stages, including proton transport, gas diffusion, electrochemical reactions, and thermal management. These models not only elucidate the microcosmic workings of the cell but also quantitatively analyze the specific impacts of key parameters such as operating temperature, gas pressure, catalyst loading, and membrane thickness on cell performance, thereby furnishing a scientific foundation for designing more efficient and durable fuel cell stacks.

In recent years, with the leap in computational capabilities and the maturation of simulation techniques, PEMFC modeling research has encountered unprecedented development opportunities. High-precision Computational Fluid Dynamics (CFD) simulations, multiphysics field coupling analyses, and the application of machine learning algorithms enable researchers to predict and optimize fuel cell performance with hitherto unattained precision and efficiency, encompassing enhancements in energy conversion efficiency, lifespan extension, cost reduction, and environmental adaptability. These advanced modeling methodologies accelerate the verification process of new concepts and materials, fostering interdisciplinary collaboration that tightly intertwines materials science, electrochemistry, thermodynamics, and engineering design, collectively propelling PEMFC technology towards deeper commercial applications.

PEMFC modeling research stands as a frontier for theoretical exploration and an indispensable technical pillar for engineering practices. It not only deepens academia's understanding of the fundamental principles of fuel cells but also provides invaluable strategic guidance to the industry, securing PEMFC technology's leading position in the realm of new energy and paving the way for a greener, more sustainable energy future.

6.2 Empirical Model

The 0D model is used to establish the relationship between the input current density and output cell voltage. The difficulty lies in accurately describing the cell voltage loss in different current density regions, which consists of activation loss, ohmic loss, and concentration loss. The following 0D model is proposed to predict the cell voltage [1]:

$$E = E_0 - b \log i - Ri - m \exp(ni) \tag{6.1}$$

where $E_0 = E_r + b \log i_0$, E_r is the reversible potential of the cell, i_0 and b are the Tafel parameters for oxygen reduction, R is the ohmic resistance that results in a linear variation of E with i, and m and n account for the mass transport overpotential as a function of current density. The ohmic resistance R primarily consists of the charge transfer resistance in the proton exchange membrane, the charge transfer resistance of the hydrogen oxidation reaction, the electric resistance in the fuel cell testing facility, and the mass transport resistance in the intermediate current density region. All the parameters can be obtained by curve fitting from an experimental current–voltage curve.

An empirical model is established [2] to describe the electrode processes (activation, ohmic, and mass-transfer) of a PEMFC stack over the entire current range. The potential-current and power-current curves of the strip PEMFC stack were fitted with the empirical equation under a variety of experimental humidity, temperature, and stack length conditions.

6.3 Mechanistic Model

6.3.1 A Basic Model

The mechanistic model means that it considers the internal physicochemical processes inside the cell. The modeling domains usually consist of anode and cathode bipolar plates, flow fields, gas diffusion layers, microporous layers, catalyst layers, and proton exchange membrane. A mechanistic model includes the model assumption, a series of conservation equations, and boundary conditions. It is worth noting that the accuracy of the numerical simulation strongly depends on the adopted transport and electrochemical parameters. Here, the two-fluid model-based PEMFC model is shown below [3].

1. Conservation equations

The conservation equations of the fuel cell model include mass, momentum, energy, charge, and species transfer. The basic assumptions of the model include: (1) the gases or gas mixture are ideal; (2) the gas flow is incompressible laminar flow; (3) the physical properties of the material are isotropic and homogeneous.

(1) Mass transfer in channels and porous media

Mass conservation equation:

$$\frac{\partial}{\partial t}(\varepsilon(1-s)\rho_g) + \nabla(\rho_g \vec{u}_g) = S_m \tag{6.2}$$

Momentum conservation equation:

$$\begin{aligned}&\frac{\partial}{\partial t}\left(\frac{\rho_g \vec{u}_g}{\varepsilon(1-s)}\right) + \nabla\cdot\left(\frac{\rho_g \vec{u}_g \vec{u}_g}{\varepsilon^2(1-s)^2}\right)\\ &= -\nabla p_g + \frac{\mu_g}{\varepsilon(1-s)}\nabla\cdot\left(\nabla\vec{u}_g + (\nabla\vec{u}_g)^T - \frac{2}{3}(\nabla\cdot\vec{u}_g)\mathbf{I}\right) + S_u\end{aligned} \tag{6.3}$$

where ρ_g is the gas mixture density, $\vec{u}_g$ is the gas velocity, μ_g is the gas mixture dynamic viscosity, ε is the porosity of porous media, p_g is the gas pressure, S_m is the mass source term, and S_u is the viscous force source term of the porous media.

(2) Species transfer in channel and porous media

Species conservation equation (i: H_2, O_2, vapor):

$$\frac{\partial}{\partial t}(\varepsilon(1-s)\rho_g C_i) + \nabla\cdot(\rho_g \vec{u}_g C_i) = \nabla\cdot(\rho_g D_i^{eff}\nabla C_i) + S_i \tag{6.4}$$

where C_i is the concentration of species i, D_i^{eff} is, the effective diffusivity of species i and S_i is the source term of species i.

(3) Heat transfer in all domains

Energy conservation equation:

$$\frac{\partial}{\partial t}\left((\rho C_p)^{eff}T\right) + \nabla\cdot\left((\rho C_p)^{eff}\vec{u}_g T\right) + \nabla\cdot(-k^{eff}\nabla T) = S_T \tag{6.5}$$

where C_p is the specific heat at constant pressure, T is the temperature, k^{eff} is the effective heat conductivity coefficient, S_T is the heat source term.

(4) Charge transfer in the solid phase and electrolyte

Electron and proton conservation equation:

$$\nabla\left(-\sigma_{\text{e}}^{\text{eff}}\nabla\phi_{\text{e}}\right) = S_{\text{e}} \tag{6.6}$$

$$\nabla\left(-\sigma_{\text{p}}^{\text{eff}}\nabla\phi_{\text{p}}\right) = S_{\text{p}} \tag{6.7}$$

where $\sigma_{\text{e}}^{\text{eff}}$ is the effective electric conductivity, $\sigma_{\text{p}}^{\text{eff}}$ is the effective protonic conductivity, ϕ_{e} is the electrode phase potential, ϕ_{p} is the electrolyte phase potential, S_{e} is the electron source term and S_{p} is the proton source term.

(5) Liquid water transfer

Liquid pressure conservation equation:

$$\frac{\partial}{\partial t}(\rho_l \varepsilon s) - \nabla\left(\rho_l \frac{K k_{rl}}{\mu_l}\nabla p_l\right) = S_l \tag{6.8}$$

Liquid saturation conservation equation:

$$\frac{\partial}{\partial t}(\rho_l s) + \nabla\left(\rho_l \vec{u}_{\text{g}} s\right) = S_l \tag{6.9}$$

where s is the liquid saturation, K is the intrinsic permeability, k_{rl} is the relative permeability, $p_l = p_g - p_c$ is the liquid pressure, μ_l is the viscosity of water, ρ_l is the density of water and S_l is the source term. The capillary pressure p_c is given by the Leverett-J function with s the liquid saturation:

$$p_c = \sigma_{\text{st}} \cos\theta_c (\varepsilon/K)^{0.5} J(s) \tag{6.10}$$

$$J(s) = \begin{cases} 1.417(1-s) - 2.120(1-s)^2 + 1.263(1-s)^3 & 0 < \theta_c < 90^\circ \\ 1.417s - 2.120s^2 + 1.263s^3 & 90^\circ < \theta_c < 180^\circ \end{cases} \tag{6.11}$$

(6) Dissolved water transfer in the membrane and ionomer

Dissolved water transport equation:

$$\frac{\partial}{\partial t}\left(\frac{\lambda \varepsilon_{ion}}{V_{\text{ion}}}\right) - \nabla\cdot\left(D_{\text{dw}}\nabla\frac{\lambda}{V_{\text{ion}}}\right) - \frac{\xi}{F}\sigma_{\text{p}}^{\text{eff}}\nabla\phi_{\text{p}} = S_\lambda \tag{6.12}$$

where ε_{ion} is the volume fraction of ionomer, λ is the dissolved water content, V_{ion} is the equivalent volume of membrane, D_{dw} is the diffusivity of dissolved water, ξ is the electro-osmotic drag coefficient, F is the Faraday's constant and S_λ is the source term of dissolved water.

The above-mentioned source terms are summarized in Table 6.1.

The concentration-dependent Butler-Volmer equation is used for describing hydrogen oxidation reaction (HOR) in anode and oxygen reduction reaction (ORR) in cathode. The transfer current density for anode and cathode are expressed as:

$$j_a = i_{0,a} \cdot a_{Pt} \cdot \left(\frac{C_{H_2}^{g}}{C_{H_2}^{ref}}\right)^{0.5} \cdot \left[\exp\left(\frac{\alpha_a F \eta_a}{RT}\right) - \exp\left(\frac{-\alpha_c F \eta_a}{RT}\right)\right] \tag{6.13}$$

Table 6.1 Source terms of conservation equations

Source term	Unit
$S_m = S_{H_2} + S_{O_2} + S_v$	Kg m^{-3} s^{-1}
$S_u = -\frac{\mu_g}{K k_{rg}} \vec{u}_g$ GDL, CL, cf-FF	kg m^{-2} s^{-2}
$S_{H_2} = -\frac{j_a}{2F} M_{H_2}$ ACL	kg m^{-3} s^{-1}
$S_{O_2} = -\frac{j_c}{4F} M_{O_2}$ CCL	kg m^{-3} s^{-1}
$S_v = -S_{ad} M_{H_2O} - S_l M_{H_2O}$	kg m^{-3} s^{-1}
$S_\lambda = \begin{cases} S_{ad} M_{H_2O} & \text{ACL} \\ S_{ad} M_{H_2O} + \frac{j_c}{2F} M_{H_2O} & \text{CCL} \end{cases}$	kg m^{-3} s^{-1}
$S_{ad} = \begin{cases} \frac{k_a}{\delta_{CL} V_{ion}} (\lambda_{eq} - \lambda) \text{ if } \lambda_{eq} > \lambda \text{ (absorption)} \\ \frac{k_d}{\delta_{CL} V_{ion}} (\lambda_{eq} - \lambda) \text{ if } \lambda_{eq} < \lambda \text{ (desorption)} \end{cases}$	mol m^{-3} s^{-1}
$S_l = \begin{cases} \gamma_{eva} \varepsilon s \frac{p}{RT} (x_{H_2O} - x_{sat}) & x_{H_2O} < x_{sat} \text{ (evaporation)} \\ \gamma_{con} \varepsilon (1-s) \frac{p}{RT} (x_{H_2O} - x_{sat}) & x_{H_2O} > x_{sat} \text{ (condensation)} \end{cases}$	mol m^{-3} s^{-1}
$S_e = \begin{cases} -j_a & \text{ACL} \\ j_c & \text{CCL} \end{cases}$	A m^{-3}
$S_p = \begin{cases} j_a & \text{ACL} \\ -j_c & \text{CCL} \end{cases}$	A m^{-3}
$S_T = S_{T,ad} + S_{T,l} + S_{T,e} + S_{T,p} + S_{T,r}$	W m^{-3}
$S_{T,ad} = H_{ad} S_{ad}$ CL	W m^{-3}
$S_{T,l} = H_{ec} S_l$ GDL, CL	W m^{-3}
$S_{T,e} = \sigma_e^{eff} (\nabla \phi_e)^2$ BP, cf-FF, GDL, CL	W m^{-3}
$S_{T,p} = \sigma_p^{eff} (\nabla \phi_p)^2$ Mem, CL	W m^{-3}
$S_{T,r} = \begin{cases} j_a \eta_a + \frac{j_a}{2F} T \Delta S_{HOR} \text{ in ACL} \\ j_c \eta_c + \frac{j_c}{2F} T \Delta S_{ORR} \text{ in CCL} \end{cases}$	W m^{-3}

$$j_c = (1 - s) i_{0,c} \cdot a_{Pt} \cdot \frac{C_{O_2}^{Pt}}{C_{O_2}^{ref}} \cdot \left[\exp\left(\frac{\alpha_a F \eta_c}{RT} \right) - \exp\left(\frac{-\alpha_c F \eta_c}{RT} \right) \right] \tag{6.14}$$

where $i_{0,a}$ and $i_{0,c}$ are the anode and cathode exchange current densities, respectively. $C_{H_2}^{ref}$ and $C_{O_2}^{ref}$ are respectively the reference hydrogen and oxygen concentrations. a_{Pt} is the active volumetric surface area of Pt, $C_{H_2}^{g}$ is the hydrogen concentration, $C_{O_2}^{Pt}$ is the oxygen concentration, α_a and α_c are the transfer coefficient. The activation overpotentials η_a and η_c are defined as [4]:

$$\eta_{a,\,c} = \phi_e - \phi_p - \begin{cases} -\frac{T\Delta S_{HOR}}{2F} - \frac{RT}{2F} \ln\left(\frac{p_{H_2}}{101325}\right) & \text{ACL} \\ \frac{T\Delta S_{ORR}}{2F} + \frac{RT}{4F} \ln\left(\frac{p_{O_2}}{101325}\right) & \text{CCL} \end{cases} \tag{6.15}$$

where ΔS_{HOR} is the hydrogen oxidation reaction entropy, ΔS_{ORR} is the oxygen reduction reaction entropy, p_{H_2} and p_{O_2} are the hydrogen and oxygen partial pressures, respectively.

2. Boundary conditions

The anode and cathode inlets are set as mass flow inlet conditions:

$$m_a = \frac{\rho_g^a \zeta_a I_{ref} A_{act}}{2FC_{H_2}} \tag{6.16}$$

$$m_a = \frac{\rho_g^a \zeta_a I_{ref} A_{act}}{2FC_{H_2}} \tag{6.17}$$

$$m_c = \frac{\rho_g^c \zeta_c I_{ref} A_{act}}{4FC_{O_2}} \tag{6.18}$$

$$m_c = \frac{\rho_g^c \zeta_c I_{ref} A_{act}}{4FC_{O_2}} \tag{6.19}$$

where ζ_a and ζ_c are respectively the anode and cathode stoichiometry ratios. I_{ref} is the reference current density, A_{act} is the active reaction area of the catalyst layer, C_{H_2} and C_{O_2} are the anode and cathode molar concentrations of hydrogen and oxygen, respectively. The C_{H_2} and C_{O_2} in anode and cathode inlets are determined by:

$$C_{H_2} = \frac{p_g^a - RH_a p_{sat}}{RT} \tag{6.20}$$

$$C_{O_2} = \frac{0.21\left(p_g^c - RH_c p_{sat}\right)}{RT} \tag{6.21}$$

where RH_a and RH_c are the relative humidity at anodic and cathodic inlets, respectively, p_{sat} is the saturated vapor pressure.

The electric potential outside the surfaces of the anode and cathode BPs are defined as:

$$\phi_{\mathrm{e}}^{\mathrm{a}} = 0 \tag{6.22}$$

$$\phi_{\mathrm{e}}^{\mathrm{c}} = E_{\mathrm{cell}} \tag{6.23}$$

The temperatures of the external walls are set as the operating temperature.

6.3.2 *Two-Phase Fluid Research*

The source of water in a fuel cell consists of two parts. On the one hand, to ensure the higher ionic conductivity in the fuel cell, gas humidification is necessary to ensure the membrane hydration. On the other hand, the cathodic oxygen reduction reaction also produces a large amount of water. The fuel cell usually works at a lower temperature (60–90 °C), therefore, affected by the gas–liquid phase equilibrium of water vapor, the water inside the cell may exist in liquid form, and a gas–liquid two-phase flow occurs inside the cell. In the past few decades, experimental techniques to study gas–liquid two-phase flow inside the fuel cell have been reported, including optical visualization, magnetic resonance imaging, neutron imaging, and X-ray radiography [5]. However, conducting such experiments is complex and expensive. For example, optical visualization experiments require the use of transparent materials, and these materials have certain differences in thermoelectric properties and surface properties (such as roughness, wettability) from those actually used in fuel cells. Other non-optical visualization techniques require different designs from real fuel cells. Therefore, numerical modeling methods to determine the two-phase flow in fuel cells and their effects on fuel cell performance have gradually been widely developed. Two-phase flow modeling methods in fuel cells include multiphase mixture model and two-fluid model that do not track the gas–liquid interface, and VOF (volume of fluid) model and level set method that track the interface.

The M2 (Multiphase Mixture) and two-fluid models can consider the two-phase transport with both liquid and gas phases and can predict the amount of liquid water (as liquid water saturation) present in a PEMFC and its effects on cell performance. However, these models cannot identify the presence of liquid water as liquid droplets or water films locally concentrated, dispersed, or accumulated in the gas channels. The VOF and level set model can trace the formation and motion of liquid water. However, the computing resource for these models is huge, because the time step is usually 10^{-6}–10^{-3} s while the liquid formation in channels may take several minutes. It is still impractical to couple the VOF model with the other transport equations to detailly describe the physicochemical process inside the fuel cell. Most researches are limited to investigating the two-phase flow only without the electrochemical reactions.

1. The M2 model

The M2 model is used to describe two-phase multicomponent transport in channels and porous media, which captures pressure, liquid saturation, and reactant composition distribution. The so-called multiphase mixture in this model is distinguished from the conventional multicomponent mixture, which is treated as a single fluid with smoothly varying phasic compositions. Liquid saturation is taken to represent the volume fraction of liquid water. The M2 model considered the gas and liquid phases as constituents of a two-phase mixture with distinct phase velocities. The mass and momentum conservation of the two-phase mixture can be written as follows:

Continuity equation:

$$\frac{\partial}{\partial t}(\varepsilon\rho) + \nabla(\rho\vec{u}) = 0 \tag{6.24}$$

Momentum conservation:

$$\rho\vec{u} = -\frac{K}{\nu}\left(\nabla P - \gamma_\rho \rho \vec{g}\right) \tag{6.25}$$

The physical properties of the two-phase mixture are defined as:

$$\rho = s\rho_l + (1 - s)\rho_g \tag{6.26}$$

$$\rho\vec{u} = \vec{u}_l\rho_l + \vec{u}_g\rho_g \tag{6.27}$$

$$\nu = \left(\frac{k_{rl}}{\nu_l} + \frac{k_{rg}}{\nu_g}\right)^{-1} \tag{6.28}$$

The liquid saturation can be determined by the mixture water concentration:

$$s = \frac{C_{\mathrm{H_2O}} - C_{sat}}{\rho_l / M_{\mathrm{H_2O}} - C_{sat}} \tag{6.29}$$

The density correction factor γ_ρ is a sole function of phase saturations and can be regarded as the property of the two-phase mixture:

$$\gamma_\rho = \frac{\rho_l\lambda_l + \rho_g\lambda_g}{s\rho_l + (1 - s)\rho_g} \tag{6.30}$$

The phase mobility λ_l and λ_g are defined as:

$$\lambda_l = \frac{k_{rl}/\nu_l}{k_{rl}/\nu_l + k_{rg}/\nu_g} \tag{6.31}$$

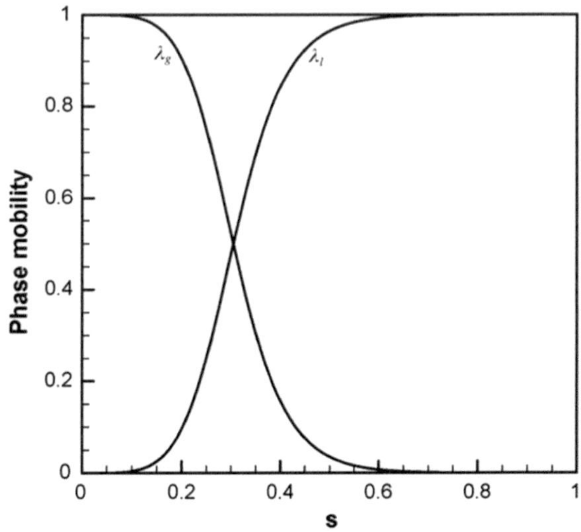

Fig. 6.1 Phase mobility versus liquid water saturation [6]

$$\lambda_g = 1 - \lambda_l \tag{6.32}$$

The relationship between the phase mobility and liquid saturation is plotted in Fig. 6.1.

The relative permeability k_{rl} and k_{rg} are defined as the ratio of the intrinsic permeability of the liquid/gas phase to the total permeability of a porous medium.

$$k_{rl} = \left(\frac{s - s_{ir}}{1 - s_{ir}}\right)^{n_k} \tag{6.33}$$

$$k_{rg} = \left(1 - \frac{s - s_{ir}}{1 - s_{ir}}\right)^{n_k} \tag{6.34}$$

where n_k is set to 4 in their work, s_{ir} is the irreducible residual saturation.

2. The two-fluid model

The multiphase flow PEMFC model describes condensation and evaporation by calculating liquid water separately and taking into account the equilibrium of phase transfer between the liquid and gas phases. The liquid saturation is calculated using Darcy's law. To avoid the liquid water saturation jump at the layer interface (e.g., CL and MPL, MPL and GDL) due to the difference in contact angle, it is better to calculate the liquid pressure in porous media. And the liquid saturation in porous media can be obtained by Leverett-J function and capillary pressure function.

Liquid pressure conservation equation:

$$\frac{\partial}{\partial t}(\rho_l \varepsilon s) - \nabla\left(\rho_l \frac{K k_{rl}}{\mu_l} \nabla p_l\right) = S_l \tag{6.35}$$

Liquid saturation conservation equation:

$$\frac{\partial}{\partial t}(\rho_l s) + \nabla\left(\rho_l \vec{u}_g s\right) = S_l \tag{6.36}$$

Capillary pressure function:

$$p_c = \sigma_{\mathrm{st}} \cos\theta_c (\varepsilon/K)^{0.5} J(s) \tag{6.37}$$

Leverett-J function:

$$J(s) = \begin{cases} 1.417(1-s) - 2.120(1-s)^2 + 1.263(1-s)^3 & 0 < \theta_c < 90^\circ \\ 1.417s - 2.120s^2 + 1.263s^3 & 90^\circ < \theta_c < 180^\circ \end{cases} \tag{6.38}$$

3. VOF model

The VOF method is based on earlier Marker-And-Cell (MAC) methods [5]. First reports of what is now called the VOF method were given by Noh and Woodward, who presented the simple line interface calculation (SLIC) algorithm. However, the first publication of the VOF method in a journal was made by Hirt and Nichols and to whom the authorship of the method is given. In these two early approaches, the interface was represented by a piecewise-constant line in each two-fluid cell, either vertically or horizontally. However, the first implementations of the VOF method using this approach showed some imperfections in the interface description. This was overcome by Youngs, who introduced the piecewise-linear interface calculation (PLIC) scheme. The method of Youngs was shown to be very robust and efficient, and the use of the VOF method with the PLIC scheme is a standard procedure currently applied in various computer codes.

The VOF formulation assumes that two or more phases (fluids) are not interpenetrating. Each volume fraction corresponds to a phase in the computational domain. In each control volume, the sum of the volume fractions of all phases is equal to one. The fields for all the variables and properties are shared by the phases and represent volume-averaged values, as long as the volume fraction of each phase is known at each location. Thus, the variables and properties in any given cell are either purely representative of one of the phases or representative of a mixture of the phases, depending upon the volume fraction values. This means that, if the volume fraction of the fluid q in the cell is α_q, the following three situations are possible:

(1) $\alpha_q = 0$, the cell is empty of the fluid q;
(2) $\alpha_q = 1$, the cell is full with the fluid q;
(3) $0 < \alpha_q < 1$, the cell contains the fluid q and other fluids.

Based on the local value of α_q, the properties and variables are assigned to each control volume within the domain. In order to track the interface(s) between the phases, a continuity equation for the volume fraction of one (or more) of the phases is solved. For the phase q, the continuity equation is given by:

$$\frac{1}{\rho_q}\left[\frac{\partial}{\partial t}\left(\alpha_q\rho_q\right)+\nabla\cdot\left(\alpha_q\rho_q\vec{u}_q\right)=S_{\alpha_q}+\sum_{p=1}^{n}\left(\dot{m}_{pq}-\dot{m}_{qp}\right)\right] \tag{6.39}$$

where ρ_q is the density of the phase q, $\vec{u}_q$ is the velocity of the phase q, S_{α_q} is the mass source term for phase q, $\dot{m}_{pq}$ is the mass transfer from phase p to phase q, and $\dot{m}_{qp}$ is the mass transfer from phase q to phase p.

The properties present in the transport equations are determined by the presence of the component phases in each control volume. For example, for an n-phase system, the volume-fraction-averaged density is given by:

$$\rho=\sum_{q=1}^{n}\alpha_q\rho_q \tag{6.40}$$

All other properties (e.g., viscosity) are calculated in the same way. In the specific case of a two-phase system (case of PEM fuel cells), for example, if the phases are represented by the subscripts i and j, and if the volume fraction of phase j is being tracked, the density and viscosity in each cell are calculated by:

$$\rho=\alpha_j\rho_j+\left(1-\alpha_j\right)\rho_i \tag{6.41}$$

$$\mu=\alpha_j\mu_j+\left(1-\alpha_j\right)\mu_i \tag{6.42}$$

A single momentum equation is solved throughout the domain, and the resulting velocity field is shared among the phases. The momentum equation is dependent on the volume fractions of all phases through the properties ρ and μ:

$$\frac{\partial}{\partial t}(\rho\vec{u})+\nabla\cdot(\rho\vec{u}\vec{u})=-\nabla p+\mu\nabla\cdot\left(\nabla\vec{u}+(\nabla\vec{u})^{\mathrm{T}}\right)+\rho\vec{g}+\vec{F} \tag{6.43}$$

where $\vec{g}$ corresponds to the gravitational acceleration and $\vec{F}$ is a source term which accounts for surface tension effects. The calculation of the term $\vec{F}$ is explained further ahead in this section.

The energy equation, which is also shared by the phases, is given by:

$$\frac{\partial}{\partial t}(\rho E)+\nabla\cdot(\vec{u}(\rho E+p))=\nabla\cdot\left(k_{eff}\nabla T\right)+S_h \tag{6.44}$$

where the energy, E, and temperature, T, are treated as mass-averaged variables:

$$E=\frac{\sum_{q=1}^{n}\alpha_q\rho_qE_q}{\sum_{q=1}^{n}\alpha_q\rho_q} \tag{6.45}$$

where E_q for each phase is based on the specific heat of that phase and the shared temperature. The properties ρ and k_{eff} (effective thermal conductivity) are shared by the phases. The source term in Eq. (6.41), S_h, contains contributions from radiation as well as any other volumetric heat sources.

A key feature of the VOF method is the ability to capture the effects of surface tension, which is an important and sometimes dominant force in microchannel flows. Two main surface tension models are used in the VOF method: the continuum surface force (CSF) model and the continuum surface stress (CSS) model.

The CSF model, proposed by Brackbill et al. [7], includes the surface tension effects to the VOF calculation by adding a source term in the momentum equation [term $\vec{F}$ in Eq. (6.43)]. The surface tension can be written in terms of pressure jump across the surface. The force at the surface can be expressed as a volume force using the divergence theorem, and it is given by:

$$F(CSF) = \sum_{pairsij,\, i<j} \sigma_{ij} \frac{\alpha_i \rho_i k_j \nabla \alpha_j + \alpha_j \rho_j k_i \nabla \alpha_i}{\frac{1}{2}\left(\rho_i + \rho_j\right)} \tag{6.46}$$

where σ_{ij} stands for the surface tension between phases i and j, and k is the curvature defined in terms of the divergence of the unit normal, $\hat{n}$:

$$k = \nabla \cdot \hat{n} \tag{6.47}$$

where

$$\hat{n} = \frac{n}{|n|} \tag{6.48}$$

n corresponds to the surface normal, defined as the gradient of the volume fraction of the phase q:

$$n = \nabla \alpha_q \tag{6.49}$$

Equation (6.46) allows for a smooth superposition of forces near cells where more than two phases are present. If only two phases are present in a cell, then $k_i = -k_j$ and $\nabla \alpha_i = -\nabla \alpha_j$, then the Eq. (6.46) takes the following form:

$$F(CSF) = \sigma_{ij} \frac{\rho k_i \nabla \alpha_i}{\frac{1}{2}\left(\rho_i + \rho_j\right)} \tag{6.50}$$

The CSS method models surface tension conservatively, contrarily to the non-conservative formulation of the CSF model. CSS avoids the explicit calculation of curvature and could be represented as an anisotropic variant of modeling capillary forces based on surface stresses. In this scheme, the surface tension force is represented as:

$$F(CSS) = \nabla \cdot \left[\sigma \left(|\nabla \alpha| \mathbf{I} - \frac{\nabla \alpha \otimes \nabla \alpha}{|\nabla \alpha|} \right) \right] \tag{6.51}$$

where $\mathbf{I}$ is the unit tensor and $\otimes$ is the tensor product of two vectors (the original normal and the transformed normal).

The CSS method has a few added advantages over the CSF method, especially for cases involving variable solution surface tension. For variable surface tension, the CSF model requires modeling an additional term in the tangential direction to the interface based on the surface tension gradient, whereas the CSS model does not require any additional term due to its conservative formulation. Moreover, the CSS method does not require any explicit calculation for the curvature. Therefore, it performs physically in under-resolved regions, such as sharp corners.

Wall adhesion effects are also taken into account in the VOF method. This is accomplished by using the contact angle that the fluid is assumed to make with the wall to adjust the surface normal in cells near the wall. This so-called dynamic boundary condition results in the adjustment of the curvature of the surface near the wall. Considering θ_w as the contact angle at the wall, then the surface normal at the cell next to the wall can be obtained by:

$$\hat{n} = \hat{n}_w \cos \theta_w + \hat{t}_w \sin \theta_w \tag{6.52}$$

where $\hat{n}_w$ and $\hat{t}_w$ are the unit vectors normal and tangential to the wall, respectively.

6.3.3 Oxygen Transportation Research

To promote the commercialization of PEMFC, researchers have been working to reduce the Pt catalyst loading. However, the experimental results show that there is a significant reduction in cell performance under low and ultra-low Pt loading conditions [8]. Researchers suggested that this unexpected change in performance was caused by the increased local mass transport resistance to the active Pt surface [9]. Considering the fast reaction kinetics and high concentration at the anode, the cathode is more susceptible to mass transport problems. Therefore, the modeling of the cathode mass transport process is presented next.

As shown in Fig. 6.2, the oxygen first flows into the cell through the cathode channel and then diffuses through the gas diffusion layer to the catalyst layers by concentration gradient. Subsequently, the oxygen reacts at the active surface to produce water. Depending on the different cell components, the total oxygen mass transport resistance (R_{Total}) of the cathode can be divided into the oxygen mass transport resistance in the flow channel (R_{Channel}), the oxygen mass transport resistance in the gas diffusion layer (R_{GDL}) and the oxygen mass transport resistance in the catalyst layer (R_{CCL}) [10].

$$R_{\text{Total}} = R_{\text{Channel}} + R_{\text{GDL}} + R_{\text{CCL}} \tag{6.53}$$

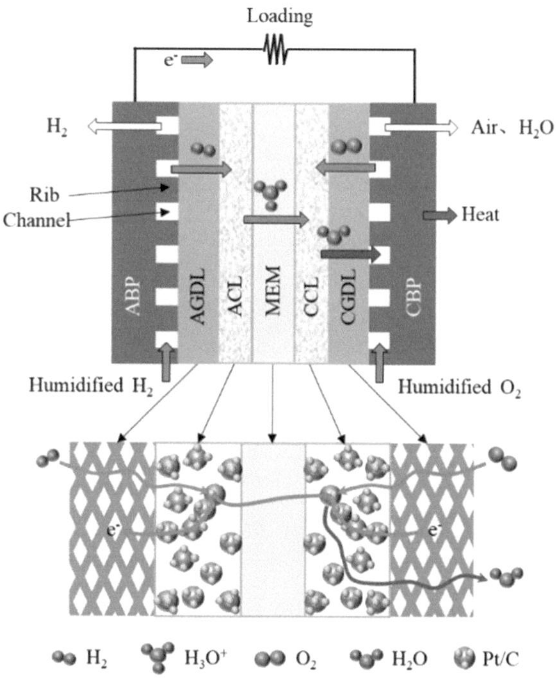

Fig. 6.2 Schematic of mass transport in PEMFC

where the unit of R_{Total}, R_{Channel}, R_{GDL} and R_{CCL} is S cm^{-1}. R_{Channel} and R_{GDL} are relatively simple, not affected by the catalyst loading, and can be measured experimentally. R_{Channel} is mainly determined by the flow field structure and fluid velocity. R_{GDL} is mainly determined by porous structure, saturation, and diffusion layer thickness. When the above factors do not change, their mass transport resistance is constant.

R_{CCL} is the most complex mass transport resistance. As shown in Figure 6.3 [10], the catalyst consists of carbon particles and their supported Pt or PT-M alloy particles, and the surface of the catalyst is covered with an ultrathin film of perfluorinated sulfonic acid (PFSA). The catalyst layer is a porous structure formed by the agglomeration and accumulation of a large number of such particles. Therefore, oxygen first diffuses between particle pores and then penetrates through the ultrathin ionomer film on the catalyst surface to the Pt surface. In this process, the mass transport resistance of the pore diffusion process is called bulk oxygen transport resistance (R_{Bulk}). The mass transport resistance of permeating through PFSA ionomer films is called local oxygen transport resistance (R_{Local}).

Many experimental studies have been carried out on local mass transport resistance, and a large number of fitting formulas have been proposed. In recent years, researchers have attempted to further explain the local mass transport resistance from the microscopic perspective. According to the "adsorption-diffusion" theory, the local mass transport processes can be divided into six steps [11]: (1) O_2 dissolution from gas pores into water film, (2) O_2 diffusion inside water film, (3) O_2 dissolution from water into ionomer, (4) O_2 adsorption into the ionomer, (5) O_2

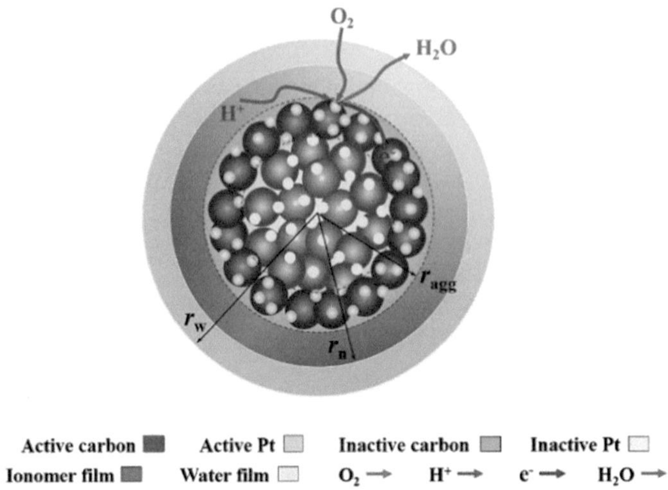

Fig. 6.3 Representation of the CCL structure and two types of mass transport behavior [10]

diffusion inside ionomer film and (6) O_2 adsorption onto the Pt surface. Figure 6.4 shows these processes [10].

In the process (1), the solubility of O_2 can be calculated by the solution balance model, not by Henry's law in previous research [12]. Therefore, based on Benson's research on dissolving oxygen in freshwater, an adaptation of a solution balance equation has been proposed [Eq. (6.54)] to clarify the dissolved behavior of oxygen on the pore/water film interface, which is completely consistent with Zhang's theoretical

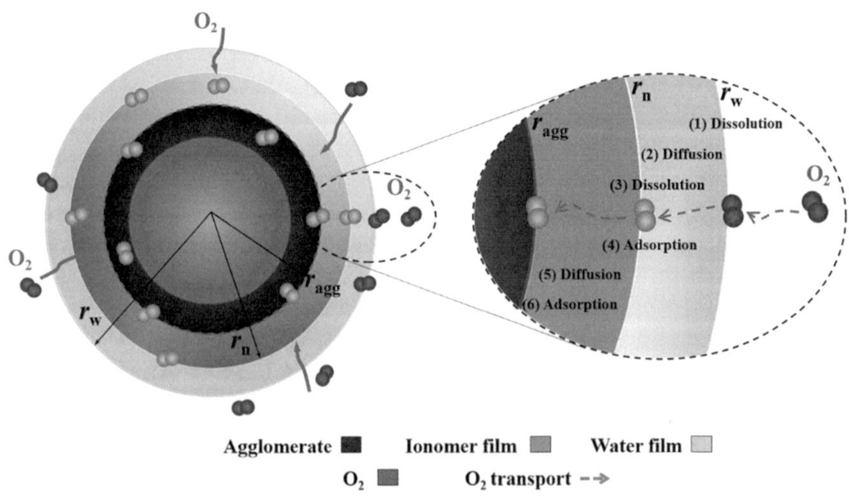

Fig. 6.4 Schematic of oxygen local transport in cathode catalyst layer [10]

analysis. In addition, the process (2)–(6) can be described as Eqs. (6.55)–(6.59), respectively. As described above, the changes in oxygen concentration from pores to active sites are shown in Fig. 6.5.

$$C_1 = 0.0214\ln\left(C_{O_2,\,g}\right) + 0.0439 \tag{6.54}$$

$$N_{O_2} = D_{O_2,w} \cdot \frac{C_1 - C_2}{\delta_w} \tag{6.55}$$

$$\frac{C_2}{C_3} = \frac{H_n C_1}{C_{O_2,\,g}} \tag{6.56}$$

$$N_{O_2} = k_n \cdot (C_3 - C_4) \tag{6.57}$$

$$N_{O_2} = D_{O_2,n} \cdot \frac{C_4 - C_5}{\delta_n} \tag{6.58}$$

$$N_{O_2} = k_{Pt} \cdot \left(C_5 - C_{O_2,r}\right) \tag{6.59}$$

where N_{O_2} and C_i represent the oxygen flux and concentration in each process, $D_{O_2,w}$ and $D_{O_2,n}$ are the diffusion coefficients in water and lifts. H_n is Henry constant in ionomer, k_n and k_{Pt} represent the constant adsorption rate at the ionomer surface and Pt surface. δ_w is the thickness of the water film, which can be obtained by [10]:

$$\delta_w = \sqrt[3]{\frac{e_v r_{agg}^3 s_{CCL}}{e_s} + r_n^3} - r_n \tag{6.60}$$

where s_{CCL} is the liquid saturation in CCL.

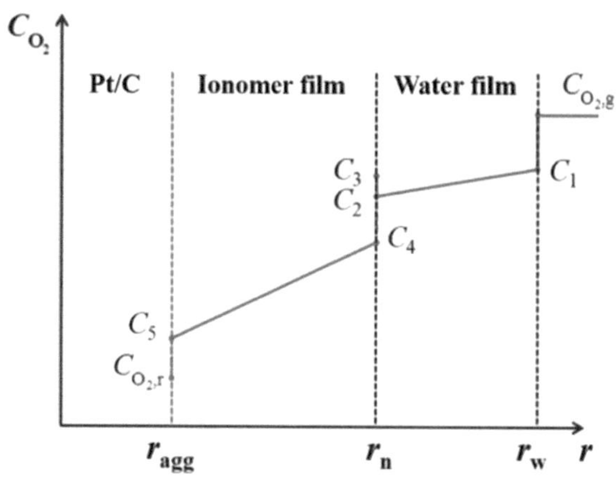

Fig. 6.5 The variable characteristics of oxygen concentration from gas pore to active sites [10]

It is worth noting that the thickness of PFSA ionomer film on the catalyst surface is only 5–10 nm, which is different from the proton exchange membrane at the micron level. In this nanoscale ultrathin film, due to the thickness of the film, the adsorption step is the dominant step in the local oxygen transport. This theory has been widely recognized and verified by the academic community. Therefore, the adsorption rate k_n of oxygen on the surface of PFSA ionomer film in Eq. (6.57) is a crucial parameter, which determines the transport of oxygen through the ionomer film. In the past, the process was usually described by Henry's law, in which the adsorption rate is a constant value, not affected by the surrounding oxygen concentration. However, the experimental results show that the oxygen adsorption rate decreases with the increase of oxygen concentration when the oxygen concentration of the cell is changed. So, the local mass transport resistance of oxygen increases with the increase of oxygen concentration. This is due to the limited adsorption sites on the surface of the ionomer film. With the increase of oxygen concentration, more adsorption sites will be occupied by oxygen. And the adsorption sites tend to be saturated, leading to the decline of the adsorption rate. Therefore, the adsorption of oxygen on the surface of the ionomer film will not show a linear relation, but will show a near-logarithmic relation.

The humidity will also have a strong impact on local oxygen transport resistance [13]. Higher humidity will weaken the adsorption of oxygen on the surface of the ionomer film, but enhance the diffusion of oxygen inside the film [14]. However, high humidity may lead to the failure discharge of the water generated by the ORR, forming a thinner ultra-thin water film between the catalyst and the ionomer film. This will increase the local mass transport resistance.

In addition to the variable operating environment, the complex structure in the catalyst layer of fuel cells also significantly affects the local mass transport of oxygen. The most obvious influence is the ionomer content. The higher the content of ionomer, the greater the local mass transport resistance. There are two reasons for this change. One is the increase of ionomer film thickness, which leads to an increase in oxygen transport resistance. The other is that the excessive ionomer leads to significant changes in the porous structure and even blocks the pore, thus significantly enhancing the local mass transport resistance.

If the model needs to accurately capture the performance variation of PEMFCs, especially for low Pt load cells, incorporating oxygen transport resistance into the model is necessary. In the early model, it is assumed that there is a local mass transfer resistance around the Pt particles, and this resistance is related to the thickness of liquid water and ionomer film on the Pt surface [15]. Compared with experimental results, it shows that the model can correctly reflect the trend of mass transport resistance with low Pt loading. The model was further optimized, and more factors, such as the oxide covering the surface of Pt, were included in the model. However, even so, the polarization curves calculated by the model will still deviate from the experimental results at high current densities, indicating that the mass transport model in low platinum electrodes is still imperfect.

What's more, some researchers hypothesized that the limited oxygen adsorption rate on the Pt surface restricted the reaction rate of the low platinum electrode, based on which a modified oxygen consumption rate model was proposed:

$$\frac{1}{R_{\mathrm{ORR}}} = \frac{1}{i_0\left(\frac{C_{\mathrm{O}_2}}{C_{\mathrm{O}_2}^{\mathrm{ref}}}\right)e^{\left(\frac{\eta}{b}\right)}} + \frac{1}{\frac{C_{\mathrm{O}_2}}{C_{\mathrm{O}_2}^{\mathrm{ref}}}i_{\mathrm{a}}} \tag{6.61}$$

where R_{ORR} is the total rate of the oxygen reduction reaction. $C_{\mathrm{O}_2}^{\mathrm{ref}}$ is the reference concentration of oxygen. The symbol b is the Tafel slope of the oxygen reduction reaction. i_{a} represents the equivalent current density of oxygen adsorption. The first term on the right of Eq. (6.61) is the oxygen reduction reaction rate, and the second term is the oxygen adsorption rate. The total reaction rate is limited by both oxygen adsorption rate and oxygen reduction rate. The above two models can improve the simulation effect, but they are still different from the experimental results.

With the further understanding of the oxygen transport process, the agglomerate model has gradually become the mainstream oxygen transport model. According to the geometric shape, the agglomerate model can be divided into three types: plate agglomeration [16], column agglomeration [17, 18], and spherical agglomeration [19]. Because the spherical agglomeration is closer to the experimentally observed geometry, it becomes the most used agglomerate structure. In addition to the standard spherical structure, some improved geometric structures have been proposed for agglomerate models, such as spherical agglomerates with primary pores [20] and geometrically overlapping spherical agglomerates [21]. The primary pores are assumed to be conical pores uniformly distributed on the agglomerate. The pores are filled with water, and the reactant gases react and diffuse synchronously in the primary pores. This geometric assumption is closer to the actual porous carbon carrier structure than the spherical assumption. The polarization curve and efficiency coefficient predicted by the model are in good agreement with the experimental results [20]. And it can be used to reveal the influencing factors of the effective coefficient of agglomerate under different current densities. The geometric overlapping agglomerate model takes two partially overlapping spheres as the object for analysis, and the overlapping geometric structure highlights the interconnection characteristics of the actual carbon carrier structure [21]. The simulation results show that when the aggregate size is small, the effect of overlapping geometry on PEMFC performance can be neglected, so such complex geometry is not needed in conventional mass transport studies.

Although the effectiveness factor equation of the agglomerate is fixed, the value of the diffusion coefficient inside the equation has been controversial. In the early studies, it was assumed that the pores inside the agglomerate were filled with water or ionomer, while in recent years, it was preferred to assume that the pores inside the agglomerate were partially occupied by water or ionomer. These assumptions correspond to different values of diffusion coefficients. Previous simulation studies have compared the two hypotheses that the agglomerate is filled with ionomer and that it is filled with water, and found that the agglomerate filled with ionomer exhibits a more uniform reaction rate distribution [22]. It should be noted that the agglomerate model itself is only a model describing the oxygen transport process in the agglomerate, and it needs to be solved together with the fuel cell model mentioned above. The most common method is to solve the material consumption in the agglomerate. The mass transfer flux in thin films is equal to the internal consumption of the agglomerate. We can use this relationship to find a solution to mass flux and concentration.

The above model can simulate the effect of Pt load and other factors on oxygen transport well, but it cannot show the details of mass transfer. In addition, a large number of empirical parameters are used in the simulation, and some parameters can only be obtained by fitting, which is difficult to explain theoretically. Besides, the parameter values of different studies vary greatly. It is therefore necessary to build models on a much smaller scale, using fewer assumptions, to reveal the details of oxygen transport. This kind of micro-scale model is called pore pore-scale model.

Pore-scale models can be roughly divided into physical rule-based solutions and governing equation-based solutions [23]. The former is to add rules to the idealized porous media network to capture the physical process, and the pore network model is a typical representative [24]. The latter method solves the governing equations in complex microstructure by different mathematical methods, including Computational fluid dynamics and Lattice Boltzmann method [25]. No matter which method is used for simulation, accurate reconstruction of porous media is a prerequisite for correct simulation, and the degree of reality of the reconstructed structure will significantly affect the simulation results. Therefore, the reconstruction method is briefly described here, and the specific micro-model will be presented in the next chapter. According to different reconstruction methods, pore-scale models can be divided into five types: regular reconstruction model, stochastic reconstruction model, process-based reconstruction model, image-based reconstruction model, and function-based reconstruction model.

The regular reconstruction model refers to the porous media model in which the geometric structure is repeated. The regular porous structure reconstructed by cubic geometry according to the steps of the literature is shown in Fig. 6.6a. The porous medium is simplified to two phases, including the solid phase for electron and ion conduction and the pore phase for reactant and product transport. Each phase is composed of the same hexahedral elements, and the geometric elements are repeated in the direction from the gas diffusion layer to the film, with the two phases occupying the grid elements alternately. This model can be used to explore the effects of pore size, thickness, and porosity on the properties of porous media.

Stochastic reconstruction is a simple method to generate disordered porous structures. A three-phase porous medium can be reconstructed as follows [26]: (1) the central coordinates of the initial carbon sphere were generated by random numbers, (2) generate random numbers to determine whether the next carbon sphere should overlap: (2a) if overlap is required, the overlap between the generated next carbon sphere and the existing carbon sphere should not be less than the specified value; (2b) if overlap is not required, the overlap between the generated next carbon sphere and the existing carbon sphere should not exceed the specified value, (3) repeat the process (2) until porosity is met or no more carbon spheres can be placed in the calculation domain, (4) all grid cells in the carbon sphere are set as carbon phase, (5) all grid cells outside the carbon sphere, but less than the radius of the ionomer film thickness plus carbon sphere from the carbon sphere center, are set as the ionomer

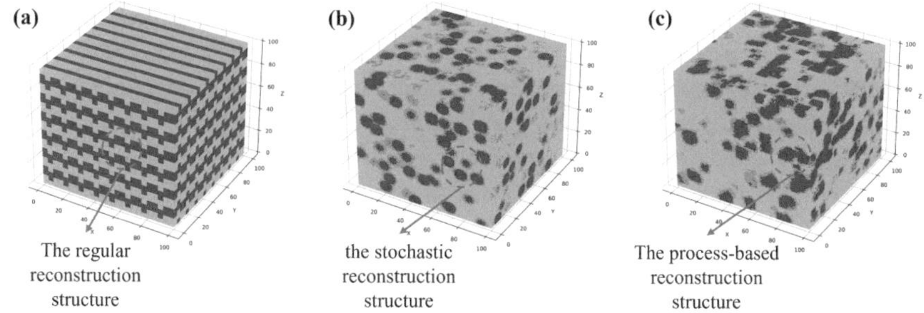

Fig. 6.6 Schematics of the reconstructed structure. **a** The regular reconstruction structure; **b** the stochastic reconstruction structure; **c** the process-based reconstruction structure

phase, (6) mark the remaining grid cells as pore phases. The radius of the carbon particles in this reconstruction method can also be set by a random number or particle size distribution law, thus generating a heterogeneous carbon support that is more consistent with the observed results. To better explore the electrochemical reaction process, porous media with four-phase substances have been developed in recent years. Pt particles are added on the surface of the carbon support, and electrochemical reactions can only occur on the surface of the Pt particles. The catalyst layer structure reconstructed by this method is shown in Fig. 6.6b.

The above method can construct the disordered catalytic layer, but the reconstructed structure is completely determined by random numbers, resulting in low reliability of simulation results. If the reconstruction area is small, the reconstruction structure is greatly affected by random numbers. That makes the effective diffusion coefficient of each calculation is different, and the numerical results are unstable [26]. If the calculation area is set too large, the reconstructed porous media is close to a uniform distribution, and the simulation results will be close to the results predicted by the homogeneous model.

Process-based reconstruction is a special kind of random reconstruction. The characteristics of this reconstruction method lie in the simulation of the agglomeration process of carbon particles "growing" around the center. The specific reconstruction methods are as follows [27]: (1) several grid cells, called "seeds", are randomly selected in the computational domain as the centers of agglomeration nucleation, (2) Set the grid cell at the "seed" as the carbon phase, and then add the grid cell adjacent to the carbon phase into the "growth set", (3) Select a unit randomly from the "growth set" and set it as the carbon phase. The new carbon phase will be adjacent to the original carbon, which symbolizes the "growth" of the original carbon in the same direction, (4) After updating the "growth set", repeat step (3) until the carbon carrier "grows" to the target volume fraction, (5) Pt and ionomers were deposited based on a similar process. The reconstruction allows the same grid cell in contact with multiple carbons and appears several times in the growth set, which increases the probability that it will be selected as the new carbon phase. This is in line with the material's tendency to reduce surface energy. The porous media reconstructed by this method is shown in Fig. 6.6c. The structure of the growth reconstruction model has better connectivity.

The pore-scale models completely reconstruct the entire porous media, and they don't rely on macroscopic empirical correlations. So, they may have higher calculation accuracy. At the same time, it can directly reflect the oxygen transport process and provide a microscopic interpretation for cell research. However, due to the limitations of microscale observation techniques and reconstruction methods, the established microstructure may not reflect the actual structure correctly. In addition, although the pore-scale model can more accurately reproduce the reaction process in porous media, it is difficult to couple the macroscopic all-cell model due to the small scale, so the homogeneous model or aggregate model is still mainly used in the all-cell simulation at present. Therefore, it is a challenge for pore-scale models to successfully build a full fuel cell model with multi-physical fields and multi-scale coupling and solve it in a reasonable computational time.

6.4 Data-Driven Model

6.4.1 Fundamentals of Machine Learning

Different from the physics-based type of mechanistic model, the data-driven model is based on huge amounts of data. The data-driven model can directly learn from a large experimental or numerical data set without explicit knowledge of the physicochemical process inside the PEMFC, and establish the relationship between the input parameters and the output parameter. The data-driven model is also referred to artificial intelligence method or machine learning method. The widely used machine learning methods in the fuel cell field include artificial neural network (ANN), support vector machine (SVM), decision tree, and so on. The data used for building a data-driven model includes experiment-derived and simulation-derived data. Due to the expensive and time-consuming nature of experiments, researchers generate a data set through performing physics-based modeling.

A neural network is an extensive, parallel, interconnected network of adaptive, simple units whose organization can simulate the interactive responses of biological nervous systems to real-world objects. The most basic component in a neural network is the neuron model. The neuron receives input signals from previous layer neurons, and these signals are transmitted through weighted connections. The total input value received by the neuron will be compared with the neuron's threshold, and then processed through the activation function to generate the output signal. Figure 6.7 illustrates the structure of the artificial neural network model. The numbers of layers and neurons depend on the complexity of the investigated problem, and can be determined by trial and error during the training process. Apart from the data set, the performance of an ANN model, including the accuracy and convenience, is strongly determined by properly selecting the number of layers and neurons. The dataset is

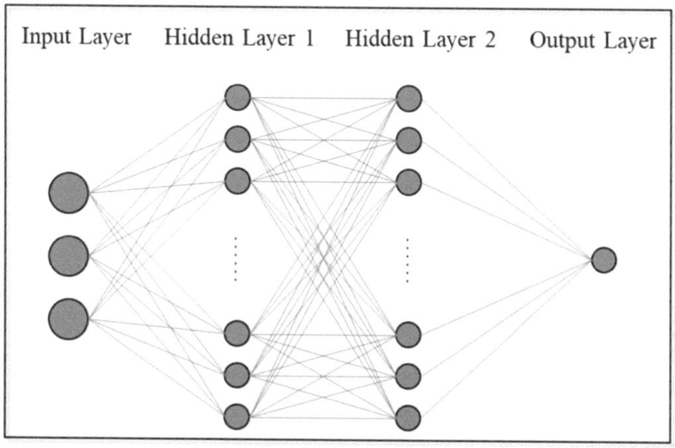

Fig. 6.7 Illustration of the artificial neural network model

usually divided into three sets for supervised learning: the training dataset, the testing and the validation dataset. The training dataset is used for the ANN learning process to optimize the weights of links among different neurons, which are constantly updated as the learning proceeds. The testing and validation datasets are applied to evaluate the performance of ANN models.

Support vector machine is a supervised learning model for analyzing data for classification and regression analysis. In support vector regression, the straight line required to fit the data is called the hyperplane. The goal of the SVM algorithm is to find a hyperplane in an n-dimensional space that unambiguously classifies the data points. The data points on both sides of the hyperplane that are closest to the hyperplane are called support vectors, which influence the position and orientation of the hyperplane. The SVM regression algorithm is called support vector regression (SVR). Support vector regression is a supervised learning algorithm used to predict discrete values. Support vector regression uses the same principles as SVM. The basic idea behind SVR is to find the line of best fit. In SVR, the line of best fit is the hyperplane with the most points. Support Vector Regression is characterized by the use of kernels, sparse solutions, and Vapnik–Chervonenkis (VC) to control the number of margins and support vectors. Although not as popular as SVM, SVR has proven to be an effective tool for real-valued function estimation. As a supervised learning method, SVR is trained using a symmetric loss function, which equally penalizes high and low error estimates. Using Vapnik's insensitive method, a flexible tube of minimum radius is formed symmetrically around the estimated function, so that absolute values of errors smaller than a certain threshold are ignored both above and below the estimated value. In this way, points outside the tube are penalized, but points inside the tube, either above or below the function, are not penalized. One of the main advantages of SVR is that its computational complexity does not depend on the dimensionality of the input space. Furthermore, it has excellent generalization ability with high prediction accuracy.

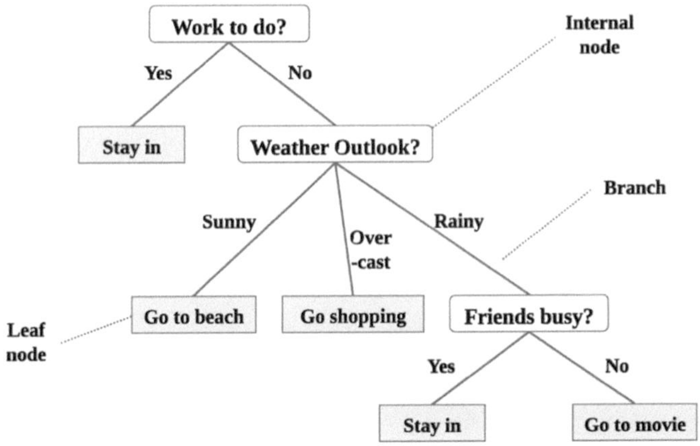

Fig. 6.8 Illustration of decision tree model [28]

A decision tree is a supervised machine learning model that learns decision rules from features to predict targets. The core idea of the decision tree regression model is that similar inputs must produce similar outputs. The purpose of the decision tree is to generate a decision tree with strong generalization ability, that is, the ability to deal with unseen examples. Decision trees are constructed by recursive splitting, starting with the root node (also called the parent node) and dividing each node into left and right children nodes, which can be further split into other child nodes. Starting from the root node, the data is split on the feature that yields the greatest information gain (IG). In the iterative process, the splitting process will continue to repeat at each child node until all leaves reach purity, that is, the samples of each node belong to the same class (Fig. 6.8).

Numerous data-driven models have been applied for PEMFC modeling.

An ANN model was developed to predict a 5 kW PEMFC stack performance [29]. The model considered 7 input (including current, air mass flow, hydrogen mass flow, nitrogen mass flow, cathode water injection, anode inlet temperature, and bulk water temperature) and 2 output (including stack voltage and cathode outlet temperature) parameters. The maximum prediction errors for the stack voltage and cathode temperature out are 9.4% and 5.6%, respectively.

Wu et al. [30]developed an advanced self-adaptive relevance vector machine (RVM) model to predict the performance degradations of a 1.2 kW and an 8 kW PEMFC stack. A comparison between single-step-ahead and multiple-step-ahead predictions of the proposed model was also given and discussed. The results showed that the predicted results by the RVM method showed a good agreement with the experimental results under different operation conditions.

Lou et al. [31]integrated the XGBoost algorithm with a PEMFC physical model to develop a data-driven model for the multi-objective prediction, examination, and optimization of the cathode catalyst layer in PEMFC. They quantified the sensitivity of features on optimization objectives, where the agglomerate radius, Pt loading, and mass ratio of Pt on carbon-supported platinum are the three significant features with average importance of 0.41, 0.24, and 0.11, respectively.

6.4.2 Examples

Combined with fuel cells, machine learning has gained impressive results in practical applications such as cell degradation prediction, safety warning, cost control, etc., proving the value of machine learning in cell performance prediction and analysis. On this basis, researchers have gradually introduced machine learning into the system components of PEMFC, which in turn better guides the actual PEMFC design optimization.

In the research of applying machine learning for PEMFC data analysis and optimization, applying machine learning algorithms to develop data-driven models based on fuel cell-related data is the current mainstream method. The data-driven model establishes a direct mapping relationship between the input parameters and the output target through machine learning algorithms based on data. The data-driven model avoids intermediate processes such as solving control equations when predicting output values under different input parameters, which can significantly improve computational efficiency. The machine learning algorithm can also capture the potential information in the data during the training process, such as parameter sensitivity. At the same time, based on the characteristics of fast prediction, data-driven models can also be coupled with search and optimization algorithms such as genetic algorithms, which in turn can achieve quantitative optimization of parameters and obtain accurate optimization parameters of the PEMFC system.

During the process of developing data-driven models, the collection of sufficient and reliable data sets is a prerequisite for research. In existing research, data sets are generated through two main methods: experimental testing and numerical simulation. Among them, the experimental test data is characterized by high reliability, but it is difficult to obtain the internal cell data information due to the limitation of test conditions. Numerical simulation takes the mathematical model as the data source, by changing the model parameters to obtain the required data set, which has greater flexibility in application and can be optimized for the analysis of parameters that are difficult to measure in the experiment.

1. Data-driven model based on experimental data

Obtaining datasets through PEMFC experimental bench testing is the most direct data collection method, and the data obtained by this method is from a reliable source with less data noise, which in turn can be trained to obtain accurate and reliable data-driven models.

Aiming at the operation degradation issue, a PEMFC long-time operation test bench was set to test the experimental parameters of three commercial fuel cells, and obtained the PEMFC operation data sets under different operating conditions [32].

They applied deep learning algorithms to develop the data-driven model by using the sliding window training method, which realized the prediction of PEMFC performance degradation and operation life under different operating conditions and operation time. The prediction errors of RMSE = 0.004 and MAPE = 0.0013 demonstrate the reliability of machine learning algorithms in fuel cell lifetime prediction.

Aiming at the formation mechanism and stability characteristics of single-atom catalysts issues, the feasibility of generating transition metal single-atom catalysts at carbon surface defects was calculated [33], using density functional theory and machine learning algorithms. The experimental results show that the prediction model developed using the machine learning algorithm is 130,000 times less than the time required by the density functional theory in terms of computation time, and the error in the prediction of the ORR reactivity is only 8.33%, which fully proves the feasibility and advantage of the machine learning approach in guiding the design optimization of single-atom catalysts.

To optimize the amplitude of wavy serpentine flow field, three wavy serpentine flow fields with different amplitudes and one straight serpentine flow field PEMFC were used for variable operation parameter tests to obtain the output currents of the four cells at different operating temperatures, fuel flow rates, and operating voltages [34]. They utilized artificial neural networks (ANNs) for data-driven model training to achieve the prediction of the output currents of PEMFCs with different flow channel types and different operating conditions. In this study, the ANN prediction error MAPE equals 2.059.

In the study of data-driven model development based on experimental data, collecting experimental data from published literature is likewise one of the methods to build a dataset. By integrating the research results in related fields and summarizing them to obtain the dataset required for data-driven modeling, it is realized to collect a large amount of data in a short time.

An optimization framework for ANN-assisted MEA analysis was designed and developed based on literature data for non-metallic catalyst PEMFC MEAs by collecting more than 10,000 sets of data from 64 publications in related fields published between 2010 and 2020 [35]. The cell performance is classified into two categories: high performance and low performance, by setting thresholds. To realize the prediction and analysis of the cell performance level under different MEA structural parameters, they developed a data-driven model for the classification problem, applying the machine learning algorithms of Decision Tree and Extreme Gradient Boosting Tree (XGBoost). After comparing 16 algorithms, ANN was chosen as the most accurate prediction algorithm for the cell maximum power density, with the coefficient of determination $R^2 = 0.9621$.

Aiming at the PEM aging issue in alkaline proton exchange membrane fuel cells, an ANN model was developed using the data collected in the literature published in the past 10 years [36]. The model input parameters include PEM physical structure parameters, chemical property parameters, and cell operating condition parameters. And the Hammett coefficient is set as the target for describing the stability of the chemical properties of the PEM. The coefficient of determination of the ANN is $R^2 = 0.9978$. Moreover, they compared the accuracy of ANN with mainstream algorithms such as XGBoost, Gaussian Process Regression (GPR), Support Vector Regression Machine (SVR), Random Forest Regression (RFR), etc., which proved the accuracy and advantages of ANN in the fuel cell proton exchange membrane decay prediction problem.

In the study of developing a data-driven model based on experimental data, ML embodies excellent prediction performance, which fully proves the feasibility of applying ML to analyze and design the PEMFC components. However, the method also suffers from the shortcomings of a long data collection time and a lack of internal data information, which limits its application in practical design optimization to a certain extent.

2. Data-driven model based on simulation data

To address the shortcomings of the long period of experimental data collection and the lack of internal data information of the cell, researchers have proposed a method to use PEMFC numerical modeling to generate datasets for data-driven model training. This is due to the advantages of numerical simulation, such as efficient generation, easy access, and the ability to compute the internal parameters of the battery.

The PEMFC models can be combined with machine learning algorithms [37]. By using a one-dimensional PEMFC agglomerate model, eight parameters, such as catalyst layer structural parameters and proton exchange membrane structural parameters, were extracted to form the dataset. An ANN was applied for training with the cell activation overpotential as the target. The training accuracy of a coefficient of determination of $R^2 = 0.8$. The effect of each parameter on the cell activation overpotential was quantitatively computed using a statistical method based on mean and variance. However, their study also suffers from the defects of low training accuracy of the agent model, overly simple parameter analysis, and optimization.

Aiming at optimizing the cell anode and cathode catalyst layer structure, a 3D PEMFC model was selected as the data-driven source to generate a dataset [38], which includes six parameters of the cathode and anode catalyst layer. The SVR algorithm was applied to train and generate the agent model, and the agent model was combined with the genetic algorithm for parameter optimization. The cell maximum power density is set as the optimization objective. In their work, the SVM algorithm demonstrated excellent training performance, with a training percentage error of only 1.395% and a coefficient of determination of $R^2 = 0.9908$. The introduction of the agent model resulted in a significant reduction in the model computation time, which made the introduction of a genetic algorithm for a global optimization solution possible.

To optimize the flow channels in PEMFC, a 3D PEMFC model was used as a data-driven source [39], and the Bagging integrated neural network method was applied to develop a data-driven model to predict the polarization curves of fuel cells under different channel heights and widths. The Bagging integrated model demonstrated excellent training accuracy and strong robustness during the training process, and could be used to predict the cell polarization curves under different fuel runner structures. The I-V curves under different channel structures are accurately predicted with the prediction errors of MSE = 1.23×10^{-5} and MAPE = 16%, which proves the feasibility of using machine learning algorithms to predict the cell performance under different channel structures.

To optimize the cell operating conditions, the operating voltage, inlet temperature, cathode and anode inlet relative humidity were selected as input parameters, and the maximum power density as the objective, using multilayer ANN to generate an agent model [40]. And the agent model was then combined with the genetic algorithm to carry out model optimization. The multilayer ANN model showed excellent performance in agent model generation with an absolute percentage error of 1.5%. The genetic algorithm optimization result showed that the maximum power density of 0.7715 W cm^{-2} could be obtained at the cell operating voltage of 0.558 V, temperature of 368.8 K, and cathode and anode inlet relative humidity of 0.78 and 0.99, respectively.

To optimize the optimization parameters, the deep belief network (DBN) is applied to construct the mapping relationship between the operating voltage, fuel pressure, fuel flow rate, and the operating current density [41]. Furthermore, a multi-objective optimization method is used, combining machine learning and PEMFC model to comprehensively optimize the cell performance from multiple perspectives [42]. For the design of fuel cell structure and operating conditions, they extracted 11 parameters, including structural parameters and operating parameters, and chose three indices of current density, system efficiency, and oxygen distribution uniformity as the objectives. The ANN was applied for agent model training. The variance-based parameter sensitivity analysis was applied for parameter screening to simplify the number of parameters to six. The nondominated sorting genetic algorithm was applied for multi-objective optimization to achieve the comprehensive performance improvement of PEMFC.

6.5 Summary and Outlook

Numerical simulation is an important method for PEMFC research. Compared with experimental tests, numerical simulation is more cost-effective. The modeling methods for PEMFC can be broadly classified into empirical [43, 44], mechanistic [45–48], and data-driven models [49–51], and the difference between these models lies in whether or not they take into account the transport and electrochemical phenomena inside the cell. Both empirical and data-driven models do not consider the physicochemical processes inside the battery, and they are based on fitting equations for inputs and outputs that have no physical meaning. Mechanistic models, on the other hand, employ a series of governing equations to describe the physicochemical phenomena inside the battery, including thermodynamics, hydrodynamics, and electrochemical reactions. As a result, mechanistic models can provide accurate results with strong physical meaning. However, mechanistic modeling is usually difficult and computationally intensive. The development of data-driven models is based on data, and data-driven models are also known as "black box models". The development of data-driven models is usually easier than the development of mechanistic models, and in addition, when the data-driven model is well trained, very little computation is required to make predictions. However, the accuracy of data-driven models deteriorates when the predictions are outside the range of the sampled data, or when the sampled data are not well distributed. Empirical models are easy to implement, fast to compute, and are more commonly used for real-time control of batteries. Overall, each of the three models has its advantages and disadvantages and has its unique applications, which require flexible model selection based on the specific research background and application scenarios.

References

1. Kim J, Lee S M, Srinivasan S, et al. Modeling of Proton Exchange Membrane Fuel Cell Performance with an Empirical Equation [J]. Journal of The Electrochemical Society, 1995, 142(8): 2670.
2. Chu D, Jiang R, Walker C. Analysis of PEM fuel cell stacks using an empirical current–voltage equation [J]. Journal of Applied Electrochemistry, 2000, 30(3): 365–70.
3. Deng S, Li Y. A porous-rib flow field for performance enhancement in proton exchange membrane fuel cells [J]. Energy Conversion and Management, 2022, 263: 115707.
4. Vetter R, Schumacher J O. Free open reference implementation of a two-phase PEM fuel cell model [J]. Computer Physics Communications, 2019, 234: 223–34.
5. Ferreira r B, Falcão D S, Oliveira V B, et al. Numerical simulations of two-phase flow in proton exchange membrane fuel cells using the volume of fluid method – A review [J]. Journal of Power Sources, 2015, 277: 329–42.
6. Wang Y, Basu S, Wang C-Y. Modeling two-phase flow in PEM fuel cell channels [J]. Journal of Power Sources, 2008, 179(2): 603–17.
7. Brackbill J U, Kothe D B, Zemach C. A continuum method for modeling surface tension [J]. Journal of Computational Physics, 1992, 100(2): 335–54.
8. Wang C, Cheng X, Lu J, et al. The Experimental Measurement of Local and Bulk Oxygen Transport Resistances in the Catalyst Layer of Proton Exchange Membrane Fuel Cells [J]. The Journal of Physical Chemistry Letters, 2017, 8(23): 5848–52.
9. Chávez-Ramírez A U, Muñoz-Guerrero R, Durón-Torres S M, et al. High power fuel cell simulator based on artificial neural network [J]. International Journal of Hydrogen Energy, 2010, 35: 12125–33.
10. Liang J, Li Y, Wang R, et al. Cross-dimensional model of the oxygen transport behavior in low-Pt proton exchange membrane fuel cells [J]. Chemical Engineering Journal, 2020, 400: 125796.
11. Jiang J, Li Y, Liang J, et al. Modeling of high-efficient direct methanol fuel cells with order-structured catalyst layer [J]. Applied Energy, 2019, 252: 113431.
12. Shen S, Cheng X, Wang C, et al. Exploration of significant influences of the operating conditions on the local O_2 transport in proton exchange membrane fuel cells (PEMFCs) [J]. Physical Chemistry Chemical Physics, 2017, 19(38): 26221–9.
13. Kudo K, Jinnouchi R, Morimoto Y. Humidity and Temperature Dependences of Oxygen Transport Resistance of Nafion Thin Film on Platinum Electrode [J]. Electrochimica Acta, 2016, 209: 682–90.
14. Novitski D, Holdcroft S. Determination of O_2 Mass Transport at the Pt | PFSA Ionomer Interface under Reduced Relative Humidity [J]. ACS Applied Materials & Interfaces, 2015, 7(49): 27314–23.

15. Gwak G, Lee J, Ghasemi M, et al. Analyzing oxygen transport resistance and Pt particle growth effect in the cathode catalyst layer of polymer electrolyte fuel cells [J]. International Journal of Hydrogen Energy, 2020, 45(24): 13414–27.
16. Gloaguen F, Convert P, Gamburzev S, et al. An evaluation of the macro-homogeneous and agglomerate model for oxygen reduction in PEMFCs [J]. Electrochimica Acta, 1998, 43(24): 3767–72.
17. Giner J, Hunter C. The Mechanism of Operation of the Teflon-Bonded Gas Diffusion Electrode: A Mathematical Model [J]. Journal of The Electrochemical Society, 1969, 116(8): 1124.
18. Lin G, He W, Van Nguyen T. Modeling Liquid Water Effects in the Gas Diffusion and Catalyst Layers of the Cathode of a PEM Fuel Cell [J]. Journal of The Electrochemical Society, 2004, 151(12): A1999.
19. Celiker H, Al-Saleh M A, Gultekin S, et al. A Mathematical Model for the Performance of Raney Metal Gas Diffusion Electrodes [J]. Journal of The Electrochemical Society, 1991, 138(6): 1671.
20. Sadeghi E, Putz A, Eikerling M. Hierarchical Model of Reaction Rate Distributions and Effectiveness Factors in Catalyst Layers of Polymer Electrolyte Fuel Cells [J]. Journal of The Electrochemical Society, 2013, 160(10): F1159.
21. Kamarajugadda S, Mazumder S. Generalized flooded agglomerate model for the cathode catalyst layer of a polymer electrolyte membrane fuel cell [J]. Journal of Power Sources, 2012, 208: 328–39.
22. Wang Q, Eikerling M, Song D, et al. Structure and performance of different types of agglomerates in cathode catalyst layers of PEM fuel cells [J]. Journal of Electroanalytical Chemistry, 2004, 573(1): 61–9.
23. Mukherjee P P, Kang Q, Wang C-Y. Pore-scale modeling of two-phase transport in polymer electrolyte fuel cells—progress and perspective [J]. Energy & Environmental Science, 2011, 4(2): 346–69.
24. El Hannach M, Pauchet J, Prat M. Pore network modeling: Application to multiphase transport inside the cathode catalyst layer of proton exchange membrane fuel cell [J]. Electrochimica Acta, 2011, 56(28): 10796–808.
25. Mukherjee P P, Wang C-Y. Direct Numerical Simulation Modeling of Bilayer Cathode Catalyst Layers in Polymer Electrolyte Fuel Cells [J]. Journal of The Electrochemical Society, 2007, 154(11): B1121.
26. Lange K J, Sui P-C, Djilali N. Pore Scale Simulation of Transport and Electrochemical Reactions in Reconstructed PEMFC Catalyst Layers [J]. Journal of The Electrochemical Society, 2010, 157(10): B1434.
27. Siddique N A, Liu F. Process based reconstruction and simulation of a three-dimensional fuel cell catalyst layer [J]. Electrochimica Acta, 2010, 55(19): 5357–66.
28. Zhang X, Cao G, Zhang X, et al. Event attendance prediction: Learning methods [J]. Event Attendance Prediction in Social Networks, 2021: 25–31.
29. Chávez-Ramírez A U, Muñoz-Guerrero R, Durón-Torres S M, et al. High power fuel cell simulator based on artificial neural network [J]. International Journal of Hydrogen Energy, 2010, 35(21): 12125–33.
30. Wu Y, Breaz E, Gao F, et al. Nonlinear Performance Degradation Prediction of Proton Exchange Membrane Fuel Cells Using Relevance Vector Machine [J]. IEEE Transactions on Energy Conversion, 2016, 31(4): 1570–82.
31. Lou Y, Hao M, Li Y. Machine-learning-assisted insight into the cathode catalyst layer in proton exchange membrane fuel cells [J]. Journal of Power Sources, 2022, 543: 231827.
32. Ma R, Yang T, Breaz E, et al. Data-driven proton exchange membrane fuel cell degradation predication through deep learning method [J]. Applied Energy, 2018, 231: 102–15.
33. Wu L, Guo T, Li T. Rational design of transition metal single-atom electrocatalysts: a simulation-based, machine learning-accelerated study [J]. Journal of Materials Chemistry A, 2020, 8(37): 19290–9.
34. Seyhan M, Akansu Y E, Murat M, et al. Performance prediction of PEM fuel cell with wavy serpentine flow channel by using artificial neural network [J]. International Journal of Hydrogen Energy, 2017, 42(40): 25619–29.

35. Ding R, Wang R, Ding Y, et al. Designing AI-Aided Analysis and Prediction Models for Nonprecious Metal Electrocatalyst-Based Proton-Exchange Membrane Fuel Cells [J]. Angewandte Chemie International Edition, 2020, 59(43): 19175–83.
36. Zou X, Pan J, Sun Z, et al. Machine learning analysis and prediction models of alkaline anion exchange membranes for fuel cells [J]. Energy & Environmental Science, 2021, 14(7): 3965–75.
37. Khajeh-Hosseini-Dalasm N, Ahadian S, Fushinobu K, et al. Prediction and analysis of the cathode catalyst layer performance of proton exchange membrane fuel cells using artificial neural network and statistical methods [J]. Journal of Power Sources, 2011, 196(8): 3750–6.
38. Wang B, Xie B, Xuan J, et al. AI-based optimization of PEM fuel cell catalyst layers for maximum power density via data-driven surrogate modeling [J]. Energy Conversion and Management, 2020, 205: 112460.
39. Yan F, Li W, Yang W, et al. Prediction of fuel cell performance based on Bagging neural network ensemble model [J]. SCIENTIA SINICA Technologica, 2019, 49(4): 391–401.
40. Tian P, Liu X, Luo K, et al. Deep learning from three-dimensional multiphysics simulation in operational optimization and control of polymer electrolyte membrane fuel cell for maximum power [J]. Applied Energy, 2021, 288: 116632.
41. Li H-W, Xu B-S, Du C-H, et al. Performance prediction and power density maximization of a proton exchange membrane fuel cell based on deep belief network [J]. Journal of Power Sources, 2020, 461: 228154.
42. Li H, Xu B, Lu G, et al. Multi-objective optimization of PEM fuel cell by coupled significant variables recognition, surrogate models and a multi-objective genetic algorithm [J]. Energy Conversion and Management, 2021, 236: 114063.
43. Bayat M, Özalp M. Effects of leak current density and doping level on energetic, exergetic and ecological performance of a high-temperature PEM fuel cell [J]. International Journal of Hydrogen Energy, 2023, 48(60): 23212–29.
44. Hong J, Yang H, Liang F, et al. State of health prediction for proton exchange membrane fuel cells combining semi-empirical model and machine learning [J]. Energy, 2024, 291: 130364.
45. González-Morán L, Suárez C, Iranzo A, et al. A numerical study on heat transfer for serpentine-type cooling channels in a PEM fuel cell stack [J]. Energy, 2024, 307: 132634.
46. Huo W, Wu P, Xie B, et al. Elucidating non-uniform assembling effect in large-scale PEM fuel cell by coupling mechanics and performance models [J]. Energy Conversion and Management, 2023, 277: 116668.
47. Sun F, Su D, Li P, et al. A novel 3D fine-mesh flow field design and performance analysis for proton exchange membrane fuel cells [J]. Journal of Power Sources, 2023, 584: 233572.
48. Zhao J, Zhou C, Han J, et al. Numerical analysis of improved water management of open-cathode proton exchange membrane fuel cells with a dead-ended anode by pulsating flow [J]. International Journal of Hydrogen Energy, 2023, 48(44): 16858–74.
49. Wang Q, Li Z, Xian L, et al. A current density distribution reconstruction method in proton exchange membrane fuel cell from external magnetic field based on convolutional neural network [J]. Energy Conversion and Management, 2024, 312: 118533.
50. Zhou X, Zhang J, Feng K, et al. Machine learning-assisted design of flow fields for proton exchange membrane fuel cells [J]. Journal of Power Sources, 2025, 626: 235753.
51. Zhu L, Liu Y, Zeng Y, et al. Energy management strategy for fuel cell hybrid ships based on deep reinforcement learning with multi-optimization objectives [J]. International Journal of Hydrogen Energy, 2024, 93: 1258–67.

Chapter 7
Proton Exchange Membrane Fuel Cells: Micro-model

7.1 Introduction

Proton exchange membrane fuel cell (PEMFC) is regarded as an important tool for building a clean energy society due to its advantages of high power density, low operating temperature, high efficiency, low emissions, and low noise, and has received extensive attention in the past decades. The flow and mass transport of different species in the gas diffusion layer (GDL), the microporous layer (MPL) and the porous catalyst layer (CL) is an important link to understanding the working mechanism of PEMFC, which is crucial to improving the performance of the cells, and has become the focus for many institutions and researchers.

Figure 7.1 shows the available models for the modeling and simulation of PEMFC porous structures, spanning from the macroscopic level to the atomic level. These models can be divided into continuum models and particle-based models. The continuum model describes transportation based on continuum space and is a natural choice for engineering applications. The continuum model can be further divided into volume-averaged type and geometric resolved type, where the volume-averaged type performs spatial homogenization within the controlled volume, and the geometric resolved type includes information on the actual microstructure of the porous medium in the model. These particle-based models describe the transport process of individual molecules or molecular clusters. Particle-based models are always time-dependent and require some time/volume averaging to extract useful information. Because these particle-based models handle the motion of molecules or particle clusters, the length scale that the models can handle is limited to submicrometers or below, and typically smaller, with time scales typically subnanoseconds or shorter. Based on the principles of motion adopted, namely, first principles, or ab initio methods following established physical laws, particle-based models can be divided into two groups without making special assumptions or involving fitting parameters. Non-first principles methods are an emerging type of modeling that simulates the aggregation

Y. Li, *Fuel Cell Fundamentals and Applications*,
https://doi.org/10.1007/978-981-96-9845-5_7

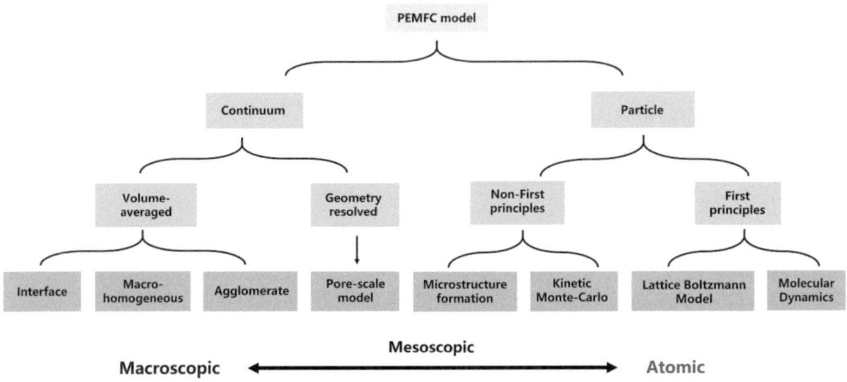

Fig. 7.1 Classification of models for the PEMFC catalyst layer

process of particles suspended in colloidal phases and the formation of microstructure in porous media by considering the interaction between forces using methods similar to particle dynamics.

In the literature, there are many computational methods for solving micro-scale transport equations, such as explicit numerical simulation (ENS) [1] and direct numerical simulation (DNS) [2]. Recently, DNS has also been referred to as pore-scale models (PSM) [3–6]. PSM has been developed to solve complex problems such as fluid flow in porous media [1], drying of porous media [3, 5, 7], conjugated heat and mass transfer of reactions [8–11], and microstructure formation [12]. The first pore-scale model of the catalyst layer in PEMFC can be traced back to Ref. [13], which studies the concentration distribution of discrete catalyst particles by using a set of diffusion reaction equations in a two-dimensional domain. The three-dimensional reconstruction of the CL was first attempted in Ref. [14] and later expanded in Ref. [15], and a series of subsequent papers [2, 6, 16, 17] reported on the comprehensive development of pore-scale modeling and simulation. Recently, [4, 18] reported on the systematic research and development of PEMFC CL PSM.

In this chapter, we review recent modeling studies on the transport of substances with spatially resolved microstructures. Due to the heterogeneous nature of the microstructure and the need to determine the average transport characteristics for macroscopic models (such as aggregate models), direct simulation of transport and reactions within the microstructure is considered to be the most effective methodology. Generally speaking, such research can be roughly divided into two steps. The first step is to reconstruct the area to be studied. According to the true appearance of porous materials under the electron microscope, the morphology of porous media is reconstructed using certain methods. Then, the pore-scale model is used to analyze the mass transport and fluid flow in the region of the reconstructed microstructure. This chapter will introduce some current research content and methods in the field of microstructure reconstruction and pore-scale modeling.

7.2 Microstructure Reconstruction

In most microscopic simulations of PEMFCs, microstructure reconstruction is required to realize the three-dimensional geometry of porous media in GDL, MPL, or CL. However, in some microscale simulations, the use of realistic pore microstructure does not play a key role, and pore geometry can be simply defined by some idealized assumptions.

Most of the microstructures applied to pore-scale simulation of fluid flow in porous components are reconstructed by random generation techniques or imaging combinations such as X-ray computed tomography (CT) or focused ion beam/scanning electron microscopy (FIB/SEM). The former reconstructs the microstructure of porous components based on the statistical information of the microstructure. The latter obtains continuous images of porous media through X-ray or scanning electron microscope (SEM), and then integrates these images to reconstruct the three-dimensional microstructure. The stochastic generation method is the preferred one in many cases due to its low cost and ease of implementation. However, the obtained microstructure is based on some assumptions, which may lead to some deviations from the actual situation.

Stochastic microstructure reconstruction is generally carried out according to the following steps: first, obtain two-dimensional (2D) microscopic images of some sections. Then, by adjusting the model parameters, a geometric model that conforms to the microstructure is randomly generated. To achieve this goal, it is necessary to determine the geometric characteristics of the microstructure and select the model parameters correctly representing the characteristic properties of the microstructure (such as porosity or permeability).

7.2.1 Gas Diffusion Layer Reconstruction

For the study of PEMFC at the macro level, GDL is usually regarded as a homogeneous, isotropic porous material, ignoring the influence of the microstructure of the porous material on the flow and mass transport of species. Different from the macroscopic model, the pore-scale model study focuses more on how the microstructure of the porous material affects fluid flow and mass transport in the fuel cell.

Most GDLs are carbon paper, carbon felt (also known as woven carbon paper), or carbon cloth. Non-woven carbon paper consists of carbon fibers with a certain diameter. Randomly oriented fibers exhibit anisotropy in both through-plane and in-plane directions. In woven carbon paper, the fibers bend more like spaghetti. Carbon cloth is woven from fiber bundles, the main component of which is fixed-diameter carbon fibers. One of the earliest random methods for GDL reconstruction was proposed by Schulz et al.

To find a simple model of carbon paper GDL with as few parameters as possible, the following information about the material was used.

(i) The fibers are long and have negligible curvature relative to the sample size.
(ii) There is no interaction between the fibers.
(iii) Due to the manufacturing process, the fibers are uniformly and isotropically placed on the material (x–y) plane.

These observations led to the adoption of a stationary Poisson's line process with a single parameter directional distribution, where the parameters represent the degree to which carbon paper is compressed. The resulting line is considered to be the axis of a cylindrical fiber. To construct a cylinder, the diameter and shape distribution can be obtained from the carbon paper manufacturer. Throughout the process, the distribution of the cross-section diameter and the direction of the line are considered independent. Employing polar coordinates, directional density—a function of altitude θ and longitude φ is used for directional distribution characterization.

$$p(\theta, \varphi) = \frac{1}{4\pi} \frac{\beta \sin \varphi}{\left(1 + \left(\beta^2 - 1\right)\cos^2\varphi\right)^{\frac{3}{2}}} \tag{7.1}$$

where β is named the anisotropy parameter. When $\beta = 0$ all fibers will be aligned vertically while by increasing β, the fibers tend to be more and more parallel to the x–y plane (the material plane) and for $\beta \to \infty$ all fibers will be aligned completely horizontally.

Anisotropy parameter values β of 10,000 and 100 for Toray090 and SGL10BA carbon papers were chose in simulation [19], respectively. Reconstructed microstructures of the Toray090 and SGL10BA GDLs are shown in Fig. 7.2.

The 3D reconstruction of GDL was performed via a relatively simple method based on the statistical information of carbon paper GDL [20]. The carbon fibers are straight with fixed diameter and randomly distributed in the in-plane direction. Then, the reconstruction procedure is implemented by piling up several carbon fiber layers. Each of the carbon fiber layers can be generated by locating fibers with random

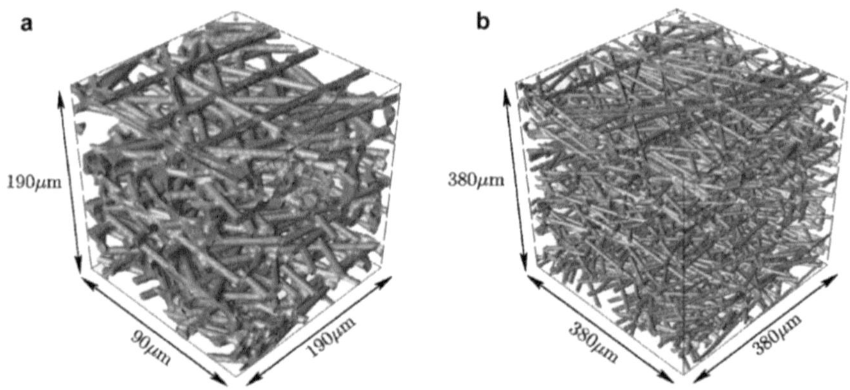

Fig. 7.2 Visualized cutout of the reconstructed **a** Toray090 and **b** SGL10BA [19]

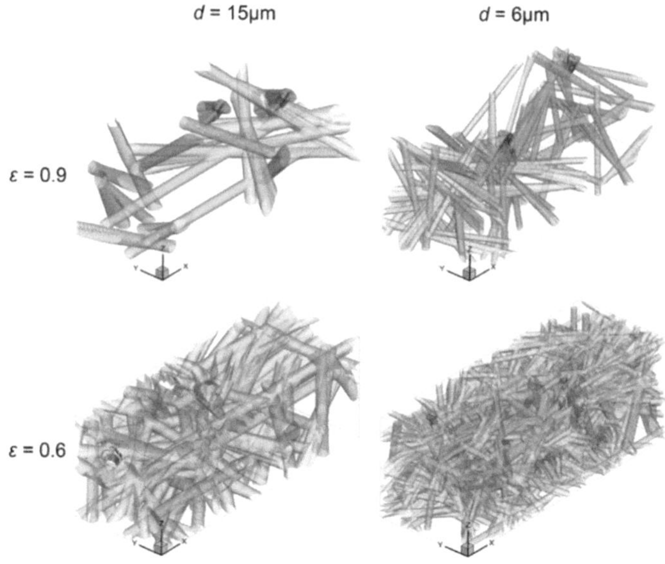

Fig. 7.3 Reconstructed GDL structure [20]

positions and angles until the prescribed porosity in this layer is satisfied. Figure 7.3 shows four of the reconstructed structures with different porosity and fiber diameter values. As shown in Fig. 7.3, with a smaller diameter, more fibers are required to achieve the same porosity, leading to much more complicated porous structures of the electrodes, which will obviously affect the macroscopic transport properties of and the reaction processes in the electrodes, which will be studied in detail in the following sections 400, 150 and 150 mm. In this work, the anisotropy parameter values are set as ∞ for all fibers are orientated in the in-plane direction.

7.2.2 Microporous Layer Reconstruction

Usually, the MPL forms a coating on the GDL based on particle aggregates comprised of tens and hundreds of 10–50 nm sized carbon black primary particles. Introducing an MPL between cathode CL and GDL is believed to be able to significantly improve PEMFC performance, durability, and stability due to three possible roles: decreasing contact resistance, preventing the membrane from being punctured by carbon fibers in GDL, and improving water management.

However, stochastic reconstruction of MPLs is more challenging compared with stochastic reconstruction of GDLs due to their multi-scale geometry. Carbon particles, agglomerates, and cracks in a real MPL sample have three different orders of magnitude. It has been concluded that the actual MPL porosity is between 35 and 60%.

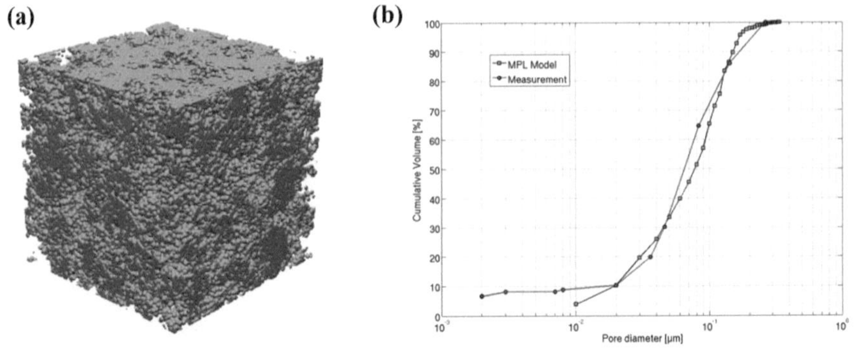

Fig. 7.4 **a** MPL structure model with side lengths of 2.5 μm, 3D view. **b** Comparison of pore size distributions: model data are averaged over three random realizations of the MPL layer [23]

Based on this material characterization, an algorithm was designed [21] to construct a virtual 3D model. As the MPL layer consists of particle agglomerates, the model is constructed with agglomerates of spheres. Each agglomerate consists of N spheres, whose centers fulfill the conditions that:

(a) Any two centers have at least a distance d_{min} from each other.
(b) For each center, there exists another center that has a distance less than d_{max}.

In this way, a connected agglomerate of slightly overlapping spheres is constructed by choosing $d_{min} < d_{max} < d$, where d is the diameter of a single sphere.

To be close to carbons that are typically used for fuel cell MPL, like acetylene black, and to match the desired pore size distributions, $N = 200$, $d_{min} = 30$ nm, $d_{max} = 50$ nm, and $d = 52$ nm. Agglomerates created with these parameters were added to the model until the desired porosity of 55% was reached. Figure 7.4 shows a model created with these parameters and the comparison of pore size distributions (model data are averaged over three random realizations of the MPL layer).

7.2.3 *Catalyst Layer Reconstruction*

The CL is the thin layer underneath the MPL where chemical reactions take place in PEMFC. For the electrochemical reaction to occur, the state-of-the-art CL of a PEMFC is a three-phase composite comprising: (1) ionomer, which is typically Nafion, provides a passage for protons to be transported in or out, (2) Pt catalysts supported on carbon, which are the electronic phase for electron conduction, and (3) pores, which provide channels for the reactant to be transferred in and product water out.

In the present study, the CL is delineated as a two-phase structure consisting of the gas phase (i.e., the void space) and a mixed electrolyte/electronic phase (i.e., the solid matrix). Compared to the electron conduction through the electronic (C/Pt) phase within CL, the assumption of a mixed electrolyte/electronic phase is reasonable from the perspective of ion transport as a limiting mechanism in the electrolyte phase, hence, it is referred to as the "electrolyte" phase. Pore-scale models can be roughly divided into two categories: solving based on physical rules and solving based on control equations [22]. The former involves adding rules to an idealized porous media network to capture physical processes, with a typical representative being the pore network model [24–28]. The latter solves the control equations in complex microstructures through different mathematical methods, including computational fluid dynamics [29], lattice Boltzmann method [30–35], and so on. Regardless of the method used for simulation, accurate reconstruction of porous media is a prerequisite for correct simulation, and the degree of authenticity of the reconstructed structure will significantly affect the simulation results. Therefore, this article divides the models into ordered reconstruction model, random reconstruction model, growth reconstruction model, image reconstruction model, and function reconstruction model according to different CL reconstruction methods, mainly introducing typical reconstruction methods in various models.

The ordered reconstruction model refers to a CL model that is repeated by geometric structures according to fixed rules. CL is simplified into two phases, including the solid phase for electron and ion conduction, and the pore phase for reactant and product transport. Each phase is composed of the same hexahedral elements, and the geometric elements are continuously repeated from the gas diffusion layer to the membrane, with the two phases alternately occupying the grid elements. This model can be used to explore the effects of pore size, thickness, and porosity on CL performance. In addition, there is a CL model reconstructed using spheres [13]. The sphere represents the solid phase composition in the catalyst layer, which is orderly arranged throughout the entire CL. The model is mainly used to study the reaction–diffusion process of gases in this structure. In recent years, there have also been CL models that use agglomerates as unit structures for reconstruction [36, 37], further increasing the complexity of reconstructed structures. The model is used to study the changes in CL pore size distribution and porosity under different pressures. It should be noted that although the ordered reconstruction method is relatively simple and takes a short time to reconstruct, the reconstructed structure differs significantly from the disordered and irregular structures observed in experiments, and needs to be fully considered before use.

Random reconstruction [38] is a simple method for generating disordered CL structures. Early random reconstructions also only had porous and solid phases. Any point in space could be defined by a discrete binary phase function:

$$Z(\vec{r}) = \begin{cases} 0, & \vec{r} \textit{ is void phase} \\ 1, & \vec{r} \textit{ is solid phase} \end{cases} \tag{7.2}$$

The reconstruction rule requires that the statistical average of the phase function is equal to the porosity, which can be measured experimentally.

The solid phase reconstructed by the above method does not have a spherical structure, which does not conform to the actual morphology, and lacks consideration for the three-phase interface. For this reason, a three-phase CL model was later developed using spherical carbon particles for reconstruction. The reconstruction strategy is as follows [4]: (1) Use random numbers to generate the center coordinates of the initial carbon balls; (2) Generate a random number to determine whether the next carbon ball needs to overlap. If it needs to overlap, the overlap between the generated next carbon ball and the existing carbon ball shall not be less than the specified value. If it does not need to overlap, the overlap between the generated next carbon ball and the existing carbon ball shall not exceed the specified value; (3) Repeat the above process (2) until the porosity is met, or until more carbon spheres cannot be placed in the computational domain; (4) Set all grid cells inside the carbon sphere as carbon phase; (5) Set all grid cells outside the carbon sphere, but close to the surface of the carbon sphere as the ionomer phase; (6) Mark the remaining grid cells as porous phases.

The radius of carbon particles in this reconstruction method can also be determined by random numbers [39] or particle size distribution patterns [40], thereby generating an uneven carbon carrier structure that is more in line with the observed results. To better explore the electrochemical reaction process, a CL model with four-phase substances has also been developed in recent years. Pt particles have been added to the surface of the carbon carrier, and the electrochemical reaction can only occur on the surface of Pt particles [41]. The above method can construct a disordered catalyst layer, but the reconstruction structure is entirely determined by random numbers, resulting in low reliability of simulation results. If the reconstruction area is set small, the reconstruction structure is greatly affected by random numbers, and the effective diffusion coefficient calculated each time is different, resulting in unstable numerical results [4]. If the calculation area is set too large, the reconstructed CL microstructure approaches a uniform distribution, and the CL performance tends towards the performance predicted by the homogeneous model. In addition, previous studies have found that the permeability of CL reconstructed by the above random methods is much lower than the actual value [42], indicating that the reconstructed structure cannot accurately reflect the characteristics of the catalyst layer structure.

The Gaussian random field method is also a popular method for catalyst layer reconstruction, which has been reported in many studies. The Gaussian random field method was applied for CL reconstruction [6]. The stochastic reconstruction method is based on the idea that an arbitrarily complex porous structure can be described by a binary phase function, which assumes a value of 0 in the pore space and 1 in the solid matrix. The intrinsic randomness of the phase function can be adequately qualified by the low-order statistical moments, namely porosity and the two-point autocorrelation function.

Figure 7.5a shows a 2-D TEM image of a typical catalyst-coated membrane (CCM) CL. Figure 7.5b is the reconstructed CL microstructure using the 2-D TEM image as the input, with the yellow representing the solid part of the CL.

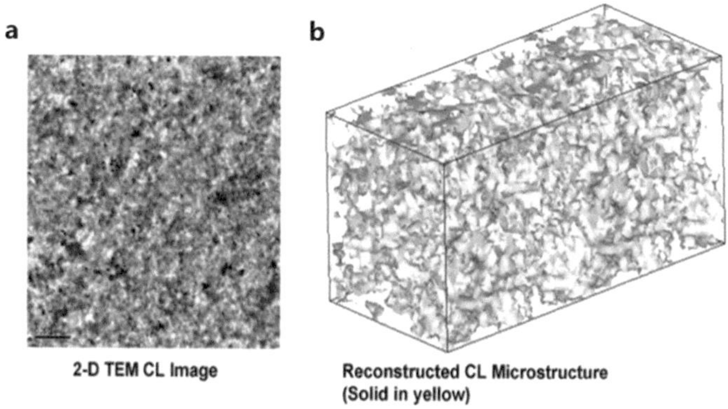

Fig. 7.5 **a** 2-D TEM image of a typical catalyst-coated membrane (CCM) CL; **b** reconstructed catalyst layer microstructure [6]

A Pt CL reconstruction was performed using the Quartet structure generation set (QSGS) method [43]. It has been demonstrated to be capable of generating morphological features closely resembling those of many real porous media. The whole domain consists of 4 types of cells generated by a set of procedures: carbon cells, Pt cells, ionomer cells, and empty cells. The procedure of the QSGS method is as follows: (1) Randomly distribute carbon seeds in the domain based on a distribution probability. (2) Grow existing carbon seeds to the neighboring 7 cells. Growth means the assignment of more carbon cells based on initial seeds. If a neighboring cell is empty and also adjacent to another carbon cell, then assign this neighboring cell as a new carbon cell. If the neighboring cell is empty but does not border with any other existing carbon cell, then assign this neighboring cell as a new carbon cell based on a certain probability. (3) Repeat (1) and (2) until the volume fraction of the carbon phase is achieved.

As for the Pt phase, a random number is assigned to each cell adjacent to the carbon phase in the domain. If the number assigned to a cell is less than a prescribed probability, the value of which is quite lower than the volume fraction of Pt, the cell is assigned as a Pt phase cell. The generation of ionomer is similar to the carbon phase.

Figure 7.6a shows the generated 3D structures of a Pt CL using the QSGS reconstruction method with porosity $\varepsilon = 0.4$. Figure 7.6b shows the detailed distributions of carbon black (top), C/Pt (middle), and ionomer on C/Pt (bottom) for $200 < x < 500$ (lattice units) at $z = 250$ (lattice units). The non-uniform coverage of ionomer on the surface of C/Pt can be clearly observed.

Figure 7.6c shows the pore size distribution of the reconstructed CL structure. The sizes of all pores are in the range of 5–130 nm, representing the nanoscale characteristics of the CL. The maximum pore size is located at around 50 nm, in good agreement with the experimental results.

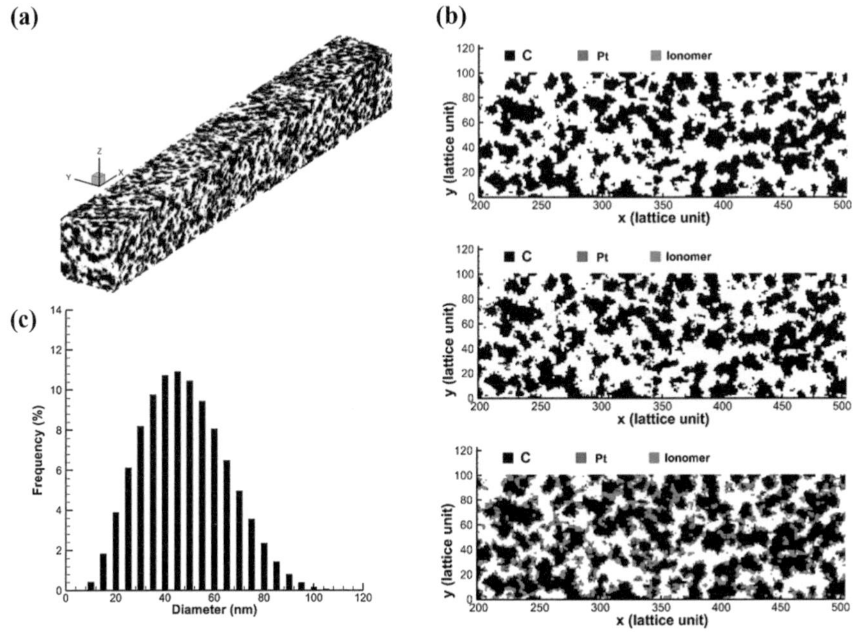

Fig. 7.6 **a** 3D structures of a Pt CL using QSGS method **b** 2-D detailed distributions of carbon black (top), C/Pt (middle), and Nafion covered on C/Pt (bottom) **c** pore size distribution of the reconstructed CL structure [43]

The image reconstruction model is a three-dimensional CL model directly reconstructed from microstructure images captured in experiments, without simplifying assumptions about the CL structure and with the highest degree of restoration. A commonly used image reconstruction technique is nano-scale X-ray computed tomography (Nano CT), which utilizes the different radiation absorption abilities of pores and solid materials to obtain X-ray images that can restore solid morphology based on changes in light and dark. Many researchers use this method to reconstruct 3D models of materials [44–47], but due to resolution limitations (about 50 nm), it is not possible to distinguish between Pt particles and ionomer films, and only the morphology of aggregates and pore channels can be reconstructed. In addition, the raw image data of Nano CT needs to be denoised and binarized before it can be used for reconstruction, and the threshold size set in the processing will affect the morphology of the reconstructed structure. Moreover, due to the presence of primary pores inside carbon particles that cannot be photographed, the porosity of the reconstructed structure will ultimately be lower than the theoretically calculated value.

Focused ion beam scanning electron microscopy (FIB-SEM) is also a commonly used imaging technique. It can reconstruct CL with a resolution of 5 nm and can be used to study the pore size distribution of the catalyst layer, but it cannot distinguish between polymer films. To model and calculate, a fixed thickness of ionomer film needs to be added numerically to the reconstructed surface of the aggregate [48]. FIB-SEM and Nano CT technologies each have their characteristics in capturing the microstructure and transport characteristics of CL [49]. After segmenting the dataset with the same porosity as the target, the models reconstructed by the two techniques have similar pore connectivity, curvature, and diffusion coefficient. However, the FIB-SEM model measures a smaller pore size distribution and exhibits significant anisotropy. When automatically segmenting the dataset, the porosity of FIB-SEM will significantly deviate from the experimental measurement results. From the raw data obtained by two methods, FIB-SEM data can only represent the 15% thickness area of the outer layer of CL, and there may be more erroneous data and noise due to the slicing operation and visible internal structure during the processing. However, compared to Nano CT, FIB-SEM has a faster data acquisition speed and has the potential for further improvement and development.

In addition, the high-angle annular dark field scanning transmission electron microscope (HITEM) has also been used by researchers to obtain CL data, in which the ionomer film surrounding the carbon carrier can be observed [50]. In recent years, a hybrid reconstruction method combining multiple image techniques has been developed, which can reconstruct the structure in CL at a higher resolution (1 nm) [51].

Although the above image reconstruction techniques can generate reconstruction models that are very close to the real CL, a threshold is required to distinguish between solid and pore phases in the image data, and the setting of the threshold brings uncertainty to the CL structure. In addition, the resolution of most images is too low to reflect the details of the carbon carrier surface. Moreover, if the size of the material used for imaging is too small, the reconstructed structure will not be physically representative. In contrast, if the size of the imaging material is too large, the reconstruction time will be too long. In summary, current image reconstruction models still have shortcomings and enormous development potential.

7.3 Micro-model Research

7.3.1 Lattice Boltzmann Method

It has been 20 years since the Lattice Boltzmann (LB) method was first put forward. In the past 20 years, it has achieved rapid development in both theoretical and applied research. The LB method is a first-principle-based numerical approach. Unlike the conventional Navier–Stokes solvers based on the discretization of the macroscopic continuum equations, the lattice Boltzmann method considers flows to be composed of a collection of pseudo-particles residing on the nodes of an underlying lattice structure, which interact according to a velocity distribution function.

In recent years, the LBM has emerged as a powerful alternative CFD (computational fluid dynamics) method for studying fluid flow and transport processes. It is a mesoscopic method for the description of fluids. In the LB method, the motion of the fluid is described by a set of particle distribution functions. Based on the simple and popular Bhatnagar–Gross–Krook (BGK) collision operator, the standard LB equation with a force term can be expressed as follows:

$$\mathrm{f}_{\sigma,\alpha}(\mathbf{x}+\mathrm{c}\mathbf{e}_\alpha\Delta\mathrm{t},\mathrm{t}+\Delta\mathrm{t})-\mathrm{f}_{\sigma,\alpha}(\mathbf{x},\mathrm{t})=-\frac{1}{\tau_{\sigma,\nu}}\left(\mathrm{f}_{\sigma,\alpha}(\mathbf{x},\mathrm{t})-\mathrm{f}^{\mathrm{eq}}_{\sigma,\alpha}(\mathbf{x},\mathrm{t})\right)+\mathrm{F}_{\sigma,\alpha}(\mathbf{x},\mathrm{t}) \tag{7.3}$$

where $\mathrm{f}_{\sigma,\alpha}(\mathbf{x},\mathrm{t})$ is the density distribution function of the component σ at the lattice site x and time t, f^{eq} is the equilibrium distribution function. $\mathrm{c}=\Delta x/\Delta t$ is the lattice with Δx and Δt as the lattice spacing and time step (both Δx and Δt equal to 1 in the lattice system), respectively, and $\tau_{\sigma,\nu}$ is the dimensionless relaxation time.

The discrete velocities $\mathbf{e}_\alpha$ depend on the particular velocity model, and the D2Q9 (two-dimensional nine velocity) square and D3Q19 (three-dimensional 19 velocity) cubic models are commonly adopted. The two-dimensional space with the D2Q9 lattice structure and the three-dimensional D3Q19 lattice structure, are shown in Fig. 7.7 [6].

For the D2Q9 model, $\mathbf{e}_\alpha$ are given by:

$$\mathbf{e}_\alpha=\begin{cases}0, & \alpha=0\\ \left(\cos\left[\frac{(\alpha-1)\pi}{2}\right],\sin\left[\frac{(\alpha-1)\pi}{2}\right]\right), & \alpha=1,2,3,4\\ \sqrt{2}\left(\cos\left[\frac{(\alpha-5)\pi}{2}+\frac{\pi}{4}\right],\sin\left[\frac{(\alpha-5)\pi}{2}+\frac{\pi}{4}\right]\right), & \alpha=5,6,7,8\end{cases} \tag{7.4}$$

For the D3Q19 model, $\mathbf{e}_\alpha$ are given by:

$$\mathbf{e}_\alpha=\begin{cases}(0,0,0), & \alpha=0\\ (\pm1,0,0),(0,\pm1,0),(0,0,\pm1), & \alpha=1,2,3,4\\ (\pm1,\pm1,0),(\pm1,0,\pm1),(0,\pm1,\pm1), & \alpha=5,6,7,8\end{cases} \tag{7.5}$$

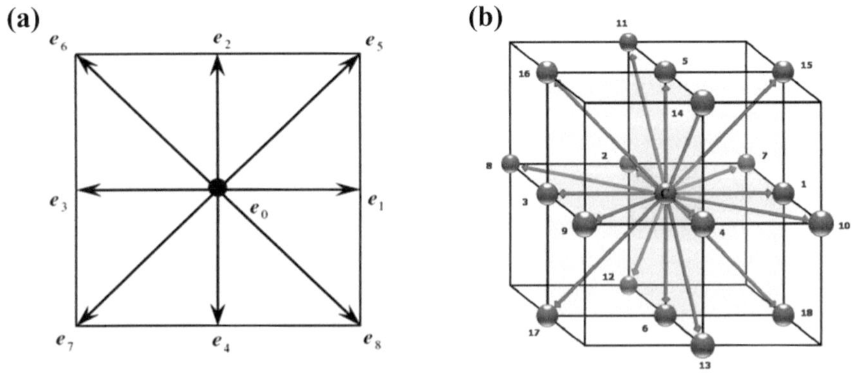

Fig. 7.7 Schematic of **a** D2Q9 lattice structure **b** D3Q19 lattice structure [6]

The equilibrium distribution functions f^{eq} for the D2Q9 lattice are of the form

$$f^{eq}_{\sigma,\alpha} = \omega_\alpha \rho_\sigma \left[1 + \frac{3}{c^2}(\mathbf{e}_\alpha \cdot \mathbf{u}) + \frac{9}{2c^4}(\mathbf{e}_\alpha \cdot \mathbf{u})^2 - \frac{3}{2c^2}\mathbf{u}^2 \right] \tag{7.6}$$

where the weight factors ω_α are given by $\omega_0 = 4/9$, $\omega_{1-4} = 1/9$, and $\omega_{5-8} = 1/36$. The fluid density ρ_σ and fluid velocity $\mathbf{u}_\sigma$ can be obtained from the following equation.

$$\rho_\sigma = \sum_\alpha f_{\sigma,\alpha} \tag{7.7}$$

$$\rho_\sigma \mathbf{u}_\sigma = \sum_\alpha f_{\sigma,\alpha} \mathbf{e}_\alpha \tag{7.8}$$

The viscosity in the lattice unit is related to the relaxation time by

$$\nu_\sigma = c_s^2 \left(\tau_{\sigma,v} - 0.5 \right) \Delta t \tag{7.9}$$

where c_s is the lattice sound speed.

Previous work based on macroscopic continuum models, while with the advantage of numerical efficiency, suffers the disadvantage of neglecting the influence of actual structures and related heterogeneous characteristics of microstructure, such as GDL. Pore-network model and lattice Boltzmann method (LBM) are useful numerical tools in numerical research considering the complicated microstructure.

Figure 7.8 shows a multi-scale modeling numerical case combining the finite volume method (FVM) and lattice Boltzmann method LBM in PEMFC [52]. The irregular GDL area uses LBM, while the computation in the gas channel (GC) is based on FVM. Figure 7.8a demonstrates the computational domain in this case. The two-dimensional computational domain representing the PEMFC cathode in the along-the-channel direction is shown, including a rectangle GC, a porous GDL, and a CL with idealized microstructures. The black obstacles in the GDL describe the carbon fibers, and the porosity of the GDL is about 0.7. Figure 7.8b is the velocity distribution of the whole computational domain. Due to the complex structure of GDL, the flow of gas mixture was blocked by carbon fibers, which present a complex flow velocity distribution as it is shown in Fig. 7.8c. Similar to the gas flow, species transport in GDL is also influenced by the complex structure of GDL consists of carbon fibers. It can be seen in Fig. 7.8d–f that the oxygen mole fraction shows an inhomogeneous distribution compared to the rather homogeneous distribution in the GC area, with the overpotential increasing from 0.2 to 0.8 V. The inhomogeneous distribution of oxygen mole fraction also resulted from the impediment of long carbon fibers.

The numerical result from macroscopic continuum models [53], on the other hand, suffers from the disadvantage of neglecting the influence of actual structures and related heterogeneous characteristics of GDL, often showing a heterogeneous distribution, whose accuracy is questionable compared to microscopic models like LBM. Figure 7.9 shows the oxygen mole fraction distribution and the local current density

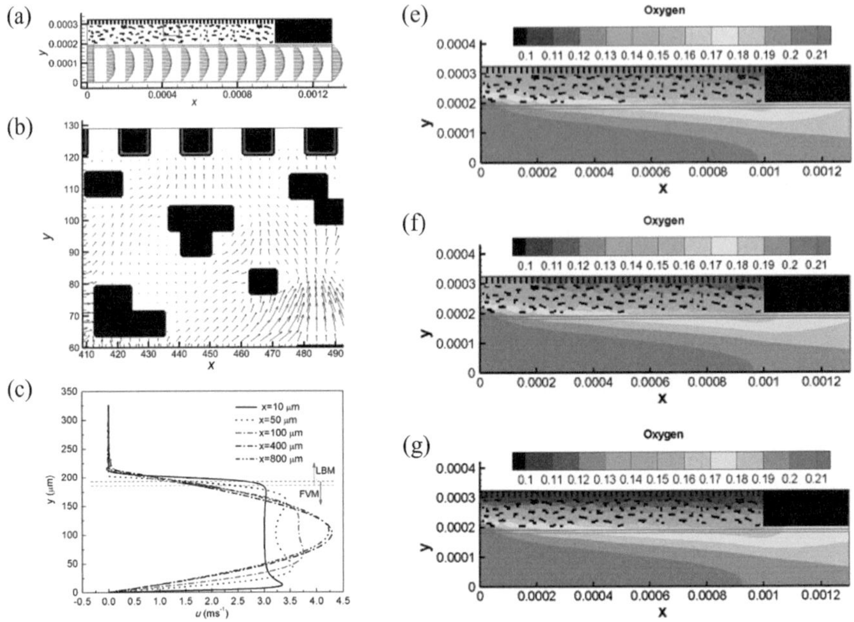

Fig. 7.8 **a** Velocity distribution in the GDL computational domain. **b** Local current density distribution under different over-potentials. **c** Oxygen mole fraction distribution under different over-potentials: (top) 0.3 V, (middle) 0.4 V, (bottom) 0.6 V [52]

distribution using macroscopic continuum models in PEMFC numerical simulation. It can be seen that the oxygen concentration distribution is heterogeneous along the flow direction, for the impediment of long carbon fiber was neglected, which leads to a heterogeneous local current density distribution, where the sudden drop of current density is missing compared to the microscopic model result.

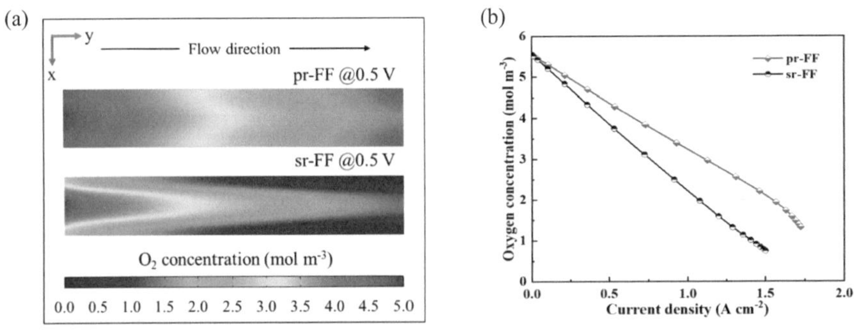

Fig. 7.9 Macroscopic continuum model numerical results in analyzing the effect of different flow fields **a** oxygen concentration distribution; **b** local current density distribution [53]

7.3.2 Lattice Boltzmann Method Two-Phase Flow

The main strength of the LBM is that it behaves like a solver for the conservation equations, such as the Navier–Stokes equation, as well as the advantages of boundary processing and programming. These characteristics of LBM make it a perfect method for dealing with complex and variable two-phase flow and phase transition problems, and it can more accurately capture the topology changes of complex interface motions in multiphase flow. Currently, various lattice Boltzmann multiphase flow models have been established and used to simulate various two-phase flow problems. Overall, these models can be divided into four major categories: pseudopotential model, free-energy model, phase-field model, and color-gradient model.

The pseudopotential model was first proposed by Shan and Chen in 1993. To introduce nonlocal interaction among particles, the force experienced by the particles of component σ at x from the particles $\overline{\sigma}$ at $\mathbf{x}'$ was defined as follows:

$$\mathrm{F}(\mathbf{x},\mathbf{x}') = -\mathrm{G}(|\mathbf{x}-\mathbf{x}'|)\psi_{\sigma}(\mathbf{x})\psi_{\overline{\sigma}}(\mathbf{x})(\mathbf{x}-\mathbf{x}') \tag{7.10}$$

where G is a Green's function and ψ is an effective mass depending on the local density. The components σ and $\overline{\sigma}$ can be different or the same, representing a single component and multi component system respectively.

The original form of effective mass in the work of Shan and Chen is

$$\psi = \rho_0\left(1 - \exp\left(-\frac{\rho}{\rho_0}\right)\right) \tag{7.11}$$

where ρ_0 is a normalization constant, which is usually chosen as 1.

A major application of the pseudopotential model is to simulate two-phase flows in the GC and porous components such as GDL and CL of PEMFC. Successful water management in the PEMFC, particularly in the cathode side, is crucial to the cell performance, as poor water management results in either dehydration of the membrane or flooding issues. The pseudopotential model is the most widely used multiphase model for studying two-phase flow in porous components of PEMFC.

The MCMP pseudopotential model was applied to study the dynamic behaviors of a droplet subjected to relatively strong air flow in a GDL under the interdigitated GC [54]. Figure 7.10 shows the dynamic behaviors of a droplet in GDL. The liquid water droplets are initially placed under the ribs because of the effect of the inlet airflow, the liquid droplets begin to move towards the outlet. In Fig. 7.10c and d, a fraction of the liquid water (in Circle 1) is trapped in a dead-end pore, and the bulk of the liquid water keeps advancing. The bulk reaches the state shown in Fig. 7.10e after a long time of creeping, with several residues left behind. These residues either are trapped by the dead-ended pores (in Circle 1) or dragged behind the bulk due to their relatively lower velocity (in Circles 2–5). In the outlet GC, the bulk is severely torn into several strips and is deflected towards the land under the influence of gas flow as shown in Fig. 7.10f.

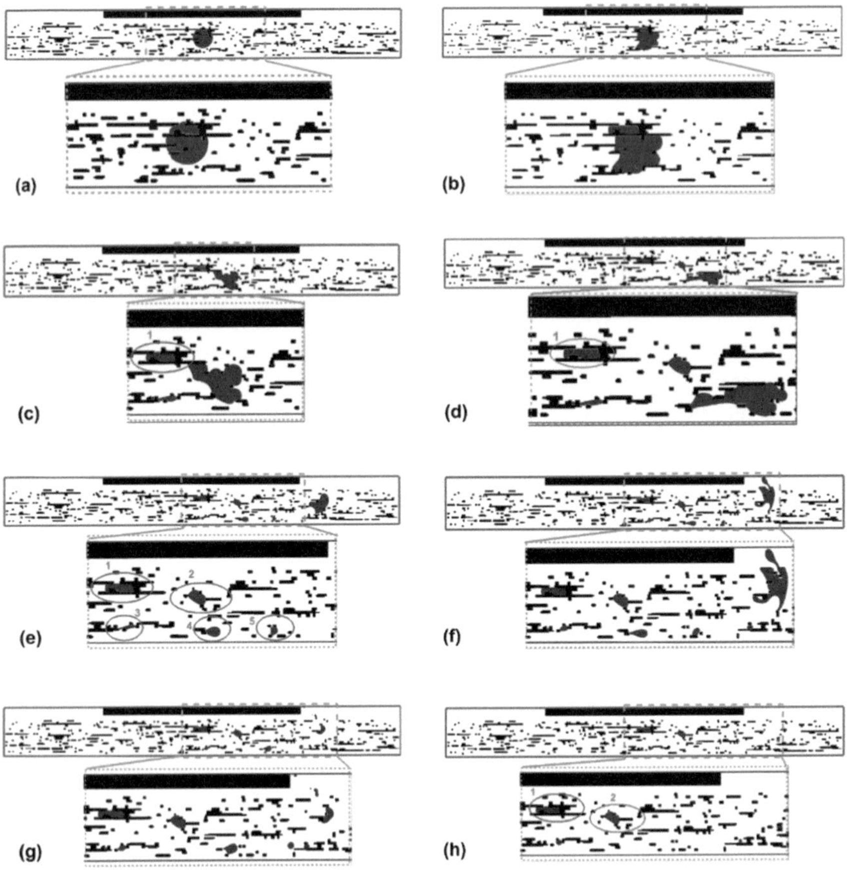

Fig. 7.10 Pore-scale behaviors of liquid water inside a GDL of interdigitated PEMFC (black: carbon fibers or land; white: void space; blue: liquid water) [54]

A 3-D MCMP pseudopotential model was applied to study the influence of the pore structure, surface wettability, and interfacial dynamics in the PEMFC catalyst layer and gas diffusion layer.

Figure 7.11 [6] displays the steady-state invading liquid water fronts corresponding to increasing capillary pressures from the primary drainage simulation in the reconstructed CL microstructure characterized by slightly hydrophobic wetting characteristics with a static contact angle of 100°. The pattern of liquid water invading changes with the increase of the capillary number, as it is shown in Fig. 7.11b. At lower capillary numbers, the invading liquid water fronts exhibit a finger-like pattern, which is similar to the displacement pattern observed typically in the capillary fingering regime. As the capillary number increases, the invading liquid water fronts tend to change to a stable flat advancing front pattern.

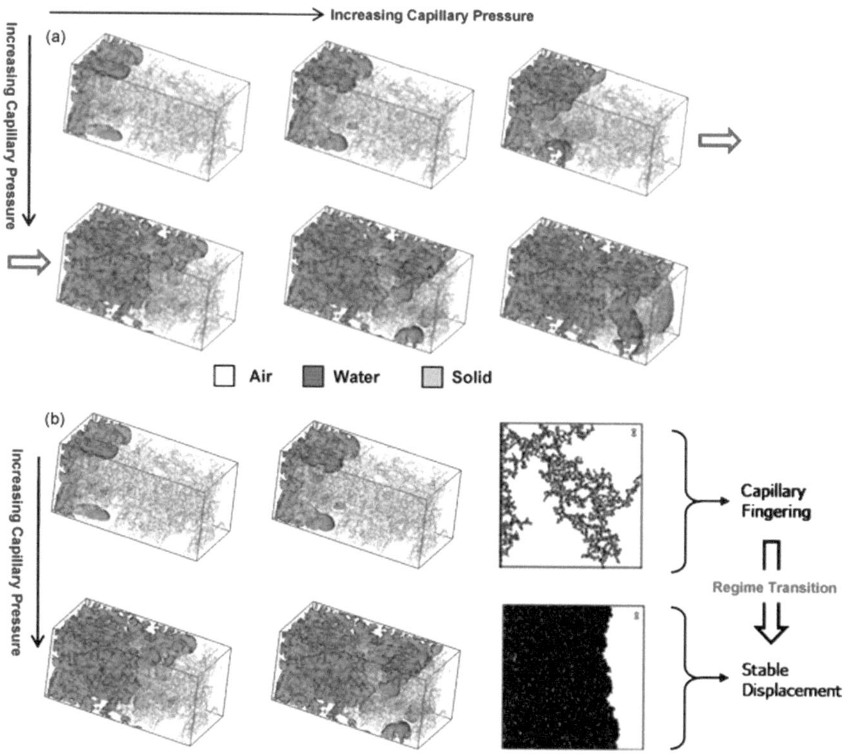

Fig. 7.11 Advancing liquid water front with increasing capillary pressure through the initially air-saturated reconstructed CL microstructure from the primary drainage simulation [6]

7.3.3 Kinetic Monte Carlo Method

The Monte Carlo method is also known as the statistical simulation method and the statistical test method. It is a numerical simulation method that takes stochastic phenomena as the research object. The inspiration for the Monte Carlo method was triggered in the late 1940s after the first US electronic computer was built.

The fundamental of the Monte Carlo method for solving problems consists of the following parts: first, establish a stochastic model or stochastic process of the problem; then calculate parameters or numerical characteristics through observation or sampling test of the stochastic process, and finally obtain the approximate value of the solution. The accuracy of the solution is represented by the standard error of the estimated value.

The Monte Carlo method also has been extensively studied in the field of PEMFC [55–58] to gain a representation of the structure of porous media and the associated effective properties and to simulate the liquid water flows in the microstructure of the fuel cell.

The stochastic model used in PEMFC is a grid-based model where the computational domain consists of voxels typically with a size of 5 μm × 5 μm × 5 μm. The lattice sites can be allocated with three types of voxels: fiber material, graphite or PTFE; liquid water, and "nothing" (gas phase not explicitly included). The iteration process of liquid water voxels is as follows. During an MC iteration, each water voxel in the whole computational domain is chosen once in a random order for a tentative move, referred to as a single MC step in the following. The acceptance of each single MC step movement will be determined by an acceptance probability calculated based on the surface energy difference between states after a certain movement.

The energy of a given state is calculated considering the interactions with all the occupied neighboring voxels. The value of interaction strength is determined by surface energy. The typical interaction values between three basic component voxels are as follows: Water-water interaction: − 72 mJ/m^2; Water-PTFE interaction: + 30 mJ/m^2; Water-graphite interaction: − 95 mJ/m^2.

Usually, the total energy difference of the possible new energy to the old state determines the acceptance probability for the MC step via the Boltzmann factor:

$$\exp\left(-\frac{\Delta E}{RT}\right) \tag{7.12}$$

Hereby, T is the absolute temperature and R denotes the universal gas constant. If the energy difference ΔE has a negative value, the factor is bigger than 1, and the MC move is accepted. If ΔE is positive, the exponent is smaller than 1, and a random number between 0 and 1 is calculated. If the Boltzmann factor is larger than this random number, the MC move is accepted. Otherwise, the old state of the system is preserved.

The liquid water behavior in GDL was studied [57], and the results presented in this work show a clear influence of the PTFE coverage upon the amount of water residing within the GDL fiber structure and the structural properties of the water phase observed. First, with decreasing PTFE coverage, more and larger water clusters are formed, but upon reaching a critical value, the behavior changes to the formation of a comparatively tall cluster geometrically connecting all GDL regions.

Figure 7.12b and c shows an 85% PTFE coverage and a 55% PTFE coverage for the given backbone structure. For all cases in the study, the number of MC iterations after reaching the steady state is about 3 × 10^6. The converged source and sink rates are reaching the same value in all cases.

Snapshots of the resulting water distribution are shown in Fig. 7.13a and b for PTFE contents of 85% and 55%. In the case of 85% PTFE content, one can only find a few water drops (represented by blue color) within the GDL. Some water can also be found on the channel walls and within the channel. In contrast, regarding the snapshot of the GDL featuring a PTFE coverage of 55%, all free pores of the GDL seem to be filled with water, and the number of water particles on the channel walls and within the channel has increased. Water drops levitating within the channel should not be interpreted as such. Although they can be observed in the simulation snapshots, they correspond to energetically unfavorable configurations. Under real

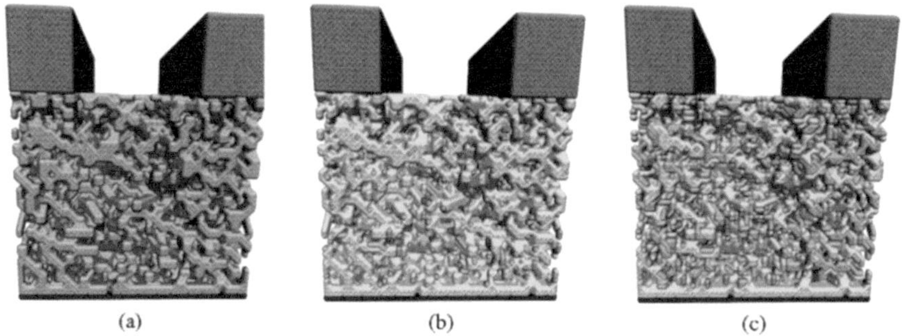

Fig. 7.12 Pure carbon 'backbone' structure (**a**) and two different PTFE loadings **b** 85% PTFE, and **c** 55% PTFE. Grey represents the graphite fibres and the flowfield ribs consisting of graphite. PTFE-covered voxels are displayed in green color [59] (color figure online)

two-phase flow conditions, water would be immediately removed by the gas flow within the channel.

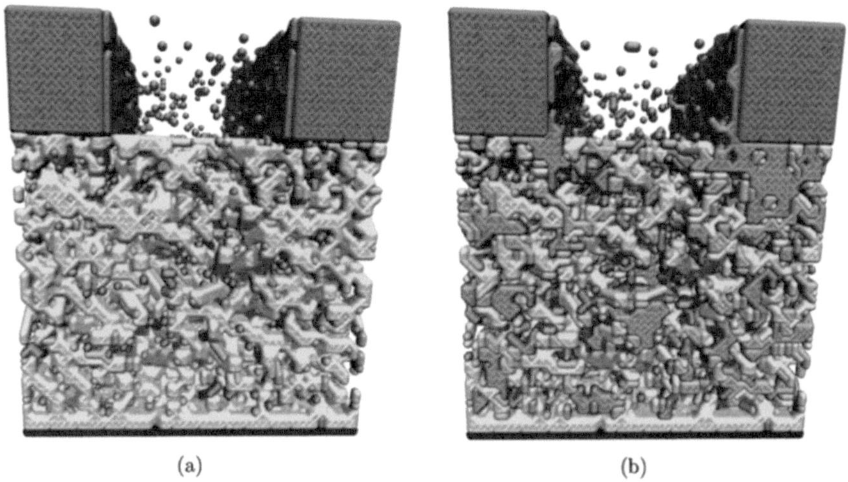

Fig. 7.13 Water distribution for different PTFE loadings. Blue represents the water, grey the graphite fibers, the flow field ribs, and green the PTFE. **a** 85% PTFE coverage; **b** 55% PTFE coverage [59] (color figure online)

7.4 Micro-model

7.4.1 Density Functional Theory

The microscopic modeling study of fuel cells can go deeper into the electrochemical reaction process on the catalyst surface, and a commonly used technique is Density Functional Theory (DFT). Density refers to the number density of electrons. DFT is a method to study the electronic structure of multi-electron systems through electron density, which is not limited to the study of microscopic models in the field of fuel cells. It has a wide range of applications in both physics and chemistry, especially for studying the properties of molecules and condensed states, and is one of the most commonly used methods in condensed matter physics and computational chemistry.

The wave function in quantum mechanics contains all the information about a system in a given state, providing us with a principled possibility to model arbitrary systems. However, there is a gap between the ideal and the reality. As the number of electrons N increases, the wave function of a system becomes very difficult to solve. As an example, a multi-electron atomic system is used here to observe the difficulties in solving it. First, use the Born–Oppenheimer approximation and ignore the motion of the nucleus. Then the Schrödinger equation can be written in the following form:

$$\hat{H}\Psi = [\hat{T} + \hat{V} + \hat{U}]\Psi = \left[\sum\nolimits_i^N \left(-\frac{\hbar^2}{2m_i}\nabla_i^2\right) + \sum\nolimits_i^N V(\boldsymbol{r}_i) + \sum\nolimits_{i<j}^N U(\boldsymbol{r}_i, \boldsymbol{r}_j)\right]\Psi \tag{7.13}$$

$$\hat{H}\Psi = [\hat{T} + \hat{V} + \hat{U}]\Psi = \left[\sum\nolimits_i^N (-)\right] \tag{7.14}$$

where, H is Hamiltonian; T in the middle is the kinetic energy operator; V is the potential energy operator for the external electric field (the electric field of the nucleus); U is the electron–electron interaction operator. For an N-electron system, each electron has three spatial coordinates, so the Schrödinger equation then contains 3N variables. For some systems with a high degree of symmetry, such as hydrogen-like particles, it is possible to derive an analytical solution for the system. However, for most of the Schrödinger equations, no analytical solution can be derived. And problems are faced in solving numerical solutions for equations with 3N variables, which is very difficult. It can be seen from the equations that the complexity of the multi-electron problem lies in U. If the solution of U can be simplified, then the computational difficulty can be drastically reduced. DFT transforms the multi-electron problem containing U into a single-electron problem, which becomes an effective method for solving such problems. In density functional theory, the most crucial variable is the electron density n. The Hohenberg–Kohn theorem is the theoretical basis for the system that can be described using the electron density. The theorem states that the ground state electron density distribution of a system has a one-to-one correspondence with the potential field in which the system is located. The wave function and energy, as well as a host of other quantum mechanical properties, are functions of the electron density. The electron density can be written as:

$$n = n(r) = N \int dr_1 \cdots \int dr_N \Psi^*(r, r_2, \ldots, r_N)\Psi(r, r_2, \ldots, r_N) \tag{7.15}$$

The total system energy can be written as:

$$E(n) = \Psi^*(n)\big(\hat{T} + \hat{V} + \hat{U}\big)\Psi(\mathrm{n}) \tag{7.16}$$

Now, the degrees of freedom of the equations have been substantially reduced, analyzing larger systems as much as possible. DFT further approximates and simplifies the kinetic energy term, the external potential energy term, and the interaction term in the equations, which can lead to an approximation of the energy:

$$E_{\text{approx}}(n) = T_s(n) + U_H(n) + V(n) \tag{7.17}$$

The difference between this energy and the exact total energy of the system is called the exchange–correlation energy and is expressed as follows:

$$\varepsilon = (T - T_s) + (U - U_H) = E_{\text{XC}} \tag{7.18}$$

The core idea of the DFT is to arrive at an easily computable but inaccurate result by approximation, and to collect all errors into a separate item to be analyzed. The exact total energy of the system can be obtained by adding the exchange–correlation energy to the approximation. Because of the complexity of exchange–correlation energies, a number of different approximations have been developed. Local Density Approximation (LDA) is the most parsimonious approximation of exchange–correlation generalization and was proposed almost along with DFT. LDA assumes that E_{XC} at a point in space is related only to the (local) electron density at that point. The exchange and correlation terms are generally discussed separately, and the E_{XC} can be expressed in the following form:

$$E_{\text{XC}}^{\text{LDA}} = E_{\text{X}}^{\text{LDA}} + E_{\text{c}}^{\text{LDA}} \tag{7.19}$$

The specific solution can be found in books on computational chemistry and will not be described here. LDA was proposed very early and has been widely used in materials research. Generally speaking, LDA is relatively accurate and can give a good estimation of structural and elastic properties. However, it has the following drawbacks: overestimation of the binding energy, underestimation of the reaction activation energy, excessive preference for high-spin structures, and misestimation of the phase stability, etc. Hence, LDA does not perform well in systems where the electron density changes rapidly. It is easy to think of a way to improve it by including the first-order gradient. This gives us the Generalized Gradient Approximation (GGA). Most of the GGA functionals are based on modifications of some LDA functionals, which introduce a first-order gradient that can be interpreted as the "speed" of the evolution of the electron cloud. GGA has several advantages over LDA: more accurate atomic and molecular energies, corrected overbinding, and more accurate reaction activation energies (which are still usually too low). However, GGA is not always superior to LDA, and the GGA functionals usually give low bond energies and high lattice parameters. Subsequent studies have also been continuously improving the approximation of the exchange association energies, proposing methods including meta-GGA and Hybrid functionals. The relationship between some of the approximate generalized functions can be represented in Fig. 7.14.

Generally, we can think that the generalized functions appearing in the picture are progressively more complex and more accurate from bottom to top. However, this does not mean that higher-order generalized functions are necessarily more accurate than lower-order generalized functions in all problems. Choosing the appropriate generalized function according to the research problem is more important.

In fuel cells, DFT is often used to calculate the catalytic properties of materials to design catalysts that are more suitable for cell reactions. For example, the high price of precious metal catalysts restricts the large-scale development of fuel cells, and transition metals have the potential to become new catalytic materials due to their lower price. Figure 7.15 shows the ability of several single transition metal

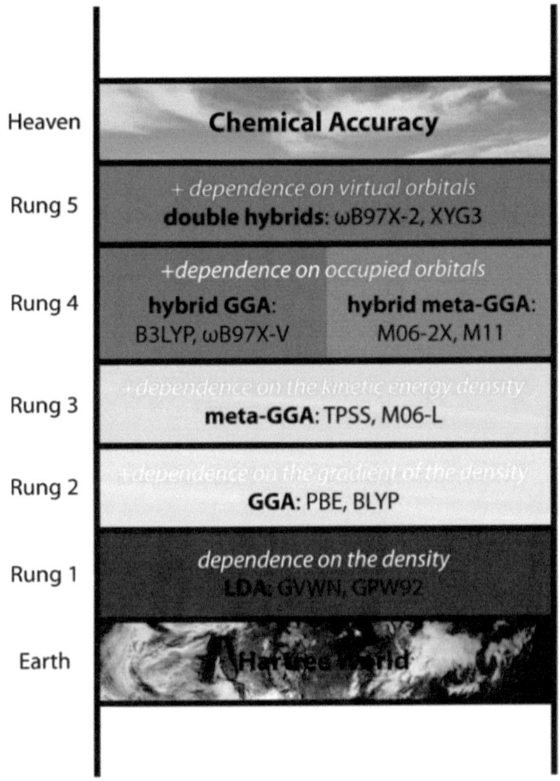

Fig. 7.14 Jacob's ladder [60]

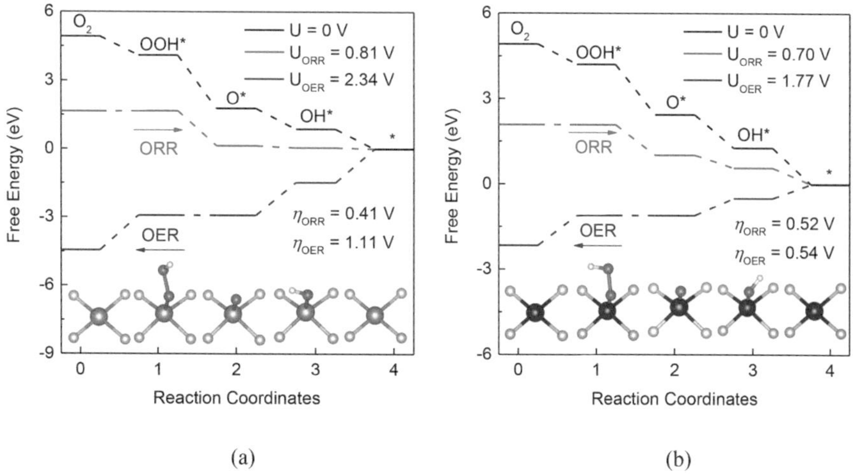

Fig. 7.15 Free energy diagrams of ORR and OER on **a** Fe-N4 and **b** Co-N4 active sites [61]

atoms on 2D BC3 to catalyze the OER and ORR reactions as obtained by DFT calculations. The exchange–correlation interactions in the simulation are described by the Perdew-Burke-Ernzerhof (PBE) GGA generalized function.

Figure 7.15 shows the free energy during the reaction of Fe-N4 and Co-N4. From the picture, we can see that Fe-N4 shows better ORR catalytic activity. Its η_{ORR} is only 0.41 V. The potential-determining step is the protonation of O_2 to OOH*. Co-N4, on the other hand, exhibits better OER catalytic activity and its η_{OER} is only 0.54 V. The potential-determining step is the hydroxylation of O* to OOH*. Figure 7.16 demonstrates the simulated adsorption free energies of oxygen and intermediates for all transition metals.

Figure 7.16a illustrates the adsorption energy of oxygen molecules on the active sites of transition metal atoms. It can be seen in the figure that the adsorption energy increases as the number of transition metal atoms increases, indicating that the interaction between the oxygen molecule and the metal sites is weakening. On the transition metals from Ti to Fe, both oxygen atoms can bond. None of the transition metal atoms after that can bond, which also means that the interaction between the oxygen and the active site is so weak that these transition metals have little catalytic ability to start the ORR reaction. Figure 7.16b illustrates the adsorption free energies of the reaction intermediates. The general trend remains that the adsorption free energy increases with the increase in the number of atoms. When reaching the Zn atoms, the change rule appears to be perverse, and all the adsorption free energies of Zn on the reaction intermediates show a significant decrease.

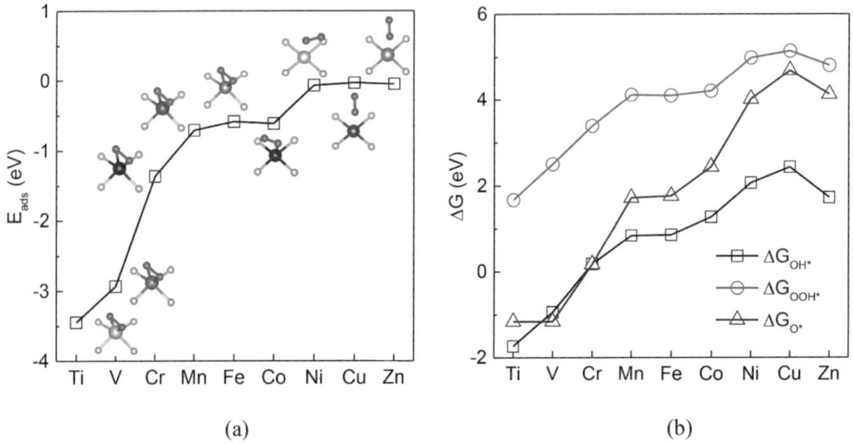

Fig. 7.16 **a** The adsorption energies of the oxygen molecule on the TM-N4 active site. **b** The adsorption free energies of intermediate species on the TM-N4 active site [61]

7.4.2 Molecular Dynamics

Molecular dynamics (MD) simulation is a method used to calculate the equilibrium and transfer properties of a multi-system. The MD method is based on Newton's laws of motion and the integration of the classical Newtonian equations of motion to obtain a system in space defined by the coordinates and velocities of the particles, taking the current position and velocity of the particles as the initial state and obtaining the final state after a number of time steps. The MD simulation incorporates the kinetic energy term and the change in time. Kinetic energy is a function of temperature, so the method can simulate the variation of a molecular analogue system with time at different temperatures. It is possible to calculate not only macroscopic physical quantities, but also to obtain accurate images of microstructure, particle motion, etc. MD simulations can be performed on spatial scales up to 10^{-10} m and temporal scales up to 10^{-15} s. MD simulations also play an important role in PEMFC calculations [62–66].

In 1957, the ideal 'hard sphere' model was first proposed [67], which used the classical MD method to investigate the state equation of the gas and liquid phases. This set the precedent for the use of MD methods to study macroscopic properties. Up to now, many microscopic phenomena and conclusions that could not be obtained experimentally could be obtained quickly and accurately using the MD method. This gives the method an unparalleled advantage in the development of new materials, catalysts, membrane permeation, etc. It has revolutionized the previous relationship between theoretical calculations and experiments. Moreover, MD also offers a simple procedure, a small computational effort, and a system of calculated particles that far exceeds the first principles approach.

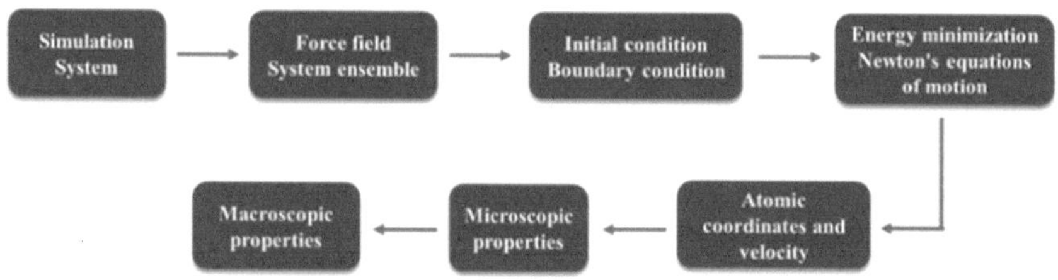

Fig. 7.17 Flow chart of the MD simulation principle

The general steps of the MD simulation can be divided into: (1) construction of the simulation system, including the construction of a reasonable system size and setting the position and number of particles, etc. (2) Selecting a suitable molecular force field and statistical synthesis based on the type of material in the system and the computational requirements. (3) Initializing the system and setting boundary conditions. (4) Calculate the energy of the atoms by procedures such as energy minimization. (5) The coordinates and velocities of the atoms are obtained by solving Newton's equations of motion. (6) The macroscopic properties of the system, such as kinetic and thermodynamic properties, are obtained by counting the microscopic properties of the atoms. The steps are shown in Fig. 7.17.

The solution process of MD follows Newton's second law, and its molecular motion can be expressed as:

$$m_i \frac{\mathrm{d}^2 \mathbf{r}_i}{\mathrm{d}t^2} = \sum_j \mathbf{F}_{ij} + \sum \mathbf{F}_k \tag{7.20}$$

where m_i is the mass of the ith molecule, $\mathbf{r}_i$ is, the molecule's position vector, F_{ij} and F_k are the forces exerted by other molecules and the body force, respectively.

Ideally, most force fields can be expressed as the gradient of the potential energy function V:

$$\mathbf{F}_k = -\nabla_{r_i} \mathbf{V} \tag{7.21}$$

Complex molecules, especially long-chain molecules, not only have bonding interactions but also have intramolecular non-bonding interactions. Therefore, some common forces involved in simulations include non-bonded and bonded interactions. Among them, there are mainly two types of nonbonding interactions: Coulomb interactions due to electrostatic effects and Van der Waals interactions due to quantum fluctuations of charges in molecules. Intramolecular hydrogen bonds are sometimes also referred to as intramolecular nonbonding interactions.

It is relatively complicated to directly calculate the charge distribution in continuous space. Therefore, some simplifications are required for the general selection of electrostatic interactions, such as the point charge method. It is generally believed that the charge is concentrated and distributed on the nucleus, and the electric field is completely formed by the expansion of point charges defined on each nucleus. Therefore, in the point charge method, we can see that the energy function describing the electrostatic potential energy is the familiar classical Coulomb form:

$$V = \sum_{ij} \frac{q_i q_j}{4\pi \varepsilon_0 r_{ij}} \tag{7.22}$$

where q_i and q_j are the signed magnitudes of two charges, r_{ij} is the distance between two particles, and ε_0 is the permittivity of free space.

Van der Waals interaction (intermolecular non-bonding interaction) includes three kinds of forces: orientation force, induction force, and dispersion force, among which the dispersion force plays a dominant role. The physical model behind the Van der Waals action is not as clear as the physical model behind the point charge method, so the main task of the Van der Waals action is to construct a reasonable physical model to describe it. A mathematically simple two-body interaction model that approximates the Van der Waals interaction was first proposed in 1924 by British mathematician Lennard–Jones [68]. Van der Waals forces have both attractive and repulsive effects. The L-J potential repulsive term has the 12th power, which describes the Pauli repulsion due to the overlapping electron orbitals, and the attractive term has the 6th power to show the dispersion force of neutral atoms.

$$u(r_{ij}) = 4\varepsilon\left[\left(\frac{\sigma}{r_{ij}}\right)^{12} - \left(\frac{\sigma}{r_{ij}}\right)^{6}\right] \tag{7.23}$$

where ε is the depth of the potential well, σ is the finite distance at which the inter-particle potential is zero, r is the distance between the particles. The relationship between L-J potential energy and distance is shown in Fig. 7.18, where U(r) is the potential energy and r is the distance between two interacting particles. For computational efficiency, the truncated L-J potential usually uses rij = 2.5σ, called the cutoff radius. Outside this radius, pair interactions are ignored.

In a molecule, there are three basic degrees of freedom, namely, bond stretch, bond bending, and rotation (torsion) that represent bonded potentials. Therefore, bonding interactions include bond stretching, bond angle, and dihedral interactions, which have energy potentials of the form:

$$V = \sum_{\text{bonds}} \frac{k_r}{2}\left(r_{ij} - r_{ij}^0\right)^2 \tag{7.24}$$

$$V = \sum_{\text{angles}} \frac{k_\theta}{2}\left(\theta_{ijk} - \theta_{ijk}^0\right)^2 \tag{7.25}$$

$$V = \sum_{\text{dihedrals}} (1 + \cos(n\phi - \phi_0))^2 \tag{7.26}$$

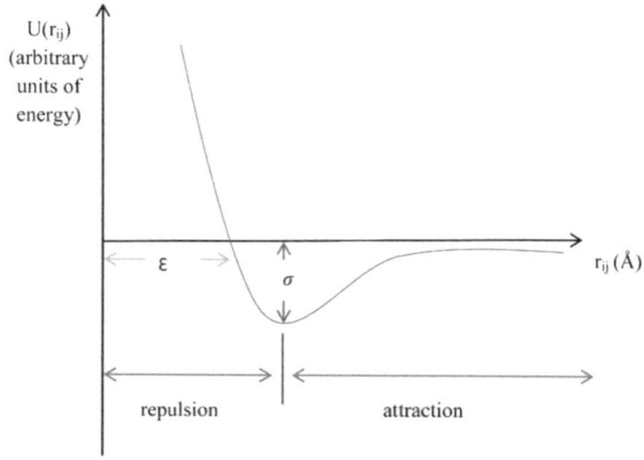

Fig. 7.18 Lennard–Jones potential represents the attractive long-range and short-range repulsive interactions [69]

The answer procedure for MD commonly requires computing the above force field equations and integrating the equations of motion. There are quite a few algorithms developed for ensembles, such as the Verlet algorithm, the velocity-Verlet algorithm, and the predictor–corrector algorithm. The preference for integration is a balance between accuracy and computational efficiency.

A major limitation of MD is its lack of ability to model processes involving chemical reactions. Nonetheless, MD methods have been widely used to analyze the morphology and structure of ionomers in PEMFCs, the diffusion behavior of protons in ionomers and membranes, the O_2 transport behavior at the electrode interface, and the electrode heat conduction process. The Balbuena research group first used molecular dynamics simulation to study the morphology of the Nafion-Pt/C interface [70]. In the simulation system, double-layer graphene is used to represent carbon loading, and Pt256 clusters are catalysts adsorbed on the surface of graphene. The formation process of the catalyst layer is simulated by gradually increasing the number of Nafion polymer chains and water molecules in the research system. The simulation results show that Nafion resin is mainly adsorbed on the Pt nanoparticles with sulfonate groups, and the sulfonate groups are surrounded by water molecules to form ion clusters, and the connectivity between adjacent ion clusters is closely related to the water content. The size of Pt nanoparticles affects the microstructure of the catalyst layer, and different Pt particle sizes lead to different morphologies of the catalyst layer [71]. By increasing the size of Pt nanoparticles, more sulfonic acid sites are dispersed close to the surface of Pt nanoparticles, and the polymer can effectively avoid the incorporation of Pt nanoparticles. The effect of ionomer content on the microstructure and heterogeneous transport properties of the cathode catalyst layer (CCL) of a polymer electrolyte membrane fuel cell was investigated by combining molecular dynamics simulations (Fig. 7.19) and single-cell experiments [72]. MD analysis shows that the Pt/C surface can be completely covered when the Nafion number exceeds 7 (29.6 wt%) in the model, at which point the maximum electrochemical surface area (ECSA) is obtained. And with increasing ionomer content

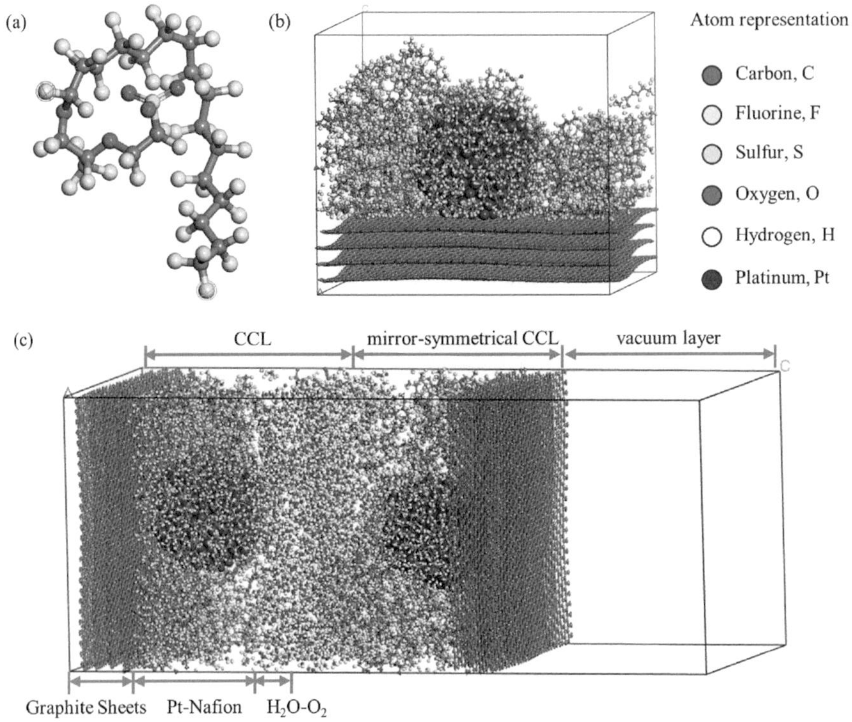

Fig. 7.19 Structure models: **a** monomer structure of ionomer chain; **b** deposition of ionomer on Pt nanoparticle-graphite sheets; **c** mirrored CCL (N7) [72]

in the CCL model, the diffusion coefficient of H_3O^+ increased by a factor of 2, while that of O_2 decreased monotonically.

Oxidation of catalysts and carbon supports is unavoidable during fuel cell operation. These processes need to be taken into account in molecular dynamics simulations to obtain more realistic information. The effect of oxidation at different water contents on the adhesion of nanoparticles on carbon substrates with different degrees of oxidation (10, 25, and 50%) was investigated using classical molecular dynamics methods [73]. The carbon support surface was chosen as epoxidized carbon and hydroxylated carbon to simulate the two most common types of carbon oxidation. Under low oxidation conditions (10%), the binding energy between the epoxidized carbon surface and the polymer electrolyte membrane increases; however, higher levels of epoxidation (25%, 50%) lead to membrane delamination. In addition, increased water content can increase the binding energy. The polymer film on the oxidized carbon surface was observed to act as a binder between the PtO nanoparticles and the oxidized surface. This work highlights the importance of carbon surface oxidation type, degree of oxidation, and water content for designing long-lived catalyst layers for PEMFC applications. In terms of the influence of catalyst oxidation and carbon support oxidation on mass transfer, the transport behavior of protons in the Pt-water film-ionomer interface under different oxide coverages on the Pt surface [74], and the O_2 transport behavior in the Pt(C)-water film-ionomer three-phase interface [75] were studied. The effects of different ratios (< 12%) of oxygen-containing functional groups loaded on the surface of amorphous carbon on the carbon support-ionomer interface structure, interfacial mass transfer, and cell performance were explored using a combination of experiments and MD calculations [76].

MD-based modeling is also a useful approach to study the properties and redox reaction mechanisms of PEMFCs. The molecular dynamics method can be used to establish the oxygen adsorption model on the Pt surface of the electrocatalyst, and the oxidation–reduction reaction model of hydrogen and oxygen on the Pt surface [77], simulate the reaction mechanism of hydrogen atoms and oxygen on the Pt surface, and then analyze the oxygen reduction reaction characteristic.

The main applications of MD in ion exchange membranes include: water transfer in membranes [78], proton transport [78–81], thermal conductivity [81], and microstructure of membranes [82]. The following is a brief description of particle transport in membranes as an example.

Take the Nafion 117 membrane as an example, using MD to study the diffusion behavior of water molecules, hydrated hydrogen ions, hydrogen, and oxygen in the polymer.

(1) Simulated system: the polymer chain length is taken as 10, which means 10 sulphonic acid groups on each polymer chain. Figures 7.20 and 7.21 show the chemical structures of the polymer chains, water molecules, hydrated hydrogen ions, oxygen, and hydrogen gas. The oxygen, hydrogen, sulphur, fluorine, and carbon atoms are shown as red, off-white, yellow, light blue, and grey balls, respectively.
(2) Force field: The COMPASS [83] force field is used, which is suitable for covalent molecular systems, including most common organic, inorganic, and polymeric substances, metal oxides, and metal halides.
(3) Non-crystalline model: Four polymer chains are inserted into the model. Forty hydrated hydrogen ions were inserted for the model to be electrically neutral as a whole. In addition, 4 hydrogen molecules, 4 oxygen molecules, and 360

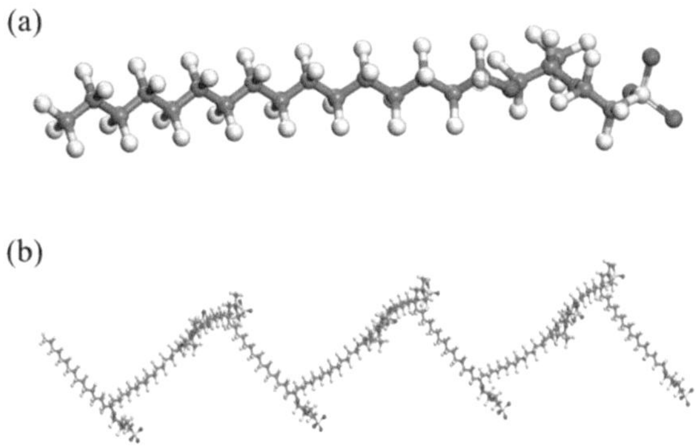

Fig. 7.20 **a** The basic repeating unit of Nafion 117; **b** a polymer chain of chain length 10

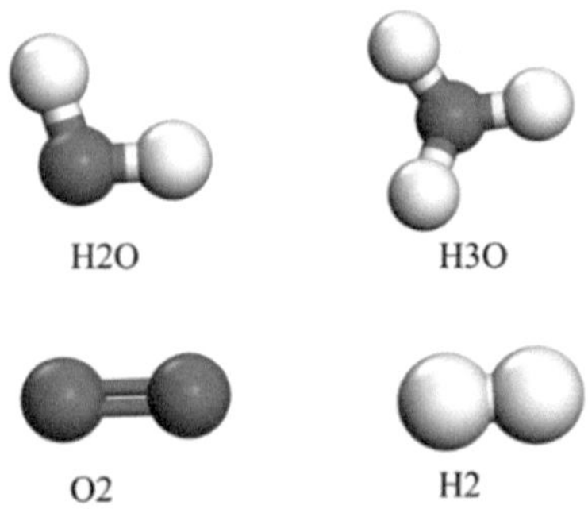

Fig. 7.21 Chemical structures of a water molecule, hydronium ion, oxygen, and hydrogen

water molecules were inserted into the model to study the diffusion coefficients of the particles in the membrane. As shown in Fig. 7.22.

(4) Energy minimization: Structural optimization is carried out first, followed by annealing.
(5) Dynamics calculation: Select the NVT ensemble and choose the current speed as the initial speed. Calculation time is 2 ns. The temperature control method is Nose, and the pressure control method is Berendsen.
(6) Calculation of the diffusion coefficient: During the molecular dynamics simulation, the particles in the system move incessantly from their initial positions, and the average of the squared particle displacements is defined as the mean squared displacement (MSD), and the equation is as follows:

$$MSD = \left\langle |\vec{r}(t) - \vec{r}(0)|^2 \right\rangle \tag{7.27}$$

where < > denotes the mean value and $\vec{r}(0)$ denotes the initial position. It can be seen from the equation that the value of MSD represents the motion activity of a particle,

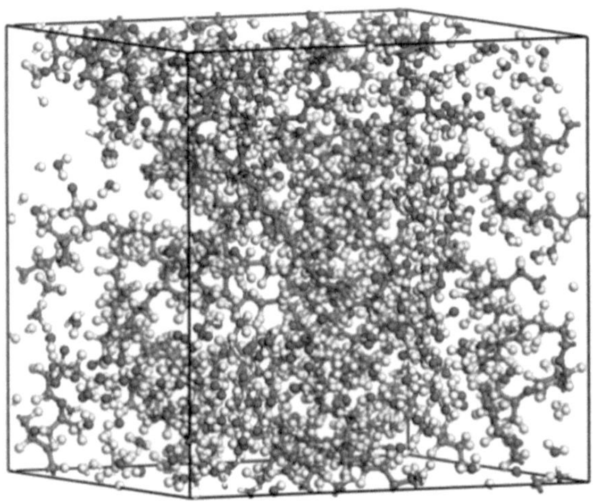

Fig. 7.22 Non-crystalline model of Nafion 117 (containing particles such as water molecules)

the larger the value of MSD, the greater the distance the particle moves in a given time.

According to Einstein's law of diffusion,

$$\lim_{t \to \infty} \left\langle |\vec{r}(t) - \vec{r}(0)|^2 \right\rangle = 6Dt \tag{7.28}$$

where D represents the self-diffusion coefficient.

The diffusion coefficient is calculated by the following equation:

$$D = \frac{1}{6N} \lim_{t \to \infty} \frac{d}{dt} \sum_{i=1}^{N} \left([r_i(t) - r_0(t)]^2 \right) \tag{7.29}$$

where N is the number of atoms in the system undergoing diffusive motion. The differential term in the equation can be replaced by the ratio of the difference in MSD with time, i.e., the slope of the MSD curve with time. The MSD has been divided by the number of diffusing atoms, so the above equation can be simplified as follows:

$$D = \frac{a}{6} \tag{7.30}$$

where a is the slope of the MSD about the time curve.

7.5 Summary and Outlook

The PEMFC model plays an important role in understanding the transmission and reaction processes within fuel cells, predicting PEMFC performance [28] and optimizing the design of fuel cell structures [84, 85], and promoting the commercialization of cells. To establish a more accurate cell model, this chapter reviews the research progress of micro models in PEMFC. A brief introduction was given to the modeling ideas and applications of various pore-scale models. Three micro models: LBM, DFT, and MD, which are commonly used in PEMFC research, are introduced in this chapter.

Unlike the continuum numerical model based on the Navier–Stokes equation, the first-principle-based method LBM considers flows to be composed of a collection of pseudo-particles residing on the nodes of an underlying lattice structure, which interact according to a velocity distribution function. This feature makes LBM a suitable approach to study microstructure in PEMFC, such as GDL and CL, which is thorny for continuum models due to its irregular computational domain. Problems in GDL and CL, such as water removal performance of GDLs [86], influence of CL structure in the kinetics of chemical reactions in CL [87], water removal performance in the flow field [88], can be reconsidered from a different perspective using LBM. Despite the outstanding ability of solving problems in irregular micro-structure regions, the application of LBM in PEMFC is still limited for its shortcoming that the computation domain size of LBM is normally limited to a small scale, which increases the computing resource demands rapidly when the computation scale rises.

DFT is a commonly used technique in fuel cells when it comes to the research of the catalyst surface, where electrochemical reaction processes take place. Focusing on calculations at the atomic level, DFT is a sprightly tool predicting catalytic properties of materials, which is critical in designing fuel cell catalysts. Catalyst design and reaction mechanism of both ORR [89, 90] and HOR [91, 92] in fuel cells are studied using DFT. Also, when it comes to the ionomer-catalyst interface [93], DFT helps to understand the mechanism of the interface behavior and improve the catalytic performance.

Also focusing on the atomic level, like DFT, MD, on the other hand, is based on Newton's laws of motion, resulting in its lack of ability to model processes that involve chemical reactions compared to DFT. Nonetheless, MD is a useful approach for studying non-chemical reaction-related processes, like water transport through the electrolyte membrane [94], oxygen transport properties in the catalyst layer [95], and ion transfer performance of the membrane [96].

References

1. Humby S J, Biggs M J, Tüzün U. Explicit numerical simulation of fluids in reconstructed porous media[J]. Chemical engineering science, 2002, 57(11): 1955–1968.
2. Wang G Q. Direct numerical simulation of porous electrodes for fuel cells and advanced batteries[D]. Ph. D. dissertation, the Pennsylvania State University, 2003.
3. Prat M. Recent advances in pore-scale models for drying of porous media[J]. Chemical engineering journal, 2002, 86(1–2): 153–164.
4. Lange K J, Sui P C, Djilali N. Pore scale simulation of transport and electrochemical reactions in reconstructed PEMFC catalyst layers[J]. Journal of The Electrochemical Society, 2010, 157(10): B1434.
5. Kohout M, Grof Z, Štěpánek F. Pore-scale modelling and tomographic visualisation of drying in granular media[J]. Journal of colloid and interface science, 2006, 299(1): 342–351.
6. Mukherjee P P, Wang C Y, Kang Q. Mesoscopic modeling of two-phase behavior and flooding phenomena in polymer electrolyte fuel cells[J]. Electrochimica Acta, 2009, 54(27): 6861–6875.
7. Kohout M, Collier A P, Štěpánek F. Mathematical modelling of solvent drying from a static particle bed[J]. Chemical engineering science, 2006, 61(11): 3674–3685.
8. Parmigiani A, Huber C, Bachmann O, et al. Pore-scale mass and reactant transport in multiphase porous media flows[J]. Journal of Fluid Mechanics, 2011, 686: 40–76.
9. Štěpánek F, Marek M, Hanika J, et al. Mesoscale modeling in multiphase catalysis[J]. Catalysis today, 2001, 66(2–4): 249–254.
10. Kočí P, Štěpánek F, Kubíček M, et al. Modelling of micro/nano-scale concentration and temperature gradients in porous supported catalysts[J]. Chemical Engineering Science, 2007, 62(18–20): 5380–5385.
11. Kočí P, Štěpánek F, Kubíček M, et al. Pore-scale modeling of non-isothermal reaction phenomena in digitally reconstructed porous catalyst[J]. Molecular Simulation, 2007, 33(4–5): 369–377.
12. Štěpánek F, Ansari M A. Computer simulation of granule microstructure formation[J]. Chemical engineering science, 2005, 60(14): 4019–4029.

13. Antoine O, Bultel Y, Durand R, et al. Electrocatalysis, diffusion and ohmic drop in PEMFC: Particle size and spatial discrete distribution effects[J]. Electrochimica acta, 1998, 43(24): 3681–3691.
14. Carcadea E, Varlam M, Marinoiu A, et al. Influence of catalyst structure on PEM fuel cell performance–A numerical investigation[J]. International Journal of Hydrogen Energy, 2019, 44(25): 12829–12841.
15. Flückiger R. Transport phenomena on the channel-rib scale of polymer electrolyte fuel cells[D]. ETH Zurich, 2009.
16. Wang G, Mukherjee P P, Wang C Y. Direct numerical simulation (DNS) modeling of PEFC electrodes: Part I. Regular microstructure[J]. Electrochimica acta, 2006, 51(15): 3139–3150.
17. Mukherjee P P, Wang C Y. Stochastic microstructure reconstruction and direct numerical simulation of the PEFC catalyst layer[J]. Journal of the Electrochemical Society, 2006, 153(5): A840.
18. Lange K J, Sui P C, Djilali N. Pore scale modeling of a proton exchange membrane fuel cell catalyst layer: Effects of water vapor and temperature[J]. Journal of Power Sources, 2011, 196(6): 3195–3203.
19. Schulz V P, Becker J, Wiegmann A, et al. Modeling of two-phase behavior in the gas diffusion medium of PEFCs via full morphology approach[J]. Journal of the electrochemical society, 2007, 154(4): B419.
20. Chen L, Luan H B, He Y L, et al. Pore-scale flow and mass transport in gas diffusion layer of proton exchange membrane fuel cell with interdigitated flow fields[J]. International Journal of Thermal Sciences, 2012, 51: 132–144.
21. Becker J, Wieser C, Fell S, et al. A multi-scale approach to material modeling of fuel cell diffusion media[J]. International Journal of Heat and Mass Transfer, 2011, 54(7–8): 1360–1368.
22. Mukherjee P P, Kang Q, Wang C Y. Pore-scale modeling of two-phase transport in polymer electrolyte fuel cells—progress and perspective[J]. Energy & Environmental Science, 2011, 4(2): 346–369.
23. El Hannach M, Pauchet J, Prat M. Pore network modeling: application to multiphase transport inside the cathode catalyst layer of proton exchange membrane fuel cell[J]. Electrochimica Acta, 2011, 56(28): 10796–10808.
24. Wu R, Liao Q, Zhu X, et al. Pore network modeling of cathode catalyst layer of proton exchange membrane fuel cell[J]. International Journal of Hydrogen Energy, 2012, 37(15): 11255–11267.
25. El Hannach M, Prat M, Pauchet J. Pore network model of the cathode catalyst layer of proton exchange membrane fuel cells: Analysis of water management and electrical performance[J]. International Journal of Hydrogen Energy, 2012, 37(24): 18996–19006.
26. Aghighi M, Gostick J. Pore network modeling of phase change in PEM fuel cell fibrous cathode[J]. Journal of Applied Electrochemistry, 2017, 47: 1323–1338.
27. Hinebaugh J, Bazylak A. Condensation in PEM fuel cell gas diffusion layers: a pore network modeling approach[J]. Journal of the Electrochemical Society, 2010, 157(10): B1382.
28. Sadeghi M A, Khan Z A, Agnaou M, et al. Predicting pemfc performance from a volumetric image of catalyst layer structure using pore network modeling[J]. Applied Energy, 2024, 353: 122004.
29. Mukherjee P P, Wang C Y. Direct numerical simulation modeling of bilayer cathode catalyst layers in polymer electrolyte fuel cells[J]. Journal of The Electrochemical Society, 2007, 154(11): B1121.
30. Chen L, Zhang R, Kang Q, et al. Pore-scale study of pore-ionomer interfacial reactive transport processes in proton exchange membrane fuel cell catalyst layer[J]. Chemical Engineering Journal, 2020, 391: 123590.
31. Park J, Li X. Multi-phase micro-scale flow simulation in the electrodes of a PEM fuel cell by lattice Boltzmann method[J]. Journal of power sources, 2008, 178(1): 248–257.
32. Molaeimanesh G R, Nazemian M. Investigation of GDL compression effects on the performance of a PEM fuel cell cathode by lattice Boltzmann method[J]. Journal of Power Sources, 2017, 359: 494–506.

33. Salah Y B, Tabe Y, Chikahisa T. Gas channel optimisation for PEM fuel cell using the lattice Boltzmann method[J]. Energy Procedia, 2012, 28: 125–133.
34. Molaeimanesh G R, Akbari M H. Agglomerate modeling of cathode catalyst layer of a PEM fuel cell by the lattice Boltzmann method[J]. International Journal of Hydrogen Energy, 2015, 40(15): 5169–5185.
35. Niu X D, Munekata T, Hyodo S A, et al. An investigation of water-gas transport processes in the gas-diffusion-layer of a PEM fuel cell by a multiphase multiple-relaxation-time lattice Boltzmann model[J]. Journal of Power Sources, 2007, 172(2): 542–552.
36. Malekian A, Salari S, Stumper J, et al. Effect of compression on pore size distribution and porosity of PEM fuel cell catalyst layers[J]. international journal of hydrogen energy, 2019, 44(41): 23396–23405.
37. Salari S, Stumper J, Bahrami M. Direct measurement and modeling relative gas diffusivity of PEMFC catalyst layers: the effect of ionomer to carbon ratio, operating temperature, porosity, and pore size distribution[J]. international journal of hydrogen energy, 2018, 43(34): 16704–16718.
38. Wang G, Mukherjee P P, Wang C Y. Optimization of polymer electrolyte fuel cell cathode catalyst layers via direct numerical simulation modeling[J]. Electrochimica Acta, 2007, 52(22): 6367–6377.
39. Chen L, Zhang R, Min T, et al. Pore-scale study of effects of macroscopic pores and their distributions on reactive transport in hierarchical porous media[J]. Chemical Engineering Journal, 2018, 349: 428–437.
40. Mu Y T, Weber A Z, Gu Z L, et al. Mesoscopic modeling of transport resistances in a polymer-electrolyte fuel-cell catalyst layer: Analysis of hydrogen limiting currents[J]. Applied Energy, 2019, 255: 113895.
41. Chen L, Kang Q, Tao W. Pore-scale numerical study of multiphase reactive transport processes in cathode catalyst layers of proton exchange membrane fuel cells[J]. International Journal of Hydrogen Energy, 2021, 46(24): 13283–13297.
42. Salomov U R, Chiavazzo E, Asinari P. Pore-scale modeling of fluid flow through gas diffusion and catalyst layers for high temperature proton exchange membrane (HT-PEM) fuel cells[J]. Computers & Mathematics with Applications, 2014, 67(2): 393–411.
43. Chen L, Wu G, Holby E F, et al. Lattice Boltzmann pore-scale investigation of coupled physical-electrochemical processes in C/Pt and non-precious metal cathode catalyst layers in proton exchange membrane fuel cells[J]. Electrochimica Acta, 2015, 158: 175–186.
44. Litster S, Epting W K, Wargo E A, et al. Morphological analyses of polymer electrolyte fuel cell electrodes with nano-scale computed tomography imaging[J]. Fuel Cells, 2013, 13(5): 935–945.
45. Epting W K, Gelb J, Litster S. Resolving the Three-Dimensional Microstructure of Polymer Electrolyte Fuel Cell Electrodes using Nanometer-Scale X-ray Computed Tomography[J]. Advanced Functional Materials, 2012, 22(3): 555–560.
46. Liu Z, Wu H. Numerical modeling of liquid–gas two-phase flow and heat transfer in reconstructed porous media at pore scale[J]. International Journal of Hydrogen Energy, 2016, 41(28): 12285–12292.
47. Cetinbas F C, Wang X, Ahluwalia R K, et al. Microstructural analysis and transport resistances of low-platinum-loaded PEFC electrodes[J]. Journal of The Electrochemical Society, 2017, 164(14): F1596.
48. Zhang X, Gao Y, Ostadi H, et al. Method to improve catalyst layer model for modelling proton exchange membrane fuel cell[J]. Journal of Power Sources, 2015, 289: 114–128.
49. Wargo E A, Kotaka T, Tabuchi Y, et al. Comparison of focused ion beam versus nano-scale X-ray computed tomography for resolving 3-D microstructures of porous fuel cell materials[J]. Journal of Power Sources, 2013, 241: 608–618.
50. Lopez-Haro M, Guétaz L, Printemps T, et al. Three-dimensional analysis of Nafion layers in fuel cell electrodes[J]. Nature communications, 2014, 5(1): 5229.
51. Cetinbas F C, Ahluwalia R K, Kariuki N N, et al. Agglomerates in polymer electrolyte fuel cell electrodes: Part I. Structural characterization[J]. Journal of The Electrochemical Society, 2018, 165(13): F1051.

52. Chen L, Feng Y L, Song C X, et al. Multi-scale modeling of proton exchange membrane fuel cell by coupling finite volume method and lattice Boltzmann method[J]. International Journal of Heat and Mass Transfer, 2013, 63: 268–283.
53. Deng S, Li Y. A porous-rib flow field for performance enhancement in proton exchange membrane fuel cells[J]. Energy Conversion and Management, 2022, 263: 115707.
54. Chen L, Kang Q, Mu Y, et al. A critical review of the pseudopotential multiphase lattice Boltzmann model: Methods and applications[J]. International journal of heat and mass transfer, 2014, 76: 210–236.
55. Seidenberger K, Wilhelm F, Haußmann J, et al. Grand canonical Monte Carlo study on water agglomerations within a polymer electrolyte membrane fuel cell gas diffusion layer[J]. Journal of power sources, 2013, 239: 628–641.
56. Pournemat A, Markötter H, Wilhelm F, et al. Nano-scale Monte Carlo study on liquid water distribution within the polymer electrolyte membrane fuel cell microporous layer, catalyst layer and their interfacial region[J]. Journal of Power Sources, 2018, 397: 271–279.
57. Bai B, Chen Y T. Simulation of the oxygen reduction reaction (ORR) inside the cathode catalyst layer (CCL) of proton exchange membrane fuel cells using the Kinetic Monte Carlo Method[J]. Energies, 2018, 11(10): 2529.
58. Pournemat A, Wilhelm F, Scholta J. Effect of Current Density on the Dynamical Evolution of Liquid Water Distribution within the Gas Diffusion Layers of PEMFC Employing a Multi-Timescale Kinetic Monte Carlo Method[J]. Journal of The Electrochemical Society, 2019, 166(4): F334.
59. Seidenberger K, Wilhelm F, Schmitt T, et al. Estimation of water distribution and degradation mechanisms in polymer electrolyte membrane fuel cell gas diffusion layers using a 3D Monte Carlo model[J]. Journal of power sources, 2011, 196(12): 5317–5324.
60. Mardirossian N, Head-Gordon M. ωB97X-V: A 10-parameter, range-separated hybrid, generalized gradient approximation density functional with nonlocal correlation, designed by a survival-of-the-fittest strategy[J]. Physical Chemistry Chemical Physics, 2014, 16(21): 9904–9924.
61. Wang Y, Zhou N, Li Y. Electrochemical catalytic mechanism of single transition metal atom embedded BC3 monolayer for oxygen reduction and evolution reactions[J]. Chemical Engineering Journal, 2021, 425: 130631
62. Srinophakun T, Martkumchan S. Ionic conductivity in a chitosan membrane for a PEM fuel cell using molecular dynamics simulation[J]. Carbohydrate polymers, 2012, 88(1): 194–200.
63. Kim D J, Park C H, Nam S Y. Molecular dynamics simulations of modified PEEK polymeric membrane for fuel cell application[J]. international journal of hydrogen energy, 2016, 41(18): 7641–7648.
64. Zheng C, Geng F, Rao Z. Proton mobility and thermal conductivities of fuel cell polymer membranes: Molecular dynamics simulation[J]. Computational Materials Science, 2017, 132: 55–61.
65. Brandell D, Karo J, Liivat A, et al. Molecular dynamics studies of the Nafion®, Dow® and Aciplex® fuel-cell polymer membrane systems[J]. Journal of molecular modeling, 2007, 13: 1039–1046.
66. Brault P, Coutanceau C, Jennings P C, et al. Molecular dynamics simulations of ternary PtxPdyAuz fuel cell nanocatalyst growth[J]. International Journal of Hydrogen Energy, 2016, 41(47): 22589–22597.
67. ALDER B J, WAINWRIGHT T E. PHASE TRANSITION FOR A HARD SPHERE SYSTEM [J]. Journal of Chemical Physics, 1957, 27(5): 1208–1209.
68. Jones J E. On the determination of molecular fields.—II. From the equation of state of a gas[J]. Proceedings of the Royal Society of London. Series A, Containing Papers of a Mathematical and Physical Character, 1924, 106(738): 463–477.
69. De Lile J R, Zhou S. Theoretical modeling of the PEMFC catalyst layer: A review of atomistic methods[J]. Electrochimica Acta, 2015, 177: 4–20.
70. Lamas E J, Balbuena P B. Molecular dynamics studies of a model polymer–catalyst–carbon interface[J]. Electrochimica acta, 2006, 51(26): 5904–5911.

71. Cheng C H, Malek K, Sui P C, et al. Effect of Pt nano-particle size on the microstructure of PEM fuel cell catalyst layers: Insights from molecular dynamics simulations[J]. Electrochimica Acta, 2010, 55(5): 1588–1597.
72. Xue Q, Zhang R, Yang D, et al. Effect of ionomer content on cathode catalyst layer for PEMFC via molecular dynamics simulations and experiments[J]. International Journal of Hydrogen Energy, 2022, 47(55): 23335–23347.
73. He Q, Joy D C, Keffer D J. Impact of oxidation on nanoparticle adhesion to carbon substrates[J]. Rsc Advances, 2013, 3(36): 15792–15804.
74. Zhao Z, Hossain M D, Xu C, et al. Tailoring a three-phase microenvironment for high-performance oxygen reduction reaction in proton exchange membrane fuel cells[J]. Matter, 2020, 3(5): 1774–1790.
75. Nouri-Khorasani A, Malek K, Malek A, et al. Molecular modeling of the proton density distribution in a water-filled slab-like nanopore bounded by Pt oxide and ionomer[J]. Catalysis Today, 2016, 262: 133–140.
76. Fernández-Alvarez V M, Malek K, Eikerling M H, et al. Molecular Dynamics Study of Reaction Conditions at Active Catalyst-Ionomer Interfaces in Polymer Electrolyte Fuel Cells[J]. Journal of The Electrochemical Society, 2022, 169(2): 024506.
77. Liu C, Wu Y H, Sun H, et al. Oxygen Reduction Reaction of PEMFC Cathode by Molecular Simulations[C]//Advanced Materials Research. Trans Tech Publications Ltd, 2010, 105: 698–700.
78. Oh K S, Kim D H, Park S. Behaviour of water molecules in Nafion 117 for polymer electrolyte membrane fuel cell by molecular dynamics simulation [J]. Molecular Simulation, 2008, 34(10–15): 1237–44.
79. Arntsen C, Savage J, Tse Y L S, et al. Simulation of Proton Transport in Proton Exchange Membranes with Reactive Molecular Dynamics [J]. Fuel Cells, 2016, 16(6): 695–703.
80. Rao Z, Zheng C, Geng F. Proton conduction of fuel cell polymer membranes: Molecular dynamics simulation [J]. Computational Materials Science, 2018, 142: 122–8.
81. Zheng C, Geng F, Rao Z. Proton mobility and thermal conductivities of fuel cell polymer membranes: Molecular dynamics simulation [J]. Computational Materials Science, 2017, 132: 55–61.
82. Brandell D, Karo J, Liivat A, et al. Molecular dynamics studies of the Nafion®, Dow® and Aciplex® fuel-cell polymer membrane systems [J]. Journal of Molecular Modeling, 2007, 13(10): 1039–46.
83. Sun H. COMPASS: An ab Initio Force-Field Optimized for Condensed-Phase ApplicationsOverview with Details on Alkane and Benzene Compounds [J]. The Journal of Physical Chemistry B, 1998, 102(38): 7338–64.
84. Wang Y, Ren H, Li C. Enhancing proton-exchange membrane fuel-cell heat transfer performance with embedded cooling channel design: A systematic numerical study[J]. Journal of Energy Engineering, 2024, 150(1): 04023052.
85. Dong E, Cao S, Zhang R, et al. A one-dimensional numerical model of carbon corrosion in catalyst layers of proton exchange membrane fuel cells[J]. Electrochimica Acta, 2024, 483: 144030.
86. Pourrahmani H, Hosseini M, Moussaoui H, et al. Quantitative measurement and comparison of breakthroughs inside the gas diffusion layer using lattice Boltzmann method and computed tomography scan[J]. Scientific Reports, 2024, 14(1): 9621..
87. Wang Y, Qin S, Liao X, et al. Lattice Boltzmann study of the effect of catalyst layer structure on oxygen reduction reaction within a PEMFC[J]. International Journal of Hydrogen Energy, 2024, 52: 1105–1114.
88. Jiang Z, Liao J, Yang G, et al. Simulation of droplet removal processes in proton exchange membrane fuel cell flow channel using lattice Boltzmann method[C]//International Conference on Optoelectronic Materials and Devices (ICOMD 2023). SPIE, 2024, 13066: 1306602.
89. Zhang X, Xia Z, Li H, et al. Factors influencing the stability of Pt-based ORR electrocatalysts in HT-PEMFCs: A theoretical investigation[J]. Computational Materials Science, 2024, 245: 113309.

90. Shin D Y, Lim D H. DFT approach for predicting the pH-potential-dependent durabilities of Pt-skinned Pt-M (M= Ni, Co, and Ir) alloys for fuel cell cathodes[J]. Applied Surface Science, 2023, 616: 156449.
91. Ren J T, Chen L, Wang H Y, et al. Hydrogen oxidation electrocatalysts for anion-exchange membrane fuel cells: activity descriptors, stability regulation, and perspectives[J]. Energy & Environmental Science, 2024.
92. Ogada J J, Ipadeola A K, Mwonga P V, et al. CeO_2 modulates the electronic states of a palladium onion-like carbon interface into a highly active and durable electrocatalyst for hydrogen oxidation in anion-exchange-membrane fuel cells[J]. ACS catalysis, 2022, 12(12): 7014–7029.
93. Gao W, Yin Q, Chen J, et al. Mechanism study of the improved catalytic activity of PEMFC catalyst layer by short-side-chain ionomer: Focusing on the ionomer/Pt interface[J]. Chemical Engineering Journal, 2024, 479: 147787.
94. Jiménez-García J C, Flores D F F R, Acosta R H, et al. Water behavior at PEMFC triple phase boundary: Exploring ionomer and catalytic layer effects via molecular dynamic simulations and NMR experiments[J]. International Journal of Hydrogen Energy, 2024, 52: 65–71.
95. Kim D, Lim J, Lee J H, et al. Investigation of Effect of Platinum Nanoparticle Shape on Oxygen Transport in PEMFC Catalyst Layer Using Molecular Dynamics Simulation[J]. ACS omega, 2023, 8(35): 31801–31810.
96. Li Y, Chu J, Su T, et al. Molecular dynamics study on impacts of different mixed metal ion contents on transport performance of proton exchange membrane for PEMFCs[J]. International Journal of Hydrogen Energy, 2024, 55: 347–356.

Chapter 8
Proton Exchange Membrane Fuel Cells: Degradation Mechanism

8.1 Introduction

Proton exchange membrane fuel cells (PEMFCs) have a series of advantages such as high efficiency, fast response speed, and zero pollution, which is considered one of the most promising power sources for future transportation, backup power systems, and stationary thermal power generation [1]. However, besides cost, degradation is another key issue for the commercialization of PEMFCs. Service life is an important parameter to measure the performance of PEMFCs for automobiles. The U.S. Department of Energy (DOE) has set a target life of PEMFCs of 5000 h by 2020 for passenger cars, 25,000 h for public transport, and 40,000 h for stationary applications. Although the definitions of PEMFC durability vary significantly across studies, it should be noted that the DOE life target for a fuel cell is defined as the time until its rated power drops by more than 10% according to the DOE durability test protocol [2].

To develop high-performing and long-lifetime PEMFCs, there is an urgent need to deeply understand the degradation mechanisms of each component and the corresponding mitigation strategies. Research on the irreversible degradation of Membrane Electrode Assembly (MEA) has mainly focused on perfluoro sulfonic acid (PFSA) membrane, platinum catalyst, carbon supports, and diffusion layer. In recent decades, numerous researchers have made great efforts on the factors affecting the degradation of Pt-based membrane electrodes. Among them, there are two main influencing factors, working conditions and operating conditions. According to harsh working conditions in vehicles, it can be divided into four categories: load changing, start-stop, high power, and idling [3]. Figure 8.1 shows the proportion of PEMFC degradation caused by different operating conditions. It can be seen that start-stop and load changes are more likely to cause degradation of key fuel cell materials than idling and high-power conditions. Moreover, operating conditions, such as operating temperature, humidity, PH, and cell potential, also have a major impact on PEMFC

Y. Li, *Fuel Cell Fundamentals and Applications*,
https://doi.org/10.1007/978-981-96-9845-5_8

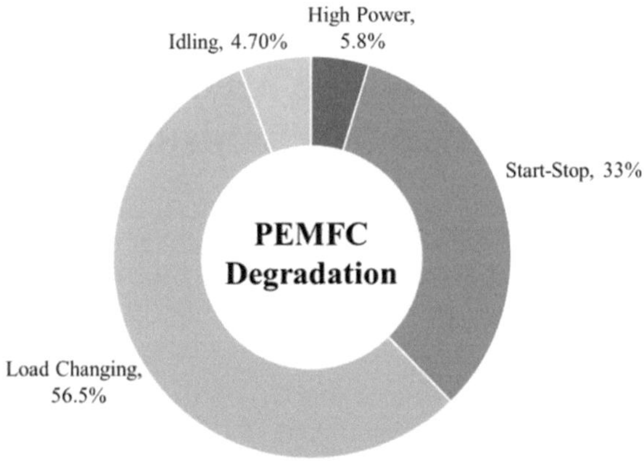

Fig. 8.1 PEMFC degradation caused by different operating conditions [3]

durability. Numerous studies have shown that higher temperature, low pH, and cell potential can lead to more severe overall performance degradation of the fuel cell.

The degradation mechanisms of fuel cells are highly intricate, with the rate and severity of degradation being strongly influenced by varying working and operating conditions. These dynamic and interdependent processes make it particularly challenging to predict and mitigate fuel cell degradation effectively. Despite decades of continuous research and technological progress in the development of core materials, cell assembly methods, and system-level optimizations, achieving the long service life required for automotive fuel cells remains a significant challenge. This issue has become a critical bottleneck in the large-scale industrialization of fuel cell vehicles. Moreover, while extensive studies have been conducted to identify the fundamental causes of degradation, there remains a notable lack of practical and universally applicable control strategies to alleviate these processes during real-world operation.

To address these challenges, it is essential to comprehensively and systematically understand the performance degradation mechanisms affecting fuel cells. Therefore, this chapter analyzes and summarizes the critical degradation mechanisms in major cell components, and the corresponding mitigation strategies and key materials development are described. It is expected to provide guidance and reference significance for understanding and improving the durability of fuel cells.

8.2 Membrane Degradation

8.2.1 Mechanism of Mechanical Degradation

Mechanical decay of the membrane means that the structure of the membrane is damaged due to the stress on the membrane, which is the main reason for early failure of the PEMs (< 1000 h) [4]. The structural damage of membranes includes delamination, tears, pinholes, wrinkles, and punctures. As shown in Fig. 8.2. There are four main sources of stress on the membrane: vertical stress from incoming gas and water vapor, stress caused by cell assembly, and stress due to membrane deformation caused by varying hygrothermal conditions. The stress generated by varying hygrothermal conditions is mainly responsible for the mechanical decay of Nafion PEMs, where the effect of humidity changes is more prominent.

Mechanical degradation occurs throughout the life of the membrane, the life path of a PEM is displayed in Fig. 8.3. It can be divided into three processes: the manufacturing process of the membrane, the assembly process of the fuel cell, and the operational process of the fuel cell stack.

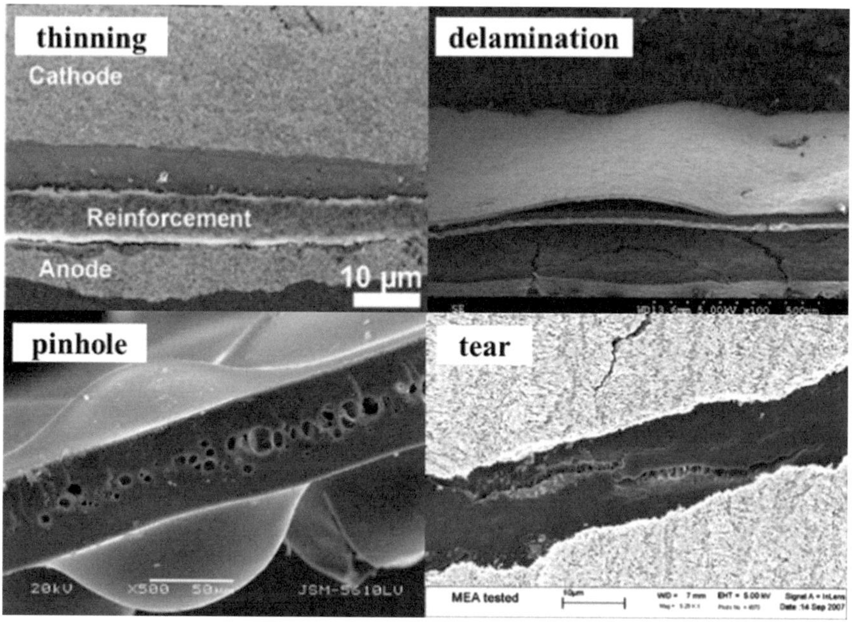

Fig. 8.2 Various mechanical failures of the membrane [4–7]

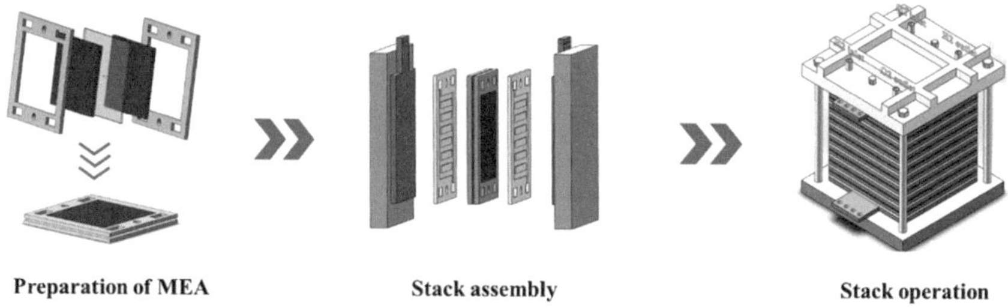

Fig. 8.3 The life path of a PEM [8]

(a) The first part refers to the mechanical decay that occurs during the manufacturing process of the membrane. As the core component of a fuel cell, the membrane plays a significant role that requires high-end qualities concerning mechanical properties. However, defects are inevitable in mass production, such as micropores, uneven thickness, and cracks, which can cause non-uniform membrane stress, the generation of local hot spots, and gas transport across the membrane. These phenomena adversely affect the durability of the membrane and the performance of the fuel cell.
(b) The second part refers to the mechanical decay that occurs during the fuel cell assembly process. Due to the presence of seals and ribs in the flow field plate, etc., the membrane is susceptible to non-uniform distribution of mechanical stress and local concentrated stress, which can cause thickness changes and tears in the membrane. Liu et al. [9] analyzed the mechanical failure mechanism of PEM in several regions and found that stress concentrations appeared on the membrane at the edge of the seal ring when the manufacturing error was greater than 0.05 mm.
(c) The third part refers to the mechanical decay that occurs during the fuel cell stack operational process. During fuel cell operation, fuel gases with a certain temperature and humidity are supplied. The constrained membrane is exposed to the pressure fluctuation of reactant gases and varying hygrothermal conditions, especially during startup or shutdown, resulting in sharp tensile and compressive stresses in the membrane due to the swelling and shrinking [10]. In addition, the freeze–thaw of a fuel cell may cause a rapid membrane failure during the short-term operation phase. During long-term operation, the membrane is subjected to varying hygrothermal conditions, causing the membrane to swell and shrink continuously, which induces fluctuating mechanical stress in the membrane. With the accumulation of time, the mechanical failure of the membrane is eventually caused by the initiation and propagation of micro-pinhole or crack, especially within the defective area [11]. Additionally, it can lead to the separation of the membrane and the catalysis layer.

It is important to note that not all changes in temperature and humidity will cause mechanical degradation. When the changes in temperature and humidity are small and the stresses generated in the membrane are less than a certain value, little mechanical degradation of the membrane will occur. Take the Nafion NR111 membrane as an example. When cyclic stresses equal to or less than 1.5 MPa, there is almost no dimensional change in the membrane. When the cyclic stress is greater than 3 MPa, significant dimensional changes occur in the membrane. When the membrane was cycled 1000 times at stresses from 0 to 4 MPa with a single cycle time of 2 min, the membrane thickness thinned, and the dimensions were elongated. When the stress increases to 6.5 MPa, the structure of the Nafion membrane undergoes significant damage, and many cracks appear on the surface. As shown in Fig. 8.4 [4]. Therefore, the Nafion NR111 membrane undergoes little mechanical degradation at cyclic stresses less than 1.5 MPa. In addition, Fig. 8.5 shows the stresses generated by the membrane under different operating conditions [4]. We can see that the cyclic changes in humidity and temperature can produce stresses less than 1.5 MPa, at which point the membrane does not undergo mechanical decay.

We can see that there are many factors affecting the mechanical decay of the membrane, both external and internal factors. The study of these influencing factors can serve well to further our understanding of mechanical degradation and suggest appropriate mitigation measures for it. Figure 8.6 summarizes the main factors influencing membrane mechanical failure in terms of three procedures.

After mechanical decay, the membranes will develop pinholes, cracks, delamination, and thinning, which can lead to degradation of fuel cell performance. In

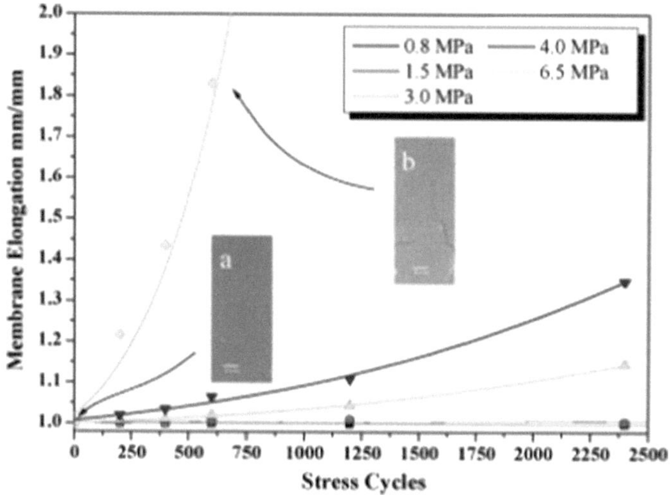

Fig. 8.4 Elongation of the membrane in the stress direction under different stress cycles. The inset pictures are the surface SEM micrographs of **a** as-received Nafion 211 membrane, **b** the Nafion 211 membrane after 540 cycles under stress switched between 0 and 6.5 MPa [4]

Fig. 8.5 Stress generated in Nafion NR111 membranes under different temperature and humidity cycles [4]

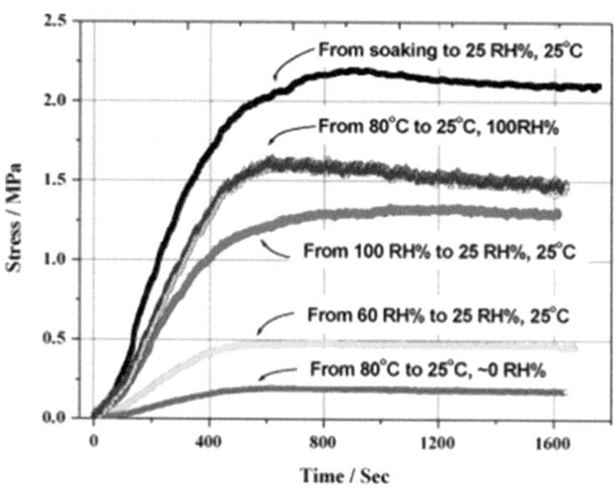

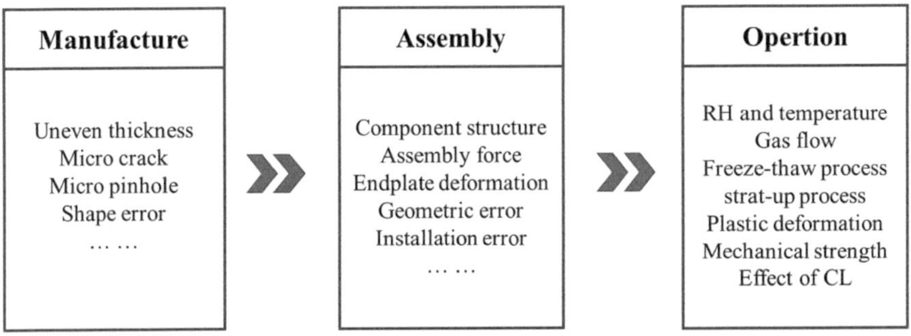

Fig. 8.6 The main factors that influence mechanical degradation during the lifespan

addition, the mechanical decay of the membrane can also exacerbate other forms of cell decay, such as chemical decay of the membrane and decay of the catalysis layer.

Mechanical failures of the membrane, such as those caused by tears or punctures, provide a pathway for hydrogen and oxygen mixing and can trigger combustion reactions and high heat spots, which can lead to transient performance degradation and failure of the fuel cell. In addition, the mixing of fuels causes additional loss of fuel, which reduces the efficiency of the fuel cell. On the other hand, the occurrence of mechanical decay exacerbates the chemical decay of the membrane while intensifying the transport of gases across the membrane, which is detailed in Sect. 8.2.2.

Membrane shrinkage and swelling over a long period of time, especially after undergoing multiple freeze–thaw processes [5], will result in delamination of the membrane and catalysis layers. The delamination of the membrane and catalysis layer will significantly reduce the proton conductivity between the two, and the ohmic resistance increases with it. This eventually leads to a significant decrease in cell performance. In addition, this phenomenon can also exacerbate the decay of the catalysis layer, such as agglomeration and migration of the catalyst.

In particular, the recent membrane of the fuel cell is designed to be particularly thin to reduce the ohmic resistance and improve proton conductivity [12]. Therefore, the mechanical integrity and durability of membranes have become a key challenge. Although mechanical degradation of membranes is inevitable, an in-depth understanding of the failure mechanisms and optimization of mitigation strategies by distinguishing the failure modes and the effects can improve the durability of the material and reduce the degradation rate [13].

8.2.2 Mitigation Strategies of Mechanical Degradation

We already know that the reasons for membrane mechanical decay can be mainly divided into three categories: (1) congenital defects in the membrane; (2) uneven distribution of mechanical stress and local concentrated stress during assembly process; (3) fluctuating mechanical stress in the membrane due to varying hygrothermal conditions during fuel cell operation, and operating conditions such as freeze–thaw, start-up and shut-down. Therefore, the mitigation strategy of membrane mechanical decay can start from the above three aspects.

In order to improve the properties of the membrane itself, some methods to change its material and structure are used. It can be divided into three main categories: reinforcement layer, nano-filling material, and membrane thickness.

(a) Reinforcement layer

The mechanical strength of the membrane can be significantly improved by adding a reinforcement layer. The main part of the reinforcement layer is porous strengthening material, such as expanded polytetrafluoroethylene (ePTFE) microporous membrane, polyvinylidene fluoride electrospinning microporous membrane, and polyvinyl alcohol microporous membrane. The reinforcement layer is located in the middle of the ionomer membrane. To ensure the continuous transmission of protons and water across the membrane thickness, the microporous reinforcement layer is filled with ionomer. With a PTFE reinforcing membrane, the mechanical degradation can be effectively mitigated. For example, a e-PTFE-reinforced membrane was fabricated with Gore Fuel Cell Technologies, exhibiting a lifetime an order of magnitude longer than a non-reinforced membrane of comparable thickness [14]. Systematic investigation of the mechanical durability of ePTFE/PFSA composite PEMs found that the composite PEMs are more durable than the homogenous PEM, which is mostly due to the water absorption rate, resulting in a small shrinkage stress. The NF-SPI-NF membranes were prepared by immersing sulfonated poly (amidic acid) (SPAA) membranes into a solution containing Nafion, followed by thermal imidization by solvent evaporation [15]. The analysis results showed that the stability of NF-SPI-NF membranes was significantly improved by adding a Nafion layer compared with native SPI membranes.

(b) Nano-filling material

In addition to adding a reinforcement layer, new substances can be added to the membrane to prepare reinforced membranes to mitigate mechanical attenuation. Some nanofillers are used to fabricate new reinforced membranes, such as carbon nanotubes, nanofibers, inorganic particles, clays, etc. [16]. Both the natural properties of the filler and the physico-chemical interaction of the filler with the Nafion matrix can affect the performance of the reinforced film. The Nafion composite membrane prepared by adding sulfonated graphene oxide (SGO) and cerium oxide showed improved mechanical properties, proton conductivity, and chemical stability [17]. The addition of SGO increased the number of sulfonic acid groups in Nafion, which enhanced the proton conductivity and tensile strength of the membrane. The addition of cerium oxide can effectively scavenge free radicals, thus enhancing the chemical stability of the membrane. Nafion-reinforced membrane prepared from titanium carbide (TiC) stabilized cerium oxide as filler has improved mechanical integrity, electrochemical durability, and hydrogen impermeability [16]. The presence of TiC enhanced the tensile strength of PEM, and cerium oxide improved the chemical stability of the membranes by scavenging free radicals.

(c) Membrane thickness

Greater thickness is effective in mitigating the mechanical decay of the membrane, but it reduces the proton conductivity. A thicker membrane has better mechanical strength, reducing the occurrence of pinholes and tears, which in turn reduces gases across the membrane. The decay of cells with four different membrane thicknesses (N117, N115, NR212, and NR211) was studied over 1000 h, and the study showed that the performance of cells with thinner membranes declined more rapidly, especially after 800 h of operation [18].

In addition to improving the performance of the membrane itself, the design of the components in the fuel cell also plays an important role in slowing down the mechanical decay of the membrane.

(a) Endplate and Bipolar plate

The endplate and bipolar plate are important structural support components in the fuel cell. Avoiding their deformation can reduce the problem of uneven stress distribution and local stress concentration caused by external mechanical stress, thereby slowing down the mechanical attenuation of the membrane. Fundamentally, the optimal design of endplates can be divided into two forms: (1) improving the flexural stiffness of endplates [19]; (2) changing the loading methods of end plates, for example, by applying forces in bands [20, 21]. For bipolar plates, using the hydroforming process instead of a stamping process is an effective way to reduce manufacturing errors [22].

(b) Flow field plate

To reduce the effects of uneven stress distribution and local concentrated stress due to dynamic hygrothermal conditions during the fuel cell operational process, the MEA and flow field structure must be carefully designed to avoid local drying of the membrane [23, 24], especially at the reactant inlet area [23]. Reducing the channel width and increasing the fillet size of the flow channel can help promote uniformity of mechanical stress in the membrane, although this may increase ohmic losses in the fuel cell [25].

(c) GDL

The 4D in-situ visualization method was used to study the effect of the roughness of the gas diffusion layer on the mechanical degradation of the membrane [26], and the results showed that using a diffusion layer with low surface roughness can suppress the buckling deformation of MEA, which has a significant mitigating effect on the mechanical decay of the membrane.

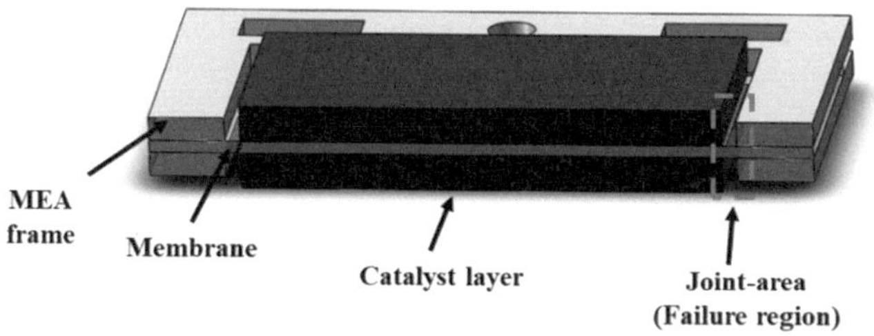

Fig. 8.7 Schematic drawing of the joint-area membrane in the PEMFC

(d) MEA

The structural design of MEA plays an important role in improving the mechanical durability of the membrane due to the strong mechanical stresses at the joints of the frame and membrane. The joints of the frame and membrane are shown in Fig. 8.7. A two-dimensional finite element model was employed to analyze the mechanical stresses in the membrane generated during the cell assembly process for different MEA frame materials, frame structures, and contact behaviors [27]. The results showed that there were regions of strong inhomogeneous stresses in the membrane below the end edges of the frame/membrane. Good initial contact of the membrane with the CL also contributes to mitigating the mechanical decay of the membrane. The contact between the membrane and the electrode can be effectively improved by utilizing the hot pressing process on the CCM. Although it may make the electrodes less porous, this method has been shown to improve cell performance and durability [28].

8.2.3 *Mechanism of Thermal Degradation*

To ensure adequate PFSA membrane hydration, the optimal operating temperature of a proton exchange membrane fuel cell is typically 60–80 °C. Thermal degradation is the change in performance of a proton exchange membrane when it is subjected to high temperatures, sub-freezing temperatures, or temperature cycling [29]. The three conditions are discussed separately below.

When serious gas permeation, fuel starvation [30], or air starvation [31] occurs in the PEMFC, a local high temperature will appear inside the PEMFC, causing thermal degradation of the proton exchange membrane. PFSA membrane will tear severely at high temperatures due to the glass transition temperature of PFSA polymers being around 80 °C [13]. Many studies have shown that the side chains with sulfonic acid groups will decompose when the temperature exceeds 280 °C. The thermal decay of PFSA resin was investigated using thermogravimetric and differential thermal analysis, and found that the sulfonate groups in PFSA resin started to fall off at temperatures above 280 °C [32]. The thermogravimetric analysis was performed to study the thermal decay of MEA and found that the decomposition of Nafion in the catalysis layer started at about 320 °C, which was lower than the degradation temperature of PFSA proton exchange membrane, due to the activation of Pt particles [33]. The thermal decay mechanism of H^+ and K^+ proton exchange membranes was investigated, and found that the H^+ proton exchange membrane lost 16% of its mass after heating at 375 °C for 2 h [34]. The analysis showed that SO_2, CO_2, CO, and SiF_4 gases were generated. At high temperature, the C–S bond in PFSA is broken first, and SO_2 and HO· radicals are formed, leaving carbon radicals to be degraded at higher temperature, and the thermal decay mechanism of H^+ type proton exchange membrane was developed. As shown in Fig. 8.8, a similar phenomenon was reported where the strength of the S–O related bond decreased significantly at 300 °C [35]. Small amounts of sulphur dioxide were measured up to 400 °C [36].

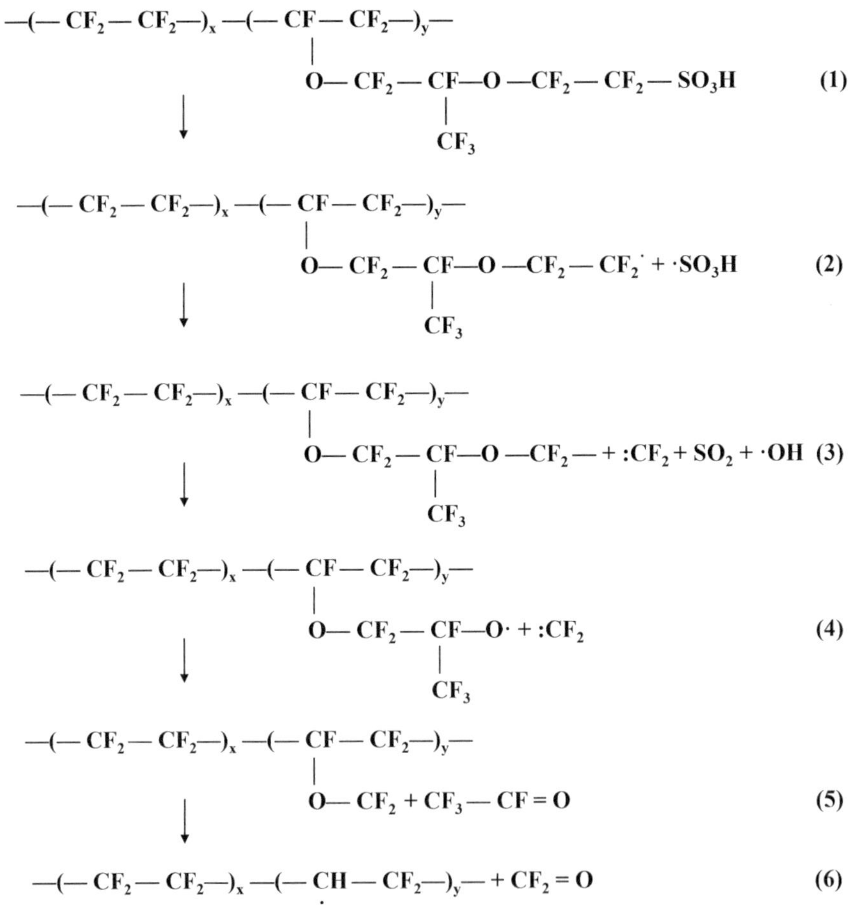

Fig. 8.8 Mechanism of thermal degradation under high temperature

Fuel cells are inevitably subjected to sub-zero temperatures and temperature cycling in practical applications, especially in transportation. Several studies have been done on these.

There are three different states of water in the membrane, and those frozen at sub-zero temperatures are "free water" that is not tightly bound to the polymer chains [37]. Similarly, studies found that only a part of the water in Nafion underwent freezing [38, 39]. Freezing conditions can create additional mechanical stresses that can change the shape of the membrane, which in turn can lead to pinholes, cracks, and delamination. In addition, the occurrence of freezing can lead to an increase in contact resistance between the cell components. The decay of the membrane due to freezing can also be categorized as mechanical attenuation.

The membrane will undergo a significant degradation of its relevant properties after thermal cycling, especially when freeze–thaw conditions occur. For example, the contact resistance between the membrane and the electrodes increased after the membrane underwent thermal cycling [40]. Studies [41] found that the proton conductivity, gas impermeability, and mechanical strength of Nafion membranes were severely damaged after 385 freeze–thaw cycles (− 40 to 80 °C).

8.2.4 Mitigation Strategies of Thermal Degradation

Mitigation strategies for thermal decay are divided into two categories according to the temperature at which they occur: mitigation strategies at high temperatures (> 100 °C) and mitigation strategies at low temperatures (< 0 °C).

(a) Mitigation strategies at high temperatures (> 100 °C)

The main reason for the thermal decay of the membrane under normal operation of the PEMFC (power stack) is the development of micro-pores or cracks in the membrane, which leads to severe gas crossover and results in the formation of a high-temperature hot spot locally. Phenomena such as micro-pores and cracks in the membrane are caused by the combined effect of chemical and mechanical decay occurring in membranes. Therefore, the mitigation measures for chemical and mechanical decay of membranes are also measures for thermal decay, as described in detail in Sects. 8.2.2 and 8.2.6.

In addition, the preparation of novel high-temperature resistant proton exchange membranes can also be effective in mitigating the thermal decay of the membrane. Many studies have been devoted to improving the durability of membranes in environments above 100 °C. Currently, high-temperature resistant proton exchange membranes are divided into three main categories [13]: (1) modified PFSA membranes, which are swelled with nonvolatile solvents or incorporate hydrophilic oxides and solid inorganic proton conductors; (2) alternative sulfonated polymers and their composite membranes, such as SPSF, SPEEK, PBI, and PVDF; and (3) acid–base polymer membranes, such as phosphoric acid-doped Nafion®–PBI composite membranes.

(b) Mitigation strategies at low temperatures (< 0 °C)

In essence, mitigation strategies fall into two categories: (i) providing heat and reducing heat loss to keep the fuel cell warm. Fuel cell stacks can be kept warm by adding insulation or by running the cell intermittently in a low-power mode. During startup, heat can be provided to the cell by catalytic combustion of hydrogen or preheated reaction gases. (ii) Keeping the fuel cell dry or replacing water with a non-freezing liquid to prevent icing during non-operating conditions. After the cell system stops operating, the fuel cell is dried by dry reaction gases (H_2, air) or dry nitrogen. All these strategies are aimed at avoiding/minimizing icing, which may lead to loss of fuel cell performance.

8.2.5 *Mechanism of Chemical Degradation*

Chemical degradation is the chain decomposition of the main and side chains of the membrane under the attack of free radicals such as ·OH and ·OOH, resulting in structural damage and performance reduction of the membrane. The whole process of chemical decay can be divided into four steps: (1) transport of gases across the membrane; (2) generation of hydrogen peroxide; (3) generation of free radicals in the presence of metal cations such as Fe^{2+}; (4) unchaining of the main and side chains. The chemical decay itself has a positive feedback effect, as the chemical decay proceeds, the decay rate will keep increasing. These steps are a coupled process, as shown in Fig. 8.9. In addition, the mechanical decay of the membranes also promotes the chemical decay.

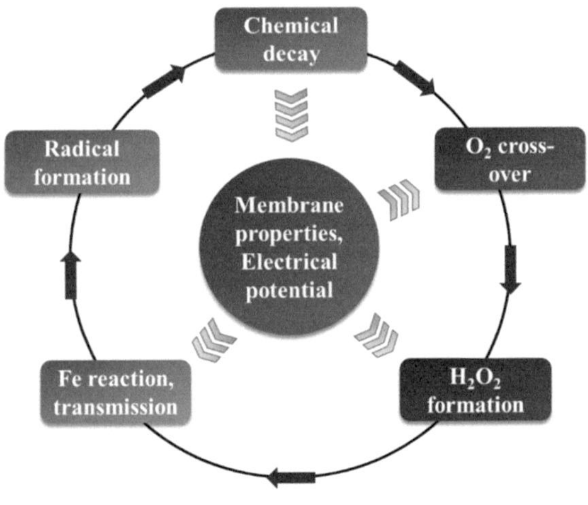

Fig. 8.9 Mechanisms represented in the chemical membrane degradation

(a) Gases across the membrane

Theoretically, the reaction gases of the fuel cell can not pass through the proton exchange membrane, while in practice, both hydrogen and oxygen can pass through the membrane to the other side of the electrode, i.e., gas permeation phenomenon. The gas crossover is a function of pressure, temperature, and membrane hydration [42]. The hydrated membrane resembles a two-phase substance, and the gas passes through the membrane via transport in the gas phase and transport in the liquid phase [43]. The permeability of oxygen in the gas and liquid phases, respectively, is [43]

$$\psi_v^{O_2} = \left(1.9 \times 10^{-11}\phi_{H_2O} + 1.1 \times 10^{-12}\right)\exp\left[\frac{22000}{R}\left(\frac{1}{T_{ref}} - \frac{1}{T}\right)\right] \times 10^{-3} \tag{8.1}$$

and

$$\psi_l^{O_2} = 1.2 \times 10^{-14}\exp\left[\frac{20000}{R}\left(\frac{1}{T_{ref}} - \frac{1}{T}\right)\right] \tag{8.2}$$

where ϕ_{H_2O} denotes the volume fraction of water in the membrane.

Hydrogen cross measurements were performed under open-circuit conditions [44] and found that gas crossover increases with increasing cell temperature, humidity, and hydrogen pressure. In particular, the effect of gas pressure is most significant.

(b) Generation of hydrogen peroxide

Most current studies suggest that hydrogen peroxide is produced as a result of a two-electron oxygen reduction reaction (ORR) [45, 46], as shown in Eq. 8.3. Oxygen crosses the membrane from the cathode to the anode and reacts with the hydrogen adsorbed on the Pt catalyst. The reaction kinetics of H_2O_2 generation depends on the catalyst loading [47], the weight ratio of catalyst to carbon support [47], and the nature of the catalyst [48].

$$O_2 + 2H^+ + 2e^- \rightarrow H_2O_2 \quad E^\circ = 0.68\ \mathrm{V} \tag{8.3}$$

Hydrogen peroxide is mainly generated at the anode, while its production at the cathode is negligible because of [49], (1) the absence of faradaic current, and (2) the high cathode potential of > 0.9 V. The destination of hydrogen peroxide can be roughly divided into three kinds: (1) diffusion into the membrane [42]; (2) diffusion with water into the gas diffusion layer and eventually out of the fuel cell [44]; (3) decomposition into oxygen and water on the Pt catalyst surface [50]. The rate of hydrogen peroxide production was studied using a rotating disk electrode and showed that the rate of H_2O_2 formation was linearly correlated with oxygen concentration and squared with water activity [45].

(c) Generation of free radicals

The generation of free radicals is similar to the process that occurs in Fenton's reagent (H_2O_2/Fe^{2+}) [51]. During the generation of free radicals, transition metal ions (Fe^{2+}, Co^{2+}, and Cu^{2+}) are involved in a series of reactions. The free radical generation reactions are shown in Table 8.1. The source of the transition metal ions may be introduced during the fabrication of the electrodes and stacks [42, 52] or may be released from the metal bipolar plates after use [53, 54]. This is also the reason why stacks with metal bipolar plates are more prone to decay than those with graphite bipolar plates.

Ghelichi et al. [66], with $C_{F_e^{2+}}$ ranging from 1 to 100 ppm and $C_{H_2O_2} = 1$ M, summarized the relative contribution of each basic reaction to the formation of HO·, HOO·, Fe^{2+}, and Fe^{3+}, which is shown in Fig. 8.10.

In addition to acting through trace transition metal ions, the reaction of hydrogen peroxide decomposition into free radicals can also be decomposed in the presence of heat, which is shown in the Eq. 8.4.

$$H_2O_2 \rightarrow 2HO\cdot \text{ (by heat)} \tag{8.4}$$

(d) Polymer structure

The molecular structures of different types of proton exchange membranes differ, mainly in the lengths of the main and side chains, while the types of functional

Table 8.1 The free radical generation reactions

Nr.	Reaction	References
1	$Fe^{2+} + H_2O_2 + H^+ \rightarrow Fe^{3+} + HO\cdot\cdot + H_2O$	[55]
2	$Fe^{3+} + H_2O_2 + H^+ \rightarrow Fe^{2+} + HOO\cdot\cdot + H^+$	[56]
3	$Fe^{2+} + HOO\cdot + H^+ \rightarrow Fe^{3+} + H_2O_2$	[57, 58]
4	$Fe^{3+} + HOO\cdot \rightarrow Fe^{2+} + O_2 + H^+$	[59]
5	$Fe^{2+} + HO\cdot + H^+ \rightarrow Fe^{3+} + H_2O$	[60]
6	$HO\cdot + H_2O_2 \rightarrow HOO\cdot + H_2O$	[61, 62]
7	$HOO\cdot + H_2O_2 \rightarrow HO\cdot + H_2O + O_2$	[63, 64]
8	$2HOO\cdot \rightarrow H_2O_2 + O_2$	[57, 65]
9	$HO\cdot + H_2 \rightarrow H\cdot + H_2O$	[61]
10	$H\cdot + O_2 \rightarrow HOO\cdot$	[61]

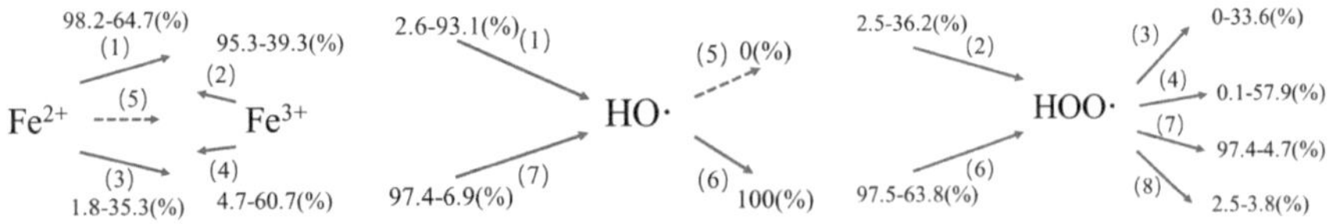

Fig. 8.10 Contribution of each reaction to the formation and consumption of the involved species in a system of Fenton's reagents [66]

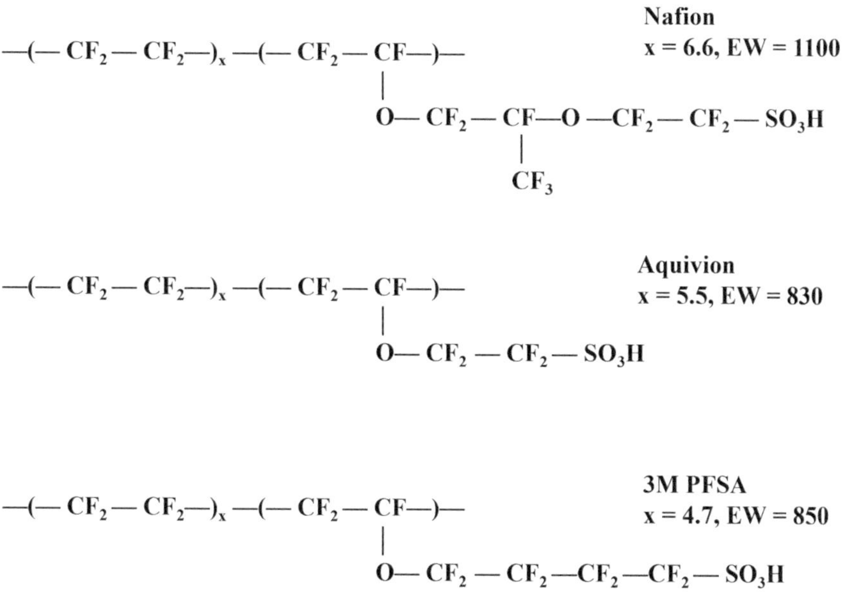

Fig. 8.11 Chemical structure of Nafion, Aquivion, and 3 M PFSA ionomers

groups contained in them are the same. Figure 8.11 depicts the chemical structures of Nafion, Aquivion, and 3 M PFSA. As we can see from Fig. 8.11, all three membrane structures have sulfonic acid groups at the end of the side chains, which serve to transfer protons.

(e) Radical attack

The chemical decay of PFSA mainly refers to the decomposition of the structure of the membrane when it is attacked by free radicals, which eventually causes membrane thinning, surface roughness, cracks, and pinholes etc. These decays are expressed in the membrane properties as a decrease in proton conductivity, an increase in gas crossover rate, and a decrease in mechanical strength. From Fig. 8.11, we can find that the basic molecular structure of a proton exchange membrane consists of a main chain and a side chain, where the main chain determines the mechanical properties of the membrane and the side chain determines the proton conductivity through the terminal sulfonic acid group [67]. The main mechanisms of unchaining in chemical decay are introduced using the widely used Nafion membrane as an example, as shown in Fig. 8.12. The HO·· plays a very important role in the chemical decay of the membrane. Gubler et al. [56] investigated the attack of free radicals on membranes using a kinetic framework based on the hydrogen peroxide reaction and found that the ranking of the oxidation strength of each species was HO· > H· > HOO· > H_2O_2. Figure 8.12b illustrates that the breakdown of the backbone is mainly due to the attack of free radicals on the carboxyl groups at the ends [68, 69], leading to the continuous unchaining and shortening of the backbone. The process occurs with the release of HF, CO_2, and H_2O. The carboxyl group at the end of the main chain is mainly introduced during the manufacturing process [70]. The attack of HO· on the membrane was investigated by ^{19}F magic angle spinning nuclear magnetic resonance and found that the main chain of the Nafion 211 membrane with negligible carboxyl group content was almost intact [67]. However, the side chain unlinking is attacked by free radicals at weak bonds, including α-OCF_2, β-OCF_2, and C-S bonds. These bonds make key contributions to membrane functionality in terms of proton conductivity [70]. As Fig. 8.12e introduces the mechanism of unchaining of C-S bonds by HO· [69], which eventually leads to the shedding of sulfonic acid groups. This is the direct cause of the decrease in membrane conductivity. The C–O bond on the side chain is also fragile, and the C–O bond on the long side chain can be divided into α-OCF_2 and β-OCF_2. There is a sensitivity difference between α-OCF_2 and β-OCF_2, as the former is attacked before the latter [71]. The mechanism of their unlinking is presented in Fig. 8.12c and d. Experiment confirmed that β-OCF_2 damage lagged behind α-OCF_2 and that β-OCF_2 was attacked less frequently [67]. The carbon at the junction of the side chain and the main chain is called the tertiary carbon, which is mainly attacked by H· radical [56]. Figure 8.12a presents the mechanism of unchaining by H· attack at the tertiary carbon. Hydrofluoric acid is produced during all of the chemical decay processes described above, which makes fluorine emissions a reliable indicator of membrane stability.

(f) Positive Feedback

The chemical decay of the membrane has a positive feedback effect. As the chemical decay proceeds, the membrane appears to thin, tear, and form a pinhole. These problems lead to an increase in the rate of gas crossing the membrane, which is the source of the chemical decay of the membrane. Thus, the end of chemical decay will in turn contribute to the source of chemical decay, causing the chemical decay of the membrane to increase dramatically in the later stages of cell operation. Inaba et al. [44]. measured the hydrogen transmembrane rate, as shown in Fig. 8.13, indirectly demonstrating the positive feedback of chemical decay.

It has been shown that chemical decay of the membrane only takes place in the presence of hydrogen, oxygen, and a catalyst at the same time [72]. The decay rate of the membrane is greater in a hydrogen-rich atmosphere than in an oxygen-rich environment, as shown in Fig. 8.14.

8.2.6 Mitigation Strategies of Chemical Degradation

Based on the mechanism of membrane chemical decay, the mitigation of membrane chemical decay can be considered from the following three aspects: (1) reducing

(a) tertiary carbon : $-CF_2-CF-(-OCF_2-R_f-)-CF_2- + \cdot H \rightarrow -CF_2-\dot{C}-(-OCF_2-Rf-)-CF_2- + HF$

(b) carboxyl groups : $R_f-CF_2COOH + \cdot OH \rightarrow R_f-CF_2\cdot + CO_2 + H_2O$

$R_f-CF_2\cdot + \cdot OH \rightarrow R_f-CF_2OH \rightarrow R_f-COF + HF$

$R_f-COF + H_2O \rightarrow R_f-COOH + HF$

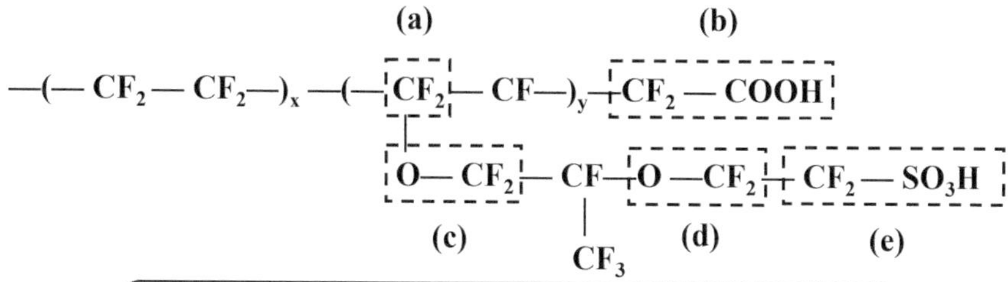

(c) β-OCF_2: $R_{f2}-O-CF_2CF(CF_3)O\cdot + 3\cdot OH \rightarrow R_{f2}-O\cdot + CF_3\cdot + 3HF + 2CO_2$

$R_{f2}-O\cdot + 2H_2O + \cdot OH \rightarrow 2-(-CF_2-)_x-COOH + 3HF$

(d) α-OCF_2: $R_{f1}-O-CF_2CF_2SO_3H + \cdot OH \rightarrow R_{f1}-O\cdot + HOCF_2CF_2SO_3H$

(e) C-S bonds: $R_f-CF_2SO_3H + \cdot OH \rightarrow R_f-CF_2\cdot + SO_3 + H_2O$

$R_f-CF_2\cdot + \cdot OH \rightarrow R_f-CF_2OH \rightarrow R_f-COF + HF$

$R_f-COF + H_2O \rightarrow R_f-COOH + HF$

Fig. 8.12 Molecular structure of a PFSA membrane with long side chains. Chemical degradation mechanisms of the PFSA membrane via free radical attack at **a** the tertiary carbon branch point between the main chain and the side chain, **b** the carboxyl group at the terminal of the main chain, **c**, **d** α-OCF2 and β-OCF2 within the side chain, and **e** the C-S bond at the terminal of a side chain

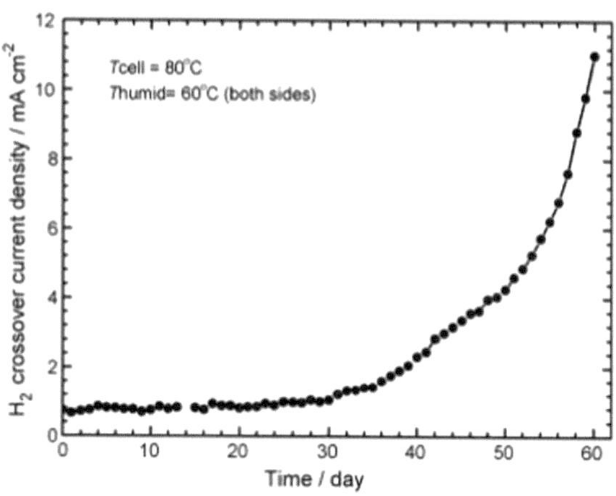

Fig. 8.13 Variation of H_2 crossover current density during OCV durability test at 80 °C [44]

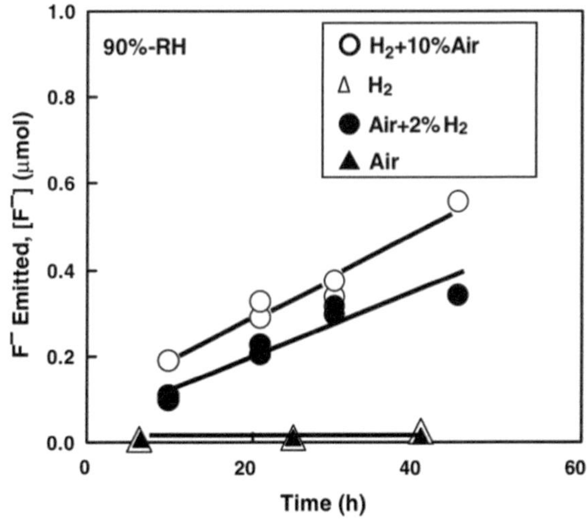

Fig. 8.14 Release of fluoride ions under different gas atmospheres [72]

the rate of gas crossover, (2) managing hydrogen peroxide and free radicals, and (3) reducing the attack sites of free radicals.

(a) Reducing the rate of gas crossover

Add a reinforcement layer in the middle of the membrane, which is a reinforced support layer based on microporous material and filled with ionomer. The transfer of substances in the reinforcement layer is similar to the transfer in porous media. Compared to normal Nafion membranes, the effective diffusion coefficient of substances in the reinforcement layer is reduced, which effectively reduces the rate of gas crossover and slows down the chemical decay of the membrane. In addition, increasing the thickness of the membrane can also effectively reduce the transport of gases across the membrane. However, it is required to note that the increase in membrane thickness is accompanied by a corresponding increase in proton transport resistance. Therefore, the thickness of the membrane should not be increased uncontrollably.

(b) Management of hydrogen peroxide and free radicals

The attack of free radicals on membranes is the direct cause of the chemical decay of membranes. Limiting the generation of free radicals in membranes or scavenging them is an effective measure to mitigate membrane decay. Composite membranes can effectively achieve these goals, which are prepared by adding fillers to the membranes. The specific filler can decompose hydrogen peroxide or scavenge free radicals from the membrane. It was shown that the degradation rate of the membrane was reduced by an order of magnitude after adding oxides such as TiO_2, CeO_2, MnO_2, and ZrO_2 to the membrane [73, 74]. Hydrogen peroxide decomposes on the oxide surface, generating water and oxygen, thus inhibiting the generation of free radicals. Meanwhile, some metal cations also have a scavenging effect on free radicals. Currently, cerium compounds are the most widely used as fillers. Take CeO_2 as an example to introduce the relevant reactions that occur, as shown in Eqs. 5.5–5.7. The membrane with CeO_2 scavenger has been commercialized, such as Nafion XL [75]. In addition to the metal oxide, the filler can be organic, such as a heteropoly acid, which has been proven to moderate or eliminate membrane deterioration due to peroxide. However, the advantage of this approach would be partially counteracted by a decrease in membrane stability and conductivity caused by the mixture of the catalysts. There is also a class of fillers which does not convert free radicals but consume them as a sacrifice, thus reducing the attack of free radicals on the membrane. Such as a radiation-grafted FEP-g-polystyrene membrane, in which polystyrene was used as a sacrificial material owing to its low resistance to radicals. Free-radical stabilizers and inhibitors, such as hindered amines or antioxidants, also have the potential to be mingled during membrane fabrication.

In addition to suppressing the amount of free radicals produced, a portion of the research mitigates the chemical decay of membranes by reducing the diffusion coefficient of free radicals, which is mainly achieved by cross-linking. Three-dimensional cross-linked networks can limit the movement of polymer molecular chains, restrict the diffusion of free radicals in PEM, keep its structure in a dry state, mitigate the stress effects caused by membrane expansion/contraction during moisture cycling, and improve the mechanical and chemical stability of the membrane [76]. For example, a composite membrane was fabricated [77], in which a thin recast Nafion membrane was bonded with a polystyrene sulfonic acid (PSSA) membrane, when positioned at the cathode of the cell, could successfully prevent oxidation degradation of the PSSA membrane.

$$H_2O_2 + 2CeO_2 \rightarrow Ce_2O_3 + O_2 + H_2O \tag{8.5}$$

$$HO \cdot + Ce^{3+} + H^+ \rightarrow Ce^{4+} + H_2O \tag{8.6}$$

$$Ce^{4+} + HOO\cdot \rightarrow Ce^{3+} + O_2 + H^+ \tag{8.7}$$

(c) Reducing the attack sites of free radicals

Increasing chemical stability can also be achieved by modifying the structure of existing membranes. Chemical decomposition of membranes is a chain reaction, and reducing the sites in membranes that are vulnerable to free radical attack can effectively mitigate chemical decay. The majority of studies suggest that free radical attack on the residual H-containing terminal bonds in the backbone of the PFSA membrane is the main degradation mechanism [78, 79]. The peroxide radical attacks the hydrogen atoms at the carboxylic acid end, triggering a chain oxidation reaction of the system, which decomposes into carbon dioxide and hydrogen fluoride. By eliminating the unstable end group, chemical stability was significantly enhanced.

Last but not least, the development and implementation of new metal coatings with improved corrosion resistance and of catalysts that produce less hydrogen peroxide are long-term goals for membrane durability enhancement.

8.2.7 Correlation Between the Three Degradations of Membrane

Mechanical decay, thermal decay, and chemical decay have a mutually reinforcing effect on each other, as shown in Fig. 8.15. Mechanical decay and thermal decay lead to an increase in the gas crossover rate, which exacerbates chemical decay. Chemical decay and thermal decay reduce the mechanical strength of the film and increase the mechanical decay. Chemical decay and mechanical decay can intensify the internal combustion of the fuel, creating localized high-temperature hot spots, which can increase thermal decay. Likewise, while the mitigation strategies corresponding to each of the three decay types are a direct mitigation of their degradation, they also mitigate the other two types of decay to some degree. Therefore, the three types of membrane degradation cannot be treated in complete isolation.

Chemical and mechanical degradation have partially similar effects on membranes, both of which damage the membrane structure. However, due to the different sources of decay between them, chemical decay is mainly due to the attack of free radicals on the membrane, and mechanical decay is mainly due to the stresses caused by temperature and humidity changes. The structural damage caused by the two is different. Take the example of the crack produced. In the case of pure mechanical decay, the crack shape is "I" type. When mechanical and chemical decay are combined, "Y" shaped cracks will be formed. As shown in Fig. 8.16 [80]. However, when purely chemical decay acts, the change in the membrane structure is mainly a thinning of the thickness, and almost no cracks are generated on the surface.

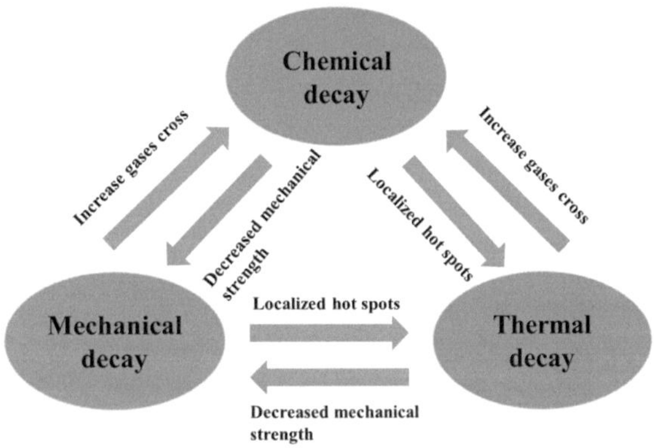

Fig. 8.15 The mutual reinforcement between the three degradations of the membrane

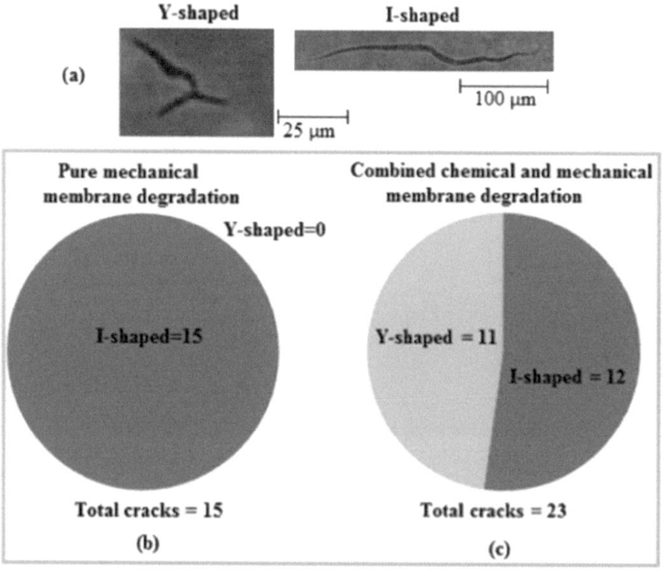

Fig. 8.16 **a** Planar membrane view of the main shapes of membrane cracks formed during chemical and mechanical degradation. Distribution of the number of type I and type Y through-thickness membrane cracks at the end of life after **b** pure mechanical membrane degradation and **c** combined chemical degradation [80]

8.3 Catalyst Degradation

As mentioned above, the catalytic layer is the most important component of the fuel cell, and the degradation of its performance plays a crucial role in the durability of the entire fuel cell. Its losses are mainly due to Pt catalyst degradation and carbon support corrosion, which are continuous, irreversible processes that can significantly shorten fuel cell life [81]. Different degradation modes at one single Pt/Vulcan aggregate under simulated start–stop conditions can be observed in Fig. 8.17 [82]. It should be pointed out that under normal working conditions, due to the H_2 oxidation reaction in the anode catalytic layer, the relative potential of the catalyst is zero, and the electrochemical dissolution process of Pt will not occur. At the same time, the oxidation corrosion process of the carbon support does not easily occur in the normal H_2 atmosphere, so the main research focus in this section is on the degradation mechanism of the cathode catalytic layer of fuel cells.

8.3.1 Platinum Degradation Mechanisms

As the catalyst surface is where the electrochemical reactions occur, changes in its ORR catalytic activity have the most direct impact on fuel cell performance, thus

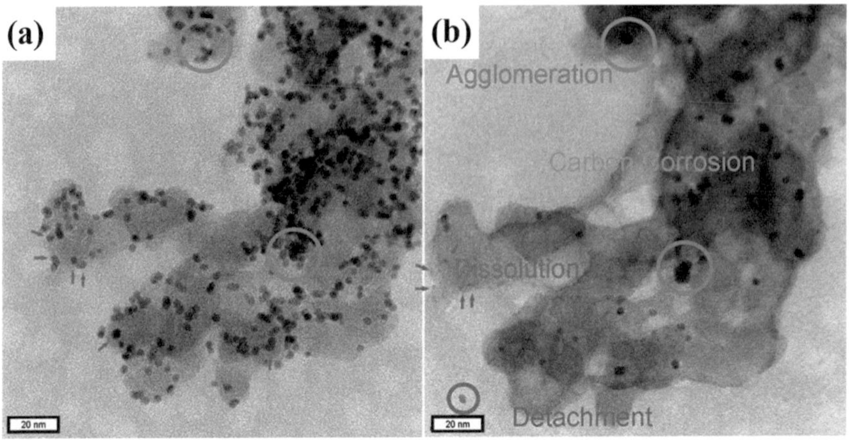

Fig. 8.17 TEM images of different degradation modes at one single Pt/Vulcan aggregate under simulated start-stop conditions [82]

making catalyst decay the most widespread concern. Typical operating conditions for fuel cells: low pH, high temperature, and potential cycling, which are generally believed to promote platinum dissolution [83]. Various Pt-based catalyst decay mechanisms have been proposed, but they can be summarized as follows [13, 84]: (1) Ostwald ripening and Pt particle agglomeration, (2) Pt loss and detachment, (3) Mechanism of catalyst active site poisoning caused by contaminants. All the above-mentioned mechanisms lead to the loss of electrochemical catalytic active sites in the catalytic layer and the loss of conductive electron paths in contact with the carbon support, which in turn leads to a significant loss of ECSA and ultimately reduces the output power of the entire PEMFC system.

8.3.1.1 Ostwald Ripening and Pt Particles Agglomeration

The Ostwald ripening process of Pt particles and the agglomeration of nanoparticles on the carbon support surface have been confirmed to be the most dominant mechanisms leading to the growth of Pt-based catalysts in PEMFC during loading cycling [85]. The schematic diagram of the process is shown in Fig. 8.18a and b. Pt Ostwald ripening is the dissolution and reprecipitation process of Pt, and the main driving force comes from the difference in electrochemical potential of different particle sizes. Indeed, according to the Gibbs–Thomson energy equation, the chemical potential depends particularly on the size of the Pt particles: the smaller the particles, the bigger the chemical potential, so smaller particles are more soluble than larger particles due to their larger chemical potential. The electrochemical Ostwald ripening process involves both the diffusion of Pt ions and the conduction of electrons: led by the difference in the chemical potential of two close Pt particles, the dissolved Pt ions from the smaller particles diffuse to the surface of the larger one through the ionomer, while the electrons from the smaller particles reach the larger one through the conductive carbon support. Then reprecipitation occurs on the surface of large particles to reach a more stable state [86]. Therefore, the strength of Ostwald ripening is determined by the particle size distribution, the conductivity of the carbon support, and the ionic conductivity of the ionomer in the catalytic layer. Eventually, two adjacent particles of different sizes generate a voltage. Existing studies have experimentally demonstrated that this voltage decreases with fuel cell operating time, indicating an increase in the average particle size of the catalyst. For the dissolution process of Pt, Meyers and Darling [87] et al. pointed out that there are two main paths, Pt direct electrochemical dissolution or chemical dissolution of PtO, and the relevant chemical reactions are as follows:

Pt direct electrochemical dissolution:

$$Pt \rightleftharpoons Pt^{2+} + 2e^- \ E_0 = 1.188\ \text{V versus SHE} \tag{8.8}$$

Pt oxide formation:

$$Pt + H_2O - 2e^- \rightleftharpoons PtO + 2H^+ \ E_0 = 0.98\ \text{V versus SHE} \tag{8.9}$$

Pt oxidation chemical dissolution:

$$PtO + 2H^+ \rightleftharpoons Pt^{2+} + H_2O \tag{8.10}$$

It is worth noting that the product after platinum oxidation is not only PtO, but also a small amount of PtO_2 [88] and $Pt(OH)_2$ [87]. Studies have shown that the dominant effects of the above two pathways on the dissolution of Pt are determined according to the cathode operating potential: when the potential is less than 1 V, the direct electrochemical dissolution of Pt is dominated, while when the potential is greater than 1.1 V, the chemical Dissolution dominates [89]. This is because at low potentials, the coverage of platinum oxide is smaller, resulting in a much higher rate of direct dissolution than chemical dissolution. However, as the potential increases, the surface of the particles is gradually covered by platinum oxide, and the coverage can even exceed 100% at 1.1 V [90]. At this time, platinum is completely covered and protected by platinum oxide without direct dissolution, so the chemical dissolution of platinum oxide takes the leading role.

To improve the utilization of Pt, the average particle size of pure Pt catalyst particles in PEMFC usually needs to be controlled at 2–5 nm, so that it exhibits the unique size-dependent properties of nanostructured elements [91]. This poses a problem of insufficient stability: nano-sized particles often have very large surface energies, making them prone to agglomeration to reduce high surface energies by moving on the surface of the carbon support [92]. And the smaller the particle, the larger its surface energy, that is, the easier it is to agglomerate. Therefore, during operation, as the Pt nanoparticles continuously migrate, agglomerate, merge, and then become larger on the surface of the carbon support, the ECSA of the catalyst begins to

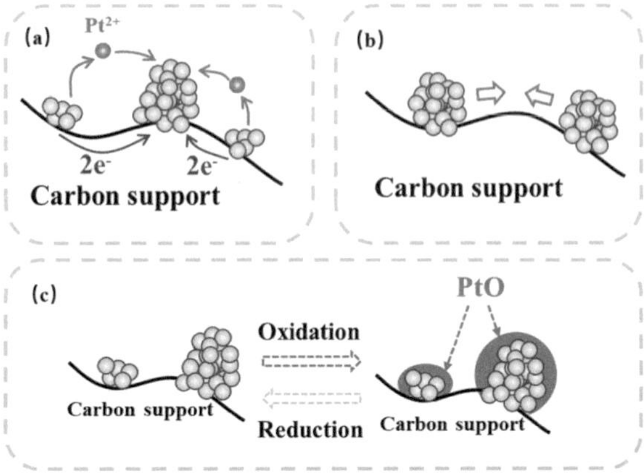

Fig. 8.18 Schematic diagram of Pt catalyst degradation: **a** Ostwald ripening process of Pt particles with different particle sizes, **b** agglomeration of nanoparticles on the carbon support surface, **c** Pt oxidation and reduction

gradually decrease, resulting in a decrease in the catalyst activity. The agglomeration mechanism is exacerbated when liquid water is present, as it promotes particle-to-particle collisions. In addition, the open-circuit idling condition also promotes the agglomeration of Pt particles: under this condition, more hydrogen permeates to the cathode through transmembrane transport to react with oxygen directly, resulting in hot spots aggravating the agglomeration and sintering of Pt particles [93].

The agglomeration mechanism has been investigated as the main reason for the growth of Pt particles at low potentials, in which Ostwald ripening is considered negligible [94]. However, although a great deal of research has been done on the Pt particle growth mechanism, it is still difficult to distinguish the contribution of these two mechanisms to the particle growth during the experiment. The sintering and agglomeration process can be divided into three stages, and the role of the two mechanisms in each stage was analyzed [85]. The results showed that Ostwald ripening was the main reason for the loss of catalyst activity in the first stage, where particles are very small, as shown in Fig. 8.19. Obvious Pt particle growth was observed after a 400 h accelerated stress test held at 1.2 V [95], due to the agglomeration and sintering of Pt particles. Figure 8.20 shows the comparison of STEM images of Pt/C before and after the test.

8.3.1.2 Platinum Detachment and Migration

During operation, the mass loss of Pt is another source of catalyst decay. In a 200-h stack experiment, the Pt/C content in the catalytic layer was measured [96]. The results showed that the Pt content decreased from 20 to 13.5% after the test, indicating that there is a serious loss of Pt mass during battery operation. The mass loss of Pt mainly originates from two processes as shown in Fig. 8.21: (1) Pt^{2+} diffuses in the electrode and then is reductively deposited by hydrogen permeation in the ionomer or the membrane which is called Pt migration, (2) due to corrosion of the carbon support catalyst particles detach from the carbon surface and migrate into nearby ionomers.

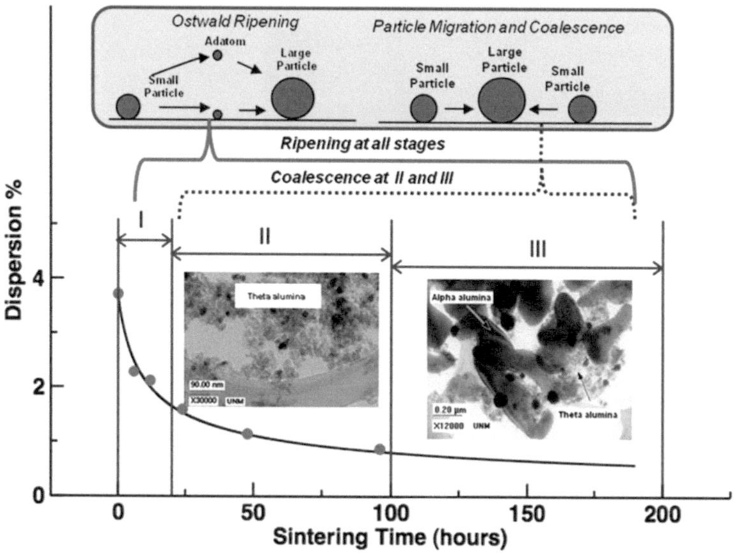

Fig. 8.19 Distribution curves of Ostwald ripening and Pt migration with sintering time [85]

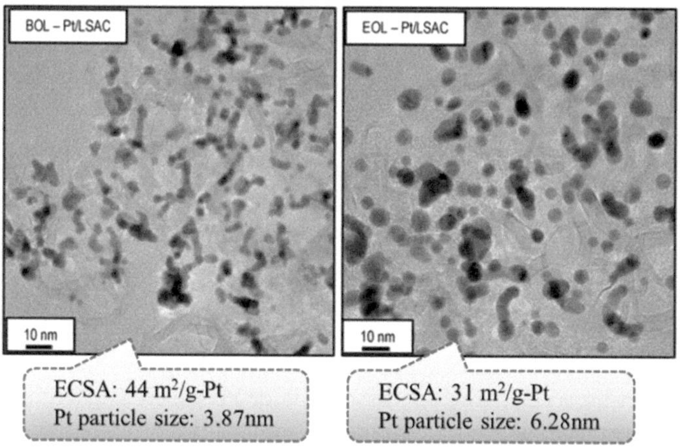

Fig. 8.20 The STEM images of Pt/C before and after the test of 400 h AST hold at 1.2 V [95]

Both cases mean that the Pt catalyst is separated from the carbon support and will lose its catalytic activity as it does not have a path to the conducting electrons. The latter is prominent mainly in the case of fuel cell start-stop cycles and fuel starvation leading to corrosion of the carbon support, which will be explained in detail in the next section. The former Pt migration mainly plays a major role during normal operation, and we will discuss it in detail next.

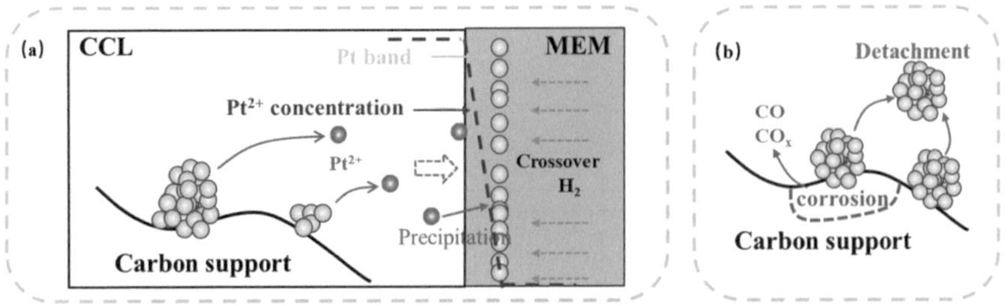

Fig. 8.21 Schematic diagram of mass loss of Pt: **a** Pt migration results in platinum band formation in membranes, **b** Pt particles detach from the carbon surface due to corrosion of the carbon support

The most discussed result of platinum migration is the formation of Pt bands in the membrane [97]. The platinum dissolution mainly occurs at the cathode at high potential, resulting in a large platinum ion concentration gradient between the cathode and the anode, which drives the platinum ions from the cathode to diffuse to the anode through the ionomer. There is also the driving effect of electric potential gradient and even water flow convection, but its influence is negligible. The process of platinum band formation can be divided into two steps: the initial nucleation stage and particle growth stage [98]. First of all, free platinum ions are reduced in ionomers or membranes to form platinum atoms due to the presence of hydrogen, which has crossed over from the anode to the cathode, represented through the reactions:

$$Pt^{2+} + H_2 \rightleftharpoons Pt + 2H^+ \tag{8.11}$$

Then, Pt atoms agglomerate and nucleate to form platinum crystal nuclei, driven by the minimized surface energy, and subsequently, through the reduction of hydrogen, Pt ions begin to continuously precipitate on the surface of the nuclei, resulting in the growth of nuclei to form platinum particles in the membrane. As this process continues to accumulate, so-called Pt bands are formed, indicating that the Pt ions are mainly precipitated near the fixed location in the membrane. It is widely believed that both open-circuit/idling conditions and load-cycling conditions promote the migration of Pt by accelerating the dissolution of Pt, which leads to higher Pt ion solubility in the electrode [99]. Pt migration dominates the catalyst decay near the cathode-membrane interface in the cathode thickness direction [100], as shown in Fig. 8.22, because the closer to the membrane side means the easier loss of Pt ions. In addition, it can also be seen from the figure that Pt migration and Ostwald ripening account for a similar contribution to the loss of Pt active area.

The position of the platinum band is directly related to the oxygen partial pressure in the cathode and the hydrogen partial pressure in the anode. Utilizing experimental studies of the position of the Pt band under different cathode and anode atmospheres, the results showed that the Pt band moved toward the cathode as the oxygen partial pressure decreased and the hydrogen partial pressure increased. Under N_2/N_2 conditions, in the absence of hydrogen, Pt ions would pass through the membrane to

Fig. 8.22 Pt surface area degradation distribution in the cathode thickness direction after potential cycling, resulting from Ostwald ripening (the upper gray region) and Pt migration (the white region) [100]

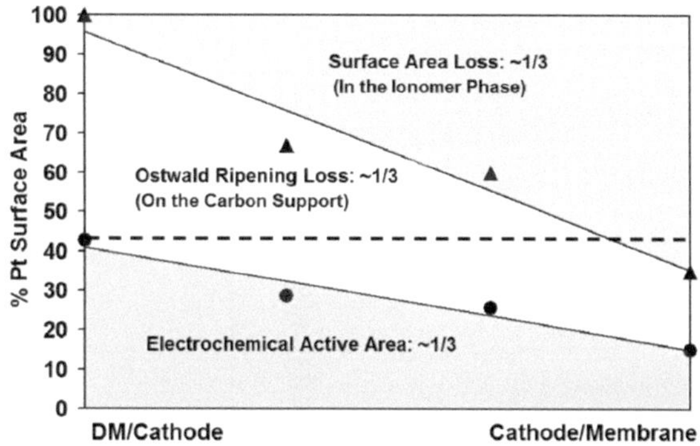

deposit on the anode catalyst; under H_2/N_2 conditions, in the absence of oxygen, Pt ions is deposited on the interface between the cathode catalytic layer and the proton exchange membrane in Fig. 8.23a [101]; while under normal hydrogen–oxygen operating conditions, platinum band can be observed in the membrane near the cathode in the H_2/AIR case in Fig. 8.23b [102], in the center of the membrane in the H_2/O_2 case in Fig. 8.23c [102].

A simple 1D model was proposed to describe Pt precipitation and hydrogen–oxygen permeation in the membrane [103], and obtained the relationship between Pt band position and hydrogen–oxygen partial pressure, which is given by:

$$\delta_{\mathrm{b}} = \frac{2c^0_{\mathrm{O_2}} D_{\mathrm{O_2}}}{c^0_{\mathrm{H_2}} D_{\mathrm{H_2}} + 2c^0_{\mathrm{O_2}} D_{\mathrm{O_2}}} \delta_{\mathrm{m}} = \frac{2D_{\mathrm{O_2}}}{D_{\mathrm{H_2}}\left(\frac{P^0_{\mathrm{H_2}} H_{\mathrm{O_2}}}{P^0_{\mathrm{O_2}} H_{\mathrm{H_2}}}\right) + 2D_{\mathrm{O_2}}} \delta_{\mathrm{m}} \quad (8.12)$$

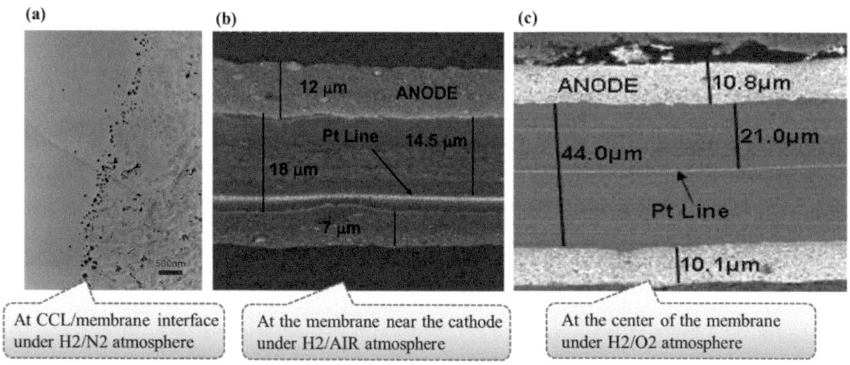

Fig. 8.23 TEM image of the Pt band in the proton exchange membrane after degradation: **a** CCL/membrane interface under H_2/N_2 atmosphere [101], MEA cross-section under H_2/AIR atmosphere **b**, and H_2/O_2 atmosphere **c** [102]

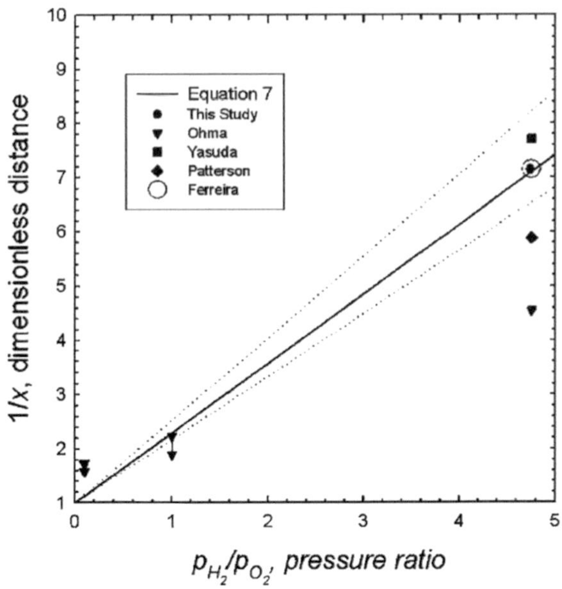

Fig. 8.24 Comparison of predicted Pt band positions with experimental data [103]

where δ_b is the distance from the platinum band to the cathode, δ_m is the thickness of the membrane, $c^0_{O_2}$ and $P^0_{O_2}$ are the oxygen concentration and oxygen partial pressure at the interface of the cathode and membrane respectively, $c^0_{H_2}$, $P^0_{H_2}$ are the hydrogen concentration and hydrogen partial pressure at the interface of the anode and membrane respectively, H_{O_2}, H_{H_2} are the Henry constants of oxygen and hydrogen in the membrane respectively. The predicted Pt band position according to this equation is in good agreement with the experimental data, as shown in Fig. 8.24.

8.3.1.3 Active Sites Contamination

The loss of ORR activity of the catalyst due to the adsorption of contamination to the active sites of the catalyst is another important factor leading to the degradation of the activity of Pt-based catalysts [104]. First, the types of contamination in PEMFC can be divided into two types according to their sources: the first is contamination from air and fuel (such as CO, H_2S, CH_4, NO_2, NO, CO_2, SO_3, SO_2, O_3 and NH_3), the second is from the system itself, such as Fe^{2+} introduced by corrosion of metal bipolar plates and F^- caused by corrosion of sealing gaskets. Moreover, for different contamination species, the poisonous characteristics leading to the performance degradation of membrane electrodes or catalysts are also different according to their physicochemical properties. In addition, for different gas impurities, whether the decay process is reversible will also be affected according to the strength of chemical adsorption between the contamination and Pt, and whether it can be oxidized and desorbed through potential cycling [105].

Among all pollutants, the most discussed and studied pollutant is CO, and its poisoning mechanism is through preferential adsorption on the active sites of Pt catalysts, making H_2 unable to enter these sites for the HOR. Even trace amounts of CO may reduce catalyst activity, resulting in higher polarization losses, ultimately leading to reduced performance of PEMFCs, especially if long-term operation is required [106]. In addition, sulfur-containing contaminants can damage the activity of Pt catalysts, and this effect is irreversible. Experiments demonstrated that catalyst poisoning by H_2S was highly irreversible and cumulative [107]. When 1 ppm H_2S was added to the anode hydrogen fuel, through the 0.5 V potentiostatic experiment for 21 h, a sharp deterioration of the polarization curve representing the cell performance was observed. Degradation of cell performance caused by certain pollutants can be recovered, such as NO_2. The performance of the cell was recovered by reintroducing clean and pure air after poisoning the cell cathode with 5 ppm NO_2 [105]. The changing process of the polarization curve is shown in Fig. 8.25. Based on the IV curves before and after poisoning, they deduced that the poisoning mechanism of NO_2 has nothing to do with the catalyst, but may be related to the ionomer in the catalytic layer. The poisoning of catalysts by Cl^- from the catalyst preparation process or fuel impurities has also been reported in the relevant literature. The degradation of cell performance caused by Cl^- was observed [108], and studies reflected that its poisoning mechanism was to promote the dissolution of Pt at the cathode to generate $[PtCl_4]^{2-}$ and $[PtCl_6]^{2-}$, which were finally reduced by permeating hydrogen in the membrane and precipitated, resulting in Formation of Pt bands.

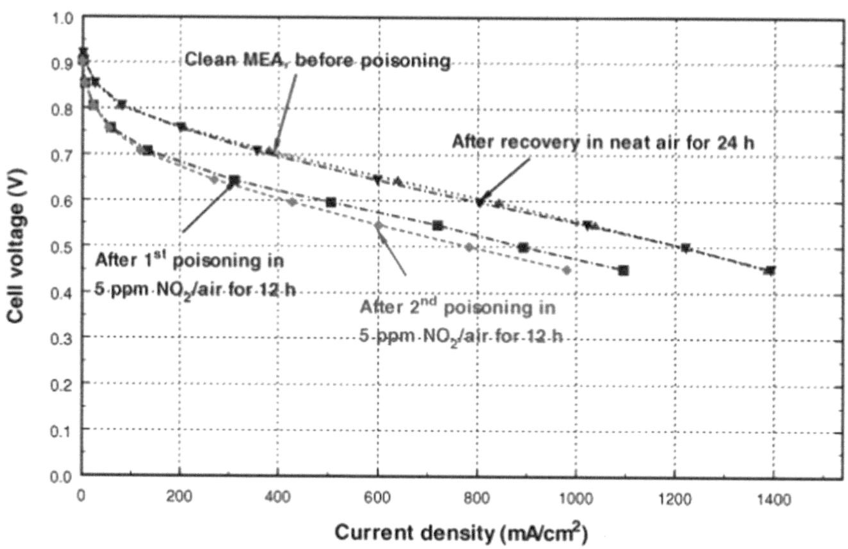

Fig. 8.25 Polarization for steady state performance showing the effects of 5 ppm NO_2/air at different times, including before poisoning, after poisoning, and after recovery [105]

8.3.2 Mitigation Strategies

Due to the existence of the above decay mechanism, it has been widely proved experimentally that the ECSA will first decrease rapidly and then tend to stabilize during the decay process, which is not conducive to the long-term stable and efficient operation of the fuel cell. Therefore, how to alleviate the decay of Pt-based catalysts to improve the durability of PEMFCs has received more extensive attention and research. First, following an appropriate control strategy can effectively reduce the degradation of Pt. For example, by way of controlling the operating parameters such as temperature, humidity, pressure, operating potential, and formulating a reasonable loading strategy to minimize the aging rate of Pt. Reduce the working potential during normal operation, and reduce the residence time of the system under UPL, etc., to balance the contribution of the various decay mechanisms of the Pt-based catalyst to achieve the optimization of catalyst durability. The effect of operating conditions and operating parameters on Pt degradation has been discussed in detail in Sect. 8.1. Secondly, the selection of appropriate catalysts, support materials, and membrane electrode structures has also been reported in a large number of literature studies as effective methods to alleviate Pt decay. Specific measures will be described in detail below.

(1) Preparation of highly durable Pt-based catalysts

The preparation and synthesis of Pt-based catalysts with higher durability can effectively avoid the severe activation loss that results from Pt dissolution, loss, migration, and Ostwald ripening in the catalyst. First, although adding a second noble metal element such as Au, Ag, Ru, Pd and Ir to Pt can not only maintain the high activity characteristics of Pt, but also improve its Stability and resistance to CO poisoning, but is not suitable for practical applications due to its high cost [109]. Therefore, transition metals such as Ni, Co, Fe, and Cu are the most attractive elements for alloying with Pt due to their lower cost and superior ORR activity [110]. However, considering that the presence of Fe and Cu ions will accelerate the degradation of ionomers and membranes, the currently commercialized alloy catalysts are dominated by Pt-Ni and Pt–Co. Corrosion experiments were performed in 0.5 M H_2SO_4 solution to compare the durability of PtCo/C and Pt/C catalysts [111]. It is observed that the amount of Pt ions dissolved in PtCo/C is smaller than that of Pt/C. The durability and stability of Pt–Co alloy catalysts were significantly improved compared to pure Pt catalysts under the condition of potential cycling [112]. In addition to balancing the effects of noble metal and non-noble metal elements on the performance and durability of Pt-based catalysts, more and more researchers have also prepared various Pt-based multicomponent catalysts, such as PbNi@Pt catalysts [113]. Second, the modulation of the structure and surface morphology of Pt-based catalysts also has great potential to improve the catalyst lifetime. Compared with Pt-M alloy catalysts with nanoporous structure, core–shell catalysts generally have higher durability and are often considered to be the most promising catalysts for next-generation commercialization. It should be noted that to improve its durability while maintaining the superior ORR activity of surface Pt, the Pt shell thickness of the catalyst needs to be controlled and optimized. A synthetic electrochemical strategy was proposed to prepare Pt–Cu alloy catalysts with a core–shell structure [114], and the experimental results showed that the dealloyed catalysts had significantly improved durability under potential cycling compared with pure Pt electrocatalysts. Third, a large number of studies have shown that appropriate elemental modification on the surface of Pt-based catalysts can reduce the dissolution of Pt during operation and improve the stability of the catalysts. Gold nanoclusters was used to modify the surface of Pt/C catalyst [115], and results showed that it can greatly improve the stability of Pt without reducing its ORR activity; The use of N element to dope Pt the nanoparticle surface can greatly improve the durability of the Pt catalyst [116], and the ORR activity only loses 3.7% after 20,000 cycles between 0.6 and 1.1 V.

(2) Enhanced Pt–C interaction force

In the fuel cell catalytic layer, Pt nanoparticles are distributed on the surface of the conductive carbon support. This arrangement shows that the degree to which the catalyst particles are stably immobilized on the carbon support depends on the interaction force between the particles and the surface of the carbon support. Their interaction forces range from weak electrostatic attraction to strong connections through surface chemical bonds or the formation of capping layers on supports [117]. Improving the interaction force between the Pt-based catalyst and the carbon support can effectively improve the stability of the catalyst particles, thereby reducing the number of catalyst particles detached from the surface of the carbon support and reducing the agglomeration of the catalyst particles. However, the stability of carbon supports under high potential and oxygen-rich, harsh cathode environments is insufficient. Therefore, the selection and improvement of supports such as Pt-C interaction force, stability, and electrical conductivity have great potential to improve catalyst durability and have been widely studied. The methods are mainly divided into the following two categories: First, using a new type of carbon material as the catalyst carrier. With the continuous production of new carbon materials, researchers have also begun to try to use them as carbon support materials for fuel cells to replace the commonly used carbon black materials. The effect of graphene oxide (GO), partially reduced graphene oxide (GO-r), and fully reduced GO (rGO) carbon materials was compared as supports on the ORR activity and durability of Pt catalysts [118]. It was found that Pt/rGO had the best durability, regardless of activity. Second, Doping modification of carbon support. Carbon material doping has attracted much attention and research as the most effective and efficient method to improve the Pt-C interaction force. Deposition of a small amount of metals or metal oxides such as CeO_2 [119], TiO_2 [120], and Se on the surface of carbon support has been shown to make Pt-based catalysts stronger rivets on the carbon support by strengthening the interaction force between Pt-based catalysts and metal oxides.

(3) Membrane electrode structure optimization

In previous studies, it has been generally found that there has been an inhomogeneous decay in the cathode catalytic layer, such that the catalyst ECSA decay in the membrane side would be more severe than that in the GDL side, which leads to little or almost complete depletion of Pt in the membrane side region after a long period of operation. This is mainly caused by the diffusion of Pt^{2+} in the cathode and then reduced deposition by hydrogen permeation in the proton exchange membrane. As a result, the Pt^{2+} concentration on the membrane side will always be lower than that on the GDL side, so the catalyst on the membrane side is more likely to be dissolved and lost. However, compared with uniform catalyst decay, when the catalyst decay degree is uniform, the non-uniform decay will lead to more severe membrane electrode performance degradation, because the non-uniform decay will not only lead to the loss of catalyst ORR activity, but also comes with very serious O_2 mass transmission loss and ohmic loss. Therefore, it is extremely important to optimize the structure of membrane electrodes by gradient design. A gradient of Pt loading was designed in the thickness direction and in-plane direction of the catalytic layer [121], as shown in Fig. 8.26, and the accurate modeling results show that it helps to improve the performance of the fuel cell and the utilization of the catalyst. Furthermore, it has been shown that increasing the Pt loading, particle size or ionomer content in the catalytic layer on the membrane side can improve the cell performance after decay and increase the durability of the membrane electrode. The particle size gradient and the Pt loading gradient can be controled in the direction of the catalytic layer to maintain a high ECSA after potential cycling for high durability and stability of PEMFCs [122], as shown in Fig. 8.27.

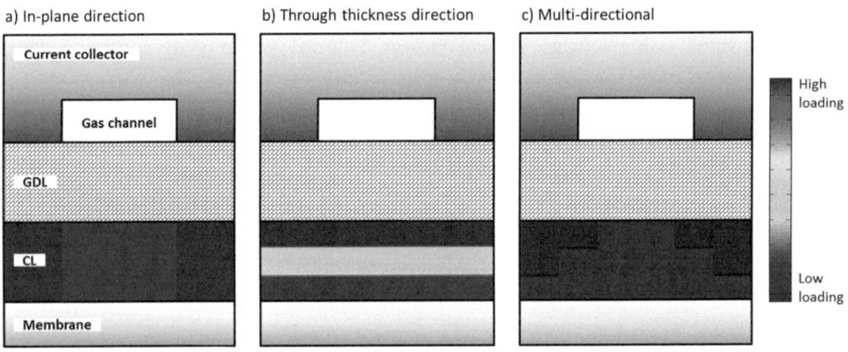

Fig. 8.26 Schematic design of Pt loading gradient in the direction of catalytic layer thickness and in-plane [121] (Color figure online)

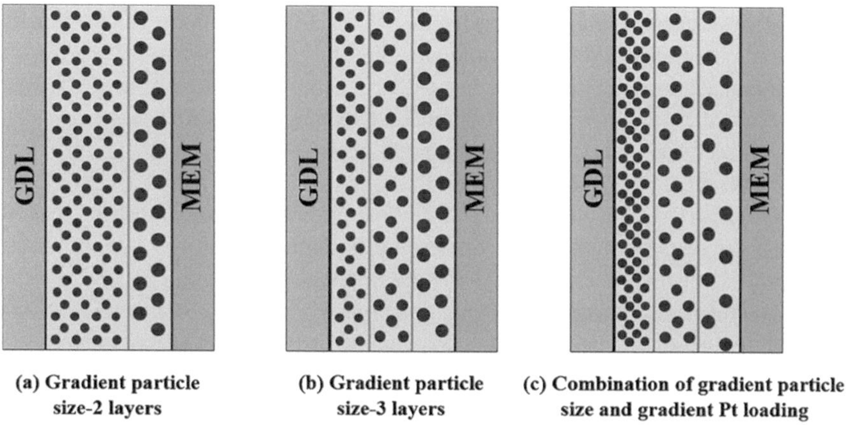

Fig. 8.27 Schematic diagram of gradient catalytic layer structure [122]

8.4 Carbon Support Degradation

8.4.1 Carbon Corrosion Reaction and Theoretical Analysis

For the catalytic layer, the nanoparticles are dispersed uniformly onto the surface of the support, thus increasing the utilization of Pt and the active area. Carbon materials are widely used as catalyst carriers for fuel cells due to their high electrical and thermal conductivity and low cost. However, due to the harsh environment with high temperature, high water content, strong acid, strong oxidation, high potential and the presence of Pt-based catalysts, the carbon cathode support inevitably faces chemical or electrochemical corrosion [106]. The electrochemical oxidation of carbon can be accelerated by increasing the potential. When the electrode interface potential is greater than 0.207 V, the carbon oxidation reaction shown in Eq. 8.13 occurs. Also, when the PEMFC is at a higher voltage, the carbon support will undergo further oxidation as shown in Eq. 8.14 to form CO, and the reaction product CO may poison the catalyst.

$$C + 2H_2O \rightarrow CO_2 + 4H^+ + 4e^-(+0.207V_{RHE}, 25\,^\circ C) \tag{8.13}$$

$$C + H_2O \rightarrow CO + 2H^+ + 2e^-(+0.518V_{RHE}, 25\,^\circ C) \tag{8.14}$$

Indeed, the CO and CO_2 emissions from the anode and cathode has been experimentally monitored [123]. Carbon corrosion rates are affected by several factors, such as temperature, relative humidity, electrical potential, and carbon surface area. It increases with operating temperature and decreases with relative humidity [124]. Of these factors, potential has the greatest impact. The mass loss of carbon on the membrane electrode was measured after the decay test [125] and concluded that the corrosion rate of carbon may increase by about an order of magnitude for every 0.1 V increase in potential when the potential is higher than 1.0 V.

In general, carbon support corrosion may occur in both cathode and anode catalytic layers. From thermodynamic analysis, at the cathode, carbon corrosion may exist even at normal operating potentials (cathode potentials are typically greater than 0.6 V), whereas at the anode there is no significant carbon support degradation [126]. Therefore, during normal steady state operation, carbon corrosion is more severe on the cathode than on the anode. Anode carbon corrosion mainly occurs in the process of local lack of H_2 or insufficient fuel supply and PEMFC startup-shutdown condition, because the anode potential will increase to exceed the COR reversible potential, at this time the corrosion of the cathode carbon support will also be particularly significant, and its potential can be as high as 1.5 V. It is worth mentioning that below 1.0 V, the reaction kinetics are so slow that they are usually ignored [127]. Therefore, it is generally accepted that potentials above 1.2 V lead to severe carbon corrosion. In addition, the interface between the catalyst and the carbon support is relatively more prone to carbon corrosion because the presence of the catalyst reduces its reaction potential [128].

8.4.2 Fuel Starvation and Startup-Shutdown Condition

The carbon support bears the brunt of the relatively high electrode interface potential under fuel starvation and start-stop conditions, which is caused by the reverse current generated at the hydrogen-air interface. This section will describe in detail how these two conditions lead to carbon support corrosion.

(1) Fuel starvation

The distribution of reactants is inherently non-uniform due to the complexity of mass transfer in the fuel cell itself, such as in different locations of the stack, between the inlet and outlet, and between the flow channel region and the rib region of a single cell. Moreover, the unreasonable design of the flow channel and the blockage of the channel by liquid water or ice will also lead to an insufficient local supply of the reaction gas. During the load-up process, the reactant supply lags the current increase [129], which is insufficient to provide the expected current for the PEMFC, which in turn causes the reactant inhomogeneity to be amplified, increasing the risk of fuel starvation and air starvation in the fuel cell. Studies have shown that the effect of cell degradation due to air starvation is much less than that of fuel starvation [130], so we mainly discuss fuel starvation. Also, depending on the degree, it can be divided into local fuel starvation and overall fuel starvation.

When the overall supply of fuel cell H_2 is sufficient, but local fuel starvation occurs due to uneven distribution, the pressure in the local starvation region decreases, resulting in enhanced air permeation through the membrane, as shown in Fig. 8.28a. As a result, an air/H_2 boundary appears in the anode, which generates reverse current

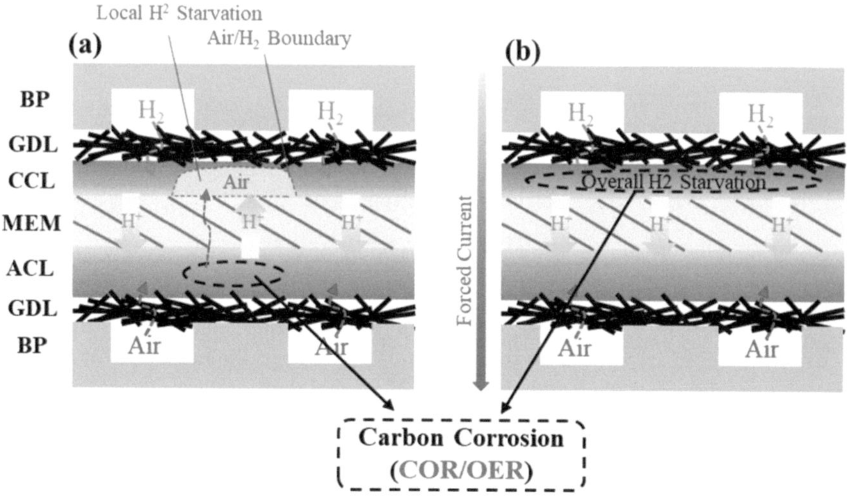

Fig. 8.28 Schematic diagram of carbon corrosion caused by local and global fuel starvation

and high interfacial potential that exacerbate carbon corrosion in the cathode, especially in the cathode local area directly facing the local starvation area. This mechanism will be discussed in detail in Sect. 8.4.3. Kynar impregnation was used to prevent mass transfer of H_2, creating a localized fuel starvation area, and observed that the localized area of the cathode facing the local Kynar impregnation of the anode suffered severe carbon corrosion after 100 h of decay test [131]. Severe Pt decay in the air inlet region was observed, as it corresponds to a locally fuel-starved H_2 outlet [132].

Furthermore, when the fuel supply is severely undersupplied, the hydrogen stoichiometric ratio is less than one, which will lead to overall fuel starvation. At the single-cell level, with no external energy supply, the cell voltage drops to zero, and both electrodes will develop an interfacial potential close to the cathode potential under OCV, resulting in a decay process similar to that of Pt and carbon support under OCV [81]. At the stack level, the fuel cell stack is usually operated in constant current mode, and the individual cells are all connected in series, but the overall fuel starvation cell cannot provide the total amount of current required by the hydrogen–oxygen reaction. Therefore, other normally operating cells will make it operate in electrolytic mode by providing the required energy [133], as shown in Fig. 8.28b. Carbon oxidation reactions (COR) and oxygen precipitation reactions (OER, Eq. 8.15) are initiated in the anode to provide protons and maintain the load current. As a result, the anode potential increases rapidly while the cell voltage decreases and even drives the cell into reverse operation, where the anode potential is higher than the cathode potential. The overall fuel starvation process was experimentally investigated by changing the stoichiometric ratio of hydrogen [134] and found that when the hydrogen stoichiometry was set to 0.55, O_2 and CO_2 accounted for 48.7% and 12% of the anode exhaust gas, respectively.

$$H_2O = \frac{1}{2}O_2 + 2H^+ + 2e^-(+1.229V_{RHE}, 25\,°C) \tag{8.15}$$

From the above analysis, it can be seen that local and global starvation affect durability in completely different ways. It is worth mentioning that they can be mitigated or even eliminated by optimally controlling the gas supply, reactant amount, gas flow field, and loading speed in the load-up process.

(2) Startup-shutdown condition

The startup and shutdown processes are the two most important dynamic processes that are unavoidable for PEMFCs in automotive applications. According to the statistics of urban road conditions in the United States, within the target life of 5500 h, the vehicle starts and stops a total of 38,500 times, with an average of 7 times/h. If each start-stop process is 10 s, the cathode exposure time above 1.2 V can reach 100 h [135]. The formation of the hydrogen-air interface is the main reason for the battery decay caused by the start-stop condition, which causes a very high interfacial potential difference in the cathode region corresponding to the air side of the anode [136], which is called the reverse current mechanism. It will be described in detail in the next section.

Since the air in the outside atmosphere will diffuse through the anode exhaust pipe or the membrane electrode, it will enter the anode flow channel of the battery after the fuel cell engine is stopped. Therefore, there is air in the anode flow field of the cell before the fuel cell engine is started the next time. Then, when starting, the hydrogen/air interface is formed the moment oxygen is introduced into the anode. As the air in the anode flow field is continuously driven from the inlet to the outlet by the continuously introduced hydrogen, a flowing hydrogen-air interface is formed, which moves continuously from the inlet to the outlet with the flow channel. When the fuel cell is turned off, due to the concentration gradient between the cathode and the anode, oxygen from the cathode diffuses to the anode through the proton exchange membrane, resulting in the formation of a hydrogen-air interface. What's more, the outside air will infiltrate the inside of the cell through the anode's exhaust pipe, which is a very slow gas diffusion process. Therefore, compared with the start-up process of the fuel cell, the process of forming the hydrogen-air interface after shutdown is very slow, and the existence time of the hydrogen-air interface inside the anode flow field is also relatively long. The specific processes above are schematically shown in Fig. 8.29 [137].

8.4.3 *Reverse Current Mechanism*

Currently, the reverse current decay mechanism is widely considered to underlie the degradation of PEMFCs during start-stop cycling and localized fuel starvation. This mechanism explains the degradation mechanism of the carbon support in the fuel cell under the hydrogen-air interface. The existence of a hydrogen–oxygen interface

Fig. 8.29 Schematic diagram of the formation process of hydrogen-air interface under Startup-shutdown condition [137]

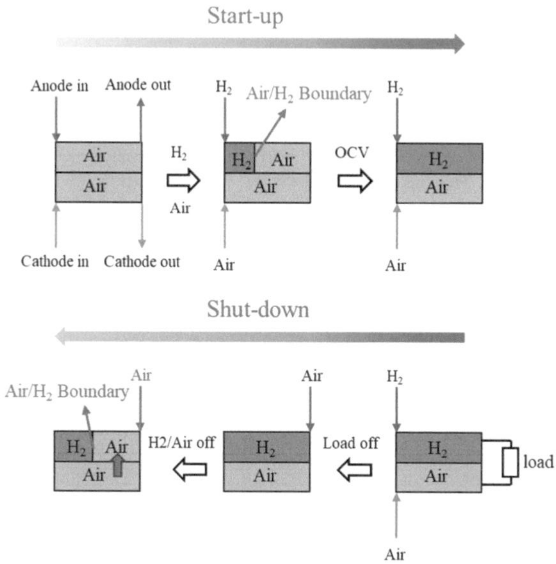

was first proposed by United Technologies Corporation (UTC) in 2000. In 2005, Reiser et al. [138] first proposed the reverse current mechanism to explain how the hydrogen-air interface leads to a high potential in the cathode, which in turn leads to corrosion of the carbon support. They propose a one-dimensional steady-state model to simulate the changes in the electrolyte phase potential along the flow path when a hydrogen-air interface is created, as shown in Fig. 8.30. In the model, the fuel cell is divided into two areas. In the left area, the two electrodes work normally: HOR at the anode and ORR at the cathode. Due to the good electrical conductivity of the bipolar plate and the diffusion layer, the metal potential drop in the direction of the flow channel is ignored, assuming an equal metal potential across the cathode and anode. The presence of anodic oxygen in the right region reduces the electrolyte potential to maintain the interfacial potential difference expected for the ORR reaction. As a result, the interfacial potential difference of the anode in the right region reaches 1.443 V, which enhances the COR and OER in the cathode. The protons generated by COR and OER in the right region are transferred from the cathode to the anode, creating a so-called reverse current. Meanwhile, it is particularly emphasized in the article that the electrolyte potential transition is very steep and occurs in the hydrogen frontier region, which indicates that the reverse current phenomenon may occur even in a small region with local hydrogen deficiency during normal operation.

The model reasonably ignores the transport of protons in the membranes of the two regions, because the very high resistance to proton conduction is attributed to the small membrane thickness. Therefore, with the hydrogen-air interface as the boundary, the entire fuel cell can be divided into a fuel cell and an electrolyzer [138], corresponding to the left and right regions, respectively, as shown in Fig. 8.31. Conductive bipolar plates and diffusion layers can be assumed to be external leads

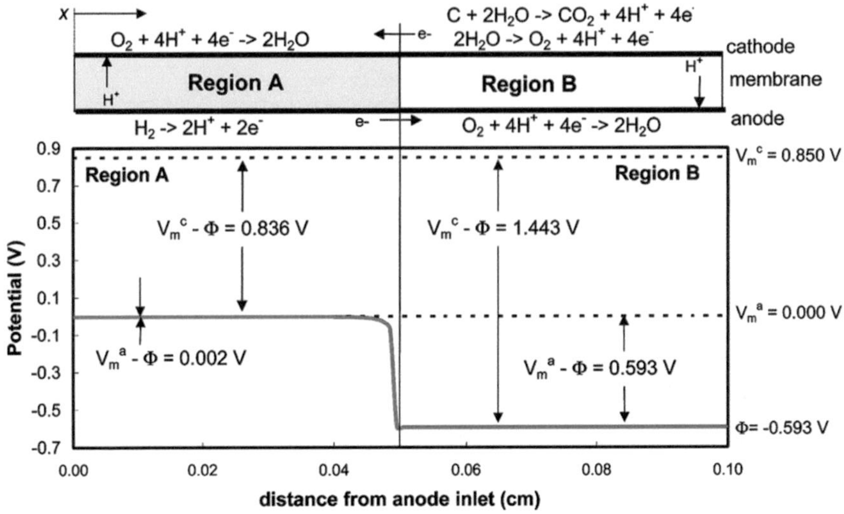

Fig. 8.30 Potential distributions along anode flow path during reverse current conditions [138]

connecting the two cells. This is called a dual-cell mechanism, equivalent to a reverse current mechanism, which has also been experimentally verified. A dual-cell experiment was conducted based on this mechanism, and monitored the current between the two cells during the process of replacing hydrogen with air [139].

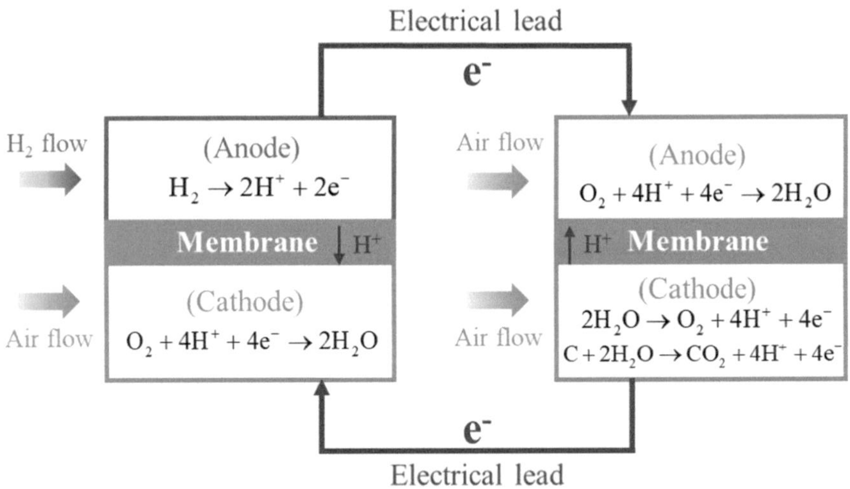

Fig. 8.31 Schematic of dual-cell mechanism [138]

8.4.4 Consequences of Carbon Corrosion

Corrosion of carbon supports is considered to be a key problem restricting the long-term service life of PEMFCs, which promotes cell decay through various mechanisms in Fig. 8.32, mainly in the following aspects:

(1) Weaken the adhesion between the catalyst and the carbon support surface and make the catalyst easily agglomerate or even detach from the carbon support [106]. In addition, the CO produced along with the carbon corrosion process also poisons the Pt reactive sites, resulting in a significant reduction in ECSA.
(2) The damage and collapse of the carbon skeleton redistributes the ionomer in the catalytic layer [140], which leads to the occurrence of uncovered and over-covered Pt particles as shown in Fig. 8.32a. The former will lose its activity due to the loss of conductive proton paths, while the latter will increase the

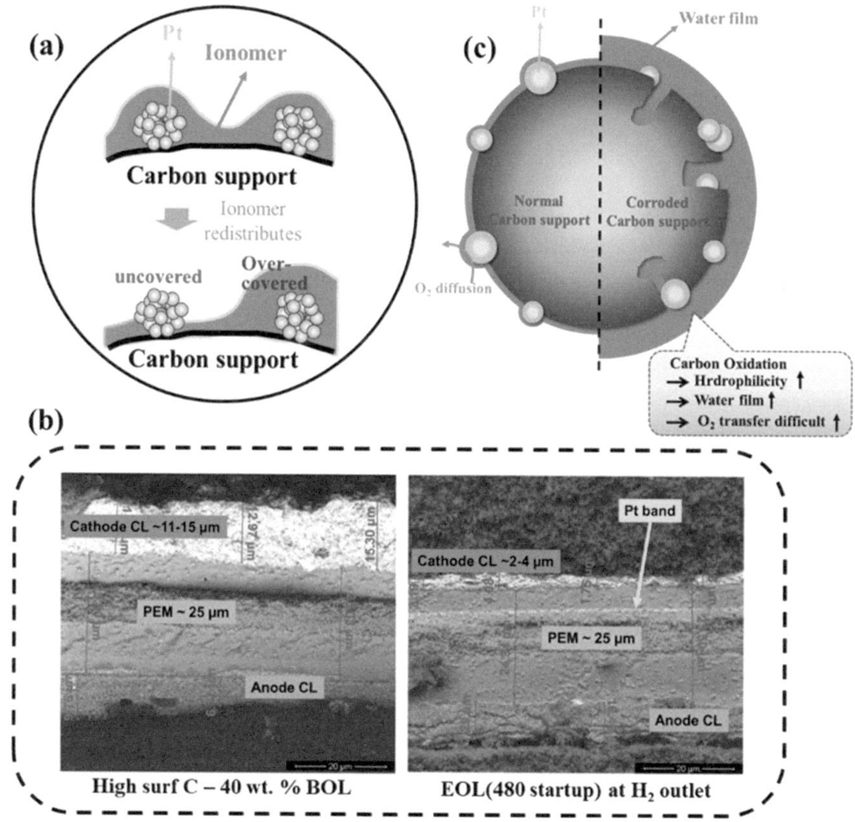

Fig. 8.32 Multi-aspects of consequences induced by carbon corrosion in cathode catalyst layer: **a** ionomer redistribution, **b** the thinning of the cathode thickness [141], **c** surface hydrophilization of the carbon support

mass transfer resistance on the particle surface. Moreover, it can also adversely affect the water management performance of the cathode, thereby increasing the likelihood of flooding. Here, it should be noted that agglomeration and redistribution of ionomers is a frequent occurrence since a uniform and highly dispersed distribution is inherently unstable.

(3) Carbon corrosion destroys the cathode catalytic layer framework. The reduction in carbon particle size, the pores on the support surface, and the thinning of the cathode thickness are widely observed in experiments in Fig. 8.32b [141], causing the porosity in the catalytic layer to decay. The porosity under the flow channel increased from 52 to 55% [142], while the porosity in the rib region decreased significantly. The collapse of the carbon skeleton is often accompanied by the shrinkage of the pore size and the severing of connected pores [143], resulting in the formation of closed pores, which in turn leads to an increase in mass transfer resistance.

(4) The oxidation process of carbon support will generate some hydrophilic oxygen-containing groups and attach to the surface, which reduces the surface hydrophobicity of the carbon skeleton [144]. This will make the water management performance worse, that is to say, the liquid water will not be removed in time and will accumulate in the catalytic layer, which will not only increase the risk of flooding, but also cause the water film of the agglomerate to thicken, thereby making mass transfer difficult as show in Fig. 8.32c.

Furthermore, carbon corrosion in the cathode is non-uniform due to differences in local gas atmospheres. Furthermore, carbon corrosion in the cathode is non-uniform due to differences in local gas atmospheres. First, from the analysis in Sect. 8.4.2, it can be seen that during the H_2 flow into the anode at start-up, the hydrogen-air interface moves from the anode inlet to the anode outlet, so that the anode outlet is always exposed to a high potential, resulting in more severe carbon corrosion to the corresponding cathode region. Conversely, the purging of air during shutdown also causes the anode inlet area to remain at high potential for a longer period. In experiments, H_2 and air were alternately fed into the anode to study the differences in carbon corrosion at the anode inlet, center, and outlet [145]. Figure 8.33 shows the cathode holes caused by carbon corrosion, indicating that the corresponding cathode regions at the anode inlet and outlet suffer from more severe corrosion. Second, carbon corrosion also showed non-uniformity in the catalytic layer under the channel and land. CL under the channel faces more severe carbon corrosion, attributed to the lower water vapor concentration under land, which acts as a reactant for carbon corrosion [146]. However, the opposite results were also reported [147] when the local current densities were measured under the channel and land during startup and shutdown. As shown in Fig. 8.34, negative currents indicate carbon corrosion leading to reverse currents. Therefore, the CL under land and the channel suffer more severe carbon corrosion at startup and shutdown, respectively. They believe that due to the more time required for the transport of gas in the catalytic layer under the land, there will be the formation of a hydrogen-air interface between the catalytic layer under the

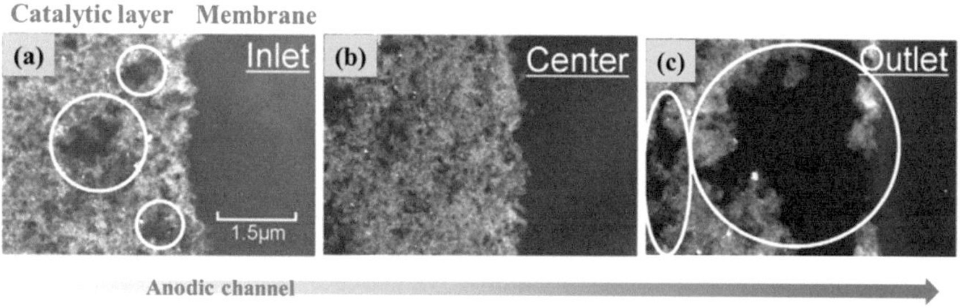

Fig. 8.33 STEM images of the MEA cross-section near the **a** anode inlet, **b** anode center, and **c** anode outlet, after 500 gas-exchange cycles in the anode [145]

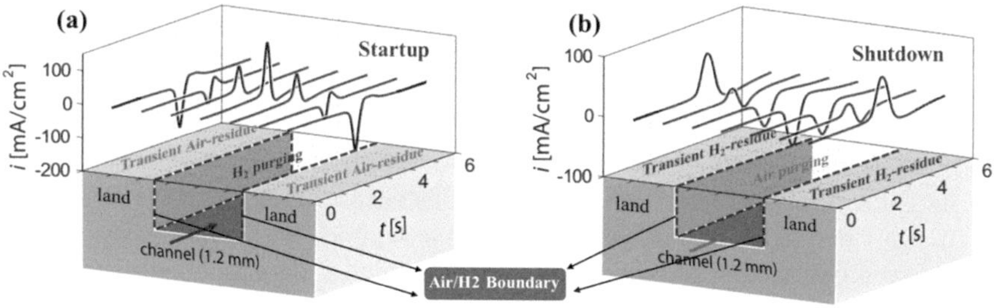

Fig. 8.34 Transient local current density corresponding to the channel and land in the **a** startup process (H_2 purging) and **b** shutdown process (air purging) [147]

channel and the land during the gas purging process, results in more severe carbon corrosion in the local cathode facing the anode air region.

8.4.5 Mitigation Strategies

To minimize the harm of carbon corrosion to fuel cells, how to mitigate the series of decays caused by carbon corrosion has become a major issue in the commercialization of fuel cells, and researchers have done a lot of work on this. At present, customized solutions are mainly from two aspects: carbon support improvement and innovation, and system control strategy optimization.

(1) Carbon support improvement and innovation

At present, the main research direction is to develop carbon supports with high oxidation resistance. The main methods can be roughly divided into the following three categories:

1. Doping modification of carbon materials. Since the chemical bonding between metals can effectively improve the stability of carbon, carbides can be used as catalyst supports. For example, titanium carbide (TiC) [148] and tungsten

carbide (WC) [149] show good stability in acidic media. Also, in recent years, a series of doped carbon support materials have been developed, such as N-doped NRGO [150], B-doped BDD [151], and P-doped P-CNT [152], by increasing the chemical bond between the support and Pt, and then obtaining Higher catalytic activity and durability

2. A new type of carbon material is used as the catalyst support. Studies have shown that the electrochemical stability of carbon materials is related to the degree of graphitization, and the catalyst stability can be improved by appropriate heat treatment of carbon supports. Using a more graphitized carbonaceous material can reduce the MEA degradation rate by 80% compared to conventional carbon supports [153]. Novel carbon support materials such as carbon nanotubes (CNTs), graphitic nanofibers, mesoporous carbons, carbon aerogels, and graphene are widely regarded as promising fuel cell catalyst support materials due to their high aspect ratios, high electronic conductivity, and specific interactions with Pt. The stability of CNTs and Vulcan-XC72 was compared by measuring the candle decay current [154]. It is found that CNTs have stronger corrosion resistance, and their candle decay current is 30% lower than that of Vulcan-XC72 under the same conditions. Carbon aerogels [155] have a controlled network structure, which can enhance the stability of catalyst supports and improve the contact between Pt and ion-conducting polymer electrolytes, showing great potential as an alternative catalyst support for proton exchange membrane fuel cells.
3. Preparation of new non-carbon materials as supports. Currently, the non-carbon materials used alone as catalyst supports in fuel cells are mainly TiO_2 [156], SnO_2 [157], and their modified materials. A Pt/TiO_2/C composite catalyst was prepared [156]. After the accelerated decay time, the Pt/TiO_2/C single cell voltage decreased by only 18 mV at 1000 mA cm^{-2}, while the Pt/C single cell voltage decreased by 88 mV. It can be seen that the Pt/TiO_2/C composite catalyst has better corrosion resistance. However, it should be pointed out that there are still many problems in the use of non-carbon materials as the carrier, such as electrical conductivity, and its attenuation mechanism is unknown.

(2) System control strategy optimization

In addition to more stable and oxidation-resistant catalyst support materials, it is also important to develop better control strategies for fuel cell systems, especially as the research and preparation of new materials is still limited. Many automotive companies, such as UTC, GM, Ford, Toyota, Nissan, and DaimlerChrysler, have come up with sound system strategies. For example, try to avoid operating the battery under extreme conditions (high potential, high temperature and low humidity, etc.) or reduce its operating time under such extreme conditions [158], reduce the number of PEMFC starts and stop during operation, and improve the system loading strategy to avoid large fluctuations in fuel cell load. From the analysis of the previous mechanism, it can be seen that the reverse current and high potential generated by the hydrogen-air interface in the start-stop condition are the main reasons for the corrosion of the carbon carrier. Therefore, some mitigation strategies are specially proposed as follows:

1. Gas purging is an effective method to avoid or reduce the existence time of the hydrogen-air interface. There are many possible strategies involving gas purging, such as anode nitrogen and air purging, cathode hydrogen purging, and cathode exhaust cleaning cycles. Several startup and shutdown processes were compared using different strategies [159] and showed that nitrogen purging during startup can effectively avoid the formation of hydrogen/air interfaces, but nitrogen purging during shutdown is not always effective. Also, the duration of the hydrogen/air interface is affected by the hydrogen and air flow rates. Hydrogen purging helped reduce the degradation rate by an order of magnitude and was easy to implement without the need for additional components [160]. The disadvantage, however, is that it can lead to anode degradation, especially when platinum alloys are used as catalysts, as hydrogen atoms can cause dissolution of the alloy components.
2. Introduction of dummy load. Gas purging can effectively prevent or mitigate the formation of anode hydrogen/air interface during startup and shutdown, but excess gas remains in the gas diffusion layer and catalytic layer [139]. It has been reported that dummy loads can be very successful in eliminating residual H_2 in the anode and residual O_2 in the cathode through electrochemical reactions. The relationship between auxiliary load access and degradation rate was studied in the start-stop cycle [161], and the results showed that auxiliary load can effectively delay the decline of ECSA and thus delay the degradation. Polarization curves, EIS, and piecewise current density were used to analyze the effect of auxiliary loading strategies [162]. The results show that the auxiliary load can effectively shorten the duration of the hydrogen/air interface during shutdown, thereby eliminating the generation of reverse current. Serious corrosion occurred on the cathode outlet side of the unprotected group, but no obvious carbon corrosion was found in the protected group.

In summary, when addressing the impact of dynamic operations such as load changes and start-stop cycles on battery degradation, it is difficult to take a single improvement measure to make the durability of dynamic operations meet the commercial requirements. Future work may focus on combining multiple strategies, or system strategies are integrated with novel corrosion-resistant carbon supports to achieve long lifetimes and meet the durability requirements of proton exchange membrane fuel cells.

8.5 Gas Diffusion Layer Degradation

The GDL is one of the key components of the fuel cell, which is usually composed of the substrate layer (MPS) and the microporous layer (MPL). It not only plays the role of supporting CL and thermal and electrical conduction, but also is responsible for the water and gas regulation of the fuel cell. An important function of GDL is to allow the reaction gas to diffuse evenly into the CL to participate in the electrochemical reaction, and at the same time, discharge excess water to prevent flooding. On the other hand, it is also necessary to avoid excessive drying of the membrane and CL caused by excessive water loss, thereby reducing the ionic conductivity of the fuel cell and increasing the internal resistance of the fuel cell [163]. Therefore, the properties and design of GDLs greatly contribute to the performance of cells and the durability of their components. However, the durability of GDL material properties, including the core capabilities of GDL such as hydrophobicity, electrical conductivity, and mechanical strength, decreases during operation which is a key factor in the degradation of fuel cell performance [164].

As can be seen from the previous content, the predecessors have done a lot of research on the proton exchange membrane and catalytic layer in the membrane electrode, the core component of PEMFC. However, current research summaries on the durability of the GDL are relatively limited, although it may be a key factor in managing mass transport and two-phase flow while mechanically supporting membrane electrode assemblies and bipolar plates. Therefore, to comprehensively and deeply investigate the GDL durability issue, this section reviews the GDL degradation process. As shown in Fig. 8.35, the degradation of GDL can be divided into mechanical degradation and physicochemical degradation. Mechanical degradation is a physical damage, including mechanical damage caused by compression, dissolution effect by water, and air flow erosion. The GDL absorbed most of the clamping force and suffered structural damage. In addition, the water brought by the gas through the humidifier and the electrochemically generated water will gradually dissolve the GDL, and the high-speed, continuous gas supply will also lead to the mechanical degradation of the GDL. Physicochemical degradation can be divided into chemical degradation and electrochemical degradation. The former refers to the occurrence of chemical carbon corrosion and PTFE loss in the presence of H_2O_2 and free radicals. The latter refers to the reaction between carbon and water under some special conditions.

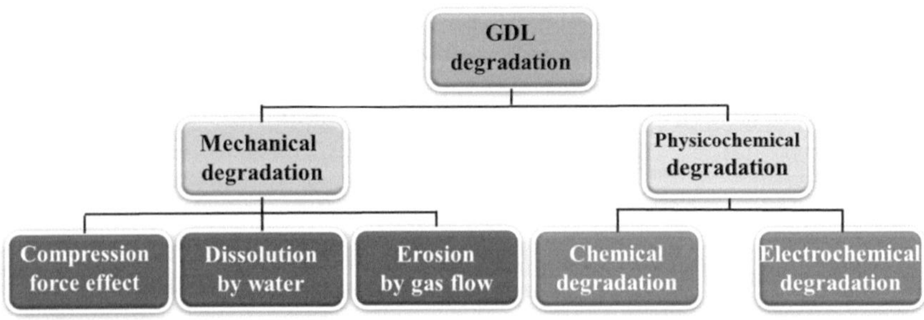

Fig. 8.35 GDL degradation mechanism

8.5.1 *Mechanical Degradation*

8.5.1.1 Effect of Compression Force

In PEMFC, a compressive force of at least 1 MPa is required to maintain contact between layers to reduce contact resistance and prevent leakage. GDL is the fuel cell component with the highest porosity, reaching 80% in some cases. This makes GDL susceptible to structural changes and even permanent degradation under mechanical stress [165].

Compression pressure strongly affects the performance of GDL, the GDL/catalyst layer interface, and the GDL/bipolar plate interface. With the increase of compressive pressure, the overall pore volume of the GDL decreases so that the water vapor permeability decreases, resulting in a higher mass transfer overpotential. The GDL permeability decreased by an order of magnitude [166] when the GDL was compressed to 65% of its initial thickness. Also, due to the ribbed channel structure of the bipolar plate (BPP), the increase in pressure increases the volume of GDL intruded into the flow channel, thereby reducing the amount of reactant gas in the flow channel and causing severe local transport differences, which in turn affect the performance of the PEMFC performance. A parallel flow channel model was established [167] and concluded that a 5% change in the amount of GDL entering the flow channel would cause a 20% decrease in the reactive gas in the flow channel. On the other hand, although compressed GDL harms gas permeability, it improves the contact between its fibers and with other components, improves the thermal/electrical conductivity of the bulk structure of GDL, and reduces GDL/CL and GDL/BPP Contact resistance. Therefore, the ohmic overpotential decreases. The electrical contact resistance was greatly reduced from the initial state of 1000–180 Ω cm^2 under an external clamping pressure of 2.5 bar [168]. In summary, compressing the GDL affects the mass transfer overpotential and ohmic overpotential, but they show opposite trends. Figure 8.36 shows the impact of GDL compression force on fuel cells. Therefore, a compromise is usually sought between gas and liquid water permeability and ohmic resistance, and an optimal compressive force range can be expected for the overall performance of the fuel cell for different GDLs [169].

In addition to the above compressive performance relationships, high compressive pressures may also lead to permanent GDL degradation:

(1) First, GDL itself will deform under high pressure, and even the internal structure will be damaged, such as cracking and falling off of the PTFE hydrophobic film covering the surface of carbon fiber [170]. The damage to the PTFE hydrophobic film will reduce the hydrophobicity of the internal pores of the GDL, causing the local accumulation of liquid water in the GDL, resulting in the obstruction of reaction gas transmission, and even more serious phenomena such as water flooding and reverse polarity in the fuel cell.
(2) In addition, the carbon fibers that form the main part of the MPS may undergo significant fractures or even displacements in the fiber alignment, which will change the morphology of the entire structure, possibly with implications for

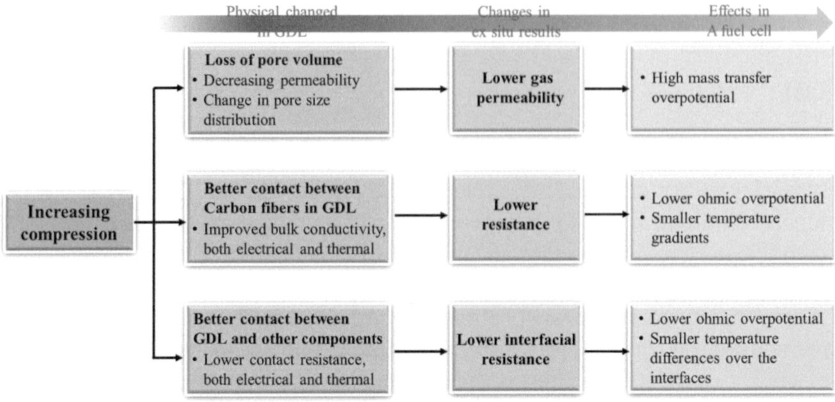

Fig. 8.36 Effects of increasing compression of the GDL

current transport and fluid permeability [167]. The uneven distribution of pressure caused by the bipolar plate flow channels and land is also more likely to cause mechanical damage to the GDL, and this uneven stress may also lead to the generation of cracks and holes in the MPL surface.

(3) Besides, the rupture or fracture of GDL carbon fibers is often a continuous regional result due to the transfer of stress, not the failure of a single carbon fiber. This results in a rougher surface of the GDL [171], which increases the contact resistance between the two interfaces and increases the ohmic polarization loss of the cell. Meanwhile, it also increases the contact thermal resistance of the entire battery, which has a certain impact on the heat of the cell and the uniformity of the current density distribution.

(4) Moreover, GDL has a significant effect on the mechanical damage of other battery components. For example, the material property transition at the GDL/CL interface provides a driving force for delamination and CL surface crack propagation [172]. The carbon fiber displacement caused by local stress may be responsible for the local damage of the ionomer membrane, causing the reaction gas to pass through the membrane more easily, thereby reducing the open-circuit voltage of the cell [173].

In conclusion, it is very important to select the clamping pressure of PEMFC, and it is necessary to study the stress–strain relationship of the GDL used, the relationship between pressure and contact resistance, and contact thermal resistance. Finally, select the appropriate clamping pressure to obtain better battery performance and longer battery life.

8.5.1.2 Dissolution Effect of GDL by Water

During the operation of PEMFCs, GDLs are exposed to water or oxidizing conditions. The GDL becomes hydrated by the water produced during the electrochemical reaction or supplied with hydrogen or oxygen gas. It can be believed that the hydrophobicity and weight loss of GDL are a result of coupled carbon resin loss, surface morphology changes, and PTFE dissolution [174]. Various experimental studies have confirmed that acidic conditions and oxygen in water lead to GDL dissolution. In the literature, dissolution is usually tested by immersing GDL in hot sulfuric acid solution or deionized (DI) water. DI water leaching experiments were conducted at 80 °C for 2160 h [175], and observed that the contact angles of both MPS and MPL decreased by 7° and tended to stabilize with time, because the material susceptible to leaching has dissolved. The aged GDL showed a high current performance loss of 4.9%/13.6% under 100%/50% relative humidity (RH) conditions, respectively. Their SEM results indicated losses in MPS carbon resin and MPL carbon particles. Deionized water was compared with acidic solutions and found that acidic environments accelerate hydrophobicity loss [176]. A similar sulfuric acid solution immersion test was performed [177]. In the SEM images of the MPL aged for 1000 h, deeper and larger cracks can be observed, as shown in Fig. 8.37. A 1200 h leaching test with H_2SO_4 was performed [178], which not only reduced the contact angle and damage to the PTFE coating but also produced larger pores, resulting in a four-fold increase in water permeability. This reduces structural integrity and negatively affects the transport of reactants and products.

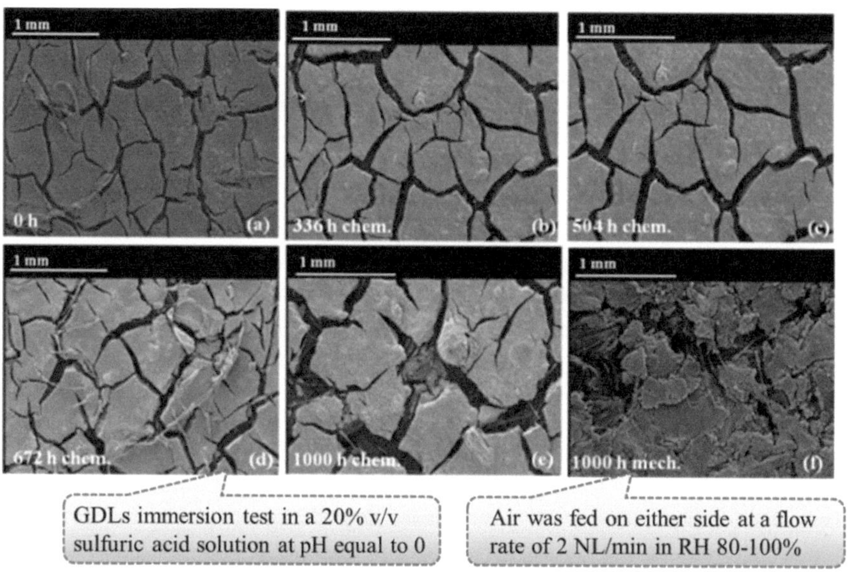

Fig. 8.37 MPL aged after **a–e** chemical degradation under immersion and **f** erosion by gas flow [177]

The dissolution of GDL is affected by several factors. First, the PTFE loss is dependent on the presence of oxygen. The thermogravimetric analysis (TGA) and SEM results [176] showed no significant loss of PTFE in deionized water or sulfuric acid solution. However, after immersing the bilayer GDL in air-enriched deionized water for 1000 h, the loss of PTFE in the NMR spectrum was observed [179]. Second, the impact of MPL appears to be controversial. While some researchers believe that the leaching test has a limited effect on MPL hydrophobicity [179] and that MPL may be beneficial for MPS durability, other researchers have found that MPL may also undergo significant degradation. In a series of studies [175, 176], MPL showed a greater loss of hydrophobicity than MPS. Third, in the case of PTFE, the amount of PTFE did not appear to affect the dissolution of GDL [179].

8.5.1.3 Erosion by Gas Flow

In PEMFC, where fuel and air are supplied to the flow channel and GDL at high velocity, when the gases come into contact with the rough surface of the MPS, a friction phenomenon occurs, which can lead to significant erosion of carbon-based materials, especially in the presence of liquid water [180]. This effect is more pronounced on the cathode side than on the anode side because the flow rates of air and hydrogen are very different under conventional conditions. After 150 h of operation, corrosion and loss of hydrophobicity may occur, causing significant mechanical degradation.

The airflow AST and leaching tests were compared [177]. Although the immersed samples experienced only 3% weight loss after 1000 h, under high-speed airflow, their weight loss was as high as 20%, and the MPL surface topography in the SEM images was severely damaged, as shown in Fig. 8.37f. Their electrochemical polarization tests also showed a sharp rise in mass transfer resistance. A gas flushing experiment on commercial SGL10BC was performed [180]. The first stage: after 14 days of flushing with a large flow of dry air at 65 °C, no cracks and small pits were found on the surface of the MPL. The second stage: based on the first stage, change to RH100% air for 14 days. The first crack appeared on the MPL surface after 1 day, and small pits also appeared. As time went on, the MPL surface cracked and the small pits became more and more. The morphological changes of MPL at each stage of the experiment are shown in Fig. 8.38. From this, it can be concluded that the erosion of the GDL is more serious in the humid gas.

The significant performance losses reported in the above studies suggest that the mechanical damage to the GDL due to gas flow may account for a large proportion of the increased mass transfer resistance. The significant performance losses reported in the above studies suggest that the mechanical damage to the GDL due to gas flow may account for a large proportion of the increased mass transfer resistance. The general explanation for the performance degradation caused by the formation of small pits and cracks on the surface of the MPL is as follows: There are two main driving forces for the discharge of water from the porous medium (GDL): (1) Capillary pressure; (2) The effect of air flow. Since the flow of gas in the MPL is very small, the discharge of water is mainly determined by the capillary pressure

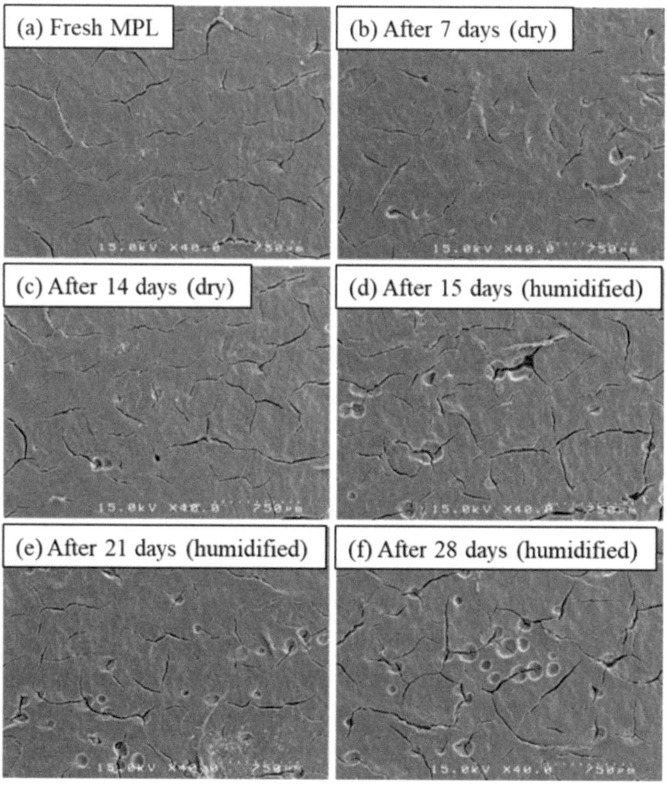

Fig. 8.38 The progression of mechanical degradation of the MPL surface: **a** Fresh MPL, **b** after 7 days of degradation, **c** after 14 days of degradation, **d** after 15 days of degradation (1 day after supplying humidified air), **e** after 21 days of degradation, **f** after 28 days of degradation [180]

inside the MPL. However, the formation of cracks and small pits on the surface of the MPL leads to a decrease in the capillary pressure difference and a decrease in the drainage capacity, which affects the gas mass transfer and leads to a decrease in the performance under high current.

To improve durability, improvements in GDL design and fabrication have been the subject of various studies. For example, by adding a polymer binder to MPL inks [180], fine carbon particles have better cohesion, resulting in a crack-free surface. This improved MPL was observed to resist the above-mentioned 14-day AST with no apparent change in the MPL surface and no significant decrease in cell performance. In addition, incorporating a certain amount of CNT into MPL can also improve the mechanical decay resistance of MPL [181]. After adding CNT, the battery performance and durability are improved.

8.5.2 Physicochemical Degradation

8.5.2.1 Chemical Dissolution by Hydrogen Peroxide

In the operation of proton exchange membrane fuel cells, hydrogen peroxide (H_2O_2) can be generated through the two-electron ORR pathway, either through the electrochemical reaction of O_2 and protons or due to gas crossover during idle [182]. Then, in the presence of transition metal ions, H_2O_2 can further form radicals, such as HO and HOO [183], which are diffused to the GDL. The mechanism of this process has been described in detail in Sect. 8.2. Hydrogen peroxide and corresponding free radicals in GDL create an oxidative environment that can lead to significant voltage loss, peak power fade, and reduced water removal capability, especially in the high current region.

The GDL performance loss caused by the presence of H_2O_2 can be attributed to two sources: chemical carbon corrosion and PTFE loss. This reduces the surface wettability of the GDL, leading to flooding under high current density and humidity conditions, significantly reducing the performance of the cell. Studies have shown that chemical carbon corrosion can lead to loss of carbon content and formation of surface oxides, and that MPL is more susceptible to oxidation. The SEM images [184, 185] show that it is difficult to observe changes in the bulk structure, so chemical carbon corrosion is mainly responsible for the formation of surface oxides, resulting in a more hydrophilic carbon surface, leading to deterioration of water management. A nonlinear relationship between CO_2 emission rate and H_2O_2 concentration was found [174], but a quantitative relationship between degradation and H_2O_2 content has not been established. Another source of GDL hydrophobicity loss is its PTFE loss and damage. The study found that acidic environments cause PTFE to dissolve, and it can be expected that the presence of H_2O_2 would further accelerate this process. Under the attack of free radicals, the C-F bond of PTFE is directly broken and decomposed. GDLs were also compared with and without PTFE coatings in immersion tests [184]. The PTFE-treated GDL showed higher CO_2 emission rates, which the authors believe indicates PTFE degradation. It should be noted that the H_2O_2 concentrations in these studies were still several orders of magnitude higher than the actual concentrations expected at ppm levels. Therefore, the relationship between these tests and actual GDL degradation still requires further investigation.

8.5.2.2 Electrochemical Degradation

As we mentioned in Sect. 8.4.2, under the start/stop, shutdown, and local hydrogen starvation conditions of PEMFC operation, the reverse current mechanism will generate a higher interfacial potential difference in the cathode region, resulting in carbon corrosion of the catalytic layer. The current study found the same mechanism for GDL [186], an external potential step was performed on conventional MEA with carbon-supported CL and MEA with platinum black. It was found that the second

MEA reduced the CO_2 emission rate by about 80% compared to the conventional MEA, indicating that 20% of the carbon corrosion originated from the GDL.

Similar to GDL degradation under radical attack, electrochemical carbon corrosion affects the transport properties of GDL by inducing structural changes and altering surface wettability. A 96 h off-line accelerated aging experiment of electrochemical corrosion was verified with GDL as the working electrode (1.0–1.4 V/SCE) in 0.5 mol/L sulfuric acid solution [187]. It was found that the number of macropores in the GDL after corrosion increased, resulting in worse mass and charge transfer. The EIS results show that the charge transfer resistance and mass transfer resistance of the GDL-assembled cells after 1.2 V constant-voltage corrosion are about 40% and 22% higher than those of the GDL-assembled cells without accelerated testing, respectively. In addition, it was found that with the increase of corrosion voltage, the corrosion phenomenon of MPL was more serious, and the cross-sectional SEM images (Fig. 8.39) showed that the corrosion degree of the inner layer was significantly greater than that of the outer layer. An accelerated carbon corrosion test of GDLs was performed [188], and by TGA analysis, found that the losses were mainly carbon materials, such as carbonized resins and carbon particles, rather than PTFE coatings. This is also confirmed by the results of Yu et al. [189]; however, they claim that the PTFE may be washed away as the supporting carbon material is washed

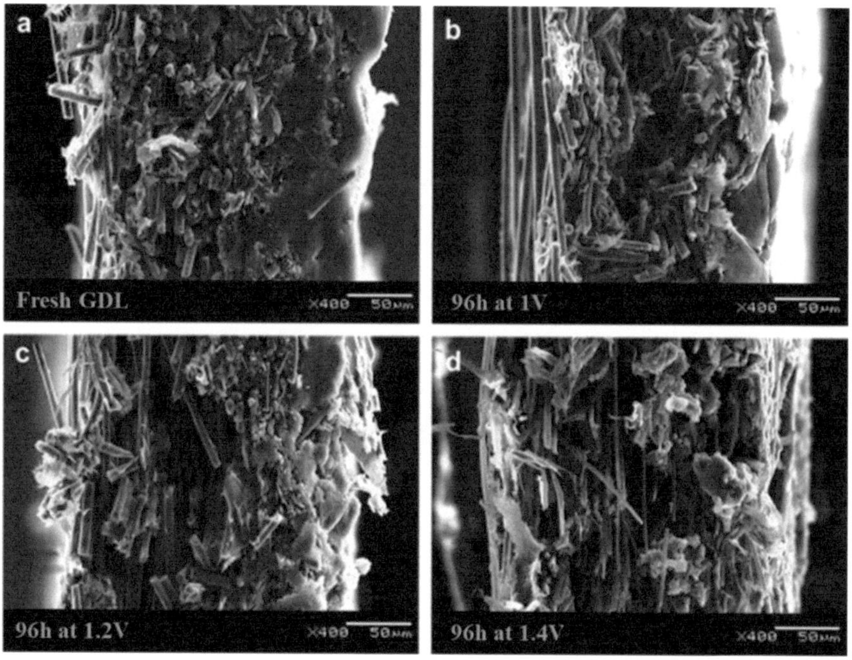

Fig. 8.39 The cross-section SEM images of **a** the fresh GDL and the GDLs after oxidation for 96 h held at **b** 1.0 V versus SCE, **c** 1.2 V versus SCE, and **d** 1.4 V versus SCE. The gas diffusion backings of the GDLs were on the left while the micro-porous layers were on the other side [187]

away, thereby reducing the hydrophobicity of the GDL. At the same time, the carbon is partially oxidized to form a relatively hydrophilic group, which also affects the discharge of water.

The effect of MPL on GDL durability has been widely investigated. The greater the penetration depths of MPL in MPS, the better durability of GDL [190], and these penetrating portions of the MPL are more susceptible to electrochemical corrosion. The cathode GDL was compared with and without MPL under the carbon corrosion accelerated stress test condition at 1 V [191] and showed that MPL is beneficial for mitigating carbon corrosion in the catalyst layer; however, MPL itself is less resistant to carbon corrosion. When the PTFE content in MPL was increased from 10 to 50% [192], the rate of performance decay was halved. Therefore, PTFE in MPL acts as an antioxidant.

8.6 Bipolar Plate Degradation

8.6.1 Mechanism of Bipolar Plate Degradation

The bipolar plate is one of the important components of the fuel cell, which serves as a separator between two adjacent cells. It conducts the electronic current between these two cells, separates the gases, and contains the flow patterns both for the reactants and for the coolant, in most cases, a liquid coolant. Its classification includes pure graphite plate, graphite-based composite bipolar plate, and metal bipolar plate. Among them, metal bipolar plate is widely used due to their high electrical conductivity, high mechanical properties, and low cost. Stainless steel is the most widely used material. However, a metal bipolar plate is more prone to corrosion in acidic and humid environments, which limits its further development. Extensive measurements have shown that the pH value of the water environment on the surface of the bipolar plate is typically 3–5. During the decay of the bipolar plate, corrosion and oxide formation occur on its surface. This has three undesirable consequences. First, the most obvious manifestation is the increase of interfacial resistance between the bipolar plate and the gas diffusion layer, which leads to the increase of voltage drop and the decrease of fuel cell efficiency. Second, the presence of corrosion products alters the hydrophobicity of the bipolar plate surface, leading to local liquid water accumulation and impeding the transport of the reaction gas. Finally, the leaching of metal cations from the bipolar plates increases the chemical decay of the membrane and decreases its conductivity, which leads to further degradation of the cell performance. The decay of the metal bipolar plate is mainly described below.

The essence of metal corrosion is the process of chemical or electrochemical reactions occurring on the surface of metals due to the interaction between the material itself and the working environment, resulting in changes to the relatively stable structure of the atoms inside the metal and further changes in the properties of the metal. As shown in Fig. 8.40, the corrosion process and corrosion mechanisms of metal bipolar

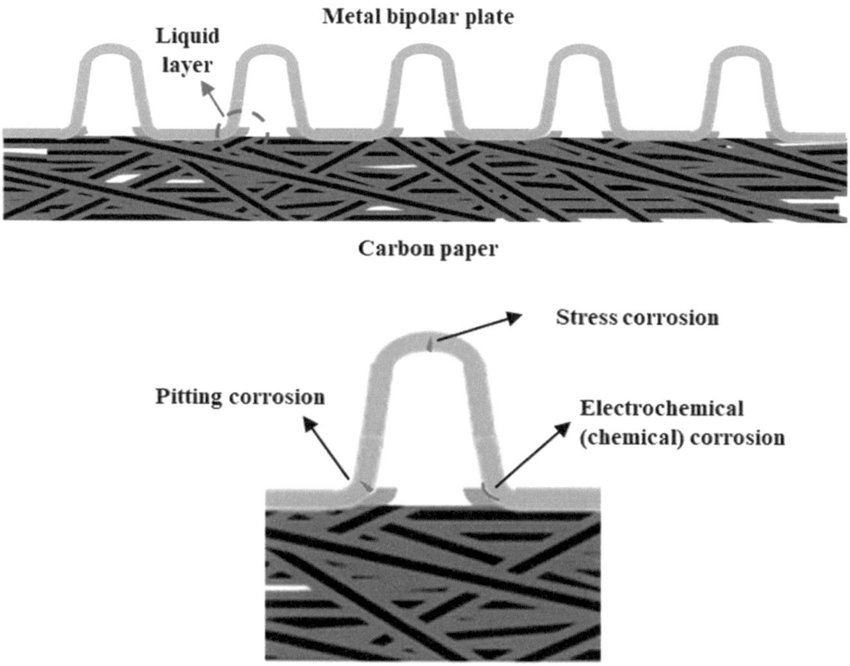

Fig. 8.40 Schematic diagram of metal bipolar plate corrosion

plates are divided into three categories [193]: chemical corrosion, electrochemical corrosion, and physical corrosion.

(a) Chemical and electrochemical corrosion

Corrosion of metal bipolar plates is propagated mainly through crevice corrosion and pitting corrosion. Pitting corrosion as a localized corrosion relatively concentrated in a small part of the metal surface, mainly occurs in aqueous solutions containing halogen ions such as Cl^{-1}, Br^{-1}, F^{-1}, etc., and is particularly serious in acidic environments. Pitting is caused by the destruction of the passivation film by reactive ions. After the bipolar plate is contaminated by halogen elements, rapid corrosion will occur.

As shown in Fig. 8.41, the pitting process can be divided into three stages [194]: (1) pitting nucleation, (2) pitting growth, and (3) repassivation of the etched holes. In the first stage, an aqueous solution containing halogen ions can enter the subsurface through the holes or loose. Both chemical corrosion and microelectrochemical corrosion will occur. In the second stage, under the combined effect of chemical and electrochemical corrosion, the etched hole will continue to develop and the breakout area will expand outward, causing the coating to peel off from the substrate. In the third stage, due to the repassivation film, the corrosion holes will stabilize and stop developing. Therefore, pitting control in the repassivation phase reduces the pitting hazard.

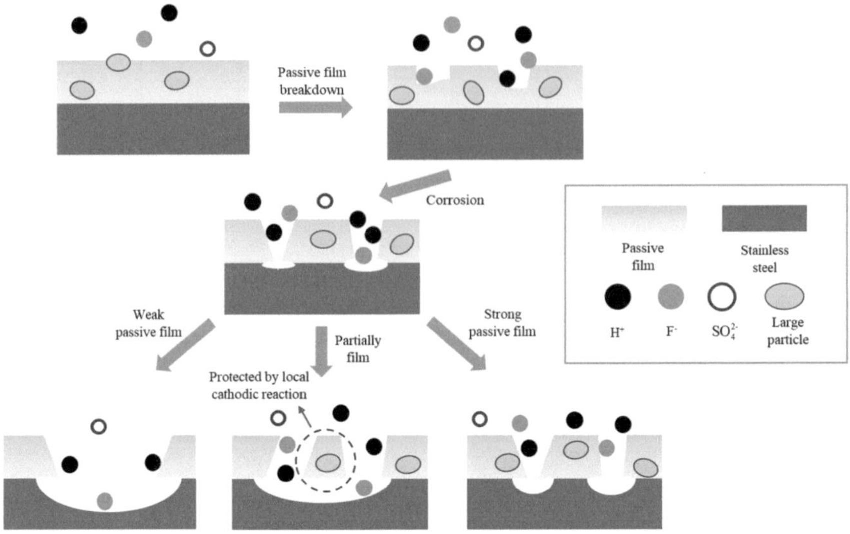

Fig. 8.41 Schematic diagram of corrosion mechanism of stainless steel passive film: **a** before passive film breakdown, **b** passive film damage begins. And three possibilities of **d** pitting corrosion: **c** weak passive film, **e** partially passive film, and **f** strong passive film

Under normal conditions, the corrosion area of the bipolar plate is concentrated at the condensate, while the corrosion area of the bipolar plate contaminated by halogen elements will occur in the whole contaminated area. This is also the reason why membrane chemical decay produces elemental fluorine, which promotes bipolar plate corrosion. At the same time, cations such as Fe^{2+} and Cr^{3+} generated by the corrosion of the metal bipolar plate itself will be transferred to the membrane through the carrying of water. This occupies sulfonic acid group sites and reduces the conductivity of the membrane. On the other hand, it also promotes chemical decay of the membrane, see Sect. 8.2.5 for details.

(b) Physical corrosion

Metal bipolar plates are generally manufactured by cutting, stamping, and etching. The surface of the bipolar plate will have various defects induced by the internal stresses generated by stamping and cutting, and the acidic environment of etching, which inevitably lead to various defects such as vacancies, grain boundaries, dislocations, and secondary defects, even with coating. These microdefects will be the source of cathode corrosion and the reduction of cell power. These defects may capture hydrogen atoms in the cathode and bias functional groups or atoms around them, resulting in an uneven distribution of hydrogen in the material. Stainless steel is prone to "hydrogen embrittlement" in long-term use in hydrogen-containing environments, resulting in irreversible hydrogen damage such as internal cracks and hydrogen bubbles. At the same time, during the assembly of the bipolar plate, the external stress is enhanced, and the hydrogen in the material interacts with the stress field, resulting in hydrogen embrittlement near the microscopic defects. This can cause an increase in contact resistance.

8.6.2 *Mitigation Strategies for Bipolar Plate Degradation*

The methods to mitigate the decay of the bipolar plate can be divided into three main types:

(a) Changing the chemical composition of the material

Stainless steel has been widely used in bipolar plates. The corrosion resistance of bipolar plates depends largely on the chemical composition of the material, as the type and content of alloying elements determine the composition of the surface passivation film, which in turn affects the corrosion resistance of bipolar plates. Eleven different alloys of stainless steel were exposed to sulfuric acid solution to study the contact resistance of bipolar plates in a PEMFC [195], and found that the chromium and molybdenum content play a decisive role in reducing contact resistance. The Cr content was corrected [196] with the corrosion and interfacial contact resistance of the material, and found that the higher the Cr content, the higher the corrosion resistance. The corrosion and interfacial resistance of 304, 310, and 316L stainless steels and four different nickel-based alloys were studied in a PEMFC environment [197], noting that oxide film composition has a strong effect on conductivity. Untreated stainless steel always lacks corrosion resistance, resulting in increasingly poor performance in fuel cell environments when used as bipolar plates. Therefore, it is mandatory that some surface modification method or protective coating be applied to stainless steels to improve their performance against corrosion [198].

In addition, the materials selected should avoid the appearance of mixtures. Because the presence of mixtures will lead to different electrochemical potentials, making it constitute a cell, and then electrochemical corrosion occurs, accelerating the corrosion of the bipolar plate.

(b) Performing surface treatment

The corrosion resistance of chromized 316L was evaluated in PEMFC simulating conditions [199]. The chromium layer was produced by a pack cementation process. The treatment was effective concerning the improvement of corrosion resistance of bare 316L. A marked improvement was realized in the corrosion and contact resistances of 316L stainless steel after a chromizing surface treatment [200]. A chromizing treatment was combined with electron discharge machining (EDM) to enhance the corrosion resistance and decrease the contact resistance of low-carbon steel bipolar plates [201]. A high corrosion and low contact resistance for chromium electroplated 316L stainless steel bipolar plates was achieved after low-temperature plasma nitriding [202].

Ion implantation was used to change the composition of the passivation layer [203, 204], which in turn improved the performance of 316L stainless steel. They used two methods: injection of nickel ions only [203] and simultaneous injection of nickel and chromium ions [204]. Due to the enrichment of nickel and chromium in the passivation film, the corrosion resistance is enhanced, and the interfacial contact resistance is reduced.

(c) Surface coating

The use of corrosion-resistant and highly conductive coatings is another surface treatment to solve the corrosion problem of metal bipolar plates. Using a physical vapor deposition (PVD) method, titanium nitride (TiN) coatings were deposited on 316L stainless steel bipolar plates [205] as an alternative to protect the material from the corrosive electrolyte of PEM fuel cells. Potentiodynamic tests were conducted [206] with TiN-coated 316L stainless steel specimens and verified a drastic decrease in the corrosion current density. In addition, they evaluated TiN-coated martensitic stainless steel bipolar plates and found that the corrosion rate was reduced by two orders of magnitude compared to the uncoated material [207]. Fluorine-doped SnO_2 coatings on austenitic stainless steel bipolar plates were tested [208] and achieved an order of magnitude reduction in corrosion current density compared to uncoated substrates.

A different approach was developed by producing a conductive composite layer consisting of a carbon black/epoxy resin mixture and coating it on a 304 stainless steel substrate [209]. It was found that the material maintained its corrosion resistance for 2000 h at 95% RH (relative humidity) and 85 °C in humidified air and 1200 h at 120 °C in 0.1 M H_2SO_4.

8.7 Impurity Effects

Impurities are an important external parameter that affects the durability of fuel cells. Impurities can harm the performance and durability of fuel cells. For example, some impurities can adsorb to the catalyst surface, thereby inhibiting the electrode charge transfer process. Cationic impurities can interact with the sulfonic acid groups of the membrane side chains, reducing proton conduction in the membrane and leading to increased ohmic losses. Some impurities can also change the transport behavior of water or gas in the GDL and increase the mass transport resistance in the cell.

Performance loss due to impurities is also divided into reversible and irreversible losses. The performance loss of some poisoned species can be offset by simply stopping the introduction of impurities or performing recovery procedures. Methods used to detect loss reversibility include exposure to high potential or repetitive cyclic voltammetry (CV) scans (for catalyst poisons, e.g., S) and operation at high currents (for external cations, e.g., NH^{4+}). Different impurities degrade the fuel cell by different mechanisms, depending on the type and chemical composition of the impurities. The types of impurities and their effects on fuel and air are described below, respectively.

8.7.1 Impurities in Fuel

(a) CO

Carbon monoxide is inevitably present in the hydrogen generated by reforming hydrocarbon fuels. The main decay mechanism of carbon monoxide on fuel cells is the poisoning of the catalyst. Many studies have been done in this area [210–212]. Figure 8.42 summarizes the poisoning of the Pt catalyst by CO in the fuel. CO will compete with H for activation sites on Pt. Because the heat of adsorption of CO on Pt (134 kJ/mol) is greater than that of H on Pt (87.9 kJ/mol) [212], CO is more likely to adsorb on the Pt surface than H. The reaction area used for the hydrogen oxidation reaction is reduced, causing a decrease in the cell reaction rate and an increase in the overpotential. The area covered by CO on Pt is determined by the adsorption/reabsorption of CO and the oxidation removal reaction of CO, as shown in Eqs. 8.16–8.18.

$$\mathrm{nPt} + \mathrm{CO} \leftrightarrow \mathrm{Pt_n} - \mathrm{CO_{ads}} \tag{8.16}$$

$$2(\mathrm{Pt} - \mathrm{H_{ads}}) + 2\mathrm{CO} \leftrightarrow 2(\mathrm{Pt} - \mathrm{CO_{ads}}) + \mathrm{H_2} \tag{8.17}$$

$$\mathrm{Pt} - \mathrm{CO_{ads}} + \mathrm{Pt} - \mathrm{H_2O_{ads}} \leftrightarrow 2\mathrm{Pt} + \mathrm{CO_2} + 2\mathrm{H^+} + 2\mathrm{e^-} \tag{8.18}$$

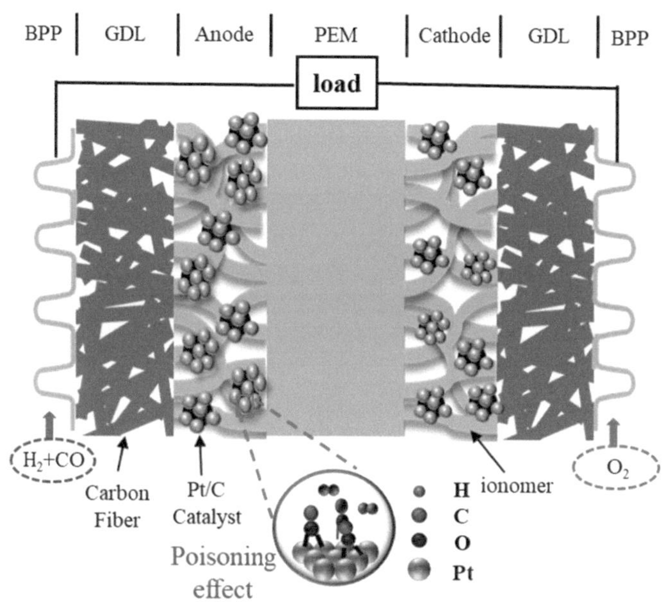

Fig. 8.42 Schematic representation of the poisoning effect of CO in fuel on Pt catalyst

Mitigation strategies for the poisoning effect of carbon monoxide on fuel cells fall into three groups: (i) the pre-treatment of reformate; (ii) 'on board' CO removal, and (iii) in-operando strategies.

(b) NH_3

NH_3 is also produced during hydrocarbon reforming to produce hydrogen. Significant degradation of cell performance can occur after NH_3 concentrations reach 13 ppm. The short-term effects of NH_3 are reversible, but the longer the exposure time, the more irreversible the negative effects on the cell will be. The cell performance was tested when the NH_3 concentration was 30 ppm, and the results are shown in Fig. 8.43 [104]. The high-frequency resistance test reveals that the resistance increases from 0.1 to 0.25 Ω cm^2 after 15 h. No significant NH_3 adsorption is found when the cell anode is tested using cyclic voltammetry. It indicates that the mechanism of NH_3 decay in the cell may be the decrease in proton conductivity. For example, NH_3 reacts with H^+, producing NH_4^+, which reduces the H^+ conduction. This negative effect occurs first in the catalysis layer of the anode and gradually transmits to the membrane.

(c) H_2S

Most of the fossil fuels contain sulfur compounds, and hydrogen sulfide is produced during the reforming process to produce hydrogen. H_2S usually has a strong affinity for metals, such as Pt catalysts. The decay effect of H_2S on the fuel cell is related to the concentration and exposure time. The polarization curves of the fuel cell at different moments for an H_2S concentration of 1 ppm are shown in Fig. 8.44 [104]. After 4 h of exposure, the loss of cell performance was evident (curve b). The cell almost stopped working after 21 h (curve c). After electrochemical cleaning, the polarization curve returned to the initial state (curve d), demonstrating that the decay of H_2S is relatively reversible.

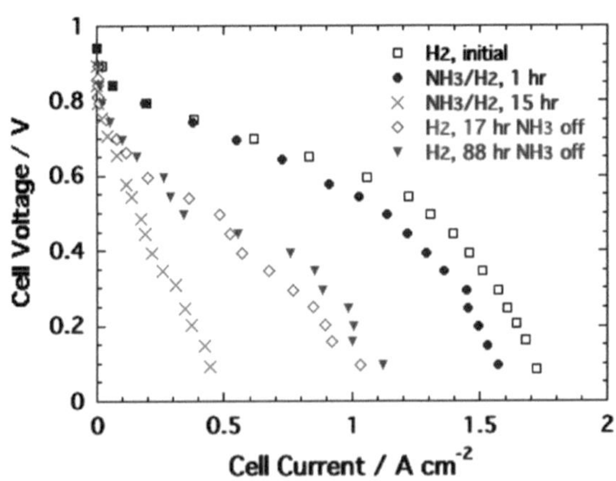

Fig. 8.43 Polarization curves of a H_2/air fuel cell showing the effect of 30 ppm NH_3 injected into the fuel stream for a total of 15 h. After this period, the cell continued operating on neat hydrogen for an additional 88 h at a constant voltage of 0.5 V. T = 80 °C [104] (Color figure online)

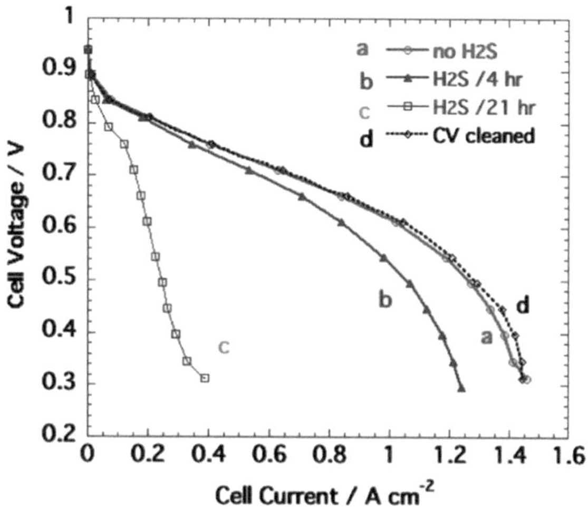

Fig. 8.44 Polarization curves of the cell at different moments at 1 ppm concentration of H_2S, and after electrochemical cleaning [104]

8.7.2 *Impurities in Air*

The cathode supply gas of the fuel cell is air, and impurities are common pollutants and aerosols in the air. Impurities in the cathode include: sulfides, nitrogen oxides, hydrocarbons, ozone, particles, and aerosol salts. Sulfur dioxide, a common air pollutant from fossil fuel combustion, causes degradation of cell performance by adsorption on the platinum catalyst surface.

For nitric oxide, strong but reversible effects have been observed in most studies. In in situ PEMFC experiments, gas supply concentrations of 1–25 ppm and injection times of 1–24 h are typically used and result in a loss of performance of 3–60%. In addition to gas-phase contaminants, salt (mainly from marine fog and road deicers) can also contaminate the cathode air supply. In the case of NaCl, Na ions can replace H^+ in the ionomer (catalysis layer and membrane), leading to reduced conductivity. High concentrations of salt can reduce the hydrophobicity of the gas diffusion layer, leading to the accumulation of liquid water in the cell and increasing the resistance of the cell to material transport. Salt also accelerates the decay of metallic components in the cell, for example, it accelerates the corrosion rate of metallic bipolar plates.

For a general air-side contaminant P, a series of electrode and surface reactions are as follows:

$$\mathrm{Pt}-\mathrm{P}\underset{k_{p1b}}{\overset{k_{p1f}}{\rightleftarrows}}\mathrm{Pt_n}-\mathrm{P} \tag{8.19}$$

$$\mathrm{P}+\mathrm{nPt}-\mathrm{O_2}\underset{k_{p2b}}{\overset{k_{p2f}}{\rightleftarrows}}\mathrm{Pt_n}-\mathrm{P}+\mathrm{nO_2} \tag{8.20}$$

Most of the contaminants present in the cathode air decay the fuel cell by chemical adsorption on the surface of the platinum catalyst layer, reducing the effective reaction area of the cell. Three main mitigation strategies are described below: (1) Use of tolerant catalysts, referring to those catalysts that are more active and selective for oxygen than impurities. This reduces the adsorption of airborne contaminants to the catalyst and uses the oxygen reduction reaction as its primary reaction path. For example, Pt uses graphene (G) as a carrier compared to C and has better durability against sulfur dioxide. After the PEMFC was exposed to a gas stream containing 1 ppm sulfur dioxide for about 3 h, the Pt/C catalyst lost 88% of its starting current density, while the Pt/G catalyst decreased by 31% under the same conditions [213]. As shown in Fig. 8.45 [213]. (2) Using an air filter is another way to address the contaminants. Filtration of gaseous contaminants by chemisorption or physical adsorption to prevent them from reaching the catalyst site, which may be more practical and cost-effective. For example, SO_2 removal by adsorption through MnO_x/Al_2O_3, NO_2 removal by adsorption through PICA activated carbon, and NO removal by oxidation through $KMnO_4/Al_2O_3$. (3) Using ozone for treatment. It has been shown that the addition of a few percent of O_3 can restore cell performance to the cathode of a sulfur-poisoned fuel cell in a short time [13]. The advantages of this strategy are: a fast cleaning process (10–20 min); it can occur at room temperature; and no special equipment or modifications need to be added to the fuel cell system. It is worth noting that excess O_3 will negatively affect the fuel cell as a new impurity.

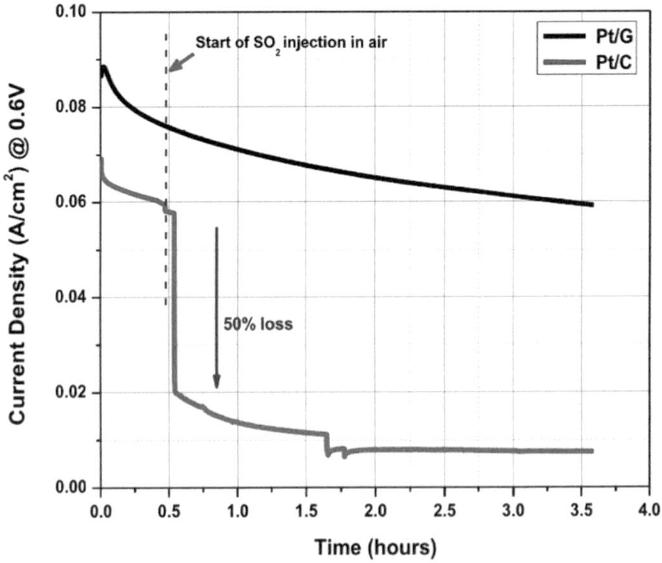

Fig. 8.45 Comparison of transient current density of Pt on carbon and Pt on graphene exposed to 1 ppm SO_2 [213]

8.8 Summary and Outlook

Due to the complex working environment and harsh operating conditions, automotive fuel cells can be severely degraded. Different components under different operating conditions accelerate fuel cell aging in different ways, including component aging, morphological changes, parameter attenuation, and ultimately leading to performance degradation. Understanding the degradation mechanisms of various components is of great interest.

The decay of membranes can be divided into: (1) Mechanical degradation. This is mainly due to stress changes on the membrane, resulting in mechanical damage such as tears and voids in the membrane. (2) Thermal degradation. PFSA membrane will tear severely at high temperatures due to the glass transition temperature of PFSA polymers being around 80 °C. Many studies have shown that the side chains with sulfonic acid groups will decompose when the temperature exceeds 280 °C. (3) Chemical degradation. The main chain and side chain structures of the membrane undergo unchaining under the attack of free radicals such as HO·. Free radicals such as hydroxyl radicals are mainly derived from the decomposition of hydrogen peroxide. Oxygen transport across the membrane from the cathode to the anode is the main cause of hydrogen peroxide production. The platinum catalysts undergo complex degradation processes in the fuel cell, including dissolution and redeposition, Ostwald ripening and agglomeration, detachment and migration, and active site poisoning. Carbon support corrosion is accelerated at high interfacial potential, which is attributed to the hydrogen-air interface and reverse current mechanism. The ECSA decreases during these processes. Material improvement, innovation, and system control strategy optimization are considered effective mitigation measures. The degradation of GDL will lead to structural destruction and surface wetness changes, which can be divided into the mechanical damage caused by compression, dissolution effect by water, and air flow erosion, and the physicochemical degradation caused by carbon corrosion and PTFE loss. In addition, the attenuation of metal bipolar plates is mainly a corrosion effect. This is mainly due to the high humidity and acidic environment in which the bipolar plates are exposed. Therefore, surface treatment or coating is required.

Fuel cell degradation during long-term operation under common conditions is a complex process that includes many mechanisms that occur in parallel, so it must be based on a comprehensive study that considers both the actual fuel cell operation and the interaction between different materials. In addition, Automotive fuel cell stacks, typically comprising hundreds of cells with large active areas, experience significant local environmental variations, leading to inhomogeneous degradation and performance decline. Addressing this challenge requires a holistic approach integrating experimental studies, advanced modeling, and system-level optimizations. Recent advancements, such as antioxidant strategies to mitigate oxidative stress [214], high-temperature proton exchange membranes with enhanced durability [215], and metal-NC catalysts resistant to corrosion and dissolution during the oxygen reduction reaction [216], have provided effective pathways to enhance fuel cell lifespan. Moreover, system-level innovations, including improved thermal and water management and advanced coatings for bipolar plates, have been shown to reduce failure rates in high-temperature operating conditions [217]. Future research should focus on integrating in situ diagnostics and machine learning techniques to optimize material design, operational strategies, and predictive maintenance, paving the way for the development of long-life fuel cell systems to meet the demands of sustainable energy and transportation applications.

References

1. Guerrero Moreno N, Cisneros Molina M, Gervasio D, et al. Approaches to polymer electrolyte membrane fuel cells (PEMFCs) and their cost [J]. Renewable and Sustainable Energy Reviews, 2015, 52: 897–906.
2. Satyapal S. U.S. Department of Energy Hydrogen and Fuel Cells Program: 2019 Annual Merit Review and Peer Evaluation Report; April 29 - May 1, 2019, Arlington, Virginia [R]. United States, 2020.
3. Pei P, Chang Q, Tang T. A quick evaluating method for automotive fuel cell lifetime [J]. International Journal of Hydrogen Energy, 2008, 33(14): 3829–36.
4. Tang H, Peikang S, Jiang S P, et al. A degradation study of Nafion proton exchange membrane of PEM fuel cells [J]. Journal of Power Sources, 2007, 170(1): 85–92.
5. Kim S, Mench M M. Physical degradation of membrane electrode assemblies undergoing freeze/thaw cycling: Micro-structure effects [J]. Journal of Power Sources, 2007, 174(1): 206–20.
6. Pestrak M, Li Y, Case S W, et al. The Effect of Mechanical Fatigue on the Lifetimes of Membrane Electrode Assemblies [J]. Journal of Fuel Cell Science and Technology, 2010, 7(4).
7. Chandesris M, Vincent R, Guetaz L, et al. Membrane degradation in PEM fuel cells: From experimental results to semi-empirical degradation laws [J]. International Journal of Hydrogen Energy, 2017, 42(12): 8139–49.
8. Qiu D, Peng L, Lai X, et al. Mechanical failure and mitigation strategies for the membrane in a proton exchange membrane fuel cell [J]. Renewable and Sustainable Energy Reviews, 2019, 113: 109289.
9. Liu W, Qiu D, Peng L, et al. Mechanical degradation of proton exchange membrane during assembly and running processes in proton exchange membrane fuel cells with metallic bipolar plates [J]. International Journal of Energy Research, 2020, 44(11): 8622–34.
10. Lai Y H, Dillard D A. Mechanical durability characterization and modeling of ionomeric membranes [M]. 2010.
11. Khorasany R M H, Kjeang E, Wang G G, et al. Simulation of ionomer membrane fatigue under mechanical and hygrothermal loading conditions [J]. Journal of Power Sources, 2015, 279: 55–63.
12. Ye D, Tu Z, Yu Y, et al. Hydrogen permeation across super-thin membrane and the burning limitation in low-temperature proton exchange membrane fuel cell [J]. International Journal of Energy Research, 2014, 38(9): 1181–91.
13. Wu J, Yuan X Z, Martin J J, et al. A review of PEM fuel cell durability: Degradation mechanisms and mitigation strategies [J]. Journal of Power Sources, 2008, 184(1): 104–19.
14. Liu W, Ruth K, Rusch G. Membrane durability in PEM fuel cells [J]. Journal of New Materials for Electrochemical Systems, 2001, 4.
15. Lin C-C, Lien W-F, Wang Y-Z, et al. Preparation and performance of sulfonated polyimide/Nafion multilayer membrane for proton exchange membrane fuel cell [J]. Journal of Power Sources, 2012, 200: 1–7.

16. Xing Y, Li H, Avgouropoulos G. Research Progress of Proton Exchange Membrane Failure and Mitigation Strategies [J]. Materials, 2021, 14(10): 2591.
17. Seo D C, Jeon I, Jeong E S, et al. Mechanical Properties and Chemical Durability of Nafion/Sulfonated Graphene Oxide/Cerium Oxide Composite Membranes for Fuel-Cell Applications [J]. Polymers, 2020, 12(6): 1375.
18. Yuan X-Z, Zhang S, Wang H, et al. Degradation of a polymer exchange membrane fuel cell stack with Nafion® membranes of different thicknesses: Part I. In situ diagnosis [J]. Journal of Power Sources, 2010, 195(22): 7594–9.
19. Chang I, Park T, Lee J, et al. Flexible fuel cell using stiffness-controlled endplate [J]. International Journal of Hydrogen Energy, 2016, 41(14): 6013–9.
20. Wang X, Song Y, Zhang B. Experimental study on clamping pressure distribution in PEM fuel cells [J]. Journal of Power Sources, 2008, 179(1): 305–9.
21. Wozniczka B, Fletcher N J, Gibb P R. Electrochemical fuel cell stack, F, 1997 [C].
22. Peng L, Yi P, Lai X. Design and manufacturing of stainless steel bipolar plates for proton exchange membrane fuel cells [J]. International Journal of Hydrogen Energy, 2014, 39(36): 21127–53.
23. Vielstich W, Lamm A, Gasteiger H. Handbook of fuel cells. Fundamentals, technology, applications [M]. United Kingdom: John Wiley and Sons Ltd., Chichester (United Kingdom), 2003.
24. Yan W-M, Mei S-C, Soong C-Y, et al. Experimental study on the performance of PEM fuel cells with interdigitated flow channels [J]. Journal of Power Sources, 2006, 160(1): 116–22.
25. Qiu D, Peng L, Yi P, et al. Contact behavior modelling and its size effect on proton exchange membrane fuel cell [J]. Journal of Power Sources, 2017, 365: 190–200.
26. Ramani D, Khattra N S, Singh Y, et al. Mitigation of mechanical membrane degradation in fuel cells – Part 1: Gas diffusion layers with low surface roughness [J]. Journal of Power Sources, 2021, 512: 230446.
27. Ye D H, Zhan Z G, Lee Y J, et al. Effects of Frame Materials and Structures on Stress Concentration of Membrane Electrode Assembly of PEMFCs [J]. Fuel Cells, 2013, 13(6): 1205–12.
28. Zhang J, Yin G-P, Wang Z-B, et al. Effects of hot pressing conditions on the performances of MEAs for direct methanol fuel cells [J]. Journal of Power Sources, 2007, 165(1): 73–81.
29. El-kharouf A, Chandan A, Hattenberger M, et al. Proton exchange membrane fuel cell degradation and testing: review [J]. Journal of the Energy Institute, 2012, 85(4): 188–200.
30. Taniguchi A, Akita T, Yasuda K, et al. Analysis of electrocatalyst degradation in PEMFC caused by cell reversal during fuel starvation [J]. Journal of Power Sources, 2004, 130(1): 42–9.
31. Ralph T R, Hogarth M P. Catalysis for Low Temperature Fuel Cells [J]. Platinum Metals Review, 2002, 46(3): 117–35.
32. Surowiec J, Bogoczek R. Studies on the thermal stability of the perfluorinated cation-exchange membrane Nafion-417 [J]. Journal of thermal analysis, 1988, 33(4): 1097–102.
33. Lee H-J, Cho M K, Jo Y Y, et al. Application of TGA techniques to analyze the compositional and structural degradation of PEMFC MEAs [J]. Polymer Degradation and Stability, 2012, 97(6): 1010–6.
34. Wilkie C A, Thomsen J R, Mittleman M L. Interaction of poly(methyl methacrylate) and nafions [J]. Journal of Applied Polymer Science, 1991, 42(4): 901–9.
35. Chu D, Gervasio D, Razaq M, et al. Infrared reflectance absorption spectroscopy (IRRAS). Study of the thermal stability of perfluorinated sulphonic acid ionomers on Pt [J]. Journal of Applied Electrochemistry, 1990, 20(1): 157–62.
36. Deng Q, Wilkie C A, Moore R B, et al. TGA–FTi.r. investigation of the thermal degradation of Nafion® and Nafion®/ [silicon oxide]-based nanocomposites [J]. Polymer, 1998, 39(24): 5961–72.
37. Kim Y S, Dong L, Hickner M A, et al. State of Water in Disulfonated Poly(arylene ether sulfone) Copolymers and a Perfluorosulfonic Acid Copolymer (Nafion) and Its Effect on Physical and Electrochemical Properties [J]. Macromolecules, 2003, 36(17): 6281–5.

38. Cappadonia M, Erning J W, Stimming U. Proton conduction of Nafion® 117 membrane between 140 K and room temperature [J]. Journal of Electroanalytical Chemistry, 1994, 376(1): 189–93.
39. Sivashinsky N, Tanny G B. The state of water in swollen ionomers containing sulfonic acid salts [J]. Journal of Applied Polymer Science, 1981, 26(8): 2625–37.
40. Cho E, Ko J-J, Ha H Y, et al. Effects of Water Removal on the Performance Degradation of PEMFCs Repetitively Brought to <0°C [J]. Journal of The Electrochemical Society, 2004, 151(5): A661.
41. Mcdonald R C, Mittelsteadt C K, Thompson E L. Effects of Deep Temperature Cycling on Nafion® 112 Membranes and Membrane Electrode Assemblies [J]. Fuel Cells, 2004, 4(3): 208–13.
42. Futter G A, Latz A, Jahnke T. Physical modeling of chemical membrane degradation in polymer electrolyte membrane fuel cells: Influence of pressure, relative humidity and cell voltage [J]. Journal of Power Sources, 2019, 410–411: 78–90.
43. Weber A Z, Newman J. Transport in Polymer-Electrolyte Membranes [J]. Journal of The Electrochemical Society, 2004, 151(2): A311.
44. Inaba M, Kinumoto T, Kiriake M, et al. Gas crossover and membrane degradation in polymer electrolyte fuel cells [J]. Electrochimica Acta, 2006, 51(26): 5746–53.
45. James P. Hoare J W. The electrochemistry of oxygen. [J]. Journal of Molecular Structure, 1968, 6: 338–9.
46. Sethuraman V A, Weidner J W, Haug A T, et al. Hydrogen Peroxide Formation Rates in a PEMFC Anode and Cathode [J]. Journal of The Electrochemical Society, 2008, 155(1): B50.
47. Inaba M, Yamada H, Tokunaga J, et al. Effect of Agglomeration of Pt/C Catalyst on Hydrogen Peroxide Formation [J]. Electrochemical and Solid State Letters, 2004, 7.
48. Paulus U A, Wokaun A, Scherer G G, et al. Oxygen Reduction on Carbon-Supported Pt−Ni and Pt−Co Alloy Catalysts [J]. The Journal of Physical Chemistry B, 2002, 106(16): 4181–91.
49. Chen C, Fuller T F. The effect of humidity on the degradation of Nafion® membrane [J]. Polymer Degradation and Stability, 2009, 94(9): 1436–47.
50. R. GERISCHER H G. Über die katalytische Zersetzung von Wasserstoffsuperoxyd an metallischem Platin [J]. Zeitschrift für Physikalische Chemie, 1956, 6(3_4): 178–200.
51. Wang F, Tang H, Pan M, et al. Ex situ investigation of the proton exchange membrane chemical decomposition [J]. International Journal of Hydrogen Energy, 2008, 33(9): 2283–8.
52. Pozio A, Silva R F, De FRANCESCO M, et al. Nafion degradation in PEFCs from end plate iron contamination [J]. Electrochimica Acta, 2003, 48(11): 1543–9.
53. Yi P, Li X, Yao L, et al. A lifetime prediction model for coated metallic bipolar plates in proton exchange membrane fuel cells [J]. Energy Conversion and Management, 2019, 183: 65–72.
54. Wilberforce T, Ijaodola O, Ogungbemi E, et al. Technical evaluation of proton exchange membrane (PEM) fuel cell performance – A review of the effects of bipolar plates coating [J]. Renewable and Sustainable Energy Reviews, 2019, 113: 109286.
55. Rigg T, Taylor W, Weiss J. The rate constant of the bimolecular reaction between hydrogen peroxide and ferrous ion [J]. Experientia, 1954, 10(5): 202–3.
56. Gubler L, Dockheer S M, Koppenol W H. Radical (HO_•, H_• and HOO_•) Formation and Ionomer Degradation in Polymer Electrolyte Fuel Cells [J]. Journal of The Electrochemical Society, 2011, 158(7): B755-B69.
57. Bielski B H J. Reevaluation Of The Spectral And Kinetic Properties Of Ho2 And O2- Free Radicals [J]. Photochemistry and Photobiology, 1978, 28(4–5): 645–9.
58. Rush J D, Bielski B H. Pulse radiolysis studies of alkaline iron(III) and iron(VI) solutions. Observation of transient iron complexes with intermediate oxidation states [J]. Journal of the American Chemical Society, 1986, 108(3): 523–5.
59. Lee C, Yoon J. Temperature dependence of hydroxyl radical formation in the hv/Fe3+/H2O2 and Fe3+/H2O2 systems [J]. Chemosphere, 2004, 56(10): 923–34.
60. Lundström T, Christensen H, Sehested K. Reactions of the HO2 radical with OH, H, Fe2+ and Cu2+ at elevated temperatures [J]. Radiation Physics and Chemistry, 2004, 69(3): 211–6.

61. Buxton G V, Greenstock C L, Helman W P, et al. Critical Review of rate constants for reactions of hydrated electrons, hydrogen atoms and hydroxyl radicals (·OH/·O− in Aqueous Solution [J]. Journal of Physical and Chemical Reference Data, 1988, 17(2): 513–886.
62. Christensen H, Sehested K, Corfitzen H. Reactions of Hydroxyl Radicals with Hydrogen Peroxide at Ambient and Elevated Temperatures [J]. The Journal of Physical Chemistry, 1982, 86: 1588–90.
63. Koppenol W H, Butler J, Leeuwen J W V. THE HABER-WEISS CYCLE [J]. Photochemistry and Photobiology, 1978, 28(4–5): 655–8.
64. Dainton F S, Rowbottom J. The primary radical yield in water. A comparison of the photolysis and radiolysis of solutions of hydrogen peroxide [J]. Transactions of the Faraday Society, 1953, 49(0): 1160–73.
65. Lundström T, Christensen H, Sehested K. The reaction of hydrogen atoms with hydrogen peroxide as a function of temperature [J]. Radiation Physics and Chemistry, 2001, 61(2): 109–13.
66. Ghelichi M, Melchy P-É A, Eikerling M H. Radically Coarse-Grained Approach to the Modeling of Chemical Degradation in Fuel Cell Ionomers [J]. The Journal of Physical Chemistry B, 2014, 118(38): 11375–86.
67. Ghassemzadeh L, Holdcroft S. Quantifying the Structural Changes of Perfluorosulfonated Acid Ionomer upon Reaction with Hydroxyl Radicals [J]. Journal of the American Chemical Society, 2013, 135(22): 8181–4.
68. Zatoń M, ROZIèRE J, Jones D J. Current understanding of chemical degradation mechanisms of perfluorosulfonic acid membranes and their mitigation strategies: a review [J]. Sustainable Energy & Fuels, 2017, 1(3): 409–38.
69. Ghassemzadeh L, Kreuer K-D, Maier J, et al. Chemical Degradation of Nafion Membranes under Mimic Fuel Cell Conditions as Investigated by Solid-State NMR Spectroscopy [J]. The Journal of Physical Chemistry C, 2010, 114(34): 14635–45.
70. Ren P, Pei P, Li Y, et al. Degradation mechanisms of proton exchange membrane fuel cell under typical automotive operating conditions [J]. Progress in Energy and Combustion Science, 2020, 80: 100859.
71. Wong K H, Kjeang E. Mitigation of Chemical Membrane Degradation in Fuel Cells: Understanding the Effect of Cell Voltage and Iron Ion Redox Cycle [J]. ChemSusChem, 2015, 8(6): 1072–82.
72. Aoki M, Uchida H, Watanabe M. Decomposition mechanism of perfluorosulfonic acid electrolyte in polymer electrolyte fuel cells [J]. Electrochemistry Communications, 2006, 8(9): 1509–13.
73. Zhao D, Yi B L, Zhang H M, et al. MnO2/SiO2–SO3H nanocomposite as hydrogen peroxide scavenger for durability improvement in proton exchange membranes [J]. Journal of Membrane Science, 2010, 346(1): 143–51.
74. Trogadas P, Parrondo J, Ramani V. Degradation Mitigation in Polymer Electrolyte Membranes Using Cerium Oxide as a Regenerative Free-Radical Scavenger [J]. Electrochemical and Solid-State Letters, 2008, 11(7): B113.
75. Baker A M, Torraco D, Judge E J, et al. Cerium Migration during PEM Fuel Cell Assembly and Operation [J]. ECS Transactions, 2015, 69(17): 1009.
76. Hou H, Di VONA M L, Knauth P. Building bridges: Crosslinking of sulfonated aromatic polymers—A review [J]. Journal of Membrane Science, 2012, 423–424: 113–27.
77. Yu J, Yi B, Xing D, et al. Degradation mechanism of polystyrene sulfonic acid membrane and application of its composite membranes in fuel cells [J]. Physical Chemistry Chemical Physics, 2003, 5(3): 611–5.
78. Curtin D E, Lousenberg R D, Henry T J, et al. Advanced materials for improved PEMFC performance and life [J]. Journal of Power Sources, 2004, 131(1): 41–8.
79. Schiraldi D A. Perfluorinated Polymer Electrolyte Membrane Durability [J]. Journal of Macromolecular Science, Part C, 2006, 46(3): 315–27.
80. Singh Y, Orfino F P, Dutta M, et al. 3D Failure Analysis of Pure Mechanical and Pure Chemical Degradation in Fuel Cell Membranes [J]. Journal of The Electrochemical Society, 2017, 164(13): F1331.

81. Yousfi-steiner N, MOçOTéGUY P, Candusso D, et al. A review on polymer electrolyte membrane fuel cell catalyst degradation and starvation issues: Causes, consequences and diagnostic for mitigation [J]. Journal of Power Sources, 2009, 194(1): 130–45.
82. Meier J C, Galeano C, Katsounaros I, et al. Degradation Mechanisms of Pt/C Fuel Cell Catalysts under Simulated Start–Stop Conditions [J]. ACS Catalysis, 2012, 2(5): 832–43.
83. Shao Y, Yin G, Gao Y. Understanding and approaches for the durability issues of Pt-based catalysts for PEM fuel cell [J]. Journal of Power Sources, 2007, 171(2): 558–66.
84. Dubau L, Castanheira L, Maillard F, et al. A review of PEM fuel cell durability: materials degradation, local heterogeneities of aging and possible mitigation strategies [J]. WIREs Energy and Environment, 2014, 3(6): 540–60.
85. Hansen T W, Delariva A T, Challa S R, et al. Sintering of Catalytic Nanoparticles: Particle Migration or Ostwald Ripening? [J]. Accounts of Chemical Research, 2013, 46(8): 1720–30.
86. Meier J C, Galeano C, Katsounaros I, et al. Design criteria for stable Pt/C fuel cell catalysts [J]. Beilstein Journal of Nanotechnology, 2014, 5: 44–67.
87. Darling R M, Meyers J P. Kinetic Model of Platinum Dissolution in PEMFCs [J]. Journal of The Electrochemical Society, 2003, 150(11): A1523.
88. Kawano A, Imabayashi S-I. Influence of Oxygen Atmosphere on Dissolution of Platinum under Potential Cycling [J]. Journal of The Electrochemical Society, 2014, 161(1): F67.
89. Zana A, Speder J, Roefzaad M, et al. Probing Degradation by IL-TEM: The Influence of Stress Test Conditions on the Degradation Mechanism [J]. Journal of The Electrochemical Society, 2013, 160(6): F608.
90. Ahluwalia R K, Arisetty S, Wang X, et al. Thermodynamics and Kinetics of Platinum Dissolution from Carbon-Supported Electrocatalysts in Aqueous Media under Potentiostatic and Potentiodynamic Conditions [J]. Journal of The Electrochemical Society, 2013, 160(4): F447.
91. Dingreville R, Qu J, Mohammed C. Surface free energy and its effect on the elastic behavior of nano-sized particles, wires and films [J]. Journal of the Mechanics and Physics of Solids, 2005, 53(8): 1827–54.
92. Shao M, Peles A, Shoemaker K. Electrocatalysis on Platinum Nanoparticles: Particle Size Effect on Oxygen Reduction Reaction Activity [J]. Nano Letters, 2011, 11(9): 3714–9.
93. Myers D J, Wang X, Smith M C, et al. Potentiostatic and Potential Cycling Dissolution of Polycrystalline Platinum and Platinum Nano-Particle Fuel Cell Catalysts [J]. Journal of The Electrochemical Society, 2018, 165(6): F3178.
94. Ous T, Arcoumanis C. Degradation aspects of water formation and transport in Proton Exchange Membrane Fuel Cell: A review [J]. Journal of Power Sources, 2013, 240: 558–82.
95. Macauley N, Papadias D D, Fairweather J, et al. Carbon Corrosion in PEM Fuel Cells and the Development of Accelerated Stress Tests [J]. Journal of The Electrochemical Society, 2018, 165(6): F3148.
96. Luo Z, Li D, Tang H, et al. Degradation behavior of membrane–electrode-assembly materials in 10-cell PEMFC stack [J]. International Journal of Hydrogen Energy, 2006, 31(13): 1831–7.
97. Péron J, Nedellec Y, Jones D J, et al. The effect of dissolution, migration and precipitation of platinum in Nafion®-based membrane electrode assemblies during fuel cell operation at high potential [J]. Journal of Power Sources, 2008, 185(2): 1209–17.
98. Burlatsky S F, Gummalla M, Atrazhev V V, et al. The Dynamics of Platinum Precipitation in an Ion Exchange Membrane [J]. Journal of The Electrochemical Society, 2011, 158(3): B322.
99. Ioroi T, Siroma Z, Yamazaki S-I, et al. Electrocatalysts for PEM Fuel Cells [J]. Advanced Energy Materials, 2019, 9(23): 1801284.
100. Shao-horn Y, Ferreira P, La O G J, et al. Coarsening of Pt Nanoparticles in Proton Exchange Membrane Fuel Cells upon Potential Cycling [J]. ECS Transactions, 2006, 1(8): 185.
101. Yasuda K, Taniguchi A, Akita T, et al. Platinum dissolution and deposition in the polymer electrolyte membrane of a PEM fuel cell as studied by potential cycling [J]. Physical Chemistry Chemical Physics, 2006, 8(6): 746–52.
102. Zhang J, Litteer B A, Gu W, et al. Effect of Hydrogen and Oxygen Partial Pressure on Pt Precipitation within the Membrane of PEMFCs [J]. Journal of The Electrochemical Society, 2007, 154(10): B1006.

103. Bi W, Gray G E, Fuller T F. PEM Fuel Cell Pt / C Dissolution and Deposition in Nafion Electrolyte [J]. Electrochemical and Solid-State Letters, 2007, 10(5): B101.
104. Borup R, Meyers J, Pivovar B, et al. Scientific Aspects of Polymer Electrolyte Fuel Cell Durability and Degradation [J]. Chemical Reviews, 2007, 107(10): 3904–51.
105. Mohtadi R, Lee W K, Van ZEE J W. Assessing durability of cathodes exposed to common air impurities [J]. Journal of Power Sources, 2004, 138(1): 216–25.
106. Zhang S, Yuan X-Z, Hin J N C, et al. A review of platinum-based catalyst layer degradation in proton exchange membrane fuel cells [J]. Journal of Power Sources, 2009, 194(2): 588–600.
107. Urdampilleta I, Uribe F, Rockward T, et al. PEMFC Poisoning with H2S: Dependence on Operating Conditions [J]. ECS Transactions, 2007, 11(1): 831.
108. Matsuoka K, Sakamoto S, Nakato K, et al. Degradation of polymer electrolyte fuel cells under the existence of anion species [J]. Journal of Power Sources, 2008, 179(2): 560–5.
109. Yang X, Roling L T, Vara M, et al. Synthesis and Characterization of Pt–Ag Alloy Nanocages with Enhanced Activity and Durability toward Oxygen Reduction [J]. Nano Letters, 2016, 16(10): 6644–9.
110. Ren X, Lv Q, Liu L, et al. Current progress of Pt and Pt-based electrocatalysts used for fuel cells [J]. Sustainable Energy & Fuels, 2020, 4(1): 15–30.
111. Tarasevich M R, Bogdanovskaya V A, Loubnin E N, et al. Comparative study of the corrosion behavior of platinum-based nanosized cathodic catalysts for fuel cells [J]. Protection of Metals, 2007, 43(7): 689–93.
112. Yu P, Pemberton M, Plasse P. PtCo/C cathode catalyst for improved durability in PEMFCs [J]. Journal of Power Sources, 2005, 144(1): 11–20.
113. Luo L, Zhu F, Tian R, et al. Composition-Graded PdxNi1–x Nanospheres with Pt Monolayer Shells as High-Performance Electrocatalysts for Oxygen Reduction Reaction [J]. ACS Catalysis, 2017, 7(8): 5420–30.
114. Koh S, Strasser P. Electrocatalysis on Bimetallic Surfaces: Modifying Catalytic Reactivity for Oxygen Reduction by Voltammetric Surface Dealloying [J]. Journal of the American Chemical Society, 2007, 129(42): 12624–5.
115. Zhang J, Sasaki K, Sutter E, et al. Stabilization of Platinum Oxygen-Reduction Electrocatalysts Using Gold Clusters [J]. Science, 2007, 315(5809): 220–2.
116. Xiong Y, Ma Y, Zou L, et al. N-doping induced tensile-strained Pt nanoparticles ensuring an excellent durability of the oxygen reduction reaction [J]. Journal of Catalysis, 2020, 382: 247–55.
117. Feng Q, Yuan X Z, Liu G, et al. A review of proton exchange membrane water electrolysis on degradation mechanisms and mitigation strategies [J]. Journal of Power Sources, 2017, 366: 33–55.
118. Kim J, Kim S-I, Jo S G, et al. Enhanced activity and durability of Pt nanoparticles supported on reduced graphene oxide for oxygen reduction catalysts of proton exchange membrane fuel cells [J]. Catalysis Today, 2020, 352: 10–7.
119. Xu F, Wang D, Sa B, et al. One-pot synthesis of Pt/CeO2/C catalyst for improving the ORR activity and durability of PEMFC [J]. International Journal of Hydrogen Energy, 2017, 42(18): 13011–9.
120. Ando F, Tanabe T, Gunji T, et al. Improvement of ORR Activity and Durability of Pt Electrocatalyst Nanoparticles Anchored on TiO2/Cup-Stacked Carbon Nanotube in Acidic Aqueous Media [J]. Electrochimica Acta, 2017, 232: 404–13.
121. Cetinbas F C, Advani S G, Prasad A K. Investigation of a polymer electrolyte membrane fuel cell catalyst layer with bidirectionally-graded composition [J]. Journal of Power Sources, 2014, 270: 594–602.
122. Zheng Z, Yang F, Lin C, et al. Design of gradient cathode catalyst layer (CCL) structure for mitigating Pt degradation in proton exchange membrane fuel cells (PEMFCs) using mathematical method [J]. Journal of Power Sources, 2020, 451: 227729.
123. Baumgartner W R R, Baumgartner W R R, WALLNöFER E, et al. Electrocatalytic Corrosion of Carbon Support in PEMFC at Fuel Starvation [J]. ECS Transactions, 2006, 3(1): 811.

124. Borup R L, Davey J R, Garzon F H, et al. PEM fuel cell electrocatalyst durability measurements [J]. Journal of Power Sources, 2006, 163(1): 76–81.
125. Makharia R, Kocha S, Yu P, et al. Durable PEM Fuel Cell Electrode Materials: Requirements and Benchmarking Methodologies [J]. ECS Transactions, 2006, 1(8): 3.
126. Takagi Y, Sato Y, Wang Z. Effect of Anode Catalyst Support on MEA Degradation Caused by Hydrogen-Starved Operation of a PEFC [J]. ECS Transactions, 2006, 3(1): 827.
127. Castanheira L, Silva W O, Lima F H B, et al. Carbon Corrosion in Proton-Exchange Membrane Fuel Cells: Effect of the Carbon Structure, the Degradation Protocol, and the Gas Atmosphere [J]. ACS Catalysis, 2015, 5(4): 2184–94.
128. Roen L M, Paik C H, Jarvi T D. Electrocatalytic Corrosion of Carbon Support in PEMFC Cathodes [J]. Electrochemical and Solid-State Letters, 2004, 7(1): A19.
129. Chen H, Zhao X, Zhang T, et al. The reactant starvation of the proton exchange membrane fuel cells for vehicular applications: A review [J]. Energy Conversion and Management, 2019, 182: 282–98.
130. Taniguchi A, Akita T, Yasuda K, et al. Analysis of degradation in PEMFC caused by cell reversal during air starvation [J]. International Journal of Hydrogen Energy, 2008, 33(9): 2323–9.
131. Patterson T W, Darling R M. Damage to the Cathode Catalyst of a PEM Fuel Cell Caused by Localized Fuel Starvation [J]. Electrochemical and Solid-State Letters, 2006, 9(4): A183.
132. GUéTAZ L, Escribano S, Sicardy O. Study by electron microscopy of proton exchange membrane fuel cell membrane-electrode assembly degradation mechanisms: Influence of local conditions [J]. Journal of Power Sources, 2012, 212: 169–78.
133. Knights S D, Colbow K M, St-pierre J, et al. Aging mechanisms and lifetime of PEFC and DMFC [J]. Journal of Power Sources, 2004, 127(1): 127–34.
134. Liang D, Shen Q, Hou M, et al. Study of the cell reversal process of large area proton exchange membrane fuel cells under fuel starvation [J]. Journal of Power Sources, 2009, 194(2): 847–53.
135. Perry M L, Darling R M, Kandoi S, et al. Operating Requirements for Durable Polymer-Electrolyte Fuel Cell Stacks [M]//BüCHI F N, INABA M, SCHMIDT T J. Polymer Electrolyte Fuel Cell Durability. New York, NY; Springer New York. 2009: 399–417.
136. Kim J, Lee J, Tak Y. Relationship between carbon corrosion and positive electrode potential in a proton-exchange membrane fuel cell during start/stop operation [J]. Journal of Power Sources, 2009, 192(2): 674–8.
137. Yu Y, Li H, Wang H, et al. A review on performance degradation of proton exchange membrane fuel cells during startup and shutdown processes: Causes, consequences, and mitigation strategies [J]. Journal of Power Sources, 2012, 205: 10–23.
138. Reiser C A, Bregoli L, Patterson T W, et al. A Reverse-Current Decay Mechanism for Fuel Cells [J]. Electrochemical and Solid-State Letters, 2005, 8(6): A273.
139. Tang H, Qi Z, Ramani M, et al. PEM fuel cell cathode carbon corrosion due to the formation of air/fuel boundary at the anode [J]. Journal of Power Sources, 2006, 158(2): 1306–12.
140. Venkatesan S V, Dutta M, Kjeang E. Mesoscopic degradation effects of voltage cycled cathode catalyst layers in polymer electrolyte fuel cells [J]. Electrochemistry Communications, 2016, 72: 15–8.
141. Komini BABU S, Spernjak D, Dillet J, et al. Spatially resolved degradation during startup and shutdown in polymer electrolyte membrane fuel cell operation [J]. Applied Energy, 2019, 254: 113659.
142. Durst J, Lamibrac A, Charlot F, et al. Degradation heterogeneities induced by repetitive start/stop events in proton exchange membrane fuel cell: Inlet vs. outlet and channel vs. land [J]. Applied Catalysis B: Environmental, 2013, 138–139: 416–26.
143. Schulenburg H, Schwanitz B, Linse N, et al. 3D Imaging of Catalyst Support Corrosion in Polymer Electrolyte Fuel Cells [J]. The Journal of Physical Chemistry C, 2011, 115(29): 14236–43.
144. Weber A Z, Borup R L, Darling R M, et al. A Critical Review of Modeling Transport Phenomena in Polymer-Electrolyte Fuel Cells [J]. Journal of The Electrochemical Society, 2014, 161(12): F1254.

145. Ishigami Y, Takada K, Yano H, et al. Corrosion of carbon supports at cathode during hydrogen/air replacement at anode studied by visualization of oxygen partial pressures in a PEFC—Start-up/shut-down simulation [J]. Journal of Power Sources, 2011, 196(6): 3003–8.
146. Ghosh S, Ohashi H, Tabata H, et al. In-plane and through-plane non-uniform carbon corrosion of polymer electrolyte fuel cell cathode catalyst layer during extended potential cycles [J]. Journal of Power Sources, 2017, 362: 291–8.
147. Schneider I A, Von DAHLEN S. Start-Stop Phenomena in Channel and Land Areas of a Polymer Electrolyte Fuel Cell [J]. Electrochemical and Solid-State Letters, 2011, 14(2): B30.
148. Wirth S, Harnisch F, Weinmann M, et al. Comparative study of IVB–VIB transition metal compound electrocatalysts for the hydrogen evolution reaction [J]. Applied Catalysis B: Environmental, 2012, 126: 225–30.
149. Esposito D V, Hunt S T, Stottlemyer A L, et al. Low-Cost Hydrogen-Evolution Catalysts Based on Monolayer Platinum on Tungsten Monocarbide Substrates [J]. Angewandte Chemie International Edition, 2010, 49(51): 9859–62.
150. He D, Jiang Y, Lv H, et al. Nitrogen-doped reduced graphene oxide supports for noble metal catalysts with greatly enhanced activity and stability [J]. Applied Catalysis B: Environmental, 2013, 132–133: 379–88.
151. Kim D H, Mei J, Ayzner A L, et al. Sequentially solution-processed, nanostructured polymer photovoltaics using selective solvents [J]. Energy & Environmental Science, 2014, 7(3): 1103–9.
152. Liu Z, Shi Q, Zhang R, et al. Phosphorus-doped carbon nanotubes supported low Pt loading catalyst for the oxygen reduction reaction in acidic fuel cells [J]. Journal of Power Sources, 2014, 268: 171–5.
153. Yu Y, Tu Z, Zhang H, et al. Comparison of degradation behaviors for open-ended and closed proton exchange membrane fuel cells during startup and shutdown cycles [J]. Journal of Power Sources, 2011, 196(11): 5077–83.
154. Wang X, Li W, Chen Z, et al. Durability investigation of carbon nanotube as catalyst support for proton exchange membrane fuel cell [J]. Journal of Power Sources, 2006, 158(1): 154–9.
155. Job N, Marie J, Lambert S, et al. Carbon xerogels as catalyst supports for PEM fuel cell cathode [J]. Energy Conversion and Management, 2008, 49(9): 2461–70.
156. Liu X, Chen J, Liu G, et al. Enhanced long-term durability of proton exchange membrane fuel cell cathode by employing Pt/TiO2/C catalysts [J]. Journal of Power Sources, 2010, 195(13): 4098–103.
157. Dou M, Hou M, Wang F, et al. Sb-Doped SnO2 Supported Platinum Catalyst with High Stability for Proton Exchange Membrane Fuel Cells [J]. Journal of The Electrochemical Society, 2014, 161(12): F1231.
158. Perry M L, Patterson T, Reiser C. Systems Strategies to Mitigate Carbon Corrosion in Fuel Cells [J]. ECS Transactions, 2006, 3(1): 783.
159. Shen Q, Hou M, Liang D, et al. Study on the processes of start-up and shutdown in proton exchange membrane fuel cells [J]. Journal of Power Sources, 2009, 189(2): 1114–9.
160. Oyarce A, Zakrisson E, Ivity M, et al. Comparing shut-down strategies for proton exchange membrane fuel cells [J]. Journal of Power Sources, 2014, 254: 232–40.
161. Jo Y Y, Cho E A, Kim J H, et al. Effects of a hydrogen and air supply procedure on the performance degradation of PEMFCs [J]. International Journal of Hydrogen Energy, 2010, 35(23): 13118–24.
162. Yang Y, Li W, Lin R, et al. Impact of dummy load shut-down strategy on performance and durability of proton exchange membrane fuel cell stack [J]. Journal of Power Sources, 2018, 404: 126–34.
163. Lapicque F, Belhadj M, Bonnet C, et al. A critical review on gas diffusion micro and macroporous layers degradations for improved membrane fuel cell durability [J]. Journal of Power Sources, 2016, 336: 40–53.
164. Kang M, Sim J, Min K. Analysis of surface and interior degradation of gas diffusion layer with accelerated stress tests for polymer electrolyte membrane fuel cell [J]. International Journal of Hydrogen Energy, 2022, 47(68): 29467–80.

165. Dafalla A M, Jiang F. Stresses and their impacts on proton exchange membrane fuel cells: A review [J]. International Journal of Hydrogen Energy, 2018, 43(4): 2327–48.
166. Nitta I, Hottinen T, Himanen O, et al. Inhomogeneous compression of PEMFC gas diffusion layer: Part I. Experimental [J]. Journal of Power Sources, 2007, 171(1): 26–36.
167. Lai Y-H, Rapaport P A, Ji C, et al. Channel intrusion of gas diffusion media and the effect on fuel cell performance [J]. Journal of Power Sources, 2008, 184(1): 120–8.
168. Chang W R, Hwang J J, Weng F B, et al. Effect of clamping pressure on the performance of a PEM fuel cell [J]. Journal of Power Sources, 2007, 166(1): 149–54.
169. Zhang Z, He P, Dai Y-J, et al. Study of the mechanical behavior of paper-type GDL in PEMFC based on microstructure morphology [J]. International Journal of Hydrogen Energy, 2020, 45(53): 29379–94.
170. Bazylak A. Liquid water visualization in PEM fuel cells: A review [J]. International Journal of Hydrogen Energy, 2009, 34(9): 3845–57.
171. Radhakrishnan V, Haridoss P. Effect of cyclic compression on structure and properties of a Gas Diffusion Layer used in PEM fuel cells [J]. International Journal of Hydrogen Energy, 2010, 35(20): 11107–18.
172. Poornesh K K, Cho C D, Lee G B, et al. Gradation of mechanical properties in gas-diffusion electrode. Part 2: Heterogeneous carbon fiber and damage evolution in cell layers [J]. Journal of Power Sources, 2010, 195(9): 2718–30.
173. Baik K D, Kim S I, Hong B K, et al. Effects of gas diffusion layer structure on the open circuit voltage and hydrogen crossover of polymer electrolyte membrane fuel cells [J]. International Journal of Hydrogen Energy, 2011, 36(16): 9916–25.
174. Chlistunoff J, Davey J R, Rau K C, et al. PEMFC Gas Diffusion Media Degradation Determined by Acid-Base Titrations [J]. ECS Transactions, 2013, 50(2): 521.
175. Cho J, Ha T, Park J, et al. Analysis of transient response of a unit proton-exchange membrane fuel cell with a degraded gas diffusion layer [J]. International Journal of Hydrogen Energy, 2011, 36(10): 6090–8.
176. Ha T, Cho J, Park J, et al. Experimental study of the effect of dissolution on the gas diffusion layer in polymer electrolyte membrane fuel cells [J]. International Journal of Hydrogen Energy, 2011, 36(19): 12427–35.
177. Latorrata S, Gallo STAMPINO P, Cristiani C, et al. Development of an optimal gas diffusion medium for polymer electrolyte membrane fuel cells and assessment of its degradation mechanisms [J]. International Journal of Hydrogen Energy, 2015, 40(42): 14596–608.
178. Yu S, Li X, Li J, et al. Study on hydrophobicity degradation of gas diffusion layer in proton exchange membrane fuel cells [J]. Energy Conversion and Management, 2013, 76: 301–6.
179. Burheim O S, Ellila G, Fairweather J D, et al. Ageing and thermal conductivity of Porous Transport Layers used for PEM Fuel Cells [J]. Journal of Power Sources, 2013, 221: 356–65.
180. Chun J H, Jo D H, Kim S G, et al. Improvement of the mechanical durability of micro porous layer in a proton exchange membrane fuel cell by elimination of surface cracks [J]. Renewable Energy, 2012, 48: 35–41.
181. Pasaogullari U, Wang C Y. Liquid Water Transport in Gas Diffusion Layer of Polymer Electrolyte Fuel Cells [J]. Journal of The Electrochemical Society, 2004, 151(3): A399.
182. Han M, Shul Y-G, Lee H, et al. Accelerated testing of polymer electrolyte membranes under open-circuit voltage conditions for durable proton exchange membrane fuel cells [J]. International Journal of Hydrogen Energy, 2017, 42(52): 30787–91.
183. Singh R, Sui P C, Wong K H, et al. Modeling the Effect of Chemical Membrane Degradation on PEMFC Performance [J]. Journal of The Electrochemical Society, 2018, 165(6): F3328.
184. Liu H, George M G, Banerjee R, et al. Accelerated Degradation of Polymer Electrolyte Membrane Fuel Cell Gas Diffusion Layers [J]. Journal of The Electrochemical Society, 2017, 164(7): F704.
185. George M G, Liu H, Muirhead D, et al. Accelerated Degradation of Polymer Electrolyte Membrane Fuel Cell Gas Diffusion Layers [J]. Journal of The Electrochemical Society, 2017, 164(7): F714.

186. Fairweather J D, Li B, Mukundan R, et al. In Situ and Ex Situ Characterization of Carbon Corrosion in PEMFCs [J]. ECS Transactions, 2010, 33(1): 433.
187. Chen G, Zhang H, Ma H, et al. Electrochemical durability of gas diffusion layer under simulated proton exchange membrane fuel cell conditions [J]. International Journal of Hydrogen Energy, 2009, 34(19): 8185–92.
188. Ha T, Cho J, Park J, et al. Experimental study on carbon corrosion of the gas diffusion layer in polymer electrolyte membrane fuel cells [J]. International Journal of Hydrogen Energy, 2011, 36(19): 12436–43.
189. Yu S, Li X, Liu S, et al. Study on hydrophobicity loss of the gas diffusion layer in PEMFCs by electrochemical oxidation [J]. RSC Advances, 2014, 4(8): 3852–6.
190. Cho J, Park J, Oh H, et al. Analysis of the transient response and durability characteristics of a proton exchange membrane fuel cell with different micro-porous layer penetration thicknesses [J]. Applied Energy, 2013, 111: 300–9.
191. Spernjak D, Fairweather J, Mukundan R, et al. Influence of the microporous layer on carbon corrosion in the catalyst layer of a polymer electrolyte membrane fuel cell [J]. Journal of Power Sources, 2012, 214: 386–98.
192. Perry M L, Patterson T, Madden T. GDL Degradation in PEFC [J]. ECS Transactions, 2010, 33(1): 1081.
193. Antunes R A, Oliveira M C L, Ett G, et al. Corrosion of metal bipolar plates for PEM fuel cells: A review [J]. International Journal of Hydrogen Energy, 2010, 35(8): 3632–47.
194. Jirarungsatian C, Prateepasen A. Pitting and uniform corrosion source recognition using acoustic emission parameters [J]. Corrosion Science, 2010, 52(1): 187–97.
195. Kim J S, Peelen W H A, Hemmes K, et al. Effect of alloying elements on the contact resistance and the passivation behaviour of stainless steels [J]. Corrosion Science, 2002, 44(4): 635–55.
196. Wang H, Sweikart M A, Turner J A. Stainless steel as bipolar plate material for polymer electrolyte membrane fuel cells [J]. Journal of Power Sources, 2003, 115(2): 243–51.
197. Silva R F, Franchi D, Leone A, et al. Surface conductivity and stability of metallic bipolar plate materials for polymer electrolyte fuel cells [J]. Electrochimica Acta, 2006, 51(17): 3592–8.
198. Li M C, Zeng C L, Luo S Z, et al. Electrochemical corrosion characteristics of type 316 stainless steel in simulated anode environment for PEMFC [J]. Electrochimica Acta, 2003, 48(12): 1735–41.
199. Cho K H, Lee W G, Lee S B, et al. Corrosion resistance of chromized 316L stainless steel for PEMFC bipolar plates [J]. Journal of Power Sources, 2008, 178(2): 671–6.
200. Yang L, Yu H, Jiang L, et al. Improved anticorrosion properties and electrical conductivity of 316L stainless steel as bipolar plate for proton exchange membrane fuel cell by lower temperature chromizing treatment [J]. Journal of Power Sources, 2010, 195(9): 2810–4.
201. Bai C-Y, Ger M-D, Wu M-S. Corrosion behaviors and contact resistances of the low-carbon steel bipolar plate with a chromized coating containing carbides and nitrides [J]. International Journal of Hydrogen Energy, 2009, 34(16): 6778–89.
202. Han D H, Hong W H, Choi H S, et al. Inductively coupled plasma nitriding of chromium electroplated AISI 316L stainless steel for PEMFC bipolar plate [J]. International Journal of Hydrogen Energy, 2009, 34(5): 2387–95.
203. Feng K, Shen Y, Mai J, et al. An investigation into nickel implanted 316L stainless steel as a bipolar plate for PEM fuel cell [J]. Journal of Power Sources, 2008, 182(1): 145–52.
204. Feng K, Shen Y, Liu D, et al. Ni–Cr Co-implanted 316L stainless steel as bipolar plate in polymer electrolyte membrane fuel cells [J]. International Journal of Hydrogen Energy, 2010, 35(2): 690–700.
205. Li M, Luo S, Zeng C, et al. Corrosion behavior of TiN coated type 316 stainless steel in simulated PEMFC environments [J]. Corrosion Science, 2004, 46(6): 1369–80.
206. Wang Y, Northwood D O. An investigation into TiN-coated 316L stainless steel as a bipolar plate material for PEM fuel cells [J]. Journal of Power Sources, 2007, 165(1): 293–8.
207. Wang Y, Northwood D O. An investigation of the electrochemical properties of PVD TiN-coated SS410 in simulated PEM fuel cell environments [J]. International Journal of Hydrogen Energy, 2007, 32(7): 895–902.

208. Wang H, Turner J A, Li X, et al. SnO2:F coated austenite stainless steels for PEM fuel cell bipolar plates [J]. Journal of Power Sources, 2007, 171(2): 567–74.
209. Kitta S, Uchida H, Watanabe M. Metal separators coated with carbon/resin composite layers for PEFCs [J]. Electrochimica Acta, 2007, 53(4): 2025–33.
210. Baschuk J J, Li X. Carbon monoxide poisoning of proton exchange membrane fuel cells [J]. International Journal of Energy Research, 2001, 25(8): 695–713.
211. Wang L, Zhou Y, Yang Y, et al. Suppression of Carbon Monoxide Poisoning in Proton Exchange Membrane Fuel Cells via Gold Nanoparticle/Titania Ultrathin Film Heterogeneous Catalysts [J]. ACS Applied Energy Materials, 2019, 2(5): 3479–87.
212. VALDéS-LóPEZ V F, Mason T, Shearing P R, et al. Carbon monoxide poisoning and mitigation strategies for polymer electrolyte membrane fuel cells – A review [J]. Progress in Energy and Combustion Science, 2020, 79: 100842.
213. Jayaraj P, Karthika P, Rajalakshmi N, et al. Mitigation studies of sulfur contaminated electrodes for PEMFC [J]. International Journal of Hydrogen Energy, 2014, 39(23): 12045–51.
214. Yu X, Bai S, Li Q, et al. Performance Optimization by Antioxidant Strategies for Proton Exchange Membrane Fuel Cells: Recent Progress and Future [J]. EnergyChem, 2024: 100142.
215. Song J, Zhao W, Zhou L, et al. Rational Materials and Structure Design for Improving the Performance and Durability of High Temperature Proton Exchange Membranes (HT-PEMs) [J]. Advanced Science, 2023, 10(30): 2303969.
216. Kumar K, Dubau L, Jaouen F, et al. Review on the degradation mechanisms of metal-NC catalysts for the oxygen reduction reaction in acid electrolyte: current understanding and mitigation approaches [J]. Chemical Reviews, 2023, 123(15): 9265–9326.
217. Wang S, Zheng Y, Xv C, et al. Performance failure mechanisms and mitigation strategies of high-temperature proton exchange membrane fuel cells [J]. Progress in Materials Science, 2024: 101389.

Chapter 9
Proton Exchange Membrane Fuel Cells: Degradation Experiments and Numerical Simulations

9.1 Introduction

Experiments and numerical simulations are two important tools for studying fuel cell degradation, where experiments provide direct physical evidence for understanding the operation and performance of PEMFCs, while simulation models provide a theoretical basis for in-depth analysis of system behavior and optimal design. Much research has been carried out in this field, and much progress has been achieved. Therefore, it is necessary to introduce the experimental and numerical simulation methods to provide a reference for new researchers in related fields.

Before starting the description of the experiments and simulations, it is necessary to introduce the main degradation indices of fuel cells. Degradation indexes are parameters or characteristics that change regularly with aging time and operating conditions, which can indicate the decay state of the cell. To measure and predict the decay of PEMFCs, we need to characterize the aging information of the cell through degradation indexes [1]. The ideal degradation indexes should be able to accurately reflect the overall, multi-component, and multi-scale aging state of PEMFCs in real time and can be used online in industrial applications. Currently, the most widely used degradation indices are stack voltage, power, and impedance, which reflect the aging information of PEMFCs at macroscopic scales. However, to more accurately determine the decay of individual components in a cell, we need to measure and predict the decay of individual components. The order of influence of the decay of each component on the overall performance of PEMFCs is: PEM, electrode, BPP, GDL, and sealing gasket. Here, we will review and analyze the degradation index of each component in this order.

(a) Degradation indexes for PEM

The most common indices of PEM decay include: PEM thickness, FRR, gas permeability, and ionomer concentration. Researchers often use multiple metrics to characterize the aging of PEM. The change process of H_2 permeability and FRR with the

Y. Li, *Fuel Cell Fundamentals and Applications*,
https://doi.org/10.1007/978-981-96-9845-5_9

aging of PEM was analyzed by the durability test of PEM [2]. The effect of relative humidity on the aging rate and performance of Nafion 212 membranes was investigated using H_2 permeability and FRR as the decay indices of PEM [3]. In addition, a comparative experiment of the OCV aging test was conducted for 1000 h on four different thicknesses of Nafion films [4], including N117, N115, NR212, and NR211. The test results showed that the PEMFCs assembled with higher films had lower OCV due to higher gas permeability, where PEM thickness, high frequency resistance, low frequency resistance, OCV, and H_2 permeability were used to characterize the aging of PEM.

(b) Degradation indexes for the electrode

The most common indices of electrode decay are Pt particle size, ECSA, and carbon mass loss. A new model for Pt catalyst degradation was proposed [5]. The model can estimate the mass of Pt catalyst, the size of Pt particles, and ECSA. Another Pt catalyst degradation model was developed considering Ostwald ripening and agglomeration processes [6], which analyzed the growth of Pt particle radius and the decay of ECSA during the degradation process.

(c) Degradation indexes for BPP

The contact resistance at the interface of GDL and BPP is the most common degradation index of BPP, which is also a decay index for GDL. A contact resistance empirical model was established based on ex-situ accelerated testing to predict the service life of coated metal BPPs [7]. An improved contact resistance model considering the bending behavior and anisotropic properties of carbon fibers in GDL was developed [8], and the relationship between contact pressure and contact resistance was studied using a multiple regression method.

(d) Degradation indexes for GDL

The three most commonly used degradation indexes for GDL are hydrophobic loss, water content, and carbon mass loss, respectively. The loss of hydrophobicity is mainly caused by the degradation of PTFE. The PTFE coverage has a significant effect on the water content and water cluster structure in GDL. As the PTFE coverage decreases, larger water clusters are more likely to form in the GDL, and when a critical value is reached, water clusters connecting all vacant regions of the GDL are formed, which will further accelerate the decay of the GDL. The performance decay of PEMFCs caused by the loss of hydrophobicity of the GDL was investigated based on a multiscale hybrid modeling approach [9]. Similar to the hydrophobicity loss index, the use of water content as a decay index for GDL presupposes an accurate model to estimate the water content in GDL for different current loading conditions. The effect of GDL aging on liquid water transport in PEMFCs was investigated by synchrotron X-ray imaging and modeled the liquid water saturation at low current densities [10]. The results showed that the aged GDL exhibited higher liquid water content at low current densities. Carbon corrosion can occur in both electrodes and the GDL. The carbon mass loss was used as a decay index to characterize the aging of

GDL [10]. In addition to the above decay indexes, the porosity of GDL, the thickness of GDL, and the conductivity of GDL are also suitable as decay indexes for GDL.

(e) Degradation indexes for sealing gasket

PEMFCs typically use silicone rubber limbs, fluoroelastomers, and ethylene propylene diene monomers as the material for sealing gaskets. The two common decay indexes for sealing gaskets are the master curve and the sealing force. The stress relaxation behavior of liquid silicone rubber (LSR) was investigated in deionized water and ambient air [11], generated a master curve by time–temperature superposition, and predicted the service life of the seal material in PEMFCs from the master curve. The same authors also used experimental methods to study and quantify the changes in material stiffness caused by high temperature aging [12]. An improved Maxwell model for viscoelastic materials was developed and used to predict the change in sealing force under temperature cycling. The results show that the sealing force is closely related to the length of time and the temperature variation range of PEMFCs operation.

It is worth noting that the membrane electrode is the core component where the electrochemical reactions take place, which determines the overall performance of the cell. Therefore, in the subsequent sections, it is inevitable that more emphasis will be placed on the content related to the degradation of the membrane electrode assembly (MEA).

9.2 Experiments

Experimental methods are an important tool for investigating fuel cell decay. However, fuel cells can operate for thousands of hours under normal conditions, and it is impractical and costly to test their full process by experimental means. To reduce the required experimental time, the accelerated test method is preferred. The principle is to use different adverse operating conditions as a source of accelerated stress for film decay, such as load cycling [13], start–stop cycles [14], low relative humidity (RH) [15], or RH cycling [16], freeze–thaw cycles [17–20], and fuel [21] or air starvation [22]. Open circuit (OC) operation has also been recognized as an effective stressor for accelerated testing, and the accelerated effects of OC operation on the degradation of PEM fuel cell components, including the PEM and catalyst layers, have been investigated [3]. The US Department of Energy has established accelerated stress test (AST) protocols for cell components, including electrocatalysts, catalyst supports, membranes, and membrane electrode assemblies, in an attempt to provide a standard set of test conditions and operating procedures to evaluate new cell component materials and structures. Several diagnostic tools are already available for detecting fuel cell degradation, including in situ and ex-situ tools [23].

9.2.1 Steady-State Durability Test

Steady-state durability testing is one of the methods to evaluate the durability of a fuel cell, which can determine the lifetime of the fuel cell under certain operating conditions. The operating conditions for steady-state durability testing of fuel cells are generally constant-current discharge, while monitoring and analyzing the change in performance over time [24, 25]. The experiments related to steady-state durability testing are summarized in Table 9.1 [26]. The end-of-life (EOL) of a fuel cell can be defined by the percentage reduction in cell voltage, current density, or power rating, or by the complete failure of a critical component, and there is no uniform standard. For a given fuel cell, variations in the definition of life may result in completely different measured EOLs. Currently, the rate of voltage degradation (μV/h) at the reference current density and total test time is mostly used to express the durability index of fuel cells.

A typical steady-state durability test protocol is uninterrupted operation at constant current. However, running a fuel cell for an extended period is almost impossible. Most tests require periodic interruptions to maintain the cell to eliminate reversible degradation [27]. In addition, in-situ characterization of the test cell also requires interrupting cell operation, such as electrochemical impedance spectroscopy and polarization curves.

While steady-state durability testing is simple, this method is impractical for large-scale applications. It is impossible to test each cell for thousands of hours, and the time and economic costs are enormous. In addition, steady-state durability tests cannot be performed perfectly for fuel cells operating under multiple operating conditions, such as vehicle-mounted PEMFCs. Current fuel cell systems and their applications are rapidly developing, and a method for rapid testing of cell decay is necessary. The AST provides a method for rapid detection of fuel cell decay, which is described in detail in the next section.

9.2.2 Accelerated Stress Test

Traditional steady-state durability testing of fuel cells has obvious drawbacks: long test cycles and high costs. In addition, the fuel cell stack is a complex system consisting of various components, and we need to understand the degradation mechanisms of individual components, the interactions between components, and the effects of operating conditions. The AST method can effectively address these issues. The AST can be an accelerated test of the entire fuel cell or individual components. The confounding effect of the components can be minimized. For example, there are different AST protocols for chemical decay, mechanical decay, and thermal decay of the membrane. So that different decay mechanisms can be investigated. AST can effectively reduce the test cycle time and cost. It is important to test the decay mechanism of individual components and to study the influence of operating conditions.

Table 9.1 Steady-state durability tests of PEMFC in literature [27]

Cell number in stack	Membrane	Pt An/Ca (mg/cm^{-2})	T (°C)	RH An/Ca	Reaction gas An/Ca	Pressure An/Ca (bar)	Voltage (V)	Current density (A/cm^{-2})	Test time (h)	Decay date (10^{-6} V/h)
15	GoreSelect	0.4/0.4	80	100%/100%	H_2/air	1.5/1.5	0.72–0.96	0–0.2	2000	20–25
1	GoreSelect	0.45/0.6	70	100%/100%	H_2/O_2	1/1	0.5–0.65	~ 0.8	26,300	~ 4.2
1	Nafion 112	0.5/0.5	75	80%/80%	H_2/air	1/1	0.61–0.68	~ 0.3	2700	~ 30
50	GoreSelect	0.3/0.4	55	0%/5.7%	H_2/air	1.3/1	~ 0.55	~ 0.3	2500	~ 20
1	Nafion 112	0.5/0.5	60	100%/100%	$(H_2 + CO_2)$/air	1/1	~ 0.6	~ 0.4	4000	~ 3
4	Nafion 117	0.25/0.55	80	Internal humidifier	$(H_2 + CO_2)$/air	3/3	~ 0.68	~ 0.4	5000	~ 4
1	Nafion 112	0.26/1.46	60	6%/4.3%	$(H_2 + CO_2)$/air	1/1	~ 0.6	~ 0.4	5100	~ 6
8	Nafion 117	–	75–85	Internal humidifier	H_2/air	3/3	~ 0.62	~ 1	6000	~ 1

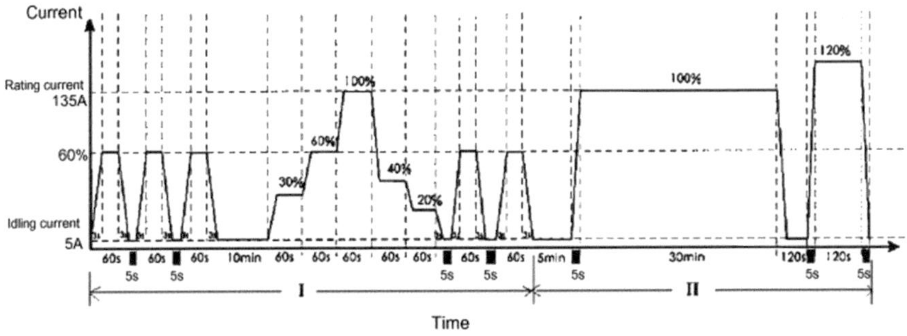

Fig. 9.1 AST protocol for hybrid PEMFC buses developed based on typical bus driving cycles in China [28]

The AST of the overall fuel cell and different components such as proton exchange membrane, catalyst, carbon carrier, and diffusion layer are described below.

9.2.2.1 Overall Fuel Cell

The AST of an overall fuel cell is a complex process with many changing operating conditions, and the automotive fuel cell is one of the most common examples. The AST cycle protocol of automotive fuel cells includes cold start, idle, acceleration, and overload conditions. Figure 9.1 shows the fuel cell AST protocol based on a typical bus driving cycle in China [28]. After 640 h of AST cycle testing, the average voltage of a single cell dropped from ~ 0.738 to ~ 0.675 V. The United Nations Economic Commission for Europe (UNECE) developed the New European Driving Cycle (NEDC), which has four repeated urban (low-speed) driving cycles (starting 0–780 s) followed by a high-speed driving cycle (followed by 780–1200 s), as shown in Fig. 9.2 [29]. In addition, the AST protocol for the automotive fuel cell can add humidity cycles, as shown in Fig. 9.3 [29]. The relative humidity was 50% for the first 780 s and 30% for the last 420 s. The temperature is kept constant at 80 °C.

9.2.2.2 The AST for PEM Degradation

The forms of membrane degradation can be divided into three categories: chemical degradation, mechanical degradation, and thermal degradation. There are different AST methods for different forms of degradation.

(a) AST for chemical decay

The main AST method for studying membrane chemical decay is the Fenton experiment [17, 30]. The main process of membrane chemical decay is the generation of free radicals by hydrogen peroxide, which attack some sites of the membrane chemical structure. Fenton solutions typically consist of 30% hydrogen peroxide

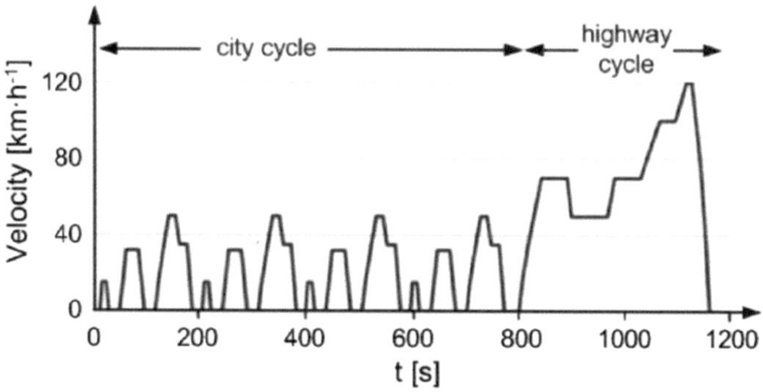

Fig. 9.2 AST protocol for new European driving cycle [29]

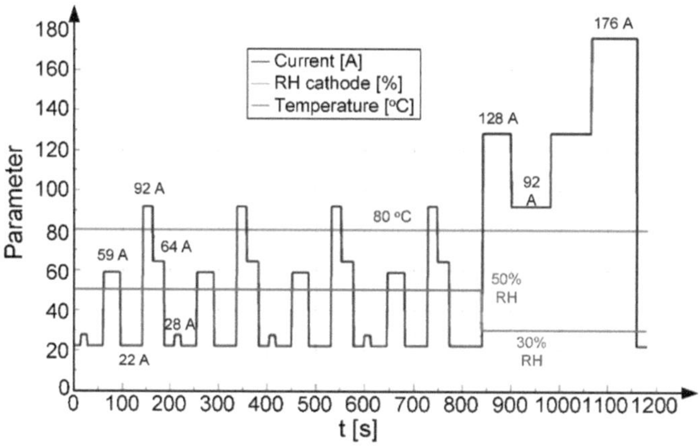

Fig. 9.3 AST protocol for the new European driving cycle after adding the humidity cycle [29]

and 20 ppm of ferrous ions. The membrane is immersed in the Fenton solution, and the solution is renewed at regular intervals, thus simulating the environment in which the membrane is exposed during chemical decay. And due to its high concentration effect, the chemical decay process is accelerated. At the end of the experiments, FTIR, NMR, and SEM are commonly used for the analysis of the chemical structure. FTIR and NMR are used to analyze the changes in the chemical composition of the membranes, as shown in Fig. 9.4 [17]. The components of CF_2, S–O, and C–O–C in the membranes decreased continuously with increasing immersion time (Fig. 9.1a). However, no change in the relative intensity of each group was observed in the NMR spectrum (Fig. 9.4b). Considering that FTIR is an area reflection and NMR is a statistical analysis of polymer contact powder, the comparison of reflection-FTIR and NMR suggests that the degradation of PEMs comes from the decomposition of repeating units on the polymer backbone. SEM is used to investigate the change in the membrane microstructure, as shown in Fig. 9.5 [30]. From the Figure, we can see that microporous structures are formed in the membrane as the chemical decay proceeds. The formation of micropores may be because the chemical decay will proceed locally and continuously, such as the chain decomposition of the membrane chemical structure.

The Fenton experiment is an ex-situ method for testing the chemical decay of membranes. In addition, there are in situ measurements, which investigate the chemical decay of the membrane in a complete fuel cell with the AST method. The chemical decay is accelerated mainly by accelerating the rate of hydrogen peroxide production [31]. Under OCV conditions, the oxygen concentration at the cathode is greater, and thus, more oxygen reaches the anode across the membrane. The rate of hydrogen peroxide production is then increased, thus accelerating the chemical decay of the membrane. The temperature setting is higher (80–95 °C), thus accelerating the decay-related reaction rate. In addition to the rate of voltage drop, the fluoride release rate (FRR) is an important indicator of the chemical decay rate. The exhaust water from the anode and cathode is collected by a condenser and is detected using an ion chromatograph. The FRR was calculated from the concentration of the anion and the flow rate of the water, and finally normalized according to the geometric active surface of the MEA. For example, the FRR can be used to study the effect of gas humidity, pressure, and current density on the chemical decay rate, as shown in Fig. 9.6 [32]. It can be found that pressure has a strong positive correlation with chemical decay, but the effect of humidity is not significant. The smaller the current density, the greater the decay rate, and the decay rate reaches a maximum under OCV conditions. It is important to note that although the graph shows that the FRR is greater at the cathode than at the anode, it does not indicate that the chemical decay rate of the membrane is greater near the cathode. This is because the water in the membrane undergoes transfer from the anode to the cathode with the transfer of protons when the cell is operating. In this process, some of the decaying fragments of the membrane may also be transferred with the water molecules.

(b) AST for mechanical decay

The mechanical decay of the membrane is mainly due to stress variations or non-uniformity in the membrane. The generation and variation of stresses are mainly due to temperature and humidity variations. Therefore, in the AST of membrane mechanical decay, the mechanical decay of the membrane is mainly accelerated by cyclic changes in the operating temperature of the fuel cell and the humidity of the supply gas. The accelerated mechanical stress test (AMST) protocol is mostly: cyclic variation of the humidity of the supply gas at a certain temperature. For example, the cell temperature is constant at 80 °C, and the duration of a single humidity cycle is 4 min, with 2 min for each of the wet and dry phases [33]. The wet phase indicates that the RH of the inlet gas is 90%, and the dry phase indicates that the RH of the inlet gas is 0%. As shown in Fig. 9.7, before the AMST cycle, the cell is purged for 20 h using nitrogen at 100% RH to ensure hydration of the proton exchange membrane. The membrane failure criterion is internal gas leakage > 10 sccm or 20,000 AST cycles, whichever occurs first. To avoid chemical decay of the membrane that could affect the test, the gas supplied to both cathode and anode during the AMST cycle

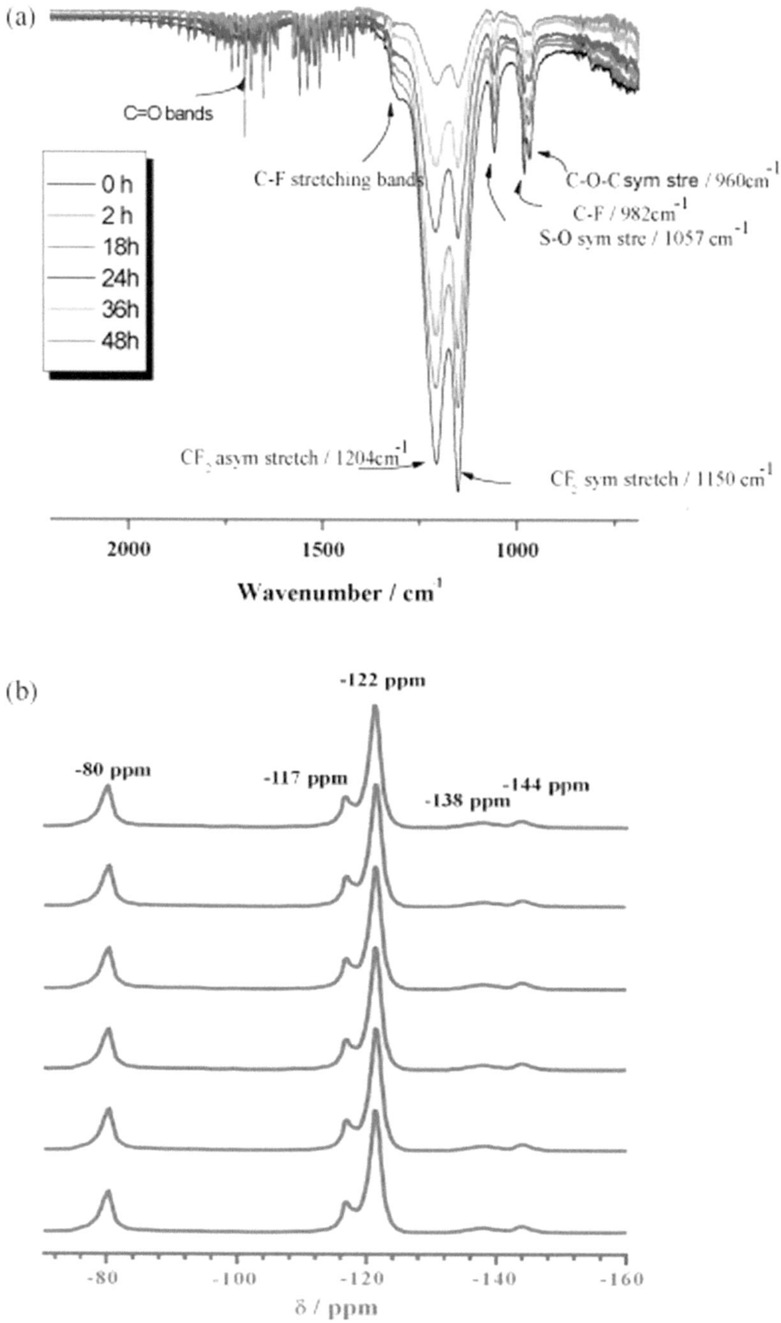

Fig. 9.4 FTIR spectrum (**a**) and NMR spectrum (**b**) of the membrane after immersion in Fenton's solution [17]

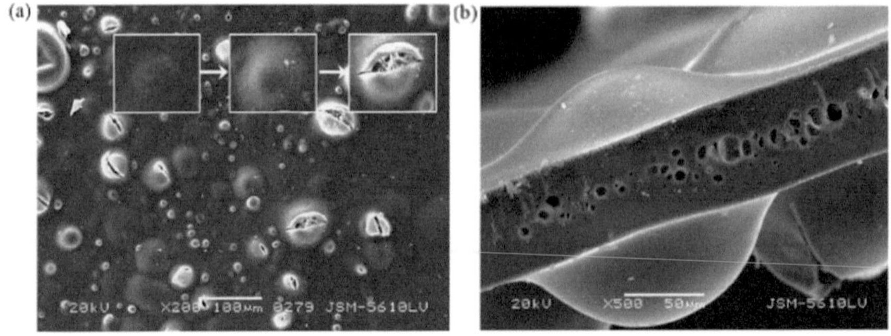

Fig. 9.5 SEM images of membrane surface (**a**) and cross section (**b**) after immersion in Fenton's solution [30]

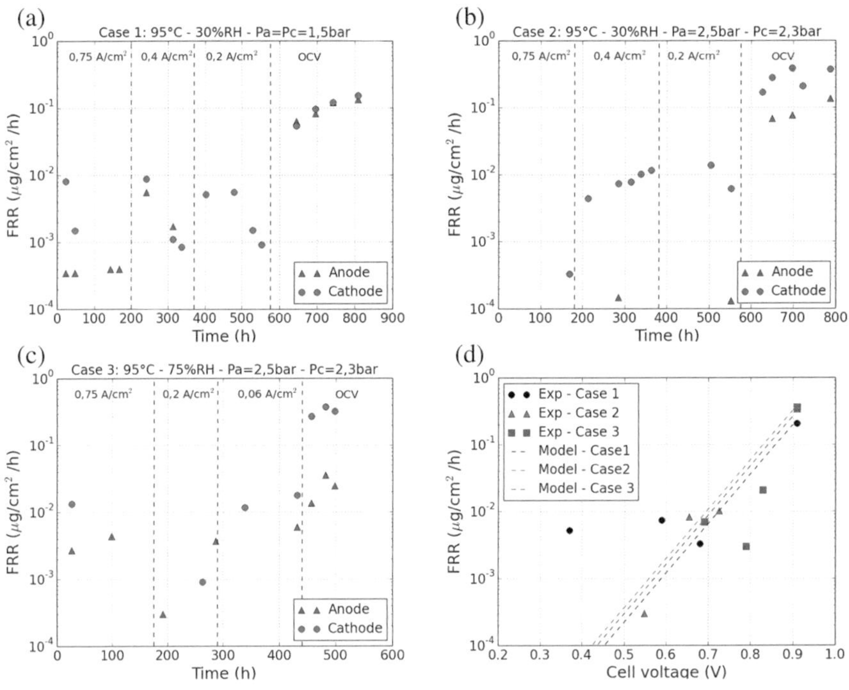

Fig. 9.6 FFR under different gas humidity, pressure, and current density conditions [32]

is nitrogen. In some in-situ tests, the gases supplied to the cathode and anode are air and hydrogen, respectively [31]. Thus, the AST of mechanical decay was performed with the cell running.

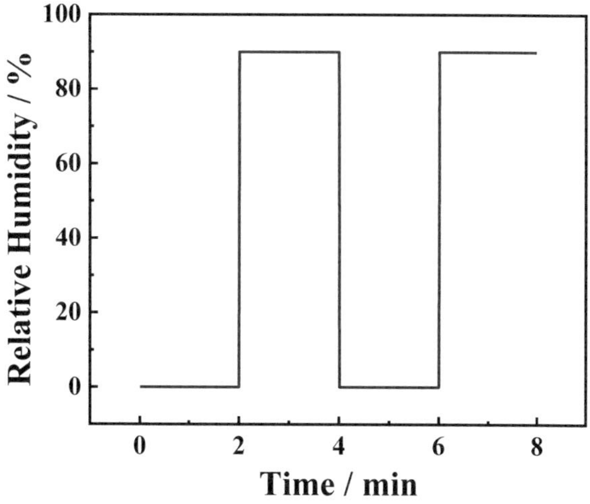

Fig. 9.7 Schematic diagram of the cyclic variation of the relative humidity for supply gases

(c) AST for thermal decay

The thermal decay of membranes is divided into decay at high temperatures and decay at sub-zero temperatures. Thermal decay at high temperatures is mainly due to the direct reaction of reactive gases at pinholes in the membrane, which generates a large amount of heat to cause thermal decomposition of the membrane structure. The AST protocol corresponding to this case is to create pinhole defects in the membrane, thus causing a rapid thermal decay in the membrane. Techniques for creating artificial pinholes are generally divided into three categories [34]: (i) electron beam lithography (EBL), (ii) focused ion beam (FIB), and (iii) glassy micro-needles. The former two methods can reproducibly implement pinholes with small deviations in pinhole size and shape. However, when using focused ion beam milling, gallium contamination in the vicinity of the pinhole cannot be excluded, even if gallium is not detected by EDX analysis. In principle, gallium ions can accelerate the decomposition of chemical polymers. Glass micro-needles are made by thermoforming and subsequent breaking. The tip of the micro-needle is conical, and therefore, the artificial pinhole formed has a similar conical shape. The artificial pinhole prepared by this method also has a large error in its diameter, which can reach 50%. The artificial pinholes in Nafion can be characterized using X-ray tomography microscopy and infrared spectroscopy. Figure 9.8 shows the characterization of a standard 10 μm diameter artificial pinhole fabricated using EBL, FIB, and glass micro-needles [34]. Cylindrical pinholes produced by EBL and FIB have an average pinhole diameter of 8.3 μm and 9.8 μm, respectively. The average diameter of pinholes made by micro-needles ranges from 10 to 20 μm.

Thermal decay at sub-zero temperatures is mostly freeze–thaw cycling and sub-zero initiation. An ex-situ freeze–thaw cycling approach can be used to understand membrane durability under thermal cycling conditions at sub-zero temperatures. As an example, Nafion 112 membrane shows minimal changes in ionic conductivity and

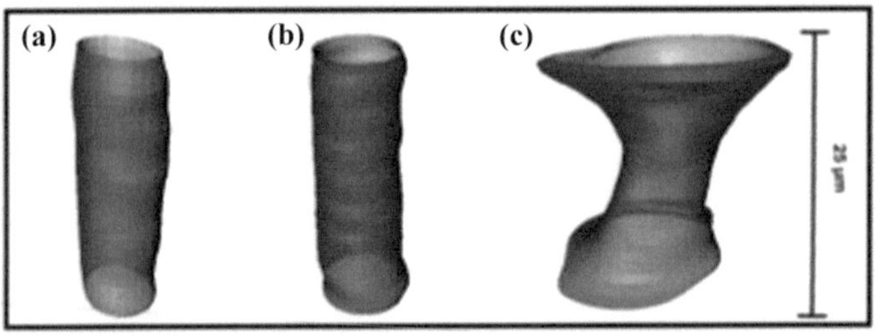

Fig. 9.8 3D visualization of artificially made pinholes using X-ray tomographic microscopy. Pinholes with a nominal diameter of 10 μm are implemented in Nafion NR 211 membranes with **a** electron beam lithography, **b** focused ion beam, and **c** micro-needles [34]

mechanical properties after 385 freeze–thaw cycles between − 40 and 80 °C [35]. However, the anisotropy of the tensile strength, oxygen permeability, and water solubilization behavior of the films decreased after dry freeze–thaw cycling, indicating some degree of rearrangement of the ionomer at the molecular level. In addition to this, there are in situ tests of freeze–thaw cycling of AST. The cold start of an automotive fuel cell is presented as an example. Figure 9.9 shows a simplified description of the entire test process cycle, namely the cryogenic cycle [36]. The SOP phase, which is held in constant potential mode for 2400 s to ensure the same initial conditions, also serves as a recovery phase for the system. The preprocessing process consists of two main phases. In the first stage, the cathode side is prepared for a low-temperature stop. The cathode and anode supply dry gas to remove water from the cell. The membrane water content is measured by a high-frequency resistance (HFR). When the HFR reaches the desired value, the bypass is closed and sealed off from the environment. In the second stage, the remaining liquid water is removed from the anode side. This liquid water is condensed out when the cell is cooled to a temperature of 10 °C. At the moment, the anode bypass is opened and the anode is flushed for 60 s using a gas mixture (90% H_2 and 10% N_2). In the low temperature phase, the temperature is lowered again to − 10 °C and held for 1200 s. After this, the start-up phase is performed. Open the cell bypass and feed dry air and hydrogen gas to the cathode and anode, respectively. Two seconds after reaching the OCV, a current density of 0.05 A/cm^2 was set for 20 s. After 20 s, start heating the cell and set the temperature to 80 °C. During the heating phase, the output current is increased continuously until the lower voltage limit is reached. After a fixed number of cycles, the cell performance is characterized. This includes polarization curves, cyclic voltammetry, and cell impedance measurements.

(d) AST for membrane-integrated decay

There are not only AST protocols for single forms of membrane decay, but also AST protocols for combined membrane decay. The integrated decay of membranes takes chemical decay, mechanical decay, and thermal decay into account simultaneously. To accelerate the integrated decay of the membrane, the AST protocol will accelerate all three forms of membrane decay simultaneously. Keeping the cell in open circuit

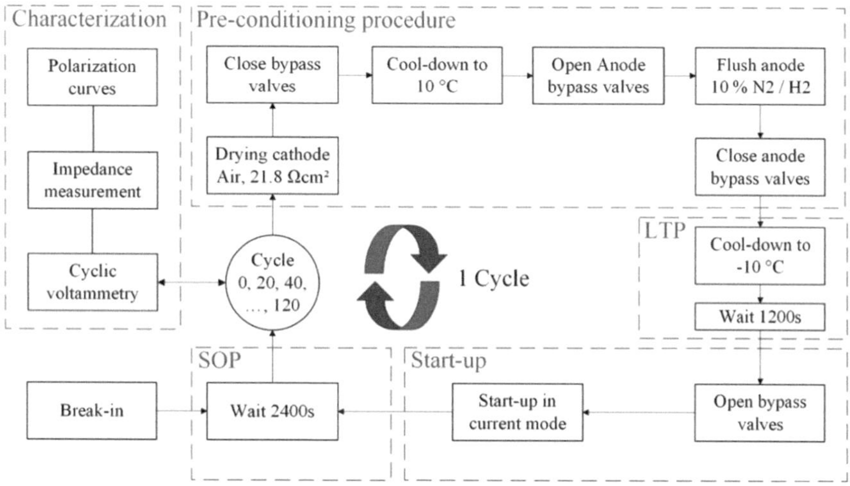

Fig. 9.9 Simplified representation of the test procedure with a process loop consisting of the following main parts (shown in blue): stable operating point (SOP), preconditioning procedure, low-temperature part (LTP), startup, and the characterization part (polarization curves, impedance measurement, and cyclic voltammetry) [36] (color figure online)

Table 9.2 Experimental set points of accelerated stress tests focusing on chemical, mechanical, and combined degradation at artificially implemented membrane defects

	Mechanical	Chemical	Combined
Current density (mA cm^{-2})	$0(N_2/N_2)$	$0(H_2/O_2)$	$0(H_2/O_2)$
Relative humidity (%)	0–70	70	0–70
Cell temperature (°C)	80	80	80
Anode/cathode pressure (kPa)	150.5/150	150.5/150	150.5/150
Cycle holding time (dry/wet/min)	10/230	None	10/230

promotes chemical decay of the membrane. Cyclically varying the humidity of the supply gas to promote mechanical decay of the membrane. Combining the above two conditions promotes both chemical and mechanical decay of the membrane, as shown in Table 9.2 [37]. Finally, artificial pinholes are fabricated in the membrane to promote thermal degradation of the membrane at the same time. Thus, the AST protocol for integrated membrane decay is achieved.

9.2.2.3 AST for Pt Catalyst Degradation

To evaluate fuel cell Pt catalyst durability, most researchers prefer to use AST for long-running lifetime tests. In AST, different accelerated stressors are applied to in situ or ex-situ fuel cell testing to determine the durability of the entire cell or specific components. These accelerated stressors include most of the factors that significantly affect the degradation of Pt catalysts, including temperature, humidity, potential, contamination, and vibration, among others.

In AST of the whole cell or stack in situ, a more comprehensive view of cell degradation can be obtained, including the degradation modes of different components. However, it is difficult to determine the main cause of degradation of individual components by in situ AST alone. Ex-situ testing is often used for specific battery components, with the advantage of using more simplified equipment and isolating the catalyst from other adjacent components, which avoids potentially confusing effects during degradation assessment, with the disadvantage of being difficult to simulate Realistic fuel cell operating conditions [26, 38].

(1) Temperature

The operating temperature has a decisive influence on the performance of a PEM fuel cell, and the cell temperature will always fluctuate during actual operation due to start-stop cycles and load variations. Experimental studies of the effect of temperature on catalyst durability have shown that the growth rate of Pt particles increases as the operating temperature of the cell increases [39]. The effect of temperature on Pt dissolution was quantified by dissolution experiments of Pt thin films in 1M $HClO_4$ [40]. The results showed that at temperatures as low as 40 °C, Pt dissolution could not be measured because the dissolved amount was too small. However, when the temperature increased from 60 to 80 °C, the Pt dissolution rate increased from 0.87 ng h^{-1} cm^2 to 1.58 μg h^{-1} cm^2. Experimental test results using square-wave potential cycling at different temperatures of 40, 60, and 80 °C showed that higher temperatures accelerated the degradation of the catalyst ECSA and the dissolution of Pt [41], as shown in Fig. 9.10. Growth of cathode Pt particles after degradation was also observed, compared with the initial particle size of 3.3 nm, and higher operating temperature resulted in more severe growth. Also, ex-situ characterization results showed that Pt bands were found to form in the membrane, and more Pt nanoparticles were found to migrate into the membrane as the test temperature increased.

(2) Humidity

Due to local climate conditions or different operating conditions, the relative humidity in the PEMFC channels and electrodes will often change. The effect of RH cycling on the long-term performance of Pt catalysts has attracted extensive research. The experimental results found that the degradation process of the cathode Pt catalyst was accelerated with the increase of relative humidity. At 60 °C, when the relative humidity was decreased from 100 to 50%, the loss rate of Pt mass and catalyst active surface area decreased by about threefold and twofold, respectively, as shown in Fig. 9.11a [42]. However, the Pt dissolution rate was only slightly affected by the humidity change in Fig. 9.11b, and the accelerated catalyst degradation was mainly attributed to the more abundant and larger water channel network within the polymer electrolyte that facilitated the transport of Pt ions. Therefore, the diffusion coefficient of Pt ions under fully humidified conditions is three times that under 50% RH. Furthermore, the loss of ECSA at high humidity was mainly attributed

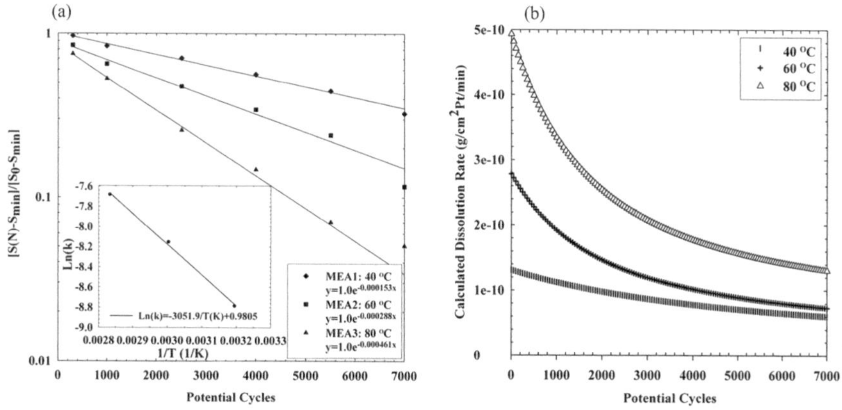

Fig. 9.10 **a** Electrochemically active surface area losses and **b** calculated Pt dissolution rates during potential cycling (0.87–1.2 V) for MEA1: 40 °C, MEA2: 60 °C, and MEA3: 80 °C [41]

to catalyst agglomeration and Ostwald ripening, which resulted in the growth of Pt particles [39, 43]. High-humidity environment will lead to more severe growth of Pt particles in Fig. 9.12. The study of CL decay behavior at various humidity is based on the automotive protocol, and various characterization means revealed that the four degradation mechanisms of Pt dissolution, Pt coarsening, carbon corrosion, and ionomer degradation were different under high and low humidity conditions. Higher water content may accelerate Pt coarsening and carbon corrosion, resulting in greater performance loss at RH 64%. The ionomer degradation and Pt dissolution were enhanced at a low humidity of RH 10%. Combined with CL and PEM attenuation, a 64% to 40% RH range is recommended for use in automotive protocols [44]. Furthermore, humidity cycling is the main cause of the microstructural change of CL, which is mainly attributed to the expansion and contraction of the ionomer during the humidity cycling. These structural changes will result in a reduction in surface area.

(3) Potential

The U.S. Department of Energy's (DOE) Fuel Cell Technologies Office (FCTO) has set a 5000-h durability goal for PEMFCs used in transportation applications. To achieve the automotive goals, the US Department of Energy-FCTO and the US DRIVE Fuel Cell Technology Team (FCTT) jointly developed a 5000-h drive cycle durability protocol to simulate the different degradation mechanisms experienced by fuel cells under normal transportation operating conditions, as shown in Fig. 9.13. Driving cycles vary operating conditions, including potential, load, and humidity cycling. The first half of the cycle (0–5 min) is run at high humidity to accelerate catalyst and support degradation. The second half of the cycle (7–17 min) is in low humidity to accelerate membrane degradation. Therefore, this driving cycle protocol is often used as a reference object for the acceleration effect of the AST protocol.

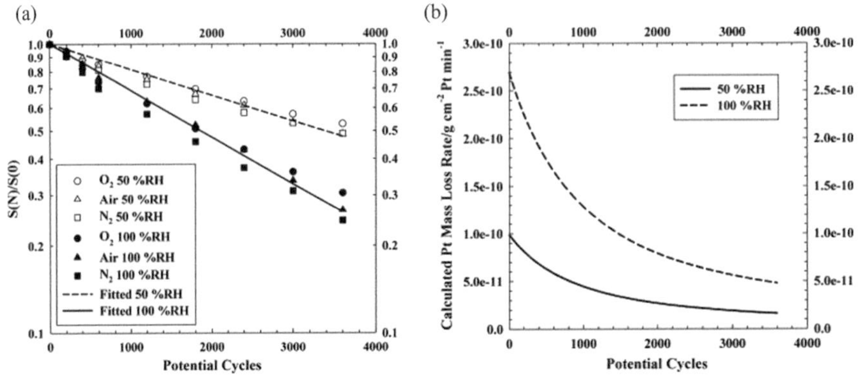

Fig. 9.11 **a** ECSA losses and **b** calculated Pt mass loss rates during potential cycling (0.87–1.2 V) for RH: 100% and RH: 50% [42]

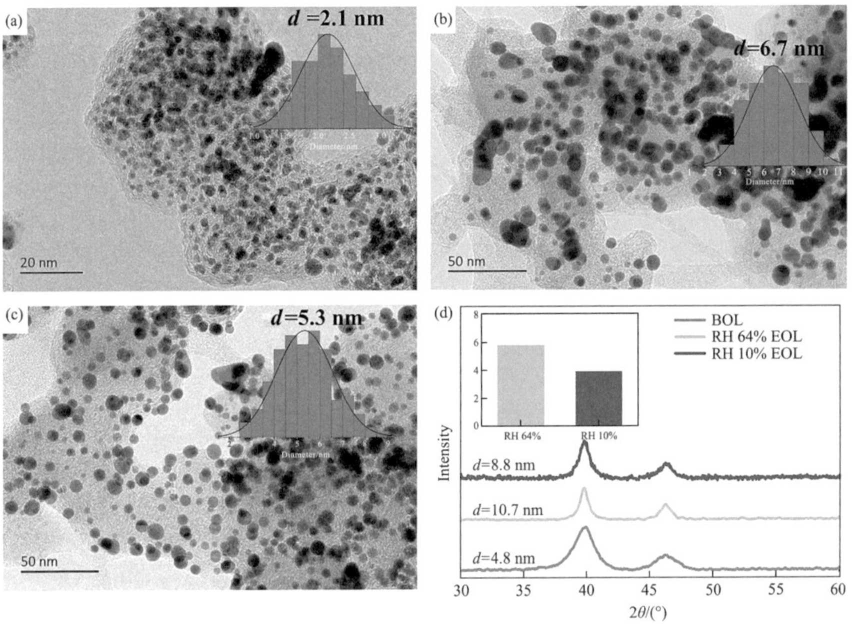

Fig. 9.12 TEM images and particle size histograms of the cathode CL **a** before of life, after automotive protocol in **b** RH 64%, **c** RH 10%, **d** XRD patterns of the cathode CL before and after simulating automotive protocol in these two RHs [44]

One of the most widespread accelerating stressors used to study catalyst degradation is potential. This includes: (1) potential cycling from low potential to high potential, (2) potential control of a specific waveform (or triangle, trapezoidal), and (3) steady-state potential control at a specified voltage value [46]. Moreover, Control

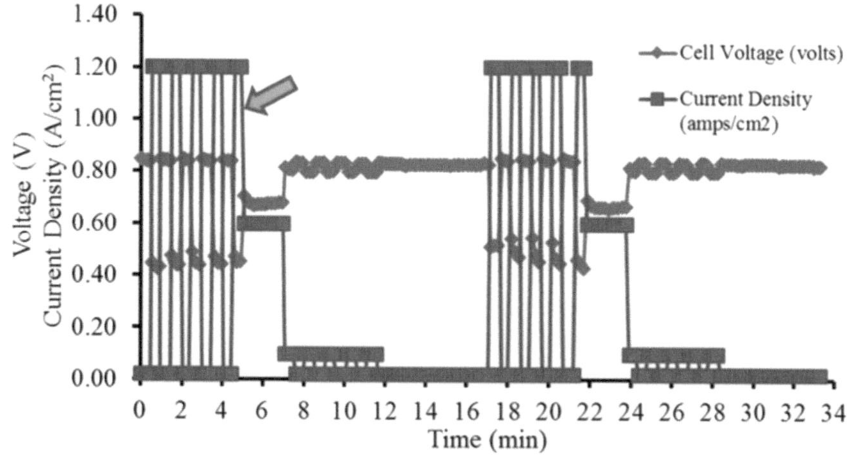

Fig. 9.13 U.S. DRIVE-FCTT drive cycle durability protocol (2 cycles shown) [45]

analysis can also be performed on the potential scan rate and the potential upper/lower limits (UPL, LPL) of the scan. Several organizations, including the U.S. Department of Energy, the former U.S. Fuel Cell Council (USFCC), the Japan Automotive Research Institute (JARI), the Fuel Cell Commercialization Conference of Japan (FCCJ), and Nissan, have developed different potential cycling protocols, as shown in Fig. 9.14 [47]. It can be seen that their difference lies in the potential range and sweep curve of AST, and most of these ranges are centered around staying below 1 V, which is to exclude the influence of carbon corrosion.

The dissolution rate of Pt has a strong dependence on the cathode potential, and higher potential can accelerate the dissolution of Pt. Pt dissolution increases exponentially with increasing steady-state potential or above the upper limit of potential cycling of 0.85 V [48]. Ex-situ experimental results show that the Pt dissolution rate increases by about 54 times as the potential increases from 0.85 to 1.15 V, and the dissolution rate reaches saturation when the potential is higher than 1.15 V, which is due to the formation of a Platinum oxide that protects Pt from dissolution [40].

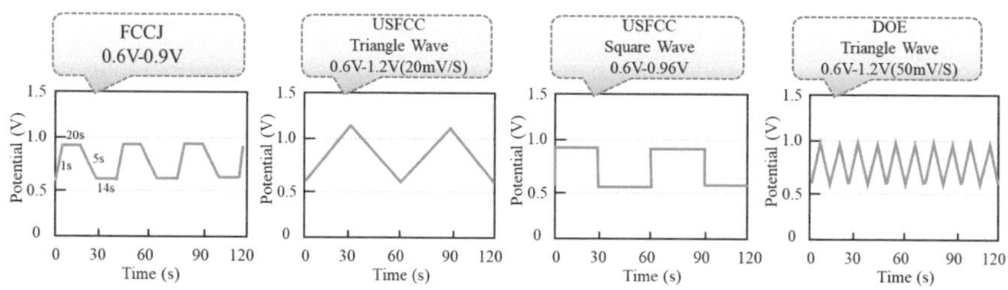

Fig. 9.14 ASTs used by FCCJ, U.S. FCC, and U.S. DOE, which focus on the potential range and type, and duration of wave [47]

A TEM image and Pt particle size distribution before and after potential cycling tests based on three different AST protocols are shown in Fig. 9.15 [49]. The Pt nanoparticles in pristine Pt/C samples have a narrow size distribution, implying uniform particle sizes. After AST, an increase in the average particle size and a broadening of the particle size distribution range can be observed. And as can be seen from the image, the Pt growth pattern is mainly due to Pt agglomeration (green circles). The ECSA loss caused by the three AST protocols (square wave, triangle wave, and wet driving cycle) was compared [47], and the TEM images and particle size distribution changes are shown in Fig. 9.16. Subjecting an MEA with a 4.5 nm PtCo alloy cathode catalyst to wet driving cycles (75,000 cycles over 1200 h) resulted in significantly higher cathode catalyst particle growth (7.4 nm) than that observed after 30,000 cycles of the standard AST protocol (5.0 and 5.1 nm). However, when averaged over time by ECSA decay rate, the square wave AST protocol shows a time acceleration factor of 5 times higher than the triangular wave AST protocol and 25 times higher than the modified wet driving cycle protocol, resulting in a significantly shorter testing time. Furthermore, these ASTs are independent of the carbon type used in the MEA, confirming the ability of these ASTs to evaluate the durability of electrocatalysts independently of the support.

Due to the differences in the local environment across the catalyst layer in a polymer electrolyte fuel cell, the Pt catalyst will undergo inhomogeneous degradation during operation. Aging of PEFC single cells using voltage cycling as an accelerated stress test. The AST was performed using a square wave potential profile from 0.60 V to open circuit potential (OCV) with 3 s holds at each potential for 30,000 cycles at 80 °C in H_2/air environment at 100% relative humidity (RH). Visualization of degradation heterogeneity at the micrometer length scale by plotting Pt catalyst particle sizes after voltage cycling in Fig. 9.17a [50]. The growth of Pt catalyst particle size follows the flow field geometry. The Pt particle size growth is larger in the region below the land and smaller in the region below the flow field channel. In addition, the Pt particle size in the outlet region increases faster than that in the inlet region. This is mainly due to the high humidity in the outlet and subterranean areas, and the high humidity environment will promote the dissolution of Pt. Similar results can be observed under the conditions of the start-up cycle AST protocol, according to Fig. 9.17b, which displays the distribution of ECSA loss along the flow channel direction, the outlet region showing more severe ECSA degradation [51]. But the reasons for these two phenomena are different, the latter is due to the severe carbon corrosion at the outlet, which leads to the easier agglomeration and detachment of Pt to lose its activity.

In the process of performing fuel cell catalyst AST, some diagnostic tools and characterization methods need to be combined to provide the surface or cross-section of CL, the size distribution of catalyst particles, element content and distribution, the atomic structure of CL local particles, and the electrochemical properties of CL, and other valuable information. The most commonly used tools can be divided into two categories: topographic observation (TEM, TEM, and XRD) and electrochemical diagnosis (IV, CV, EIS). With the help of TEM, the morphological changes of particles at the micro- and nano-scales can be observed after in-situ and ex-situ

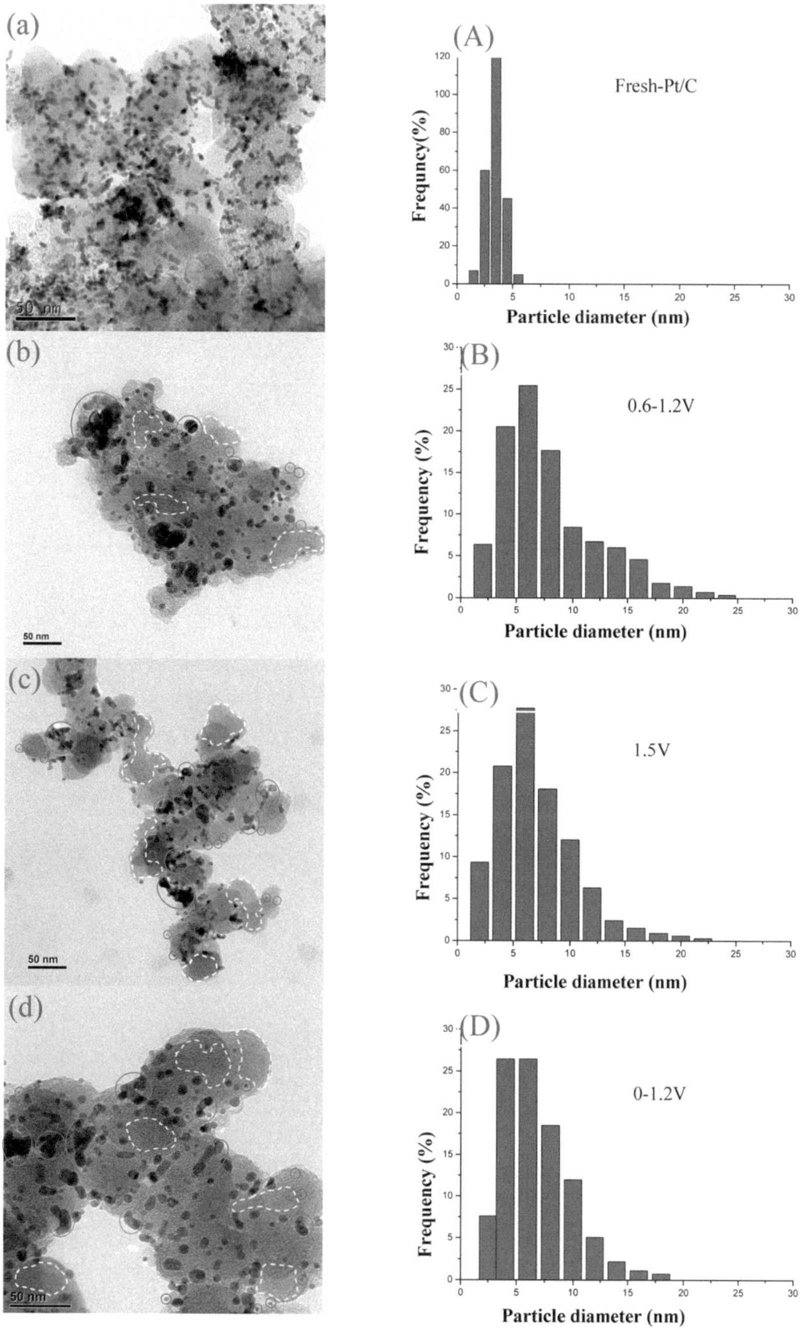

Fig. 9.15 TEM images and Pt particle-distribution histograms of **a**, **A** fresh Pt/C catalyst, **b**, **B** aged-Pt/C catalyst by potential cycling between 0.6 and 1.2 V for 4000 cycles, **c**, **C** aged-Pt/C catalyst by constant potential at 1.5 V for 26.67 h, **d**, **D** aged-Pt/C catalyst by potential cycling between 0 and 1.2 V for 2000 cycles [49]

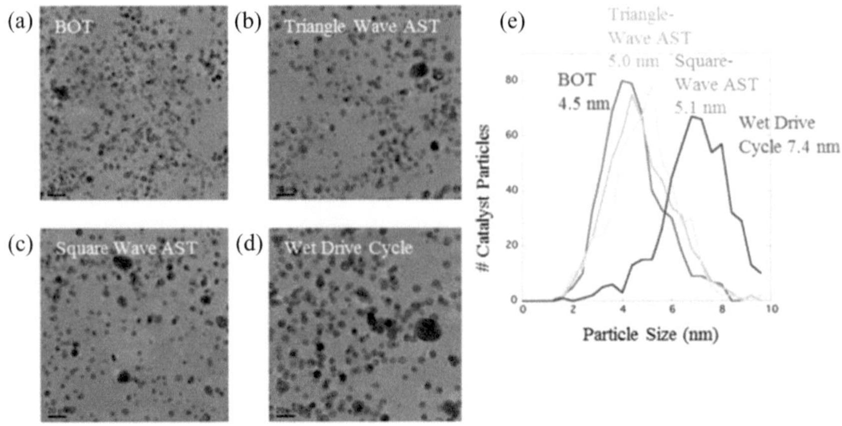

Fig. 9.16 S/TEM images of the SOA Pt–Co/HSAC particle size at **a** BOT and **b–d** EOT conditions and **e** their particle size distributions [47]

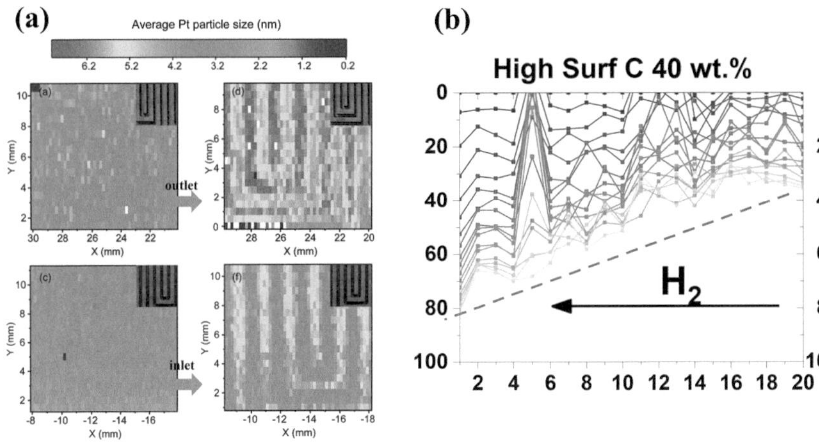

Fig. 9.17 **a** Pt catalyst nanoparticle size mapping of a 1 cm × 1 cm area near the air outlet and inlet of the pristine and aged MEA [50], **b** ECSA loss along the channel segments after start-up cycle AST protocol [51]

aging processes [52]. Scanning electron microscopy (SEM) can be used to examine the surface condition, thickness, and interfacial changes of CLs. X-ray diffraction (XRD) can obtain the average size of Pt particles [53]. Polarization curves (IV) are the most commonly used tool for characterizing fuel cell performance [54]. The electrochemical surface area (ECSA) can be obtained by integrating the CV curve [55]. Catalyst surface area, catalyst loading, and catalyst utilization can all be characterized using data obtained from EIS measurements [56].

9.2.2.4 AST for Carbon Corrosion

(1) Fuel starvation and start-stop cycling

The carbon corrosion reaction occurring at the anode electrode due to severe fuel shortage should be the direct cause of electrocatalyst degradation. Using TEM, EDX, and electrochemical methods, the degradation of the electrocatalyst due to fuel starvation was characterized, revealing a severe loss of electrocatalyst surface area and a decrease in cell performance due to corrosion of the carbon support after cell reversal [21]. The start-stop cycle of the car can cause carbon corrosion in the cathode. According to the experimental results of the start-stop cycle [57], the most obvious damage occurs in the first 30 cycles of the corrosion process, when air and fuel are alternately supplied to the anode. After 80 such cycles, the cathodic ECSA lost up to 70%. This is also evidenced by the fact that the cathodic layer is generally thinner than the anode layer after start-stop cycle aging [14].

Aging heterogeneity in PEMFCs during operation under startup and shutdown conditions has also been extensively studied [58]. A 30 cm^2 fuel cell was segmented into 20 similar regions along the flow field in Fig. 9.18a, and the MEA was analyzed after aging using physical and chemical techniques. Particular attention was paid to identifying various aging heterogeneities within a single MEA, i.e., inlet versus outlet and channel versus land. After 136 SU/SD cycles, locally resolved SEM observations of the aged MEA revealed that the active layer had aged inhomogeneously throughout the length of the MEA in Fig. 9.18b. Close to the cathode exit region, the MEA looks very similar to its pristine state. A significant thinning of the cathode thickness is observed close to the cathode inlet region, implying a severe carbon corrosion rate. Furthermore, based on SEM images taken close to the cathode entrance in Fig. 9.18c, the cathode appears to be more degraded under the land than under the channel: the remaining cathode thickness is smaller. The parts of the anode facing the cathode land area are indeed expected to suffer longer fuel starvation events than those facing the cathode channel area due to the time it takes for hydrogen to diffuse beneath the land.

(2) Potential control

The rate of the carbon corrosion reaction is strongly dependent on the potential, and corrosion occurs at high potentials (> 1.2 V). Carbon corrosion rates increase with potential [59] and are accelerated by rapid changes in potential [60]. The CL thickness of the Toyota Mirai battery after driving for 300 h was 10–11 μm. However, after the catalyst AST (0.6–0.95 V), it exhibited a thinning of 8–9 μm. The carbon-supported AST will lead to a more severe thinning of CL (4 μm), which will lead to a sharp decrease in electrode porosity. Carbon oxidation or corrosion weakens the adhesion of Pt particles on the carbon surface, eventually leading to structural collapse and the detachment of Pt particles from the carbon support, resulting in severe agglomeration and performance degradation of Pt during long-term operation. Figure 9.19a shows the 3D reconstruction of three MEAs after carbon corrosion of AST (BOT, after 500 cycles, after 1000 cycles), where the agglomeration and growth of Pt particles can be

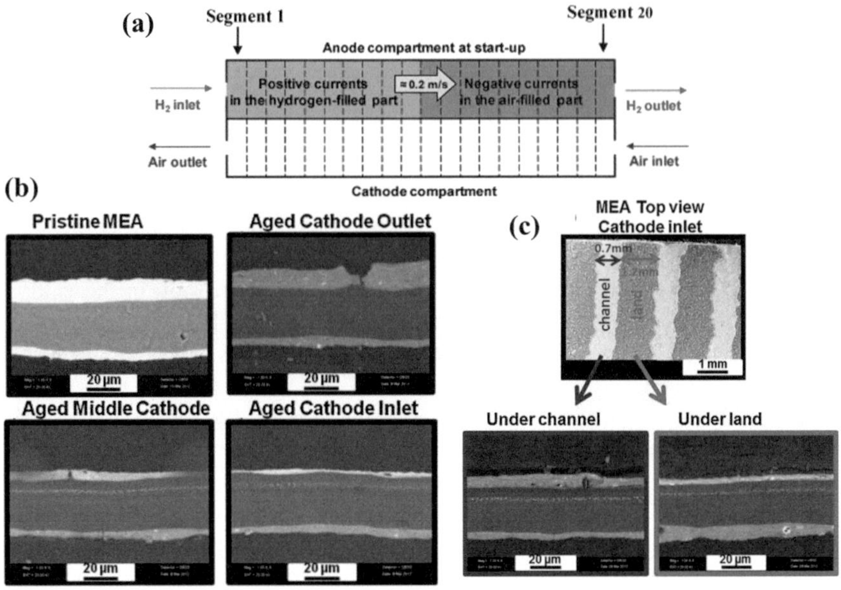

Fig. 9.18 **a** Schematic diagram of a segmented fuel cell. **b** SEM images of a pristine MEA and aged under repeated 136 SU/SD cycles, taken at different locations, i.e., from the cathode outlet (segment #1) to the cathode inlet (segment #20). The upper electrodes are the cathodes. **c** SEM images of the top view of the aged MEA and the cross-sections of the two distinct regions in segment #18: under a channel and a land, respectively. The upper electrodes are the cathodes [58]

significantly observed [47]. In addition, the CCL thickness before and after carbon corrosion of AST also changed significantly, from 7 μm in BOT to 6 microns after 500 cycles, and greatly reduced to 3 μm after 1000 cycles in Fig. 9.19b.

9.2.2.5 AST for BBP Degradation

The decay of the bipolar plate is mainly carried out by corrosion. Therefore, the AST protocol for bipolar plates promotes the decay of bipolar plates by facilitating their corrosion. The corrosion of bipolar plates is mainly influenced by the acidic environment they are exposed to and the voltage. The PEMFC environment includes ions such as SO_4^{2-}, Cl^-, and F^- as well as cathode and anode reactant gases [61]. To simulate the PEMFC medium, 0.01 M HCl + 0.01 M Na_2SO_4 was used as the test solution. Fluoride ion is not introduced due to their low content and low corrosive effect on the bipolar plate. The solution in the glass cell was fed with pure oxygen and pure hydrogen at 80 °C to simulate the cathodic and anodic environments in the PEMFC, respectively. Pre-bubbling is performed approximately 2 h before each experiment. The exit gas is passed through an air-cooling duct and a water lock to prevent water from flowing out and air from entering. AST of bipolar plates is mostly

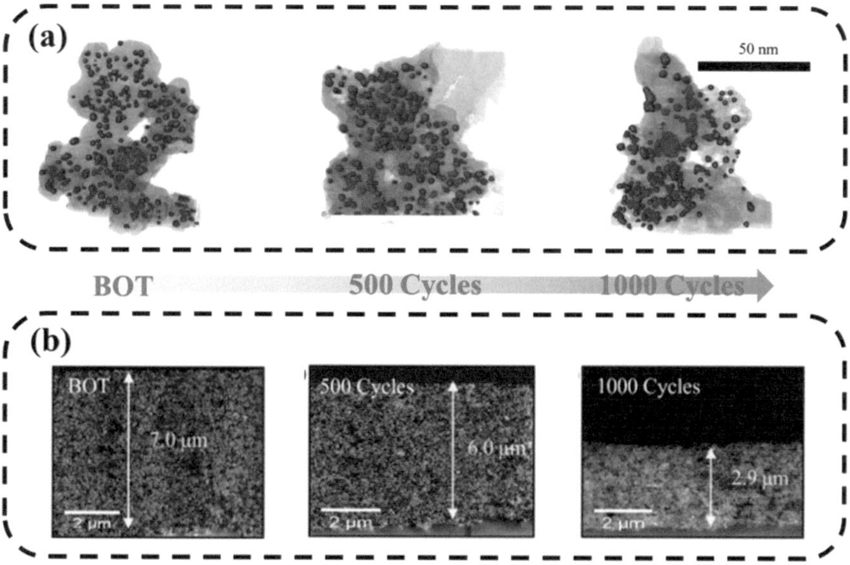

Fig. 9.19 **a** The 3D reconstruction of three MEAs after carbon corrosion of AST (BOT, after 500 cycles, after 1000 cycles), **b** TEM images of catalyst layer thickness: BOT, after 500 cycles, after 1000 cycles [47]

used to study new corrosion protection measures. The effect of corrosion protection measures can be quickly analyzed using the AST method.

9.2.3 Diagnostic Tools

Several diagnostic tools are already available for detecting fuel cell degradation, including in situ and ex-situ tools [23].

(a) In situ tools

Polarization curve (IV), electrochemical impedance spectroscopy (EIS), and linear sweep voltammetry (LSV).

(b) Ex-situ tools

- Atomic force microscopy (AFM)—Nanoscale Nafion membrane and GDL degradation [62]
- Infrared (IR) imaging—Membrane thinning and pinhole detection [63]
- X-ray diffraction (XRD)—Particle size change [64] and loss of surface area [65]
- Scanning electron microscopy (SEM)—Morphological changes [66] and degradation of membrane [67], CL [68], and GDL [69]

- Transmission electron microscopy (TEM)—Diagnosis of electrocatalyst degradation [70]
- Fourier transform infrared spectroscopy (FTIR)—Polymer alterations induced by degradation [71]
- X-ray photoelectron spectroscopy (XPS)—Surface component loss [71]
- Ion chromatography (IC)—Membrane and ionomer degradation through analysis of dissolved fluoride ions [72].

9.3 Model Predictions

9.3.1 Classification of Predictive Models

Most of the current research on PEMFCs is focused on solving two bottlenecks: cost reduction and service life extension [1]. In addition to developing new high-performance, low-cost materials and designing new PEMFCs, prognostics and health management (PHM) techniques can be used to effectively extend the service life of PEMFCs and reduce their cost of use. The main idea of PHM is to monitor the health of the system in real time throughout the life cycle of the target system and take appropriate maintenance measures at the right time to improve the reliability and durability of the target system. PHM consists of seven steps: data acquisition, data processing, condition assessment, diagnosis, prognostics, decision making, and human–machine interface, as shown in Fig. 9.20. Prognostics is one of the most critical steps in PHM. Predicting the trend of decay as well as the time and risk of failure of a target system is a key technique to transform system maintenance from the traditional "fail and fix" to "predict and prevent".

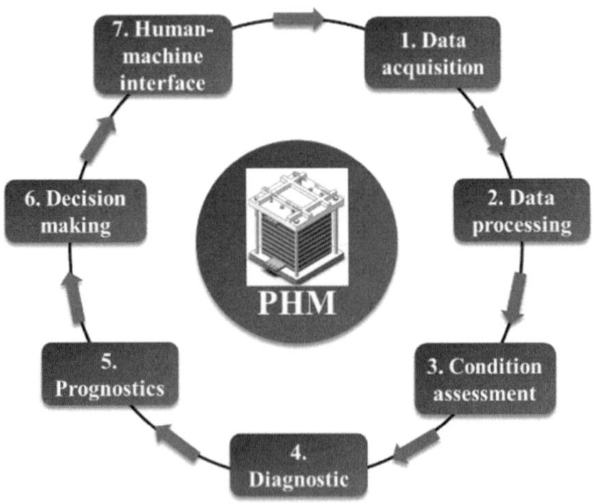

Fig. 9.20 PHM structure principle diagram

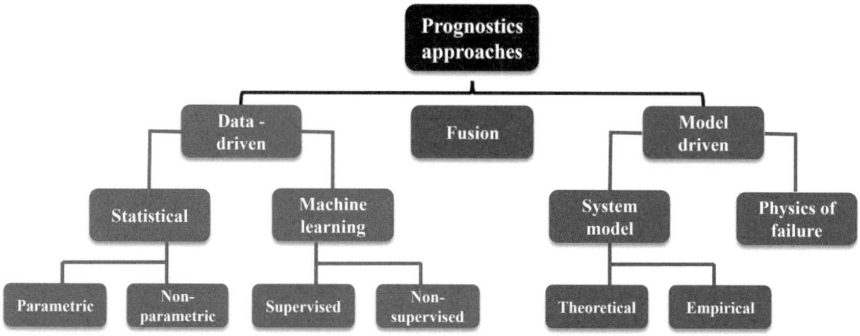

Fig. 9.21 Classification of prognostics approaches

As shown in Fig. 9.21, prognostics methods for PEMFCs are mainly divided into three categories, which are (i) data-driven, (ii) model-driven, and (iii) fusion approach.

(a) Data-driven

Data-driven approaches typically use machine learning, neural networks, or signal processing algorithms for decay prediction and remaining life estimation. These methods have the advantage of simplicity of implementation and do not require an in-depth understanding of the aging mechanism of PEMFCs. The disadvantage is that it requires large and comprehensive training data, which requires a significant investment of time and cost. It is also less versatile, and typically, the aging data used for training and testing must come from the same type and manufacturer of PEMFCs to ensure the accuracy and reliability of the prediction results, which further increases the cost of using data-driven methods. In addition, the data-driven approach suffers from the black-box problem of knowing the changes in key parameters and states during the aging process of PEMFCs. Therefore, users are inevitably limited by the lack of critical state information when making decisions and maintenance, which will eventually affect the effectiveness of maintenance measures.

There are several studies on data-driven research. The regime-switching vector autoregressive was used to achieve accurate estimation of the decay state of PEMFCs [73]. A novel artificial intelligence framework was developed by combining a data-driven surrogate model and a stochastic optimization algorithm to achieve multi-variable global optimization for improving the maximum power density of PEMFCs [74]. A data-driven prediction method was investigated based on the summation wavelet-extreme learning machine algorithm [75], and integrated learning and incremental learning methods were used to improve the accuracy and reliability of the prediction results.

(b) Model-driven

Model-driven approaches can be further classified as mechanistic, empirical, semi-mechanistic, or semi-empirical models.

Advantages of the mechanism model: less data required, high accuracy, and generalizability. In addition, the changes of internal states and key parameters during the decay of PEMFCs can be obtained based on the model parameters. Its main drawback is that the establishment of accurate mechanism models is very complicated, which requires not only an in-depth understanding of the complex aging mechanisms of PEMFCs, but also extensive modeling experience and strong modeling ability. However, some decay mechanisms of PEMFCs are still unclear. A failure-prediction-oriented model of Pt catalyst dissolution mechanism was developed [76] and used a trace-free Kalman filter algorithm to estimate the relative distribution of Pt catalyst particle radius and electrochemical surface area (ECSA) with decay time. The ECSA decay model based on the Pt catalyst dissolution mechanism was embedded into a voltage decay semi-mechanical model by a model inversion approach to achieve the long-term decay prediction of PEMFCs [77]. Based on the Pt catalyst dissolution and Ostwald ripening mechanism model [5], an improved ECSA decay mechanism model was developed [78].

Due to the complexity and computational overload of mechanistic models, researchers have developed empirical models to avoid these problems. Empirical models have the advantages of simple implementation, low computational complexity, and easy online use. However, compared with the mechanistic model, the empirical model also has disadvantages such as low accuracy, poor generality, and more experimental data are needed to identify the model parameters. An empirical model was developed to predict the lifetime of PEMFCs in fuel cell vehicle applications by calculating the voltage decay of PEMFCs under different operating conditions (including start-stop, idle, general load, and high power load) to estimate the decay state [79]. A failure prediction method for PEMFCs was developed based on the particle filter algorithm [80]. The authors compare the performance of linear decay empirical models, exponential decay empirical models, and logarithmic decay empirical models of stack power in decay prediction. The test results show that the logarithmic empirical model has the best predictive performance.

Considering the respective advantages and disadvantages of the mechanistic and empirical models, the researchers compromised between the two models and developed a semi-mechanistic and semi-empirical model. The semi-mechanical model means that the main structural framework of the model is established based on the aging mechanism of PEMFCs, while some parameter values in the model are determined based on experimental data or expert experience. A typical semi-mechanical model of PEMFCs was developed [81]. In this model, the structural framework of the model is established based on the mechanism of polarization losses (including activation losses, ohmic losses, permeation losses, and concentration losses) of PEMFC voltages, while the key parameter values are determined based on expert experience and fitting to the polarization curve data. Semi-empirical modeling refers to designing the structural framework of the model based on experimental data or expert experience, and then determining the values of some parameters in the model based on the decay mechanism of PEMFCs. A stack power model based on experimental data and experience was developed, and the parameter values of the model were set based on the results of the mechanistic analysis of the EIS equivalent circuit model [82].

(c) Fusion method

The researchers proposed a fusion prediction method based on the advantages and disadvantages of data-driven and model-driven methods. According to different hybrid forms, fusion prediction methods can be divided into three categories: (1) The decay features are extracted by a model-driven approach, and then a data-driven approach is used to achieve decay prediction and remaining life estimation. As in Refs. [83, 84]. (2) Fitting attenuation data and predicting attenuation trends by data-driven methods, and then estimating remaining lifetime by model-driven methods. As in Refs. [85, 86]. (3) Two or more model-driven and data-driven methods are used to achieve attenuation trend prediction and remaining life estimation. The results of each method are weighted and summed to obtain the final results. As in Refs. [87, 88].

9.3.2 Mechanism Model for Proton Exchange Membrane Degradation

Currently, the decay prediction models for membranes are mainly for chemical decay. The chemical decay prediction model of PEM, i.e., adding gas transport across the membrane, hydrogen peroxide generation and decomposition, metal cation (Fe^{2+}) transport, free radical generation and transport, and free radical attack on the membrane structure to the simple hydrogen–oxygen fuel cell model. The main material transfer and related physical fields in the model are shown in Fig. 9.22. The model should also have a positive feedback effect: the decay rate increases as the chemical decay proceeds. This is achieved mainly by correcting for the rate of gas transport across the membrane. The details of the model construction are described below.

(a) Gas flows across over membrane

Since the main chain structure of the PEM membrane is hydrophobic, the sulfonic acid group of the side chain is hydrophilic. After the membrane absorbs water, the water will gather in the membrane to form water clusters, and as the water content of the membrane increases, the water clusters are connected to form water channels. Therefore, the PEM resembles a porous structure of two-phase material after hydration, and the gas transport through the membrane is divided into the transport of gas phase and liquid phase. The controlling equation for gas transport across the membrane is [89]:

$$\frac{\partial \phi_{\text{ion}} c^j}{\partial t} + \nabla \cdot \left[S_{\text{ch}} \left(-\psi_{\text{l}}^j \nabla p^j \right) + (1 - S_{\text{ch}}) \left(-\psi_{\text{v}}^j \nabla p^j \right) \right] = 0 \tag{9.1}$$

where ϕ_{ion}, c^j and S_{ch} denote the ionomer volume fraction in the PEM, the molar concentration of gas species j, the fraction of expanded water channels in the membrane, and the gas species partial pressure, respectively. The symbols ψ_l^j

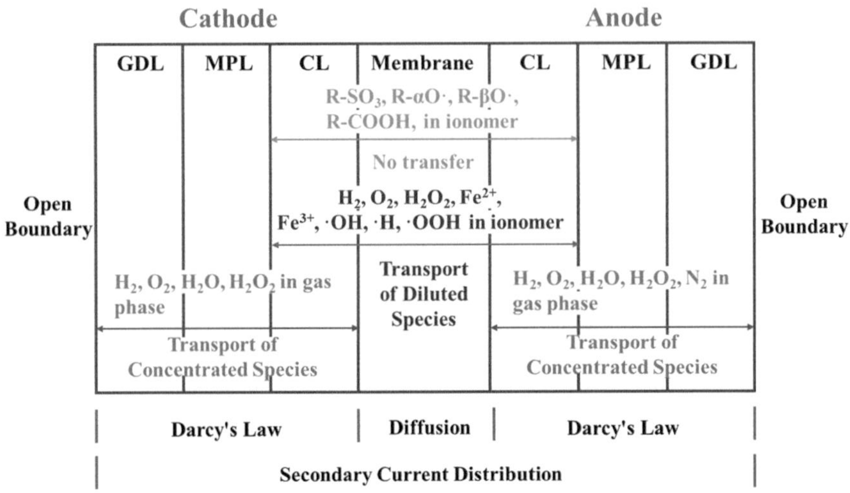

Fig. 9.22 The main material diffusion and related physical fields in the decay model

and ψ_v^j represent the permeation coefficients for the liquid- and vapor-equilibrated membrane.

(b) Generation and decomposition of hydrogen peroxide

The reactions related to hydrogen peroxide have been described in Sect. 8.2.5. The source term settings for the generation and decomposition of hydrogen peroxide are [90]:

$$r^{H_2O_2,f} = A_{eff} i^0 \frac{c^{O_2}}{c_{ref}^{O_2}} \chi \exp\left(\frac{-F\eta}{RT}\right) \tag{9.2}$$

$$r^{H_2O_2,r} = 2A_{eff} i^0 \frac{c^{H_2O_2}}{c_{ref}^{H_2O_2}} \sinh\left(\frac{\alpha^{H_2O_2,r} F\eta}{RT}\right) \tag{9.3}$$

where A_{eff}, i^0, χ and $\alpha^{H_2O_2,r}$ denote the effective reaction area, the exchange current density, the selectivity of hydrogen peroxide, and the activity of hydrogen peroxide, respectively.

(c) Reaction and transport of iron ions

The source of the transition metal ions may be introduced during the fabrication of the electrodes and stacks, or they may be released from the metal bipolar plates. This is the reason why stacks with metal bipolar plates are more prone to decay than stacks with graphite bipolar plates. A redox reaction occurs at the electrode with iron ions. The reaction rate equation is [90]:

$$r^{\mathrm{Fe}} = A_{\mathrm{eff}}\, k^{\mathrm{Fe}}\mathrm{F}\sqrt{c^{\mathrm{Fe2+}}c^{\mathrm{Fe3+}}} \times \left[\exp\left(\frac{\alpha^{\mathrm{Fe}}\mathrm{F}\eta^{\mathrm{Fe}}}{RT}\right) - \exp\left(\frac{-\alpha^{\mathrm{Fe}}\mathrm{F}\eta^{\mathrm{Fe}}}{RT}\right)\right] \tag{9.4}$$

The total amount of ferrous ions in the electrode and PEM is considered to be constant. Ferrous and ferric ions are constantly transformed into each other through redox reactions. Their transport in the electrode and PEM is divided into two parts: diffusion and electromigration, and the controlling equation is as follows [90]:

$$\Psi^j = -D^j\nabla c^j - u^j z^j c^j \mathrm{F}\nabla\Phi_{\mathrm{ion}} \tag{9.5}$$

where D^j, u^j, z^j and Φ_{ion} denote the diffusion coefficient, the electromobility, the charge number, and the electrolyte potential, respectively.

(d) Generation of free radicals

The generation of free radicals is similar to the process occurring in Fenton's reagent (H_2O_2/Fe^{2+}). It is described in Sect. 8.1.5. The reaction rate is calculated by the following equation [91].

$$r^i = \phi_{ion}k^i\prod_j \left(c^j\right)^{v^{i,j}} \tag{9.6}$$

where k^i, c^j and $v^{i,j}$ denote the reaction rate constant, the concentration of substances involved in the reaction, and the stoichiometric number, respectively.

(e) Free radical attack on the membrane structure

It is considered that the attack of free radical HO· on the membrane structure is divided into three parts: (i) the shedding of sulfonic acid groups at the side chains; (ii) the chain decomposition reaction of the side chains that lose functional groups; and (iii) the chain decomposition reaction of the main chain after the loss of side chains. The reaction rate equations are shown below.

$$r^{11} = \phi_{ion}k^{11}c^{\mathrm{HO}} \cdot c^{\mathrm{h}} \tag{9.7}$$

$$r^{12} = \phi_{ion}\frac{k^{12}}{N^{\mathrm{t}}}c^{\mathrm{HO}} \cdot c^{\mathrm{ta}} \tag{9.8}$$

$$r^{13} = \phi_{ion}\frac{k^{13}}{N^{\mathrm{bb}}}c^{\mathrm{HO}} \cdot c^{\mathrm{bba}} \tag{9.9}$$

where c^{HO}, c^{h}, c^{ta} and c^{bba} denote the HO• concentration, the sulfonic acid group concentration, the concentration of side chains that have lost their sulfonic acid groups, and the concentration of main chains that have lost their side chains, respectively. The symbols N^{t} and N^{bb} denote he number of decompositions of the side and main chains that undergo chain reactions, respectively.

(f) Feedback effect of decay

The feedback effect of decay means that: membrane decay leads to changes in membrane properties, and these changes further affect the rate of chemical decay. The effect of chemical decay on membrane performance can be divided into two main areas: proton conductivity (electrical conductivity) and the rate of gas transport across the membrane.

As described in the previous sections, there are interconnected water channels in the hydrated membrane, and the whole membrane resembles a porous structure with water filling it. The conduction of protons in the membrane is similar to the transfer of protons in the porous structure. The transport mechanisms of protons in membranes broadly include three types: (1) proton transfer by jumping through sulfonic acid groups; (2) on-board mechanism, where protons jump between different water molecules by combining with water to form hydrated hydrogen ions; and (3) diffusion mechanism, where protons diffuse through the pores in the form of hydrated hydrogen ions. The equation for the conductivity of the membrane is as follows [92].

$$\sigma_{H^+} = \frac{\varepsilon}{\tau}\left[\frac{F^2}{RT}\left(D_{H^+}^{\Sigma} \cdot C_{H^+}^{\Sigma} + D_{H^+}^{G} \cdot C_{H^+} + D_{H^+}^{E} \cdot C_{H^+}\right)\right] \tag{9.10}$$

where ε and τ denote porosity and curvature, respectively. The symbols $D_{H^+}^{\Sigma}$, $D_{H^+}^{G}$ and $D_{H^+}^{E}$ denote the diffusion coefficients of protons through the three modes of transport, respectively. The symbols $C_{H^+}^{\Sigma}$ and C_{H^+} denote the concentration of protons at the pore surface and the middle of the pore, respectively.

The equations for porosity and curvature are as follows [92]:

$$\varepsilon = \frac{\lambda}{\lambda + \frac{\overline{V}_{\text{Nafion}}}{\overline{V}_{H_2o}}} \tag{9.11}$$

$$\tau = \frac{2(1-\varepsilon) + 2\varepsilon\ln\varepsilon - 0.5\varepsilon(\ln\varepsilon)^2}{\varepsilon(1-\varepsilon) + \varepsilon^2\ln\varepsilon} \tag{9.12}$$

where $\overline{V}_{\text{Nafion}}$, $\overline{V}_{H_2o}$ and λ denote the volume of ionomer per mole of sulfonic acid group, the molar volume of water, and the number of water molecules per mole of sulfonic acid group are expressed respectively. The equation of membrane water content is as follows [93].

$$\lambda = 0.043 + 17.81a - 39.85a^2 + 36.0a^3 \tag{9.13}$$

where a denotes the water activity.

The correction for gas transport across the membrane consists of two main aspects: (i) the effect of membrane microstructure changes on gas transport, and (ii) the effect of membrane thickness changes on gas transport. The effect of microstructure is expressed through Eqs. 9.11 and 9.12, where the water volume fraction, the number of integrals of the ionomer, and the proportion of water channels change accordingly

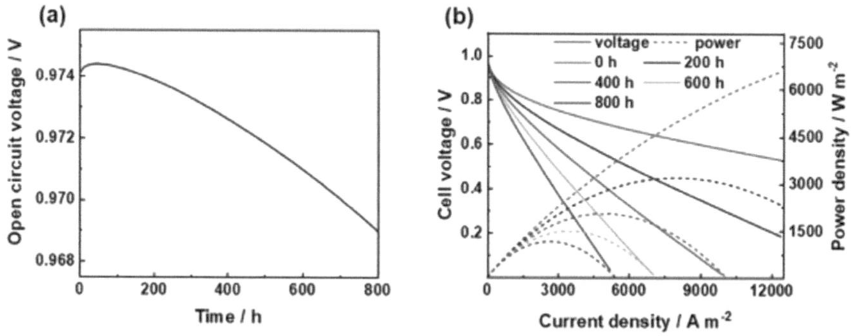

Fig. 9.23 Evolution of PEMFC performance: **a** OCV, **b** polarization and power density curves [94]

during the decay process. The calculation equation for the membrane thickness in the decay process is:

$$l = \overline{V}_{\mathrm{Nafion}} \times n_{\mathrm{SO3}} \div S \tag{9.14}$$

where n_{SO3} and S denote the amount of the substance of the sulfonic acid group and the cross-sectional area of the membrane, respectively.

A bidirectionally coupled physical model was proposed [94], which combines the above control equations with a conventional fuel cell model, to investigate the process and effect of PEM degradation. Using this model, they found that the degradation leads to a slight increase in the OCV at first and then a continuous decrease. At the same time, the ohmic loss of the PEMFC increases rapidly due to the decrease in the effective proton conductivity. The proposed model has been proven to be useful for studying the changes of PEM under different cell operating conditions and for proposing mitigation strategies (Fig. 9.23).

9.3.3 Mechanism Model for Pt Degradation

In addition to the method of performing decay experiments, numerical simulations are also widely used in the study of cathode catalyst degradation in PEMFC. Compared with the experimental method, the numerical simulation is not only more efficient and faster, saves manpower and material resources, but also can obtain the evolution of the particle size distribution and ECSA during the catalyst degradation process in real time, and analyze the non-uniform degradation distribution in the direction of the catalyst layer and the relative contribution between each decay mechanism, to provide a more comprehensive and in-depth understanding of the degradation mechanism of the cathode catalyst and the performance degradation of the catalyst layer. This section mainly reviews the research history and current status of the cathode catalyst degradation model, and introduces the construction process of the mechanism-based cathode catalyst degradation model in detail.

9.3.3.1 Development History

In fact, for the various decay mechanisms of Pt catalysts in PEMFC that we have learned before, several models have been developed to describe them, and a lot of numerical simulation research work has been carried out. In 2003, the first mechanism-based Pt catalyst degradation kinetic model was proposed, which included the electrochemical dissolution of Pt, the formation of PtO, and the chemical dissolution of PtO, and provided its kinetic parameters [95]. The evolution of oxide coverage on the catalyst surface and the dissolution kinetics of cathode Pt during potential cycling were analyzed, and the model was claimed to be in good agreement with the experimental data. Subsequently, the original author added the transport model of Pt ions in the membrane electrode based on this model, established a one-dimensional transient degradation model covering electrochemical reactions, and used this model to study the Pt catalyst with double particle size distribution ECSA change during degradation [96]. Later, other researchers optimized the model by considering the Pt mass loss from the cathode to the PEM and the formation of a Pt band. The dissolution of Pt and the precipitation of Pt ions demonstrate the mass exchange between large particles and small particles, thereby achieving the growth of overall Pt particles, which well describes the Ostwald ripening mechanism mentioned above [5]. Some models consider the chemical dissolution process of PtO and believe that chemical dissolution plays a major role in the dissolution process of Pt [97]. Based on previous modeling work on the Pt degradation process, a mathematical-physical model for the coupling of the degradation kinetics of Pt nanoparticles with the operating conditions and particle size distribution was proposed [98]. The authors extensively modified the kinetic equation for Pt oxidative dissolution using a Butler-Volmer equation-like form, using a lognormal distribution function to fit hundreds of particle size groups to represent a nearly continuous PSD. The results show that the particle size distribution predicted by the model is in good qualitative agreement with the experiment. This mathematical model was widely adopted in the subsequent Pt degradation research work.

The models discussed above only consider pure Pt catalysts. However, it is also important to understand the degradation mechanism of Pt-based bimetallic catalysts. Existing research work has proposed a transient multiscale model of Pt degradation in the cathode electrode using a multilayer model of the electrochemical interface, which was used to predict the performance and degradation rate of Pt–Co catalysts, and the results showed a good agreement with the experimental data [99]. Platinum oxide coverage has a significant effect on Pt dissolution. The thermodynamic and kinetic models of Pt dissolution based on solid solution theory show that the coverage of Pt oxide protects Pt from being dissolved at high potentials. The Pt dissolution rate predicted by the model results is consistent with the experimental results [6]. Although platinum oxides have been studied for more than 40 years, accurate mathematical models describing oxide formation and reduction under transient operating conditions remain a challenge, both experimentally and theoretically. From the previous content, we know that the Ostwald ripening and coalescence mechanisms are the two main reasons for the coarsening of Pt particles. Numerically, however, modeling the coalescence mechanism is more challenging than the Ostwald ripening process because the coalescence mechanism requires solving the integral–differential equations for the PSD. In this regard, some studies have proposed a coalescence thermodynamic model that is only related to the particle size of adjacent particles to simulate the coalescence mechanism of Pt particles, and the results show that the particle growth observed in experiments can also be reproduced using a model without coalescence [100]. Similarly, the radius-dependent Smoluchowski particle formation and disappearance equations are also used to describe the coalescence kinetics of particles, and for simplicity, a size-independent kinetic model is constructed to express the process of particles detaching from carbon supports [101]. However, the numerical methods still do not allow a clear distinction between Ostwald ripening and coalescence mechanisms. The relevance of the coalescence mechanism to the overall Pt degradation in PEMFCs remains unclear, as experimental measurements of ECSA losses also do not allow the distinction between Ostwald ripening and coalescence mechanisms. Therefore, most existing Pt catalyst degradation models also consider Ostwald ripening and coalescence as processes.

9.3.3.2 Mechanism Model Construction

In summary, in PEMFCs, the decay model of pure Pt catalysts at the cathode has been developed relatively comprehensively, which involves all the decay mechanisms of Pt catalysts. Considering that most of the time, the vehicle PEMFCs are mainly under the load cycle condition of below 1 V. In this case, the corrosion rate of the carbon carrier is very slow, so the detachment of Pt particles caused by carbon corrosion is not considered. At the same time, the chemical dissolution of PtO was also neglected because of its very small attenuation effect on ECSA. As mentioned above, the coalescence process was also considered as a part of the Ostwald ripening to be studied. Therefore, most existing degradation models for pure Pt catalysts mainly consist of three parts: (1) Electrochemical dissolution and reprecipitation of Pt; (2) Surface oxidation and reduction of Pt; (3) Pt ions are transported in the membrane electrode and then reduced by hydrogen permeation in the proton exchange membrane. In this section, we will elaborate on the process of constructing a mathematical-physical model for each degradation mechanism of cathode Pt/C catalysts in detail, and review some research results based on this model.

For the physical model of Pt catalyst degradation in the catalyst layer, most studies use a one-dimensional transient degradation model, and there are few reports on zero-dimensional [102] and two-dimensional [103]. As shown in Fig. 9.24 [104], where L represents the thickness of the cathode catalyst layer, the two sides are the CCL/GDL interface and the CCL/MEM interface, respectively. The entire cathode catalyst layer is evenly divided into N control volumes. In any control volume *i*, all Pt nanoparticles are assumed to be hemispherical, and the initial Pt particle size distribution (PSD) is the same, generally using a normal distribution. PSD is discretized into M particle size groups. The more particle size groups, the more continuous the simulated degradation process, and the more in line with the experimental results. Therefore, the diameter

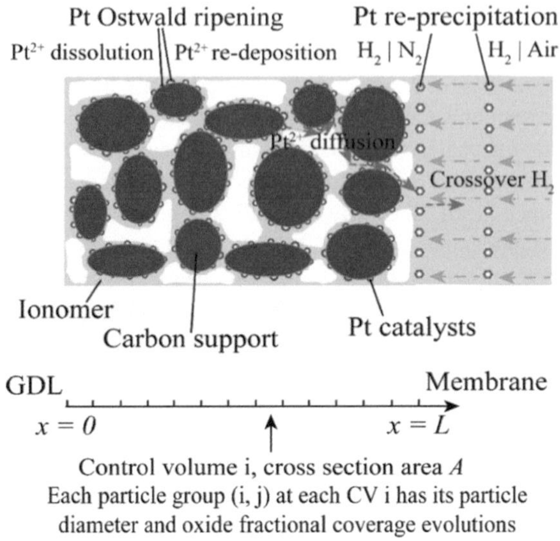

Fig. 9.24 Schematic of 1D Pt degradation model in a PEMFC cathode [104]

of any particle size group j is expressed as d_{ij}, and the corresponding number of particles and PtO coverage are expressed as Num_{ij} and θ_{ij}.

The dissolution and redeposition process of Pt is the core factor that triggers the Ostwald ripening mechanism of the catalyst, and it is also the cause of the loss of Pt mass in the cathode. The associated thermodynamic forward and reverse reaction rates are based on a modified form of Butler-Volmer kinetics, to describe the size dependence of Pt particle stability. The Gibbs–Thomson approximation is employed, which considers the curvature energetics expressed as a function of the surface tension, molar volume, and diameter. The dissolution and redeposition rate expressions of the Pt catalyst are as follows [98]:

$$
\begin{aligned}
r_{\mathrm{net,Pt^{2+}}} = {} & \nu_1 \Gamma \exp\left[\frac{-\overline{H_1}}{RT}\right]\left(1-\min\left(1,\theta_{i,j}\right)\right) \\
& \times \left(\begin{array}{c} \exp\left[\frac{n_1 \mathrm{F}(1-\beta_1)}{RT}\left(U_1^{\mathrm{eq}} - \frac{4\Omega_{\mathrm{Pt}}\gamma_{\mathrm{Pt}}}{d_{i,j} n_1 \mathrm{F}} - V\right)\right] \\ -\frac{\nu_2}{\nu_1}\left(\frac{c_{\mathrm{Pt^{2+}}}}{c_{\mathrm{Pt^{2+}}}^{\mathrm{ref}}}\right)\exp\left[\frac{n_1 \mathrm{F}(\beta_1)}{RT}\left(U_1^{\mathrm{eq}} - \frac{4Q_{\mathrm{Pt}}\gamma_{\mathrm{Pt}}}{d_{i,j} n_1 \mathrm{F}} - V\right)\right] \end{array}\right) \\
& \times \mathrm{RH}^{1.7}
\end{aligned}
\tag{9.15}
$$

In the above expression, v_1 and v_2 are the forward and reverse reaction rate constants, respectively; $\overline{H_1}$ is the molar activation enthalpy of the dissolution reaction; U_1^{eq} is the Pt dissolution equilibrium potential; V is the interface voltage; γ_{Pt} is the Pt surface tension; $c_{\mathrm{Pt^{2+}}}^{\mathrm{ref}}$ is the reference Pt ion concentration. From the expression, we can see that the PtO coverage directly affects the dissolution rate of the Pt catalyst, and high PtO coverage will prevent the dissolution of Pt.

Studies have shown that the coverage of PtO on the particle surface helps to slow down the dissolution of surface Pt atoms [105]. The oxidation process of Pt on the particle surface also plays an important role in the decay process of Pt-based catalysts. Similarly, the expressions for Pt oxidation and reduction are [98]:

$$r_{net,oxide} = \nu_1^* \Gamma \exp\left[\frac{-1}{RT}\left(\overline{H}_2 + \lambda\theta_{i,j}\right)\right] \times \left(\begin{array}{l} \left(1 - \frac{\theta_{i,j}}{2}\right)\exp\left[-\frac{n_2\mathrm{F}(1-\beta_2)}{RT}\left(U_2 + \frac{\omega\theta_{i,j}}{n_2\mathrm{F}} - V\right)\right] \\ -\frac{\nu_2^*}{\nu_1^*}\left(10^{-2pH}\right)\frac{\theta_{i,j}}{2}\exp\left[\frac{n_2\mathrm{F}\beta_2}{RT}\left(U_2 + \frac{\omega\theta_{i,j}}{n_2\mathrm{F}} - V\right)\right] \end{array}\right) \tag{9.16}$$

In the above expression: ν_1^* and ν_2^* are the oxidation and reduction reaction rate constants, respectively; $\overline{H}_2$ is the oxidation reaction activation enthalpy on the Pt surface at zero coverage, which is assumed to be linear with the platinum oxide coverage; λ is the kinetic energy barrier constant related to PtO; ω is the oxide-oxide interaction energy. At the same time, U_2 represents the equilibrium oxidation potential of Pt on the surface of the catalyst, which is related to the particle size, and can be expressed as:

$$U_2 = U_{\mathrm{ftt}} + \frac{1}{\mathrm{F}}\left[\frac{\gamma_{\mathrm{PtO}}\Omega_{\mathrm{PtO}}}{d_{i,j}} - \frac{\gamma_{\mathrm{Pt}}\Omega_{\mathrm{Pt}}}{d_{i,j}}\right] \tag{9.17}$$

where U_{fit} is the equilibrium oxidation potential of the bulk Pt surface (a constant independent of particle size). which is generally taken as 0.98 V.

Next, the governing equations and boundary conditions of the model are introduced. The process of dissolution and reprecipitation of Pt particles is bound to be accompanied by the change of particle size. According to material conservation, the rate of change of Pt particle diameter with time can be expressed by the following formula [95]:

$$\frac{\mathrm{d}(d)}{\mathrm{d}t} = -R_{\mathrm{diss}}\Omega_{\mathrm{Pt}} \tag{9.18}$$

In the same way, the oxidation and reduction of Pt surface and the change of particle size will also lead to the change of surface PtO coverage, and its change rate with time can be expressed as:

$$\frac{\mathrm{d}(\theta)}{\mathrm{d}t} = \frac{R_{\mathrm{oxide}}}{\Gamma} - \frac{2\theta}{d}\frac{\mathrm{d}(d)}{\mathrm{d}t} \tag{9.19}$$

For the transport process of Pt ions in the catalyst layer, it is driven by three mechanisms in the practical application process, namely convective diffusion, electromigration and concentration diffusion. Considering the assumption that the ionomer in the CCL is always in a saturated state, there is no obvious convective process of liquid water at this time, so the convective diffusion can be ignored. At the same time, the transport process of Pt ions is obvious only under high Pt ion concentration in high potential region, and the current density at this time is very small, resulting in a small electrolyte potential gradient in the catalyst layer. Studies have shown that the electromigration term of Pt^{2+} at this time is only about 1/12 of the diffusion term [5]. To simplify the model, the electromigration process was also ignored, although some studies have considered this process [96]. Therefore, this model only considers the concentration diffusion process of Pt ions in the catalyst layer, and its governing equation is expressed as:

$$\varepsilon\frac{\partial c_{\mathrm{Pt}^{2+}}}{\partial t} = \nabla\cdot(\varepsilon^{1.5}D_{\mathrm{Pt}^{2+}}\nabla c_{\mathrm{Pt}^{2+}}) + S_{\mathrm{Pt}^{2+}} \tag{9.20}$$

In the expression, $S_{\mathrm{Pt}^{2+}}$ represents the Pt^{2+} source term produced by Pt dissolution and reprecipitation in the control body, which can be calculated by the dissolution rate of Pt, as shown in the following formula:

$$S_{\mathrm{Pt}^{2+}} = \sum_{i}^{N}\frac{\pi}{2}(d_i)^2 Num_i R_{\mathrm{diss}} \tag{9.21}$$

In addition, in order to be able to solve the Pt ion diffusion model, it is also necessary to set boundary conditions. For the GDL/CCL boundary, because Pt^{2+} has not been detected in the water discharged from the cathode channel in the current experiment, we can think that Pt^{2+} will not diffuse out through the GDL/CCL boundary, so it can be expressed as a zero-flux boundary condition [96]:

$$\left.\frac{\partial c_{\mathrm{Pt}^{2+}}}{\partial x}\right|_{x=0} = 0 \tag{9.22}$$

However, for the CCL/MEM boundary condition, it is related to the fuel cell inlet condition. Under the condition of H_2/N_2, the Pt band is exactly located on the CCL/MEM interface [105], which means that Pt^{2+} will be reduced by H_2 transported across the membrane by the anode as soon as it reaches this interface, so this interface is usually set as a zero concentration boundary condition.

Under H_2/O_2 (AIR) conditions, the CCL/MEM interface was set as a flux boundary condition. The flux is related to the position of the Pt band in the membrane, which is determined by the partial pressure of hydrogen and oxygen. The distance of the Pt band from the CCL/MEM interface can be expressed as [106]:

$$\delta_{\mathrm{Pt}} = \frac{2K_{\mathrm{O}_2}p_{\mathrm{O}_2}\delta_{\mathrm{M}}}{K_{\mathrm{H}_2}p_{\mathrm{H}_2} + 2K_{\mathrm{O}_2}p_{\mathrm{O}_2}}$$

In the expression, δ_M represents the thickness of the proton exchange membrane; K and p represent the permeability and partial pressure, respectively. In general, the Pt^{2+} concentration at the Pt band site is assumed to be zero. According to the

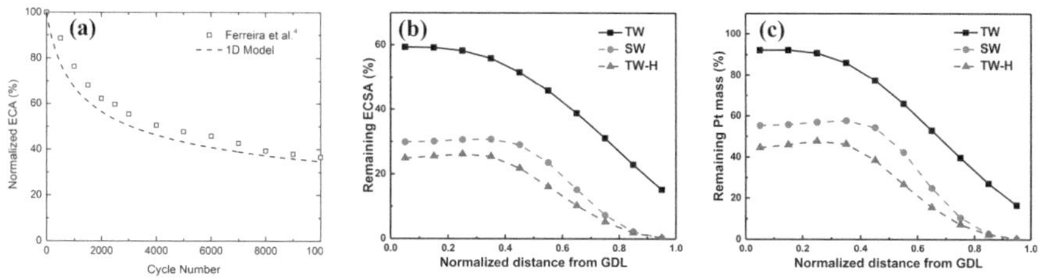

Fig. 9.25 **a** Model predicts ECSA evolution with load cycles (0.6–1.0 V) during Pt degradation [104], **b** ECSA and **c** Pt remaining mass distribution along the direction of the catalyst layer after AST with different loading waveforms [106]

diffusion theory, under the condition of the passive term, the Pt^{2+} concentration should increase linearly at the CCL/MEM interface. From this, the Pt^{2+} flux at the CCL/MEM interface can be obtained, and its boundary conditions can be expressed as [104]:

$$\varepsilon^{1.5}\frac{\partial c_{Pt^{2+}}}{\partial x}\bigg|_{x=L} = -\frac{c_{Pt^{2+}}}{\delta_{Pt}}$$

So far, the one-dimensional transient Pt degradation model has been constructed. Pt degradation affects fuel cell performance through degradation of ECSA. The model can accurately predict the evolution of the cathode catalyst layer ECSA and agrees well with the experimental data [104], as seen in Fig. 9.25a. The ECSA faced rapid degradation at the beginning, and gradually stabilized as the degradation progressed. This is because the properties of Pt particles are more stable as they are coarser. The degradation of ECSA in the direction of the catalyst layer is uneven, and the side near the membrane will face more serious ECSA degradation due to the serious loss of Pt mass [107], as shown in Fig. 9.25b and c. The decrease in ECSA is closely linked to performance degradation, it not only increases the activation overpotential by reducing the sites for oxygen reduction reaction (ORR) kinetics, but also increases the mass transfer loss by causing additional microscale oxygen transport resistance in the ionomer film surrounding the Pt particles, especially at high current densities. As shown in Fig. 9.26, the ionomer effective permeation area available for oxygen transport decreases with decreasing ECSA. The results show that more than half of the performance degradation at high current densities originates from the increase in the transport overpotential [108].

9.3.3.3 Research Status

Based on the mechanical model established above, we can conduct a more in-depth study on the Pt degradation process. From the previous content, we know that temperature, relative humidity, intake conditions, operating voltage, load cycle mode, and particle size have intricate effects on the durability of PEMFC. The predecessors

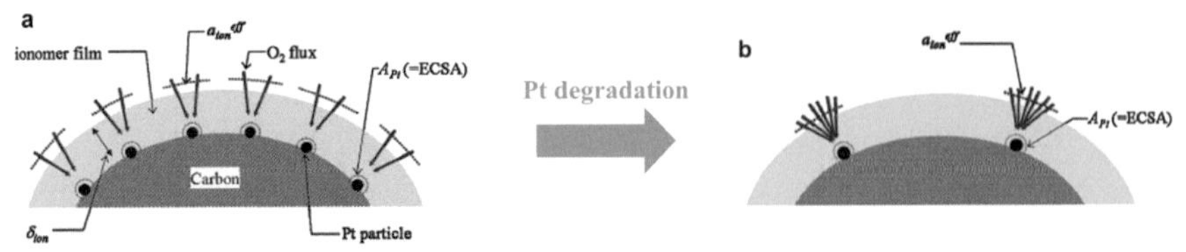

Fig. 9.26 Schematic diagram of the increase of microscale oxygen transport resistance in the ionomer film caused by Pt degradation [108]

have done a lot of research for this by the method of establishing a Pt degradation mechanical model. The simulation results show that both the increase in temperature and relative humidity accelerate the degradation of the cathode catalyst ECSA [104, 109], as shown in Fig. 9.27. Under high temperature and high humidity conditions, the degradation of Pt catalysts is relatively serious. High temperature will accelerate the kinetics of the Pt degradation reaction, and high humidity will promote the dissolution and diffusion of Pt. Also, it can be seen from Fig. 9.27c that ECSA degradation is more sensitive to changes in humidity. Voltage cycling under different cycle curves is also critical to the degradation of Pt catalysts [107]. Comparing the simulation results of triangular wave (TW), square wave (SW), and trapezoidal wave (TW-H), it is evident that longer dwell time at higher potential during SW and TW-H cycling will promote Pt dissolution and Pt oxide covering formation. With longer residence time at high Pt^{2+} concentration, ECSA loss increases significantly due to mass loss of Pt, which then leads to more severe catalyst degradation, as shown in Fig. 9.28. In addition, Fig. 9.28a shows the Pt degradation rate under different periods, which means different scan rates. The scan rate has little effect on the total ECSA loss during the cycle time, mainly because the Ostwald ripening effect on the catalyst surface is slowed down while the cathodic Pt mass loss intensifies as the scan rate decreases. A longer residence time under UPL would accelerate the degradation of catalysts in Fig. 9.28b at different positions of the CCL, especially within the catalyst layer on the membrane side. Because it will lead to more serious Pt mass loss and Ostwald ripening. In addition, for the contribution ratio of these two Pt degradation mechanisms, the simulation results under most working conditions show that the mass loss of Pt plays a leading role in the degradation of ECSA in Fig. 9.28c [107, 109]. Studies have pointed out that the loss of Pt mass is mainly determined by the residence time at high potential, while the Ostwald ripening effect on the catalyst surface is mainly affected by the number of cycles or the frequency of potential oscillation during the potential cycle.

Catalyst layer microstructural variables, like catalyst particle size distribution, also have a significant impact on degradation. Comparing the ECSA evolution results after 10,000 potential cycles under three average diameters of 2, 2.8, and 3.7 nm, as shown in Fig. 9.29a, we can see that the smaller average diameter leads to more serious ECSA degradation, and this relationship is highly sensitive [109]. A slight increase in particle size beyond the commonly used value of 2–3 nm can greatly improve the stability of Pt catalysts, however, the use of large particle sizes can

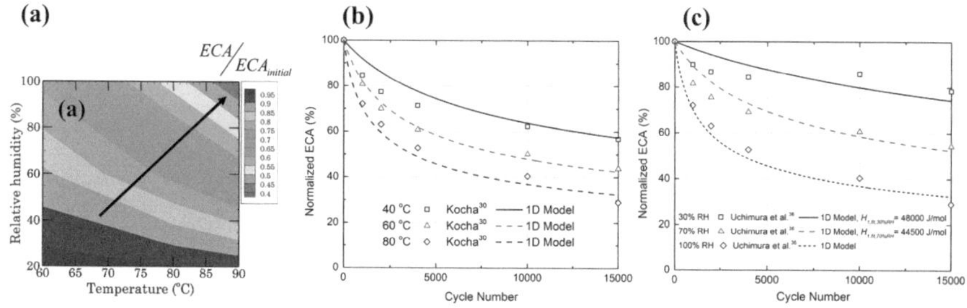

Fig. 9.27 Effect of **a** temperature and **b** humidity on the degradation rate of Pt after load cycle AST (0.6–1.0 V) [104, 109]

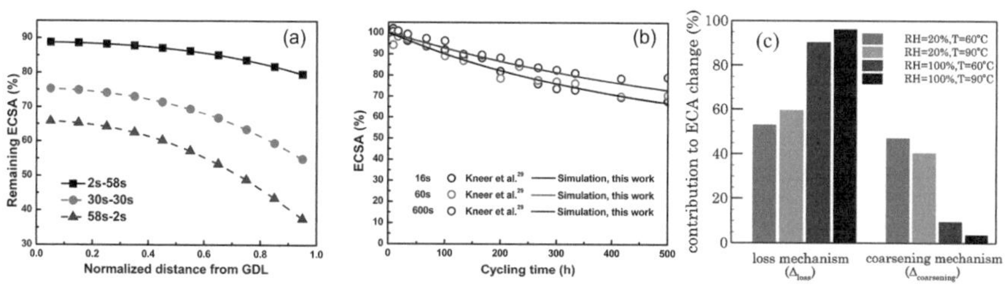

Fig. 9.28 Effect of **a** period and **b** residence time under UPL on the degradation rate of Pt after load cycle AST (0.6–1.0 V) [107], **c** contribution ratio of degradation mechanism (Ostwald ripening and Pt mass loss) under different temperature and humidity conditions [109]

also reduce the initial specific activity, resulting in substandard performance of fuel cells, thus, there is a trade-off. The relationship between performance and durability. The 1D transient Pt degradation model can also be coupled as a sub-model to the transient 2D full cell model [103]. Figure 9.29b shows the polarization curve after 5000 cycles of the current cycle test. It can be seen that the voltage loss is serious under high current density. The non-uniform degradation of the entire 2D catalyst layer, including the channel-land direction and the through-plane direction, can be obtained using the coupled mathematical model, as shown in Fig. 9.30, which demonstrates the remaining ECSA distribution of the catalyst layer after the current cycling test. From Fig. 9.30a, it can be observed that under the land and near the CCL/MEM interface, more severe Pt degradation occurs, which is due to the higher relative humidity below the land due to the longer water transport path, and the higher RH leads to faster Pt degradation. Also, the non-uniformity of Pt degradation in the channel-land direction is slightly smaller than that in the through-plane direction. Non-uniform Pt degradation can lead to changes in the current density distribution in cycled CCLs. Figure 9.30b shows the local current density distribution in the initial and cycled CCL at an average current density of 2.0 A cm^{-2}. Shifts in the current density distribution can be found in both through-plane and channel-land directions. Because of the amplified effect of ECSA loss on the local oxygen transport resistance at high current densities, lower ECA remains under the land in the cycled CCL,

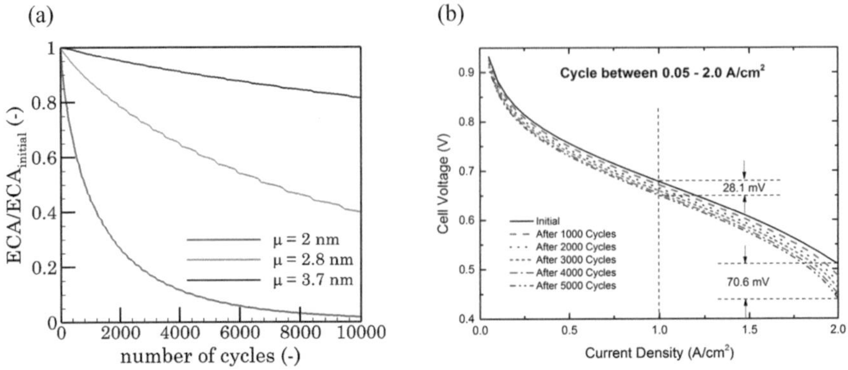

Fig. 9.29 **a** Effect of average particle size (2, 2.8, 3.7 nm) of catalyst layer on degradation rate [109], **b** polarization curve after 5000 cycles of current cycle test (0.05–2 A cm^{-2}) [103]

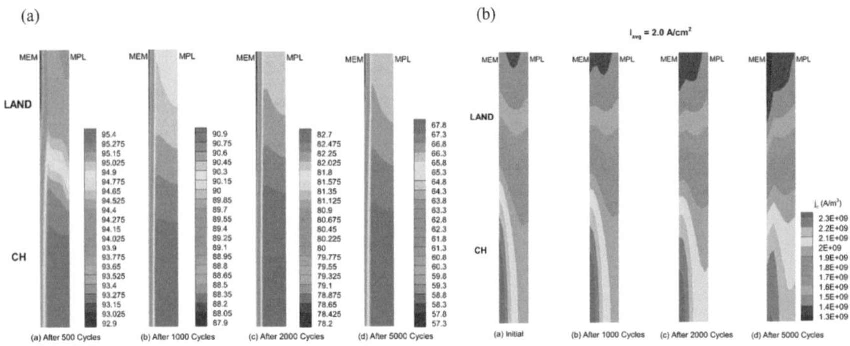

Fig. 9.30 **a** Remaining ECSA distribution and **b** local current density distribution of the entire 2D catalyst layer after 5000 cycles of current cycle test (0.05–2 A cm^{-2}) [103]

which will result in a lower reaction current density. Recently, Hao et al. [110] also developed a two-dimensional Pt aging model, which is used to reveal the reason and influence of the non-uniform degradation (Fig. 9.31). On this basis, a variety of regionally designed MEAs are proposed to provide a possible strategy for extending the lifetime of a fuel cell.

9.3.4 Mechanism Model for Carbon Corrosion

To date, most of the research and development on PEMFC durability has focused on proton exchange membranes and catalysts, with less attention paid to carbon corrosion issues. However, corrosion of the catalyst's support carbon is also a major challenge to the long-term durability of fuel cells. The corrosion of the carbon support

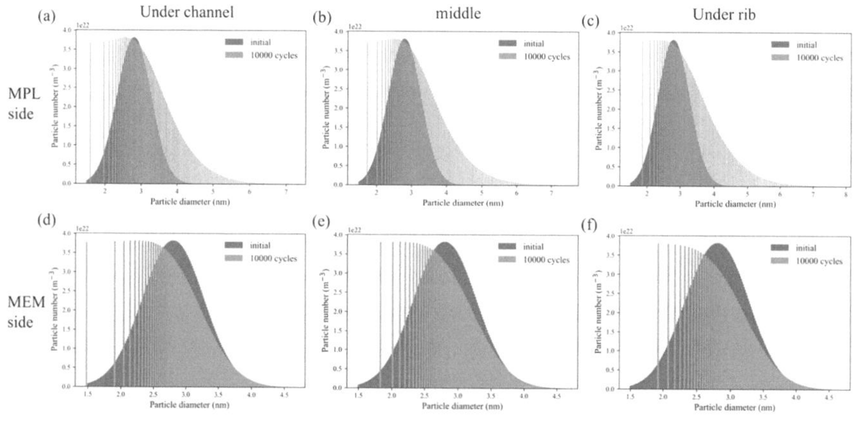

Fig. 9.31 Evolution of particle size distribution at various positions: initial particle size distribution (blue) and simulated particle size distribution after test (orange) [110]

will weaken the binding force between the Pt catalyst and the support, which will lead to the migration and agglomeration of the catalyst, or even fall off. In addition, carbon corrosion will inevitably lead to the collapse and compaction of the carbon skeleton of the catalyst layer, thereby increasing the oxygen transport resistance of the cathode. At the same time, the hydrophilic groups produced by carbon corrosion will also increase the hydrophilicity of the catalyst layer, which increases the risk of water flooding [111]. The foregoing has detailed the mechanism of carbon corrosion, which mainly occurs during fuel cell start-stop cycles and local fuel starvation. At this time, due to the existence of the anode hydrogen-air interface, a reverse current is caused, which leads to a higher interface potential corresponding to the cathode, thereby promoting the occurrence of carbon corrosion reaction [112]. Based on the understanding of the mechanism of carbon corrosion, some studies have reported related modeling work to provide a deeper understanding of carbon corrosion and guide the development of some effective mitigation strategies.

The establishment of a carbon corrosion model requires the use of different models and coupling according to different physical phenomena, mainly including kinetic models based on electrochemical reactions and transport models based on physics [113]. The earliest report on modeling the carbon corrosion process of the cathode catalyst layer can be traced back to 2006, which was developed by Meyers and Darling et al. [114]. Interestingly, they were also the first to develop a model for Pt catalyst degradation. The model is a one-dimensional single-cell model along the channel direction, including the distribution of reactant gas concentration, electrode and electrolyte potential distribution, and local current density. The Butler-Volmer equation was used to describe the hydrogen and oxygen reactions, while the Tafel expression with a Langmuir adsorption term was used for the carbon corrosion reaction. The results show that partial flooding or uneven distribution of hydrogen will form a reverse current, thereby accelerating carbon corrosion. Furthermore, the effect of start/stop cycles on carbon corrosion was investigated. Based on this model, a start-up procedure for mitigating carbon corrosion is proposed. However, this model ignores the in-plane currents in the membrane and thus fails to describe the physics of the hydrogen-air interface.

A multiscale model coupled with electrocatalysis was developed to study the competition between the carbon oxidation reaction and the electrocatalytic mechanism [115]. The model features detailed electrochemical characterization and explicitly describes the instantaneous feedback between performance and the corrosion process of carbon supports. The simulation results show that carbon corrosion can lead to the reduction of cathode thickness and the degradation of cell voltage, and in addition, carbon corrosion is beneficial to platinum coarsening. A two-dimensional steady-state carbon corrosion model was developed along the gas channel direction and the MEA thickness direction [116]. The effects of cell voltage, ionomer proton conductivity, fuel-starved region size, and oxygen transmembrane transport speed on carbon corrosion rate were studied, and the conditions that can significantly accelerate carbon corrosion were proposed. Later, the model was extended to simulate transient processes with quasi-steady-state assumptions, resulting in the electrolyte potential distribution in the membrane as shown in Fig. 9.32a [117, 118]. In a normal cell, the electrolyte potential drops uniformly and linearly across the membrane, but in the absence of hydrogen from the anode, the drop of electrolyte potential can be seen. As a result, the corresponding cathode generates a large interface potential, which promotes the carbon corrosion reaction in this region, so a significant carbon corrosion current is observed in Fig. 9.32b. Moreover, the equipotential lines bend almost perpendicular to the electrodes and then form an arc. Most of the proton current flows directly along a rough semicircle on the anode from the normal area on the left to the hydrogen-depleted area on the right, and the electron current flows through the anode, effectively shorting out the cell. In addition, the effect of MEA design parameters on carbon corrosion, such as oxygen reduction kinetics and membrane thickness, was investigated. The results show that the carbon corrosion current depends on the oxygen partial pressure on the anode, and as the oxygen reduction kinetics increases, the current from carbon corrosion increases sharply in Fig. 9.33a, which means that carbon corrosion can be mitigated by reducing the catalyst activity of the anode. Membrane thickness also affects carbon corrosion. Thicker membrane reduces carbon corrosion, while thinner films cause large carbon corrosion currents in Fig. 9.33b. The effect of cell voltage was also investigated, as a higher electrode potential would cause a larger overpotential and thus a larger carbon corrosion current in Fig. 9.33c.

A fully two-dimensional model considering multiple electrochemical and chemical reactions and coupled transport of charged and uncharged components was developed to study carbon corrosion caused by localized fuel starvation and start-up/shutdown processes [119]. In the simulation, a part of the GDL is set to have an extremely low gas diffusivity to simulate a situation where the local GDL is submerged by liquid water, resulting in a local fuel starvation. The distributions of the electrolyte potential, the potential difference between the electrolyte and the electrodes, and the current density at the interface of CCL and CGDL along the channel

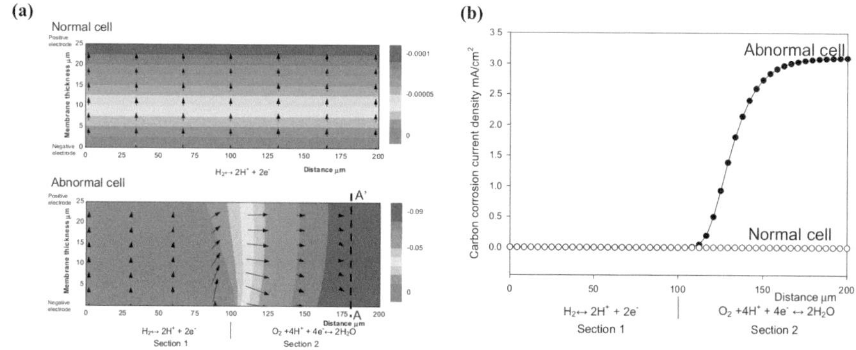

Fig. 9.32 **a** The electrolyte potential distribution in the membrane for normal cell and abnormal cell; **b** carbon corrosion current density distribution on the positive electrode in the normal cell and the abnormal cell. The oxygen replaces the hydrogen in the right part of the negative electrode [118]

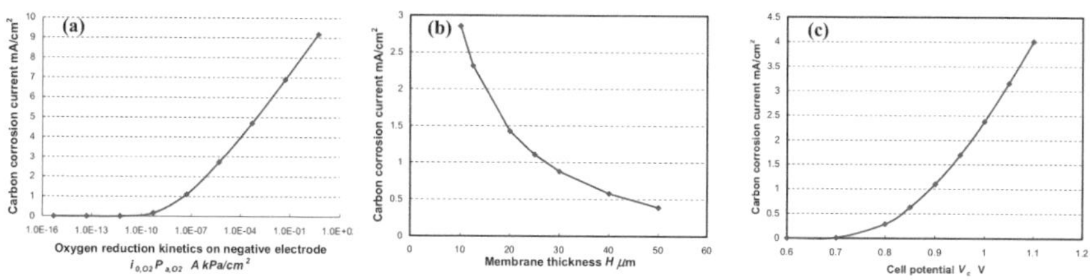

Fig. 9.33 Effect of MEA design parameters on carbon corrosion, **a** oxygen reduction kinetics and **b** membrane thickness; **c** the influence of the cell potentials for carbon corrosion [118]

direction are obtained, as shown in Fig. 9.34a. At the interface between the normal region and the starved region, the electrolyte potential suddenly drops, and the current reversal behavior can be captured, with a reverse current of about 250 mA m^{-2}. For the start/shutdown procedure, it is achieved by setting a part of the anode channel to be filled with air, similar results are shown in Fig. 9.34b. A sudden drop in electrolyte potential is seen at the H_2/Air interface and is more pronounced than in the case of local fuel starvation. In addition, the current reversal is also amplified, about 4000 mA m^{-2}, which is much larger than the local fuel shortage of 250 mA m^{-2}. It was found that under fuel-starved conditions, the rate of carbon corrosion on the cathode side strongly depends on the diffusion of oxygen through the membrane. Experimentally, carbon corrosion has been observed to be associated with cracked membrane [120], which is in good agreement with the hypothesis that O_2 crossover is the root cause of carbon corrosion. The use of OER-favorable catalysts and low O_2 permeable membranes is proposed to be two effective techniques for mitigating carbon corrosion.

CFD (Computational Fluid Dynamics) based models have also been developed, incorporating electrode dynamics and transport, for 2D steady-state simulations [121, 122] as well as 2D and 3D transient simulations [123]. Furthermore, attempts

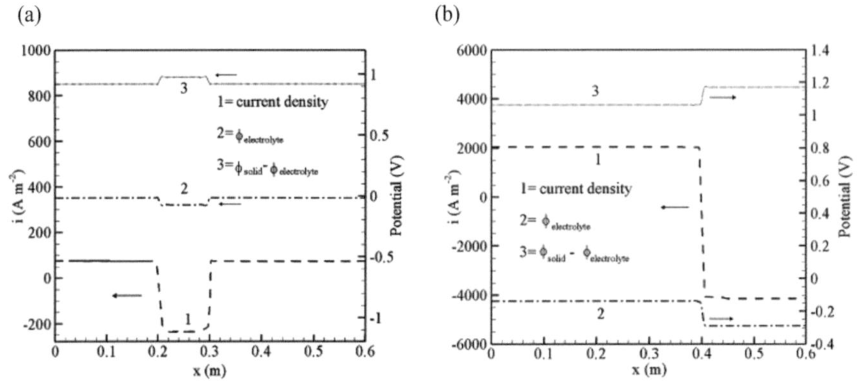

Fig. 9.34 Electrolyte potential, potential difference, and current density profiles on the CCL/CGDL interface for **a** cell voltage = 0.9 V and **b** start/shutdown procedure [119]

have been made to develop multiscale modeling, including carbon corrosion, which includes a description of the electrochemical double layer and considers multiple reaction steps, linking carbon support corrosion to voltage degradation [124]. By combining microstructure data obtained from coarse-grained molecular dynamics (CGMD) simulations with a detailed description of the nanoscale elementary kinetic processes and electrochemical double-layer effects, it is demonstrated that the structural properties of Nafion can affect the carbon corrosion kinetics in the cathode catalyst layer. Nafion adsorbed on carbon supports was shown to reduce carbon corrosion due to the hydrophobicity of the Nafion backbone [125]. Most recently, the significant effect of humidity on the carbon corrosion rate has been investigated by combining experiments with single-cell PEMFCs and a simple SUSD model using the RH-dependent carbon oxidation reaction (COR) rate kinetics [126]. It was found that when the RH was decreased from 100 to 25%, the COR rate decreased by about 3 times during the start-stop cycle at 80 °C. A 3D model considering the multi-step reaction of carbon corrosion was developed to study the carbon corrosion characteristics under different operating conditions [127]. The results show that after 5000 h of operation in the non-flooded area, the carbon carrier degradation rate is only about 4%. However, the degradation of carbon support in the flooded area was as high as about 97.76% after 100 h. This means that good water management is essential to mitigate carbon corrosion. The study also pointed out that increasing the pressure and relative humidity will promote carbon corrosion, while increasing the temperature will only increase the carbon corrosion rate in the non-flooded area, and have a mitigation effect on the carbon corrosion in the flooded area. The study also pointed out that increasing the pressure and relative humidity will promote carbon corrosion, while increasing the temperature will only increase the carbon corrosion rate in the non-flooded area, and have a mitigation effect on the carbon corrosion in the flooded area. This is consistent with the conclusions drawn from the experiment [128].

9.4 Summary and Outlook

Fuel cell durability test methods can be broadly divided into two categories: steady-state durability test and AST. Among them, the traditional steady state test is less used because of the long testing time and high cost, and AST has become the mainstream. In AST, the cell operating voltage will change regularly according to a certain AST protocol, which will cause some components to experience significant performance degradation [129]. By carefully designing different AST protocols, it is possible to reduce test costs and to investigate the degradation of specific components while minimizing the impacts of other components. Therefore, AST has become the mainstream experimental method for fuel cell degradation studies.

In addition to experiments, simulation is another tool to study fuel cell degradation, especially playing an important role in fuel cell health management [130, 131]. The current fuel cell degradation models can be roughly divided into three categories: data-driven models, mechanism models, and fusion models [132]. Data-driven models use deep learning algorithms to predict remaining lifetime and belong to the black-box model, which has the simplicity of no requirement for knowledge of fuel cell degradation mechanisms [133]. However, the model lacks generalization, and it is difficult to obtain the internal details of the fuel cell. A large amount of measured data is required in the model construction, and only specific parameter variations can be output. Mechanism models, on the other hand, can output the detailed information of the fuel cell during degradation, but the establishment process is very complicated, which requires not only an in-depth understanding of the complex aging mechanisms but also extensive modeling experience and strong modeling ability [134]. In addition, the current research on the cell degradation mechanism is still insufficient, and it is difficult to construct a complete mechanism model. Now, there are also some researchers trying to combine these two models, using the advantages of both types to achieve faster and more accurate degradation prediction [135].

In general, the current understanding of fuel cell aging is still not enough. During durability tests, researchers can only perform some limited electrochemical characterization [136], and there is no way to directly characterize the fuel cell degradation process in situ. Therefore, it is necessary to design more test methods that can directly observe the degradation process, which helps to understand the aging mechanism. As for the simulation, it can be said that it is still at an early stage. Taking the mechanism model as an example, a few three-dimensional mechanism models have been proposed, while experimental tests have proved the variability of the degradation, indicating the need for three-dimensional models. Furthermore, most of the mechanism models only couple one degradation mechanism (For example, the Pt degradation model only couples the Pt degradation mechanism), whereas multiple degradation processes exist simultaneously in a fuel cell [137]. Hence, there is a need to develop models that couple more degradation mechanisms to achieve accurate predictions. Finally, the combination with deep learning is also promising, and it requires continuous efforts from researchers to better integrate deep learning algorithms with degradation mechanism models.

References

1. Wang Y, Chen K S, Mishler J, et al. A review of polymer electrolyte membrane fuel cells: Technology, applications, and needs on fundamental research [J]. Applied Energy, 2011, 88(4): 981–1007.
2. Inaba M, Yamada H, Tokunaga J, et al. Effect of Agglomeration of Pt/C Catalyst on Hydrogen Peroxide Formation [J]. Electrochemical and Solid State Letters, 2004, 7.
3. Xu H, Boroup R, Brosha E, et al. Effect of Relative Humidity on Membrane Degradation Rate and Mechanism in PEM Fuel Cells [J]. ECS Transactions, 2007, 6(13): 51–62.
4. Yuan X-Z, Zhang S, Wang H, et al. Degradation of a polymer exchange membrane fuel cell stack with Nafion® membranes of different thicknesses: Part I. In situ diagnosis [J]. Journal of Power Sources, 2010, 195(22): 7594–9.
5. Bi W, Fuller T F. Modeling of PEM fuel cell Pt/C catalyst degradation [J]. Journal of Power Sources, 2008, 178(1): 188–96.
6. Ahluwalia R K, Arisetty S, Wang X, et al. Thermodynamics and Kinetics of Platinum Dissolution from Carbon-Supported Electrocatalysts in Aqueous Media under Potentiostatic and Potentiodynamic Conditions [J]. Journal of The Electrochemical Society, 2013, 160(4): F447-F55.
7. Yi P, Li X, Yao L, et al. A lifetime prediction model for coated metallic bipolar plates in proton exchange membrane fuel cells [J]. Energy Conversion and Management, 2019, 183: 65–72.
8. Wu Z, Zhou Y, Lin G, et al. An improved model for predicting electrical contact resistance between bipolar plate and gas diffusion layer in proton exchange membrane fuel cells [J]. Journal of Power Sources, 2008, 182(1): 265–9.
9. Pauchet J, Prat M, Schott P, et al. Performance loss of proton exchange membrane fuel cell due to hydrophobicity loss in gas diffusion layer: Analysis by multiscale approach combining pore network and performance modelling [J]. International Journal of Hydrogen Energy, 2012, 37(2): 1628–41.
10. Liu H, George M G, Banerjee R, et al. Accelerated Degradation of Polymer Electrolyte Membrane Fuel Cell Gas Diffusion Layers [J]. Journal of The Electrochemical Society, 2017, 164(7): F704-F13.
11. Cui T, Lin C W, Chien C H, et al. Service life estimation of liquid silicone rubber seals in polymer electrolyte membrane fuel cell environment [J]. Journal of Power Sources, 2011, 196(3): 1216–21.
12. Cui T, Chao Y J, Van Zee J W. Sealing force prediction of elastomeric seal material for PEM fuel cell under temperature cycling [J]. International Journal of Hydrogen Energy, 2014, 39(3): 1430–8.
13. Kusoglu A, Karlsson A M, Santare M H, et al. Mechanical response of fuel cell membranes subjected to a hygro-thermal cycle [J]. Journal of Power Sources, 2006, 161(2): 987–96.
14. Takagi Y, Takakuwa Y. Effect of Shutoff Sequence of Hydrogen and Air on Performance Degradation in PEFC [J]. ECS Transactions, 2006, 3(1): 855–60.
15. Yu J, Matsuura T, Yoshikawa Y, et al. Lifetime behavior of a PEM fuel cell with low humidification of feed stream [J]. Physical Chemistry Chemical Physics, 2005, 7(2): 373–8.
16. Lai Y H, Dillard D A. Mechanical durability characterization and modeling of ionomeric membranes [M]. 2010.
17. Tang H, Peikang S, Jiang S P, et al. A degradation study of Nafion proton exchange membrane of PEM fuel cells [J]. Journal of Power Sources, 2007, 170(1): 85–92.

18. Kim S, Mench M M. Physical degradation of membrane electrode assemblies undergoing freeze/thaw cycling: Micro-structure effects [J]. Journal of Power Sources, 2007, 174(1): 206–20.
19. Pestrak M, Li Y, Case S W, et al. The Effect of Mechanical Fatigue on the Lifetimes of Membrane Electrode Assemblies [J]. Journal of Fuel Cell Science and Technology, 2010, 7(4).
20. Chandesris M, Vincent R, Guetaz L, et al. Membrane degradation in PEM fuel cells: From experimental results to semi-empirical degradation laws [J]. International Journal of Hydrogen Energy, 2017, 42(12): 8139–49.
21. Taniguchi A, Akita T, Yasuda K, et al. Analysis of electrocatalyst degradation in PEMFC caused by cell reversal during fuel starvation [J]. Journal of Power Sources, 2004, 130(1): 42–9.
22. Ralph T R, Hogarth M P. Catalysis for Low Temperature Fuel Cells [J]. Platinum Metals Review, 2002, 46(3): 117–35.
23. Xing Y, Li H, Avgouropoulos G. Research Progress of Proton Exchange Membrane Failure and Mitigation Strategies [J]. Materials, 2021, 14(10): 2591.
24. Wu J, Yuan X Z, Martin J J, et al. A review of PEM fuel cell durability: Degradation mechanisms and mitigation strategies [J]. Journal of Power Sources, 2008, 184(1): 104–19.
25. Schmittinger W, Vahidi A. A review of the main parameters influencing long-term performance and durability of PEM fuel cells [J]. Journal of Power Sources, 2008, 180(1): 1–14.
26. Zhao J, Li X. A review of polymer electrolyte membrane fuel cell durability for vehicular applications: Degradation modes and experimental techniques [J]. Energy Conversion and Management, 2019, 199: 112022.
27. Cleghorn S J C, Mayfield D K, Moore D A, et al. A polymer electrolyte fuel cell life test: 3 years of continuous operation [J]. Journal of Power Sources, 2006, 158(1): 446–54.
28. Lu L, Ouyang M, Huang H, et al. A semi-empirical voltage degradation model for a low-pressure proton exchange membrane fuel cell stack under bus city driving cycles [J]. Journal of Power Sources, 2007, 164(1): 306–14.
29. Nandjou F, Poirot-crouvezier J P, Chandesris M, et al. Impact of heat and water management on proton exchange membrane fuel cells degradation in automotive application [J]. Journal of Power Sources, 2016, 326: 182–92.
30. Fernandes A C, Ticianelli E A. A performance and degradation study of Nafion 212 membrane for proton exchange membrane fuel cells [J]. Journal of Power Sources, 2009, 193: 547–54.
31. Huang X, Solasi R, Zou Y, et al. Mechanical endurance of polymer electrolyte membrane and PEM fuel cell durability [J]. Journal of Polymer Science Part B: Polymer Physics, 2006, 44: 2346–57.
32. Chandesris M, Vincent R Q, Guetaz L, et al. Membrane degradation in PEM fuel cells: From experimental results to semi-empirical degradation laws [J]. International Journal of Hydrogen Energy, 2017, 42: 8139–49.
33. Singh Y, Orfino F P, Dutta M, et al. 3D Failure Analysis of Pure Mechanical and Pure Chemical Degradation in Fuel Cell Membranes [J]. Journal of The Electrochemical Society, 2017, 164.
34. Kreitmeier S, MichiardI M, Wokaun A, et al. Factors determining the gas crossover through pinholes in polymer electrolyte fuel cell membranes [J]. Electrochimica Acta, 2012, 80: 240–7.
35. Mcdonald R C, Mittelsteadt C K, Thompson E L. Effects of Deep Temperature Cycling on Nafion® 112 Membranes and Membrane Electrode Assemblies [J]. Fuel Cells, 2004, 4(3): 208–13.
36. Sabawa J P, Bandarenka A S. Investigation of degradation mechanisms in PEM fuel cells caused by low-temperature cycles [J]. International Journal of Hydrogen Energy, 2021, 46(29): 15951–64.
37. Kreitmeier S, Lerch P, Wokaun A, et al. Local Degradation at Membrane Defects in Polymer Electrolyte Fuel Cells [J]. Journal of The Electrochemical Society, 2013, 160.
38. Zhang S, Yuan X-Z, Hin J N C, et al. A review of platinum-based catalyst layer degradation in proton exchange membrane fuel cells [J]. Journal of Power Sources, 2009, 194(2): 588–600.

39. Borup R L, Davey J R, Garzon F H, et al. PEM fuel cell electrocatalyst durability measurements [J]. Journal of Power Sources, 2006, 163(1): 76–81.
40. Dam V A T, De Bruijn F A. The Stability of PEMFC Electrodes: Platinum Dissolution vs Potential and Temperature Investigated by Quartz Crystal Microbalance [J]. Journal of The Electrochemical Society, 2007, 154(5): B494.
41. Bi W, Fuller T. Temperature Effects on PEM Fuel Cells Pt/C Catalyst Degradation [J]. ECS Transactions, 2007, 11(1): 1235.
42. Bi W, Sun Q, Deng Y, et al. The effect of humidity and oxygen partial pressure on degradation of Pt/C catalyst in PEM fuel cell [J]. Electrochimica Acta, 2009, 54(6): 1826–33.
43. Xie J, Wood D L, More K L, et al. Microstructural Changes of Membrane Electrode Assemblies during PEFC Durability Testing at High Humidity Conditions [J]. Journal of The Electrochemical Society, 2005, 152(5): A1011.
44. Wang J, Yang W, Geng J, et al. Experimental investigation on degradation mechanism of membrane electrode assembly at different humidity under automotive protocol [J]. Chinese Journal of Chemical Engineering, 2023, 56: 70–9.
45. Stariha S, Macauley N, Sneed B T, et al. Recent advances in catalyst accelerated stress tests for polymer electrolyte membrane fuel cells[J]. Journal of The Electrochemical Society, 2018, 165(7): F492-F501.
46. Mitsushima S, Kawahara S, Ota K-I, et al. Consumption Rate of Pt under Potential Cycling [J]. Journal of The Electrochemical Society, 2007, 154(2): B153.
47. Stariha S, Macauley N, Sneed B T, et al. Recent Advances in Catalyst Accelerated Stress Tests for Polymer Electrolyte Membrane Fuel Cells [J]. Journal of The Electrochemical Society, 2018, 165(7): F492.
48. Pavlišič A, Jovanovič P, Šelih V S, et al. Platinum Dissolution and Redeposition from Pt/C Fuel Cell Electrocatalyst at Potential Cycling [J]. Journal of The Electrochemical Society, 2018, 165(6): F3161.
49. Zhang Y, Chen S, Wang Y, et al. Study of the degradation mechanisms of carbon-supported platinum fuel cells catalyst via different accelerated stress test [J]. Journal of Power Sources, 2015, 273: 62–9.
50. Cheng L, Khedekar K, Rezaei Talarposhti M, et al. Mapping of Heterogeneous Catalyst Degradation in Polymer Electrolyte Fuel Cells [J]. Advanced Energy Materials, 2020, 10(28): 2000623.
51. Komini Babu S, Spernjak D, Dillet J, et al. Spatially resolved degradation during startup and shutdown in polymer electrolyte membrane fuel cell operation [J]. Applied Energy, 2019, 254: 113659.
52. Matsuoka K, Sakamoto S, Nakato K, et al. Degradation of polymer electrolyte fuel cells under the existence of anion species [J]. Journal of Power Sources, 2008, 179(2): 560–5.
53. Yamada H, Shimoda D, Matsuzawa K, et al. Stability of Pt-Ru/C Catalysts: Effects of Ru Content [J]. ECS Transactions, 2007, 11(1): 325.
54. Yoda T, Uchida H, Watanabe M. Effects of operating potential and temperature on degradation of electrocatalyst layer for PEFCs [J]. Electrochimica Acta, 2007, 52(19): 5997–6005.
55. Mayrhofer K J J, Meier J C, Ashton S J, et al. Fuel cell catalyst degradation on the nanoscale [J]. Electrochemistry Communications, 2008, 10(8): 1144–7.
56. Yamazaki O, Oomori Y, Shintaku H, et al. Study on Degradation of the Pt-Ru Anode of PEFC for Residential Application (2): Analyses of the MEAs Degraded in the Anode Performance [J]. ECS Transactions, 2007, 11(1): 287.
57. Tang H, Qi Z, Ramani M, et al. PEM fuel cell cathode carbon corrosion due to the formation of air/fuel boundary at the anode [J]. Journal of Power Sources, 2006, 158(2): 1306–12.
58. Durst J, Lamibrac A, Charlot F, et al. Degradation heterogeneities induced by repetitive start/stop events in proton exchange membrane fuel cell: Inlet vs. outlet and channel vs. land [J]. Applied Catalysis B: Environmental, 2013, 138–139: 416–26.
59. Makharia R, Kocha S, Yu P, et al. Durable PEM Fuel Cell Electrode Materials: Requirements and Benchmarking Methodologies [J]. ECS Transactions, 2006, 1(8): 3.

60. Macauley N, Papadias D D, Fairweather J, et al. Carbon Corrosion in PEM Fuel Cells and the Development of Accelerated Stress Tests [J]. Journal of The Electrochemical Society, 2018, 165(6): F3148.
61. Hornung R, Kappelt G. Bipolar plate materials development using Fe-based alloys for solid polymer fuel cells [J]. Journal of Power Sources, 1998, 72(1): 20–1.
62. Zhang S, Yuan X-Z, Hiesgen R, et al. Effect of open circuit voltage on degradation of a short proton exchange membrane fuel cell stack with bilayer membrane configurations [J]. Journal of Power Sources, 2012, 205: 290–300.
63. Yuan X-Z, Zhang S, Ban S, et al. Degradation of a PEM fuel cell stack with Nafion® membranes of different thicknesses. Part II: Ex situ diagnosis [J]. Journal of Power Sources, 2012, 205: 324–34.
64. Uvarov V, Popov I. Metrological characterization of X-ray diffraction methods for determination of crystallite size in nano-scale materials [J]. Materials Characterization, 2007, 58(10): 883–91.
65. Ferreira P, Shao-Horn Y, Morgan D, et al. Instability of Pt/C Electrocatalysts in Proton Exchange Membrane Fuel Cells [J]. Journal of The Electrochemical Society - J ELECTROCHEM SOC, 2005, 152.
66. Lin R, Gülzow E, Schulze M, et al. Investigation of Membrane Pinhole Effects in Polymer Electrolyte Fuel Cells by Locally Resolved Current Density [J]. Journal of The Electrochemical Society, 2011, 158(1): B11.
67. Fernandes A C, Ticianelli E A. A performance and degradation study of Nafion 212 membrane for proton exchange membrane fuel cells [J]. Journal of Power Sources, 2009, 193(2): 547–54.
68. Young A P, Stumper J, Gyenge E. Characterizing the Structural Degradation in a PEMFC Cathode Catalyst Layer: Carbon Corrosion [J]. Journal of The Electrochemical Society, 2009, 156(8): B913.
69. Wood D L, Borup R L. Durability Aspects of Gas-Diffusion and Microporous Layers [M]// BüCHI F N, INABA M, SCHMIDT T J. Polymer Electrolyte Fuel Cell Durability. New York, NY; Springer New York. 2009: 159–95.
70. Lin R, Li B, Hou Y P, et al. Investigation of dynamic driving cycle effect on performance degradation and micro-structure change of PEM fuel cell [J]. International Journal of Hydrogen Energy, 2009, 34(5): 2369–76.
71. Yuan X-Z. PEM fuel cell diagnostic tools [M]. 2011.
72. Kundu S, Fowler M W, Simon L C, et al. Open circuit voltage durability study and model of catalyst coated membranes at different humidification levels [J]. Journal of Power Sources, 2010, 195(21): 7323–31.
73. Hochstein A, Ahn H-I, Leung Y T, et al. Switching vector autoregressive models with higher-order regime dynamics Application to prognostics and health management [J]. 2014 International Conference on Prognostics and Health Management, 2014: 1–10.
74. Wang B, Xie B, Xuan J, et al. AI-based optimization of PEM fuel cell catalyst layers for maximum power density via data-driven surrogate modeling [J]. Energy Conversion and Management, 2020, 205: 112460.
75. Javed K, Gouriveau R, Zerhouni N, et al. Prognostics of Proton Exchange Membrane Fuel Cells stack using an ensemble of constraints based connectionist networks [J]. Journal of Power Sources, 2016, 324: 745–57.
76. Zhang X, Pisu P. Prognostic-oriented Fuel Cell Catalyst Aging Modeling and Its Application to Health-Monitoring and Prognostics of a PEM Fuel Cell [J]. International Journal of Prognostics and Health Management, 2014, 5.
77. Robin C, Gérard M, Quinaud M, et al. Proton exchange membrane fuel cell model for aging predictions: Simulated equivalent active surface area loss and comparisons with durability tests [J]. Journal of Power Sources, 2016, 326: 417–27.
78. Polverino P, Pianese C. Model-based prognostic algorithm for online RUL estimation of PEMFCs; proceedings of the 2016 3rd Conference on Control and Fault-Tolerant Systems (SysTol), F 7–9 Sept. 2016, 2016 [C].

79. Chen H, Pei P, Song M. Lifetime prediction and the economic lifetime of Proton Exchange Membrane fuel cells [J]. Applied Energy, 2015, 142: 154–63.
80. Jouin M, Gouriveau R, Hissel D, et al. Prognostics of PEM fuel cell in a particle filtering framework [J]. International Journal of Hydrogen Energy, 2014, 39(1): 481–94.
81. Bressel M, Hilairet M, Hissel D, et al. Extended Kalman Filter for prognostic of Proton Exchange Membrane Fuel Cell [J]. Applied Energy, 2016, 164: 220–7.
82. Zhang D, Cadet C, Bérenguer C, et al. Some Improvements of Particle Filtering Based Prognosis for PEM Fuel Cells [J]. IFAC-PapersOnLine, 2016, 49(28): 162–7.
83. Zhou D, Al-Durra A, Zhang K, et al. Online remaining useful lifetime prediction of proton exchange membrane fuel cells using a novel robust methodology [J]. Journal of Power Sources, 2018, 399: 314–28.
84. Li Z, Jemei S, Gouriveau R, et al. Remaining Useful Life Estimation for PEMFC in Dynamic Operating Conditions; proceedings of the 2016 IEEE Vehicle Power and Propulsion Conference (VPPC), F 17–20 Oct. 2016, 2016 [C].
85. Yang C, Li Z, Liang B, et al. A novel fusion strategy for failure prognostic of proton exchange membrane fuel cell stack; proceedings of the 2017 36th Chinese Control Conference (CCC), F 26–28 July 2017, 2017 [C].
86. Cheng Y, Zerhouni N, Lu C. A hybrid remaining useful life prognostic method for proton exchange membrane fuel cell [J]. International Journal of Hydrogen Energy, 2018, 43(27): 12314–27.
87. Zhou D, Gao F, Breaz E, et al. Degradation prediction of PEM fuel cell using a moving window based hybrid prognostic approach [J]. Energy, 2017, 138: 1175–86.
88. Wu X, Ye Q, Wang J. A hybrid prognostic model applied to SOFC prognostics [J]. International Journal of Hydrogen Energy, 2017, 42(39): 25008–20.
89. Weber A Z, Newman J. Transport in Polymer-Electrolyte Membranes : II. Mathematical Model [J]. Journal of The Electrochemical Society, 2004, 151(2): A311.
90. Wong K H, Kjeang E. Mitigation of chemical membrane degradation in fuel cells: understanding the effect of cell voltage and iron ion redox cycle [J]. ChemSusChem, 2015, 8(6): 1072–82.
91. Futter G A, Latz A, Jahnke T. Physical modeling of chemical membrane degradation in polymer electrolyte membrane fuel cells: Influence of pressure, relative humidity and cell voltage [J]. Journal of Power Sources, 2019, 410–411: 78–90.
92. Choi P, Jalani N H, Datta R. Thermodynamics and Proton Transport in Nafion: II. Proton Diffusion Mechanisms and Conductivity [J]. Journal of The Electrochemical Society, 2005, 152(3): E123.
93. Jiao K, Li X. Water transport in polymer electrolyte membrane fuel cells [J]. Progress in Energy and Combustion Science, 2011, 37(3): 221–91.
94. Chen S, Hao M, Hu Y, et al. Insight into the evolution of membrane chemical degradation in proton exchange membrane fuel cells: From theoretical analysis to model developing[J]. Journal of Power Sources, 2024, 599: 234238.
95. Darling R M, Meyers J P. Kinetic Model of Platinum Dissolution in PEMFCs [J]. Journal of The Electrochemical Society, 2003, 150(11): A1523.
96. Darling R M, Meyers J P. Mathematical Model of Platinum Movement in PEM Fuel Cells [J]. Journal of The Electrochemical Society, 2005, 152(1): A242.
97. Rinaldo S G, Stumper J, Eikerling M. Physical Theory of Platinum Nanoparticle Dissolution in Polymer Electrolyte Fuel Cells [J]. The Journal of Physical Chemistry C, 2010, 114(13): 5773–85.
98. Holby E F, Morgan D. Application of Pt Nanoparticle Dissolution and Oxidation Modeling to Understanding Degradation in PEM Fuel Cells [J]. Journal of The Electrochemical Society, 2012, 159(5): B578.
99. Franco A A, Passot S, Fugier P, et al. Ptx Coy Catalysts Degradation in PEFC Environments: Mechanistic Insights: I. Multiscale Modeling [J]. Journal of The Electrochemical Society, 2009, 156(3): B410.

100. Ahluwalia R K, Arisetty S, Peng J-K, et al. Dynamics of Particle Growth and Electrochemical Surface Area Loss due to Platinum Dissolution [J]. Journal of The Electrochemical Society, 2014, 161(3): F291.
101. Urchaga P, Kadyk T, Rinaldo S G, et al. Catalyst Degradation in Fuel Cell Electrodes: Accelerated Stress Tests and Model-based Analysis [J]. Electrochimica Acta, 2015, 176: 1500–10.
102. Schneider P, Sadeler C, Scherzer A-C, et al. Fast and Reliable State-of-Health Model of a PEM Cathode Catalyst Layer [J]. Journal of The Electrochemical Society, 2019, 166(4): F322.
103. Li Y, Wang C Y. Modeling of Transient Platinum Degradation in a Low Pt-Loading PEFC under Current Cycling [J]. Journal of The Electrochemical Society, 2017, 164(2): F171.
104. Li Y, Moriyama K, Gu W, et al. A One-Dimensional Pt Degradation Model for Polymer Electrolyte Fuel Cells [J]. Journal of The Electrochemical Society, 2015, 162(8): F834.
105. Redmond E L, Trogadas P, Alamgir F M, et al. The effect of Platinum Oxide Growth on Platinum Stability in PEMFCs [J]. ECS Transactions, 2013, 50(2): 1369.
106. Zhang J, Litteer B A, Gu W, et al. Effect of Hydrogen and Oxygen Partial Pressure on Pt Precipitation within the Membrane of PEMFCs [J]. Journal of The Electrochemical Society, 2007, 154(10): B1006.
107. Zheng Z, Yang F, Lin C, et al. Voltage Cycling-Induced Pt Degradation in Proton Exchange Membrane Fuel Cells: Effect of Cycle Profiles [J]. ACS Applied Materials & Interfaces, 2020, 12(31): 35088–97.
108. Jomori S, Nonoyama N, Yoshida T. Analysis and modeling of PEMFC degradation: Effect on oxygen transport [J]. Journal of Power Sources, 2012, 215: 18–27.
109. Goswami N, Grunewald J B, Fuller T F, et al. Mechanistic interactions in polymer electrolyte fuel cell catalyst layer degradation [J]. Journal of Materials Chemistry A, 2022, 10(28): 15101–15.
110. Hao M, Hu Y, Chen S, et al. A scale-bridging model for proton exchange membrane fuel cells: Understanding interactions among multi-physics transports, electrochemical reactions and heterogeneous aging[J]. Nano Energy, 2024, 128: 109957.
111. Promislow K, Chang P, Haas H, et al. Two-Phase Unit Cell Model for Slow Transients in Polymer Electrolyte Membrane Fuel Cells [J]. Journal of The Electrochemical Society, 2008, 155(7): A494.
112. Reiser C A, Bregoli L, Patterson T W, et al. A Reverse-Current Decay Mechanism for Fuel Cells [J]. Electrochemical and Solid-State Letters, 2005, 8(6): A273.
113. Gu W, Yu P T, Carter R N, et al. Modeling of Membrane-Electrode-Assembly Degradation in Proton-Exchange-Membrane Fuel Cells – Local H2 Starvation and Start–Stop Induced Carbon-Support Corrosion [M]//WANG C-Y, PASAOGULLARI U. Modeling and Diagnostics of Polymer Electrolyte Fuel Cells. New York, NY; Springer New York. 2010: 45–87.
114. Meyers J P, Darling R M. Model of Carbon Corrosion in PEM Fuel Cells [J]. Journal of The Electrochemical Society, 2006, 153(8): A1432.
115. Franco A A, Gerard M. Multiscale Model of Carbon Corrosion in a PEFC: Coupling with Electrocatalysis and Impact on Performance Degradation [J]. Journal of The Electrochemical Society, 2008, 155(4): B367.
116. Fuller T, Gray G. Carbon Corrosion Induced by Partial Hydrogen Coverage [J]. ECS Transactions, 2006, 1(8): 345.
117. Takeuchi N, Fuller T. Modeling of Transient State Carbon Corrosion for PEMFC Electrodes [J]. ECS Transactions, 2007, 11(1): 1021.
118. Takeuchi N, Fuller T F. Modeling and Investigation of Design Factors and Their Impact on Carbon Corrosion of PEMFC Electrodes [J]. Journal of The Electrochemical Society, 2008, 155(7): B770.
119. Hu J, Sui P C, Kumar S, et al. Modelling and simulations of carbon corrosion during operation of a Polymer Electrolyte Membrane fuel cell [J]. Electrochimica Acta, 2009, 54(23): 5583–92.
120. Guétaz L, Escribano S, Sicardy O. Study by electron microscopy of proton exchange membrane fuel cell membrane-electrode assembly degradation mechanisms: Influence of local conditions [J]. Journal of Power Sources, 2012, 212: 169–78.

121. Hu J, Sui P C, Kumar S, et al. Modelling of Carbon Corrosion in a PEMFC caused by Local Fuel Starvation [J]. ECS Transactions, 2007, 11(1): 1031.
122. Hu J, Sui P C, Djilali N, et al. Modelling and Simulations on Mitigation Techniques for Carbon Oxidation Reaction Caused by Local Fuel Starvation in a PEMFC [J]. ECS Transactions, 2008, 16(2): 1313.
123. Jain K, Gidwani A, Kumar S, et al. CFD Study of Carbon Corrosion in PEM Fuel Cells [J]. ECS Transactions, 2008, 16(2): 1323.
124. Franco A A. A multiscale modeling framework for the transient analysis of PEM Fuel Cells - From the fundamentals to the engineering practice [D]; Université Claude Bernard - Lyon I, 2010.
125. Malek K, Franco A A. Microstructure-Based Modeling of Aging Mechanisms in Catalyst Layers of Polymer Electrolyte Fuel Cells [J]. The Journal of Physical Chemistry B, 2011, 115(25): 8088–101.
126. Mittermeier T, Weiß A, Hasché F, et al. PEM Fuel Cell Start-Up/Shut-Down Losses vs Relative Humidity: The Impact of Water in the Electrode Layer on Carbon Corrosion [J]. Journal of The Electrochemical Society, 2018, 165(16): F1349.
127. Zhao J, Huang X, Tu Z, et al. Water distribution and carbon corrosion under high current density in a proton exchange membrane fuel cell [J]. International Journal of Energy Research, 2022, 46(3): 3044–56.
128. Zhao J, Huang X, Chang H, et al. Effects of operating temperature on the carbon corrosion in a proton exchange membrane fuel cell under high current density [J]. Energy Conversion and Management: X, 2021, 10: 100087.
129. Xu K, Di Q, Sun F, et al. Degradation mechanism analysis of substrate and microporous layer of gas diffusion layer in proton exchange membrane fuel cell[J]. Fuel, 2024, 358: 130198.
130. Shojayian M, Kjeang E. Simulation of cathode catalyst durability under fuel cell vehicle operation–Effects of stack size and temperature[J]. Journal of Power Sources, 2024, 591: 233820.
131. Zhao Y, Luo M, Yang J, et al. Numerical analysis of PEMFC stack performance degradation using an empirical approach[J]. International Journal of Hydrogen Energy, 2024, 56: 147–163.
132. Izadi M J, Hassani P, Raeesi M, et al. A novel WaveNet-GRU deep learning model for PEM fuel cells degradation prediction based on transfer learning[J]. Energy, 2024, 293: 130602.
133. Jia C, He H, Zhou J, et al. A performance degradation prediction model for PEMFC based on bi-directional long short-term memory and multi-head self-attention mechanism[J]. International Journal of Hydrogen Energy, 2024, 60: 133–146.
134. Dong E, Zhao H, Zhang R, et al. Effects of gradient structures of cathode catalyst layers on performance and durability of proton exchange membrane fuel cells[J]. Electrochimica Acta, 2024, 477: 143772.
135. Hong J, Yang H, Liang F, et al. State of health prediction for proton exchange membrane fuel cells combining semi-empirical model and machine learning[J]. Energy, 2024, 291: 130364.
136. Zhang X, Huang L, Jiang Y, et al. Investigation of nonlinear accelerated degradation mechanism in fuel cell stack under dynamic driving cycles from polarization processes[J]. Applied Energy, 2024, 355: 122286.
137. Liu Z, Cai S, Tu Z, et al. Recent development in degradation mechanisms of proton exchange membrane fuel cells for vehicle applications: problems, progress, and perspectives[J]. Energy Storage and Saving, 2024.

Chapter 10
Anion Exchange Membrane Fuel Cells

10.1 Introduction

In the early stage of fuel cell development, alkaline fuel cells (AFCs) received much attention. AFCs have the highest efficiency (up to 70%) compared to other types of fuel cells [1]. They can provide electricity for the equipment on the one hand, and the reaction products can also provide drinking water on the other. So, in 1960, AFCs were introduced into NASA's space program. But AFCs used liquid electrolytes that were sensitive to carbon dioxide, making them prone to side effects. When air is used as an oxidant, hydroxide ions react with carbon dioxide to form carbonate deposits, which can degrade cell performance or even block the fuel cell. In addition, liquid electrolytes may leak, endangering equipment safety. Therefore, AFCs were mainly used in the space program and not commercialized on a large scale. After the 1980s, PEMFCs using solid electrolytes completely solved the problem of electrolyte leakage and carbonate deposition, making AFCs no longer the focus of research. In recent years, with the emergence of anion exchange membranes, Anion Exchange Membrane Fuel Cells (AEMFCs) have become more and more popular in fuel cell technology and is gradually replacing the traditional AFCs technology. The emergence of Anion Exchange Membrane (AEM) also solves the problem of overflow leakage caused by the use of liquid electrolytes in traditional AFCs [2]. In addition, the AEM has better carbon dioxide tolerance than liquid alkaline systems. The conductive material is now in a stationary solid polymer. Due to the reaction of alkali with CO_2, some carbonates will be produced. But because there is no moving cation, no solid crystal of metal carbonate will be formed to block the gas diffusion electrode [3]. Compared with PEMFCs, the alkaline environment of AEMFCs provide additional advantages. For example, the cathode has faster reaction kinetics, so non-noble metal catalysts can be used. This is important to reduce cell costs and further drive commercialization. Besides, AEMFCs have a higher open-circuit voltage and higher efficiency [4]. All the above characteristics have led to the attention of researchers since the emergence of AEMFCs.

Y. Li, *Fuel Cell Fundamentals and Applications*,
https://doi.org/10.1007/978-981-96-9845-5_10

However, there are still many issues that need to be addressed before AEMFCs become mainstream, including performance, stability, durability, mechanical strength, and low-cost production methods. After years of efforts, AEMFCs have been able to achieve a power density of 0.8W/cm^2, which is close to the power density of PEMFCs. But the lifetime of AEMFCs still faces significant challenges, especially when operating at higher temperatures. This may be due to the long-term stability of the anion film being limited. Up to now, AEMFCs can be prepared to operate stably for hundreds of hours under limited cell temperature, but next-generation AEM technology is still needed to achieve higher stability and durability.

The AEMFCs have a component structure and working mode similar to those of the PEMFCs. Fuel reacts at the anode, while air or an oxidant reacts at the cathode of the AEMFCs. Electrons are transferred through external circuits to generate electrical energy. The main difference is that the ion exchange membrane in AEMFCs is an anion exchange membrane rather than a cation exchange membrane. So, a slightly different chemical reaction happens. Figure 10.1 shows a comparison of the two cells.

In AEMFCs, oxygen reacts at the cathode to form hydroxide ions. The transport of hydroxide ions across the membrane from the cell cathode to the anode is in contrast to the conduction of hydrogen ions in PEMFCs. The hydroxide ions that arrive at the anode react with hydrogen to form water. The equation of the reaction is given below.

$$1/2O_2 + H_2O + 2e^- \rightarrow 2OH^- \quad (10.1)$$

$$2OH^- + H_2 \rightarrow 2H_2O + 2e^- \quad (10.2)$$

Despite these differences, the components of the two fuel cells are still very similar. AEMFCs also consist of a flow channel, a diffusion layer, a catalytic layer, and a

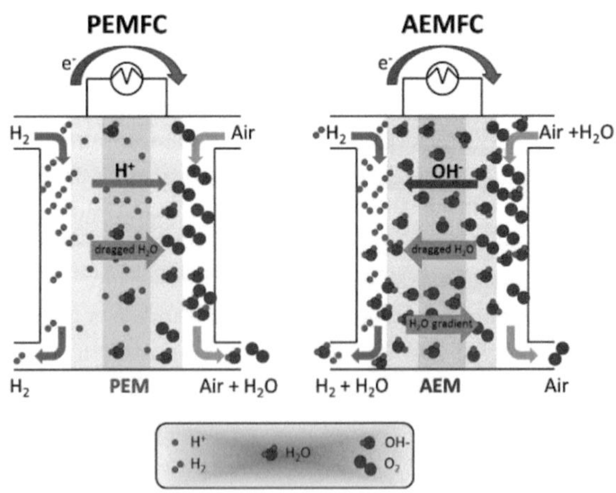

Fig. 10.1 Schematic of the PEMFC and the AEMFC [5]

membrane. Therefore, AEMFCs can use the same fuel cell components as PEMFCs and only needs to replace the ion-exchange membrane and reaction solution. In addition, because the fuel cell operates in an alkaline environment, AEMFCs can use much lower-cost components as cell support materials. For example, stainless steel can be used instead of a high-purity graphite flow field. This is forbidden in the extremely acidic environment of proton exchange membrane fuel cells.

In addition to hydrogen, AEMFCs can also use formate, methanol, and ethanol as fuel. The type of catalyst used for the anode varies depending on the fuel used. Therefore, the fuel cell performance varies with different fuels. Air or pure oxygen is usually used as an oxidant, which is similar to PEMFCs. Thus, the cathode catalyst will not be introduced here. In the following sections, the relevant studies of AEM will be introduced first, and then the anodic catalysts in AEMFCs will be introduced according to the different fuels. Finally, the AEMFCs technology is summarized.

10.2 Anion Exchange Membrane

10.2.1 Overview of Anion Exchange Membrane

Since the twentieth century, PEMFCs have been intensively studied and are widely used in automotive transportation, portable electronic devices, military equipment, and other fields. However, due to durability and cost issues, the widespread commercialization of PEMFCs still face great challenges [1–3]. On the one hand, PEMFCs have a high dependence on precious metal catalysts; on the other hand, hydrogen as a fuel is not only expensive to produce, but also presents huge challenges in transporting, storing, and handling gaseous hydrogen [1, 4]. Today, because of the many advantages of faster electrochemical kinetics, lower catalyst cost, and weaker corrosion compared to PEMFCs [4, 5], AEMFCs are getting more and more attention. As shown in Fig. 10.2, they have a similar cell structure between PEMFCs and AEMFCs, with the most important difference being that an acidic polymer electrolyte membrane is used in PEMFCs to conduct H^+, while AEMFCs use a basic polymer electrolyte membrane to conduct OH^- [2]. Because AEM contains positively charged head groups, it allows only anions to pass and rejects cations, and AEM can be divided into AEM and Alkaline Anion Exchange Membrane (AAEM) depending on the type of anion it can pass through. AEM can pass non-basic anions (such as Cl^-, SO_4^{2-}, PO_4^{3-}), while AAEM can pass basic anions (such as OH^-, HCO_3^-, CO_3^{2-}) [7, 8].

As a core component of a fuel cell, the AEM functions to transport OH^- from the cathode to the anode and acts as a barrier for the electrons and gases generated by the electrochemical reaction. The AEM consists of a polymer backbone coupled with an anion-exchange functional group. Typically, the polymer backbone is polysulfone (PSF) or polystyrene cross-linked with divinylbenzene (DVB), and the ion exchange groups are $-NH_3^+$, $-RNH_2^+$, $-RN^+$, $-R_2N^+$, $-R_3P^+$, $-R_2S^+$ and quaternary

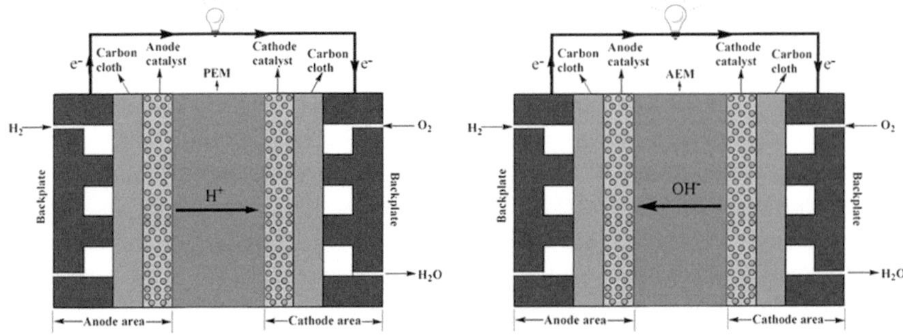

Anode:	$2H_2 \rightarrow 4H^+ + 4e^-$	$E_a^0 = 0V$	**Anode:**	$2H_2 + 4OH^- \rightarrow 4H_2O + 4e^-$	$E_a^0 = -0.83V$
Cathode:	$O_2 + 4H^+ + 4e^- \rightarrow 2H_2O$	$E_c^0 = 1.23V$	**Cathode:**	$O_2 + 2H_2O + 4e^- \rightarrow 4OH^-$	$E_c^0 = 0.40V$
Overall:	$2H_2 + O_2 \rightarrow 2H_2O$	$E_{cell} = 1.23V$	**Overall:**	$2H_2 + O_2 \rightarrow 2H_2O$	$E_{cell} = 1.23V$

Fig. 10.2 PEMFC (left) and AEMFC (right) both filled with H_2 and O_2 [2]

ammonium salts [9]. Desirable properties of highly efficient AEMs include high mechanical, thermal, and chemical stability, ionic conductivity, and potential barrier effects on electrons and gases. The polymer backbone is responsible for mechanical and thermal stability, the functional groups are responsible for ion exchange capacity, ionic conductivity, and transport coefficient, and the polymer matrix and functional groups together determine chemical stability.

The efficiency of the fuel cell is affected by many parameters, including the operating temperature, pressure, and relative humidity of the gas stream. However, the AEM is one of the core components, and to ensure the performance of the fuel cell, the anionic membrane must meet stringent mechanical, chemical, and thermal requirements. It must also ensure that it can make good contact with the irregular electrodes. All the required properties are summarized as follows:

(1) It must ensure efficient transfer of hydroxyl ions from one electrode to the other, requiring high ionic conductivity.
(2) The membrane must act as a barrier between the cathode and anode fuels, because the diffusion between fuels will reduce the efficiency and performance of the fuel cell, and even gases can explode when they mix.
(3) It should be chemically stable because it acts in strongly acidic or alkaline media.
(4) Ideally, the durability of the performance is required regardless of the hydration state and temperature of the membranes. The amount of water in fuel cells is difficult to quantify because it is continuously generated. on the one hand, the membranes in these fuel cells need to be highly hydrophilic and their performance highly dependent on the hydration state; on the other hand, they must remain water-insoluble and only soluble in suitable organic solvents to treat the membranes by casting. Finally, these properties possessed by the membranes should remain essentially constant at different operating temperatures [10, 11].

AEMFCs are currently paid much attention because they have the potential to offer more significant performance advantages over PEMFCs: Good stability at high pH values, the use of non-precious catalysts, high electrocatalytic activity at alkaline conditions (especially for organic fuels such as methanol or ethanol), reduced fuel cross rates and improved water management (caused by proton opposite direction of electroosmotic resistance and hydroxide ion flux in proton exchange membrane fuel cells) [2, 12, 13]. Despite the many advantages, there are still certain challenges to commercialization, among which, improving CO_2 tolerance, reducing carbonate precipitation, enhancing chemical stability of AEM in harsh environments, and improving ionic conductivity of AEM have become urgent problems to be solved.

10.2.2 Properties and Characterization of Anion Exchange Membrane

10.2.2.1 Characterization of Polymer Backbones and Cationic Functional Groups

AEMs are usually solid polymer electrolytes consisting of a hydrophobic polymer backbone with positively charged hydrated ions and surrounding water molecules [14]. Their main function is to transport anions from the cathode to the anode while preventing the penetration of neutral molecules or cations. Based on the research so far, the main concerns of AEMs are the anion conductivity, chemical stability, and dimensional stability. To date, a great deal of research work has been invested in this field, which can be divided into two main categories. The first category involves the synthesis of new anionic conducting groups (ACG), while the second category involves the design of specific polymer structures [15]. The traditional AEM head group is the quaternary ammonium (QA) ion; other newly studied head groups include tertiary diamine, phosphonium, sulfoxide, and metal cations [8, 15, 16]. QA groups are more thermally and chemically stable than tertiary ammonium (TA) and primary ammonium (PA), and therefore are most commonly used as hydroxide ion transfer carriers [17]. Table 10.1 demonstrates the common anion-conducting cations in the AEM headgroups [15].

However, the drawbacks of low anionic conductivity and insufficient tolerance to highly alkaline environments hindered the development and commercialization of

Table 10.1 Common anion-conducting cations in AEM head groups [15]

Nitrogen-containing groups	Nitrogen-free groups
Quaternary ammonium/tertiary diamines	Phosphonium
(Benz)Imidazolium	Sulphonium
Guanidinium	Metal cation (ruthenium, nickel, cobalt)
Pyridinium	

AEMs, and to address this problem, tertiary amines with different chemical structures were investigated as precursors for ACGs (as shown in Fig. 10.3) [15].

Along with the enhancement of AEM head groups, research needs to focus on polymer structures to improve ion exchange capacity (IEC) and chemical stability. AEM polymer research focuses on increasing polymer cross-linking and forming ion channels in polymers with different hydrophilic and hydrophobic regions [15]. The ideal polymer for AEMs must have excellent OH^-conductivity, chemical and thermal stability, flexibility, low permeability, low water resistance, low cost, and good availability. Typically, the implementation strategy for anion-conductive structures (Fig. 10.3b) is to design a hydrophobic side chain in the original structure of the membrane, where the cation (usually quaternary ammonium (-NR_3^+)) is tightly attached to the polymer backbone and the anion (OH^-) dissociates in the aqueous phase. The membrane conducts the anion based on the interaction between the hydrophilic and positively charged functional group and the negatively charged OH^-. When hydrated ions and surrounding water molecules (ion clusters) in the membrane are dispersed on a hydrophobic substrate (Fig. 10.4a), the surrounding hydrophobic walls inhibit ion transport, although small ion channels can be formed through

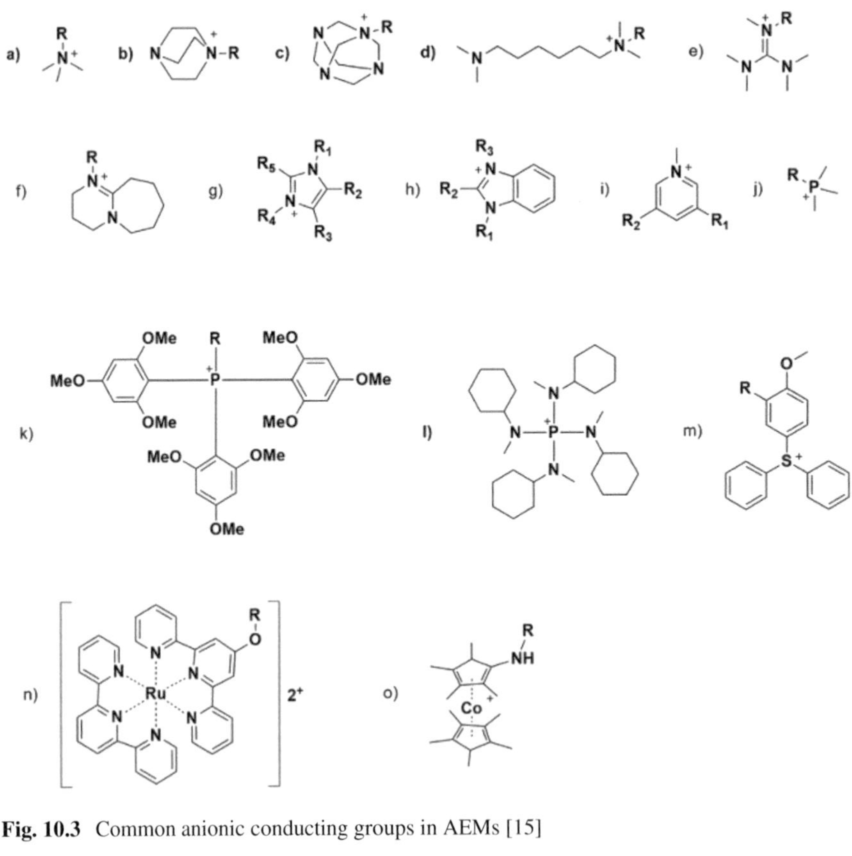

Fig. 10.3 Common anionic conducting groups in AEMs [15]

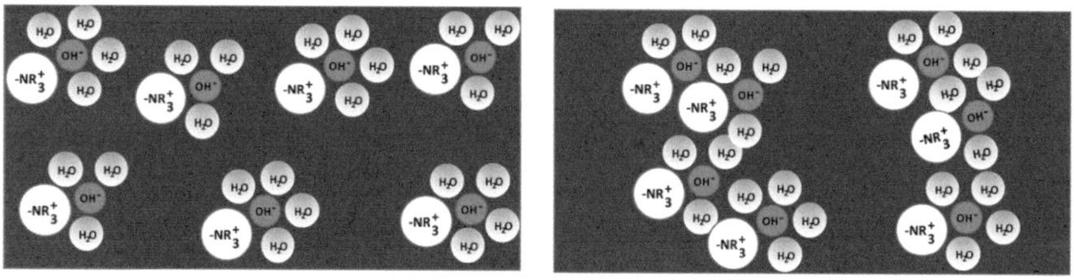

Fig. 10.4 Schematic diagram of ion channels in alkaline polymer membranes [5, 18]

the dynamic assembly of ion clusters. Aggregation of ion clusters by increasing the content of hydrophobic structures results in larger ion clusters to facilitate the formation of interconnected ion channels (Fig. 10.4b).

(a) Ionic clusters are uniformly dispersed in hydrophobic matrices.
(b) By introducing additional hydrophobic structures, ion clusters aggregate, which contribute to the formation of interconnected ion channels.

The polymer backbone is usually a polysulfone or fluorinated polymer [10, 11]. Figure 10.5 shows the common polymer backbone degradation pathways for polysulfone and fluorinated polymers. Due to hydroxide attack, polysulfone is susceptible to ether hydrolysis and quaternary carbon hydrolysis, while fluorinated polymers are susceptible to defluorination. However, the chemical stability of AEMs is influenced by the polymer backbone design in addition to the alkaline degradation of the AEM headgroups [7]. In these types of polymers, chemical modification allows the design and selection of polymer backbones with higher thermal and chemical stability [19–21]. For Nafion, more than just QA is attached to the polymer backbone. The Nafion polymer is formed from regularly spaced flexible side chains containing one or more QA groups [15, 22]. It can be seen that changing the polymer backbone to a low-polar polymer slightly improves the stability; however, the greatest stability improvement is achieved by linking the QA group through the long aliphatic side chain. Hydrophobic (polymer backbone) and hydrophilic (polymer side chain) regions have been developed by grafting multiple QA groups on the side chains, which have been shown to improve IEC and chemical stability [23].

10.2.2.2 Ion Conductivity and Electrical Resistance

The ionic conductivity of AEM directly affects the cell resistance and is one of the key parameters for the development of high-performance AEMFCs, however, the ionic conductivity of AEM is lower compared to that of PEM [24, 25]. Figure 10.6 compares the water uptake and ionic conductivity of PEMs and AEMs under the action of different IECs [26], where the water uptake is a measure of how the

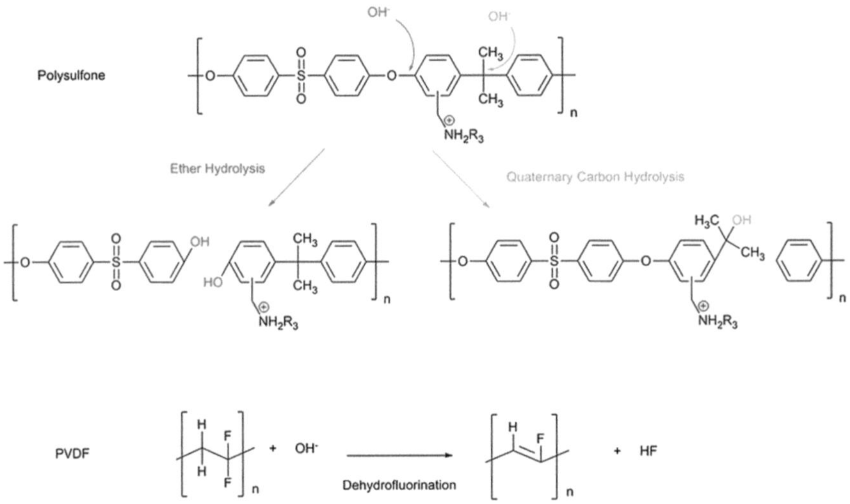

Fig. 10.5 Polymer degradation pathways for polysulfone (top) and poly(vinylidene fluoride), PVDF (bottom) [7]

membrane mass changes when exposed to water [7]. Although the water absorption rates of PEMs and AEMs are similar for a given IEC, the hydroxide conductivity of AEMs is reported to be 2–8 times lower than that of PEMs. The large difference in Fig. 10.6 is apparently due to the relatively wide range of AEM materials covered and the non-standard testing conditions. There are three recognized reasons for the lower hydroxide conductivity in AEMs compared to that in proton exchange membranes. (1) AEMs are polyhydrocarbon-based membranes, therefore, the formation of a solvent-based ionic phase capable of ion permeation requires a higher water content compared to perfluorinated PEMs. (2) AEMs may be partially carbonized when in contact with CO_2 from the air, thus reflecting a carbonate anion with a conductivity four times lower than the hydroxide anion. (3) The conductivity of hydroxide ions in dilute aqueous solutions is two times lower than that of protons. As can be seen from the report in Fig. 10.6b, there is a large conductivity gap between PEMs and AEMs, which is mainly due to the higher degree of nano-hydrogen phase agglomeration in the PEM materials than in the AEM materials for some given water content.

Several methods to increase the conductivity of AEM hydroxides have been demonstrated [27–29]. The most common approach is to increase the IEC, however, it usually leads to loss of mechanical properties by promoting excessive water uptake. Therefore, strict control of membrane morphology is required to improve mechanical properties, and loss or degradation of membrane mechanical properties will have a profound negative impact on fuel cell performance when the membrane is overhydrated or embrittlement due to intense drying [10]. Another approach is to change the polymer molecular structure to enhance the hydrophilic/hydrophobic phase contrast and improve hydrophilic connectivity [26]. Various AEMs with flexible side chains

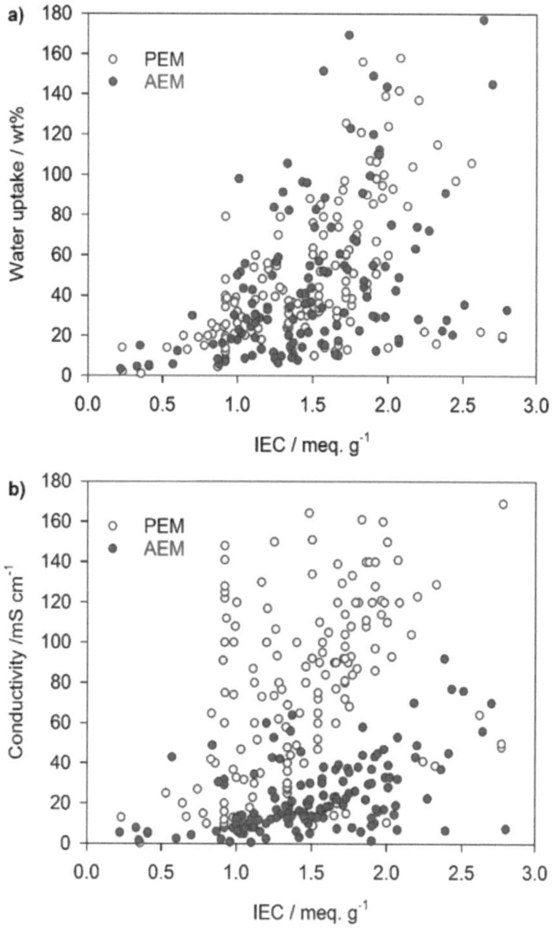

Fig. 10.6 Comparing PEMs and AEMs **a** water absorption rate **b** ionic conductivity [26]

[30], polycationic groups [22], multi-block polymeric backbones [31], comb-like polymer structure [32], and Laminar Master Chain [33] can be designed today to improve the hydroxide conductivity under conditions controlled by membrane water uptake. To take full advantage of this approach, a better understanding of the transport mechanisms and the relationship between ionomer structure and ion transport is still needed. In addition, the choice of cationic groups is more critical, as it will directly affect the chemical stability of the membrane. AEM is less chemically stable in alkaline media due to vapor generation from the attack of cationic groups by hydroxide. The result of this degradation is an important loss in the number of anion exchange groups, as well as a decrease in ionic conductivity. An important challenge in the preparation of efficient AEMs is the ionic conductivity and mechanical stability, therefore, the ionic conductivity must be sufficiently high ($\geq 100 \times 10^{-3}$ S/cm) to ensure high current with minimal resistive losses.

When certain specific hydroxide conductivities are given, the membrane-related ohmic losses in AEMFCs can be minimized by reducing the membrane thickness. However, good mechanical properties are required to fabricate thinner membranes, which in turn requires high molecular weight and good membrane flexibility. Due to the brittle nature of these polymers, the thickness of Polycyclic aromatic hydrocarbons (PAH) AEMs is usually limited to about 20 mm [34]. However, other types of AEMs or enhanced AEMs can be thinner, less than 15 mm thick [35]. For thinner membranes, reactant gas permeation will be higher, but for most AEMs, the cell resistance benefit of reducing the AEM thickness to 10 mm outweighs the performance loss due to gas crossover.

10.2.2.3 Water Transport and Management Strategies

It is well known that commercial PGM-based catalysts (Pt/C for the cathode and PtRu/C for the anode) are resistant to chemical and electrochemical degradation in alkaline media for thousands of hours (at least under potential or intensity static conditions). Why do many of the reported AEMFCs not survive even 100 h? One of the variables responsible for this is the imbalance of water transport in the cell [36]. For example, in Fig. 10.7, there is a difference of six water molecules in the water consumption/production (per oxygen) resulting from the reaction in this AEMFC, while there is a difference of two water molecules at the cathode and anode in a PEMFC, where water is only produced at the cathode. In addition, it is estimated that in the highly hydrated state, each hydroxide that migrates from the cathode to the anode to complete the electrochemical circuit can bring up to eight H_2O molecules through electroosmotic resistance (compared to 1–2 water molecules per H^+ in a proton exchange membrane fuel cell) [37]. Although HOR produces a large amount of water, the diffusion of water through the AEMFC anode is slow due to the adsorption of hydroxide cation water on the HOR catalyst.

On the one hand, this means that the water in the AEMFC anode is not uniformly distributed, and the AEMFC cathode requires high porosity for fast water transport but is more prone to flooding. On the other hand, the water in the cathode is consumed to generate current, which means that a large amount of reaction water in the cathode is supplied from the anode under high current density conditions [38, 39]. The implications of this are twofold. First, if the water diffusivity of a given AEM is too low, the mass transport limiting process in the cell is the diffusion of water through the AEM, and the cell has no chance to achieve high performance at all. This helps explain why the highest performing cells in the literature also use very thin AEMs (5–25 mm). This is not to say that thickness is the only important variable, as the physicochemistry of the alkaline ionomer can play an important role in its ionic conductivity, water mobility, and hydration state [40].

Due to the role of water imbalance in causing reversible performance loss (and even limiting initial cell performance) in AEMFCs, it is important to identify methods to mitigate water imbalance. Since the most pressing problem in cells is anode overflow, operating conditions that promote convective evaporation of the reaction gas

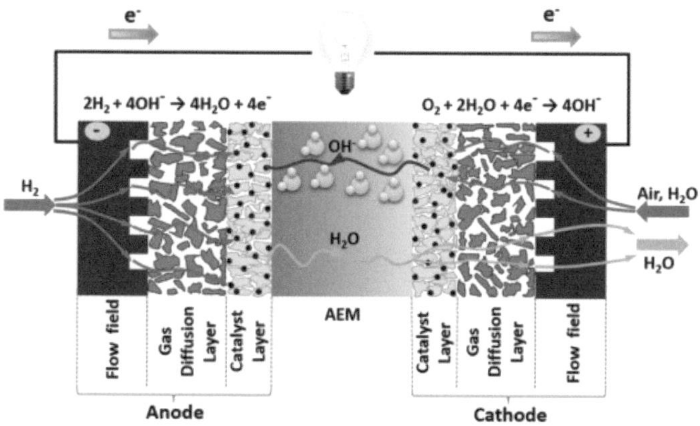

Fig. 10.7 Description of the hydrodynamics in AEMFC operation [36]

from the anode catalyst layer to the flow through the anode have become a generally accepted strategy. On the one hand, a practical approach to balancing cell water must rely heavily on the reverse diffusion shown in Fig. 10.7, where rapid reverse diffusion would allow the AEMs to absorb most of the water and transport it to the cathode. Here, a portion of the diffused water will react, and the rest of the water can be discharged from the cell via the cathode reaction gas, where a higher flow rate (higher stoichiometric ratio) is more desirable because its effect on the overall efficiency is lower than the anode reaction gas flow rate. Another way to reduce the degree of anode flooding is to redesign the electrodes themselves. In general, it is advantageous to avoid extremely thin electrodes. This is because the anode ionomer provides a water bath for absorbing liquid water. In addition, it is helpful to increase the hydrophobicity of the anode, which helps the anode layer to drain the liquid water formed [41]. Water conditions at the cathode are also important in determining the operational stability of the cell [42].

10.2.3 Membrane Stability and Degradation

10.2.3.1 Mechanism of Degradation on Cationic Functional Groups and Polymer Backbones

An important challenge for alkaline AEMs is their stability at high pH. Chemical degradation of AEMs may occur in the cationic functional groups and polymer backbones, including the polymer-cationic group linkage. Degradation of cationic groups increases the ohmic resistance, while degradation of the polymer backbone will reduce mechanical properties. To develop more stable AEMs, there is a need

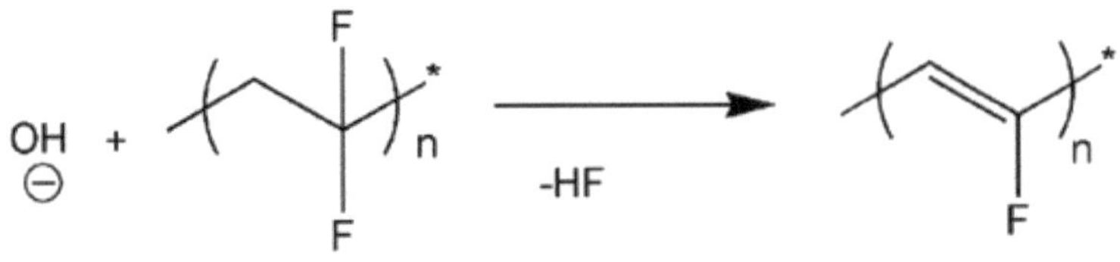

Fig. 10.8 Degradation of PVDF in alkaline media [10]

to investigate the degradation mechanisms of the polymer backbone and cationic groups that occur in alkaline media [8, 43, 44].

Degradation of polymer backbone stability is usually the result of oxygen [45] or hydroxyl attack [46]. Some polymers exhibit better resistivity, such as polysulfone and fluorocarbon-based membranes. Many of these AEMs are developed through radiation-induced grafting. However, polyvinylidene fluoride (PVDF) and vinylidene fluoride (VDF) copolymers are susceptible to degrade rapidly in high pH environments (as shown in Fig. 10.8), which can lead to membrane darkening and loss of mechanical properties [10]. Therefore, VDF and PVDF membranes are not recommended as starting materials for AFCs. The stability of the positive charges is equally important, and these fixed charges must not only be stable at higher temperatures and pH, but also need to resist the attack of hydroxyl ions.

Although the polymer backbone is less stable, cationic groups are more sensitive to alkaline media than the polymer backbone. Therefore, the degradation mechanisms of AEMs are almost always studied on cationic groups, especially quaternary ammonium groups, which usually follow two main degradation pathways [11, 47]. One is Hoffman degradation or E2 elimination, which is the preferred decomposition pathway for ammonium cations containing β-hydrogen. The other is a nucleophilic substitution reaction, which is caused by the hydroxyl attack of the α-hydrogen on the ammonium cation. As shown in Fig. 10.9, R1 can be a hydrogen atom, an alkyl group, or a polymer chain. Reaction 10.1 is an elimination (Hoffman elimination), while reactions 10.2 and 10.3 are nucleophilic substitutions. To solve the problems faced by AEMs with quaternary ammonium groups, such as easy degradation and excessive swelling, many researchers began to explore AEMs with functional groups different from quaternary ammonium, and the reported anionic functional groups different from quaternary ammonium of AEMs are shown in Fig. 10.10 [2].

10.2.3.2 Thermal Stability and Mechanical Strength

The thermal stability of AEMs depends mainly on the polymer backbone of AEMs [14]. Although running fuel cells at high temperature facilitates the performance improvement, it reduces the thermodynamic voltage loss due to pH differences across the AEMs and improves the electrochemistry. However, when used in portable electronic devices, they should operate below 60 °C [48]. Thermal stability can be determined by thermogravimetric/differential scanning calorimetry (TGA/DSC), usually

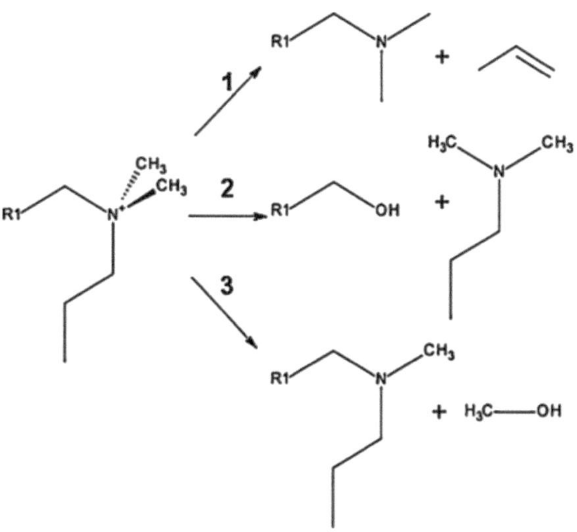

Fig. 10.9 Major degradation reactions of quaternary ammonium at high temperature in alkaline medium [11]

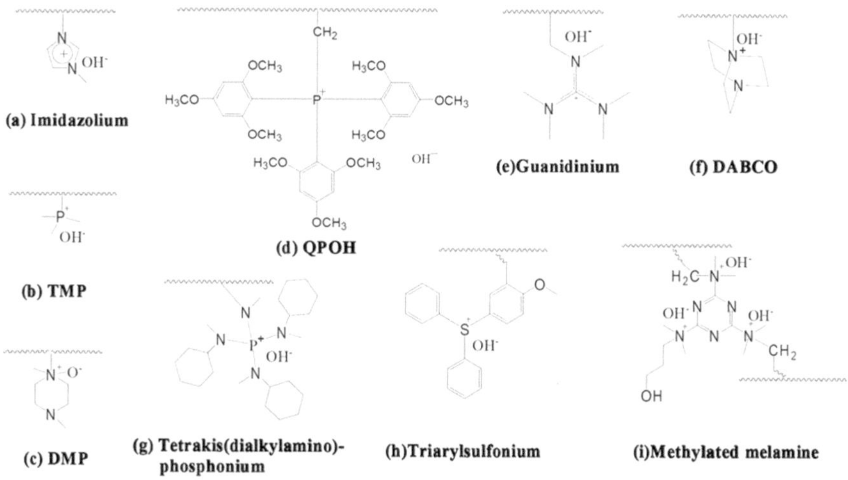

Fig. 10.10 Chemical structures of several anion exchange groups [2]

at a heating rate of 10 °C/min ramp rate under N_2, with a general temperature span of 900 °C.

The mechanical toughness of AEMs plays an important role in the AEMFC properties. Figure 10.11 shows the relationship between the molecular weight, mechanical properties, and AEMFC properties of quaternized poly(phenylene) AEMs, and due to the strong dependence of the mechanical properties of the polymer on the degree of polymer chain entanglement [49], AEMs with high molecular weight

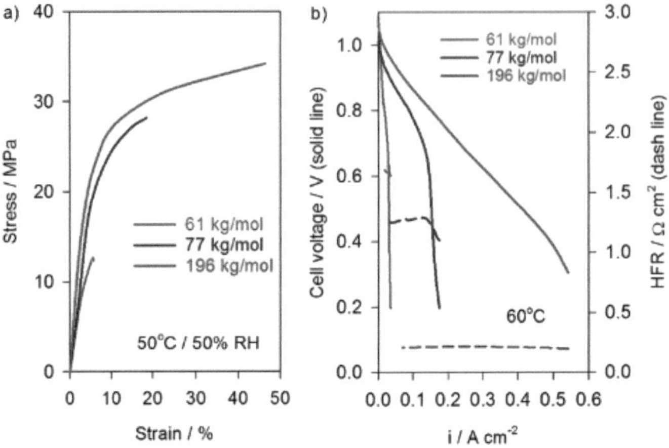

Fig. 10.11 Relationship between molecular weight, mechanical properties, and AEMFC properties of quaternized poly(phenylene) AEMs [49]

have better mechanical properties. Experimental studies have shown that organic–inorganic hybrid films with inorganic components introduced into organic polymer matrices can improve mechanical, thermal and chemical stability [50]: The tensile strength of the membrane can reach 65 MPa, the elongation at break is higher than 30%, the thermal degradation temperature in air is within the range of 250–300 °C, and the alkaline resistance of Fenton reagent is strong, the resistance time can up to 12 h.

10.2.3.3 Chemical Stability of Anion Exchange Membrane

Most of the AEMs studied so far demonstrate the alkaline stability of AEMs in dilute NaOH or KOH solutions, where the cationic groups and the chemical structure of the polymer backbone play a major role, and Fig. 10.12 shows the possible degradation pathways of AEMs under high pH conditions [37]. Currently, most of the work has been done to prepare stable AEMs by changing the chemical structure of the cationic groups. Most AEMs contain quaternary ammonium functional groups, which have good stability in alkaline media, and it has been observed that the benzyl trimethyl ammonium exchange site is particularly stable. Among other things, the instability of AEMs in alkaline media can be attributed to the substitution of the ammonium group by the OH^-, since OH^- is a very good nucleophile [6]. Some studies [51] compared the alkaline stability of various organic cations and similar cation-based AEMs, as well as the chemical degradation mechanisms of quaternary ammonium cations, imidazole, phosphonium, guanidine, benzimidazole, and metal cations.

However, the selection of a suitable polymer backbone structure is as important to ensure the stability of AEMs as the selection of cationic functional groups, and

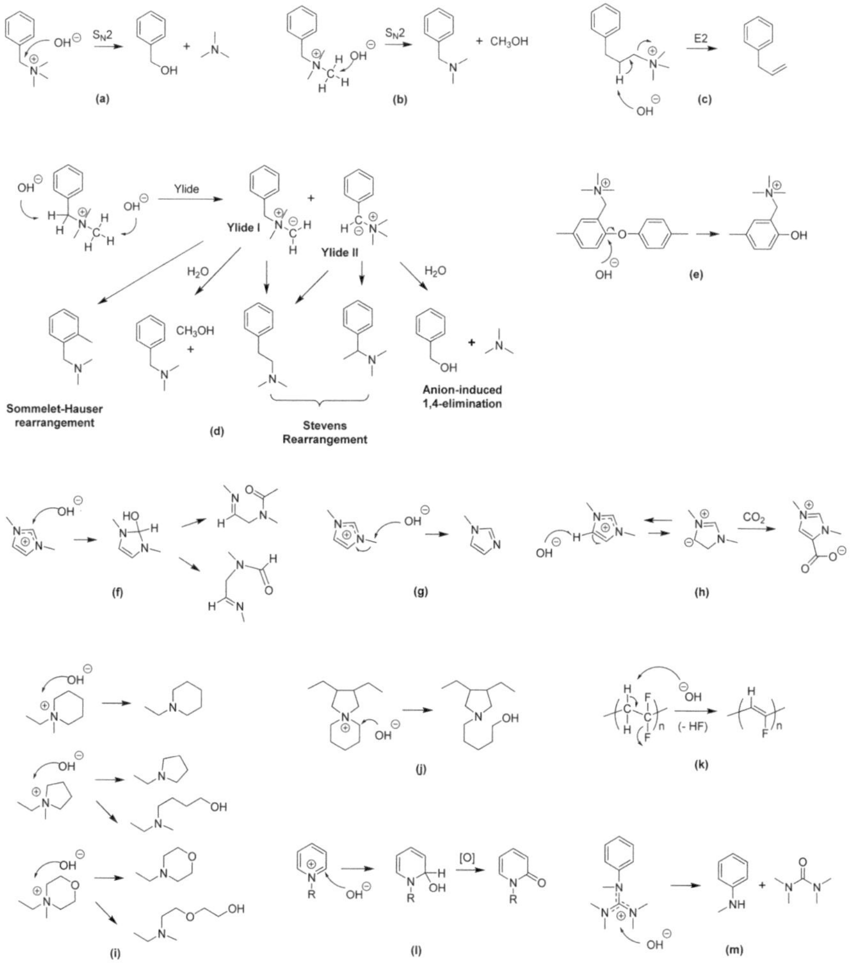

Fig. 10.12 Various degradation pathways of AEMs [37]

several polymer backbone structures have been investigated for AEMFC applications. Polystyrene is stable at high pH conditions; however, due to the brittle nature of polystyrene, benzyl trimethyl ammonium groups are usually introduced by copolymerization with flexible olefin segments [52, 53]. Good alkaline polymer backbone stability was also observed in flexible, non-fluorinated or perfluorinated polyolefins [54]. In contrast, the degradation of the polymer backbone of partially fluorinated polyolefins is due to defluorination caused by hydroxide attack, which produces carbon-carbon double bonds and leads to chain breakage [26].

10.2.3.4 Nonchemical Degradation: Impact of Carbonation in Anion Membrane

In addition to the degradation of AEMs, carbonation is one of the culprits affecting the performance of the cell. When the cell is exposed to the environment or when the cathode fuel is air, it causes the OH^- on the membrane to react with CO_2 to form the carbonate ion, and the reaction process is shown in Eqs. (10.3) and (10.4). Compared to hydroxide, the conductivity of this ion is four to five times lower [7]. This increases the film resistance and causes the carbonate to adsorb onto the anode catalyst layer, greatly reducing the performance of the cell.

$$OH^- + CO_2 \rightarrow HCO_3^- \tag{10.3}$$

$$HCO_3^- + OH^- \rightarrow CO_3^{2-} + H_2O \tag{10.4}$$

The current approach to mitigate carbonation is to use the self-purification mechanism in the AEMFC to improve CO_2 tolerance by operating at higher currents, thereby reversing the deleterious effects of CO_2 entry when operating at lower currents. In AEMFCs, hydroxides are generated at the cathode and transported through the AEM to the anode. Under high current conditions, the hydroxide generation rate exceeds the carbonate generation rate, resulting in excess hydroxides in the AEM transporting carbonate ions (HCO^{3-}, $CO_3{}^{2-}$) from the membrane to the anode, thus ensuring the presence of large amounts of OH^- ions in the AEM and maintaining high conductivity [55]. Recent work [56] investigated the addition of a gas mixture (e.g., ammonia-hydrogen) at the anode to promote the $HCO_3{}^-$ depletion reaction and to improve the CO_2 tolerance of AEMFC. They concluded that the low CO_2 tolerance of AEMFC is mainly due to the adsorption of carbonate onto the hydroxide reaction catalyst at the anode, and therefore, by promoting $HCO_3{}^-$ depletion at the anode, it removes the adsorbed species and releases the catalyst to perform its function, thus maintaining AEMFC performance. CO_2 tolerance in AEMFCs is being researched and developed, and understanding the mechanisms of carbonation and mitigation strategies holds great research potential [7].

Although CO_2 in the air is the main contributor to carbonation, carbonation ions from the water supply also have an effect. Another important source is the carbon oxidation reaction of the cathode material, as commonly used electrocatalyst carrier materials, ionomers, GDLs, and graphite plates contain various forms of carbon, which are converted to CO_2, hydroxyl, carbonyl, and carboxyl groups on the carbon surface under high cathode potential according to the following reactions:

$$C + 2H_2O \rightarrow CO_2 + 4H^+ + 4e^- \tag{10.5}$$

$$C + H_2O \rightarrow CO + 2H^+ + 2e^- \tag{10.6}$$

$$C + 2H_2O \rightarrow HCOOH + 2H^+ + 2e^- \tag{10.7}$$

The negative effects of electrochemical oxidation (corrosion) of carbon carrier materials are not only in the reduction of the intrinsic activity of ORR catalysts, catalyst particle migration, but also in the reduction of electron conductivity of the cathode and electrode overflow. The potential contribution of carbonation to long-term performance degradation plays an important role in recoverable performance changes and may amplify existing non-recoverable degradation mechanisms. Of course, the most obvious mitigation strategy is to pre-purge the air before feeding CO_2 into the cathode. Finally, since the transport number, the number of charge groups, and the interaction with water are determined by the AEM, the backbone and functional groups of the AEM can also be manipulated to reduce the effects of carbonation [37].

10.2.4 Applications and Perspectives

Over the last decades, anion exchange membranes have been extensively studied and are expected to become an alternative to PEMFCs. Of course, this is due to its many advantages, such as the ability to operate in alkaline environments improves the electrode reaction kinetics and allows the use of a wider variety of liquid fuels (ammonia and hydrazine, etc.). In addition, non-precious metal catalysts can be used, which effectively reduces the high costs associated with metal catalysts. Moreover, the application of anion exchange membrane is not only limited to fuel cells, but it is also widely used in wastewater treatment and water electrolysis.

Nevertheless, the AEM still has a long way to go to achieve full commercialization, and currently faces huge challenges such as chemical and mechanical stability under high temperature and high pH conditions, the problem of anionic membrane carbonation, and the loss of system efficiency due to low ionic conductivity, based on which the AEM needs to continue to improve in the following areas in the future:

(1) AEMs with higher hydroxide conductivity must be developed to ensure better fuel cell performance at high current densities.
(2) Research on the degradation mechanism of AEM polymers and ionic groups should be enhanced to select the appropriate number of cation exchange groups and sites on the polymer and control the membrane morphology to improve mechanical properties, scalability, and stability.
(3) Optimization of the polymer structure of the AEM head groups allows the combination of valuable properties obtained by the head group, which has improved chemical stability. It also takes advantage of polymer structures with ion channels and improved membrane properties using porous carriers or nanoparticles.

(4) The tolerance of AEM to CO_2 should be improved, the blockage of the catalyst surface by carbonate precipitation should be reduced, and the durability of AEMs should be improved.
(5) Realizing the application of AEMFCs in portable devices, it is necessary to reduce the cost on the one hand and improve its power density as much as possible on the other hand.

10.3 Anode Electrocatalysts

10.3.1 Catalysts for Methanol Oxidation

AEMFCs may be more cost-effective and scalable than PEMFCs. The fuels that can be used in AEMFCs are hydrogen, ethanol, methanol, formate, etc. Some catalysts suitable for hydrogen oxidation have been described previously, and electrocatalysts suitable for several other fuels will be further described below

10.3.1.1 Methanol Oxidation Reaction

Direct Methanol Fuel Cells (DMFCs), as an energy conversion device that directly converts the chemical energy in methanol into electrical energy, are one of the most promising alternative energy sources for portable devices, the military, and low-power vehicles [57–59]. Unlike the hydrogen fuel used in PEMFCs, the safety of methanol storage and transport can be guaranteed, and its simpler process system and mild operating conditions make DMFCs one of the strong contenders for sustainable energy conversion and storage devices [60]. It can be expected that in the future, due to its high energy density (6000 Wh kg^{-1}), DMFCs will be one of the most promising products to address the energy and environmental crisis. In the last two decades, commercial DMFCs have developed rapidly from 25 to 1000 W (Fig. 10.13).

In DMFCs, methanol and oxygen are consumed as reactants at the anode and cathode, respectively, to produce carbon dioxide and water (Fig. 10.14).

The chemical reaction process of DMFCs in acidic and alkaline environments is shown below:

In acid:

Chemical reaction at the anode:

$$CH_3OH + H_2O \rightarrow CO_2 + 6H^+ + 6e^- \quad (10.8)$$

Chemical reaction at the cathode:

$$3/2O_2 + 6H^+ + 6e^- \rightarrow 3H_2O \quad (10.9)$$

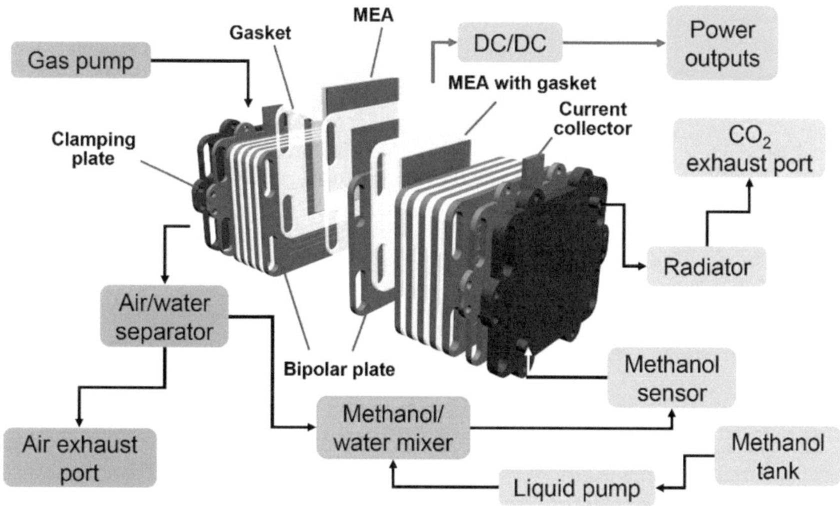

Fig. 10.13 Schematic of a typical DMFC stack and system [61]

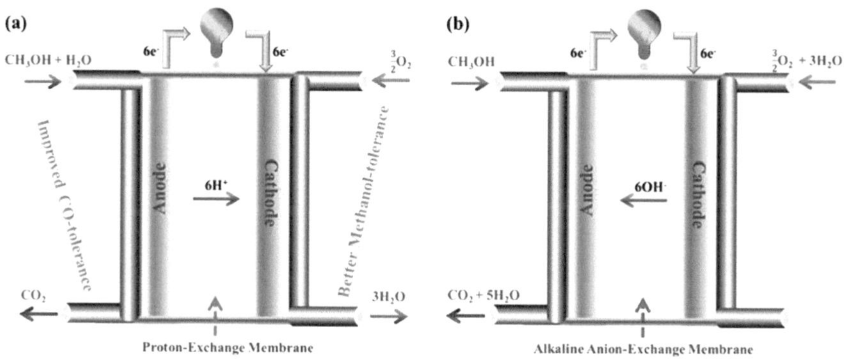

Fig. 10.14 Principles of the operation of **a** acid and **b** alkaline based DMFCs [62]

In alkaline:
Chemical reaction at the anode:

$$CH_3OH + 6OH^- \rightarrow CO_2 + 5H_2O + 6e^- \tag{10.10}$$

Chemical reaction at the cathode:

$$3/2O_2 + 6H_2O + 6e^- \rightarrow 6OH^- \tag{10.11}$$

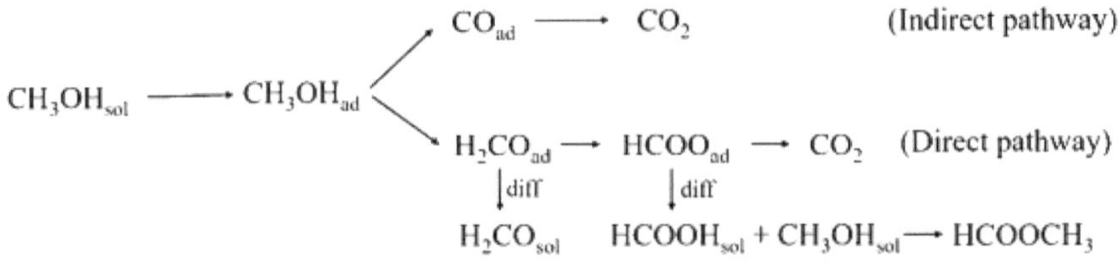

Fig. 10.15 Dual pathways proposed for MOR on Pt electrodes in acidic solutions [64]

Overall chemical reaction:

$$CH_3OH + H_2O \rightarrow CO_2 + 6H^+ + 6e^- \tag{10.12}$$

To develop efficient Pt-based catalysts, it is essential to understand the associated methanol oxidation reaction (MOR) mechanism. So far, researchers have summarised the dual-pathway methanol oxidation reaction mechanism based on a large number of experiments [63]. Among them, the indirect pathway is dominant for Pt in acidic solution, while the direct pathway is secondary (Fig. 10.15).

The theoretical voltage of DMFC at 25 °C is 1.21 V. However, due to various polarisations within the cell, the actual voltage of a DMFC single cell will be much lower than the theoretical value. The reaction rate of the MOR, which occurs at the anode, is several orders of magnitude slower than that of the hydroxide reaction (HOR), resulting in a loss of kinetics [65]. It is necessary to further study the catalytic mechanism, develop high-performance catalysts, improve reaction activity, and reduce activation loss [66].

10.3.1.2 Catalysts for MOR

It has been mentioned above that kinetic losses due to sluggish methanol oxidation reactions dominate the overpotential of DMFCs. In addition, methanol molecules can permeate through the PFSA membrane to the cathode, and the methanol crossover phenomenon can seriously affect the performance of the cell, whereas a higher Pt loading can effectively inhibit the methanol crossover [67]. Therefore, the most widely used electrocatalysts in MOR are still Pt and Pt alloys [68]. Platinum metal seems to be essential for DMFCs, but considering the scarcity of Pt precious metals, the production cost of Pt-based DMFCs severely limits their further commercialization [69]. Moreover, DMFCs with low Pt content are also susceptible to methanol poisoning, reducing the lifetime of the cell.

To weigh the cost and performance and accelerate the commercialization of DMFCs, a large number of researchers have invested in the optimized design of MOR catalysts, such as reducing the particle diameter to the nanoscale (< 10 nm) to improve the atomic utilization; alloying with other metals to reduce the Pt content and mitigate the poisoning; and selecting suitable carriers to make the catalyst particles uniformly dispersed while improving the stability. Existing MOR catalysts can be broadly classified into three categories: monometallic Pt catalysts, platinum-based catalysts, and platinum-free catalysts [69].

Monometal Pt Catalyst

For the time being, the mechanism of MOR is mostly derived from the study of Pt or Pt alloys as catalysts, which can probably include the following steps (M refers to another metal) [68]:

$$CH_3OH + Pt \rightarrow Pt{\cdot}CH_2OH^* + H^+ + e^- \tag{10.13}$$

$$Pt{\cdot}CH_2OH^* + Pt \rightarrow Pt_2{\cdot}CHOH^* + H^+ + e^- \tag{10.14}$$

$$Pt_2{\cdot}CHOH^* + Pt \rightarrow Pt_3{\cdot}COH^* + H^+ + e^- \tag{10.15}$$

$$Pt_3{\cdot}COH^* \rightarrow Pt{\cdot}CO^* + 2Pt + H^+ + e^- \tag{10.16}$$

$$Pt(M) + H_2O \rightarrow Pt(M){\cdot}OH^* + H^+ + e^- \tag{10.17}$$

Since Muller [70] first used Pt as a catalyst to study the methanol electro-oxidation process in 1922, many researchers have started to study the characteristics of MOR on the Pt surface in a large number of studies [71–73]. The catalytic activity of Pt is closely related to the particle size, morphology, structure, and surface state. Usually, the utilisation factor and specific activity of Pt increase with decreasing Pt nanoparticle size because the proportion of Pt atoms exposed on the surface increases. However, some studies have demonstrated that too small a particle size is detrimental to the promotion of activity. The specific activity of Pt/C catalyst particles with particle sizes in the range of 1.2–4.5 nm decreases as the particle size gets smaller [74], whereas the specific activity stays the same when the particle size is larger than 4.5 nm. This may be due to the smaller Pt particles being more susceptible to oxidation and forming Pt-OH at lower potentials. A significant size effect was observed for Pt particle sizes in the range of 1–4 nm [75]. The above size effect can be attributed to the fact that the loading nucleates rather than forming monolayers on very small Pt particles due to the ultra-high surface energy. At the same potential, coalesced CO is more difficult to oxidize than monolayer CO. Therefore, the overpotential for CO oxidation is higher on smaller Pt particles.

The dependence of the activity of Pt catalysts on their morphology and size is very obvious. Structural design at the nanoscale is often considered a strategy that can effectively enhance the performance of catalysts [76]. A large number of nanostructures, such as nanoflowers, nanotubes, and nanosheets, have been successively reported in the literature in the last two decades due to their excellent physicochemical properties, among which 3D Pt nanostructures have a great potential for development due to their high catalytic activity and higher yield [77]. However, single Pt atoms are poorly stabilized and generally require inert carriers (e.g., TiN [78], θ-Al_2O_3 [79], and FeO_x [80]) for stabilization to mitigate severe aggregation and coarsening during the catalytic reaction. In addition, the inability to achieve mass production and the low active site density of single Pt atom catalysts are also serious problems. Therefore, these catalysts are still in the development stage, and it is necessary to further investigate their reaction mechanisms.

Pt-Based Catalysts

Given the scarcity of Pt and the nature of Pt poisoning, other metals/non-metals must be added to the anode catalyst to reduce the Pt content and inhibit poisoning. So far, many transition metals such as Ru, Rh, Pd, Ir, Cu, etc. have been used to synergize with Pt to test their activity as MOR catalysts, among which Ru is the most effective complementary metal. The properties of these catalysts are shown in Table 10.2.

Currently, there are three main types of PtRu catalysts: the PtRu alloys [81], the Ru@Pt core-shell structures [82], and the Ru atom or cluster modified Pt [83]. Compared with PtRu alloys, the optimized Pt@Ru core-shell structure (shown in Fig. 10.16) not only effectively protects Ru and reduces Ru dissolution, but also increases the density of surface active sites, which is more promising for application in DMFC. Deposition of Ru atoms or their clusters onto Pt nanoparticles is also an important method to improve catalyst utilization as well as inhibit poisoning. Despite many advantages of Pt-based catalysts, most of the metals alloyed with them are also precious metals, and the cost is still unacceptable.

Table 10.2 The comparison of catalyst performance according to different literature [69]

Catalyst	ECSA ($m^2\ g^{-1}$)	Mass activity ($mA\ mg_{Pt}^{-1}$)	Specific activity ($mA\ cm^{-2}$)	CO oxidation peak potentials
Pt/CeO_2 NWS	30.2	438.2	1.45	– 0.005 V versus Hg Hg_2SO_4
PtCu NWNs	82	1287	1.87	0.488 V versus SCE
PtRu nano sponge	48.4	410	–	0.251 V versus SCE
PtRuCuW	26	467.1	1.8	0.47 V versus SCE
Pt-Cu/GNs-CD	44.6	1360	–	0.57 V versus Ag/AgCl
Pt-Ni/GNs-CD	26.5	960	–	0.58 V versus Ag/AgCl
Pt-Co/GNs-CD	30.2	840	–	0.60 V versus Ag/AgCl
Pt/GNs-CD	20.4	500		0.61 V versus Ag/AgCl
FePtSn/C	40.7	1077	1.27	0.45 V versus SCE
PtRu/N-GA	–	668	–	0.44 V versus Ag/AgCl
PtFe/C	–	575	1.6	0.53 V versus SCE

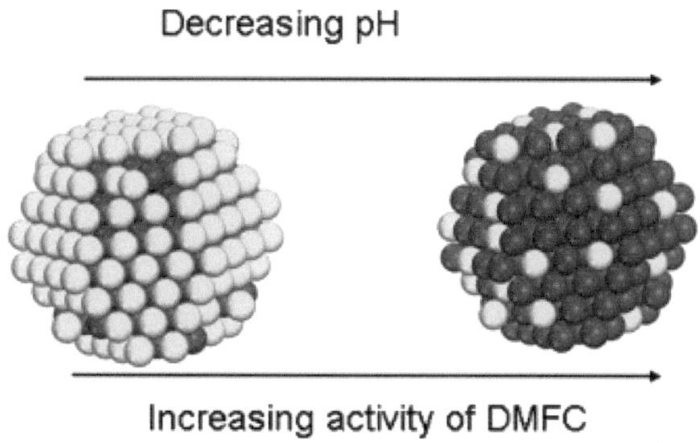

Fig. 10.16 Ru@Pt core–shell structure [84]

Pt-Free Catalysts

For platinum-free catalysts, the methanol oxidation reaction has to be catalyzed in an alkaline medium since methanol needs to be activated by an alkaline solution [85]. Statistically, the number of studies on platinum-free catalysts for DMFC is much less than that of platinum and platinum-based catalysts, and only a few have been reported, e.g., WC/rGO [86], CoCu alloy [87], and ZnO@C [85]. Platinum-free catalysts come at the cost of sacrificing part of their catalytic activity, and even though some platinum-free catalysts have better catalytic activity than Pt, there are no matching alkaline electrolyte ion exchange membranes yet, resulting in their further application being hindered [88].

10.3.1.3 Conclusion

Methanol electrooxidation over electrocatalysts is a non-homogeneous catalytic process, an interfacial reaction, which is mainly determined by the nature of the catalyst and the interfacial structure is highly active in catalyzing MOR, but its catalytic mechanism still needs to be further explored. If the exact relationship between the electronic states, crystal parameters, and surface morphology of the catalysts and the catalytic activity can be thoroughly understood in future studies, more advanced preparation methods can be developed to simulate the catalytic performance of Pt with non-Pt elements.

High dispersibility, small and narrow particle sizes, and favorable crystalline surfaces are essential elements of high-performance catalysts. The strength of the bond with the carrier and the resistance to poisoning determine the lifetime of the catalyst under fuel cell operating conditions. From an application point of view, stability and durability are even more important than activity. In fact, many highly active catalysts tested in conventional three-electrode systems proved to be less efficient in in situ fuel cell tests. In addition, cost is another key factor for catalysts. The synthesis process of catalyst systems should be simplified using inexpensive agents and methods suitable for mass production.

10.3.2 *Catalysts for Ethanol Oxidation*

10.3.2.1 Reaction Pathways for Ethanol Oxidation Reaction (EOR)

For the mechanism of ethanol oxidation, the Dual-Pathway mechanism is more generally accepted at present. According to this mechanism, there are two parallel and competing reaction pathways in the oxidation process of ethanol. On the one hand, the ethanol molecule generates CH_x and CO intermediates through C-C bond breaking, and these C1 intermediates are further oxidized to CO_2, achieving the complete oxidation of ethanol, with an electron transfer number of 12 in the cell. The oxidation pathway is named as C1 pathway. On the other hand, the ethanol molecule undergoes a series of dehydrogenation reactions to generate intermediates, such as CH_3CHO and acetyl, which are further oxidized to CH_3COOH by the C2 intermediates. Since it is very difficult to further oxidize CH_3COOH, ethanol is only partially oxidized, and the number of electrons transferred is 4, which is named the C2 pathway. In the actual process of oxidizing ethanol, the generation of the fully oxidized product, CO_2, during the C1 pathway, is hindered by the extreme kinetics due to the difficulty in breaking the C-C bond in the molecule (Fig. 10.17).

In contrast, the C2 Pathway has a greater thermodynamic advantage in generating the by-product CH_3COOH [90]. Therefore, among the two parallel reaction pathways of the Dual-Pathway mechanism, the C2 pathway has both kinetic and thermodynamic advantages, which means that the selectivity of the complete oxidation product CO_2 is very low. The generation of the by-product CH_3COOH not only reduces the number of electrons transferred in the cell, but also drastically lowers the efficiency of fuel ethanol utilization, which makes the output power of the cell much lower than the theoretical value, and the performance of the cell decreases significantly. In addition, the generation of CH_3COOH will also cause corrosion to the internal components of the cell and shorten the service life of the cell [91]. Therefore, developing efficient anode catalysts to promote the C-C bond breakage in ethanol molecules and inhibit the generation of CH_3COOH as a by-product has become the focus of the current research on ethanol fuel cells.

The following is the reaction path mechanism for ethanol oxidation on the anode [92]:

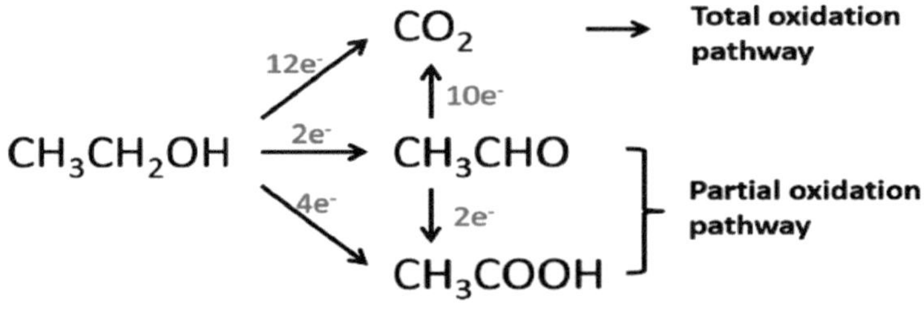

Fig. 10.17 Schematic diagram of the parallel pathways during the ethanol oxidation reaction [89]

C1 Pathway

$$CH_3 - CH_2OH + 3H_2O \rightarrow 2CO_2 + 12H^+ + 12e^- \quad (10.18)$$

$$CH_3 - CH_2OH + 5H_2O \rightarrow 2HCO_3^- + 14H^+ + 12e^- \quad (10.19)$$

$$CH_3 - CH_2OH + 5H_2O \rightarrow 2CO_3^{2-} + 16H^+ + 12e^- \quad (10.20)$$

C2 Pathway

$$CH_3 - CH_2OH + H_2O \rightarrow 2CH_3 - COOH + 4H^+ + 4e^- \quad (10.21)$$

$$CH_3 - CH_2OH \rightarrow CH_3 - CHO + 2H^+ + 2e^- \quad (10.22)$$

10.3.2.2 Catalysts for EOR

Suitable electrode catalysts are of great significance to improve the efficiency of the cell, however, the current inactive intermediate products tend to poison the catalyst, reduce the reaction rate, and cause various problems such as shorter lifetime [93]. Therefore, to reduce the poisoning of intermediate products to a certain extent, it is important to design and synthesize a new type of high-efficiency anode catalyst by breaking the C-C bond at lower temperatures and lower potential [90]. The most widely researched catalysts are Pt and Pd [94–96].

Pt-Based Catalysts

Most of the chemical energy of ethanol is locked in the C–C bond, and the transfer of $12e^-$ is required to achieve complete energy release, but the reaction kinetics of this process is extremely slow, which greatly affects the commercialization of ethanol fuel cells [97]. In the actual ethanol oxidation process, in most cases, ethanol is oxidized along the $4e^-$ C2 pathway, and its final products are acetaldehyde, acetic acid, etc. Obviously, the C1 pathway, in which the complete oxidation of the final product is CO_2, is more conducive to the complete conversion of ethanol energy [98, 99]. Therefore, the development of electrocatalysts with high catalytic activity and high C1 path selectivity for the ethanol oxidation reaction is the core of direct ethanol fuel cell research. EOR is a structure-sensitive reaction in which Pt (110) effectively promotes C–C bond breaking, and our previous work has shown that the noble metal Ir facilitates the C–C bond breaking in the ethanol oxidation reaction (EOR) process. The efficiency of ethanol oxidation is limited due to the low selectivity of the C1 pathway. To address this issue, researchers developed PtPb@PtIr1 HNPs, which are single-atom hexagonal Ir (Ir1) hcp-PtPb/fcc-Pt nanoplatelets that were encapsulated on a Pt (110) surface with a tensile strength of 7.2% [100]. By utilizing low-ligand

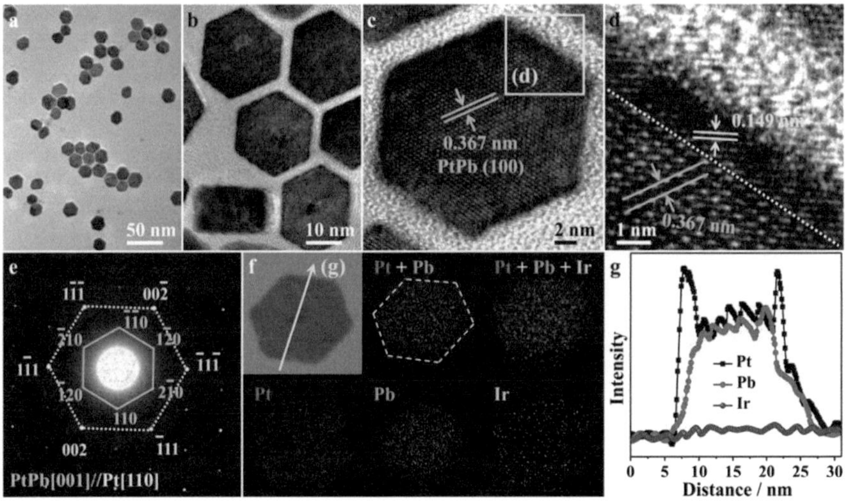

Fig. 10.18 **a**, **b** TEM images of the PtPb@PtIr$_1$HNPs, **c** HRTEM image of one single HNP, **d** an HRTEM image from the selected area in **c**, **e** SAED of the single HNP in **c**, **f** HAADF–STEM image and elemental mappings of one single HNP: Pt mapping in red, Pb mapping in cyan, Ir mapping in yellow, and integrated mapping of (Pt + Pb) and (Pt + Pb + Ir) are shown, and **g** Line-scan analysis across the yellow arrow in the HAADF–STEM image [100]

active sites on nanocrystals with controlled shape (Fig. 10.18), the researchers were able to modulate the selectivity and activity of the catalytic reaction, resulting in complete electro-oxidation of ethanol. In the field of catalysis, the strain on the catalyst surface has also been shown to improve the adsorption energy of the catalyst by modulating the electronic structure of the catalyst surface, thus affecting the catalytic performance of the catalyst, and also single-atom catalysis has been considered as an effective strategy due to the interfacial nature of the catalytic reactions.

The design synthesizes Pt (110) surface-covered materials with optimal activity in acid and lattice expansion, which is a tensile strain structure that can modulate the electronic structure of the catalyst surface and further optimize its adsorption energy, and the introduction of Ir into the Pt system can enhance the C-C bond breaking ability for ethanol oxidation. The design contains the above three advantages can synergistically increase the catalytic effect and optimize the selectivity of the EOR reaction. Pt-based bimetallic alloy structures are an effective strategy to accelerate the EOR process [101, 102]. Bimetallic PtNi composite nanocatalysts were prepared for accelerated EOR by reduction in a hydrothermal reaction. The results of microscopic morphology tests showed that the diameter range of PtNix nanoparticles was 6–13 nm, and the PtNix nanoparticles were firmly bound to the surface of the carbon carrier [103]. The electrochemical test results showed that the PtNi3/MWNTs nanocatalysts had a certain EOR stability. The alloy and catalyst have higher EOR catalytic activity than the Pt/MWNTs nanocatalyst.

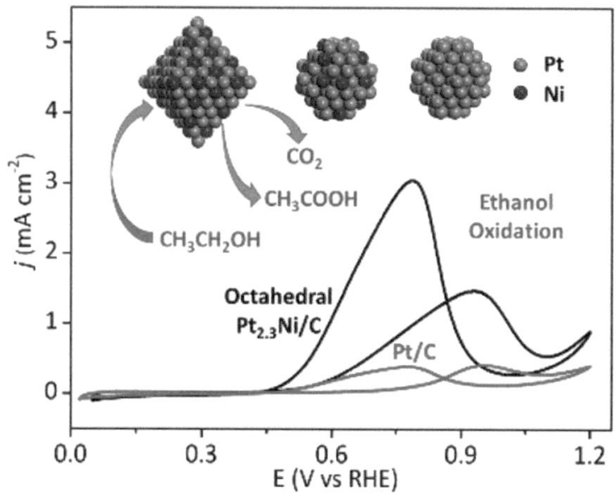

Fig. 10.19 CVs in an Ar-saturated 0.1 M $HClO_4$ solution at a scan rate of 50 mV s^{-1} for Pt/C and octahedral $Pt_{2.3}Ni/C$ [101]

The EOR catalytic activity of the binary Pt-based alloy has been greatly improved, but its stability and the current density in acidic electrolytes still need to be improved [104]. Alloying Pt electrocatalysts with other transition metals (Ni, Co, and Fe) is an effective strategy to reduce the cost of the catalysts and to improve their CO-resistant performance on the anodes of direct ethanol fuel elements [105]. Octahedral Pt-Ni/C nanocrystals with uniform open (111) faces and average 10 nm edges were synthesized (Fig. 10.19). The activities of octahedral Pt-Ni/C nanocatalysts were 4.6 and 7.7 times higher than those of conventional Pt-Ni/C catalysts and commercial Pt/C catalysts, respectively. The results showed that ethanol oxidation on octahedral Pt-Ni/C produced more acetic acid than on other surfaces. The synergistic effect of electrons and surfaces may be the reason why octahedral Pt-Ni/C has better reactivity for ethanol oxidation.

Platinum-based nanocrystals are the most effective electrocatalysts for accelerating chemical conversion at the anode of direct alcohol fuel cells. Most electrocatalysts for direct ethanol fuel cells suffer from severe CO poisoning and low selectivity for C-C bond cleavage [105]. Here, researchers report a class of sub-nanometer platinum-rhodium nanowires (Pt-Rh NWs) as efficient electrocatalysts for EOR. The prepared Pt-Rh nanowires have ultrafine characteristics with an average diameter of only 1.2 nm, which greatly improves the efficiency of Pt utilization. Combining structural and synergistic effects, the optimized PtRh nanowires/C showed nearly threefold increase in EOR activity and improved stability compared with commercial Pt/C. Detailed mechanistic studies show that the introduction of Rh significantly improves the anti-poisoning ability, and the PtRh NWs/C possesses stronger C-C bond cleavage ability and significant C1 pathway selectivity, which together improve the electrocatalytic performance of EOR [106].

Pd-Based Catalysts

A class of one-dimensional (1D) ultrathin Pt3Sn nanofibers (NFs) with controllable aspect ratios and intermetallic structures is designed as efficient EOR electrocatalysts.

The optimized ordered structure of Pt3Sn NFs with a maximum aspect ratio (Pt3Sn NFs-L) showed the highest activity and stability compared to commercial Pt/C and other similar products. In addition to the inherent structural advantages, the binding strength of the reaction intermediates can be effectively reduced by doping Sn on Pt3Sn NFs-L, and the electrooxidation of the intermediates can be further promoted by providing abundant oxygenophilic substances, which largely enhances the EOR activity and stability.

Exploring Pd-based bimetallic catalysts has become a hot topic for long-term stable high-performance electrocatalyst applications in fuel cells [107–111]. The ethanol oxidation reactions of Pd_2Ru/C and Pd/C catalysts in alkaline media were comprehensively investigated. The results show that Pd_2Ru/C favors the formation of acetaldehyde and hinders its oxidation compared to Pd/C. Ru promotes a larger electron vacancy in the Pd 4d band and thus is expected to favor the formation of adsorbed ethoxylates on Pd_2Ru/C, which are subsequently oxidized to acetaldehyde, facilitated by the oxygenated species supplied by Ru. In contrast, the oxidation of acetaldehyde was more difficult on Pd_2Ru/C than on Pd/C, probably because of the increased adsorption energy of the active species [112]. Under the same test conditions, the Pd_2Ru/C anode initially performed better in an alkaline direct ethanol fuel cell (ADEFC) but degraded much faster than the Pd/C anode (Fig. 10.20). The degradation is due to the accumulation of large amounts of acetaldehyde, which forms dimers through aldol condensation reactions in alkaline media.

Pd-based catalysts have been the main catalyst system investigated due to their unique structural features and their superior activity in electrocatalytic alcohol oxidation reactions [113, 114]. The alloyed PdAg NPs were optimized by solvothermal synthesis. The optimized PdAg NPs exhibited excellent current activity in EOR in an alkaline ethanol solution, which is 3.3 times higher than that of commercial Pd/C NPs, and this is closely related to the changes in the electronic structure and the active sites. In addition, the PdAg NPs showed better antitoxicity, lower reaction

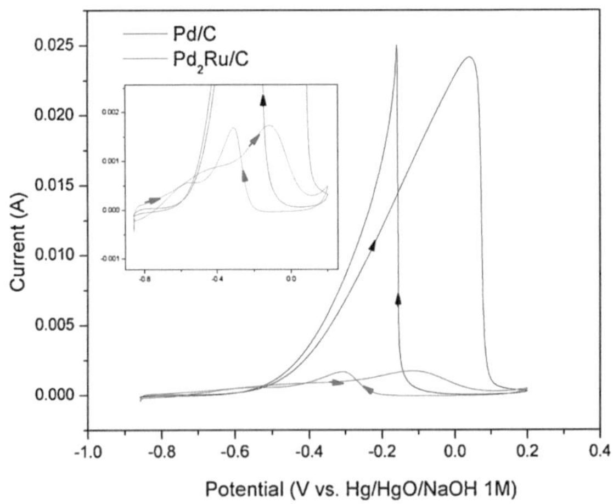

Fig. 10.20 CV curves of Pd/C and Pd_2Ru/C catalysts in 1.0 M NaOH + 1.0 M ethanol. Scan rate: 20 mV s^{-1} [112]

potential, and better long-term stability than other PdAg NPs and commercial Pd/C catalysts [115]. Therefore, higher catalytic temperature, higher pH, and higher ethanol concentration are favorable for ethanol oxidation.

The rational design and synthesis of efficient nanocatalysts for electrochemical EOR are of great significance for the practical application of direct ethanol fuel cells [116, 117]. Nano-engineering of the nanostructures and compositions of the EOR electrocatalysts is of considerable interest because these parameters can improve the electrocatalytic kinetics and efficiency of EOR. The synergistic EOR catalyst based on 3 nm thick bimetallic single-crystal palladium-silver nanowires showed a significant improvement in the electrochemical performance of the catalyst [118]. Kinetically stable but thermodynamically unfavorable ultrathin single-crystal PdAg nanowires were grown by in situ epitaxial growth on nano-hexagonal mesophases directly self-assembled from amphiphilic chloride surfactant under optimal synthesis conditions in aqueous solution. The PdAgNWs exhibited higher mass activity and stability in PNP due to the ultrathin, ultra-long nanostructures, single crystals with a high density of low ligand atomic order, high palladium utilization, and more pro-oxidative Ag attached to palladium. Kinetic studies showed that the significant increase in EOR efficiency could be attributed to the synergistic electronic and bifunctional effects of the ultrafine PdAg sNWs.

10.3.2.3 Conclusion

The EOR process that occurs at the anode of a direct ethanol fuel cell is complex, with difficult C-C bond breaking and slow kinetics. Currently, palladium (Pd) is the most widely studied anode catalyst for direct ethanol fuel cells, but Pd is a rare precious metal, and its introduction into a wide range of applications will lead to an irreversible increase in its market price. Therefore, the construction of efficient anode catalysts for direct ethanol fuel cells based on Pd is of great research value. It is necessary to improve the catalytic performance of ethanol oxidation by introducing a second element to form a multifunctional catalyst, changing the morphological composition and structure of the catalyst, and constructing defects.

10.3.3 Catalysts for Formate Oxidation

10.3.3.1 Overview of Formate Oxidation Mechanism

Compared with alcohol fuels, direct formate fuel cells (DFFCs) have the advantages of safety, high voltage, and high energy conversion efficiency. At the same time, formate can be stored in solid form and has great advantages in transportation safety. Therefore, formate fuel cells are considered a potential candidate for transportation and stationary power generation equipment. However, the output power density of DFFCs is limited by the slow reaction kinetics of the anode formate oxidation reaction

(FOR), which limits its development and practical application [119, 120]. Therefore, developing and preparing a high electrocatalytic performance and low price FOR catalyst is crucial for the wide application of DFFCs.

Shedding light on the electrocatalytic mechanism of formate on a certain catalyst will improve our understanding of formate oxidation, which promotes the development of high-activity catalysts toward formate oxidation. As compared to extensive studies of formic acid oxidation in acid media, formate oxidation in alkaline media has not received much attention yet, with only a few works focusing on Pd and platinum (Pt). Compared with the electrooxidation reaction of formate in alkaline media, in acidic media, the catalyst is not easily corroded or poisoned. In addition, the dependence on precious metals in acidic media leads to the high cost of the catalyst, thus, the formate electrooxidation reaction in alkaline media has development prospects. Moreover, formate fuel cells are widely regarded as ideal direct liquid fuel cells. The basic principle of their operation is that on the anode, the fuel solution flowing into the anode channel (typically, formate and alkali dissolve in water) is transferred through the anode diffusion layer (DL) to the anode catalyst layer (CL), where formate ions are oxidized to produce electrons, water, and carbonate ions. The formate electrooxidation reaction of the anode is as follows [121]:

$$HCOO^- + 3OH^- \rightarrow CO_3^{2-} + 2H_2O + 2e^- \quad E^0 = -1.05\,V \tag{10.23}$$

After the anodic reaction, the water generated by the FOR, along with the water in the solution, diffuses to the cathode CL through AEM, and the generated electrons are transported to the cathode through an external load. On the cathode, the oxygen diffuses to the cathode CL through the cathode DL, and the oxygen reduction reaction (ORR) occurs on the cathode CL, losing electrons and generating hydroxide ions according to [122]:

$$\frac{1}{2}O_2 + H_2O + 2e^- \rightarrow 2OH^- \quad E^0 = 0.40\,V \tag{10.24}$$

Combining the FOR and the ORR given by Eqs. (10.23) and (10.24), the overall reaction of the fuel cell system can be obtained:

$$HCOO^- + \frac{1}{2}O_2 + OH^- \rightarrow CO_3^{2-} + H_2O \quad E^0 = 1.45\,V \tag{10.25}$$

This theoretical voltage shows an increase as compared with that of fuel cells running methanol (1.23 V) and ethanol (1.14 V) [123, 124]. Research on the reaction pathway of formate electrooxidation mainly focuses on Pt and Pd catalysts. The following is a brief introduction to the research progress of the reaction mechanism.

Palladium

The most probable mechanisms for the FOR can be related to the H_{ad} and COO_{ad}^- which adsorbs on the Pd surface [125]. Potential scanning and other methods were used to study the electrocatalytic oxidation of formate on Pd. They suggested that formate adsorbed on the Pd surface was decomposed into two fragments of H_{ad} and COO_{ad}^- according to:

$$\begin{aligned} &HCOO^- \rightarrow H_{ad} + COO_{ad}^- \\ &H_{ad} + OH^- \rightarrow H_2O + e^- \\ &COO_{ad}^- + 2OH^- \rightarrow CO_3^{2-} + H_2O + e^- \end{aligned} \tag{10.26}$$

The net reaction of the anodic oxidation of formate is described as:

$$HCOO^- + \frac{1}{2}O_2 + OH^- \rightarrow CO_3^{2-} + H_2O \quad E^0 = 1.45\,V \tag{10.27}$$

They also showed that the electrocatalytic oxidation of formate did not involve a strongly bound poisoning intermediate on the Pd surface.

Platinum

The initial research on the mechanism of formate electrooxidation reaction was mainly focused on Pt catalysts. The mechanism of the formate oxidation on the polycrystalline Pt was studied in a highly basic medium (pH = 14) [8]. Based on the experimental results, they proposed a dual-pathway mechanism on the Pt surface for the formate oxidation, which is shown in Fig. 10.21 [126]. Interestingly, this dual pathway mechanism is similar to the electrooxidation reaction of formic acid in acidic media. The dual pathway mechanism suggests that formate ions are first slowly oxidized to adsorbates in the primary and secondary pathways. The adsorbates can be directly oxidized to CO_2 (direct oxidation pathway) or transformed into adsorbed CO_{ad} and further oxidized to CO_2 (indirect pathway) by binding with OH^- at higher potentials. There are other similar reaction pathways. A three-way mechanism was proposed by studying the electrocatalytic oxidation of formate on Pt black in alkaline medium [127]. Unlike the dual pathway mechanism, the third pathway mechanism includes an independent third pathway, which involves the oxidation of less active adsorbates to CO_2. The in-situ FTIR technology shows that the adsorbed hydroxyl groups play an important role in controlling the reaction mechanism and participating in the oxidation of intermediates [128]. In addition, the presence of adsorbed hydroxyl groups prevents part of the surface from being blocked by adsorbed intermediate species (such as CO_{ad}) and promotes the adsorption of active O-bonded species.

In summary, due to the slow kinetics of the formate electrooxidation reaction, the problem of high overpotential and inability to spontaneously react is crucial. Research on catalysts is mainly focused on the design and optimization of Pt-based and Pd-based catalysts.

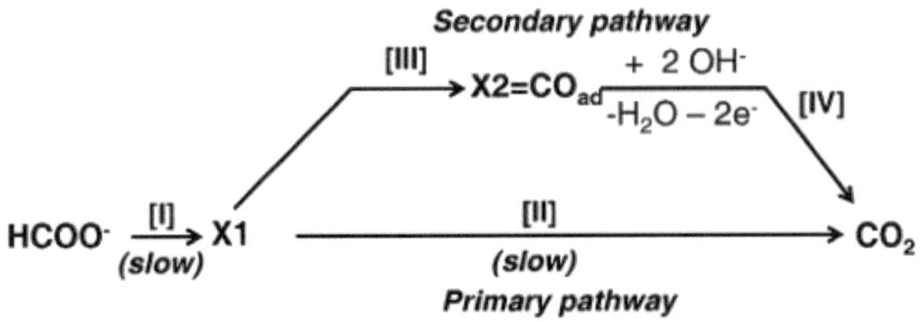

Fig. 10.21 Proposed mechanism for the formate oxidation on Pt [125]

10.3.3.2 Pt-Based Catalysts

Due to the poisoning effect of intermediate products on Pt catalysts, the efficiency of formate electrooxidation in alkaline media on Pt catalysts is much lower than that on Pd catalysts. Therefore, there is relatively little research on Pt-based catalysts, but the initial catalyst research still started with Pt catalysts. The oxidation of formate on a platinum black modified electrode was studied under the conditions of microbial fuel cell by CV, chronoamperometry, and potentiostatic coulometry [129]. It has been proven that the kinetic constant of formate electrooxidation is ten times higher than that of ethanol oxidation. The experimental results indicate that the current density of the formate oxidation reaction could reach 6 mA cm^{-2} at a 0.2 V oxidation potential. Similarly, to improve the performance of the catalyst, it is common to dope other metal elements based on Pt to form binary alloy catalysts. A fuel cell using ammonium formate as fuel was proposed by The electrocatalytic activity of Pt and Pt iridium (Ir) on Ammonium formate and Pt iridium (Ir) was studied by CV [130]. It was found that the Pt-Ir catalysts could effectively oxidize ammonium formate, as they could effectively oxidize the CO layer to CO_2. In addition, the studies about the oxidation of ammonium formate show that the peak current of the Pt–Pd catalyst is much higher than that of Pt [130]. Results also confirmed that among various catalysts (Pt, Pt-Ir, Pt–Pd, Pt-Ru), the Pt–Pd catalysts resulted in the highest current density for the electrocatalytic oxidation of ammonium formate (0.1 M), as presented in Fig. 10.22. Afterwards, research on Pd-based catalysts gradually took a dominant position.

10.3.3.3 Pd-Based Catalysts

Pd-based catalysts are considered the most promising formate electrooxidation catalysts because their activity is much higher than that of Pt-based catalysts. On the other hand, palladium-based catalysts not only have higher activity but are less prone to deactivation due to carbon monoxide under alkaline conditions compared to Pt-based catalysts. As a result, palladium and palladium-based catalysts have been widely studied as electrocatalytic materials for the oxidation of formate. The Pd catalysts for the electrocatalytic oxidation of formate were investigated in alkaline

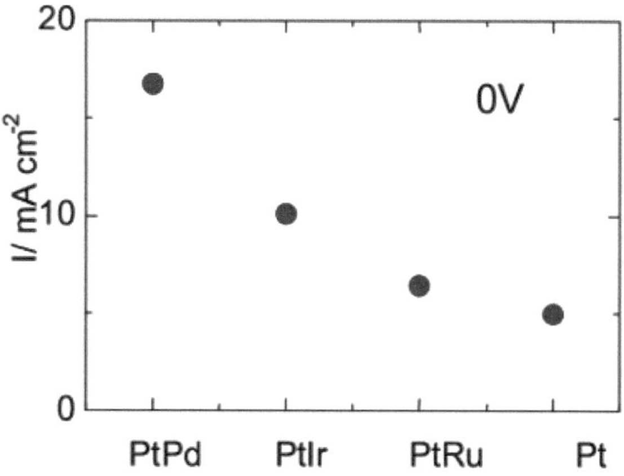

Fig. 10.22 Current density for a series of Pt-based catalysts (Pt–Pd, Pt-Ir, Pt-Ru, Pt) in 0.1 M $HCOONH_4$ by CV [130]

media [131]. The results show that compared with the Pt catalyst, formate can be oxidized better on the surface of Pd, and there is no strongly bound intermediate on the surface of Pd. They demonstrated that using the Pd catalysts enabled this type of fuel cell to operate with the air or even without a supporting electrolyte such as KOH. However, other liquid fuels, such as ethanol, have to be mixed with the alkali in the fuel solution to achieve high performance, which is the unique advantage of formate electrooxidation. In addition, they also studied the effect of the formate salt type (HCOOK/HCOONa) on the cell performance and showed that the peak power density of the fuel cell running on HCOOK was 144 mW cm^{-2}, while the fuel cell running on HCOONa resulted in a peak power density of 125 mW cm^{-2} at 60 °C. To improve the performance of the catalyst, research on FOR catalysts is mainly based on palladium-based nanoalloys [133–135]. Recently, a 3D flower-like PdCu nanosheet (NS) catalyst was synthesized (Figure 10.23) via a method of oil bath reduction under mild conditions as a bifunctional electrocatalyst for DFFCs [132]. As compared to Pd/C, the kinetics of the formate oxidation was enhanced by using the PdCu alloy nanosheet catalyst in alkaline media, which could be explained by the high exposed facets of Pd and the richness of the surface area. Moreover, the electronic effect of charge transfer from Pd to Cu, which plays an important role in formate oxidation, further improves performance. When using the PdCu nanosheet catalyst as a bifunctional catalyst for DFFCs, the power density of 53 mW cm^{-2} and the ultimate current density of 500 mA cm^{-2} can be achieved at room temperature. In addition, Pd-Au/C catalysts with different atomic ratios (90:10, 80:20, 70:30, and 50:50) were prepared by the metal salt reduction method with sodium borohydride for the electrocatalytic oxidation of formate in alkaline media [133]. The mean particle size of the Pd-Au/C catalysts was between 4 nm and 10 nm. The use of Pd-Au/C catalysts (90:10 and 80:20) in DFFCs resulted in about 15% higher peak power density than the Pd/C catalysts did. Similarly, surface reconstruction is a key to improving the performance of alloy catalysts. The AgPd-Ag_2O and AgPd-AgF heterointerface catalysts were synthesized through a self-limiting reconstruction strategy to illustrate the improvement of the mechanism for formate electrooxidation reaction [134]. The

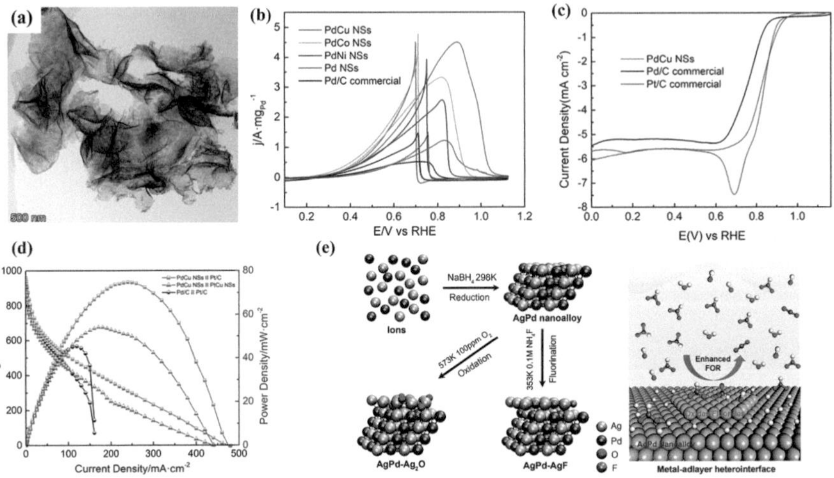

Fig. 10.23 **a** TEM image **b** CV curves for formate oxidation reaction **c** LSV for ORR **d** performance of DFFC of PdCu nanosheet catalyst. **e** Surface reconstruction of AgPd catalyst [135]

formation of AgOx and AgFx layers after adsorption of O and F atoms was demonstrated through molecular dynamics simulations, which facilitates the kinetics and thermodynamics compared to smooth metal surfaces.

In summary, the discussion has been focused on two types of electrocatalysts, i.e., Pt/Pt-based and Pd/Pd-based electrocatalysts. In the future, non-precious metals should also be considered as a good candidate for electrocatalysts toward formate oxidation in alkaline media.

The fuel cell performance depends not only on the electrocatalytic and membrane materials, but also on the structural design parameters that determine the transport rates of mass, ions, and electrons through the fuel cell structure, as well as the operating parameters that directly link to the fuel cell power output. As summarized in Table 10.3, although the DFFC is still an emerging energy technology, the power density has been substantially boosted from ~50 mW cm^{-2} to 591 mW cm^{-2} in recent years.

Table 10.3 The DFFC performance at 60 °C in the literature

Fuel/oxidant	Anode catalyst	Cathode catalyst	Membrane	Power density ($mW\ cm^{-2}$)	References
2 $mL\ min^{-1}$ 1 M HCOONa + 2 M NaOH/10 $mL\ min^{-1}$ 1 M H_2O_2 + 1 M H_2SO_4	PdAu/Ni foam	PdAu/Ni foam	N 115	331	[121]
1 $mL\ min^{-1}$ 2 M HCOOK + 2 M KOH/100 sccm O_2	Pd/C	Pt/C	A 201	144	[136]
1 $mL\ min^{-1}$ 2 M HCOOK + 2 M KOH/400 sccm air	Pd/C	Pt/C	A 201	109	[136]
1 $mL\ min^{-1}$ 2 M HCOOK + 2 M NaOH/100 sccm O_2	Pd/C	Pt/C	A 201	125	[136]
1 $mL\ min^{-1}$ 2 M HCOOK + 2 M KOH/100 sccm O_2	Pd/C	Pt/C	A 201	144	[137]
1 $mL\ min^{-1}$ 2 M HCOOK + 1 M KOH/100 sccm O_2	Pd/C	Pt/C	A 201	125	[137]
1 $mL\ min^{-1}$ 2 M HCOOK + 2 M KOH/100 sccm O_2	Pd/C	Pt/C	A 201	123	[137]
1 $mL\ min^{-1}$ 1 M HCOONa + 2 M KOH/100 sccm O_2	Pd/C	Pt/C	A 201	125	[137]
1 $mL\ min^{-1}$ 1 M HCOOK + 2 M KOH/400 sccm air	Pd/C	Pt/C	A 201	110	[137]
2 $mL\ min^{-1}$ 1 M HCOONa + 2 M NaOH/10 $mL\ min^{-1}$ 1 M H_2O_2 + 1 M H_2SO_4	Pd/C	Pt/C	N 115	407	[138]
2 $mL\ min^{-1}$ 1 M HCOONa + 2 M NaOH/10 $mL\ min^{-1}$ 1 M H_2O_2 + 1 M H_2SO_4	Pd/C	Pt/C	N 115	442	[138]
2 $mL\ min^{-1}$ 1 M HCOONa + 2 M NaOH/10 $mL\ min^{-1}$ 1 M H_2O_2 + 1 M H_2SO_4	Pd/C	Pt/C	N 115	412	[138]
2 $mL\ min^{-1}$ 1 M HCOONa + 2 M NaOH/10 $mL\ min^{-1}$ 1 M H_2O_2 + 1 M H_2SO_4	Pd/C	Pt/C	N 115	~ 400	[138]
2 $mL\ min^{-1}$ 1 M HCOONa + 2 M NaOH/10 $mL\ min^{-1}$ 1 M H_2O_2 + 1 M H_2SO_4	Pd/C	Pt/C	N 115	519	[138]
2 $mL\ min^{-1}$ 1 M HCOONa + 2 M NaOH/10 $mL\ min^{-1}$ 1 M H_2O_2 + 1 M H_2SO_4	Pd/C	Pt/C	N 115	591	[138]
2 $mL\ min^{-1}$ 1 M HCOONa + 2 M NaOH/10 $mL\ min^{-1}$ 1 M H_2O_2 + 1 M H_2SO_4	Pd/C	Pt/C	N 115	550	[138]

(continued)

Table 10.3 (continued)

Fuel/oxidant	Anode catalyst	Cathode catalyst	Membrane	Power density (mW cm^{-2})	References
2 mL min^{-1} 1 M HCOONa + 2 M NaOH/10 mL min^{-1} 1 M H_2O_2 + 1 M H_2SO_4	Pd/C	Pt/C	N 115	520	[138]
2 mL min^{-1} 1 M HCOONa + 2 M NaOH/10 mL min^{-1} 1 M H_2O_2 + 1 M H_2SO_4	Pd/C	Pt/C	N 115	410	[138]

10.4 Summary and Outlook

As a substitute for PEMFC, AEMFC has a good prospect in terms of performance and cost, and has attracted much attention at present. Compared with PEMFCs, AEMFCs under alkaline conditions has the following advantages: (1) it is easier to carry out ORR, and can use non-platinum group metal (PGM) catalyst with lower cost and higher stability. (2) The electrooxidation kinetics of many liquid fuels (such as sodium borohydride, methanol, ethanol, etc.) can be enhanced in an alkaline environment. (3) The electroosmotic resistance associated with ion transport can ease the penetration of liquid fuels in AEMFCs, thus enabling the use of more concentrated liquid fuels, which is an advantage for portable fuel cell devices. (4) Flexibility in fuel and ORR catalyst selection also expands the parameter space for the discovery of highly selective catalysts for permeation-resistant fuels. The alkaline environment permits the usage of lower-cost non-precious metal catalysts, such as nickel, cobalt, and silver. This characteristic greatly reduces the reliance on precious metals like platinum, thereby significantly lowering the overall cost of the fuel cell [139].

In recent years, to achieve high-performance AEMFCs, the research has mainly focused on [36]:

(1) the membrane with high conductivity and high water transport under alkaline stable conditions.
(2) ensure the balance of water content in the battery to achieve good water management.
(3) electrode development and optimization.
(4) reduce the effect of carbonation.
(5) using PGM catalysts.

The water absorption, ion conductivity, and ion exchange capacity (IEC) of various polyolefin and polyaromatic AEM were compared with that of PEM and found that the conductivity of AEM may be 2–10 times that of PEM, and that AEM requires a higher water content than PEM to obtain conductivity [140]. To increase the conductivity, the IEC of the film is increased, which has a certain effect on the mechanical stability of the film. It has been found that adding an ionomer layer between GDE and the film can reduce the interface contact resistance and improve the performance. The performance comparison between the catalyst-coated electrode method, GDE method, and GDE method with ionomer interlayer shows that GDE with ionomer interlayer can be close to the catalyst-coated film method due to the decrease of interface resistance [141].

Nevertheless, AEMFCs still face several challenges along the way [140, 142]:

(1) In recent years, some progress has been made in the synthesis of new ORR catalysts to replace Pt catalysts, but there are still some problems that have not been solved, such as high overpotential and low reactivity, and durability and stability problems still need to be solved.
(2) As a key component of MEA, one of the primary challenges is the performance and stability of anion exchange membranes. These membranes need to possess high ion conductivity, excellent chemical stability, and mechanical robustness simultaneously. However, many existing membrane materials are prone to degradation in alkaline environments, leading to significant declines in conductivity and service life. This adversely affects the overall performance and reliability of the fuel cell. In addition, water management in alkaline environments poses another major technical hurdle. AEMFCs must maintain appropriate water content within the cell to ensure efficient OH^- ion conduction.
(3) The battery cathode is usually a high flow rate of air, which will cause the cathode to have a strong local drying tendency, and the lack of water in the cathode will not only affect the electrochemical reaction, but also reduce the conductivity of AEM.

To address the above problems, it is necessary to maximize the water absorption, electrical conductivity, stability, and tolerance of the membrane, further develop non-PGM catalysts, and optimize the catalyst layer composition and preparation technology, including ionomer dispersion control, to obtain higher performance and stability. In addition, it is also necessary to design the hydrophilicity of the electrode and optimize the water management strategy, and the optimized electrode must also have appropriate thickness, hydrophobicity, and porosity. Of course, to realize the potential of anion exchange membrane fuel cells, material and system management strategies need to be studied.

In summary, anion exchange membrane fuel cells, with their unique technological advantages and extensive application potential, are emerging as a key direction in the field of fuel cell research. Although challenges related to materials, catalysts, and system integration persist, ongoing research and technological advancements are expected to overcome these hurdles. As such, AEMFCs are poised to occupy a vital position in the future clean energy landscape, contributing significantly to global energy transition and sustainable development goals.

References

1. Arges C G, Ramani V K, Pintauro P N. The Chalkboard: Anion Exchange Membrane Fuel Cells[J]. The Electrochemical Society Interface, 2010, 19(2): 31–35.
2. Zhang H-W, Chen D-Z, Xianze Y, et al. Anion-Exchange Membranes for Fuel Cells: Synthesis Strategies, Properties and Perspectives[J]. Fuel Cells, 2015, 15(6): 761–780.
3. Zhang H, Shen P K. Recent Development of Polymer Electrolyte Membranes for Fuel Cells[J]. Chemical Reviews, 2012, 112(5): 2780–2832.
4. Sun X, Li Y, Li M-J. Highly Dispersed Palladium Nanoparticles on Carbon-Decorated Porous Nickel Electrode: An Effective Strategy to Boost Direct Ethanol Fuel Cell up to 202 mW cm^{-2}[J]. ACS Sustainable Chemistry & Engineering, 2019, 7(13): 11186–11193.
5. Dekel D R. Review of cell performance in anion exchange membrane fuel cells[J]. Journal of Power Sources, 2018, 375: 158–169.
6. Hren M, Božič M, Fakin D, et al. Alkaline membrane fuel cells: anion exchange membranes and fuels[J]. Sustainable Energy & Fuels, 2021, 5(3): 604–637.
7. Hagesteijn K F L, Jiang S, Ladewig B P. A review of the synthesis and characterization of anion exchange membranes[J]. Journal of Materials Science, 2018, 53(16): 11131–11150.
8. Varcoe J R, Atanassov P, Dekel D R, et al. Anion-exchange membranes in electrochemical energy systems[J]. Energy Environ. Sci., 2014, 7(10): 3135–3191.
9. Pavel C C, Cecconi F, Emiliani C, et al. Highly Efficient Platinum Group Metal Free Based Membrane-Electrode Assembly for Anion Exchange Membrane Water Electrolysis[J]. Angewandte Chemie International Edition, 2014, 53(5): 1378–1381.
10. Merle G, Wessling M, Nijmeijer K. Anion exchange membranes for alkaline fuel cells: A review[J]. Journal of Membrane Science, 2011, 377(1–2): 1–35.
11. Couture G, Alaaeddine A, Boschet F, et al.z. Progress in Polymer Science, 2011, 36(11): 1521–1557.
12. Jiao K, He P, Du Q, et al. Three-dimensional multiphase modeling of alkaline anion exchange membrane fuel cell[J]. International Journal of Hydrogen Energy, 2014, 39(11): 5981–5995.
13. Pivovar B S, Leader F C T, Fedkiw P, et al. Alkaline membrane fuel cell workshop final report[J]. Work. Held, 2006: 13.
14. Cermenek B, Ranninger J, Hacker V. Alkaline Direct Ethanol Fuel Cell[M]. Elsevier, 2019: 383–405.
15. Ran J, Wu L, He Y, et al. Ion exchange membranes: New developments and applications[J]. Journal of Membrane Science, 2017, 522: 267–291.
16. Strathmann H, Grabowski A, Eigenberger G. Ion-Exchange Membranes in the Chemical Process Industry[J]. Industrial & Engineering Chemistry Research, 2013, 52(31): 10364–10379.
17. Pan Z F, An L, Zhao T S, et al. Advances and challenges in alkaline anion exchange membrane fuel cells[J]. Progress in Energy and Combustion Science, 2018, 66: 141–175.
18. Pan J, Chen C, Li Y, et al. Constructing ionic highway in alkaline polymer electrolytes[J]. Energy Environ. Sci., 2014, 7(1): 354–360.
19. Ponce-González J, Ouachan I, Varcoe J R, et al. Radiation-induced grafting of a butyl-spacer styrenic monomer onto ETFE: the synthesis of the most alkali stable radiation-grafted anion-exchange membrane to date[J]. Journal of Materials Chemistry A, 2018, 6(3): 823–827.
20. Han J, Peng H, Pan J, et al. Highly Stable Alkaline Polymer Electrolyte Based on a Poly(ether ether ketone) Backbone[J]. ACS Applied Materials & Interfaces, 2013, 5(24): 13405–13411.
21. Fujimoto C, Kim D-S, Hibbs M, et al. Backbone stability of quaternized polyaromatics for alkaline membrane fuel cells[J]. Journal of Membrane Science, 2012, 423–424: 438–449.
22. Zhu L, Pan J, Wang Y, et al. Multication Side Chain Anion Exchange Membranes[J]. Macromolecules, 2016, 49(3): 815–824.
23. Han J, Liu Q, Li X, et al. An Effective Approach for Alleviating Cation-Induced Backbone Degradation in Aromatic Ether-Based Alkaline Polymer Electrolytes[J]. ACS Applied Materials & Interfaces, 2015, 7(4): 2809–2816.

24. Agel E, Bouet J, Fauvarque J F. Characterization and use of anionic membranes for alkaline fuel cells[J]. Journal of Power Sources, 2001, 101(2): 267–274.
25. Anion Exchange Membrane Fuel Cells: Principles, Materials and Systems[M]. L. An, T.S. Zhao. Cham: Springer International Publishing, 2018, 63.
26. Gottesfeld S, Dekel D R, Page M, et al. Thermally Cross-Linked Anion Exchange Membranes from Solvent Processable Isoprene Containing Ionomer[J]. Journal of Power Sources, 2018, 375: 170–184.
27. Pan J, Chen C, Zhuang L, et al. Designing Advanced Alkaline Polymer Electrolytes for Fuel Cell Applications[J]. Accounts of Chemical Research, 2012, 45(3): 473–481.
28. Li N, Yan T, Li Z, et al. Comb-shaped polymers to enhance hydroxide transport in anion exchange membranes[J]. Energy & Environmental Science, 2012, 5(7): 7888.
29. Yang Z, Guo R, Malpass-Evans R, et al. Highly Conductive Anion-Exchange Membranes from Microporous Tröger's Base Polymers[J]. Angewandte Chemie International Edition, 2016, 55(38): 11499–11502.
30. Hibbs M R. Alkaline stability of poly(phenylene)-based anion exchange membranes with various cations[J]. Journal of Polymer Science Part B: Polymer Physics, 2013, 51(24): 1736–1742.
31. Tanaka M, Fukasawa K, Nishino E, et al. Anion Conductive Block Poly(arylene ether)s: Synthesis, Properties, and Application in Alkaline Fuel Cells[J]. Journal of the American Chemical Society, 2011, 133(27): 10646–10654.
32. Li N, Leng Y, Hickner M A, et al. Highly Stable, Anion Conductive, Comb-Shaped Copolymers for Alkaline Fuel Cells[J]. Journal of the American Chemical Society, 2013, 135(27): 10124–10133.
33. Lee W-H, Park E J, Han J, et al. Poly(terphenylene) Anion Exchange Membranes: The Effect of Backbone Structure on Morphology and Membrane Property[J]. ACS Macro Letters, 2017, 6(5): 566–570.
34. Lee W-H, Kim Y S, Bae C. Robust Hydroxide Ion Conducting Poly(biphenyl alkylene)s for Alkaline Fuel Cell Membranes[J]. ACS Macro Letters, 2015, 4(8): 814–818.
35. Oshiba Y, Hiura J, Suzuki Y, et al. MEA with Thin Pore-Filling Anion Exchange Membrane and Water Transport Analysis in SAFC[J]. ECS Meeting Abstracts, IOP Publishing, 2015, MA2015–02(37): 1339.
36. Mustain W E. Understanding how high-performance anion exchange membrane fuel cells were achieved: Component, interfacial, and cell-level factors[J]. Current Opinion in Electrochemistry, 2018, 12: 233–239.
37. Mustain W E, Chatenet M, Page M, et al. Durability challenges of anion exchange membrane fuel cells[J]. Energy & Environmental Science, 2020, 13(9): 2805–2838.
38. Gerhardt M R, Pant L M, Weber A Z. Along-the-Channel Impacts of Water Management and Carbon-Dioxide Contamination in Hydroxide-Exchange-Membrane Fuel Cells: A Modeling Study[J]. Journal of The Electrochemical Society, 2019: 14.
39. Mandal M, Huang G, Hassan N U, et al. The Importance of Water Transport in High Conductivity and High-Power Alkaline Fuel Cells[J]. Journal of The Electrochemical Society, 2020: 12.
40. Marino M G, Melchior J P, Wohlfarth A, et al. Hydroxide, halide and water transport in a model anion exchange membrane[J]. Journal of Membrane Science, 2014, 464: 61–71.
41. Dekel D R, Willdorf S, Ash U, et al. The critical relation between chemical stability of cations and water in anion exchange membrane fuel cells environment[J]. Journal of Power Sources, 2018, 375: 351–360.
42. Leonard D P, Maurya S, Park E J, et al. Asymmetric electrode ionomer for low relative humidity operation of anion exchange membrane fuel cells[J]. Journal of Materials Chemistry A, 2020, 8(28): 14135–14144.
43. Choe Y-K, Fujimoto C, Lee K-S, et al. Alkaline Stability of Benzyl Trimethyl Ammonium Functionalized Polyaromatics: A Computational and Experimental Study[J]. Chemistry of Materials, 2014, 26(19): 5675–5682.

44. The Chemistry of Membranes Used in Fuel Cells: Degradation and Stabilization[M]. S. Schlick. Hoboken, NJ, USA: John Wiley & Sons, Inc., 2018.
45. Neagu V, Bunia I, Plesca I. Ionic polymers VI. Chemical stability of strong base anion exchangers in aggressive media[J]. Polymer Degradation and Stability, 2000, 70(3): 463–468.
46. Sata T, Matsuo Y, Yamaguchi T, et al. Preparation and transport properties of anion-exchange membranes containing viologen moieties as anion-exchange groups in the presence or absence of photoirradiation[J]. Journal of the Chemical Society, Faraday Transactions, 1997, 93(15): 2553–2560.
47. Peng H, Li Q, Hu M, et al. Alkaline polymer electrolyte fuel cells stably working at 80 °C[J]. Journal of Power Sources, 2018, 390: 165–167.
48. Varcoe J R, Slade R C T, Lam How Yee E. An alkaline polymer electrochemical interface: a breakthrough in application of alkaline anion-exchange membranes in fuel cells[J]. Chemical Communications, 2006(13): 1428.
49. Kim Y S, Welch C F, Hjelm R P, et al. Origin of Toughness in Dispersion-Cast Nafion Membranes[J]. Macromolecules, 2015, 48(7): 2161–2172.
50. Antolini E, Gonzalez E R. Alkaline direct alcohol fuel cells[J]. Journal of Power Sources, 2010, 195(11): 3431–3450.
51. Sun Z, Lin B, Yan F. Anion-Exchange Membranes for Alkaline Fuel-Cell Applications[J]: The Effects of Cations. ChemSusChem, 2018, 11(1): 58–70.
52. Sun L, Guo J, Zhou J, et al. Novel nanostructured high-performance anion exchange ionomers for anion exchange membrane fuel cells[J]. Journal of Power Sources, 2012, 202: 70–77.
53. Tsai T-H, Ertem S P, Maes A M, et al. Thermally Cross-Linked Anion Exchange Membranes from Solvent Processable Isoprene Containing Ionomers[J]. Macromolecules, 2015, 48(3): 655–662.
54. Zhang M, Liu J, Wang Y, et al. Highly stable anion exchange membranes based on quaternized polypropylene[J]. Journal of Materials Chemistry A, 2015, 3(23): 12284–12296.
55. Ziv N, Mustain W E, Dekel D R. The Effect of Ambient Carbon Dioxide on Anion-Exchange Membrane Fuel Cells[J]. ChemSusChem, 2018, 11(7): 1136–1150.
56. Katayama Y, Yamauchi K, Hayashi K, et al. Anion-Exchange Membrane Fuel Cells with Improved CO2 Tolerance: Impact of Chemically Induced Bicarbonate Ion Consumption[J]. ACS Applied Materials & Interfaces, 2017, 9(34): 28650–28658.
57. Aricò A S, Srinivasan S, Antonucci V. DMFCs: From Fundamental Aspects to Technology Development[J]. Fuel Cells, 2001, 1(2): 133–161.
58. Kamarudin S K, Achmad F, Daud W R W. Overview on the application of direct methanol fuel cell (DMFC) for portable electronic devices[J]. International Journal of Hydrogen Energy, 2009, 34(16): 6902–6916.
59. Acres G J K. Recent advances in fuel cell technology and its applications[J]. Journal of Power Sources, 2001, 100(1): 60–66.
60. Zhao T S, Chen R, Yang W W, et al. Small direct methanol fuel cells with passive supply of reactants[J]. Journal of Power Sources, 2009, 191(2): 185–202.
61. Kimiaie N, Wedlich K, Hehemann M, et al. Results of a 20000 h lifetime test of a 7 kW direct methanol fuel cell (DMFC) hybrid system–degradation of the DMFC stack and the energy storage[J]. Energy & environmental science, 2014, 7(9): 3013–3025.
62. Kim J, Johnson T A, Miller J E, et al. Fuel production from CO 2 using solar-thermal energy: system level analysis[J]. Energy & Environmental Science, 2012, 5(9): 8417–8429.
63. Zhou Y-W, Chen Y-F, Jiang K, et al. Probing the enhanced methanol electrooxidation mechanism on platinum-metal oxide catalyst[J]. Applied Catalysis B: Environmental, 2021, 280: 119393.
64. Scott K, Taama W, Argyropoulos P, et al. The impact of mass transport and methanol crossover on the direct methanol fuel cell[J]. Journal of Power Sources, 1999, 83(1–2): 204–216.
65. Joghee P, Malik J N, Pylypenko S, et al. A review on direct methanol fuel cells–In the perspective of energy and sustainability[J]. MRS Energy & Sustainability, 2015, 2: E3.
66. Tiwari J N, Tiwari R N, Singh G, et al. Recent progress in the development of anode and cathode catalysts for direct methanol fuel cells[J]. Nano Energy, 2013, 2(5): 553–578.

67. Xia Z, Zhang X, Sun H, et al. Recent advances in multi-scale design and construction of materials for direct methanol fuel cells[J]. Nano Energy, 2019, 65: 104048.
68. Zuo Y, Sheng W, Tao W, et al. Direct methanol fuel cells system–A review of dual-role electrocatalysts for oxygen reduction and methanol oxidation[J]. Journal of Materials Science & Technology, 2022, 114: 29–41.
69. Gong L, Yang Z, Li K, et al. Recent development of methanol electrooxidation catalysts for direct methanol fuel cell[J]. Journal of energy chemistry, 2018, 27(6): 1618–1628.
70. Müller E. Die kathodische Abscheidung von Tellur und Selen aus ihren Sauerstoffsäuren und ihre elektroanalytische Bestimmung[J]. Zeitschrift für Physikalische Chemie, 1922, 100(1): 346–366.
71. Lamy C, Leger J, Clavilier J. Structural effects in the electrooxidation of methanol in alkaline medium: Comparison of platinum single crystal and polycrystalline electrodes[J]. Journal of Electroanalytical Chemistry and Interfacial Electrochemistry, 1982, 135(2): 321–328.
72. Leiva E, Giordano M. Multiple current components for methanol electrosorption and electrooxidation at platinum in acidic solutions[J]. Journal of electroanalytical chemistry and interfacial electrochemistry, 1983, 158(1): 115–130.
73. Mundy G, Potter R, Christensen P, et al. A study of the electro-oxidation of methanol on platinum in 1 M Na2SO4 by electrochemical and in-situ FTIR techniques[J]. Journal of electroanalytical chemistry and interfacial electrochemistry, 1990, 279(1–2): 257–272.
74. Frelink T, Visscher W, Van Veen J. Particle size effect of carbon-supported platinum catalysts for the electrooxidation of methanol[J]. Journal of Electroanalytical Chemistry, 1995, 382(1–2): 65–72.
75. Maillard F, Eikerling M, Cherstiouk O, et al. Size effects on reactivity of Pt nanoparticles in CO monolayer oxidation: The role of surface mobility[J]. Faraday discussions, 2004, 125: 357–377.
76. Ensafi A A, Jafari-Asl M, Rezaei B. A new strategy for the synthesis of 3-D Pt nanoparticles on reduced graphene oxide through surface functionalization, Application for methanol oxidation and oxygen reduction[J]. Electrochimica Acta, 2014, 130: 397–405.
77. Centi G, Perathoner S. The role of nanostructure in improving the performance of electrodes for energy storage and conversion[J]. European Journal of Inorganic Chemistry, 2009, 2009(26): 3851–3878.
78. Yang S, Kim J, Tak Y J, et al. Single-atom catalyst of platinum supported on titanium nitride for selective electrochemical reactions[J]. Angewandte Chemie International Edition, 2016, 55(6): 2058–2062.
79. Moses-DeBusk M, Yoon M, Allard L F, et al. CO oxidation on supported single Pt atoms: experimental and ab initio density functional studies of CO interaction with Pt atom on θ-Al2O3 (010) surface[J]. Journal of the American Chemical Society, 2013, 135(34): 12634–12645.
80. Qiao B, Wang A, Yang X, et al. Single-atom catalysis of CO oxidation using Pt1/FeOx[J]. Nature chemistry, 2011, 3(8): 634–641.
81. Li M, Zheng H, Han G, et al. Facile synthesis of binary PtRu nanoflowers for advanced electrocatalysts toward methanol oxidation[J]. Catalysis Communications, 2017, 92: 95–99.
82. Xie J, Zhang Q, Gu L, et al. Ruthenium–platinum core–shell nanocatalysts with substantially enhanced activity and durability towards methanol oxidation[J]. Nano Energy, 2016, 21: 247–257.
83. Du Y, Wang C. Preparation Ru, Bi monolayer modified Pt nanoparticles as the anode catalyst for methanol oxidation[J]. Materials Chemistry and Physics, 2009, 113(2–3): 927–932.
84. Muthuswamy N, de la Fuente J L G, Tran D T, et al. Ru@ Pt core–shell nanoparticles for methanol fuel cell catalyst: Control and effects of shell composition[J]. International journal of hydrogen energy, 2013, 38(36): 16631–16641.
85. Ghouri Z K, Al-Meer S, Barakat N A M, et al. ZnO@C (core@shell) microspheres derived from spent coffee grounds as applicable non-precious electrode material for DMFCs[J]. Scientific Reports, 2017, 7(1).

86. Lao X, Yang M, Chen J, et al. The ethanol oxidation reaction on bimetallic PdxAg1-x nanosheets in alkaline media and their mechanism study[J]. Electrochimica Acta, 2021, 374.
87. Sun L, Lv H, Wang Y, et al. Unveiling Synergistic Effects of Interstitial Boron in Palladium-Based Nanocatalysts for Ethanol Oxidation Electrocatalysis[J]. J Phys Chem Lett, 2020, 11(16): 6632–6639.
88. Liu M, Xie M, Jiang Y, et al. Core–shell nanoparticles with tensile strain enable highly efficient electrochemical ethanol oxidation[J]. Journal of Materials Chemistry A, 2021, 9(27): 15373–15380.
89. Rizo R, Pérez-Rodríguez S, García G. Well-Defined Platinum Surfaces for the Ethanol Oxidation Reaction[J]. ChemElectroChem, 2019, 6(18): 4725–4738.
90. Yaqoob L, Noor T, Iqbal N. A comprehensive and critical review of the recent progress in electrocatalysts for the ethanol oxidation reaction[J]. RSC Adv, 2021, 11(27): 16768–16804.
91. Jiang K, Wang P, Guo S, et al. Ordered PdCu-Based Nanoparticles as Bifunctional Oxygen-Reduction and Ethanol-Oxidation Electrocatalysts[J]. Angewandte Chemie, 2016, 55(31): 9030–5.
92. Zhu M, Sun G, Xin Q. Effect of alloying degree in PtSn catalyst on the catalytic behavior for ethanol electro-oxidation[J]. Electrochimica Acta, 2009, 54(5): 1511–1518.
93. Eid K, Ahmad Y H, Yu H, et al. Rational one-step synthesis of porous PtPdRu nanodendrites for ethanol oxidation reaction with a superior tolerance for CO-poisoning[J]. Nanoscale, 2017, 9(47): 18881–18889.
94. Shen S, Guo Y, Luo L, et al. Comprehensive Analysis on the Highly Active and Stable PdAu/C Electrocatalyst for Ethanol Oxidation Reaction in Alkaline Media[J]. The Journal of Physical Chemistry C, 2018, 122(3): 1604–1611.
95. Liu C, Shen Y, Zhang J, et al. Multiple Twin Boundary-Regulated Metastable Pd for Ethanol Oxidation Reaction. Advanced Energy Materials, 2022, 12(8).
96. Li L, Liu H, Qin C, et al. Ultrathin PtxSn1-x Nanowires for Methanol and Ethanol Oxidation Reactions: Tuning Performance by Varying Chemical Composition[J]. ACS Applied Nano Materials, 2018, 1(3): 1104–1115.
97. Xiao L, Li G, Yang Z, et al. Engineering of Amorphous PtOx Interface on Pt/WO3 Nanosheets for Ethanol Oxidation Electrocatalysis[J]. Advanced Functional Materials, 2021, 31(28).
98. Zhu Y, Bu L, Shao Q, et al. Subnanometer PtRh Nanowire with Alleviated Poisoning Effect and Enhanced C–C Bond Cleavage for Ethanol Oxidation Electrocatalysis[J]. ACS Catalysis, 2019, 9(8): 6607–6612.
99. Qin Y, Huang H, Yu W, et al. Porous PdWM (M = Nb, Mo and Ta) Trimetallene for High C1 Selectivity in Alkaline Ethanol Oxidation Reaction[J]. Advanced science, 2022, 9(5): e2103722.
100. Zhang G, Cao D, Guo S, et al. Tuning the Selective Ethanol Oxidation on Tensile-Trained Pt(110) Surface by Ir Single Atoms[J]. Small, 2022, 18(33): e2202587.
101. Wu R, Wang L. Activity Enhancement of PtIr Catalysts for Complete Ethanol Oxidation Reaction by Tuning C–O Coupling Abilities[J]. The Journal of Physical Chemistry C, 2022, 126(51): 21650–21666.
102. Sulaiman J E, Zhu S, Xing Z, et al. Pt–Ni Octahedra as Electrocatalysts for the Ethanol Electro-Oxidation Reaction[J]. ACS Catalysis, 2017, 7(8): 5134–5141.
103. Ding K, Zhao Y, Liu L, et al. Pt–Ni bimetallic composite nanocatalysts prepared by using multi-walled carbon nanotubes as reductants for ethanol oxidation reaction[J]. International Journal of Hydrogen Energy, 2014, 39(31): 17622–17633.
104. Zhu Y, Bu L, Shao Q, et al. Structurally Ordered Pt3Sn Nanofibers with Highlighted Antipoisoning Property as Efficient Ethanol Oxidation Electrocatalysts[J]. ACS Catalysis, 2020, 10(5): 3455–3461.
105. Yu X, Liu J, Li J, et al. Phosphorous incorporation in Pd2Sn alloys for electrocatalytic ethanol oxidation[J]. Nano Energy, 2020, 77.
106. Yan S-Y, Huang Y-R, Yang C-Y, et al. Enhanced activity of ethanol oxidation reaction on PtM (M=Au, Ag and Sn): The importance of oxophilicity and surface oxygen containing species[J]. Electrochimica Acta, 2018, 259: 733–741.

107. Lao X, Yang M, Sheng X, et al. Monodisperse PdBi Nanoparticles with a Face-Centered Cubic Structure for Highly Efficient Ethanol Oxidation[J]. ACS Applied Energy Materials, 2021, 5(1): 1282–1290.
108. Seo M H, Choi S M, Seo J K, et al. The graphene-supported palladium and palladium–yttrium nanoparticles for the oxygen reduction and ethanol oxidation reactions: Experimental measurement and computational validation[J]. Applied Catalysis B: Environmental, 2013, 129: 163–171.
109. Du W, Mackenzie K E, Milano D F, et al. Palladium–Tin Alloyed Catalysts for the Ethanol Oxidation Reaction in an Alkaline Medium[J]. ACS Catalysis, 2012, 2(2): 287–297.
110. Li Z, Chen Y, Fu G, et al. Porous PdRh nanobowls: facile synthesis and activity for alkaline ethanol oxidation[J]. Nanoscale, 2019, 11(6): 2974–2980.
111. Li X, You H, Wang C, et al. 3D Taraxacum-like porous Pd nanocages with Bi doping: High-performance non-Pt electrocatalysts for ethanol oxidation reaction[J]. Journal of colloid and interface science, 2021, 591: 203–210.
112. Guo J, Chen R, Zhu F-C, et al. New understandings of ethanol oxidation reaction mechanism on Pd/C and Pd2Ru/C catalysts in alkaline direct ethanol fuel cells[J]. Applied Catalysis B: Environmental, 2018, 224: 602–611.
113. Zhang H, Shang Y, Zhao J, et al. Enhanced Electrocatalytic Activity of Ethanol Oxidation Reaction on Palladium-Silver Nanoparticles via Removable Surface Ligands[J]. ACS applied materials & interfaces, 2017, 9(19): 16635–16643.
114. Zhang Z, Xin L, Sun K, et al. Pd–Ni electrocatalysts for efficient ethanol oxidation reaction in alkaline electrolyte[J]. International Journal of Hydrogen Energy, 2011, 36(20): 12686–12697.
115. Chen J, Yang M, Pang M, et al. Bimetallic PdAg nanoparticles for enhanced electrocatalysis of ethanol oxidation reaction[J]. Colloids and Surfaces A: Physicochemical and Engineering Aspects, 2021, 629.
116. Yang H, Yu Z, Li S, et al. Ultrafine palladium-gold-phosphorus ternary alloyed nanoparticles anchored on ionic liquids-noncovalently functionalized carbon nanotubes with excellent electrocatalytic property for ethanol oxidation reaction in alkaline media[J]. Journal of Catalysis, 2017, 353: 256–264.
117. Shen S Y, Zhao T S, Xu J B, et al. Synthesis of PdNi catalysts for the oxidation of ethanol in alkaline direct ethanol fuel cells[J]. Journal of Power Sources, 2010, 195(4): 1001–1006.
118. Lv H, Wang Y, Lopes A, et al. Ultrathin PdAg single-crystalline nanowires enhance ethanol oxidation electrocatalysis[J]. Applied Catalysis B: Environmental, 2019, 249: 116–125.
119. Pan, Bowei, et al. "PdAuAg alloy nanoparticles on nickel foam as anode for passive air-breathing formate fuel cell." Journal of The Electrochemical Society 168.6 (2021): 064519.
120. Jin T, Chen F, Guo L, et al. Pd4O3 Subsurface Oxide on Pd (111) Formed during Oxygen Adsorption-Induced Surface Reconstruction and Its Activity toward Formate Oxidation Reactions[J]. The Journal of Physical Chemistry C, 2021, 125(35): 19497–19508.
121. Li, Yinshi, Yaling He, and Weiwei Yang. "A high-performance direct formate-peroxide fuel cell with palladium–gold alloy coated foam electrodes." Journal of Power Sources 278 (2015): 569–573.
122. An, Liang, and T. S. Zhao. "An alkaline direct ethanol fuel cell with a cation exchange membrane." Energy & Environmental Science 4.6 (2011): 2213–2217.
123. Zhao, T. S., et al. "Mass transport phenomena in direct methanol fuel cells." Progress in energy and Combustion Science 35.3 (2009): 275–292.
124. An, Liang, T. S. Zhao, and Y. S. Li. "Carbon-neutral sustainable energy technology: Direct ethanol fuel cells." Renewable and Sustainable Energy Reviews 50 (2015): 1462–1468.
125. Takamura, T., and F. Mochimaru. "Adsorption and oxidation of formate on palladium in alkaline solution." Electrochimica Acta 14.1 (1969): 111–119.
126. John, Jimmy, et al. "Mechanistic studies of formate oxidation on platinum in alkaline medium." The Journal of Physical Chemistry C 116.9 (2012): 5810–5820.
127. Sun, S. G., J. Clavilier, and A. Bewick. "The mechanism of electrocatalytic oxidation of formic acid on Pt (100) and Pt (111) in sulphuric acid solution: an emirs study." Journal of electroanalytical chemistry and interfacial electrochemistry 240.1–2 (1988): 147–159.

128. Jiang, Junhua, John Scott, and Andrzej Wieckowski. "Direct evidence of a triple-path mechanism of formate electrooxidation on Pt black in alkaline media at varying temperature. Part I: The electrochemical studies." Electrochimica Acta 104 (2013): 124–133.
129. Christensen, P. A., A. Hamnett, and D. Linares-Moya. "The electro-oxidation of formate ions at a polycrystalline Pt electrode in alkaline solution: an in situ FTIR study." Physical Chemistry Chemical Physics 13.24 (2011): 11739–11747.
130. Rosenbaum, Miriam, Uwe Schröder, and Fritz Scholz. "Investigation of the electrocatalytic oxidation of formate and ethanol at platinum black under microbial fuel cell conditions." Journal of Solid State Electrochemistry 10 (2006): 872–878.
131. Aoki, Hidemitsu, et al. "Improvement of Anode Oxidation Reaction of a Fuel Cell using Ammonium Formate with Pt-Pd Catalysts." ECS Transactions 16.2 (2008): 849.
132. Bartrom, Amy M., et al. "Catalysts for alkaline direct ethanol and direct formate fuel cells." ECS Transactions 50.2 (2013): 1913.
133. Zhipeng Liu, Jinling Xue, and Yinshi Li. "Ultrathin PdCu Nanosheet as Bifunctional Electrocatalysts for Formate Oxidation Reaction and Oxygen Reduction Reaction." Small Methods (2023): 2300021.
134. da Silva, Sirlane G., et al. "PdAu/C electrocatalysts as anodes for direct formate fuel cell." Electrocatalysis 6 (2015): 442–446.
135. Guo, Longfei, et al. "Reconstruction of an AgPd nanoalloy with oxidation for formate oxidation electrocatalysis." Journal of Materials Chemistry A 10.26 (2022): 13998–14010.
136. A.M. Bartrom, G. Ognibene, J. Ta, J. Tran, J.L. Haan, Catalysts for alkaline direct ethanol and direct formate fuel cells, ECS Trans. 50 (2012) 1913–1918.
137. A.M. Bartrom, J.L. Haan, The direct formate fuel cell with an alkaline anion exchange membrane, J. Power Sources 214 (2012) 68–74.
138. Y.S. Li, H. Wu, Y.L. He, Y. Liu, L. Jin, Performance of direct formate-peroxide fuel cells, J. Power Sources 287 (2015) 75–80.
139. Li F, Chan S H, Tu Z. Recent development of anion exchange membrane fuel cells and performance optimization strategies: a review[J]. The Chemical Record, 2024, 24(1): e202300067.
140. Gottesfeld S, Dekel D R, Page M, et al. Anion exchange membrane fuel cells: Current status and remaining challenges[J]. Journal of Power Sources, 2018, 375: 170–184.
141. Ashdot A, Kattan M, Kitayev A, et al. Design strategies for alkaline exchange membrane–electrode assemblies: Optimization for fuel cells and electrolyzers[J]. Membranes, 2021, 11(9): 686.
142. Ren J T, Chen L, Wang H Y, et al. Hydrogen oxidation electrocatalysts for anion-exchange membrane fuel cells: activity descriptors, stability regulation, and perspectives[J]. Energy & Environmental Science, 2024.

Chapter 11
Solid Oxide Fuel Cells

11.1 Introduction

With the increasing global energy demand and a series of problems caused by environmental pollution and climate warming, people are forced to continuously seek new clean alternative energy sources. In recent years, fuel cells have received more and more attention and research. Fuel cells are static energy conversion devices that can convert chemical energy directly into electrical energy through an electrochemical reaction between fuel and oxidizer without the combustion process, and are not subject to the Carnot cycle [1]. Among the different fuel cells, Solid Oxide Fuel Cells (SOFCs) is one of the most efficient power generation technologies because of its flexible fuel choice, lack of noise, low CO_2 emissions, and potentially long life of 40,000–80,000 h [2].

In 1937, the concept of SOFCs was first proposed by Swiss scientists Emil Bauer and Hans Preis [3]. Subsequently, more and more researchers began to study SOFCs more intensively and systematically, and found the problem of oxygen vacancy in doped oxides [4]. In 1957, some scholars proposed the thermodynamic study of a concentrated battery based on solid electrolyte, which laid a certain foundation for the modern solid-state electrochemistry of SOFCs [5].

SOFCs also have many advantages over other types of fuel cells, such as: it is a solid electrolyte that is easy to manage and doesn't cause corrosion for the cell; it can operate at high operating temperature (400–1000 °C) and allow the use of various fuels such as hydrocarbons, hydrogen, carbon monoxide, etc. [6].

Nevertheless, SOFCs still face some difficulties and challenges: SOFCs are characterized by high operating temperature, slow start-up, high costs, and intolerance to sulfur content; they are particularly unsuitable for large fluctuations in load demand, and, therefore, the medium and large power sector is their main application area [7]. Sintering problems and catalytic poisoning of electrodes due to high operating temperatures [8]. It also faces interface problems due to the mismatch in the thermal expansion coefficient of the components, which makes the selection of the appropriate

Y. Li, *Fuel Cell Fundamentals and Applications*,
https://doi.org/10.1007/978-981-96-9845-5_11

manufacturing method an important factor in the development of each component [9]. Considering the harsh working conditions of SOFCs [10], such as high temperature, redox and thermal cycling, and toxic atmospheres, their components (cathode, electrolyte, anode, interconnect, sealant) require a variety of properties, including [11–15]:

(1) Good electrical conductivity;
(2) Good stability and maintain good compatibility with other components;
(3) Ensure that the electrode is a porous and dense electrolyte (for preventing gas mixing);
(4) High electrical conductivity, perfect gas tightness, insulation, and high resistivity for interconnected oxidation resistance, sulfation, and carbon deposition.

The SOFCs have four basic functional components: electrolyte, cathode, anode, and interconnect. The electrons required for the cathode reaction are released from the anode and reach the cathode via an external load, through which electricity is generated. The function of the electrolyte determines that it must be an electronic insulator; in addition, as a support for the cathode and anode, the mechanical and thermal properties of the electrolyte and electrodes should be well matched. At the same time, it should also have good chemical stability [16].

The working principle of SOFCs are shown in Fig. 11.1 [17]. The cathode reduces the oxygen molecules in the oxidizer gas to oxygen ions, and then the oxygen anions (O^{2-}) migrate through the electrolyte interlayer to the anode and react with H_2 at the anode to form water and produce electrons. Finally, these generated electrons are transferred to the cathode via an external circuit to produce an electric current to complete the discharge process. During the chemical reaction, the Gibbs free energy (or chemical potential energy) of the overall reaction of the fuel and oxidizer is converted into electricity and heat [18].

SOFCs allow different cell geometries to be set up, with flat and tubular designs being the most common configurations for practical applications [11]. Planar SOFCs can be in the form of discs or square plates, which have simpler and cheaper manufacturing processes, higher power density, and lower internal resistance due to their short current path. The tubular cell consists of a sandwich electrolyte and electrode array with a specific length and diameter, which has a more robust thermal cycling performance and is easier to seal [19, 20]. In the planar design, oxygen and fuel flow through the tank at an angle from different sides to reach the electrodes. The interconnections on both sides of the electrodes provide proper space for fuel movement and can be used as an anode and cathode gas separator. The main advantages of planar SOFCs over tubular SOFCs are that they can produce higher current densities and are relatively easy to fabricate. However, their limitations include unstable sealing, which leads to fuel and oxidant leakage, poor thermal cycling stability, and relatively high production costs [17].

Both cell designs require sufficient mechanical strength to withstand the operational stresses provided by the support layer. The support layer has the maximum thickness, and the other layers have the minimum thickness to avoid high internal resistance, improve cell efficiency, and reduce cost. Typically, SOFCs have a support

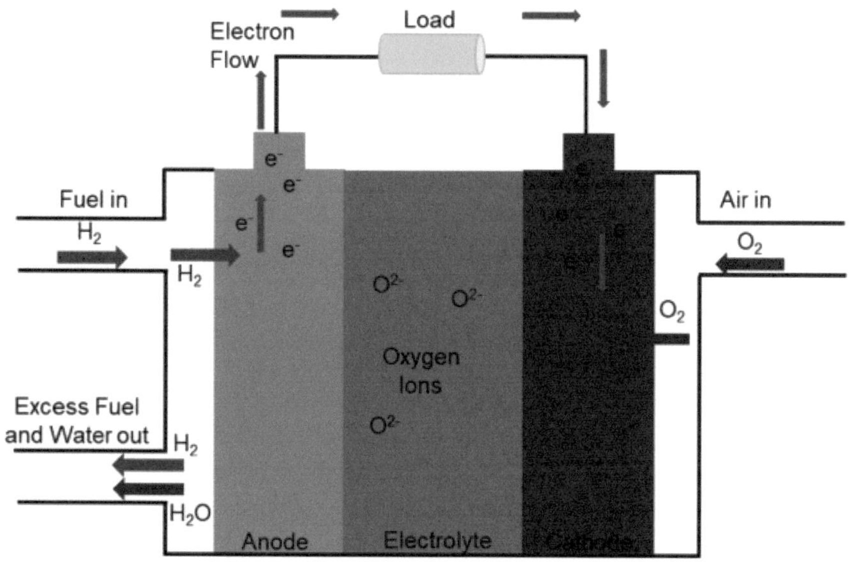

Fig. 11.1 SOFCs working principle diagram [17]

layer, and they can be designed as anode, cathode, or electrolyte support [21, 22]. The planar design mainly uses anode support, while the tubular design uses electrolyte support structure [23]. In a planar SOFCs, the reaction gas diffuses from the center to the periphery into the porous microstructure, while in a cathode-supported tubular cell, the fuel flows externally and the oxidizer flows internally. Notably, there is another geometry for SOFCs stacks, the flat tubular structure, which combines the characteristics of planar and tubular SOFCs into one design, such as high power density, good thermal stability, and easy sealing [24].

11.2 Electrolyte Materials

11.2.1 Stabilized ZrO_2

Electrolyte is the core of ion transport in the SOFCs unit. According to the different types of transported ions, the existing electrolyte systems can be divided into oxygen ion conducting electrolytes (ZrO_2, CeO_2, Bi_2O_3, etc.), proton conducting electrolytes ($BaCO_3$, $BaZrO_3$, etc.), and electrolyte materials with both oxygen ion and proton conducting capabilities. The most commonly used solid electrolyte material in SOFC is yttrium-stabilized zirconia (YSZ), and the addition of yttrium oxide can stabilize the conductivity of the cubic fluorite phase, which increases the ionic conductivity by increasing the concentration of oxygen vacancies. As shown in Fig. 11.2 [11],

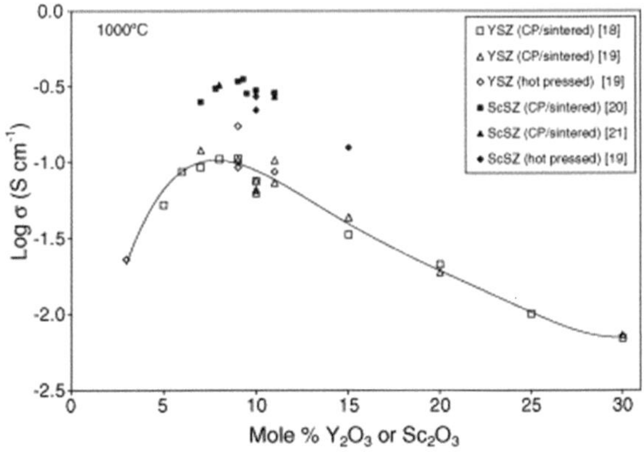

Fig. 11.2 Conductivity of yttrium oxide and scandium oxide stabilized zirconia in air at 1000 °C [11]

the electrical conductivity of YSZ increases with the addition of yttrium oxide, but decreases with increasing yttrium oxide content after reaching a certain dopant content. The decrease at higher dopant content is due to the association of point defects, which leads to a decrease in defect mobility and consequently conductivity. In addition to yttrium oxide stabilized zirconium oxide, scandium oxide is also a promising dopant with higher conductivity than yttrium oxide doping.

Pure ZrO_2 has three different crystal structures, monoclinic, tetragonal, and cubic [9], however, the crystal structure transformation process is accompanied by a change in the crystal volume, which makes the thermal stability of pure ZrO_2 materials very poor. To enhance the thermal stability, it is necessary to select suitable divalent or trivalent cations to replace some of the Zr, because when oxygen vacancies are introduced, oxygen ion conductivity is generated in the oxide lattice, which stabilizes the cubic structure over a large temperature range and can enhance its mechanical properties and thermal stability. As shown in Fig. 11.2, the two oxygen vacancies created in the lattice provide channels for oxygen ion conduction. It has been shown that the crystal conductivity significantly depends on the concentration of oxygen vacancies, which is determined by the amount of doping. It was found that at lower concentrations, the conductivity increases with increasing doping levels and reaches a maximum at 8%; however, if the doping level exceeds 8%, especially at lower temperatures, the oxygen ion conductivity decreases significantly. In addition to the temperature limitation, when using YSZ as an electrolyte, it should be ensured that the film is thin enough to reduce the ohmic resistance and get a good power output [16].

To improve the ionic conductivity of stabilized zirconia, different dopants have been extensively tested for zirconia oxides in the current study, and the results show that doping with trivalent ions (rare earth metals) is more effective than divalent ions (alkali metals), mainly because cubic fluorite has a higher propensity for defect conjugation and lower thermodynamic stability. However, a major challenge in using stable ZrO_2 as an electrolyte is the very limited choice of cathode materials. Cathodes containing Ba, La, and other elements were found to react readily with YSZ, which led to a significant increase in electrode polarization resistance and ohmic resistance, as well as poor long-term cell stability. Therefore, it required the application of a buffer layer to avoid direct contact between the YSZ electrolyte and the cathode. Currently, doped cerium oxide is widely used as a buffer layer due to its high oxygen ion conductivity and chemical inertness to the electrolyte and cathode [25].

11.2.2 Acceptor Doped CeO_2

Cerium dioxide is a common electrolyte material for SOFCs. Like zirconia, cerium oxide also forms a fluorite structure, but has better ionic conductivity and lower polarization resistance at lower operating temperatures than zirconia [14, 26]. However, unlike the multiphase nature of ZrO_2, CeO_2 can maintain its cubic fluorite structure from room temperature to its melting point. Pure CeO_2 has negligible oxygen vacancies in its crystal structure and thus has a very low oxygen ion conductivity. To improve its oxygen ion conductivity at intermediate temperatures, current research suggests that trivalent or divalent cations (such as Gd^{3+}, Sm^{3+}, Y^{3+}, La^{3+} or Ca^{2+}) can be used to partially replace Ce^{4+} in the crystal structure, leading to the creation of more oxygen vacancies that are carriers of oxygen ions, and the ionic radius of the dopant cation is closer to Ce^{4+}, the higher the expected oxygen ion conductivity [27, 28]. Although cerium oxide is less stable than zirconium oxide at low oxygen partial pressures, cerium oxide with cathode material is more chemically stable than zirconium oxide.

At moderate temperatures, the oxygen ion conductivity of doped CeO_2 materials is much higher than that of stable ZrO_2. Many studies on the conductivity of doped CeO_2 electrolytes have been conducted by many researchers over several years, but the reported conductivity varies widely. There are two main reasons for this difference: one is the relative density of the electrolyte, and the other is the impurities in the electrolyte. The sintering of the electrolyte layer is related to the nature of the initial powder and the sintering temperature. The addition of Al_2O_3, ZnO, and Fe_2O_3 can effectively improve the sintering performance of doped CeO_2, thus achieving higher oxygen ion conductivity [29–32]. The purity of the electrolyte is another important issue. A common impurity in electrolytes is silicon oxide, which is commonly found in chemicals, water, and the environment. It was found that silicon oxide is usually deposited at the grain boundaries of the sintered electrolyte, which hinders the conduction of oxygen ions at the grain boundaries [33, 34]. Therefore, the purity of the product, the silicon content of the liquid used, and the sintering environment should be strictly controlled during the preparation of the electrolyte.

11.2.3 $LaGaO_3$-Based Solid Electrolytes

Calcium titanite structure (ABO_3) oxide has ion–electron hybrid conductivity and is a low-temperature electrolyte material with good overall performance. Chalcogenides such as $LaBO_3$ exhibit high oxygen ion conductivity at medium temperatures, where $LaGaO_3$-based solid oxides are the most studied chalcogenide oxide electrolytes [35, 36]. The ionic conductivity of $LaGaO_3$ itself is very low, but by doping divalent alkaline earth cations (such as Sr_{2+}, Ba_{2+}, Ca_{2+}, etc.) at the A-site, a large number of oxygen vacancy defects are created, especially by partially replacing La^{3+} with Sr^{2+}, which has a similar ionic radius to La^{3+}, and the oxygen ionic conductivity can be substantially increased. However, the oxygen ion conductivity doesn't always increase with the increase of the doping amount. When its doping amount is greater than 10%, it will instead produce an impurity phase and affect the conduction of oxygen ions.

The high ionic conductivity $La_{1\text{-}x}Sr_xGa_{1\text{-}y}Mg_yO^{3-}{}_z$ (LSGM) electrolyte, co-doped by Sr and Mg, is one of the most studied SOFCs chalcogenide oxide electrolytes, with ionic conductivity higher than typical YSZ or CeO_2-based solid electrolytes and slightly lower than Bi_2O_3-based solid electrolytes, but with good thermodynamic stability at high temperature oxidation or fuel reduction, and high ionic conductivity. However, despite the good conductivity performance of LSGM, some disadvantages limit the application of this system electrolyte in SOFCs, such as: the high cost of crop oxide, which can't be used in large quantities and is volatile, and LSGM is difficult to densify and is easily reduced to produce impurity phases, thus reducing the conductivity of the material. In addition, LSGM can easily react with a metal anode and form low low-conductivity phase at the interface, and these challenges make LSGM not much used in SOFCs [37].

11.3 Cathodes

11.3.1 Overview of Cathode Reaction Mechanism

The electrochemical reaction at the SOFCs cathode can only take place at the three-phase interface (TPB) between the oxygen ion conductor, the electron conductor, and the gas confluence [38], as shown in Fig. 11.3 [39]. If any one of the three phases is not in contact with each other, then no oxygen reduction reaction can occur [40]. Blocking the approach of ions, electrons, or gases to the reaction sites can prevent them from reacting, so microstructure and electrode materials have an important influence on the size and distribution of TPB. There are two main ways to enlarge the area of TPB, either by choosing a mixed ion–electron conductor (MIEC) material, a cathode material that allows both ion and electron migration, so that the oxygen reduction reaction (ORR) can take place on the electrode surface and within the electrode, or by mixing electron-conducting and ion-conducting materials as cathode materials.

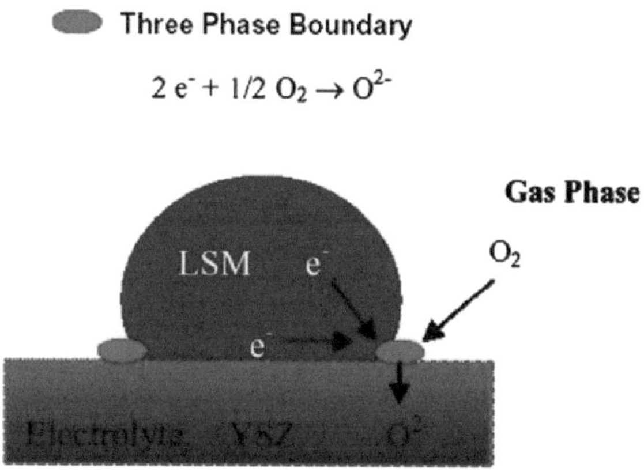

Fig. 11.3 Schematic diagram of the cathode triple-phase boundary [39]

By employing these strategies, the electrochemical reaction is enhanced by several orders of magnitude over the electron-conducting cathode material. The extent of this enhancement depends mainly on the defect conduction rate of the solid MIEC, the degree of gas coverage on the surface of the MIEC, and the catalytic activity of the interface.

Although the molecules involved in the overall electrochemical reaction at the cathode are diatomic species (O_2), they must first be converted by one or more processes to some "electroactive" intermediate form. Usually, this electrochemical kinetic step is restricted to the vicinity of the TPB region [41]. The electrochemical processes in SOFCs cathodes are very complex [42, 43], and generally, researchers consider the basic reactions as follows [44, 45]:

$$O_2 \rightarrow 2O_{ad} \tag{11.1}$$

$$O_{ad} + e^- \rightarrow O_{ad}^- \tag{11.2}$$

$$O_{ad}^- \rightarrow O_{TPB}^- \tag{11.3}$$

$$O_{TPB}^- + e^- \rightarrow O_{TPB}^{2-} \tag{11.4}$$

$$O_{TPB}^{2-} + V_O^{\cdot\cdot} \leftrightarrow O_O^{\times} \tag{11.5}$$

The oxygen reduction step (11.2) is the main factor leading to excessive cell resistance, and improving the cathode catalytic activity has an important impact on the performance of the cell [46]. There are three reaction paths for electron-conducting cathode materials [47], namely the electrode surface path, the bulk path, and the electrolyte surface path, and the reaction paths are shown in Fig. 11.4. It is worth noting that LSMs are p-type semiconductors only at higher Po_2 than the SOFC cathode

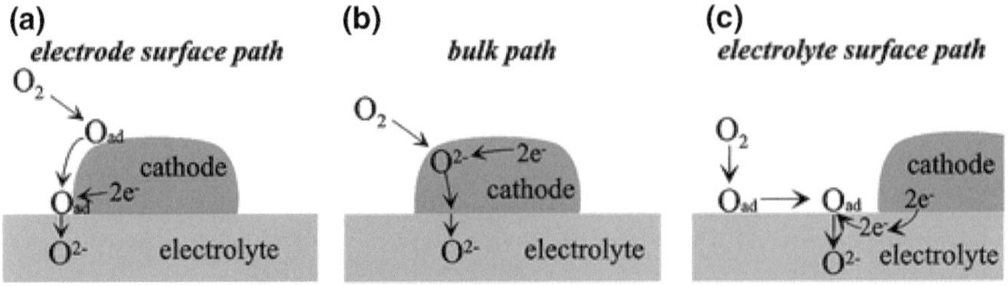

Fig. 11.4 Three reaction paths of oxygen reduction: **a** electrode surface path; **b** bulk path; **c** electrolyte surface path [48]

operating conditions; however, oxygen vacancies are formed in LSMs only as Po_2 decreases and the temperature increases. The decisive step of the oxygen reduction reaction is not always determinable, as it may also depend on local conditions, such as the operating temperature, the oxygen partial pressure, and the microstructure of the cathode. Despite the complexity of the reaction mechanism, the study of the oxygen reduction mechanism still provides important data for the study of SOFCs, especially for well-defined SOFCs systems [48]. Dense films [49, 50] and micro-electrodes with well-defined geometries [51, 52] allow an in-depth study of ORR kinetics through the quantitative relationship between the cathode geometry and the impedance of the oxygen reduction reaction.

11.3.2 Cathode Materials

The requirements of SOFCs for cathode materials mainly include [53, 54]: (1) High electron conductivity and ionic conductivity. The high electronic conductivity of the cathode, the lower the ohmic loss during electron transport, and the sufficient oxygen ion conductivity improve the surface reaction and ion diffusion ability; (2) High catalytic activity. The cathode performs the function of a catalyst, which has to open the covalent bonds of oxygen molecules, and therefore must have a sufficiently high catalytic activity for oxygen reduction, which is associated with the chemical composition of the cathode material and the cathode microstructure; (3) Good compatibility. The cathode is required at room temperature, and in the temperature range of cell operation and preparation, there should be no chemical reaction with adjacent components (electrolyte, linker, etc.). There should be no significant inter-diffusion and a similar coefficient of thermal expansion. Concerning the materials researched so far, the main cathode materials are metal, metal-ceramic composite cathode materials, and oxide materials of chalcogenide structure.

11.3.2.1 Metal and Cermet Cathode Materials

These materials are generally worked at low and medium temperatures. Pt-based precious metal is a cathode material used in early research. In addition to Pt, precious metals suitable for cathode materials also include Pd、Rh, and so on. However, these materials are expensive, so research on such materials has focused on cost reduction. Ceramic-based materials metal-$Sc_{0.10}Ce_{0.01}Zr_{0.89}O_2$ (SSZ), were prepared by the vacuum high-energy ball milling method [55], in which Pt, Pd, Rh, and Ag and their alloys were used as electron conducting phases. The material was used with the electrolyte Y_2O_3-doped ZrO_2(YSZ) and showed good cathodic activity. The interfacial conductivity of the Pt-SSZ cathode material was up to 6.7 S/cm^2 at 700 °C with a Pt content of 40 mg/cm^2. The composite obtained by replacing Pt with a certain mass ratio of Pt–Ag alloy showed improved performance, with an interfacial conductivity of 12 S/cm^2 at 700 °C. The cathodic reaction activation energy of the alloy (Pt–Ag, Pd–Ag)/SSZ material is smaller than that of the Pt/ SSZ material. Further research has been carried out on Ag composites, and it is found that the cathode material with good low and medium temperature performance can be produced by using relatively low-priced Ag compounded with Y_2O_3-doped Bi_2O_3 (YDB). The polarization resistance reaches a minimum value of 1 Ω cm^2 when the Ag content reaches about 50% [56]. If other materials can be developed to match, this is also a promising cathode material.

11.3.2.2 La-Based Perovskite Oxides

The cathode material of SOFCs generally adopts the perovskite material [57]. To better design and optimize the cathode material, it is necessary to first understand the basic structure of perovskite. ABO_3 perovskite has unique structural properties, and this material has a general structure as shown in Fig. 11.5 [58]. Where A and B represent metal cations, A cations with lower valence and larger ionic radius (eg La, Ba or, Sr etc.) coordinate with twelve oxygen anions, while B cations (transition metals such as: Mn, Fe, Co or Ni) have a higher valence state and a smaller ionic radius and coordinates with six oxygen anions. When the total valence state of the A-site and B-site cations is less than 6, oxygen vacancies are generated to allow the material to achieve charge conservation [59, 60]. From the effect of A and B elements on the material, the properties of A-La elements, and the amount of alkali earth metal substitution affect the electrochemical properties of the cathode and the catalytic activity of oxygen reduction. The reaction rate of the cathode is affected by the transition metal at the B position, and the order is Co > Mn > Fe > Cr [61]. The most representative ones are doped $LaMnO_3$ and $LaCoO_3$.

Mn-Based Perovskite Cathode

Considering the catalytic activity and compatibility with related components, lanthanum manganate (LSM) doped with strontium or calcium titanite type is the best cathode material known to match with Y_2O_3-doped ZrO_2(YSZ) [62]. Pure $LaMnO_3$ is

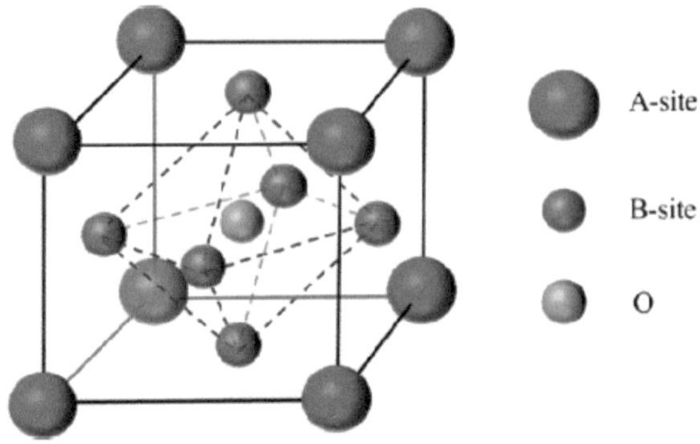

Fig. 11.5 Structure of ABO_3 type perovskites [58]

a semiconductor material with a p-type chalcogenide structure, and when Sr partially replaces the position of La, it generates small polaritons by charge compensation, and the conductivity of the cathode material is increased by this jump of small polaritons. LSM has high electronic conductivity (higher than 100 S/cm) and catalytic activity (specific resistance of LSM/YSZ interface is less than 1 Ω cm^2 at 1000 °C) and good chemical stability at high temperatures of 800–1000 °C. These advantages make LSM the cathode material with the highest degree of application at present.

In addition to lanthanum-Mn-based materials, the possibility of other Mn-based oxides as cathode materials has been investigated. Sr-doped $PrMnO_3$(PSM) has a high electronic conductivity and low cathodic overpotential [63]. The cathodic overpotential decreases with increasing Sr doping, and the conductivity increases with increasing Sr doping below 40%. PSM can obtain the same power density as $La_{0.6}Sr_{0.4}MnO_3$ (LSM) at operating temperatures below 100 °C. Moreover, PSM reacts less with YSZ than LSM, and the coefficient of thermal expansion is closer to that of YSZ, so $Pr_{0.6}Sr_{0.4}MnO_3$ is promising to replace LSM. The cathodic properties of $Ln_{l-x}Sr_xMnO_3$ (Ln = La, Pr, Nd, Sm, and Gd, $0 \leq x \leq 0.5$) system materials were studied [64, 65] and found that the conductivity of all materials was above 200 S/cm at 1000 °C. At a temperature of 1400 °C and a Sr doping of 30%, no insulating phase was produced by co-sintering $Ln_{l-x}Sr_xMnO_3$ (Ln = Pr, Nd, and Sm) with YSZ. Among this series of materials, manganates of Pr and Nd have more desirable cathode performance and have the possibility of replacing LSM as a high-temperature SOFC cathode material.

Co-Based Perovskite Cathode

Cobalt-based perovskite structure materials used as the cathode in SOFCs are materials in which cobalt is at the B-site of the ABO_3 structure. The double perovskite-structured materials can also be used as cathode materials. These materials exhibit high electro-catalytic activity along with appropriate electronic and ionic conductivity. For the identical conditions, the conductivity of $LaCoO_3$ is much larger than that of $LaMnO_3$. Cobalt-based materials also help in reducing the cathode polarisation resistance. Along with these properties, they act as good catalysts for oxygen reduction in the operations of SOFCs. As reported in the literature, cobalt-based materials have higher conductivity values than cobalt-free perovskite materials [66]. Some of the promising cathode materials are differentiated based on the A-site elements. In addition to A-site rare earth-alkaline-earth doping, co-doping with B-site transition metals has also been investigated. $Ba_{0.5}Sr_{0.5}Co_{0.8}Fe_{0.2}O_{3-\delta}$ (BSCF) was prepared by completely replacing La series elements with alkaline earth elements, showing excellent cathode properties [67]. The good catalytic activity and oxygen conductivity at medium and low temperatures make this material promising as a cathode material for medium-temperature SOFCs, but the biggest problem is that the thermal expansion coefficient is too large, which is about twice that of the electrolyte materials YSZ and Gd-Doped CeO_2 (GDC) under the same conditions. To adjust the thermal expansion coefficient of $LaCoO_3$ and improve the long-term stability, Fe replaces Co in an attempt to reduce the thermal expansion coefficient. At 100–900 °C, the thermal expansion coefficient of $La_{0.8}Sr_{0.2}Co_{0.2}Fe_{0.8}O_3$ is 15.4×10^{-6}/°C [68]. When a certain amount of vacancies is introduced at the A position, the thermal expansion coefficient can be further reduced, such as $La_{0.6}Sr_{0.2}Co_{0.2}Fe_{0.8}O_{3-\delta}$ at 700 °C. The thermal expansion coefficient at 13.8×10^{-6}/°C matches the thermal expansion coefficient (13.5×10^{-6}/°C) of the intermediate temperature electrolyte $Ce_{0.8}Gd_{0.2}O_{1.9}$ in this temperature range. This provides a way to reduce the thermal expansion coefficient approach, but the related properties still need further study.

Fe-Based Perovskite Cathode

Fe-based oxides are also the most studied cathode materials. Although the electrical conductivity of Fe-based materials is not as good as that of Co-based materials, the thermal expansion coefficient close to that of the electrolyte makes it an option to replace LSCs in the mid-temperature range. However, the conductivity of $LaFeO_3$ is not high, and it needs to be doped with metal ions (such as Sr) to increase the porosity and conductivity [69]. The electrical properties of $(La_{1-x}Sr_x)_{0.99}FeO_{3-\delta}$ ($x = 0$, 0.05, 0.15, 0.25, 0.35, 0.50, 0.70) (LSF) cathodes were studied with $Ce_{1.9}Gd_{0.1}O_{1.95}$ as electrolyte chemical properties [70]. The TEC of $(La_{1-x}Sr_x)_{0.99}FeO_{3-\delta}$ cathode increases with the increase of Sr content. The ASR of the $(La_{1-x}Sr_x)_{0.99}FeO_{3-\delta}$ cathode first decreased with the increase of the Sr content. When the Sr content was $x = 0.15$, the ASR had a minimum value, and then the ASR increased with the continuous increase of the Sr content. As shown in Fig. 11.6. In addition, the conductivity of the Sr and Ni double-doped $(La_{1-x}Sr_x)_{0.99}FeO_{3-\delta}$ cathode increased from 90 S/cm (LSFN0) to a maximum of 200 S/cm (LSFN0.2) at 700 °C, and then descended [69]. This is because the conductivity is related to the dopant content. An appropriate doping amount can increase the conductivity and reduce the impedance, while an excessive doping amount will increase the concentration and increase the concentration polarization resistance.

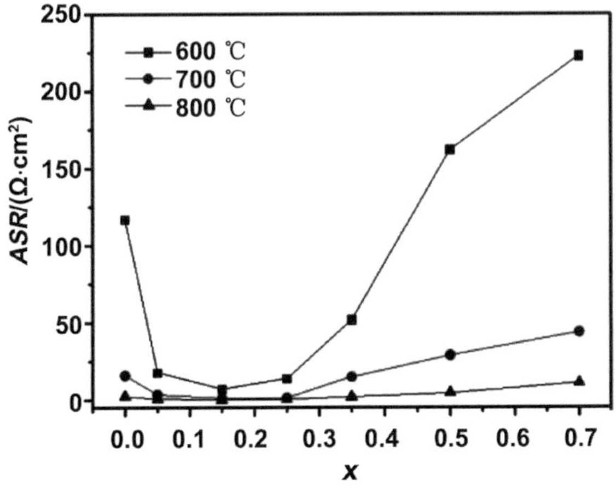

Fig. 11.6 Relationship between ASR and Sr content of $(La_{1-x}Sr_x)_{0.99}FeO_{3-\delta}$ (x = 0, 0.05, 0.15, 0.25, 0.35, 0.50, 0.70) cathode [70]

11.3.2.3 Ba-Based Perovskite Oxides

$Ba_{0.5}Sr_{0.5}Co_{0.8}Fe_{0.2}O_{3-\delta}$ (BSCF) can achieve high power density output at lower operating temperatures with very low ASR [71]. The conductivity of this system using SDC electrolyte is 4.7×10^{-3} S cm^{-1} at 500 °C, which is higher than the highest conductivity of 9.5×10^{-3} S cm^{-1} in other doped ceria [72]. Therefore, the resistance of the electrolyte can be reduced by optimizing the fabrication process or reducing the thickness to about 10 μm [73]. Although BSCF has a higher maximum power density, the oxygen flux-induced phase transition of the material needs to be investigated to evaluate the long-term stability of the material. Recently, researchers have investigated the redox properties of perovskite BSCF and LSCF by coulometric titration [74]. And the electrical conductivity of BSCF and LSCF increases with the increase of oxygen partial pressure, indicating that these two materials are *p*-type semiconductors, which are sufficient to be used as SOFCs cathode material according to their good electrical conductivity and redox stability. However, the wide applications of BSCF and LSCF are seriously hindered by the unstable crystal structure and low power density at mid-temperature, especially at the working temperature below 850 °C, where the oxygen permeation flux of BSCF increases with the working time. Under the working conditions of SOFCs, BSCF will transform from a cubic structure to a hexagonal structure in the long-term durability test. To improve the stability of the cubic structure, the Goldschmidt tolerance factor (t) of perovskite should be close to 1 [75]. Therefore, some researchers have investigated lanthanide-doped BSCF and found that lanthanide doping can significantly improve the stability and electrochemical performance of the structure.

11.3.2.4 Double Perovskite Cathode

These materials are represented by the general structure $LaBaMO_{5+\delta}$, where Ln is a lanthanide metal and M is a transition metal. In this double perovskite $AA'B_2O_6$ crystal structure, Ln and Ba occupy the A site of the perovskite structure, and M occupies the B sites. Due to the large size difference between Ba and Ln, the A-site cations are arranged in an orderly manner, the dodecahedral-coordinated LnO and BaO layers are separated by the octahedral-coordinated MO_6 layer, and the structure of $LaBaMO_{5+\delta}$ is shown in Fig. 11.7 [76].

The structural anisotropy of this class of cobaltates is the result of the ordering of oxygen vacancies, which can lead to new modes of migration [77]. Researchers have reported the oxygen exchange properties of some manganese-based perovskites. The crystallographic order between lanthanides and alkaline earth ions can allow oxygen [78], the diffusion rate in $Gd_{0.5}Ba_{0.5}MnO_{3-\delta}$ increases by an order of magnitude. Transforming simple cubic perovskites into layered $GdBaMn_2O_{5+\delta}$ reduces the strength of oxygen bonding and provides disordered channels for ion migration.

The stability of the $LnBaCo_2O_{5+\delta}$ layered perovskite structure is closely related to the ionic radius of the Ln^{3+} cation [79]. In the A-site cation-ordered $LnBaCo_2O_{5+\delta}$ oxides, the bulk diffusion of oxygen is mainly in the LnO_δ plane, while the oxygen diffusion through the CoO_2 plane or in the direction perpendicular to the LnO_δ plane is not obvious, indicating a significant change in the oxygen diffusion path, possibly from changes in the polycrystalline microstructure. The oxygen transport path may

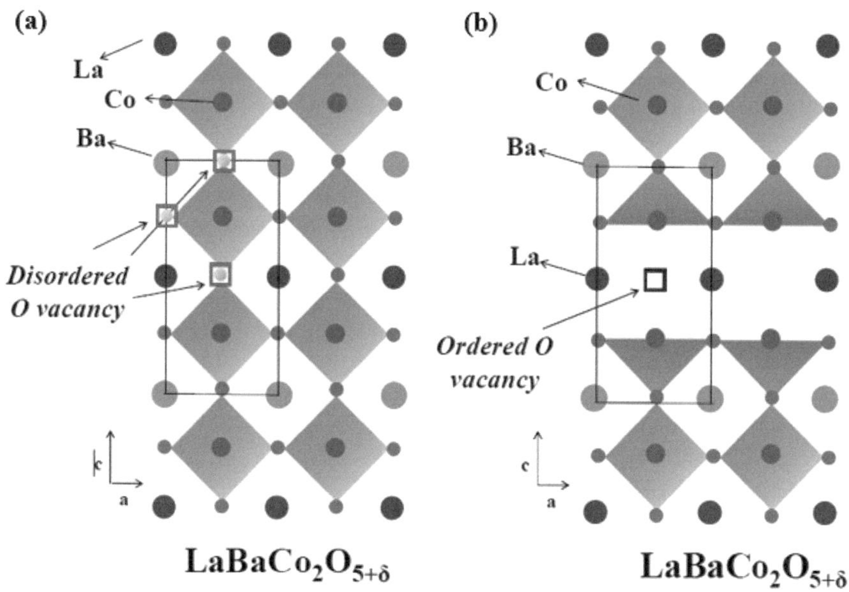

Fig. 11.7 The structure diagrams of double perovskite, **a** tetragonal P4/mmm $LaBaMO_{5+\delta}$; and **b** orthorhombic Pmmm $LaBaMO_{5+\delta}$ [76]

be much longer than the dense film thickness. This explains why the oxygen ion conductivity calculated from the transmembrane oxygen flux is lower than the actual oxygen ion conductivity. The redox performance, structural stability, and electrical conductivity of $NdBaCo_2O_{5+\delta}$/YSZ composite as a novel SOFCs cathode were also reported [80]. The redox properties, including oxidation enthalpy and entropy, were investigated by coulometric titration, and the isotherm of the cathode material was determined over a wide range of PO_2. Oxygen non-stoichiometric (δ) analysis showed that these composites rapidly reacted to a reduced state with a PO_2 value of 10^{-5} atm at 750 °C, and decomposed at 10^{-5} atm. The electrical conductivity of the NBCO/YSZ composite material can exceed 20 $S{\cdot}cm^{-1}$ in the range of 650–750 °C, and when the PO_2 is above 10^{-5} atm, it can meet the requirements of the cathode material.

The double perovskite cathode material has good electrochemical performance at low temperature, but the research that simply pursues high catalytic activity may not achieve the desired effect in the actual application process. The research of perovskite anion materials must comprehensively consider electrochemical performance, thermal expansion coefficient, estoppel, material stability, and adaptability to the operating environment.

11.3.3 New Development on Cathode Materials

In addition to traditional perovskite materials, new Ruddlesden-Popper (RP) type perovskite-like structure oxides have attracted the attention of researchers and are expected to be candidate materials for cathode materials. The general structure of R-P type material is A_{n+1}-MnO_{3n+1} (A = RE, Sr, Ba; M = Co, Ni). Its structure can be regarded as the alternating superposition of ABO_3-type perovskite structure and AO-type rock-salt structure in the c-axis direction. Since there is a large enough space in the middle of the AO layer, it can accommodate an excess amount of oxygen as an oxygen interstitial material. This interstitial oxygen has a negative charge, which can be balanced by the oxidation of the B-site rigid ion, so that the whole material exhibits electrical neutrality. Due to the high interstitial oxygen concentration, such materials have high oxygen diffusion coefficients and surface exchange coefficients, which significantly improve the activity of cathode materials. The most commonly used system is the n = 1 phase, which has a typical K_2NiF_4 structure [81]. In addition to the fast oxygen migration rate, the lattice expansion of $La_2NiO_{4+\delta}$ caused by changes in temperature and oxygen partial pressure is small, and the thermal expansion coefficient of La_2NiO_4 (11×10^{-6} to 13×10^{-6} K^{-1}). It can be compared with YSZ, CGO, and LGSM.

Similar to perovskite-structured materials, their properties can be optimized by doping La or Ni sites. Doping Sr at La site reduces the oxygen self-diffusion coefficient of $La_{1.9}Sr_{0.1}NiO_{4+\delta}$. The surface exchange and self-diffusion coefficients of $La_{2-x}Sr_xNiO_{4+\delta}$ were measured [82], and and found that the oxygen diffusion coefficient of $La_2NiO_{4+\delta}$ is higher than that of $La_{0.6}Co_{0.2}Fe_{0.8}O_{3-\delta}$, 1 order of magnitude lower than that of $La_{0.3}Sr_{0.7}CoO_{3-\delta}$. The study of different rarefied ions in the

A site shows [83] that $Pr_2NiO_{1+\delta}$ has a lower cathode ASR value. Nd instead of La can improve the chemical stability of these compounds and YSZ [84]. When $Nd_{1.05}NiO_{3+\delta}$ and $Nd_{2-x}Sr_xNiO_{4+\delta}$ are used as cathodes, they do not react with the electrolyte. Commonly used substitute ions for Ni sites are Co, Fe, Cu, etc. The ASR of the system doped with Cu at the Ni site is lower than that of $La_2NiO_{4+\delta}$ without Cu [85]. Although $La_2NiO_{4+\delta}$ has good cathode electrocatalytic performance and thermal matching performance. However, the performance of the reported SOFCs is still lower than expected. The main reason may be due to the compatibility issue with electrolytes, so the application of such materials still needs further research.

Ru oxide with pyroclase structure is a kind of cathode material with high catalytic activity. The cathode materials of $Pb_2Ru_2O_{6.5}$ and $Bi_2Ru_2O_7$ were studied [86]. Although these two materials have a larger cathode overpotential than LSM, their electrical conductivity is better than that of LSM, and they have good chemical and thermal compatibility with the electrolyte YSZ. In addition, there are spinel-type structure cathode materials. The properties of spinel (AB_2O_4) type oxide materials are mainly caused by the distribution of divalent (Mg^{2+}, Zn^{2+}, etc.) and trivalent (Al^{3+}, Co^{3+}, Cr^{3+}, etc.) ions. In the AB_2O_4 structure, oxygen ions are arranged in cubic close packing, divalent cations are located in tetrahedral voids, and trivalent cations are located in octahedral voids. Spinel materials have strong chemical stability and corrosion resistance at high temperatures. Due to the advantages of high strength and good thermal stability, they have received extensive attention as a cathode material [87, 88]. For example, $Mn_{1.5}Co_{1.5}O_4$(MCO) is an excellent SOFCs cathode material due to its good thermal and structural stability as well as high electronic conductivity [89].

11.4 Anodes

11.4.1 Overview of Anode Reaction Mechanism

Because the performance of SOFCs depends strongly on the anode structure, it is useful to consider how the anode works on a microscopic scale [90, 91]. The electrochemical reaction can only occur at the three-phase boundary (TPB), which is defined as the line at which the electrolyte, the electron-conducting metal phase, and the gas phase all come together. A cartoon of the region between the electrolyte and the anode where the TPB exists is shown in Fig. 11.8 [92, 93]. Hydrogen is oxidized at the anode to form hydrogen ions, which combine with oxygen ions transferred from the cathode through the electrolyte to form water vapor. If there is a breakdown in connectivity in any one of the three phases, the reaction cannot occur. If ions from the electrolyte cannot reach the reaction site, if gas-phase fuel molecules cannot reach the site, or if electrons cannot be removed from the site, then that site cannot contribute to the performance of the cell. While the structure and composition affect the size of the TPB, various theoretical and experimental methods have been used to estimate

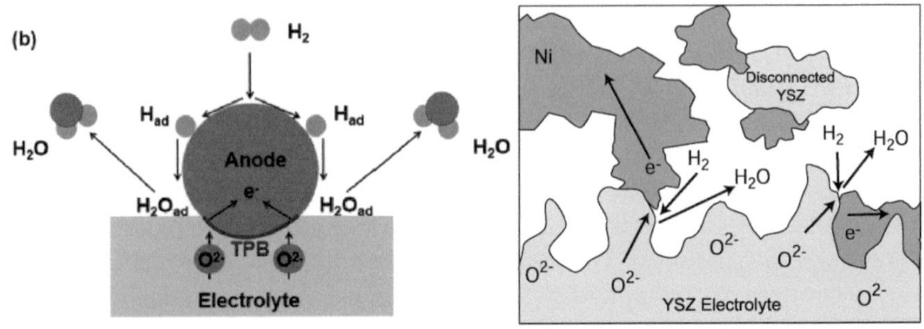

Fig. 11.8 Schematic diagram of the anode reaction process in SOFCs [82]; Schematic of Ni/YSZ anode three-phase boundary [93]

that, under normal conditions, the region in which the TPB exists extends no more than approximately 10 μm from the electrolyte into the electrode [94]. Essentially, so long as the diffusion of ions through the electrolyte partially limits the performance, the concentration of excess ions in the oxide phase of the anode will be insignificant beyond this distance. The electrochemical reaction process in SOFCs anodes is as follows:

$$H_2 \rightarrow 2H_{ad} \tag{11.6}$$

$$2H_{ad} \rightarrow 2H^+_{ad} + 2e^- \tag{11.7}$$

$$2H^+_{ad} + O^{2-} \rightarrow H_2O \tag{11.8}$$

11.4.2 Anode Materials

The anode of SOFCs is where the electrochemical reaction of oxygen ions and fuel gas occurs. Considering its operating environment and function, the anode of SOFCs should meet the following conditions [95, 96]: (1) The anode material needs to have good catalytic activity for the oxidation reaction of various fuels, thereby reducing the activation polarization during the oxidation of the anode fuel. (2) The anode material needs to have high electronic and ionic conductivity, thereby reducing the ohmic polarization caused by electron conduction in the anode and increasing the anode reaction rate. (3) The anode needs to have a good pore structure and suitable porosity to ensure that the fuel gas can quickly diffuse to the three-phase interface where the electrochemical reaction occurs, and the reaction products can be quickly transferred out, reducing the concentration polarization on the anode side. (4) Surrounding electrochemical components, such as electrolytes, have similar thermal expansion coefficients and good chemical compatibility. Avoid delamination, shedding or fracture of the cell, and avoid the formation of impurity phases with high impedance to damage the performance of the cell. (5) Under the high working temperature and strong reduction conditions of the fuel cell, it has good physical and chemical stability. At present, SOFCs anode materials mainly include Metal, metal (Ni, Cu)-based cermet anodes, and perovskite oxide anodes.

11.4.2.1 Metal

The oxygen partial pressure of the anode is very low due to the constant supply of fuel gas on the anode side of the SOFCs, which makes it possible to use metal as the anode material. The metal used as the anode material must be stable at the fuel feed port in a reducing atmosphere, and cannot be oxidized at the product discharge port in a weakly oxidizing atmosphere. Moreover, the operating temperature of SOFCs is very high, so the suitable metals are limited to Ni, Fe, Co, and some precious metals [97]. Metal Co is a good anode material, and its electrochemical catalytic activity is better than Ni, but due to the high price of Co, it is rarely used in SOFCs. Although Fe is relatively cheap, it is easily oxidized to Fe_2O_3 and loses its catalytic activity. In contrast, Ni is widely used as an anode material for SOFCs due to its cheap price, excellent catalytic performance, and relatively high chemical stability.

Because pure metal anodes cannot conduct O^{2-}, the electrochemical reaction of the fuel can only take place at the interface of the anode and the electrolyte. Not only that, the thermal expansion matching between the metal anode and the electrolyte is not good. After many cycles of heating and cooling, cracks are easily generated at the interface, resulting in electrode peeling. In addition, the metal anode is easy to sinter and vaporize at high temperatures, which will seriously affect the working performance of the cell.

Metal-based cermet anode materials

At present, the most widely used SOFCs anode material is a composite material formed by dispersing a catalytically active metal component into an electrolyte material matrix and sintering, which is referred to as a cermet. It can meet the requirements of anode to electrical conductivity, electrocatalysis, chemical stability, and thermal conductivity. The requirements of properties, microstructure, and thermal expansion coefficient can also minimize the effects of carbon deposition and sulfur poisoning. Ni is regarded as the metal of choice for cermet composites due to its electrochemical properties for catalyzing hydrogen oxidation and hydrocarbon reforming [98]. The composite material formed by it and the electrolyte material can well match the thermodynamic properties of the electrolyte, and can also prevent the agglomeration of Ni particles, thereby avoiding the reduction of the surface area and the three-phase interface area. Because of its cheap price and high stability, metal is often mixed with electrolyte yttria-stabilized zirconia to make porous cermet (M-YSZ). The skeleton structure of the YSZ electrolyte provides a carrier for the attachment of metal particles, which effectively inhibits the high-temperature sintering properties of Ni and ensures the porous structure of the anode, thereby maintaining the high electronic conductivity and catalytic activity of the metal anode, and improving the ionic conductivity of the material. In addition, the electrochemical reaction of the YSZ framework extends from the original two-phase interface to the three-phase interface, thereby increasing the area of the electrochemically active region. Among them, metal-based ceramic anode materials are mainly researched on Ni-based and Cu-based cermet materials.

Ni-YSZ cermet anode materials

Ni-based cermet anode materials have good catalytic activity for hydrogen oxidation and hydrocarbon fuels, high electronic conductivity, good chemical stability, and low price, which are mainly due to the conductive and catalytic effects of metallic nickel. At present, the most widely used anode material is Ni/YSZ [99]. The electrochemical reaction of the anode occurs at the contact position between Ni and YSZ, that is, near the three-phase interface of Ni-YSZ-fuel. The metal Ni plays the role of electronic conduction and catalytic oxidation, and YSZ mainly plays the role of oxygen ion conduction and inhibits the sintering and growth of Ni particles. Ni also has the advantages of low price, high catalytic activity, good thermal stability, etc. Under normal circumstances, the conductivity of the SOFCs anode reaching more than 100 S/cm can meet industrial applications. The conductivity of Ni/YSZ anode at 1000 °C can reach 102–104 S/cm, which can almost meet all the basic conditions required for SOFCs operation. However, the Ni/YSZ anode still needs to be further optimized, and the main problems it faces are Ni coarsening under high temperature, carbon deposition under hydrocarbon fuel, and sulfur poisoning under sulfur-containing fuel [100, 101]. The main reason for carbon deposition is that the decomposition reaction of CH_4 or the disproportionation reaction of CO is easy to proceed thermodynamically at the operating temperature [102], and the Ni-based cermet anode can not only catalyze the breaking of C-H bonds, but also catalyze bond formation of C–C at the same time. This promotes the deposition and growth of C on the Ni-based anode and further diffuses into the Ni particles, which finally causes the reduction of Ni catalytic activity and produces stress changes, resulting in the degradation of the cell performance and even the destruction of the anode structure [103]. The current solution is generally to adjust the operating conditions (such as operating temperature, fuel composition, or discharge current). In addition, metals or metal oxides can also be modified in the anode, such as Cu, Sn, Fe, MgO, BaO, GDC, etc., to reduce the catalytic activity of Ni on the formation of C–C bonds and inhibit the formation of C–C bonds. The role of carbon deposition. In addition, Ni/SDC or Ni/GDC can also be used directly as the anode of SOFCs to improve the anti-carbon deposition and electrochemical performance of the anode [104].

Cu-YSZ cermet anode materials

Cu-based cermet composites are another cermet composite anode. Compared with Ni, the catalytic activity of Cu-metallic for the formation of C–C bonds is very low, so it can effectively prevent the formation of carbon deposits. However, its catalytic activity for hydrogen and hydrocarbon fuels is also low, so the electrochemical performance is not comparable to that of Ni-based anodes. To improve the catalytic effect of Cu/YSZ composite anode on hydrocarbon fuels, oxide catalysts, such as CeO_2, are usually added to the anode [105]. Cu-CeO_2-based anodes have strong resistance to carbon deposition and sulfur poisoning, but have obvious shortcomings. Cu has relatively few active sites, a low melting point, easy agglomeration under high temperature operation, low electronic conductivity, and low catalytic ability to alkane gas.

11.4.2.2 Mixed Conductor Oxide Anode Materials

Mixed ion electron conductors (MIEC) are materials with high electronic and ionic conductivity at high temperatures and reducing environments. Using MIEC materials as anodes can expand the reaction site to the whole anode and significantly reduce its polarization resistance; some MIEC materials with chalcogenide structure also outperform metal-ceramic anodes in terms of sulfur poisoning resistance. The main reason is that in mixed conductors, both oxygen ions and electrons are mobile, and oxygen ions can directly pass to the anode particles and electrons can quickly pass to the linker, so that electrochemical reactions do not need the three-phase interface as in ceramic anode materials, much less being confined to the interface between the electrolyte and the anode as in metal anodes, but occur at the interface between the entire anode and the fuel gas, greatly increasing the effective area of the electrochemically active The effective area of the zone is greatly increased [106]. Not only that, mixed conductor oxides can also be used to catalyze electrochemical oxidation reactions of dry hydrocarbon gases such as methane, and since the oxides are not active enough to induce the formation of C–C bonds and do not react easily with sulfur, there is no carbon deposition and no sulfur poisoning occurs in this anode material, all of which show their advantages as anode materials of SOFCs [107].

CeO_2-based oxides

CeO_2 crystals have a fluorite structure (Fm3m), and perfect CeO_2 crystals have a face-centered cubic (fcc) stacking pattern of Ce and O atoms. CeO_2 has a high melting point and is not easily sintered. CeO_2 has been shown to have good catalytic activity for the oxidation of dry methane, and both doped and undoped CeO_2-based materials are capable of exhibiting mixed conductor performance at low oxygen partial pressures and are promising SOFCs anode materials [108]. The volume of CeO_2 is closely related to the air pressure, and under a reducing atmosphere, Ce^{4+} is reduced to Ce^{3+}, leading to lattice expansion, which can produce cracks on the surface of the electrode and electrolyte, and in severe cases, lead to the detachment of the electrode from the electrolyte surface. Although the doping of low-valent ions in CeO_2 can greatly reduce the lattice expansion and contraction during the redox cycle, the doping of low-valent ions will inevitably reduce the electronic conductivity of the material. In summary, the molar fraction of Gd^{3+} or similar alkaline earth metal cations doped into CeO_2 is 40–50%, which can balance the expansion and contraction of the lattice and the reduction of the electronic conductivity [109].

Perovskite Anode Materials

Oxides with perovskite structure are widely used because of their stable structure and properties over a wide oxygen partial pressure and temperature range. ABO_3-type perovskite anodes have attracted widespread attention due to their flexible composition and tunable catalytic activity. The A and B sites of perovskite can be doped with other ions in a wide range, thereby improving the intrinsic properties of ABO_3 perovskite materials. By doping low-valent cations at the A site, oxygen vacancies can be generated according to the principle of electrical neutrality, thereby improving the oxygen ion conductivity of perovskite materials. The electronic conductivity and catalytic activity of perovskite materials can be improved by doping with B-site transition metals. In addition, perovskite materials are structurally stable in a reducing atmosphere, are good mixed-ion electronic conductors, and have good resistance to carbon deposition. The above advantages make perovskite materials a very promising anode material for direct hydrocarbon fuel cells.

Although pure $LaCrO_3$ material has the advantages of high electrical conductivity and good stability in high-temperature oxidizing atmosphere, it is not an ideal anode material, because it is prone to lattice expansion in reducing atmosphere, and is a *p*-type semiconductor. Incorporation of Sr and Ti into $LaCrO_3$ material will make the material possess *n*-type semiconductor properties and improve its stability under reducing atmosphere. In addition, the A-site doped $LaCrO_3$ material has a certain catalytic activity for methane, and it is not easy to deposit carbon. Tao et al. found that the synthesized $La_{0.75}Sr_{0.25}Cr_{0.5}Mn_{0.5}O_{3-\delta}$ anode had a polarization resistance of about 0.2 Ω cm^2 at 900 °C by doping Sr and Mn at the A and B sites of $LaCrO_3$, respectively [110]. Philippe et al. improved the catalytic activity of perovskite for methane steam reforming by incorporating Ru at the B site, and no carbon deposition was observed after running for 300 h; The oxidative activity of the anode towards hydrogen is enhanced after Mn doping, which is close to that of the Ni-YSZ anode material [111]. Danilovic et al. studied $La_{0.75}Sr_{0.25}Cr_{0.5}X_{0.5}O_{3-\delta}$ (X = Co, Fe, Ti, Mn) as the anode of SOFCs by doping a series of transition metal elements with high catalytic activity at the B site, and the catalytic activity for CH_4 conversion was found to be Co > Mn > Fe > Ti. The maximum power density of the cell depends on the fuel: in CH_4, Fe > Mn > Ti; In H_2, Fe > Mn > Ti; in 0.5% H_2S/CH_4, Ti > Fe > Mn [112]. The researchers achieved the purpose of improving the conductivity of the perovskite anode and the catalytic activity of fuel gas by doping Sr and transition metal elements at the A and B sites of the $LaCrO_3$ perovskite structure, respectively.

$SrTiO_3$ is a typical perovskite anode material. At 800 °C, a small amount of Y was incorporated into its B site, and the anode conductivity of $Sr_{0.86}Y_{0.08}Ti\,O_{3-\delta}$ with A site vacancy could reach 82 S cm^{-1} [113]. The conductivity of $Sr_{0.94}Ti_{0.9}Nb_{0.1}O_{3-\delta}$ was further improved to 120 S cm^{-1} by doping Nb at the B site to form A site defects [114]. Then the $La_xSr_{1-x}TiO_3$ ($x = 0.1 \sim 0.4$) perovskite anode was studied, and found that although this material is structurally stable and satisfies the thermal stability of the anode and the matching of the thermal expansion coefficient, the catalytic activity for hydrogen oxidation is poor [115]. $La_4Sr_8Ti_{11}Mn_{0.5}Ga_{0.5}O_{37.5}$ was formed by doping Ga and Mn at the B site. Under reducing conditions, Mn^{3+} and Ti^{4+} were reduced to Mn^{2+} and Ti^{3+}, respectively, resulting in enhanced electronic conductivity. Under the conditions of hydrogen humidification and 950 °C, the formation of oxygen ion vacancies in perovskite improves the ionic conductivity, and the peak power density is close to 500 mW cm^{-2}. Then, after two days of testing with methane fuel, no carbon residue was found in the anode, showing excellent anti-carbon performance [116]. Table 11.1 presents the fuel cell performance of the LST-based anode materials formed by A-site doping of La [117]. Doping La and transition metal elements at the A-site and B-site of the $SrTiO_3$ perovskite structure, respectively, can significantly improve its electrical conductivity, enhance its catalytic activity for CH_4 fuel, and also exhibit excellent anti-carbon properties.

In addition, there have been many discoveries of anodes with double perovskite structures. The expression of the double perovskite structure is generally $AA'B_2X_6$ (or $A_2BB'X_6$). Figure 11.9 describes the formation process of the double perovskite structure. The double perovskite structure expands the composition space of the perovskite family to accommodate higher valence elements; at the same time, there is also a more complex atomic and electronic environment, showing stronger chemical stability under catalytic conditions [118]. The double perovskite structure of $Sr_2Mg_{1-x}Mn_xMoO_{6-\delta}$ ($x = 0, 0.2, 0.5, 1$)) and $Sr_2Mg_{0.9}Cr_{0.1}MoO_{6-\delta}$ were studied when in H_2 fuel without or with H_2S addition [119]. The single cell with $Sr_2MgMoO_{6-\delta}$ double perovskite anode achieves maximum power densities of 840 and 823 mW cm^{-2} at 800 °C, respectively, and greatly improves the tolerance to sulfur poisoning compared with Ni-YSZ. Doping or replacing with Ni, Fe, Co, Ti, and other elements at the B site forms anodes with double perovskite structure $Sr_2NiMoO_{6-\delta}$, $Sr_2FeMoO_{6-\delta}$, $Sr_2Fe_{1.5}Mo_{0.5}O_{6-\delta}$, $Sr_2Mg_{1-x}Co_xMoO_{6-\delta}$, and $Sr_2TiMoO_{6-\delta}$, both exhibit MIEC characteristics and have high resistance to sulfur poisoning and carbon deposition [120–123].

Table 11.1 Fuel cell performance of selected LST-based anode materials [117]

Anode composition	Electrolyte (thickness)/ cathode	Tem. (°C)	Fuel	MPD[a] (mw cm^{-2})
$La_{0.4}Sr_{0.6}TiO_3$	YSZ (160 μm)/LSM	1000	Humidified H_2	260
$La_{0.20}Sr_{0.25}Ca_{0.45}TiO + CeO_2 + Ni$	ScSZ (310 μm)/LSM	900	Humidified H_2	~ 300
$La_4Sr_8Ti_{11}Mn_{0.5}Ga_{0.5}O_{37.5}$	YSZ (2 mm)/LSM	950	Humidified H_2	500
$La_{0.2}Sr_{0.8}Ti_{0.98}Co_{0.02}O_3$ + GDC + Ni	LSGM (250 μm)/ $La_{0.6}Sr_{0.4}Co_{0.2}Fe_{0.8}O_{3-\delta}$	800	Humidified H_2	638
$La_{0.2}Sr_{0.8}TiO_3$ + NiO-SDC + NiO-YSZ	YSZ/LSM	800	Dry H_2	850

[a] *MPD*—maximal power density

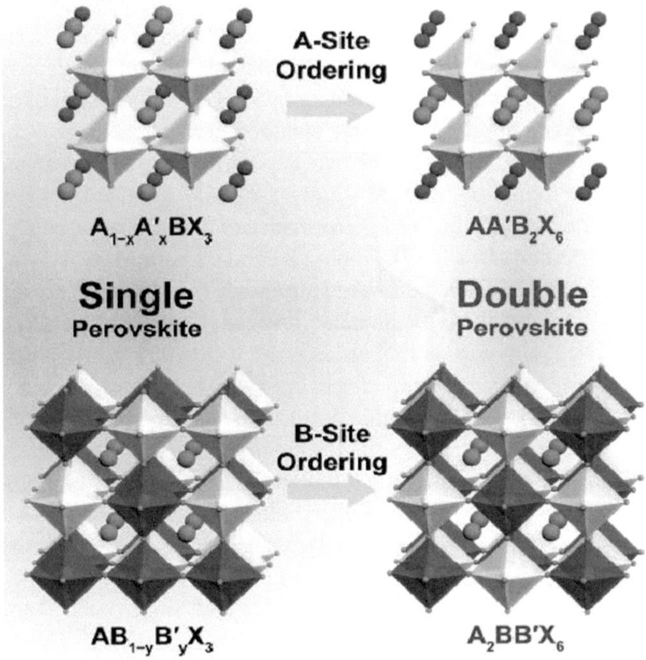

Fig. 11.9 Schematic depicting the formation of double perovskites [118]

The discovery of perovskite materials has broadened the choice of anode materials, and the complexity and diversity of perovskite anodes also ensure the development prospects of perovskite anodes. At present, although the conductivity of the anode is not comparable to that of metal-ceramic anodes such as Ni-YSZ, the MIEC characteristic of the perovskite structure has a natural TPB length advantage, and the maximum power density of a single cell at high temperature has exceeded that of a metal-ceramic anode single cell [124]. However, at low temperatures, the main factor limiting perovskite anodes is the lack of catalytic activity. The catalytic activity can be improved by different preparation processes, such as impregnation and in-situ desolubilization, which have a very good commercialization prospect.

11.4.3 New Development on Anode Materials

In recent years, some scholars have put more effort into the development of SOFCs anodes without rare-earth materials. Titanium oxycarbide samples were prepared as anode materials [125] for medium temperature solid oxide fuel cells. The titanium oxycarbide sample is made by vacuum sintering of a mixed powder of titanium oxide and titanium carbide at high temperature, which can produce more phase-pure oxycarbide than can be produced by carbothermal reduction of TiO_2 [126]. Titanium oxycarbide can be seen as a solid solution of titanium carbide (TiC) and titanium monoxide (TiO), so it can usually be represented as TiO_xC_{1-x}, where $0 \leq x \leq 1$. Notably, the material has the same face-centered cubic (FCC) rock salt structure as TiC. The experimental results show that titanium oxycarbide material can be successfully used for the SOFCs anode, which is a substitute material for the SOFC anode.

11.5 Solid Oxide Fuel Cells Systems

11.5.1 Solid Oxide Fuel Cells Subsystems

As the core of the fuel cell system, the performance of the fuel cell subsystem determines the performance of the entire SOFCs system. The fuel cell subsystem is usually referred to as the fuel cell stack. Because the voltage generated by a single cell is less than 1 V, it is necessary to stack many cells together to get a high voltage. The structure, stacking mode, and interconnection mode of the fuel cell will have an important impact on the performance of the stack.

For all stack types, electrical performance, thermal management capabilities, and mechanical strength must be considered. The ideal stack should have minimal resistance, a stable operating temperature, and sufficient mechanical strength to meet the assembly and handling requirements. In addition, the design of the stack should avoid short circuits and gas leakage. Because this will seriously reduce the fuel cell stack performance. It should be able to uniformly distribute the reactants to each area of the stack. And the reactants can reach the reaction site quickly to reduce mass transport limitations. The performance of each single cell in the stack should be close to the performance in the single-cell test. Therefore, it is necessary to review the structure design of SOFCs.

The stack structure design of SOFCs has matured over the years, starting from pressed thimbles and discs in the 1930s. The dominant designs are now flat and tubular, and other geometries have fallen out of favor, so this section focuses on these two. Some interesting designs will be introduced in the third subsection. Moreover, since the fuel cells need to be connected in series, the performance of the cell will be affected by the different interconnection modes. Cell interconnection designs will also be introduced, along with fuel supply methods and sealing measures.

11.5.1.1 Planar Type

Each cell in the planar SOFC is a flat plate, connected like that in a PEMFC stack [127]. Figure 11.10 shows a schematic diagram of the planar SOFC design.

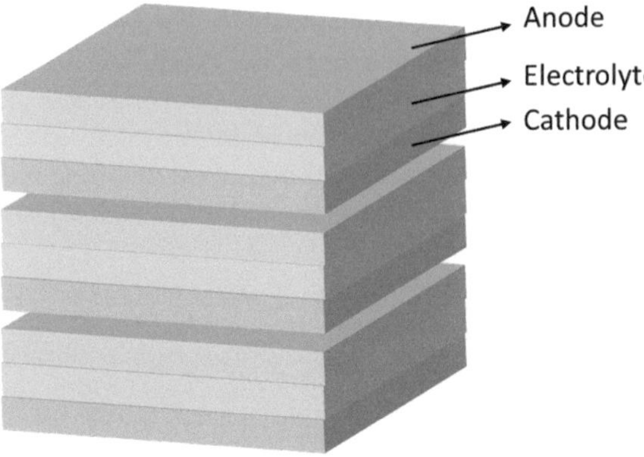

Fig. 11.10 Schematic of the planar SOFC design (Color figure online)

The planar stack design can be subdivided into two types: self-supported and externally supported. In a self-supported design, one of the cell components acts as the cell structural support and is made thicker than the other components. In the externally supported design, a porous substrate or metal serves as a support for the cell. Figure 11.11 shows the various support designs of a flat SOFC. Choosing one for all stacks is difficult because each of these designs has its advantages and disadvantages. For example, for the design of the electrolyte support, a thicker electrolyte results in a higher ohmic resistance, and to minimize the ohmic resistance, the cell must operate at a higher temperature. For anode-supported and cathode-supported configurations, the cell can operate at a lower temperature because of the thin electrolytes. The benefits of SOFCs low-temperature operation include a wider selection of materials (particularly low-cost metal materials for interconnects), longer cell life, reduced thermal stress, improved reliability, and potentially lower cell costs. However, thicker electrodes enhance mass transport resistance, and lower temperatures slow the electrochemical reaction kinetics, both of which lead to a loss in the performance of the stack. In contrast, externally supported configuration does not result in excessively thick layers in the cell and can provide good mechanical support. As a result, most designs in recent years have chosen this configuration. However, this design is the most complex and requires more considerations, such as connection oxidation, flow field design constraints, short circuits, etc.

Apart from the support configuration, the most important design features in planar SOFCs stacks are related to the gas flow configuration and gas manifolding. Gas flow configuration contains two aspects, one is flow direction selection and the other is flow field design. The air flow direction in the planar SOFCs can be arranged as cross flow, co-flow, or counter-flow. The selection of a particular flow direction has significant effects on the temperature and current distribution within the stack, depending on the specific stack design. The flow field of planar SOFCs also has a variety of choices. The flow field design used in PEMFCs can also be used in SOFCs, including Z-flow and serpentine flow field. Some combinations of these flow fields are also common in a planar SOFCs stack. Figure 11.12 shows some examples. In planar SOFCs, good flow field design can increase the uniformity of gas distribution

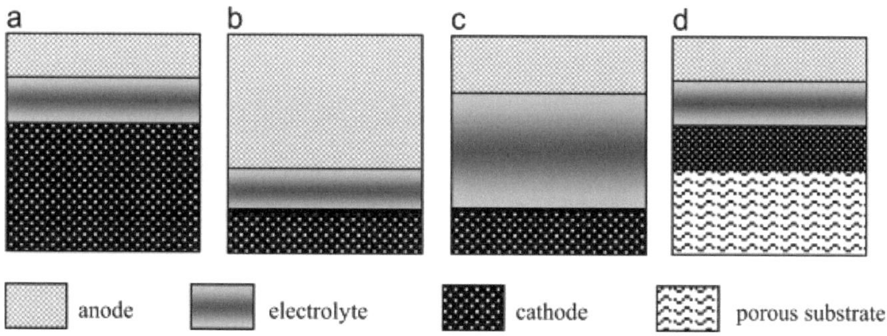

Fig. 11.11 SOFCs configurations. **a** cathode supported **b** anode supported **c** electrolyte supported, and **d** porous substrate supported [128]

and facilitate heat and mass transfer in each cell. Besides, in a planar SOFCs stack, the flow field often plays an important role in connecting different single cells, like the bipolar plates in PEMFCs. This requires that the contact area between the flow field and electrodes must also be considered in the design to minimize the contact resistance losses. Therefore, determining the flow field for fuel and oxidant flow is an important aspect in designing planar SOFCs.

Gas manifolds are gas channels used to deliver gas from a common supply point to each cell and remove unreacted gas and reaction products. The gas manifolds can be classified as external manifolds and integral manifolds. The external manifolds are constructed separately from the interconnecting components of the cell or cell stack. Figure 11.13 shows an external manifold design. The fuel and oxidant flow into the reactor from different directions, and after the reaction, they are discharged from the other side. The integral manifold is usually designed in the flow field and is part of the fuel cell. Figure 11.14 shows an integral manifold design. The holes on the left and on the right form the channels for transporting the gases inside the stack. In principle, the manifold must be designed to have a low pressure drop, relative to the individual cell drop, to provide a uniform flow distribution to the stack.

Currently, the most common fabrication method for flat-panel SOFCs is tape casting, dry pressing, and screen printing [132]. Tape casting is a process of preparing a thin film of ceramic slurry over a flat surface. It is well-suited for the fabrication of multilayer ceramic structures. Dry pressing is a powder compaction technique that is usually used to generate ceramic parts. This technique compacts dry powder with high pressure to fabricate cells. This method is widely used due to its simplicity and low cost. Screen printing is another low-cost wet ceramic processing technique, which is always used to fabricate a few components of the cell. It usually works with other technologies to fabricate the whole cells. The production of planar SOFCs typically begins with the fabrication of a support layer. The other layers are then assembled or deposited onto the support layer by the same or a different fabrication technology.

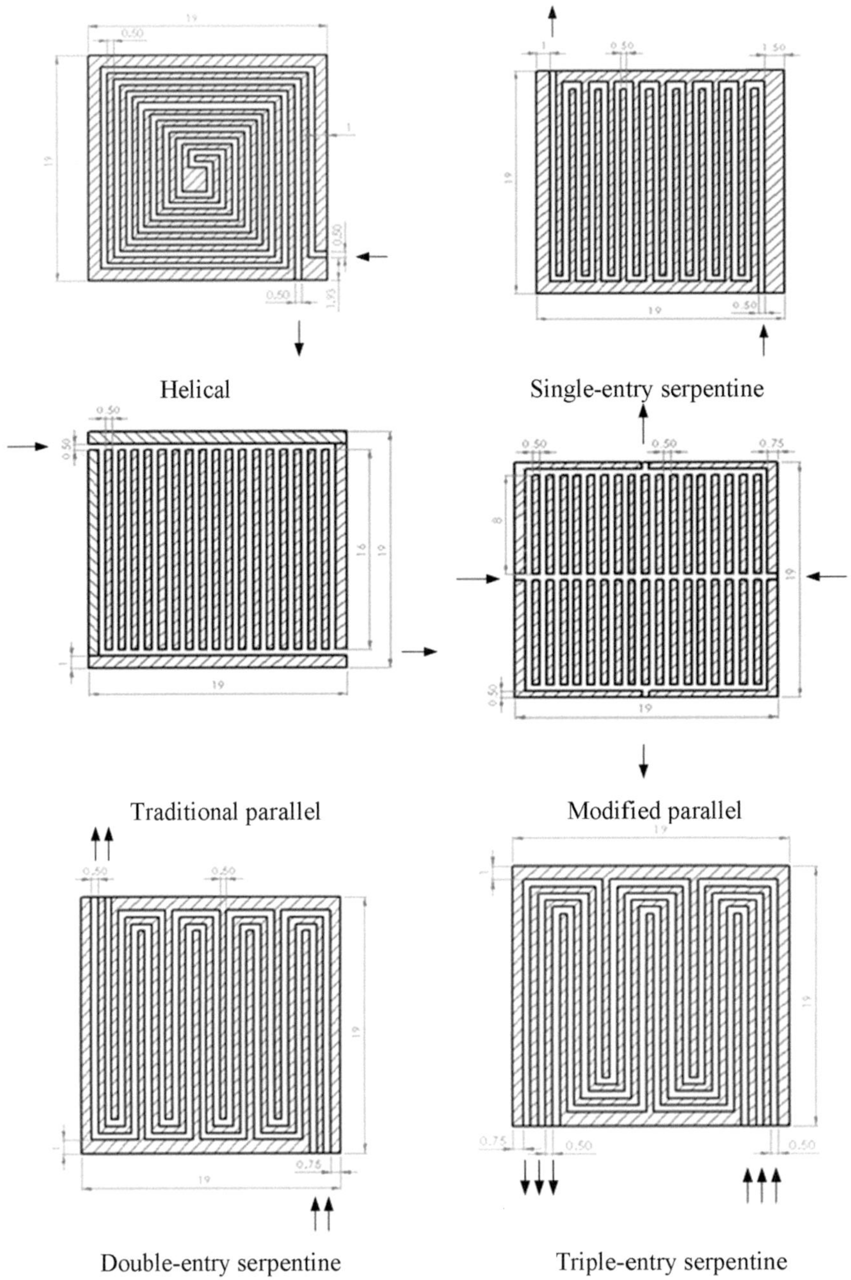

Fig. 11.12 Examples of flow fields in planar SOFCs [129]

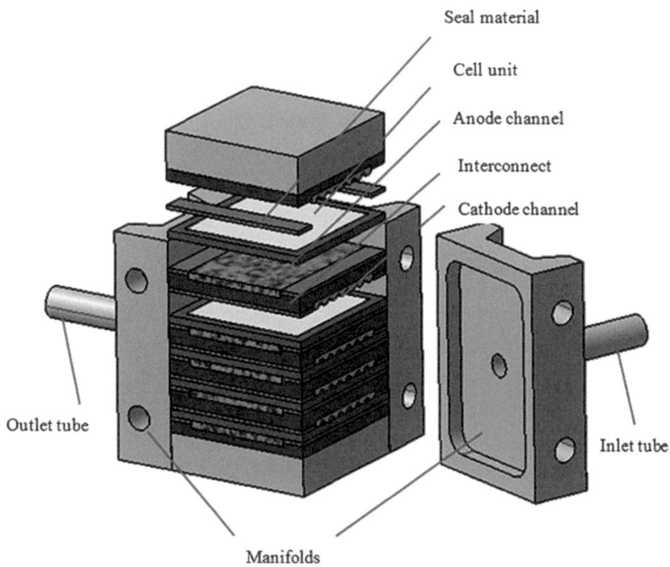

Fig. 11.13 An example of an external manifold [130]

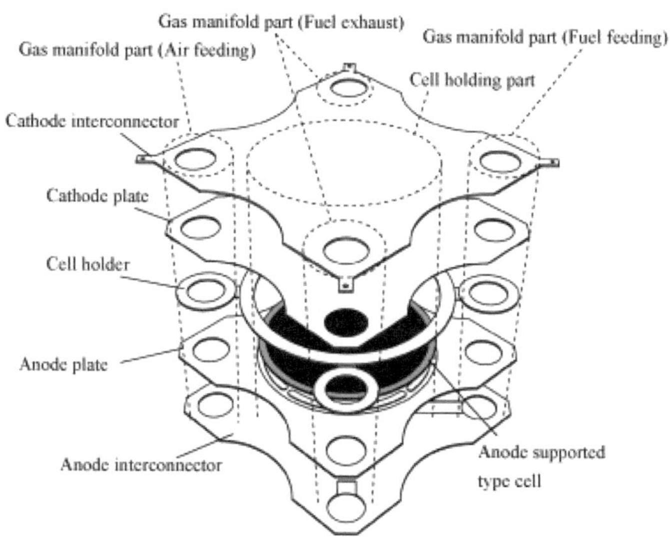

Fig. 11.14 An example of an integral manifold [131]

Once the cells are manufactured, the interconnects between the cells need to be considered. Planar SOFCs mainly has metal and ceramic two kinds of interconnects. The choice of interconnect depends on the cell operating temperature. Ceramic interconnects are usually used at 900–1000 °C, while metal interconnects are used at less than 800 °C. The most commonly used material for ceramic interconnects is doped lanthanum chromate ($LaCrO_3$). Lanthanum chromate interconnects for planar SOFCs are usually made by conventional ceramic processing methods, such as pressing or tape casting, followed by sintering [128]. The most commonly used materials for metal interconnection are chromium-based alloys and ferritic stainless steels. These materials were considered for planar SOFCs interconnects because their thermal expansion coefficients closely match the thermal expansion rates of the cell components. Chromium-based alloys are preferred for interconnect applications due to the high electrical conductivity of the chromium oxide scales that form on the surface of the alloy. However, the migration of chromium species into the cell will result in cell performance loss during SOFCs operation [133].

After the cells are assembled, the stack often needs to be sealed to prevent gas leakage or crossover. It is important to note that the sealing must be insulated to prevent an internal short circuit. Two main types of sealing methods are used in the planar SOFCs: compressive loads and high-temperature sealants. The compression seal is formed by compressing the fuel cell components. This type of sealing has the advantage of not requiring a sealant. However, since the surface of the cell is often uneven, mechanical stress can occur during compression. Gaskets are now commonly used to relieve this mechanical stress, which can also improve gas-tightness. High-temperature sealant is another popular sealing method. Commonly used high-temperature sealants include cement, glass, and glass ceramics. The sealant selected for planar SOFCs must have stability in oxidizing and reducing environments, chemical compatibility with cell components, and proper sealing and insulation properties.

In terms of performance, designs of planar SOFCs have relatively low ohmic resistance due to the short current flow path and are therefore able to provide high power densities. In addition, the planar design is easy to manufacture, so the production cost can be reduced [133]. However, the planar design also suffers from a major disadvantage. It requires the use of sealing around the edges of the cell components. In addition, the thermal stress at the interface of different materials in the cell poses a great challenge to mechanical properties. This problem is exacerbated by the thin cell components in the planar design.

11.5.1.2 Tubular Type

The tubular type of SOFCs was first developed by Westinghouse Electric Corporation (now Siemens Westinghouse Power Corporation). The original design used cylindrical tubes made of calcium-stabilized zirconia, about 1–2 mm thick. The cell components are deposited in the form of thin layers on the cylindrical tube, and an anode is deposited on the support tube first. By the early 1980s, the sequence of cell

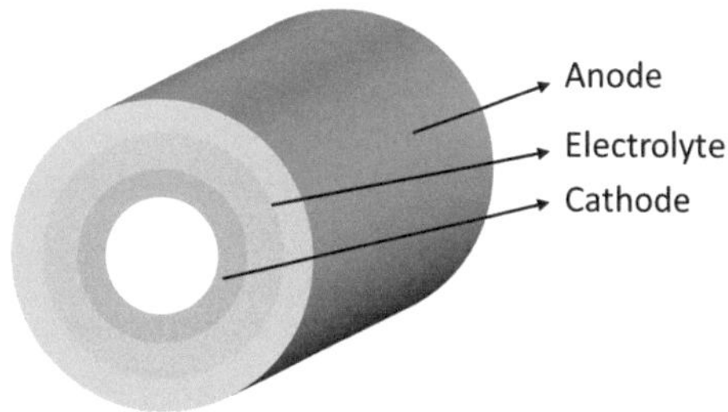

Fig. 11.15 Schematic of a tubular SOFC (Color figure online)

components was changed. The cathode is deposited in the first layer of the support tube, while the anode is the outermost layer. The experimental results showed that the performance of the cell with the cathode support tube was significantly improved, so this tube became the standard tube. In addition, to further increase the cell performance, the active length of a single cell has been increased. But increasing length also has negative effects, such as a larger pressure drop. The diameter of the cell has also been increased to alleviate the large pressure drop in the cell. Figure 11.15 shows a diagram of the typical tubular SOFC.

The most marked advantage of tubular SOFCs is that it realizes a seal-less design. The reason why high-temperature sealing materials are no longer required can be explained by their specific structure, as shown in Fig. 11.16. Each support tube is shaped like a large test tube, and the tube is sealed at one end. The fuel flows along the outer wall of the tube towards the unsealed end. Right in the middle of the tubular fuel cell is an intake pipe, through which air enters the cell. Air and fuel flow in and out of the tube separately, so there is no need to worry about mixing. For the open end of the tube, the air meets the unused fuel and burns immediately because of the high operating temperature. This causes a cell exhaust temperature of up to 1000 °C and provides heat to preheat the air. In this way, the tubular SOFCs have a built-in exhaust combustion chamber, thus eliminating the need for high-temperature sealing materials.

But tubular SOFCs still have two drawbacks. One is the high fabrication cost. The second is the lower power density than the planar type. The high fabrication cost is due to the way electrolytes and electrodes are deposited. The electrochemical vapor deposition process is used to form the electrolyte layer. Although it is possible to reduce the cost after expanding production, there are still problems such as complex technology, high investment cost, and the need to use vacuum equipment. So, we still need to develop more economical techniques to prepare the desired tubular SOFCs. The disadvantage of a lower power density is due to the larger resistance in the cell. The larger pores in the cell extend the path through which the current flows, thus weakening the performance of the cell.

The tubular cell stack usually connects many single cells in series and parallel to form a semi-fixed cell bundle, thus constituting the SOFCs subsystem. Multiple single cells are connected in series as shown in Fig. 11.17 to achieve the desired cell voltage and constitute a group of cell units. Then, several cell units connect in

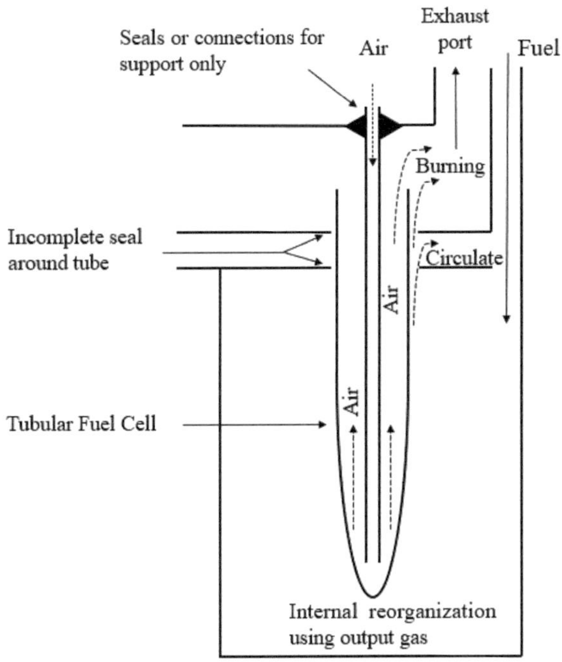

Fig. 11.16 Tubular SOFCs are constructed without sealing [133]

parallel to constitute the entire SOFCs subsystem, to meet the output power demand. Figure 11.18 shows a tubular stack structure.

Similar to planar SOFCs, the manifold design for tubular SOFCs is necessary. The distribution of fuel and air can have a serious impact on cell performance, and a large number of simulation studies have been carried out in this field. In addition to affecting the mass transport, the geometry of tubular SOFCs can also make a significant difference to the ohmic resistance. Conventional circular tube designs

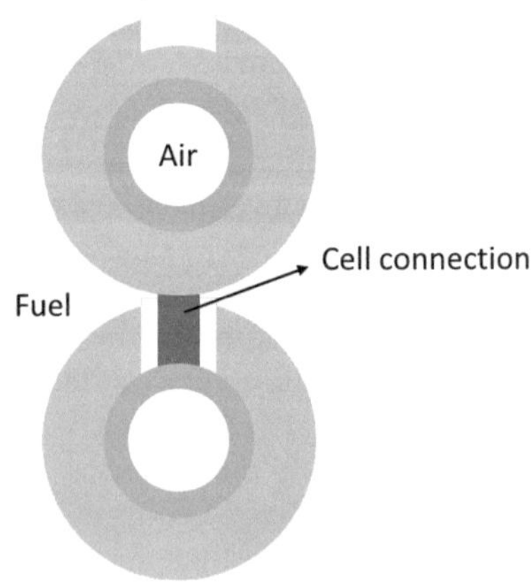

Fig. 11.17 Schematic of two tubular cells (Color figure online)

Fig. 11.18 A tubular SOFCs stack [134]

lack efficient interconnectors, and the novel structure is expected to mitigate this problem.

11.5.1.3 Other Types

There are advantages and disadvantages to each of the two designs described in the previous section. Although the planar SOFCs have a relatively high power density, the structure instability problem occurs when the temperature changes rapidly. In addition, the need for sealing leads to higher production costs. And for the tubular type, it has better thermal stability, but its relatively low power density limits its application. Therefore, most research in the past two decades has focused on solving these problems by offering novel cell designs [128].

The flat-tube design of SOFCs is a combination of tubular and planar stack designs proposed by Siemens Westinghouse. It solves the low power density problem of traditional tubular design by reducing the internal ohmic resistance. Figure 11.19 shows a flat-tube design [135].

A flat tube SOFC consists of a large flat thick plate with several gas channels inside. The air passes through these channels, while the fuel flows between the plates in the stack. A thin interconnect film is coated on one side of the plate to electrically connect the cells into a stack. By 2012, it has been reported that the power density of this cell design can reach 0.36 W/cm^2 at an operation temperature of 750 °C [136].

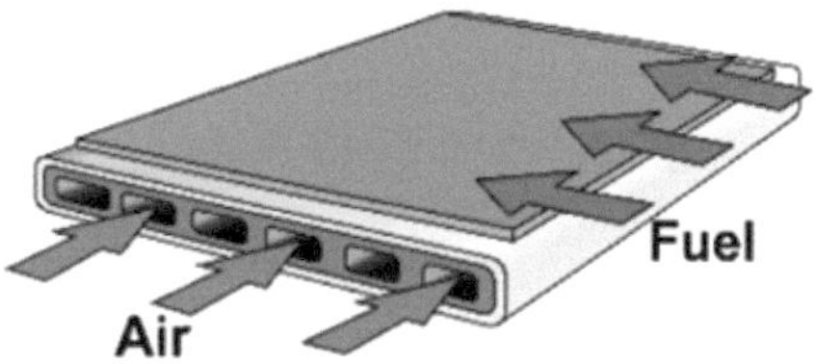

Fig. 11.19 Schematic of a flat-tube SOFC [135]

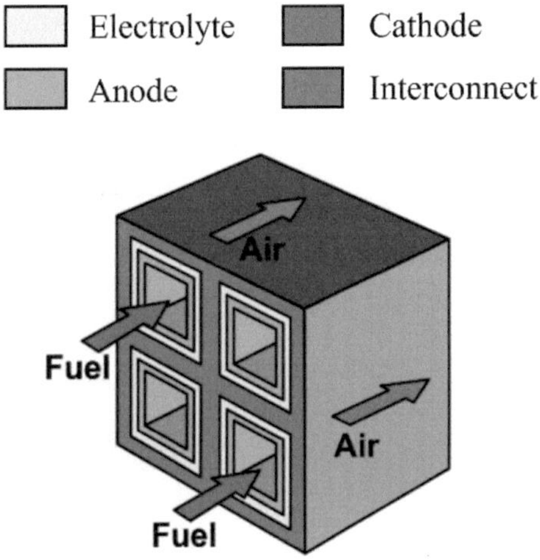

Fig. 11.20 Schematic of a honeycomb SOFC [135] (Color figure online)

Numerous simulation studies have been carried out to optimize structural designs such as the shape and number of channels.

The honeycomb design is another novel stack design that also combines the advantages of the tubular and planar SOFCs, similar to the flat-tube SOFCs design. It also has a similar geometry to the flat tube SOFCs, which looks like multiple layers of flat tubes connected directly together. It has the advantages of high mechanical strength and high volumetric power density. The structure consists of many channels. In some cases, these channels transport fuel only, as shown in Fig. 11.20, with air outside the structure. In some other cases, fuel and air flow in different channels in an alternating sequence.

The honeycomb structure can be electrolyte or electrode supported. Interconnects and electrical leads are major problems in this design. The honeycomb structure of SOFCs was patented in 1992 [137]. Initially, the power density of this stack design is only 0.1W/cm^2 because of the thick electrolyte material used as support. To improve the performance of honeycomb SOFCs, researchers have carried out a large number of studies on electrolyte materials with high ionic conductivity. In 2008, a cathode-supported honeycomb SOFC with a power density up to 1W/cm^2 was developed in the laboratory [138].

11.5.2 Ancillary Subsystems

In a fuel cell system, the cell subsystem typically accounts for 25% of the volume, with the remainder consisting of auxiliary devices, including insulation, piping, pumps, heat exchangers, heat utilization equipment, combustion processors, control systems, starting heaters, and power regulation. The auxiliary system is also an important part of the whole fuel cell system. They will ensure the proper and stable operation of the fuel cell system.

The fuel passes through the desulfurizer before entering the SOFCs stack. Because sulfur in any form is harmful to the cell stack, it poisons the anode and degrades the performance of the cell stack. The usual way to remove sulfides is to dehydrogenate them with hydrogen gas in a catalytic reactor to produce hydrogen sulfide, which is then absorbed in a separation reactor containing zinc oxide. The treated fuel is then reformed in a pre-reformer to produce methane, carbon monoxide, and hydrogen to prevent carbon deposition on the SOFCs stack. Some of the exhaust gas can also be recycled to provide steam for the pre-reformer. The maximum fuel utilization of the SOFCs stack is between 85 and 90%, and the remaining fuel is burned with excess air in the after-burner. The waste heat in the exhaust gas is used through a heat exchanger to heat the air to avoid thermal shock after air enters the SOFCs. Finally, the direct current generated by the stack is converted into alternating current. A two-level transformation is typically used. The first level is a DC voltage regulator, which converts the unstable voltage of the stack into a stable DC voltage. The second level then converts the direct current into the alternating current required by the grid. Converter efficiency is generally 94–98%.

A fan and air filter are usually present in the system to provide clean gas to the stack. Before air flows into the SOFCs stack, it also passes through a pre-heater. When the stack is just starting up or running at low load, auxiliary heating of the air is carried out. In addition to the above components, there are some subsystems in the SOFCs system that are not shown in the diagram. An example is a steam generation system, which supplies steam to the pre-reformer when the stack is just starting up. There is also a purging system, which cleans the inside of the SOFCs stack during stack start and stop.

The reactor control system is another important part of the auxiliary system design. A good stack control system can improve system efficiency and protect stack safety. For SOFCs systems such as the one described above, the main control parameters are system output power and stack temperature. The output power of the system is determined by the power of the stack, which is related to the fuel supply rate. In addition, the same power may correspond to a variety of fuel efficiencies, which leads to a more complex power control system. If the fuel efficiency is too high, it will lead to the oxidation of nickel, completely damaging the anode. If fuel efficiency is too low, a lot of fuel will be burned in the tail gas. This is a waste of fuel, and it causes a risk of overheating. Another important control is the stack temperature control. Decreasing the temperature is beneficial to the electrochemical conversion of hydrogen, and increasing the temperature is beneficial to the decrease of electrolyte ohmic resistance. Therefore, there is an optimal operating temperature in theory, and the stack control system should keep the stack temperature near this optimal temperature as much as possible by changing the gas flow rate of the air.

11.6 Applications

11.6.1 Small-Scale Solid Oxide Fuel Cells Systems

The power level of the small-scale SOFC system is watt level, and the power of the fuel cell is within a kilowatt. The main application is a portable mobile energy system. Two application cases will be introduced as follows.

Figure 11.21 demonstrates a conceptual diagram of a small SOFC system with a power of 20 W [139]. The system uses jet fuel as fuel and can output 12 V DC power with an output power of 20 W. The weight of the system is 0.6 kg, and the volume is 8.5 by 11.5 by 20 cm. Figure 11.22 shows a conceptual diagram of a 500 W SOFC system [139]. The output voltage is 28 V, the output power is about 500 W, the weight of the system is 7 kg, and the volume is 48 by 28 by 23 cm.

Table 11.2 shows some important parameters of the structure and operating conditions of the 500 W-SOFC system.

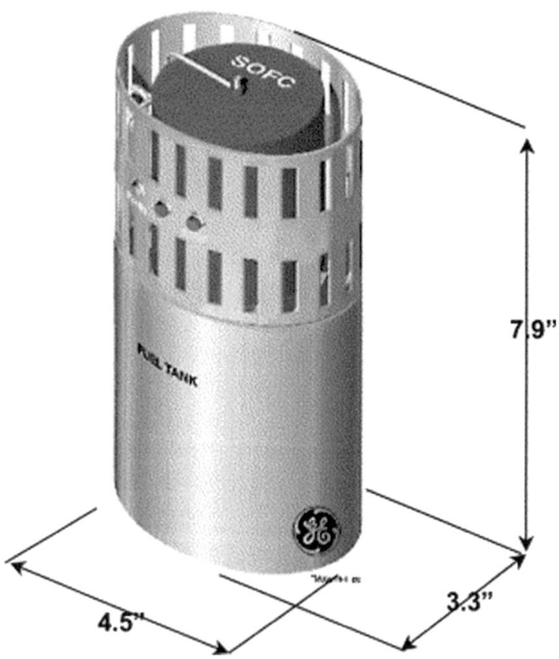

Fig. 11.21 Conceptual diagram of a small SOFC system with a power of 20 W [139] (Color figure online)

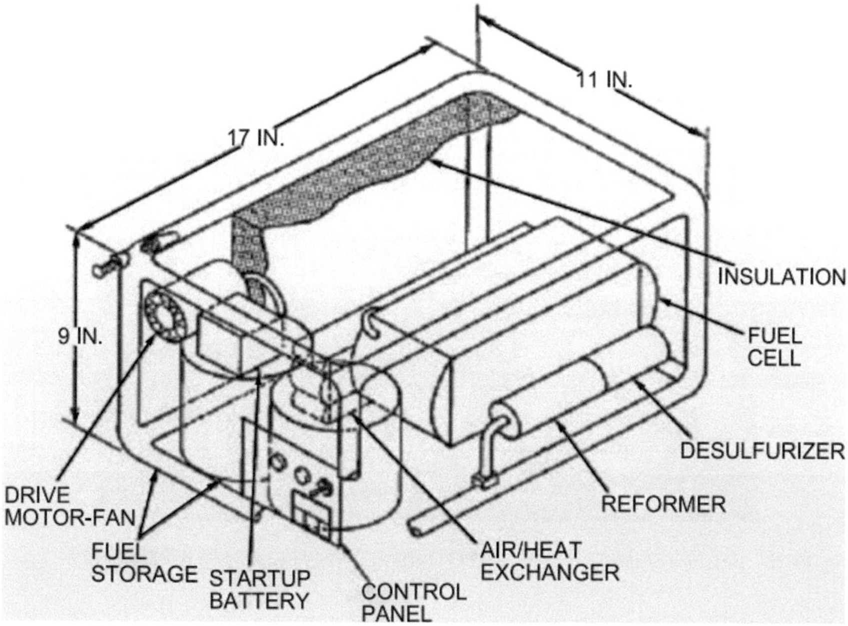

Fig. 11.22 A 500W-SOFC system [139]

Table 11.2 Structure parameters and operating conditions of the 500 W-SOFC system

SOFC module	
Voltage	0.65 V/cell
Fuel utilization	80%
Stack weight	1.7 kg
Thermal management	
Effectiveness	0.75
Weight	1.8 kg
Reactant requirement	
Air flow	18.2 kg/h
Fuel flow	0.16 kg/h

11.6.2 Solid Oxide Fuel Cells in Transportation

In the field of transportation, SOFCs also have a wide range of application scenarios. Compared with the application scenario of small-scale power at the watt level described above, the power of the system in the field of transportation is generally at the kilowatt level. Initially, in the field of transportation, SOFCs were mainly applied to the auxiliary power unit system (APU) to provide vehicles, charge batteries and drive small-scale power demands such as air-conditioning components. Delphi

Fig. 11.23 Structure of onboard SOFCs APU [140]

Automotive Systems has developed a vehicle APU based on a SOFC system [140], with a power of 5 kW. The system uses gasoline as fuel and can provide the required power for the vehicle's air conditioning system without the intervention of the automobile engine. The SOFC systems are composed of four SOFCs stacks, an energy recovery unit, a gasoline reformer, and other subsystems. Its structure is shown in Fig. 11.23.

In addition to vehicle APUs, renewable energy vehicles driven by SOFC systems also have application prospects and are widely studied by researchers. A typical SOFCs vehicle is a fuel cell hybrid vehicle. Usually, the power unit of the vehicle is composed of a SOFC system and a small capacity buffer cell. The SOFCs converts fuel into electric energy and stores it in the buffer cell. The vehicle is driven by a drive motor. A feasibility study, technical and economic analysis was conducted on a SOFC hybrid power system for vehicle traction applications [141]. Figure 11.24 is a schematic diagram of the system.

The SOFCs hybrid vehicle has the advantages of quiet operation, good low-speed acceleration performance, low emission level, and high energy conversion efficiency. Traditional fuel cell-powered vehicles use PEMFCs as the power source. The performance parameters and operating characteristics of vehicles were compared when driven by two different fuel cells [142], and the performance parameters of conventional internal combustion engine vehicles and pure electric vehicles widely used at present.

In general, compared with traditional pure electric vehicles and internal combustion engine vehicles, SOFCs hybrid vehicles have the advantages of high energy conversion efficiency and long range. At the same time, their disadvantages are a

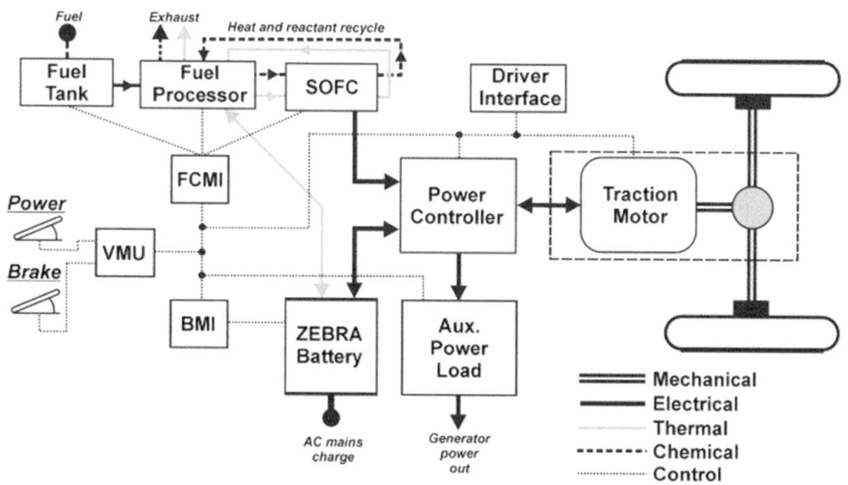

Fig. 11.24 SOFC-powered hybrid vehicle systems [141]

long start-up time and high operating temperature. Further research is still needed to solve these problems.

11.6.3 Solid Oxide Fuel Cells for Stationary Power Generation

In the field of fixed power generation with higher power levels, the SOFC system also has application scenarios. The power of the SOFC system applied to fixed power generation is generally in the order of hundreds of kilowatts to megawatts. SOFC system is usually combined with conventional power generation forms, such as gas turbine and steam turbine, to improve energy conversion efficiency. Research shows that the energy conversion efficiency of gt-sofc system can reach more than 60%. In general, the gas turbine SOFC system (GT-SOFC) is divided into direct type and indirect type. The structure of the direct GT-SOFC system is shown in Fig. 11.25 [143]. The components of the system include: compressor, recuperator, SOFCs unit, combustion chamber, and turbine. Compared with the traditional gas turbine system, the direct GT-SOFC system adds an SOFC unit. The air enters the cathode side of the SOFC after being pressurized and preheated by the compressor, and part of the methane fuel enters the anode side for reaction. The exhaust gas of the SOFC enters the combustion chamber of the gas turbine together with another part of the methane fuel for combustion and then enters the gas turbine.

In the indirect GT-SOFC system, the combustion chamber is replaced by a heat exchanger. The air enters the cathode of the fuel cell after being preheated by the exhaust gas of the turbine, and the fuel enters the anode side for reaction. The outlet

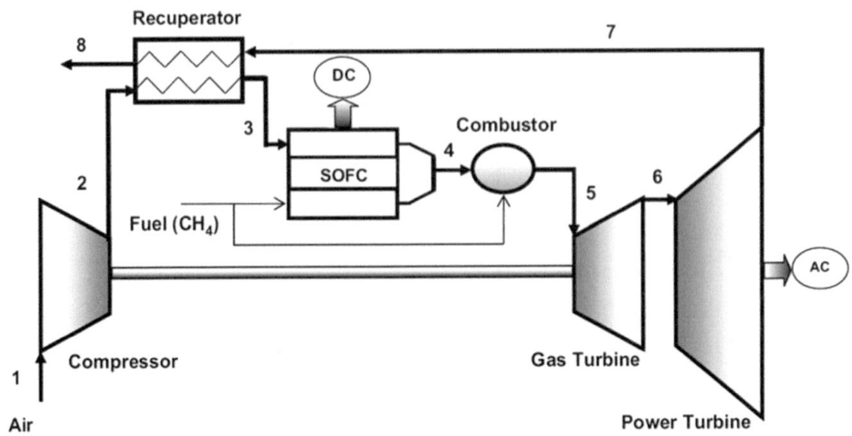

Fig. 11.25 Direct GT-SOFCs system [143]

air from the cathode is used to burn the residual hydrogen, carbon dioxide, and methane in the outlet gas of the anode. The exhaust gas after combustion enters the heat exchanger of the gas turbine to heat the air compressed by the compressor, replacing the combustion chamber in a conventional gas turbine system. The heat exchanger in the indirect GT-SOFC system needs to operate in a high-temperature and high-pressure environment, which has high requirements for its materials, thus, the system is not commonly used.

CCHP technology can combine a power generation system with a heating and cooling system to meet users' demand for heat, electricity, cooling, and other energy. The traditional CCHP system operates in the Otto cycle, using a gas turbine or steam turbine to generate electricity. The CCHP system, driven by SOFCs, benefits from many advantages of fuel cells and has a high energy conversion efficiency. Burer et al. proposed the CCHP system driven by SOFCs. The research results show that the energy conversion efficiency of the system driven by SOFC can be improved by 22% compared with the traditional CCHP system [144].

11.7 Summary and Outlook

SOFCs are a kind of electrochemical power generation device. They possess numerous advantages, including high energy conversion efficiency, not requiring catalysts, a wide range of fuel options (natural gas, methanol, biomass gas, etc., can be used), being environmentally friendly, being easy to assemble in a modular fashion, having strong adaptability, and being quiet and reliable. They have shown extremely broad application prospects in multiple fields such as power generation, transportation, and the military.

However, there is still much room for improvement in SOFCs technology, such as the need for improvement of electrode materials, relatively high costs, insufficient utilization of waste heat, and overly high operating temperatures. Nevertheless, efforts are being made in these aspects, and there has also been certain research so far. For example, the latest ones are as follows: Liang has recovered the waste heat of the SOFCs system by using D-ARC and MED-TVC technologies [145]; Solar chemical looping has been utilized to reduce costs [146]; Ni-doped Fe-based perovskites were employed to obtain SOFCs electrodes with high electrocatalytic activity [147]; Dielectric titanium-copper-based electrolytes were exploited to develop low-temperature solid oxide fuel cells [148].

Given the increasingly severe situation of the world's energy and environment, solid oxide fuel cells will surely gradually become the new focus that people pay attention to when developing and utilizing energy. As a future power source with broad application prospects, it will also definitely become one of the key technologies supporting the development of the global economy.

References

1. Kuterbekov K A, Nikonov A V, Bekmyrza K Zh, et al. Classification of Solid Oxide Fuel Cells [J]. Nanomaterials, 2022, 12(7): 1059.
2. Stambouli A B, Traversa E. Solid oxide fuel cells (SOFCs): a review of an environmentally clean and efficient source of energy [J]. Renewable and Sustainable Energy Reviews, 2002, 6(5): 433–455.
3. Über Brennstoff-Ketten mit Festleitern [J].
4. Waonr V C. l~ber den Mechanismus der elektrischen Stromleitung im Nernststift. [J].
5. Kiukkola K, Wagner C. Measurements on Galvanic Cells Involving Solid Electrolytes [J].
6. Singhal S C. Solid oxide fuel cells for power generation: Solid oxide fuel cells for power generation [J]. Wiley Interdisciplinary Reviews: Energy and Environment, 2014, 3(2): 179–194.
7. Kirubakaran A, Jain S, Nema R K. A review on fuel cell technologies and power electronic interface [J]. Renewable and Sustainable Energy Reviews, 2009, 13(9): 2430–2440.
8. Irshad M, Siraj K, Raza R, et al. A Brief Description of High Temperature Solid Oxide Fuel Cell's Operation, Materials, Design, Fabrication Technologies and Performance [J]. Applied Sciences, 2016, 6(3): 75.
9. Deepi A S, Dharani Priya S, Samson Nesaraj A, et al. Component fabrication techniques for solid oxide fuel cell (SOFC) – A comprehensive review and future prospects [J]. International Journal of Green Energy, 2022: 1–13.
10. Price R, Cassidy M, Grolig J G, et al. Upscaling of Co-Impregnated $La_{0.20}Sr_{0.25}Ca_{0.45}TiO_3$ Anodes for Solid Oxide Fuel Cells: A Progress Report on a Decade of Academic-Industrial Collaboration [J]. Advanced Energy Materials, 2021, 11(15): 2003951.
11. Zarabi Golkhatmi S, Asghar M I, Lund P D. A review on solid oxide fuel cell durability: Latest progress, mechanisms, and study tools [J]. Renewable and Sustainable Energy Reviews, 2022, 161: 112339.
12. Mahato N, Banerjee A, Gupta A, et al. Progress in material selection for solid oxide fuel cell technology: A review [J]. Progress in Materials Science, 2015, 72: 141–337.
13. Sun C, Hui R, Roller J. Cathode materials for solid oxide fuel cells: a review [J]. Journal of Solid State Electrochemistry, 2010, 14(7): 1125–1144.

14. Fergus J W. Electrolytes for solid oxide fuel cells [J]. Journal of Power Sources, 2006, 162(1): 30–40.
15. Kendall K, Kendall M. High-temperature solid oxide fuel cells for the 21st century: fundamentals, design and applications [M]. London: Academic Press is an imprint of Elsevier, 2016.
16. Shi H, Su C, Ran R, et al. Electrolyte materials for intermediate-temperature solid oxide fuel cells [J]. Progress in Natural Science: Materials International, 2020, 30(6): 764–774.
17. Singh M, Zappa D, Comini E. [J]. International Journal of Hydrogen Energy, 2021, 46(54): 27643–27674.
18. Li W, Wang Y, Liu W. A review of solid oxide fuel cell application [J]. IOP Conference Series: Earth and Environmental Science, 2020, 619(1): 012012.
19. Jamil S M, Othman M H D, Rahman M A, et al. Recent fabrication techniques for micro-tubular solid oxide fuel cell support: A review [J]. Journal of the European Ceramic Society, 2015, 35(1): 1–22.
20. Patcharavorachot Y, Arpornwichanop A, Chuachuensuk A. Electrochemical study of a planar solid oxide fuel cell: Role of support structures [J]. Journal of Power Sources, 2008, 177(2): 254–261.
21. Hsieh Y D. Effects of pressurization and temperature on power generating characteristics and impedances of anode-supported and electrolyte-supported planar solid oxide fuel cells [J]. Journal of Power Sources, 2015: 10.
22. Modeling, design, construction, and operation of power generators with solid oxide fuel cells [M]. New York, NY: Springer Berlin Heidelberg, 2018.
23. Fang X, Zhu J, Lin Z. Effects of Electrode Composition and Thickness on the Mechanical Performance of a Solid Oxide Fuel Cell [J]. 2018: 13.
24. Park B-K, Lee J-W, Lee S-B, et al. A fiat-tubular solid oxide fuel cell with a dense interconnect film coated on the porous anode support [J]. Journal of Power Sources, 2012: 5.
25. Wang G, Zhang Y, Han M. Densification of Ce0.9Gd0.1O2-δ interlayer to improve the stability of La0.6Sr0.4Co0.2Fe0.8 O3-δ/Ce0.9Gd0.1O2-δ interface and SOFC [J]. Journal of Electroanalytical Chemistry, 2020, 857: 113591.
26. Dalslet B, Blennow P, Hendriksen P V, et al. Assessment of doped ceria as electrolyte [J]. Journal of Solid State Electrochemistry, 2006, 10(8): 547–561.
27. Kim D-J. Lattice Parameters, Ionic Conductivities, and Solubility Limits in Fluorite-Structure MO2 Oxide [M = Hf4+, Zr4+, Ce4+, Th4+, U4+] Solid Solutions [J]. Journal of the American Ceramic Society, 1989, 72(8): 1415–1421.
28. Eguchi K, Setoguchi T, Inoue T, et al. Electrical properties of ceria-based oxides and their application to solid oxide fuel cells [J]. Solid State Ionics, 1992, 52(1–3): 165–172.
29. Kang Y J, Choi G M. The effect of alumina and Cu addition on the electrical properties and the SOFC performance of Gd-doped CeO2 electrolyte [J]. Solid State Ionics, 2009, 180(11–13): 886–890.
30. Villas-Boas L A, Figueiredo F M L, de Souza D P F, et al. Zn as sintering aid for ceria-based electrolytes [J]. Solid State Ionics, 2014, 262: 522–525.
31. Zheng Y, Zhou M, Ge L, et al. Effect of Fe2O3 on Sm-doped ceria system solid electrolyte for IT-SOFCs [J]. Journal of Alloys and Compounds, 2011, 509(2): 546–550.
32. Preethi S, Abhiroop M, Suresh Babu K. Low temperature densification by lithium co-doping and its effect on ionic conductivity of samarium doped ceria electrolyte [J]. Ceramics International, 2019, 45(5): 5819–5828.
33. Lane J, Neff J, Christie G. Mitigation of the deleterious effect of silicon species on the conductivity of ceria electrolytes [J]. Solid State Ionics, 2006, 177(19–25): 1911–1915.
34. Hsieh T H, Ray D T, Fu Y P. Co-precipitation synthesis and AC conductivity behavior of gadolinium-doped ceria [J]. Ceramics International, 2013, 39(7): 7967–7973.
35. Yu S, Bi H, Sun J, et al. Effect of grain size on the electrical properties of strontium and magnesium doped lanthanum gallate electrolytes [J]. Journal of Alloys and Compounds, 2019, 777: 244–251.

36. Hosoi K, Sakai T, Ida S, et al. Ce0.6Mn0.3Fe0.1O2-δ as an Alternative Cathode Material for High Temperature Steam Electrolysis Using LaGaO3-based Oxide Electrolyte [J]. Electrochimica Acta, 2016, 194: 473–479.
37. Jaiswal N, Tanwar K, Suman R, et al. A brief review on ceria based solid electrolytes for solid oxide fuel cells [J]. Journal of Alloys and Compounds, 2019, 781: 984–1005.
38. Gorte R J, Vohs J M. Novel SOFC Anodes for the Direct Electrochemical Oxidation of Hydrocarbons [J]. Journal of Catalysis, 2003, 216 (1–2): 477–486.
39. Sarikaya A, Dogan F. Effect of Various Pore Formers on the Microstructural Development of Tape-Cast Porous Ceramics [J]. Ceramics International, 2013, 39 (1): 403–413.
40. Adler S B. Factors Governing Oxygen Reduction in Solid Oxide Fuel Cell Cathodes [J]. Chemical Reviews, 2004, 104 (10): 4791–4844.
41. Antoni L. Materials for Solid Oxide Fuel Cells: the Challenge of their Stability [C] Materials Science Forum. 2004:1073–1090.
42. Gellings P J, Bouwmeester H J. Solid State Aspects of Oxidation Catalysis [J]. Catalysis Today, 2000, 58 (1): 1–53.
43. Gellings P J. Handbook of Solid State Electrochemistry [M]. CRC Press LLC, 2019: 578–580.
44. Jorgensen M J, Mogensen M. Impedance of Solid Oxide Fuel Cell LSM/YSZ Composite Cathodes [J]. Journal of the Electrochemical Society, 2001, 148 (5): 433.
45. Hassel V, Boukamp B A, Burggraaf A J. Oxygen Transfer Properties of Ion-Implanted Yttria-Stabilized Zirconia [J]. Solid State Ionics, 2017, 53: 890–903.
46. Baumann F S, Fleig J, Habermeier H-U, et al. Impedance Spectroscopic Study on Well-Defined (La, Sr)(Co, Fe)$O_{3-\delta}$ Model Electrodes [J]. Solid State Ionics, 2006, 177 (11–12): 1071–1081.
47. Fleig J. Solid Oxide Fuel Cell Cathodes: Polarization Mechanisms and Modeling of the Electrochemical Performance [J]. Annual Review of Materials Research, 2003, 33 (1): 361–382.
48. Beez A, Yin X, Menzler N H, et al. Insight into the Reaction Mechanism of $(La_{0.58}Sr_{0.40})(Co_{0.20}Fe_{0.80})O_{3-\delta}$ Cathode with Volatile Chromium Species at High Current Density in a Solid Oxide Fuel Cell Stack [J]. Journal of The Electrochemical Society, 2017, 164 (10): 3028.
49. Endo A, Fukunaga H, Wen C, et al. Cathodic Reaction Mechanism of Dense $La_{0.6}Sr_{0.4}CoO_3$ and $La_{0.81}Sr_{0.09}MnO_3$ Electrodes for Solid Oxide Fuel Cells [J]. Solid State Ionics, 2000, 135 (1–4): 353–358.
50. Brichzin V, Fleig J, Habermeier H U, et al. The Geometry Dependence of the Polarization Resistance of Sr-doped $LaMnO_3$ Microelectrodes on Yttria-Stabilized Zirconia [J]. Solid State Ionics, 2002, 152 499–507.
51. Chen D, Hu B, Ding K, et al. The Geometry Effect of Cathode/Anode Areas Ratio on Electrochemical Performance of Button Fuel Cell Using Mixed Conducting Materials [J]. Energies, 2018, 11 (7): 1875.
52. Jiang S. A Comparison of O_2 Reduction Reactions on Porous (La, Sr) MnO_3 and (La, Sr)(Co, Fe)O_3 Electrodes [J]. Solid State Ionics, 2002, 146 (1–2): 1–22.
53. Tang, H., Jin, Z., Wu, Y., Liu, W., & Bi, L. Cobalt-free nanofiber cathodes for proton conducting solid oxide fuel cells [J]. Electrochemistry Communications, 2019,100(3):108–112.
54. Yamamoto O., Solid oxide fuel cells: Fundamental aspects and prospects [J]. Electrochim Acta, 2000, 45:2423.
55. Sasaki K., Tamura J., Dokiya M.,Noble metal alloy-Zr(Sc)O_2 cermet cathode for reduced-temperature SOFCs [J]. Solid State Ionics, 2001,144:233–240.
56. Doshi R., Bichards V. L., Carter J. D.et a1. Devdopment of solidoxide fuel cells that operate at 500 °C [J]. J Electrcehem Sec, 1999, 146(4):1273–1278.
57. Skinner S. J. Recent Advances in Perovskite-Type Materials for Solid Oxide Fuel Cell cathodes [J]. International Journal of Inorganic Materials, 2001, 3 (2): 113–121.
58. Xu P., Chang D., Lu T., et al. Search for ABO_3 Type Ferroelectric Perovskites with Targeted Multi-Properties by Machine Learning Strategies [J]. 2021.

59. Tao S, Irvine J T. A Redox-Stable Efficient Anode for Solid-Oxide Fuel Cells [J]. Nature Materials, 2003, 2 (5): 320–323.
60. Boukamp B. A. The Amazing Perovskite Anode [J]. Nature Materials, 2003, 2 (5): 294–296.
61. Ngamou P. H. T., Bahlawane N., Chemical vapor deposition and electric characterization of perovskite oxides $LaMO_3$ (M=Co, Fe, Cr and Mn) thin films [J]J Solid State Chem, 2009, 182(4):849–854.
62. Wang W., Jiang S. P., A mechanistic study on the activation process of (La, Sr) MnO_3 electrodes of solid oxide fuel cells [J]. Solid State Iortics, 2006,177(15–16):1361–1369.
63. lshihara T., Kudo T., Matsuda H., et a1. Doped $PrMnO_3$ perovskite oxide as a new cathode of solid oxide fuel cells for low temperature operation [J]. J Electroehem Soc., 1995,142(5):1519–1524.
64. Sakaki Y., Takeda Y.,Kato A., et a1. $Ln_{l\text{-}x}Sr_xMnO_3$ (Ln=Pr, Nd, Sm and Gd) as the cathode material for solid oxide fuel cells [J]. Solid State lonics, 1999,118(3–4):187–194.
65. Huang X., Liu J., Lu Z., et a1. Properties of nonstoichiometric $Pr_{0.6\text{-}x}Sr_{0.4}MnO_3$ the cathodes of SOFCs [J].Solid State Ionics, 2000,130(3–4):195–201.
66. Q. Zhou, T. He, Q. He, Y. Ji, Electrochemical performances of LaBaCuFeO5+x and LaBaCuCoO5+x as potential cathode materials for intermediate-temperature solid oxide fuel cells, Electrochem. Commun. 11 (2008) 80–83.
67. Zhou W., Ran R., Shan Z. P. Progress in understanding and devdopment of $Ba_{0.5}Sr_{0.5}Co_{0.8}Fe_{0.2}O_{3\text{-}\delta}$-based cathodes for intermediate temperature Solid-oxide fuel cells: A review [J]. J Power Sources, 2009, 192(2): 231–246.
68. Kostogloudis G. C., Ftikos C., Properties of A-site-deficient $La_{0.6}Sr_{0.2}Co_{0.2}Fe_{0.8}O_{3\text{-}\delta}$-based perovskite oxides [J]. Solid Srate lonics, 1999, 126(1–2):143–151.
69. Ben Mya O., Dos Santos-Gómez L., Porras-Vázquez J. M.,et al. La1-x Srx Fe0.7 Ni0.3 O3-δ as both cathode and anode
70. materials for Solid Oxide Fuel Cells [J]. InternationalJournal of Hydrogen Energy, 2017, 42 (36): 23160–23169.
71. Hansen K. K. Evaluation of LSF based SOFC cathodes using cone-shaped electrodes and EIS [J]. Solid State Ionics, 2020, 344:115096.
72. Shao Z. P., Haile S. M., Ahn J., et al. A Thermally Self-Sustained Micro Solid-Oxide Fuel-Cell Stack with High Power Density [J]. Nature, 2005, 435 (7043): 795–798.
73. Steele B. C. Appraisal of $Ce_{1\text{-}y}Gd_yO_{2-y/2}$ Electrolytes for IT-SOFC Operation at 500 °C [J]. Solid State Ionics, 2000, 129 (1–4): 95–110.
74. Xia C., Liu M. A., Simple and Cost-Effective Approach to Fabrication of Dense Ceramic Membranes on Porous Substrates [J]. Journal of the American Ceramic Society, 2001, 84 (8): 1903–1905.
75. Jun A., Yoo S., Gwon O. H., et al. Thermodynamic and Electrical Properties of $Ba_{0.5}Sr_{0.5}Co_{0.8}Fe_{0.2}O_{3\text{-}\delta}$ and $La_{0.6}Sr_{0.4}Co_{0.2}Fe_{0.8}O_{3\text{-}\delta}$ for Intermediate-Temperature Solid Oxide Fuel Cells [J]. Electrochimica Acta, 2013, 89 372–376.
76. Bhalla A, Guo R, Roy R. The Perovskite Structure-A Review of its Role in Ceramic Science and Technology [J]. Materials Research Innovations, 2000, 4 (1): 3–26.
77. Bernuy L. C., Høydalsvik K., Einarsrud M. A., et al. Effect of A-Site Cation Ordering on Chemical Stability, Oxygen Stoichiometry and Electrical Conductivity in Layered $LaBaCo_2O_{5+\delta}$ Double Perovskite [J]. Materials, 2016, 9 (3): 154.
78. Maignan A., Martin C., Pelloquin D., et al. Structural and Magnetic Studies of Ordered Oxygen-Deficient Perovskites $LnBaCo_2O_{5+\delta}$, Closely Related to the “112” Structure [J]. Journal of Solid State Chemistry, 1999, 142 (2): 247–260.
79. Taskin A., Lavrov A., Ando Y. Achieving Fast Oxygen Diffusion in Perovskites by Cation Ordering [J]. Applied Physics Letters, 2005, 86 (9): 091910.
80. Zhang K., Ge L., Ran R., et al. Synthesis, Characterization and Evaluation of Cation-Ordered $LnBaCo_2O_{5+\delta}$ as Materials of Oxygen Permeation Membranes and Cathodes of SOFCs [J]. Acta Materialia, 2008, 56 (17): 4876–4889.
81. Yoo S., Shin J. Y., Kim G., Thermodynamic and Electrical Characteristics of $NdBaCo_2O_{5+\delta}$ at Various Oxidation and Reduction States [J]. Journal of Materials Chemistry, 2011, 21 (2): 439–443.

82. Kharton V. V., Yaremchenko A. A., Shaula A. I., et al. Transport properties and stability of Ni-contailling mixed conductors
83. Tors with perovskite- and K_2NiF_4-type structure [J]. J solid state Chem, 2004, 177: 26.
84. Skinner S. J., Kilner J. A., Oxygen diffusion and surface exchange in $La_{2-x}Sr_xNiO_{4+\delta}$ [J]. solid State Ionics, 2000, 135:709.
85. Ishihara T., Miyoshi S., Furuno T., et a1. Mixed conductivity and oxygen permeability of doped $Pr_2NiO_{1+\delta}$ based oxide [J]. Solid state Ionics, 2006, 177: 3087.
86. Mauvy F., Bassat J. M., Boehm, et a1. Oxygen electrode reaction on $Nd_2NiO_{4+\delta}$ cathode materials: Impedance spectroscopy study [J]. solid State Ionics, 2003, 158: 17.
87. Aguadem A., Alonso J. A., Escudero M. J., et a1. Evaluation of the $La_2Ni_{1-x}Cu_xO_{4+\delta}$ system as SOFC cathode material with 8YSZ and LSGM as electrolytes [J]. Solid State Ionics, 2008, 179(11-12): 393.
88. Takeda T., Kanno R., Kawamoto Y., et a1. New cathode materials for solid oxide fuel cell. Ruthenium pyrochlores and perovskites [J]. J Electrechem Soc, 1999,147(5): 1730-1733.
89. Zhen S., Sun W., Li P., et al. High performance cobalt-free $Cu_{1.4}Mn_{1.6}O_4$ spinel oxide as an intermediate temperature solid oxide fuel cell cathode [J]. Journal of Power Sources, 2016, 315:140–144.
90. Shao L., Wang Q., Fan L., et al. Copper cobalt spinel as a high performance cathode for intermediate temperature solid oxide fuel cells [J]. Chemical Communications, 2016,52(55):8615–8618.
91. Zhang L., Li D., Zhang S., One-step synthesized $Co_{1.5}Mn_{1.5}O_4$-$Ce_{0.8}Sm_{0.2}O_{1.9}$ composite cathode material for intermediate temperature solid oxide fuel cells [J]. Ceramics International, 2017, 43(2): 2859–2863.
92. Wang, X., Nakagawa, N., Kato, K. J. Electrochem. Soc. 2001,148, A565.
93. Horita, T., Yamaji, K., Sakai, N., Xiong, Y., Kato, T., Yokokawa, H., Kawada, T. J. Power Sources 2002, 106, 224.
94. Shu L., Sunarso J., Hashim S. S., et al. Advanced perovskite anodes for solid oxide fuel cells: A review [J]. International Journal of Hydrogen Energy, 2019, 44(59). 31275–31304.
95. Mc Intosh S., Gorte R..J. Direct hydrocarbon solid oxide fuel cells [J]. Chem. Rev., 2004, 104: 4845–4865.
96. Brown, M., Primdahl, S., Mogensen, M. J. Electrochem. Soc.2000, 475, 147.
97. Goodenough J. B., Huang Y. H. Alternative anode materials for solid oxide fuel cells [J]. Journal of Power Sources, 2007, 173(1): 1–10.
98. Singhal, S. C., Advances in solld oxlde fuel cell technology [J]. Solid State Ionics, 2000, 135: 305–313.
99. Minh N. Q., Ceramic fuel cell [J]. J Am Ceram soc, 1993, 76(3): 563–588.
100. Setoguchi T. Okamoto K., Eguchi K., et a1. Effects of anode material and fuel on anodic reaction of solid oxide fuel cells [J]. eletrochem. Soc., 1992, 139(10): 2875–2880.
101. Kanawka K., Othman M. H. D., Wu Z., et al. Electrochemistry Communications, 2011, 13(1), 93.
102. Grtm. Comprehensive review of methane conversion in solid oxide fuel cells: prospects for efficient electricity generation from natural gas [J]. Prog Energ Combust, 2016, 54: 61–64.
103. Boldrin P., Ruiz-trejo E., Mermelstein J., et al. Strategies for carbon and sulfur tolerant solid oxide fuel cell materials, incorporating lessons from heterogeneous catalysis [J]. Chem Rev, 2016, 116(22): 13633–13684.
104. Hanna J., Lee W. Y., Shi Y., et al. Fundamentals of electro-and thermochemistry in the anode of solid-oxide fuel cells with hydrocarbon and syngas fuels [J]. Prog Energ Combust, 2014, 40: 74–111.
105. Yang L., Choi Y., Qin W., et al. Promotion of water-mediated carbon removal by nanostructured barium oxide/nickel interfaces in solid oxide fuel cells [J]. Nat Commun, 2011(2): 357.
106. Liu S., Wang J. X., He C. R., et al. J China Ceram Soc, 2013, 41(8): 1057–1062.
107. Mcintosh S., Gorte R. J. Direct Hydrocarbon Solid Oxide Fuel cels [J]. Chem Rev, 2004, 104(10): 4845–4866.

108. Schafer W.,Koch A., Herold-Scllmidt U., Materials, Interfaces and production techniqucs for planar solid oxide fuel cells [J]. solid state Ionics, 1996, 86–88: 1235-1239.
109. Stashans A., Sanchez P., A theoretical study of La-doping in strontium titanatc [J]. Mater. Lett., 2000, 44: 153–157.
110. Ramyrez-Cabrera F. E., Atkinson A., Chadwick D., The influence of point defects on resistance of ceria to carbon depsition in hydrocarbon catalysis [J]. solid statce Ionics, 2000, 136–137: 825–831/
111. Mogensen M., Lindegaard T., Hansen U. R., et al. Physical properties of mixed conductor solid oxide fuel cell anodes doped CeO2 [J]. J Electrochem Soc, 1994, 141: 2122–2128.
112. Tao S., Irvine J. T. S., A redox-stable efficient anode for solid-oxide fuel cells [J]. Nature Materials, 2003, 2(5): 320–323.
113. Vernoux P., Djurado E., Guillodo M., Catalytic and Electrochemical Properties of Doped Lanthanum Chromites as New Anode Materials for Solid Oxide Fuel Cells [J]. Journal of the American Ceramic Society, 2001, 84(10): 2289–2295.
114. Danilovic N., Vincent A., Luo J. L., et al. Correlation of Fuel Cell Anode Electrocatalytic and ex situ Catalytic Activity of Perovskites $La_{0.75}Sr_{0.25}Cr_{0.5}X_{0.5}O_{3-\delta}$ (X = Ti, Mn, Fe, Co) [J]. Chemistry of Materials, 2010, 22(3): 957–965.
115. Hui S., Petric A., Electrical properties of yttrium-doped strontium titanate under reducing conditions [J]. Journal of the Electrochemical Society, 2002, 149(1): J1-J10.
116. Blennow P., Hansen K. K., Wallenberg L. R., et al. Electrochemical characterization and redox behavior of Nb-doped $SrTiO_3$ [J]. Solid State Ionics, 2009, 180(1): 63–70.
117. Marina O. A., Canfield N. L., Stevenson J. W., Thermal, electrical, and electrocatalytical properties of lanthanum-doped strontium titanate [J]. Solid State Ionics, 2002, 149(1): 21–28.
118. Ruiz-Morales J. C., Canales-Vázquez J., Savaniu C., et al. Disruption of extended defects in solid oxide fuel cell anodes for methane oxidation [J]. Nature, 2006, 439(7076): 568–571.
119. Zhou X, Yan N, Chuang K T, et al. Progress in La-doped $SrTiO_3$ (LST)-based anode materials for solid oxide fuel cells [J]. RSC Advances, 2014, 4(1): 118–131.
120. Xu X., Zhong Y., Shao Z., Double Perovskites in Catalysis, Electrocatalysis, and Photo(electro)catalysis [J]. Trends in Chemistry, 2019, 1(4): 410–424.
121. Huang Y. H., Dass R. I., King Z. L., et al. Double perovskites as anode meterials for solid-oxide fuel cells [J]. Science, 2006, 312(5771): 254–257.
122. Gwan M. A., Yun J. W., Carbon tolerance effects of $Sr_2NiMoO_{6-\delta}$ as an alternative anode in solid oxide fuel cell under methane fuel condition [J]. Journal of Electroceramics, 2018, 40(3): 171–179.
123. Yang X, Chen J, Panthi D, et al. Electron doping of $Sr_2FeMoO_{6-\delta}$ as high performance anode materials for solid oxide fuel cells [J]. Journal of Materials Chemistry A, 2019, 7(2): 733–743.
124. Munoz-Garcia A. B., Bugaris D. E., Pavone M., et al. Unveiling Structure-Property Relationships in $Sr_2Fe_{1.5}Mo_{0.5}O_{6-\delta}$, an Electrode Material for Symmetric Solid Oxide Fuel Cells [J]. Journal of the American Chemical Society, 2012, 134(15): 6826–6833.
125. Xie Z, Zhao H., Du Z., et al. Effects of Co doping on the electrochemical performance of double perovskite oxide $Sr_2MgMoO_{6-\delta}$ as an anode material for solid oxide fuel cells [J]. Journal of Physical Chemistry C, 2012, 116(17): 9734–9743.
126. Cowin P. I., Petit C. T. G., Lan R., et al. Recent Progress in the Development of Anode Materials for Solid Oxide Fuel Cells [J]. Advanced Energy Materials, 2011, 1(3): 314–332.
127. Sinha, A.; Miller, D. N.; Irvine, J. T. S. Development of Novel Anode Material for Intermediate Temperature SOFC (IT-SOFC). *J. Mater. Chem. A* **2016**, *4* (28), 11117–11123.
128. Maitre, A.; Lefort, P. Carbon Oxidation at High Temperature during Carbothermal Reduction of Titanium Dioxide. *Phys. Chem. Chem. Phys.* **1999**, *1* (9), 2311–2318.
129. Minh N Q. Ceramic fuel cells [J]. Journal of the American Ceramic Society, 1993, 76(3): 563–588.
130. Timurkutluk B, Timurkutluk C, Mat M D, et al. A review on cell/stack designs for high performance solid oxide fuel cells [J]. Renewable and Sustainable Energy Reviews, 2016, 56: 1101–1121.

131. Saied M, Ahmed K, Nemat-Alla M, et al. Performance study of solid oxide fuel cell with various flow field designs: numerical study [J]. international journal of hydrogen energy, 2018, 43(45): 20931–20946.
132. Zhao C, Yang J, Zhang T, et al. Numerical simulation of flow distribution for external manifold design in solid oxide fuel cell stack [J]. International Journal of Hydrogen Energy, 2017, 42(10): 7003–7013.
133. Yokoo M, Tabata Y, Yoshida Y, et al. Highly efficient and durable anode-supported SOFC stack with internal manifold structure [J]. Journal of power sources, 2008, 178(1): 59–63.
134. Singh M, Zappa D, Comini E. Solid oxide fuel cell: Decade of progress, future perspectives and challenges [J]. International Journal of Hydrogen Energy, 2021, 46(54): 27643–27674.
135. Kendall K, Ming N Q, Singhal S C. Cell and stack designs [M]//High Temperature and Solid Oxide Fuel Cells. Elsevier, 2003: 197–228.
136. Boigues-Muñoz C, Santori G, McPhail S, et al. Thermochemical model and experimental validation of a tubular SOFC cell comprised in a 1 kWel stack designed for μCHP applications [J]. International journal of hydrogen energy, 2014, 39(36): 21714–21723.
137. Kuterbekov K A, Nikonov A V, Bekmyrza K Z, et al. Classification of solid oxide fuel cells [J]. Nanomaterials, 2022, 12(7): 1059.
138. Park B K, Lee J W, Lee S B, et al. A flat-tubular solid oxide fuel cell with a dense interconnect film coated on the porous anode support [J]. Journal of Power Sources, 2012, 213: 218–222.
139. Yamaguchi T, Shimizu S, Suzuki T, et al. Development and evaluation of a cathode-supported SOFC having a honeycomb structure [J]. Electrochemical and Solid-State Letters, 2008, 11(7): B117.
140. van Gerwen R J F. Systems and applications [M]//High Temperature and Solid Oxide Fuel Cells. Elsevier Science, 2003: 363–392.
141. Minh N Q. Solid oxide fuel cell technology—features and applications [J]. Solid State Ionics, 2004, 174(1–4):271–277.
142. Singhal S.C. Solid oxide fuel cells for stationary, mobile, and military applications [J]. Solid State Ionics, 2002, 152:405–410.
143. Aguiar P, Brett DJL, Brandon NP. Feasibility study and techno economic analysis of an SOFC/battery hybrid system for vehicle applications [J]. Journal of Power Sources 2007; 171:188–197.
144. Haseli Y, Dincer I, Naterer GF. Thermodynamic analysis of a combined gas turbine power system with a solid oxide fuel cell through exergy [J]. Thermochimica Acta 2008;480:1–9.
145. Selman JR. In: Blomen LJM, Mugerwa MN, editors. Research, development,and demonstration of molten carbonate fuel cell systems in fuel cell systems [J].New York, NY: Plenum Press; 1993. p. 419.
146. Colson C, Nehrir H .Evaluating the benefits of a hybrid solid oxide fuel cell combined heat and power plant for energy sustainability and emissions avoidance [C]. 2011 IEEE Power and Energy Society General Meeting. IEEE, 2011.
147. Liang W, Han J, Ge Y, et al. Investigation on combining multi-effect distillation and double-effect absorption refrigeration cycle to recover exhaust heat of SOFC-GT system [J]. Energy Conv Manag, 2024, 301: 118054.
148. Cannone SF, Ishaq M, Lanzini A, et al. Thermodynamic analysis of a synergistic integration of solid oxide fuel cell and solar-based chemical looping methane reforming unit for solar energy storage, power production, and carbon capture [J]. Energy Conv Manag, 2024, 302: 118080.

MIX
Papier aus verantwortungsvollen Quellen
Paper from responsible sources
FSC® C105338

If you have any concerns about our products, you can contact us on
ProductSafety@springernature.com

In case Publisher is established outside the EU, the EU authorized representative is:
Springer Nature Customer Service Center GmbH
Europaplatz 3, 69115 Heidelberg, Germany

Printed by Libri Plureos GmbH
in Hamburg, Germany